Mathematical Methods in Computer Aided Geometric Design II

Mathematical Methods in Computer Aided Geometric Design II

EDITED BY

Tom Lyche
Institutt for informatikk
Universitetet i Oslo
Oslo, Norway

Larry L. Schumaker
Department of Mathematics
Vanderbilt University
Nashville, Tennessee

ACADEMIC PRESS, INC.
Harcourt Brace Jovanich, Publishers
Boston San Diego New York
London Sydney Tokyo Toronto

This book is printed on acid-free paper. ♾

ACADEMIC PRESS, INC.
1250 Sixth Avenue, San Diego, CA 92101-4311

United Kingdom Edition published by
ACADEMIC PRESS LIMITED
24–28 Oval Road, London NW1 7DX

Library of Congress Cataloging-in-Publication Data

Mathematical methods in computer aided geometric design II / edited by Tom Lyche, Larry L. Schumaker.
p. cm.
ISBN 0-12-460510-9
1. Geometry—Data processing—Congresses. I. Lyche, Tom. II. Schumaker, Larry L., Date
QA448.D38M382 1992
516'. 15'028—dc20 92-1590
CIP

Printed in the United States of America

92 93 94 95 BB 9 8 7 6 5 4 3 2 1

CONTENTS

PREFACE

During the week of June 20–25, 1991 an international conference on Curves, Surfaces, CAGD, and Image Processing was held at Biri, Norway. The conference was attended by 125 mathematicians from 22 countries. There were 8 one hour survey talks, 12 invited half hour talks, and 67 contributed research talks. This volume contains papers based on the surveys, along with 40 full length refereed research papers.

Topics discussed in the book include blossoming, cyclides, data fitting and interpolation, finding intersections of curves and surfaces, geometric continuity, geometrical optics, image and signal processing, modelling of geological structures, multiresolution analysis, NURBS, offsets, radial basis functions, rational splines, robotics, spline and Bézier methods for curve and surface modelling, subdivision, terrain modelling, wavelets, and others.

The conference was supported by grants from a number of international scientific organizations and two industrial concerns. For their support, we would like to thank Norsk Hydro, the University of Oslo, the Center for Industrial Research, the Office of Naval Research European Office, the Royal Norwegian Council for Scientific and Industrial Research (NTNF), the US Air Force European Office of Aerospace Research and Development, and Digital Equipment Corp.

We would like to thank Bjørg Alfsen, Morten Dæhlen, Sigurd Glærum, Per Ø. Hvidsten, Lise Jensen, Chantal Lyche, Evy Madsen, Knut Mørken, Lise Oppsahl, and Geir Westgaard for their help in organizing and running the conference. We are also grateful to Gerda Schumaker for her help with preparing the proceedings.

Nashville, January 17, 1992

PARTICIPANTS

Numbers in parentheses indicate pages on which authors' contributions begin.

LAURENT ALT, *Thomson Digital Image, 22 rue Hégésippe Moreau, 75018 Paris, FRANCE* [laurent%tdisys@inria.inria.fr].

ERLEND ARGE, *Universitetet i Oslo, Institutt for Informatikk, P. B. 1080 Blindern, 0316 Oslo 3, NORWAY* [erlend@ifi.uio.no].

JESUS PÈRES ARRIAGA, *Universidad de Cantabria, Dept. Electronica, 39005 Santander, SPAIN* .

KETIL ÅSTEBØL, *NORSAR, P. O. Box 51, N-2007 Kjeller, NORWAY* [ketil@lus.norsar.no].

PHILLIP BARRY (1,297), *University of Minnesota, Computer Science Dept., 4-192 EE/CSci Building, 200 Union St. SE, Minneapolis, MN 55455, USA* [barry@cs.umn.edu].

BRIAN A. BARSKY (567), *University of California, Berkeley, Computer Science Division, Elec. Eng. & Comp. Sci. Dept., Berkeley, CA 94720, USA* [barsky@beta.berkeley.edu].

RICHARD BARTELS (431), *University of Waterloo, Computer Science Department, Waterloo, Ontario N2L 3G1, CANADA* [rhbartels@watcgl.uwaterloo.ca].

B. J. C. BAXTER (9), *University of Cambridge, Dept. of App. Math. and Theor. Phys., Silver Street, Cambridge CB3 9 EW, ENGLAND* [bjcb%camnum@atm.amtp.cam.ac.uk].

MAREK BEŚKA, *Technical University of Gdańsk, Department of Mathematics, Majakowskiego 11/12, 80-952 Gdańsk, POLAND* .

LUC BIARD, *Université Joseph Fourier, LMC-IMAG, BP 53X, 38041 Grenoble, Cedex, FRANCE* [luc@aigle.imag.fr].

BIECHY, *Peugeot S.A. DITA, 62 Bd Victor Hugo, 92208 Neuilly-sur-Seine Cedex, FRANCE* .

CARL DE BOOR, *University of Wisconsin, Center for Mathematical Sciences, Madison, WI, USA* [deboor@cs.wisc.edu].

LEONARD BOS, *University of Calgary, Department of Mathematics, Calgary, Alberta T2N-1N4, CANADA* [lpbos@uncamult.bitnet].

ABDELAZIZ BOURAS, *Université Claude Bernard - Lyon I, LIGIA Bât 710/, 43 Boulevard du 11 Novembre 1918, 69622 Villeurbanne Cedex, FRANCE* [ligia@frmop11.bitnet].

KARL-HEINZ BRAKHAGE (19), *RWTH Aachen, Institut f. Geom. u. Praktische Mathematik, Templergraben 55, 5100 Aachen, GERMANY* [ez010br@dacht11.bitnet].

PERE BRUNET (29), *Universitat Politècnica de Catalunya, Departament de Llenguatges i Sistemes Informàtics, Av. Diagonal 647, 8ena Planta, 08028 Barcelona, SPAIN* [brunet@decnet.upc.es].

GUIDO BRUNNETT (43), *Department of Computer Science, Arizona State University, Tempe, AZ 85281, USA* [brunnett@asuvax.eas.asu.edu].

LEV BRUTMAN, *Haifa University, Dept. of Math. & Comp. Sc., Haifa 31905, ISRAEL* [rsma303@haifauvm.bitnet].

MARTIN BUHMANN, *University of Cambridge, Dept. of App. Math. and Theor. Phys., Silver Street, Cambridge CB3 9 EW, ENGLAND* [mdb10@phx.cam.ac.uk].

GIULIO CASCIOLA, *Universitá di Bologna, Dipartimento di Matematica, Piazza di Porta San Donato 5, 40127 Bologna, ITALY* [casciola@dm.unibo.it].

PAUL DE FAGET DE CASTELJAU (57), *Société des Automobiles Citroën, Centre Technique de Velizy, Route de Gizy chemin vicinal 2, 78140 Velizy Villacoublay, FRANCE* .

PATRICK CHENIN, *Université Joseph Fourier, LMC-IMAG, BP 53X, 38041 Grenoble, Cedex, FRANCE* [chenin@aigle.imag.fr].

ENG WEE CHIONH (101), *National University of Singapore, Department of Computer Science, Kent Ridge, Singapore 0511, SINGAPORE* [isccew@nusvm.bitnet].

VICTOR CHIU, *Structural Dynamics Research Corporation, 2000 Eastman Drive, Milford, OH 45150, USA* .

CHARLES K. CHUI (111), *Texas A&M University, Department of Mathematics, College Station, TX 77843-3368, USA* [e443cc@tamvm1.bitnet].

ELAINE COHEN (229), *University of Utah, Dept. of Computer Science, 3190 MEB, Salt Lake City, Utah 84112, USA* [cohen@cs.utah.edu].

MORTEN DÆHLEN (135), *Senter for Industriforskning, Postboks 124, Blindern, 0314 Oslo 3, NORWAY* [mortend@ifi.uio.no].

WOLFGANG DAHMEN, *Freie Universität Berlin, Institut für Mathematik I, Arnimalee 2-6, 1000 Berlin 33, GERMANY* [dahmen@math.fu-berlin.de].

MARC DANIEL (161), *Ecole Nationale Supérieure Mécanique, 1 rue de la Noë, 44072 Nantes Cedex 03, FRANCE* [daniel@cc01.ensm-nantes.fr].

WENDELIN L. F. DEGEN (171), *Universität Stuttgart, Mathematisches Institut B, Pfaffenwaldring 57, D-7000 Stuttgart 80, GERMANY* [lbaa@dsorus1i.bitnet].

VITTORIA DEMICHELIS, *Universitá di Torino, Dipartimento di Matematica, Via Carlo Alberto 10, 10123 Torino, ITALY* [demichelis@unimat.to.cnr.it].

TONY DEROSE (185), *University of Washington, Department of Computer Science, Seattle, WA 98195, USA* [derose@cs.washington.edu].

WENHUI DU (197), *Thomson Digital Image, 22 rue Hégésippe Moreau, 75018 Paris, FRANCE* [du%tdisys@inria.inria.fr].

NIRA DYN (209), *Tel-Aviv University, School of Mathematical Sciences, Ramat Aviv, Tel Aviv 69978, ISRAEL* [niradyn%taurus.bitnet@cunyvm.cuny.edu].

MATTHIAS ECK (217), *Technische Hochschule Darmstadt, Arbeitsgruppe Differentialgeometrie, Schloßgartenst. 7, 6100 Darmstadt, GERMANY* [xbr1dbok@ddathd21.bitnet].

TIMO EIROLA, *Helsinki University of Technology, Institute of Mathematics, Otakaari 1 A, SF-02150 Espoo, FINLAND* [timo@spazio.hut.fi].

PER EVENSEN, *Senter for Industriforskning, P. O. Box 124, Blindern, 0314 Oslo 3, NORWAY* [evensen@bivrost.cs.si.no].

RIDA T. FAROUKI (239), *IBM Watson Research Center, P. O. Box 218, Yorktown Heights, New York 10598, USA* [farouki@ibm.com].

MICHAEL FLOATER (261), *Senter for Industriforskning, P. O. Box 124, Blindern, 0314 Oslo 3, NORWAY* [floater@bivrost.cs.si.no].

THOMAS A. FOLEY (275), *Arizona State University, Dept. of Computer Science, Tempe, Arizona 85287, USA* [foley@asuvax.eas.asu.edu].

LARS ANDREAS FRØYLAND (287), *Norsk Hydro, Postboks 4313, Nygårdstangen, N-5028, Bergen, NORWAY* .

LARS GÆDE, *Danmarks Tekniske Højskole, Matematisk Institut, Bygning 303, DK-2800 Lyngby, DENMARK* [lars@mat.dth.dk].

Henri Giaume, *Peugeot S.A. DITA, 62 Bd Victor Hugo, 92208 Neuilly-sur-Seine Cedex, FRANCE* .

Sigurd Glærum, *Universitetet i Oslo, Institutt for Informatikk, P. B. 1080 Blindern, 0316 Oslo 3, NORWAY* [sigurdg@ifi.uio.no].

John Goldak, *Carleton University, Mechanical Engineering, Ottawa, Ontario K1S 5B6, CANADA* [goldak@mrco.carleton.ca].

Ronald Goldman (101,297,431,585), *Rice University, Dept. Computer Science, Brown School of Engineering, P.O Box 1892, Houston, Texas 77251, USA* [rng@rice.edu].

Jens Gravesen, *Danmarks Tekniske Højskole, Matematisk Institut, Bygning 303, DK-2800 Lyngby, DENMARK* [gravesen@mat.dth.dk].

Carla Guerrini, *Universitá di Bologna, Departimento di Matematica, Piazza di Porta San Donato 5, 40127 Bologna, ITALY* [guerrini@dm.unibo.it].

Lu Han, *Vanderbilt University, Dep. of Mathematics, Nashville, TN, 37235, USA* .

Erich Hartmann, *Technische Hochschule Darmstadt, Fachbereich Mathematik, Scloßgartenstr. 7, D-6100 Darmstadt, GERMANY* .

Øyvind Hjelle, *Senter for Industriforskning, P. O. Box 124, Blindern, 0314 Oslo 3, NORWAY* .

Bas Hogervorst, *University of Twente, Dept. of Applied Mathematics, P. O. Box 217, 7500 AE Enschede, THE NETHERLANDS* [hogervor@math.utwente.nl].

Josef Hoscheck (331), *Technische Hochschule Darmstadt, Fachbereich Mathematik, AG 2 - Geometrie und Algebra, Scloßgartenstr. 7, D-6100 Darmstadt, GERMANY* [xbr1d1sd@ddathd21.bitnet].

Steinar Hveberg, *Møre og Romsdal Distriktshøyskole, Postboks 308, 6401 Molde, NORWAY* [hveberg@dhmolde.no].

Per Ø. Hvidsten, *Universitetet i Oslo, Institutt for Informatikk, P. O. Box 1080, Blindern, 0316 Oslo 3, NORWAY* .

Mohammed K. Ismail (343), *Dundee University, Dept. of Mathematical Sciences, Dundee DD1 4HN, SCOTLAND* [mismail@mcs.dund.ac.uk].

Michael Kallay, *Electronic Data Systems, 13555 SE 36th St., Suite 300, Bellevue, WA 98008, USA* [kallay@c4west.eds.com].

PER ERIK KOCH (351), *Norges Tekniske Høgskole, Institutt for Matematiske Fag, 7034 Trondheim - NTH, NORWAY* [pek@imf.unit.no].

REKHA KULKARNI (367), *Université Joseph Fourier, LMC-IMAG, BP 53 X, 38041 Grenoble, FRANCE* .

BORIS I. KVASOV, *USSR Academy of Sciences, Institute of Pure & Applied Mechanics, Novosibirsk 630090, USSR* .

ARNE LAKSÅ (287), *Senter for Industriforskning, P. O. Box 124, Blindern, 0314 Oslo 3, NORWAY* .

VINCENT K. H. LAU, *Brunel University, Department of Mathematics and Statistics, Uxbridge, Middlesex, UB 8 3PH, ENGLAND* [vincent.lau@brunel.ac.uk].

PIERRE-JEAN LAURENT (367), *Université Joseph Fourier, LMC-IMAG, BP 53 X, 38041 Grenoble, FRANCE* [pjl@imag.fr].

EUGENE LEE (381), *Boeing Commercial Airplane Group, P. O. Box 3707, M/S 6A-WA, Seattle, WA 98124, USA* .

ALAIN LE MÉHAUTÉ, *Université de Nantes, Institut de Math. et d'Inf., 2 chemin de la Houssinière, 44072 Nantes-Cedex, FRANCE* [alm@cicb.fr].

MATIJA LOKAR (359), *University of Ljubljana, Institute of Mathematics, Physics and Mechanics, Jadranska 19, 61111 Ljubljana, YUGOSLAVIA* [matija.lokar@uni-lj.ac.mail.yu].

ANTONIO LÓPEZ-CARMONA, *Granada, Depto. Matematica Aplicada, Facultad de Ciencias, Avda Fuentenueva Sin., 18071 Granada, SPAIN* .

JEAN-LUC LOSCHUTZ, *Intergraph, Software Development Center, 95-101, rue des Solets, Silic 578, 94653 Rungis Cedex, FRANCE* .

BRADLEY LUCIER (391), *Purdue University, Department of Mathematics, West Lafayette, IN 47907, USA* [lucier@gauss.math.purdue.edu].

TOM LYCHE (135), *Universitetet i Oslo, Institutt for Informatikk, P. B. 1080 Blindern, 0316 Oslo 3, NORWAY* [tom@ifi.uio.no].

JOAQUIM S. MADEIRA, *Universidade de Coimbra, Dept. de Matemática, Apartado 3008, P-3000 Coimbra, PORTUGAL* .

JEAN-JAQUES MALOSSE, *Intergraph Co., IW17C3, Huntsville AL 35894-0001, USA* .

STEPHEN MANN (185), *University of Washington, Dept. of Computer Science, FR-35, Seattle, WA 98195, USA* [smann@cs.washington.edu].

CARLA MANNI (401), *Dipartimento di Energetica, c/o Fontanella, Via di S. Marta 3, 50139 Firenze, ITALY* [font1@ifiidg.bitnet].

DENIS MARCHEPOIL, *Université Joseph Fourier, LMC-IMAG, BP 53 X, 38041 Grenoble, FRANCE* [denis@aigle.imag.fr].

MARIE-LAURENCE MAZURE (367,413), *Université Joseph Fourier, LMC-IMAG, BP 53 X, 38041 Grenoble, FRANCE* .

CHARLES A. MICCHELLI (69), *IBM Watson Research Center, P. O. Box 218, Yorktown Heights, NY 10598, USA* [cam@yktvmx.bitnet].

JACOB MICHELSEN, *Danmarks Tekniske Højskole, Instituttet for skibs- og havteknik, Bygning 101 E, DK-2800 Lyngby, DENMARK* [ishjm@odin.ish.dth.dk].

MIKHAIL N. MIKHAIL, *American University in Cairo, 113 Kasr el aini st., Cairo, EGYPT* [mikhail@auc.eg.bitnet].

LAURA MONTEFUSCO, *Universitá di Bologna, Dipartimento di Matematica, Piazza di Porta San Donato 5, 40127 Bologna, ITALY* .

ANNE METTE MØLLER, *Danmarks Tekniske Højskole, Matematisk Institut, Bygning 303, DK-2800 Lyngby, DENMARK* [anne@mat.dth.dk].

DOUG MOORE, *Rice University, Dept. of Computer Science, P. O. Box 1892, Houston, TX 77254-1892, USA* [dougm@rice.edu].

KNUT MØRKEN, *Universitetet i Oslo, Institutt for Informatikk, P. B. 1080 Blindern, 0316 Oslo 3, NORWAY* [knutm@ifi.uio.no].

ARVID MYKLEBUST, *Virginia Polytechn. Inst. and State Univ., College of Engeneering, Blacksburg, VA 24061-0238, USA* .

OLAVI NEVANLINNA, *Helsinki University of Technology, Institute of Mathematics, 02150 Espoo,, FINLAND* [onevanli@vipunen.hut.fi].

GREGORY M. NIELSON, *Arizona State University, Dept. of Computer Science, Tempe, Arizona 85287, USA* [nielson@enuxva.eas.asu.edu].

BOON–HUA ONG (419), *University of Science of Malaysia, School of Math. & Computer Sciences, 11800 Penang, MALAYSIA* [obh@cs.usm.my].

JAN PAJCHEL (287), *Norsk Hydro, Postboks 4313, Nygårdstangen, N-5028, Bergen, NORWAY* .

J. F. RIVAS PENA, *Universidad de Cantabria, Dept. Electronica, 39005 Santander, SPAIN* .

A.R.M. Piah (441), *University of Dundee, Dept. of Mathematics and Comp. Science, Dundee DD1 4HN, SCOTLAND* [arahni@mcs.dund.ac.uk].

Bruce Piper, *Rensselaer Polytechnic Institute, Dept. of Mathematical Sciences, Troy, NY 12180, USA* [piperb@turing.cs.rpi.edu].

Michael J. Pratt (457), *Cranfield Institute of Technology, Dept. of Appl. Comp. And Math, Cranfield, Bedford MK43 OAL, ENGLAND* .

Hartmut Prautzsch (475), *Universität Karlsruhe, Institut für Betriebs- und Dialogsysteme, Am Fasanengarten 5, Postf. 6980, W-7500 Karlsruhe 1, GERMANY* [prautzsch@ira.uka.de].

Luigia Puccio, *Universitá di Messina, Dipartimento di Matematica, Salita Sperone, 31, Contrada Papaado, 98166 Messina, ITALY* [laura@imeuniv.bitnet].

Ewald Quak, *Texas A&M University, Center for Approximation Theory, College Station, TX 77843-3368, USA* [egq4460@tamvenus.bitnet].

Christophe Rabut (481), *INSA Toulouse, Service de Mathématiques, Avenue de Rangueil, 31077 Toulouse Cedex, FRANCE* [rabut@cix.cict.fr].

Louise A. Raphael, *Howard University, Department of Mathematics, Washington, D.C. 200059, USA* [alama38@humain.bitnet].

Richard F. Riesenfeld, *University of Utah, Dept. of Computer Science, 3190 MEB, Salt Lake City, Utah 84112, USA* [rfr@cs.utah.edu].

Dirk-Jan Ripmeester, *University of Amsterdam, Faculty of Mathematics and Computer Science, Kruislaan 403, 1098 SJ Amsterdam, THE NETHERLANDS* [dirk-jan@fwi.uva.nl].

Paul Sablonnière (497), *INSA Rennes, Laboratoire LANS, 20 Av. des Buttes de Coësmes, 35043 Rennes, Cedex, FRANCE* .

Kęstutis Šalkauskas, *University of Calgary, Dept. of Mathematics and Statistics, Calgary, Alberta, CANADA* [salkauskas@uncamult.bitnet].

Ramon F. Sarraga, *General Motors Research Laboratories, Computer Science Department, P. O. Box 9055, Warren, Michigan 48090-9055, USA* [sarraga@gmr.com].

Robert Schaback (517), *Universität Göttingen, Inst. für Num. und Angew. Math., Lotzestrasse 16-18, D-3400 Göttingen, GERMANY* [u0881@dgogwdg5.bitnet].

GERD SCHMELTZ (537), *Technische Hochschule Darmstadt, Arbeitsgruppe Differentialgeometrie, Schloßgartenst. 7, 6100 Darmstadt, GERMANY* [xbr3dc56@ddathd21.bitnet].

CAJA SCHMIDT, *FORCE-Institutes, Data-Centre, Park allé 345, DK-2605 Brønøy, DENMARK* .

LARRY L. SCHUMAKER, *Vanderbilt University, Dept. of Mathematics, Nashville, TN, 37240, USA* [schumake@ctrvax.vanderbilt.edu].

THOMAS SEDERBERG (549), *Brigham Young University, Dept. of Civil Engineering, 368 Clyde Bldg., Provo Utah 84602, USA* [tom@tws.byu.edu].

HANS-PETER SEIDEL (559), *University of Waterloo, Department of Computer Science, Waterloo, Ontario N2L 3G1, CANADA* [hpseidel@watcgl.waterloo.edu].

GADIEL SEROUSSI (567), *Hewlett-Packard Laboratories, Building 3U, P. O. Box 10490, Palo Alto, CA 94303, USA* [seroussi@hplabs.hp.com].

XI-QUAN SHI, *Freie Universität Berlin, Institut für Mathematik I, Arnimalee 2-6, 1000 Berlin 33, GERMANY* .

YOSEFA SHRAGA, *Soreo Nuclear Research Center, 19 Ben Yosef, Givat Shmuel, ISRAEL* .

YOHANES STEFANUS (585), *Rice University, Department of Computer Science, P. O. Box 1892, Houston, Texas 77251, USA* [yohanes@rice.edu].

BIRGIT STIMPEL, *Mercedes-Benz AG, Abt. KTB, Postfach 226, D-7032 Sindelfingen, GERMANY* .

KYRRE STRØM, *Universitetet i Oslo, Institutt for Informatikk, P. B. 1080 Blindern, 0316 Oslo 3, NORWAY* [kyrres@ifi.uio.no].

JOACHIM STÖCKLER, *Universität Duisburg, Fachbereich Mathematik, Lotharstr. 65, 4100 Duisburg, GERMANY* [hn277st@math.uni-duisburg.de].

GEOVAN TAVARES DOS SANTOS, *Catholic University, Dep. of Mathematics, Rua Marquês de São Vicente, 225, Rio de Janeiro RJ 22453, BRAZIL* [geta@lncc.bitnet].

CORNELIS R. TRAAS, *University of Twente, Dept. of Applied Mathematics, P. O. Box 217, 7500 AE Enschede, THE NETHERLANDS* [traas1@utwente.nl].

LEONARDO TRAVERSONI, *Universidad Autónoma Metropolitana, Division de Ciencias Basicas e Ingenieria, Dept. de Ing. de Procesos e Hidraulica, Apdo. Post. 55-532, C.P. 09340 México D.F., MÉXICO* .

WILFRIED TRUMP (475), *Universität Karlsruhe, Institut für Betriebs- und Dialogsysteme, Am Fasanengarten 5, Postf. 6980, W-7500 Karlsruhe 1, GERMANY* [trump@ira.uka.de].

KENJI UEDA (597), *Ricoh Company, Ltd., 1-1-17, Koishikawa, Bunkyo-ku, Tokyo, 112, JAPAN* [ueda@src.ricoh.co.jp].

KEITH UNSWORTH (419), *University of Dundee, Department of Mathematics and Computer Science, Dundee DD1 4HN, SCOTLAND* [unsworth@mcs.dund.ac.uk].

CHRISTINE VERCKEN, *Telecom Paris, Dpt. Inf., 46 rue Barrault, 75634 Paris Cedex 13, FRANCE* [vercken@inf.enst.fr].

ÀLVAR VINACUA (29), *Universitat Politècnica de Catalunya, Departament de Llenguatges i Sistemes Informàtics, Av. Diagonal 647, 8ena Planta, 08028 Barcelona, SPAIN* [ealvar@ebrupc51.bitnet].

GEIR WESTGAARD, *Universitetet i Oslo, Institutt for Informatikk, P. O. Box 1080, Blindern, 0316 Oslo 3, NORWAY* .

NORMAN WEYRICH (607), *Universität Rostock, FB Mathematik, Universitätsplatz 1, 0-2500 Rostock, GERMANY* .

DANIEL A. WILLIAMS, *Howard University, Department of Mathematics, Washington, D.C. 200059, USA* [alama38@humain.bitnet].

P. K. YUEN (321), *Brunel University, Department of Mathematics and Statistics, Uxbridge, Middlesex, UB8 3PHcd, ENGLAND* .

RAINER ZEIFANG (615), *Universität Stuttgart, Mathematisches Institut A, Pfaffenwaldring 57, D-7000 Stuttgart 80, GERMANY* [zeifang@merkur.mathematik.uni-stuttgart.de].

Symmetrizing Multiaffine Polynomials

Phillip J. Barry

Abstract. Polar forms are n-variate multiaffine symmetric polynomials which are intimately related to B-spline curves. Certain problems in computer aided geometric design produce (non-symmetric) multiaffine polynomials, which can be "symmetrized" to yield the polar form. Any n-variate polynomial can be symmetrized by obtaining it at all possible permutations of its n arguments, and averaging the results. However, for *multiaffine* polynomials, one may not need to use all $n!$ permutations. The question then arises: for a given n, what is the smallest set that will symmetrize any n-variate multiaffine polynomial? The main result presented here is that for $n \leq 6$, the number of elements in a minimal symmetrizing set is the least common multiple of $\{\binom{n}{k}\}_k$.

§1. Introduction and Motivation

Polar forms, or blossoms [8,9], have proved a valuable tool in the study of B-spline curves and surfaces. Let $G(t)$ be any polynomial of degree n or less. Its polar form is the unique n-variate polynomial $B_G(u_1, \ldots, u_n)$ such that (i) B_G is multiaffine, *i.e.*, each u_i has exponent no greater than 1, (ii) B_G is symmetric, *i.e.*, $B_G(u_1, \ldots, u_n) = B_G(u_{\pi(1)}, \ldots, u_{\pi(n)})$ for any permutation π, and (iii) on its diagonal, B_G reduces to G, *i.e.*, $B_G(t, \ldots, t) = G(t)$. The polar form of a piecewise polynomial $G(t)$ of degree n or less is the piecewise n-variate polynomial $B_G(u_1, \ldots, u_n)$ that has the three above properties for each of its polynomial segments. The polar form and B-spline representation of a piecewise polynomial are closely connected. Let $\{t_i\}$ be a knot sequence. Then the B-spline control vertices $\{V_i\}$ for a given curve $G(t)$ in the related spline space can be simply expressed in terms of the polar form:

$$V_i = B_G(t_{i+1}, \ldots, t_{i+n}). \tag{1}$$

The values $B_G(t_{i+1}, \ldots, t_{i+n})$ are *polar values* of the curve.

Mathematical Methods in Computer Aided Geometric Design II
Tom Lyche and Larry L. Schumaker (eds.), pp. 1–7.

ISBN 0-12-460510-9.

The usefulness of the polar form motivates considering techniques for finding either the polar form itself or certain polar values. Some problems in computer aided geometric design produce an n-variate (non-symmetric) multiaffine polynomial, which can be symmetrized to yield the polar form. We give three examples of such problems.

First, by Equation (1), one means for converting to B-spline form is to find the appropriate polar values. Suppose we have a curve scheme given by a "de Boor-like" evaluation algorithm. That is, suppose the curve scheme possesses a recursive evaluation algorithm consisting of a triangular array of computations grouped into n levels (see Figure 1). The computations begin with the (relevant) original control points, which are placed at the base of the triangle. Points in each row are affine combinations of pairs of consecutive points in the previous row; this combination is linear in t, the evaluation parameter. The point on the curve appears at the apex of the triangle. There are actually many curve schemes that possess such an evaluation algorithm, including Lagrange interpolating polynomials, "Pólya curves," Catmull-Rom splines, and generalized Ball curves [1,2,6].

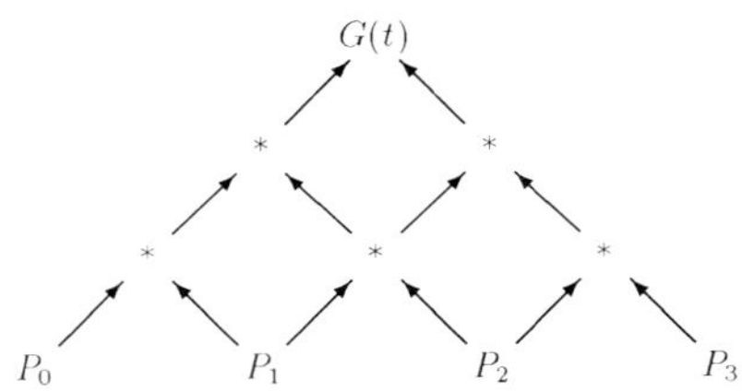

Figure 1. The form of a recursive evaluation algorithm for a cubic polynomial G with control points $P_0, \ldots, P_3$.

If one replaces t in the jth level of such algorithms with u_j, for all j, the result of the algorithm is an n-variate multiaffine polynomial $C(u_1, \ldots, u_n)$ which reduces to the original curve on the diagonal; however, C is not symmetric. One manner of finding $B_G(u_1, \ldots, u_n)$ from C is to obtain C at all possible permutations of $u_1, \ldots, u_n$ and normalize by dividing by the number of permutations (this insight is due to R. N. Goldman). That is,

$$B_G(u_1, \ldots, u_n) = \frac{1}{n!} \sum_{\pi \in S_n} C(u_{\pi(1)}, \ldots, u_{\pi(n)}), \tag{2}$$

where S_n is the group of all permutation on n objects.

For our second example, suppose we have polar forms $B_{G_1}(u_1, \ldots, u_{n_1})$ and $B_{G_2}(u_1, \ldots, u_{n_2})$ of polynomials G_1 and G_2, respectively, and wish to find the polar form $B_G(u_1, \ldots, u_{n_1+n_2})$ of $G = G_1 G_2$. Let $n = n_1 + n_2$, and

$$C(u_1, \ldots, u_n) = B_{G_1}(u_1, \ldots, u_{n_1}) B_{G_2}(u_{n_1+1}, \ldots, u_n). \tag{3}$$

C is multiaffine, n-variate, and reduces to G on the diagonal, so again Equation (2) holds.

For the final example, consider composition. Let $G = G_1 \circ G_2$, $n = n_1 n_2$, and

$$C(u_1, \ldots, u_n) = B_{G_1}(B_{G_2}(u_1, \ldots, u_{n_1}), \ldots, B_{G_2}(u_{n_1(n_2-1)+1}, \ldots, u_n)). \quad (4)$$

Since C is again multiaffine, n-variate, and reduces to G on the diagonal, Equation (2) once more yields B_G.

In each case one would like to avoid applying all $n!$ permutations. Define a *minimal symmetrizing set* $S \subseteq S_n$ to be any set of minimal cardinality such that, for *any* n-variate multiaffine polynomial C,

$$\sum_{\pi \in S} C(u_{\pi(1)}, \ldots, u_{\pi(n)}) \quad (5)$$

is symmetric. Thus, for any degree n or less polynomial G, if C is any n-variate multiaffine polynomial that reduces to G on its diagonal,

$$B_G(u_1, \ldots, u_n) = \frac{1}{|S|} \sum_{\pi \in S} C(u_{\pi(1)}, \ldots, u_{\pi(n)}). \quad (6)$$

The purpose of this paper is to discuss minimal symmetrizing sets. The main result is

Theorem. *For* $n \leq 6$, $|S| = \operatorname{lcm}\{\binom{n}{k}\}_k$, *where* lcm *denotes the least common multiple.*

The proof will follow from the discussion in the next two sections. Note that $\operatorname{lcm}\{\binom{n}{k}\}_k$ grows exponentially in n [7], so even minimal symmetrizing sets will not be useful for applications involving large n; however, in many applications, n is small.

§2. Cubics

To explore the problem further, we now look at the cubic case (the linear and quadratic cases are trivial). Consider any trivariate multiaffine polynomial $C(u_1, u_2, u_3)$. What characteristics must a minimal symmetrizing set S have? In order to symmetrize the terms of total degree 1, S must map $i \in \{1, 2, 3\}$ to every element in $\{1, 2, 3\}$, and must map it to each the same number of times. Since there are three such elements, 3 must divide $|S|$. A similar argument holds for symmetrizing the terms of total degree 2. Since there is only one term each of total degrees 0 and 3 we need not be concerned with these. Therefore 3 divides $|S|$, and $|S| \geq 3$.

For general n, a similar argument yields that $\binom{n}{k}$ divides $|S|$ for all k, and that S must have at least $\operatorname{lcm}\{\binom{n}{k}\}_k$ elements.

Note that if a set of permutations symmetrizes all terms of total degree k, it must symmetrize those of total degree $n - k$. So for the cubic case we need only find a set that works on the terms of total degree 1. We now find a set of 3 elements that does this. Let $(\alpha_1\alpha_2\cdots\alpha_k)$ indicate the permutation that maps α_1 to α_2, α_2 to α_3, ..., α_k to α_1 and every element not equal to any α_i to itself; let (1) denote the identity permutation. A permutation of the form $(\alpha_1\alpha_2\cdots\alpha_k)$ is said to be a *k-cycle*. If there exists a symmetrizing set of 3 elements, exactly one element fixes u_1. Therefore either (1) or (23) is in the set, but not both. The same argument applied to u_2 and u_3 yields that either (1) is in the set and the set contains no 2-cycles, and is therefore $\{(1),(123),(132)\}$, or the set consists only of 2-cycles; *i.e.*, it is $\{(12),(13),(23)\}$. Simple computation shows both of these work (note in general that if S is a symmetrizing set, so is $S_n\setminus S$). Therefore there exists a symmetrizing set of $3=\text{lcm}\{\binom{n}{k}\}_k$ elements for $n=3$. As our example shows, such a set need not be unique.

The above analysis indicates that the problem need not be phrased in terms of multiaffine polynomials. For example, consider any set of n objects, and let $\mathcal{P}_n$ be the power set of this set. Partition $\mathcal{P}_n$ into subsets

$$H_{i,n}:=\{A:\ A\in\mathcal{P}_n, |A|=i\} \qquad i=0,\ldots,n.$$

Then an equivalent statement of the problem is to find the smallest non-empty set S such that, for all i, for all $h,h_1,h_2\in H_{i,n}$,

$$|\{\pi:\pi\in S,\pi(h)=h_1\}|=|\{\pi:\pi\in S,\pi(h)=h_2\}|.$$

§3. Quartics, Quintics, and Sextics

In this section we will consider the cases $n=4,5$, and 6. In each case we show $|S|=\text{lcm}\{\binom{n}{k}\}_k$. Although these results can be shown by brute force, we provide arguments which illustrate some alternative approaches. As in the case $n=3$, the sets given below are not unique.

Case n = 4. The least common multiple of $\{\binom{4}{k}\}$ is 12. Consider the set of all even permutations of 4 objects:

$$\{(1),(123),(124),(134),(234),(132),(142),(143),(243),$$
$$(12)(34),(13)(24),(14)(23)\}.$$

We need consider only the terms of total degree 1 and the terms of total degree 2. Consider an arbitrary term, u_j, of total degree 1. All 3-cycles in S_4 are included; these take u_j to each term of total degree 1 twice. The elements that are products of two 2-cycles take u_j to each term of total degree 1 other than itself once, and then the identity takes u_j to itself. Therefore the terms of total degree 1 are symmetrized.

The set of terms of total degree 2 has 6 elements so each must be mapped twice to each term. Consider the effects of the 3-cycles on u_iu_j. There are four terms of the form u_iu_k or u_ju_k for $k\neq i,j$. If i or j is fixed by the 3-cycle it will go to such a term. On the other hand, if both are moved by the 3-cycle, either $i \mapsto j$ or $j \mapsto i$, so any 3-cycle maps $u_i u_j$ to a term involving either u_i or u_j, but not both. Since all eight 3-cycles are present, the set of 3-cycles map $u_i u_j$ onto these elements twice each.

The identity and one of the elements that is a product of two 2-cycles maps $u_i u_j$ onto itself. Finally, the other terms that are products of two 2-cycles maps $u_i u_j$ onto the term not involving u_i or u_j, and the result follows.

Case n = 5. The least common multiple of $\{\binom{5}{k}\}$ is 10. Consider the set

$$\{(12345), (13524), (14253), (15432), (1), (1342), (1435), (1523), (2453), (1254)\}.$$

This set is the direct product of two sets, one consisting of all powers of the 5-cycle (12345), and the other composed of (1) and (1254). (This product is taken so that elements in the latter set are applied first.) We need consider only the terms of total degree 1 and the terms of total degree 2. Note first that the set consisting of all powers of (12345) symmetrizes all terms of total degree 1. Now if a set T symmetrizes all terms of a given total degree, so does $T\pi$, for any $\pi \in S_n$. So all terms of total degree 1 are symmetrized by the set above.

Consider then the terms of total degree 2. Partition this set into subsets $\{u_1u_2, u_2u_3, u_3u_4, u_4u_5, u_1u_5\}$, and $\{u_1u_3, u_2u_4, u_3u_5, u_1u_4, u_2u_5\}$. Note that (1) maps each subset onto itself and (1254) maps each subset onto the other. Moreover, when applied to either subset, the set of powers of (12345) is bijective. Therefore the combined effect is to symmetrize the terms of total degree 2, and the result follows.

Case n = 6. In this case $\mathrm{lcm}_k\{\binom{n}{k}\} = 60$. Again we consider the direct product of two sets: the symmetrizing set above for $n = 5$, and the set of all powers of (123456). Call these sets T_1 and T_2, respectively.

Consider first the effect of this direct product on an arbitrary term of degree 1. T_2 maps this term onto every term of degree 1 once. Then the effect of the ten elements of T_1 is to map each non-u_6 term to each non-u_6 term twice and to fix u_6. Thus the original term goes to each term of total degree one exactly ten times, and the terms of total degree 1 are symmetrized.

Next consider the terms of total degree 2. T_2 maps an arbitrary term onto four terms not involving u_6, and onto two terms involving u_6. T_1 is bijective on the set of terms not involving u_6, and maps each term involving u_6 onto each term involving u_6 twice each. So the arbitrary term is mapped to each term of total degree 2 exactly 4 times, and the set of elements of total degree 2 is symmetrized.

Finally, consider the terms of total degree 3. T_2 maps an arbitrary term onto three terms not involving u_6, and onto three terms involving u_6. T_1 is bijective on the set of terms not involving u_6, and is also bijective on the set of terms involving u_6. So the arbitrary term is mapped to each term of total degree 3 exactly 3 times, and the set of elements of total degree 3 is symmetrized. Since it suffices to consider the terms of total degrees 1, 2, and 3, the result follows.

§4. Observations on Conversion to B-spline Form

Multiaffine polynomials related to spline schemes often possess special properties which may be exploited in problems involving symmetrizing the multiaffine polynomial. Consider, for example, the problem of coverting a curve given by a de Boor-like recursive evaluation algorithm to B-spline form. One possible way to do this is through computing, by Equation (6), the polar values corresponding to the B-spline control points. The following observations may apply to such a conversion technique:

1. Because of the overlap in knots associated with consecutive polar values, there is significant overlap in the associated computations.
2. Special purpose hardware has been developed [4] for rapid evaluation of curves possessing recursive evaluation algorithms. Slightly modified versions of this hardware could perhaps be coupled with the results above to perform fast conversions.
3. Some curve schemes, such as Catmull-Rom splines, already have some symmetry in their recursive evaluation algorithms (certain of the levels in their recursive evaluation algorithm are "B-spline" levels — see [1]). In these special cases, one can further reduce the number of permutations needed.
4. The results in this paper extend to include transformations involving other representations. For example, Bézier curves can be considered a special type of B-spline curve. Furthermore, other curve schemes such as the monomial form of a curve can be considered an extended type of B-spline curve [5], and a slight modification of the results above provide conversions to this form. Moreover, the class of curves possessing recursive evaluation algorithms can be enlarged to include curves such as those written in Newton form [3]; a slight modification of the results above provides conversions from these forms to B-spline form.
5. Curve schemes such as Bézier curves involve multiple knots. In these cases the effect of some elements of a symmetrizing set may be identical; also, the overlap mentioned in Observation (1) may be more prominent. In either case further simplification may be possible.

§5. Open Questions

In this paper we have discussed symmetrizing multiaffine polynomials. The results given in this paper are partial; many more questions have been raised. We conclude by listing a few.

1. Does the result that a minimal symmetrizing set has $\mathrm{lcm}\{\binom{n}{k}\}_k$ elements extend to all n?
2. For a given n, how many different minimal symmetrizing sets are there? Is there any simple way to characterize them?
3. Is there a simple algorithm for constructing minimal symmetrizing sets?
4. Can the results in this paper be used to provide *useful* algorithms for computer aided geometric design?

5. Observations such as those in Section 4 may provide more efficient algorithms when used with some symmetrizing sets than with others. Is this indeed the case, and, if so, which sets are most efficient?
6. Is there any algebraic structure to minimal symmetrizing sets? For example, minimal symmetrizing sets need not be subgroups of S_n, but are they related to any group structures?
7. Can the results in this paper be extended in any way to encompass any types of non-multiaffine polynomials?

Acknowledgments. The author would like to thank Dr. R. N. Goldman of Rice University for many discussions about this problem, as well as for helpful comments on the preparation of this paper.

References

1. Barry, P. J., and R. N. Goldman, A recursive evaluation algorithm for a class of Catmull-Rom splines, Computer Graphics — *SIGGRAPH '88* Conference Proceedings **22** (1988), 199–204.
2. Barry, P. J., and R. N. Goldman, Recursive polynomial curve schemes and computer aided geometric design, Constr. Approx. **6** (1990), 65–96.
3. Barry, P. J., and R. N. Goldman, Shape parameter deletion for Pólya curves, Numerical Algorithms **1** (1991), 121–138.
4. DeRose, T., M. Bailey, B. Barnard, R. Cypher, D. Dobrikin, C. Ebeling, S. Konstantinidou, L. McMurchie, H. Mizrahi, and B. Yost, The Apex: Two VLSI designs for generating parametric curves and surfaces, The Visual Computer **5** (1989), 264–276.
5. Goldman, R. N., and P. J. Barry, Algorithms for progressive curves: Extending B-spline and blossoming techniques to the monomial, power, and Newton dual bases, in *Knot Insertion and Deletion Techniques for B-spline Modeling*, T. Lyche and R. N. Goldman (eds.), SIAM, to appear.
6. Goodman, T. N. T., and H. B. Said, Shape preserving properties of the generalised Ball basis. Computer Aided Geom. Design **8** (1991), 115–121.
7. Shalit, J., private communication.
8. Ramshaw, L., *Blossoming: A Connect-the-Dots Approach to Splines*, Digital Equipment Technical Report 19, Palo Alto, California, 1987.
9. Ramshaw, L., Blossoms are polar forms, Computer Aided Geom. Design **6** (1989), 323–358.

Phillip J. Barry
Computer Science Department, University of Minnesota
4-192 EE/CSci Bldg., 200 Union St. SE
Minneapolis, MN 55455 USA
barry@cs.umn.edu

Norm Estimates for Inverses of Distance Matrices

B. J. C. Baxter

Abstract. A radial basis function approximation has the form $s(x) = \sum_{j=1}^{n} y_j \, \varphi(\|x - x_j\|_2)$, $\quad x \in \mathbb{R}^d$, where $\varphi: [0, \infty) \to \mathbb{R}$ is some given function, $(y_j)_1^n$ are real coefficients, and the centres $(x_j)_1^n$ are points in $\mathbb{R}^d$. For a wide class of functions φ, it is known that the interpolation matrix $A = (\varphi(\|x_j - x_k\|_2))_{j,k=1}^n$ is invertible (see [5]). Further, several recent papers have provided upper bounds on $\|A^{-1}\|_2$, where the points $(x_j)_1^n$ satisfy the condition $\|x_j - x_k\|_2 \geq \delta$, $j \neq k$, for some positive constant δ. In this paper, we show that the generalized Fourier transform of φ provides a simple and elegant tool for the derivation of such estimates.

§1. Introduction

This paper uses Fourier transform techniques to derive inequalities of the form

$$y^T A y \leq -\mu \, y^T y, \qquad y \in \mathbb{R}^n, \tag{1}$$

where μ is a positive constant and $\sum_{j=1}^{n} y_j = 0$. Here we are using the notation of the abstract. It can be shown that equation (1) implies the bound $\|A^{-1}\|_2 \leq 1/\mu$ for a wide class of functions including the multiquadric (see Section 4 of Baxter [3]). Such estimates have been derived in [2,6,7,9], using a different technique. The author submits that the derivation presented here for the Euclidean norm is more perspicuous. Further, we relate the generalized Fourier transform to the measure that occurs in an important characterization theorem for those functions φ considered here. This is useful because tables of generalized Fourier transforms are widely available, thus avoiding several of the technical calculations of [6,7,9]. Finally, we mention some recent work of the author that provides the *least upper bound* on $\|A^{-1}\|_2$ when the points $(x_j)_{j \in \mathbb{Z}^d}$ form a subset of $\mathbb{Z}^d$.

The norm $\|\cdot\|$ will always be the Euclidean norm in this paper. We shall denote the inner product of two vectors x and y by xy.

Mathematical Methods in Computer Aided Geometric Design II
Tom Lyche and Larry L. Schumaker (eds.), pp. 9–18.

ISBN 0-12-460510-9.

§2. The Univariate Case for the Euclidean Norm

Let $n \geq 2$ and let $(x_j)_1^n$ be points in $\mathbb{R}$ satisfying the condition $\|x_j - x_k\| \geq 1$ for $j \neq k$. We shall prove that

$$\Big|\sum_{j,k=1}^{n} y_j y_k \,\|x_j - x_k\|\Big| \geq \frac{1}{2}\|y\|^2,$$

whenever $\sum_{j=1}^n y_j = 0$.

We shall use the fact that the generalized Fourier transform of $\varphi(x) = |x|$ is $\hat{\varphi}(t) = -2/t^2$ in the univariate case. A proof of this may be found in Jones [4], Theorem 7.32.

Proposition 2.1. *If* $\sum_{j=1}^n y_j = 0$, *then*

$$\begin{aligned}\sum_{j,k=1}^{n} y_j y_k \,\|x_j - x_k\| &= (2\pi)^{-1}\int_{-\infty}^{\infty} (-2/t^2) \sum_{j,k=1}^{n} y_j y_k \,\exp(i(x_j - x_k)t)\,dt \\ &= -\pi^{-1}\int_{-\infty}^{\infty} |\sum_{j=1}^{n} y_j e^{ix_j t}|^2 t^{-2}\,dt.\end{aligned} \tag{2}$$

Proof: The two expressions on the righthand side above are equal because of the useful identity

$$\sum_{j,k=1}^{n} y_j y_k \,\exp(i(x_j - x_k)t) = |\sum_{j=1}^{n} y_j e^{ix_j t}|^2.$$

This identity will be used several times below. We now let

$$\hat{g}(t) = (-2t^{-2})\,|\sum_{j=1}^{n} y_j e^{ix_j t}|^2, \quad \text{for } t \in \mathbb{R}.$$

The condition $\sum_{j=1}^n y_j = 0$ implies that $\hat{g}$ is uniformly bounded. Further, since $\hat{g}(t) = \mathcal{O}(t^{-2})$ for large $|t|$, we see that $\hat{g}$ is absolutely integrable. Thus we have the equation

$$g(x) = (2\pi)^{-1}\int_{-\infty}^{\infty} \hat{g}(t)\exp(ixt)\,dt.$$

A standard result of the theory of generalized Fourier transforms (cf. [4], Theorem 7.14, pages 224ff) provides the expression

$$\sum_{j,k=1}^{n} y_j y_k \,\|x + x_j - x_k\| = (2\pi)^{-1}\int_{-\infty}^{\infty} (-2t^{-2})|\sum_{j=1}^{n} y_j e^{ix_j t}|^2 \exp(ixt)\,dt,$$

where we have used the identity stated at the beginning of this proof. We now need only set $x = 0$ in this final equation. ■

Proposition 2.2. *Let $B : \mathbb{R} \to \mathbb{R}$ be a continuous function such that supp $(B) \subset [-1, 1]$ and $0 \le \hat{B}(t) \le t^{-2}$. If $n \ge 2$, $\|x_j - x_k\| \ge 1$ for $j \ne k$, and $\sum_{j=1}^n y_j = 0$, then*

$$\sum_{j,k=1}^{n} y_j y_k \, \|x_j - x_k\| \le -2B(0)\|y\|^2.$$

Proof: By Proposition 2.1 and properties of Fourier transforms,

$$\begin{aligned}\sum_{j,k=1}^{n} y_j y_k \, \|x_j - x_k\| &\le (2\pi)^{-1} \int_{-\infty}^{\infty} (-2\hat{B}(t)) \sum_{j,k=1}^{n} y_j y_k \, \exp(i(x_j - x_k)t) dt \\ &= -2 \sum_{j,k=1}^{n} y_j y_k \, B(x_j - x_k) \\ &= -2B(0)\|y\|^2,\end{aligned}$$

where the first inequality follows from the condition $\hat{B}(t) \le t^{-2}$. The last line is a consequence of $\mathrm{supp}(B) \subset [-1, 1]$. ■

Corollary 2.3. *Let*

$$B(x) = \begin{cases} (1 - |x|)/4, & \text{if } |x| \le 1 \\ 0, & \text{otherwise.} \end{cases}$$

Then B satisfies the conditions of Proposition 2.2 and $B(0) = 1/4$.

Proof: By direct calculation, we find that

$$\hat{B}(t) = \frac{\sin^2(t/2)}{t^2} \le \frac{1}{t^2}.$$

It is clear that the other conditions of Proposition 2.2 are satisfied. ■

We have therefore shown the following theorem to be true.

Theorem 2.4. *Let $(x_j)_1^n$ be points in $\mathbb{R}$ such that $n \ge 2$ and $\|x_j - x_k\| \ge 1$ when $j \ne k$. If $\sum_{j=1}^n y_j = 0$, then*

$$\sum_{j,k=1}^{n} y_j y_k \, \|x_j - x_k\| \le -\frac{1}{2} \, \|y\|^2.$$

We see that a consequence of this result is the non-singularity of the Euclidean distance matrix when the points $(x_j)_1^n$ are distinct and $n \ge 2$. It is important to realise that the homogeneity of the Euclidean norm allows us to replace the condition "$\|x_j - x_k\| \ge 1$ if $j \ne k$" by "$\|x_j - x_k\| \ge \epsilon$ if $j \ne k$". We restate Theorem 2.4 in this form for the convenience of the reader:

Theorem 2.4b. *Choose any $\epsilon > 0$ and let $(x_j)_1^n$ be points in $\mathbb{R}$ such that $n \geq 2$ and $\|x_j - x_k\| \geq \epsilon$ when $j \neq k$. If $\sum_{j=1}^n y_j = 0$, then*

$$\sum_{j,k=1}^{n} y_j y_k \|x_j - x_k\| \leq -\frac{1}{2}\, \epsilon \|y\|^2.$$

We shall now show that this bound is optimal. Without loss of generality, we return to the case $\epsilon = 1$. We take our points to be the integers $0, 1, \ldots, n$, so that the Euclidean distance matrix, A_n say, is given by

$$A_n = \begin{pmatrix} 0 & 1 & 2 & \ldots & n \\ 1 & 0 & 1 & \ldots & n-1 \\ \vdots & \vdots & \ddots & \vdots & \\ n & n-1 & n-2 & \ldots & 0 \end{pmatrix}.$$

It is straightforward to calculate the inverse of A_n:

$$A_n^{-1} = \begin{pmatrix} (1-n)/2n & 1/2 & & & & 1/2n \\ 1/2 & -1 & 1/2 & & & \\ & 1/2 & -1 & & & \\ & & & \ddots & & \\ & & & & -1 & 1/2 \\ 1/2n & & & & 1/2 & (1-n)/2n \end{pmatrix}.$$

Proposition 2.5. *We have the inequality* $2 - (\pi^2/2n^2) \leq \|A_n^{-1}\|_2 \leq 2$.

Proof: We observe that $\|A_n^{-1}\|_2 \leq \|A_n^{-1}\|_1 = 2$, establishing the upper bound. For the lower bound, we focus attention on the $(n-1) \times (n-1)$ symmetric tridiagonal minor of A_n^{-1} formed by deleting its first and last rows and columns, which we shall denote by T_n. Thus we have

$$T_n = \begin{pmatrix} -1 & 1/2 & & & & \\ 1/2 & -1 & 1/2 & & & \\ & 1/2 & -1 & & & \\ & & & \ddots & & \\ & & & & -1 & 1/2 \\ & & & & 1/2 & -1 \end{pmatrix}.$$

Now

$$\begin{aligned} \|T_n\|_2 &= \max\{y^T A_n^{-1} y : y^T y = 1 \text{ and } y_1 = y_{n+1} = 0\} \\ &\leq \max\{y^T A_n^{-1} y : y^T y = 1\} \\ &= \|A_n^{-1}\|_2, \end{aligned}$$

so that $\|T_n\|_2 \leq \|A_n^{-1}\|_2 \leq 2$. But the eigenvalues of T_n are given by

$$\lambda_k = -1 + \cos(k\pi/n), \text{ for } k = 1, 2, \ldots, n-1.$$

Thus $\|T_n\|_2 = 1 - \cos(\pi - \pi/n) \geq 2 - \pi^2/2n^2$, where we have used an elementary inequality based on the Taylor series for the cosine function. The proposition is proved. ■

§3. The Multivariate Case for the Euclidean Norm

We first prove the multivariate versions of Propositions 2.1 and 2.2, which generalize in a very straightforward way. We shall require the fact that the generalized Fourier transform of $\varphi(x) = \|x\|$ in $\mathbb{R}^d$ is given by

$$\hat{\varphi}(t) = -c_d\|t\|^{-d-1},$$

where

$$c_d = 2^d \pi^{(d-1)/2}\Gamma((d+1)/2).$$

This may be found in [4], Theorem 7.32. We now deal with the analogue of Proposition 2.1.

Proposition 3.1. *If* $\sum_{j=1}^n y_j = 0$, *then*

$$\sum_{j,k=1}^n y_j y_k \|x_j - x_k\| = -c_d(2\pi)^{-d}\int_{\mathbb{R}^d} |\sum_{j=1}^n y_j e^{ix_jt}|^2 \|t\|^{-d-1}\,dt. \tag{3}$$

Proof: We define

$$\hat{g}(t) = -c_d\|t\|^{-d-1}|\sum_{j=1}^n y_j e^{ix_jt}|^2.$$

The condition $\sum_{j=1}^n y_j = 0$ implies this function is uniformly bounded and the decay for large argument is sufficient to ensure absolute integrability. The argument now follows the proof of Proposition 2.1, with obvious minor changes. ■

Proposition 3.2. *Let* $B : \mathbb{R}^d \to \mathbb{R}$ *be a continuous function such that* supp $(B) \subset \{x \in \mathbb{R}^d : \|x\| \le 1\}$, $0 \le \hat{B}(t) \le \|t\|^{-d-1}$ *and* $B(0) > 0$. *If* $n \ge 2$, $\|x_j - x_k\| \ge 1$ *for* $j \ne k$, *and* $\sum_{j=1}^n y_j = 0$, *then*

$$\sum_{j,k=1}^n y_j y_k \|x_j - x_k\| \le -c_d B(0)\|y\|^2.$$

Proof: The proof of Proposition 2.2 clearly generalizes to this case. ■

However, to exhibit a function B satisfying the conditions of Proposition 3.2 is harder than in the univariate case. We modify a construction of [6,9]. Let

$$B_0(x) = \begin{cases} 1, & \text{if } \|x\| \le 1/2 \\ 0, & \text{otherwise.} \end{cases}$$

Then, using [6], equation 1.10 or [9], Lemma 3.1, we find that

$$\hat{B}_0(t) = (2\|t\|)^{-d/2} J_{\frac{d}{2}}(\|t\|/2),$$

where J_k denotes the kth-order Bessel function of the first kind. Further, $\hat{B}_0$ is a radially symmetric function since B_0 is radially symmetric. We now define

$$B = B_0 * B_0,$$

so that, by the convolution theorem,

$$\begin{aligned}\hat{B}(t) &= (\hat{B}_0)^2(t)\\ &= (2\|t\|)^{-d} J^2_{\frac{d}{2}}(\|t\|/2),\end{aligned}$$

and the behaviour of J_0 for large argument provides the inequality

$$\hat{B}(t) \leq \mu_d \|t\|^{-d-1},$$

for some constant μ_d. Since the conditions of Proposition 3.2 are now easy to verify when B is scaled by μ_d^{-1}, we see that we are done .

§4. Fourier Transforms and Bessel Transforms

Here we relate our technique to the work of Ball [2] and Narcowich and Ward [6, 7].

Definition 4.1. *A real sequence $(y_j)_{j\in\mathbb{Z}^d}$ is said to be zero-summing if it is finitely supported and $\sum_{j\in\mathbb{Z}^d} y_j = 0$.*

Definition 4.2. *A function $\varphi\colon [0,\infty) \to \mathbb{R}$ will be said to be conditionally negative definite of order 1 on $\mathbb{R}^d$, hereafter shortened to CND1(d), if it is continuous and, for any points $(x_j)_{j\in\mathbb{Z}^d}$ in $\mathbb{R}^d$ and any zero-summing sequence $(y_j)_{j\in\mathbb{Z}^d}$, we have*

$$\sum_{j,k\in\mathbb{Z}^d} y_j y_k \varphi(\|x_j - x_k\|) \leq 0.$$

Such functions were characterized by von Neumann and Schoenberg (1941). For every positive integer d, let $\Omega_d\colon [0,\infty) \to \mathbb{R}$ be defined by

$$\Omega_d(r) = {\omega_{d-1}}^{-1} \int_{S^{d-1}} \cos(ryu)\, dy,$$

where u may be any unit vector in $\mathbb{R}^d$, S^{d-1} denotes the unit sphere in $\mathbb{R}^d$, and ω_{d-1} its $(d-1)$-dimensional Lebesgue measure. Thus Ω_d is essentially the Fourier transform of the normalized rotation invariant measure on the unit sphere.

Theorem 4.3. *Let $\varphi\colon [0,\infty) \to \mathbb{R}$ be a continuous function. A necessary and sufficient condition that φ be a CND1(d) function is that it have the form*

$$\varphi(r) = \varphi(0) + \int_0^\infty (1 - \Omega_d(rt))\, t^{-2} d\beta(t),$$

for every $r \geq 0$, where $\beta: [0, \infty) \to \mathbb{R}$ is a non-decreasing function such that $\int_1^\infty t^{-2}\, d\beta(t) < \infty$ and $\beta(0) = 0$. Furthermore, β is uniquely determined by φ.

Proof: The first part of this result is Theorem 7 of [8], restated in our terminology. The uniqueness of β is a consequence of Lemma 2 of that paper. ■

It is a consequence of this theorem that there exist constants A and B such that $\varphi(r) \leq Ar^2 + B$. For we have

$$\left| \int_1^\infty (1 - \Omega_d(rt))\, t^{-2}\, d\beta(t) \right| \leq 2 \int_1^\infty t^{-2}\, d\beta(t) < \infty,$$

using the fact that $|\Omega_d(r)| \leq 1$ for every $r \geq 0$. Further, we see that

$$0 \leq 1 - \Omega_d(\rho) = 2{\omega_{d-1}}^{-1} \int_{S^{d-1}} \sin^2(\rho u t/2)\, dt \leq \rho^2/2,$$

which provides the bound

$$\left| \int_0^1 (1 - \Omega_d(rt))\, t^{-2}\, d\beta(t) \right| \leq r^2 \int_0^1 \frac{1}{2}\, d\beta(t) = \frac{1}{2} r^2 \beta(1).$$

Thus $A = \beta(1)/2$ and $B = \varphi(0) + 2\int_1^\infty t^{-2}\, d\beta(t)$ suffice. Therefore the function $\{\varphi(\|x\|) : x \in \mathbb{R}^d\}$ is a tempered distribution in the sense of Schwartz (1966) and possesses a generalized Fourier transform $\{\hat{\varphi}(\|\xi\|) : \xi \in \mathbb{R}^d\}$. There is a rather simple relation between the generalized Fourier transform and the nondecreasing function of Theorem 4.3 for a certain class of functions. This is our next topic.

Definition 4.4. *A function $\varphi: [0, \infty) \to \mathbb{R}$ will be termed* admissible *if it is a continuous function of algebraic growth which satisfies the following conditions:*

1. *$\hat{\varphi}$ is a continuous func ion on $\mathbb{R}^d \setminus \{0\}$.*
2. *The limit $\lim_{\|\xi\| \to 0} \|\xi\|^{d+1} \hat{\varphi}(\|\xi\|)$ exists.*
3. *The integral $\int_{\{\|\xi\| \geq 1\}} |\hat{\varphi}(\|\xi\|)|\, d\xi$ exists.*

It is straightforward to prove the analogue of Propositions 2.1 and 3.1 for an admissible function.

Proposition 4.5. *Let $\varphi: [0, \infty) \to \mathbb{R}$ be an admissible function and let $(y_j)_{j \in \mathbb{Z}^d}$ be a zero-summing sequence. Then for any choice of points $(x_j)_{j \in \mathbb{Z}^d}$ in $\mathbb{R}^d$, we have the identity*

$$\sum_{j,k \in \mathbb{Z}^d} y_j y_k \varphi(\|x_j - x_k\|) = (2\pi)^{-d} \int_{\mathbb{R}^d} \left| \sum_{j \in \mathbb{Z}^d} y_j \exp(i x_j \xi) \right|^2 \hat{\varphi}(\|\xi\|)\, d\xi. \quad (4)$$

Proof: Let $\hat{g}\colon \mathbb{R}^d \to \mathbb{R}$ be the function defined by

$$\hat{g}(\xi) = \Big| \sum_{j\in\mathbb{Z}^d} y_j \exp(ix_j\xi)\Big|^2 \hat{\varphi}(\|\xi\|).$$

Then $\hat{g}$ is an absolutely integrable function on $\mathbb{R}^d$, because of the conditions on φ and because $(y_j)_{j\in\mathbb{Z}^d}$ is a zero-summing sequence. Thus $\hat{g}$ is the generalized transform of $\sum_{j,k} y_j y_k \varphi(\|\cdot + x_j - x_k\|)$, and by standard properties of generalized Fourier transforms we deduce that

$$\sum_{j,k} y_j y_k \varphi(\|x+x_j - x_k\|) = (2\pi)^{-d} \int_{\mathbb{R}^d} \Big| \sum_{j\in\mathbb{Z}^d} y_j \exp(ix_j\xi)\Big|^2 \hat{\varphi}(\|\xi\|) \exp(ix\xi)\, d\xi.$$

The proof is completed by setting $x = 0$. ■

Proposition 4.6. *Let $\varphi\colon [0,\infty) \to \mathbb{R}$ be an admissible CND1(d) function. Then*

$$d\beta(t) = -(2\pi)^{-d}\omega_{d-1}\hat{\varphi}(\|tu\|)t^{d+1}\, dt,$$

where u may be any unit vector in $\mathbb{R}^d$.

Proof: Let μ and ν be different integers and let $(y_j)_{j\in\mathbb{Z}^d}$ be a sequence with only two nonzero elements, namely $y_\mu = -y_\nu = 2^{-1/2}$. Choose any point $\zeta \in \mathbb{R}^d$ and set $x_\mu = 0$, $x_\nu = \zeta$, so that equation (4) provides the expression

$$\varphi(0) - \varphi(\|\zeta\|) = (2\pi)^{-d} \int_{\mathbb{R}^d} (1 - \cos(\zeta\xi))\, \hat{\varphi}(\|\xi\|)\, d\xi.$$

Employing spherical polar coordinates, this integral takes the form

$$\varphi(0) - \varphi(\|\zeta\|) = 2\pi)^{-d}\omega_{d-1} \int_0^\infty (1 - \Omega_d(t\|\zeta\|))\, \hat{\varphi}(\|tu\|)t^{d-1}\, dt,$$

where u may be any unit vector in $\mathbb{R}^d$. Setting $r = \|\zeta\|$, we have

$$\varphi(r) = \varphi(0) + \int_0^\infty (1 - \Omega_d(rt))\, \gamma(t)t^{-2}\, dt,$$

where $\gamma(t) = -(2\pi)^{-d}\omega_{d-1}\hat{\varphi}(\|tu\|)t^{d+1}$. Now Theorem 2.6 of [3] implies that $\hat{\varphi}$ is a nonpositive function. Thus there exists a nondecreasing function $\tilde{\beta}\colon [0,\infty) \to \mathbb{R}$ such that $\gamma(t)\, dt = d\tilde{\beta}(t)$, and $\int_1^\infty t^{-2}\, d\tilde{\beta}(t)$ is finite and $\tilde{\beta}(0) = 0$. But the uniqueness of the representation of Theorem 4.3 implies that $\beta = \tilde{\beta}$, that is

$$d\beta(t) = -(2\pi)^{-d}\omega_{d-1}\hat{\varphi}(\|tu\|)t^{d+1}\, dt,$$

and the proof is complete. ■

This proposition is useful if we want to calculate β for a particular function φ, since tables of generalized Fourier transforms are readily available.

Example 4.7. Let $\varphi(r) = (r^2+1)^{1/2}$. This is a non-negative CND1(d) function for all d (see [5]). When $d = 3$, the generalized Fourier transform is $\hat{\varphi}(r) = -4\pi r^{-2} K_2(r)$. Here K_2 is a modified Bessel function which is positive and smooth in $\mathbb{R}^+$, has a pole at the origin, and decays exponentially (See [1]). Consequently $\hat{\varphi}$ is a non-negative admissible function. Applying Theorem 4.7 gives the equation

$$\begin{aligned} d\beta(r) &= (2\pi)^{-3}(4\pi)r^4(4\pi r^{-2}K_2(r)) \\ &= (2r^2/\pi)K_2(r)dr, \end{aligned}$$

agreeing with [7], equation 3.12.

§5. The Least Upper Bound for Subsets of the Integer Grid

The following theorem is proved in [3] using extensions of the technique provided here.

Theorem 5.1. *Let $\varphi\colon [0,\infty) \to \mathbb{R}$ be an admissible function that is not identically zero, let $\varphi(0) \geq 0$, and let φ be CND1(d) for every positive integer d. Further, let $(x_j)_{j\in\mathbb{Z}^d}$ be any elements of $\mathbb{Z}^d$ and let $A = (\varphi(\|x_j - x_k\|))_{j,k \in kcalN}$, where $\mathcal{N}$ can be any finite subset of $\mathbb{Z}^d$. Then we have the inequality*

$$\|A^{-1}\| \leq \Big(\sum_{k\in\mathbb{Z}^d} |\hat{\varphi}(\|\pi e + 2\pi k\|)| \Big)^{-1},$$

where $e = [1,\ldots,1]^T \in \mathbb{R}^d$ and $\hat{\varphi}$ is the generalized Fourier transform of φ. Moreover, this is the least upper bound valid for all finite subsets of $\mathbb{Z}^d$.

Proof: See Section 4 of Baxter [3]. ■

References

1. Abramowitz, M., and I. A. Stegun, *Handbook of Mathematical Functions*, Dover, New York, 1970.
2. Ball, K. M., Invertibility of Euclidean distance matrices and radial basis interpolation, CAT report no. 201, Texas A & M University, College Station, 1989.
3. Baxter, B. J. C., Norm estimates and Toeplitz distance matrices, Report DAMTP 1990/NA16, University of Cambridge, 1991.
4. Jones, D. S., *The Theory of Generalised Functions*, Cambridge University Press, Cambridge, 1982.
5. Micchelli, C. A., Interpolation of scattered data: distance matrices and conditionally positive definite functions, Constr. Approx. **2** (1986), 11–22.
6. Narcowich, F. J. and J. D. Ward, Norm estimates for inverses of scattered data interpolation matrices associated with completely monotonic functions, preprint, 1990.

7. Narcowich, F. J. and J. D. Ward, Norms of inverses and condition numbers of matrices associated with scattered data, J. Approx. Th. **64** (1991), 69–94.
8. von Neumann, J. and I. J. Schoenberg, Fourier integrals and metric geometry, Trans. Amer. Math. Soc. **50** (1941), 226–251.
9. Sun, X., Norm estimates for Inverses of Euclidean Distance Matrices, preprint, 1990.

B. J. C. Baxter
DAMTP, University of Cambridge
Silver Street
Cambridge CB3 9EW, England
bjcb@damtp.cam.ac.uk

Numerical Treatment of Surface–Surface Intersection and Contouring

Karl–Heinz Brakhage

Abstract. Three Surface–Surface–Intersection (SSI) algorithms will be introduced. Surfaces may be given implicitly ($f(\mathbf{x}) = 0$) or parametically ($\mathbf{x} = \mathbf{x}(u, v)$). Thus we have to find the solution of $\mathbf{F}(\mathbf{y}) = \mathbf{0}$ with a function $\mathbf{F} : \mathbb{R}^{n+1} \to \mathbb{R}^n$ and $n \in \{1, 2, 3\}$. Contouring is a problem of the same type ($n = 1$). The first algorithm is for the special case $\mathbf{F} : \mathbb{R}^2 \to \mathbb{R}$ and the others are marching methods with different high order approximants for taking steps of variable length. They proceed in two stages. The first is to find an initial guess of the curve or a point on the curve, and the second is to refine the curve or to march along the curve respectively.

§1. Introduction

Surface–Surface–Intersection (SSI) is an important and often occurring problem in CAGD, CAD/CAM, geometric modeling, robotics, NC–milling etc. For an overview of the results up to 1984, see [10]. Some new results especially on marching methods can be found in [3] and [2]. The calculation should be robust, accurate and fast, which may be/is contradictory. Therefore it is necessary to develop SSI–Algorithms that allow the user to control certain tolerances to get a quick overview of the intersection as well as to compute the intersection very accurately.

Surfaces are assumed to be smooth and may be given parametically ($\mathbf{x} = \mathbf{x}(u, v)$) or implicitly ($f(\mathbf{x}) = 0$). (We denote vectors and vector–valued functions by bold letters.) Normally the parametric form is known. Three cases of intersections arise:

(i) implicit/parametric

$$F(u, v) := f(\mathbf{x}(u, v)) = 0$$
$$F : \mathbb{R}^2 \to \mathbb{R}$$

Mathematical Methods in Computer Aided Geometric Design II
Tom Lyche and Larry L. Schumaker (eds.), pp. 19–28.

ISBN 0-12-460510-9.

(ii) parametric/parametric

$$\mathbf{F}(u_1, v_1, u_2, v_2) := \mathbf{x}_1(u_1, v_1) - \mathbf{x}_2(u_2, v_2) = \mathbf{0}$$
$$\mathbf{F} : \mathrm{I\!R}^4 \to \mathrm{I\!R}^3$$

(iii) implicit/implicit

$$\mathbf{F}(\mathbf{x}) := \begin{pmatrix} f_1(\mathbf{x}) \\ f_2(\mathbf{x}) \end{pmatrix} = \mathbf{0}$$
$$\mathbf{F} : \mathrm{I\!R}^3 \to \mathrm{I\!R}^2.$$

All these cases result in

$$\mathbf{F}(\mathbf{y}) = \mathbf{0}$$

with a function

$$\mathbf{F} : \mathrm{I\!R}^{n+1} \to \mathrm{I\!R}^n \text{ and } n \in \{1, 2, 3\}.$$

Computing contours of a surface (not only horizontal cuts) and its profile can be handled as special cases of SSI. The cutting plane is given by:

$$\mathbf{x}(u, v) = \mathbf{x}_0 + u\,\mathbf{dx}_1 + v\,\mathbf{dx}_2$$

or

$$\mathbf{n}^T\mathbf{x} + d = 0$$

Thus we have two cases:

(i) parametric surfaces:

$$F(u, v) := \mathbf{n}^T\mathbf{x}(u, v) + d = 0$$
$$F : \mathrm{I\!R}^2 \to \mathrm{I\!R}$$

(ii) implicit surfaces:

$$F(u, v) := f(\mathbf{x}_0 + u\,\mathbf{dx}_1 + v\,\mathbf{dx}_2) = 0$$
$$F : \mathrm{I\!R}^2 \to \mathrm{I\!R}.$$

The profile of a surface under parallel or central projection is determined by the condition that the scalar product of the surface–normal and the projection ray is equal to zero. If the surface is given implicitly, we can use the *gradient* instead of the normal and $f(\mathbf{x}) = 0$ is the second equation ($\mathbf{F} : \mathrm{I\!R}^3 \to \mathrm{I\!R}^2$). In the parametric case we compute the normal as vector product of the partial derivatives of $\mathbf{x}(u, v)$ ($F : \mathrm{I\!R}^2 \to \mathrm{I\!R}$).

The intersection of two surfaces consists of isolated points, smooth curves, smooth patches, a combination of the previous situations, or is empty. Normally, it is a collection of smooth curves. Therefore, we will concentrate on this situation. The solutions are called regular in this case.

Most SSI algorithms proceed in two stages. The first is to find an initial guess of the curve or a point on the curve and the second is to refine the curve or to march along the curve, respectively. It might be necessary to sort the different parts of a curve.

Three SSI algorithms will be introduced. One for the (easy) special case $\mathbf{F} : \mathbb{R}^2 \to \mathbb{R}$, and two marching methods with different high order approximants (predictors) for taking steps of variable length. Here is a brief outline of the three algoritms:

• For the special case $n{=}1$ a grid is generated in the two dimensional domain. F will be evaluated at the grid points ($\to F_{ij}$) and an initial guess of the intersection curve is computed using only the values F_{ij}. Then this guess will be refined until a given error tolerance is reached. Methods of this kind are called *lattice evaluation* in [10].

• A regular solution curve $\mathbf{y} : \mathbb{R} \ni t \mapsto \mathbf{y}(t) \in \mathbb{R}^n$ fulfills Davidenko's differential equation. Thus, having a starting point $\mathbf{y}_0$, the marching is done via the numerical solution of an initial value problem (IVP) with (optional) Newton iteration for refinement and error control. For the IVP methods with error and step size control and dense output for high order interpolation between the evaluated curve points are known.

• The predictor is based on the the Taylor expansion of $\mathbf{y}(t)$ at $\mathbf{y}_0$ and needs the derivatives of $\mathbf{y}(t)$ at $\mathbf{y}_0$. The computation of the latter, which is done analytically in [2], appears to be rather cumbersome, but a valuable observation shows that only derivatives for special directions are needed. These are approximated numerically by differentiation formulas and extrapolation in [9]. Again a Newton iteration is used for refinement and error control.

All three SSI algorithms work with error control, they can be used to get a quick overview of the intersection as well as to compute the intersection very accurate. A lot of examples show that they are also robust.

§2. Lattice Evaluation (2D Grid–Methods)

Here we consider the (easy) case $F : \mathbb{R}^2 \to \mathbb{R}$. The problem we have to solve is

$$F(u,v) = 0 \quad , \quad \begin{cases} u_{min} \leq u \leq u_{max} \\ v_{min} \leq v \leq v_{max}. \end{cases}$$

For given N and M we define for $i = 0, 1, 2, \ldots, N$ and $j = 0, 1, 2, \ldots, M$ our initial grid and starting values:

$$\begin{aligned} u_i &:= u_{min} + i \cdot \Delta u \ , \ \Delta u := \frac{u_{max} - u_{min}}{N} \\ v_j &:= v_{min} + j \cdot \Delta v \ , \ \Delta v := \frac{v_{max} - v_{min}}{M} \\ F_{ij} &:= F(u_i, v_j). \end{aligned}$$

Using only this data, we compute an initial guess of the intersection curve by the following (simplified) algorithm: for $i = 0, 1, 2, \ldots, N-1$ and $j =$

$0, 1, 2, \ldots, M-1$:

$$
\begin{array}{l}
k := 0 \\
\text{if } \; F_{ij} \cdot F_{i+1,j} \leq 0 \\
\text{then } (u,v)_{ijk} := \left(u_i - \dfrac{F_{ij}}{\dfrac{F_{i+1,j} - F_{ij}}{\Delta u}}, v_j \right) \; ; k := k+1 \\
\text{if } \; F_{i+1,j} \cdot F_{i+1,j+1} \leq 0 \\
\text{then } \ldots \\
\text{if } \; F_{i+1,j+1} \cdot F_{i,j+1} \leq 0 \\
\text{then } \ldots \\
\text{if } \; F_{i,j+1} \cdot F_{ij} \leq 0 \\
\text{then } \ldots \\
\text{if } \; k = 2 \\
\text{then } \mathit{plot}\left((u,v)_{ij0}, (u,v)_{ij1}\right).
\end{array}
$$

If equality holds in one of the above tests, then the corresponding edge is dropped from the list of tests. The case $k = 4$ is treated below. The result of this first step for an example taken from [10] ($F(u,v) = (u-1)u(u+1)(v-1)v(v+1) + 0.05$) is shown in Fig. 1 for the parameter domain $[-2.8, 2.8] \times [-2, 2]$.

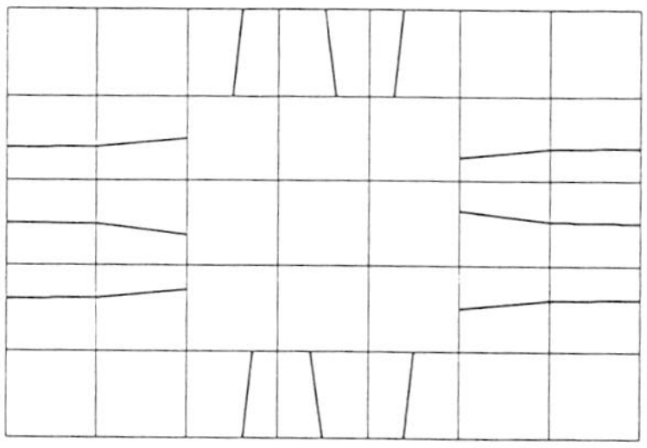

Fig. 1. Initial grid and first guess

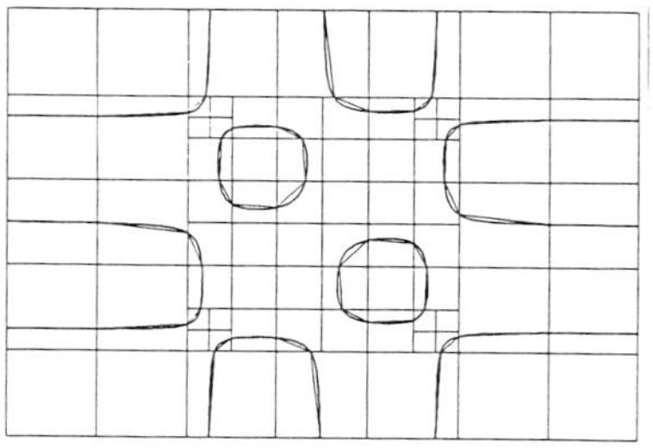

Fig. 2. Final grid and refinements

After this the curve is refined in two steps. We first devide the (rectangular) cells with $k = 4$ into four parts (recursive) until $k = 2$ or $k = 0$ or the diameter of the cell is smaller then the given error tolerance and calculate the (*exact*) zeros on the grid. The calculation is done with the method of *Anderson and Björk* (see [1]), which is a modification of the *regula falsi* (zero inclusion). It is shown in [1] that the efficency index of this method is ≈ 1.7, which is better than Newton's method ($\sqrt[2]{2}$).

In a second step we approximate the curve recursively with two secants instead of one until a given error tolerance is reached (compare Fig. 2). Let the secant be determined by $(u,v)_0, (u,v)_1$. Then (for example) we have to find the zero $\overline{t}$ of

$$g(t) := F\left(\frac{(u,v)_0 + (u,v)_1}{2} + t \cdot ((u,v)_1 - (u,v)_0)^{\perp}\right) =: F((u,v)(t)).$$

Since $g'(t) = (\operatorname{grad} F((u,v)(t)))^T ((u,v)_1 - (u,v)_0)^{\perp}$, this can be done by Newton's method, or by the following method, which doesn't use derivatives. Comparing $g(0)$ and $g(0.1)$ we get the direction of descent. Then for $k = 1, 2, \ldots$, we evaluate $g(2^k \cdot 0.1 \cdot t)$ or $g(-2^k \cdot 0.1 \cdot t)$ until we get an inclusion of the zero, or leave the actual cell. Now we can either use the method of *Anderson and Björk* again to compute this zero or have to refine the corresponding cells (compare the case $k = 4$ above). (An example for this case is given in Fig. 3 and Fig. 4 for $F(u,v) = v^2 + u^2(u+1) - 0.0005$. The parameter domain is $[-2,2] \times [-1.2,1.2]$.) The error can be estimated by $|(u,v)(\overline{t}) - (u,v)(0)|$. When computing $\mathbf{x}(u,v)$, it is a good idea to scale by an upper bound for the norm of the Jacobian of $\mathbf{x}(u,v)$ in the just mentioned error estimation.

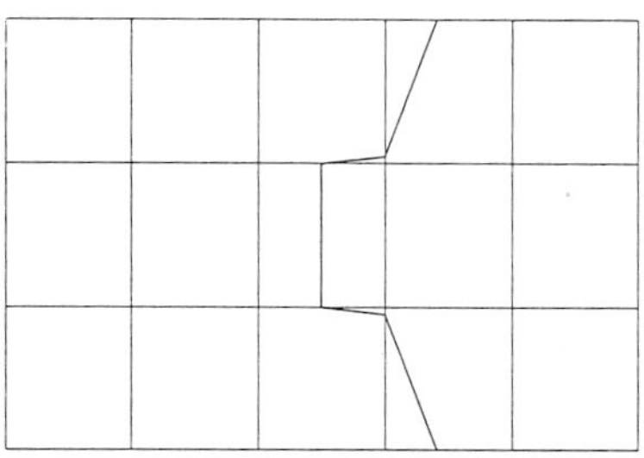

Fig. 3. Initial grid and first guess

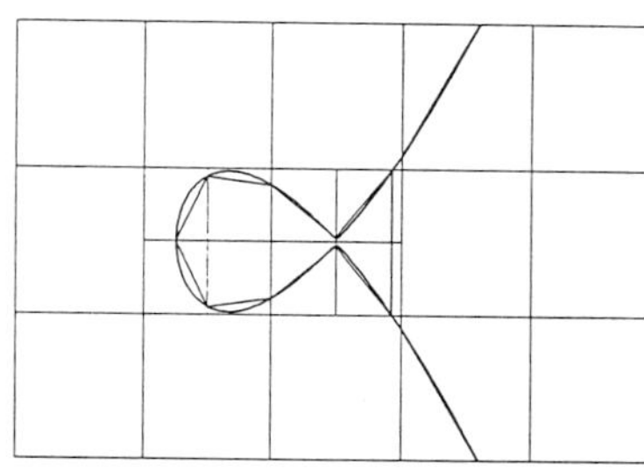

Fig. 4. Final grid and refinements

Remarks

- If a curve enters and leaves the cell by the same edge this will not be detected, but can be avoided by using a finer grid or quadratic interpolation for the initial guess.
- Small loops inside cells will be missed. Again this can be avoided by 2D–quadratic interpolation or local refinement (tangent tests).

§3. Marching Based on Davidenko's Differential Equation

Here we consider the general case

$$\mathbf{F} : \mathbb{R}^{n+1} \to \mathbb{R}^n.$$

Let $\mathbf{y}_0$ be a solution of

$$\mathbf{F}(\mathbf{y}) = \mathbf{0}.$$

A regular solution curve

$$\mathbf{y} : \mathbb{R} \ni t \mapsto \mathbf{y}(t) \in \mathbb{R}^n$$

fulfills Davidenko's differential equation ($\mathbf{T} = \mathbf{T}(y)$)

$$\mathbf{y}'(t) = \mathbf{T} \qquad \mathbf{y}(0) = \mathbf{y}_0,$$

where $\mathbf{T}$ is determined by

$$\begin{aligned} (D\,\mathbf{F}(\mathbf{y}))\,\mathbf{T} &= 0 \quad \text{(kernel property)} \\ \mathbf{T}^T\mathbf{T} &= 1 \quad \text{(quasi arc–length condition)} \\ \det\begin{pmatrix} (D\,\mathbf{F}(\mathbf{y})) \\ \mathbf{T}^T \end{pmatrix} &> 0 \quad \text{(choice of direction).} \end{aligned}$$

Thus, given the starting point $\mathbf{y}_0$, the marching is done via the numerical solution of an initial value problem (IVP) with (optional) Newton iteration for refinement and error control. For the IVP methods with error and step size control and dense output (high order interpolation between the evaluated curve points) are known. The best known *Runge–Kutta–Methods* (RKM) for tolerances about 10^{-8} are those of *Bogacki–Shampine* ([4]) and *Dormand–Prince* ([6], [7]). Both are embedded formula pairs of order 4 and 5 and use 7(6) function evaluations. A complete set of embedded formulas for order 1 to 5 was given by *Cash* and *Karp* ([5]). This method uses only 6 function evaluations and is nearly as good as that of *Dormand* and *Prince*. For more details on IVPs, see [8].

In the following, a brief description of *embedded* RKM's with step size control is given. Let

$$\begin{aligned} \mathbf{y}'(t) &= \mathbf{f}(t, \mathbf{y}(t)) \quad \text{and} \\ \mathbf{y}(a) &= \mathbf{y}_0 \end{aligned}$$

be given. For

$$a = t_0 \le t_1 \le t_2 \le \ldots \le t_n \le t_{n+1} \le \ldots$$

compute

$$\mathbf{y}_{n+1} \approx \mathbf{y}(t_{n+1})$$

with the formulas

$$\mathbf{k}_i = \mathbf{f}\left(t_n + c_i h_n, \mathbf{y}_n + h_n \sum_{j=1}^{i-1} a_{ij}\mathbf{k}_j \right), \ i = 1, 2, \ldots, s$$

$$\mathbf{y}_{n+1} = \mathbf{y}_n + h_n \sum_{i=1}^{s} b_i \mathbf{k}_i$$

and given constants c_i, b_i and a_{ij}.
Example: (Euler's method)

$$\begin{aligned} \mathbf{k}_1 &= \mathbf{f}(t_n, \mathbf{y}_n) \\ \mathbf{y}_{n+1}^E &= \mathbf{y}_n + h_n \mathbf{k}_1 \end{aligned}$$

Example: (Heun's method)

$$\begin{aligned} \mathbf{k}_1 &= \mathbf{f}(t_n, \mathbf{y}_n) \\ \mathbf{k}_2 &= \mathbf{f}(t_n + h_n, \mathbf{y}_n + h_n \mathbf{k}_1) \\ \mathbf{y}_{n+1}^H &= \mathbf{y}_n + 0.5 h_n (\mathbf{k}_1 + \mathbf{k}_2). \end{aligned}$$

Euler's method is of order one and Heun's method is of order two. Since both methods have different order and use the same $\mathbf{k}_i$ (it is an embedded formula pair), we can estimate the error without further function evaluations by

$$\varepsilon_{n+1} = \left|\mathbf{y}_{n+1}^H - \mathbf{y}_{n+1}^E\right|$$

and with

$$\varepsilon_{n+1} = c \cdot h_n^2 \quad \text{and} \quad h_n \approx h_{n+1}$$

and the desired tolerance ε we can compute the new step size

$$h_{n+1} = \alpha \sqrt[2]{\frac{\varepsilon}{\varepsilon_{n+1}}} h_n.$$

In the above formula, α is a security–factor ($\alpha \approx 0.8$) to ensure $\varepsilon_{n+2} \leq \varepsilon$. Steps are rejected and recomputed with a smaller h_n, if the estimated error is too large.

If the intersection curve has only regular points, then it can be traced only using the IVP solver. But Newton's method should be used for refinement and error control to be more efficent, especially in the case of curves with singularities. Refining $\mathbf{y}_{n+1}$ with Newton's method, for instance, means solving

$$\begin{aligned} \mathbf{F}(\overline{\mathbf{y}}_{n+1}) &= \mathbf{0} \\ \mathbf{T}^T(\mathbf{y}_{n+1})\,(\overline{\mathbf{y}}_{n+1} - \mathbf{y}_{n+1}) &= 0. \end{aligned}$$

These are $n+1$ equations for the $n+1$ unknowns in $\overline{\mathbf{y}}_{n+1}$. As starting value we choose $\mathbf{y}_{n+1}$. For low order predictors we can use the $\mathbf{k}_i$'s with c_i near 1 instead of $\mathbf{T}(\mathbf{y}_{n+1})$. Now the error can be estimated by

$$\overline{\varepsilon}_{n+1} = |\mathbf{y}_{n+1} - \overline{\mathbf{y}}_{n+1}|,$$

and we can use $\overline{\varepsilon}_{n+1}$ instead of ε_{n+1} to compute h_{n+1}. This way we have computed the *local truncation error* of the RKM, which is of order $p+1$ if the order of the RKM is p. This has to be taken into account when we compute the new step size.

Crossing a node

$$\det \begin{pmatrix} (D\,\mathbf{F}(\mathbf{y})) \\ \mathbf{T}^T \end{pmatrix}$$

changes the sign. Thus we have to take care of the sign of $\mathbf{T}$. Even more importantly, we have to treat nodes with extreme care, because at a node, $\operatorname{rank}(D\,\mathbf{F}(\mathbf{y})) = n-1$.

In *easy* cases the computation of $\mathbf{T}$ can be done as follows:

$\underline{n=1:}$

$$F(u,v) = 0$$

$$\mathbf{T}(u,v) = \pm\frac{(\operatorname{grad}F(u,v))^{\perp}}{|\operatorname{grad}F(u,v)|}$$

$\underline{n=2:}$

$$\begin{pmatrix} f_1(x,y,z) \\ f_2(x,y,z) \end{pmatrix} = \mathbf{0}$$

$$\mathbf{T}(x,y,z) = \pm\frac{\operatorname{grad}f_1(x,y,z)\times\operatorname{grad}f_2(x,y,z)}{|\operatorname{grad}f_1(x,y,z)\times\operatorname{grad}f_2(x,y,z)|}$$

$\underline{n=3:}$

$$\mathbf{x}_1(u_1,v_1) - \mathbf{x}_2(u_2,v_2) = 0$$

$$\left(\tfrac{\partial}{\partial u_1}\mathbf{x}_1 \times \tfrac{\partial}{\partial v_1}\mathbf{x}_1\right) \times \left(\tfrac{\partial}{\partial u_2}\mathbf{x}_2 \times \tfrac{\partial}{\partial v_2}\mathbf{x}_2\right) =: \overline{\mathbf{T}}$$

$$\left(\tfrac{\partial}{\partial u_1}\mathbf{x}_1 \quad \tfrac{\partial}{\partial v_1}\mathbf{x}_1\right)\begin{pmatrix} du_1 \\ dv_1 \end{pmatrix} = \overline{\mathbf{T}}$$

$$\left(\tfrac{\partial}{\partial u_2}\mathbf{x}_2 \quad \tfrac{\partial}{\partial v_2}\mathbf{x}_2\right)\begin{pmatrix} du_2 \\ dv_2 \end{pmatrix} = \overline{\mathbf{T}}$$

$$\mathbf{T} = (du_1, dv_1, du_2, dv_2)^T.$$

In the above formulas, $\mathbf{T}$ or $\overline{\mathbf{T}}$ can be normalized. The arguments of the partial derivatives are omitted.

In critical cases (for instance near singular points) we have to use the *QR*– or *Singular–Value–Decomposition* for the computation of the kernel element $\mathbf{T}$.

§4. Marching Based on Taylor Expansion

Again we consider

$$\mathbf{F}(\mathbf{y}(t)) = \mathbf{0}.$$

We now approximate the solution curve by

$$\mathbf{y}(t) \approx \mathbf{y}(0) + t\,\mathbf{y}'(0) + \frac{t^2}{2}\mathbf{y}''(0) + \frac{t^3}{6}\mathbf{y}^{(3)}(0) + \dots$$

The Taylor expansion of $\mathbf{F}(\mathbf{y}(t))$ is given by

$$\mathbf{F}(\mathbf{y}(t)) = \mathbf{F}(\mathbf{y}(0)) + t\left(D\,\mathbf{F}(0)\,\mathbf{y}^{(1)}(0)\right) + \frac{t^2}{2}\left(D\,\mathbf{F}(0)\,\mathbf{y}^{(2)}(0) + D^2\mathbf{F}(0)\,(\mathbf{y}^{(1)}(0))^2\right) + \cdots$$

To get an approximation of order 2, 3, 4, ... the terms in the first 1, 2, 3, ... brackets have to vanish. Therefore we get the determining equations

$$\begin{aligned} D\,\mathbf{F}(0)\,\mathbf{y}^{(1)}(0) &= \mathbf{0} \\ D\,\mathbf{F}(0)\,\mathbf{y}^{(2)}(0) &= -D^2\mathbf{F}(0)\,(\mathbf{y}^{(1)}(0))^2 =: \mathbf{R}_2 \\ D\,\mathbf{F}(0)\,\mathbf{y}^{(3)}(0) &= -3\,D^2\mathbf{F}(0)\,\mathbf{y}^{(1)}(0)\,\mathbf{y} - D^3\mathbf{F}(0)\,(\mathbf{y}^{(1)}(0))^3 =: \mathbf{R}_3 \\ D\,\mathbf{F}(0)\,\mathbf{y}^{(4)}(0) &= \cdots \end{aligned}$$

In [2] the right–hand sides $\mathbf{R}_2$ and $\mathbf{R}_3$ are computed explicitly with the partial derivatives (for n=2 these are totally 38 partial derivatives). *Mackens* ([9]) avoided this by the observation

$$\left(\frac{d}{ds}\right)^k \mathbf{F}\left(\mathbf{y}_0 + \sum_{i=1}^{k-1} \mathbf{y}^{(i)}(0)\frac{s^i}{i!}\right)\Bigg|_{s=0} = -\mathbf{R}_k,$$

and numerical differentiation of

$$\mathbf{F}_{k-1}(t) := \mathbf{F}\left(\mathbf{y}_0 + \sum_{i=1}^{k-1} \mathbf{y}^{(i)}(0)\frac{s^i}{i!}\right)$$

done by extrapolation methods.

Since the linear systems are underdetermined, one can choose a further condition, for example to get a parameterization with respect to arc–length.

The above technique is used to get a predictor. Then Newton's method is used for refinement, error control and step size selection as described in section 3. We only need the decomposition of $D\,\mathbf{F}(0)$ for the computation of $\mathbf{y}^{(i)}(0)$.

§5. Conclusion and Future Extensions

The methods introduced above have been implemented and tested for a number of examples. The methods with high order predictors are very robust and are efficent if the predictor has to be of small tolerance to get the correct solution curve (See Fig. 5). For *easy* cases the Euler predictor should be used for efficiency. This is due to the observation that in this case even a bad prediction will be traced to the solution curve with a few Newton steps, and for instance the Euler predictor and one Newton step with $\mathbf{T} = \mathbf{k}_1$ will produce an approximation of order four by solving two linear systems (one for the Euler step, and one for the Newton step). By a RKM step an approximation of order four can only be achieved by solving at least three linear systems. For this reason it is worth thinking of multistep and multistep multiderivative methods. This will be done by the author in a future work.

Fig. 6 shows a tangent branch–point. It is detected by the code but not described in this paper.

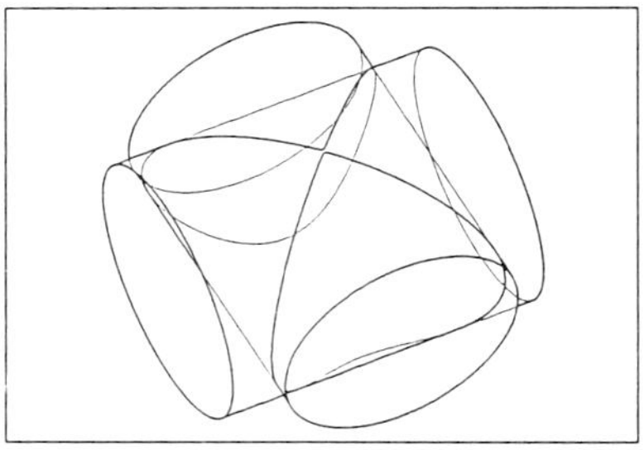

Fig. 5. Nearly a branch-point

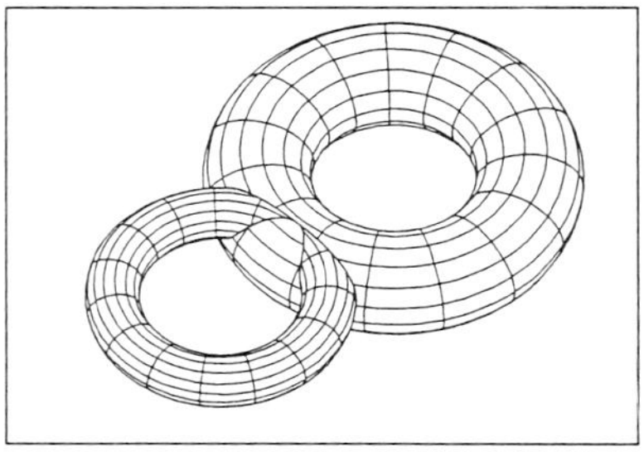

Fig. 6. Tangent branch-point

References

1. Anderson, N., and Å. Björk, A new high order method of Regula Falsi type for computing the root of an equation, BIT **13** (1973), 253–264.
2. Bajaj C. L., C. M. Hoffmann, R. E. Lynch, and J. E. H. Hopcraft, Tracing surface intersections, in Computer Aided Geom. Design **5** (1988), 285–307.
3. Barnhill, R. E., and S. E. Kersey, A marching method for parametric surface/surface intersection, Computer Aided Geom. Design **7** (1990), 257–280.
4. Bogacki, P., and L. F. Shampine, An efficent Runge–Kutta (4,5) pair, Math. Department Southern Methodist University, 1990.
5. Cash, J. R., and A. H. Karp, A variable order Runge–Kutta method for initial value problems with rapidly varying right–hand side, in ACM Trans. Math. Software **16** (1990), 201–222.
6. Dormand, J. R., and P. J. Prince, A family of embedded Runge–Kutta formulas, J. Comp. Appl. Math. **6** (1980), 19–26.
7. Dormand, J. R., and P. J. Prince, Runge–Kutta triples, in Comp. & Maths. with Appls. **12A** (1986), 1007–1017.
8. Hairer, E., S. P. Nørsett, and G. Wanner, *Solving Ordinary Differential Equations I (Nonstiff Problems)*, Springer Verlag, Springer Series in Computational Mathematics 8, 1987.
9. Mackens, W., Numerical differentiation of implicitly defined space Curves, Computing **41** (1989), 232–260.
10. Pratt, M. J., and A.D. Geisow, Surface/surface intersection problems, in *The Mathematics of Surfaces*, J. A. Gregory (ed.), Oxford Science Publications, New Series Number 6, 1986, 117–142.

Modeling Closed Surfaces: A Comparison of Existing Methods

P. Brunet and A. Vinacua

Abstract. As C^1 overall continuity is not possible in closed surfaces, modeling schemes are closely related to geometric continuity. Algorithms that generate surfaces from a set of polyhedron vertices include recursive subdivision methods, piecewise algebraic interpolation, discrete interpolation or transfinite interpolation of a closed mesh of curves. On the other hand, boolean operations between simple surfaces lead to trimmed surfaces. Volumes can also be modeled by approximate models based on the subdivision of the space. The present paper includes a comparative study of some of these algorithms discussing shape control and fairness of the final surface, geometric continuity, space complexity of the model and their performance in basic solid interrogations.

§1. Introduction

Geometric modeling has to provide efficient, interactive and powerful tools for the representation and manipulation of three-dimensional objects. However, so far there is no unified framework for modeling 3D shapes. In fact, surface models can describe either open surfaces or closed sets of patches that enclose a finite volume. Solid models, on the other hand, represent closed regions of the space in an unambiguous way and support classification tests (inside-on-outside) of points against objects. Both models have evolved independently [27] and whereas surface models are able to deal with sculptured surfaces of complex shapes, volume models are usually restricted to objects limited by simple planar or quadric surfaces.

In this context, several authors have addressed the problem of creating general solid models with non restricted surfaces, see [6]. This paper presents different existing techniques for the modeling of non-restricted closed surfaces, and discusses them in terms of the surface continuity and geometric interrogations and operation.

Mathematical Methods in Computer Aided Geometric Design II
Tom Lyche and Larry L. Schumaker (eds.), pp. 29–42.

ISBN 0-12-460510-9.

Volume models can either store the explicit geometry of the closed boundary surface or volume, or use a constructive approach via a tree of elementary solid operations that act on solid primitives and define the final solid, as in the CSG tree models. In the rest of the paper, we will concentrate on the first class of models. We can divide them into the following classes, depending on the complexity of the geometric elements used,

1.- Closed triangulation of points. A set of points in space plus a closed triangulation define a polyhedron of triangular faces enclosing a finite volume.
2.- Closed triangulation of points with normals or tangential data. This model is obtained from the last one through an estimation of tangent planes at the vertices.
3.- General 3D polyhedra.
4.- Grid of curves, with the appropriate topology.
5.- Piecewise parametric closed surfaces, either polynomial or rational, using rectangular, triangular, trimmed or general patches. Trimmed surfaces appear as a consequence of surface intersections and set operations [8], [30] (see Section 7.3).
6.- Piecewise algebraic surfaces
7.- Hierarchical space subdivision schemes like octrees.

Most of the methods for representing closed volumes start from simple schemes of the above list and generate more sophisticated representation schemes of the same list. In particular, subdivision methods refine either polyhedra (3) or triangulations of points (1) while giving more smooth polyhedra of the same kind. Discrete interpolation can generate piecewise parametric closed surfaces (5) from simple polyhedron models (2,3). While several algebraic interpolation methods start from (2) and obtain piecewise algebraic surfaces, transfinite interpolation schemes obtain piecewise parametric surfaces (5) from a grid of curves model (4). Surfaces obtained by any of the preceding algorithms can be used in turn for the generation of more complex surfaces through specific shape operations like set operations or deformations (see for example [32], [12]). Finally, octrees usually start from closed surface models (5,6) and generate a hierarchical volume representation of the internal volume. The following sections survey some of the most significative methods in each family, and finally compare their performances in modeling and operation of closed solid volumes.

§2. Subdivision Schemes

Subdivision schemes derive from the work of Chaikin [10] who proposed an algorithm for the generation of a curve from a polygon, by successively cutting out its corners. The resulting curve is the quadratic B-spline provided that every edge is cut at 25 and 75 percent of its length. Similar algorithms in 3D work by successively refining a polyhedron in a way that approaches a limit surface. If the polyhedron is composed by four sided faces although a general number of edges per vertex is permitted, the Catmull and Clark approach [9] works by dividing every face onto four subfaces at each subdivision step. On the other hand, the Doo and Sabin [14] approach applies to n-sided topologies and is based on a shrink of every polyhedron face. General n-edged vertices in the initial polyhedron are converted to n-sided faces after the first subdivision step which shrink in successive steps and tend to their centroid in the limit surface. Except for a finite number of so-called extraordinary points [14] which are singular, the surface is representable as a quadratic B-spline. Doo and Sabin show that the limit surface presents tangent plane continuity, and that the second order continuity behavior of the surface in the vicinity of the extraordinary points is regular and independent of n. A general discussion on the continuity properties of the limit surface of subdivision algorithms can be found in [21].

The biquadratic subdivision algorithm of Doo and Sabin has been extended to the generation of interpolating limit surfaces [22]. On the other hand, it is possible to incorporate tension parameters (shape factors) to every vertex of the initial polyhedron [2] in order to control the shape of the limit surface and maximize GC^2 continuity.

A different approach has been proposed by Dyn, Levin and Gregory [15]. A butterfly scheme is used for the subdivision of a general closed triangulation of the control points. Although the algorithm must start from a particular polyhedron made up of triangular faces, it generates a uniform GC^1 limit surface, without extraordinary points.

A simple analysis of the two last schemes, [2] and [15], brings us the following conclusions in addition to the requirements for the initial polyhedron and the uniformity of the limit surface:

- The extended Doo and Sabin approach [2] works by anisotropically shrinking every face of the polyhedron towards its centroid, while the butterfly scheme [15] subdivides every triangle into four by halving every edge of the triangulation at every subdivision step.
- The extended Doo and Sabin approach generates GC^1 surfaces with a finite number of singular extraordinary points. On the other hand, the surface is also GC^1 in the butterfly scheme if all vertices have degree greater than three. The limit surface in the extended Doo and Sabin scheme is a quadratic B-spline except at the singular extraordinary points.
- Both the extended Doo and Sabin approach and the butterfly scheme use scalar tension parameters in the initial vertices, for shape control of the limit surface.
- Both schemes have no known simple solution for the point classification problem — point-in-the-solid test — based on the initial data.
- Both schemes produce a similar rate of increase of the model size during the subdivision process. The increase on the complexity of the polyhedron in the extended Doo and Sabin approach is given by the equations $f_{k+1} = f_k + e_k + v_k$, $e_{k+1} = 4 * e_k$, $v_{k+1} = 4 * e_k$ where f_k, e_k and v_k are the number of faces, edges and vertices in the k-subdivision step. The eigenvalue of the subdivision transformation is 4, while the limit polyhedron has double the number of edges than the number of faces and vertices. On the other

hand, the butterfly scheme obviously applies a factor of 4 to the number of triangles at every subdivision step, $t_{k+1} = 4 * t_k$.

§3. Discrete Interpolation

Discrete interpolation methods start from a polyhedron and obtain a closed set of interpolating patches meeting at the polyhedron vertices and having the same topology of its faces. Among all proposed algorithms, only few of them are presented and briefly discussed in this section. Van Wijk's approach [33], for example, is restricted to a limited set of polyhedra and generates a surface made of rectangular patches. Nielson [24] starts from a triangulation of a 3D set of points plus normal vectors and uses a Hermite interpolation scheme; on the other hand, Piper [26] proposes a Clough Tocher scheme for the same structure of data points. Recent work includes papers from Loop and De Rose [19] and Seidel (in this volume). Both approaches support general topologies; while the first one is based on a subdivision net and uses s-patches in non-rectangular corners, Seidel's approach uses a general B-spline set of 2D basis functions on the data net.

Discrete interpolation schemes must face corner compatibility problems for patches meeting at common vertices, as will be discussed in Section 7.1.

§4. Algebraic Interpolation

Algebraic interpolation methods describe the closed surface by a set of polynomial implicit patches $f(x, y, z) = 0$. The main advantage of the representation is the simplicity of most of the solid volume interrogations, namely classification of points with respect to the solid and surface-straight line intersections, see Section 7.

Sederberg [31] proposed a scheme for the shape control of an algebraic patch restricted to a tetrahedron (simplex), by using a Bézier representation in a higher dimension. One of the extensions of Sederberg's work which is suitable for modeling solids with free form boundaries has been proposed by Dahmen, [13]. Dahmen's methods starts from a triangulation of the data points which have also associated a tangent plane (or normal). The topology of the triangulation is in some sense restricted, as it must support a set of lines called a transversal system. The transversal system consists of a line associated to every triangle, and compatibility constraints must be fulfilled between lines of neighbor triangles, [13]. The algorithm then constructs an interpolating surface built from pieces of quadrics contained in tetrahedra associated to the data triangles and stitched together with GC^1 continuity.

Dahmen's scheme generates 12 algebraic patches per triangle in the original triangulation. Geometric interrogations can be solved by two step procedures, first finding the tetrahedron which contains the point or intersects the line, and then interrogating the corresponding implicit patch. The location of the tetrahedron vertex in every line of the transversal system gives one shape handle per triangle, although the effect and shape control of these parameters has not been investigated. On the other hand, the computation of the transversal system and the topology of the final mesh of tetrahedra associated to the triangles pose certain constraints to the initial topology and tangent planes of the data which are difficult to foresee.

§5. Transfinite Interpolation

The problem of constructing a continuous surface which interpolates —filling the holes— a grid of curves appears in many applications where data are digitized along lines in model surfaces [29]. Starting from a closed grid, transfinite interpolation leads to volume models of the enclosed solid volume. Standard schemes like Coons approach require grids with strong topology limitations. In any case, the main problem of transfinite interpolation is that the user is either faced with non-rectangular regions in the grid or with vertices where a number of edges other than four meet.

Transfinite methods based on polynomial patches pose significant limitations in the initial grid or the final patches. As an example, Sarraga's approach [29] is only derived for topologies with 3,4 or 5 patches meeting at vertices and generates a surface composed of 3×3 and 6×6 Bézier patches. Jones [18] requires corner compatibility conditions, while the more general approach [25] needs grid curves meeting at any vertex compatible with a second fundamental form. On the other hand, transfinite schemes based on rational patches [11], [24] avoid these limitations at the cost of increasing the complexity of the patch representation. Both advantages and drawbacks of these transfinite algorithms will be discussed in Section 7.1.

§6. Octree and Face Octree Representations

An octree is a tree that represents solid objects through the recursive subdivision of a finite cubic universe. In this structure, each tree node is terminal or has eight descendants. The tree divides the space of the universe into cubes of different sizes. The root of the tree represents the universe, a cube with 2^N edge length. This cube is divided into eight identical cubes, called octants, with an edge length of 2^{N-1}. Each octant is represented by one of the eight descendants of the root. If an octant contains too complex a part of the solid (Grey node), it is divided into another eight identical cubes which are represented as descendants of the octant in question. The previous process is repeated recursively until valid terminal nodes are obtained, [20]; the process also stops when octants of a minimum edge length called resolution (usually $2^0 = 1$) appear. Consequently, a specific octree representation is characterized by the definition of the types of terminal nodes which are allowed. The more extensive the set of geometric cases considered in the definition of the allowed terminal nodes, the less the total number of nodes in the octree representation of a particular object.

In the simplest octree representation, called *classical octrees* (CO), only homogeneous Black and White terminal nodes are allowed. Nodes with associated cubic octants which are completely inside the object are coded as Black nodes, whereas those completely outside it are called White nodes, [20].

Octree models that incorporate new terminal nodes containing parts of the object surface can also be derived. They are sometimes called *vector octrees*, [28]. Different models can be considered, depending on the specific node types used in the representation: Face, Edge and Vertex nodes [23] can be used in addition to the classical White and Black terminal nodes. Face nodes are crossed only by a single planar face (or planar approximation) of the solid, whereas Edge nodes contain two neighboring faces and a part of their common edge, and Vertex nodes contain one vertex of the polyhedron and part of the faces and edges converging to it. Extended octrees [1], [3] or polytree representations [7] are obtained when all face, edge and vertex nodes are incorporated to the classical octree node types. On the other hand, Face octrees are octrees in which the set of terminal nodes includes White, Black and Face nodes [4]. The following definition makes precise the concept of face octrees:

Definition. *A face octree is an octree which allows White, Black, Grey and Face nodes, together with a tolerance ε. White, black and grey nodes are defined as in classical octrees. Face nodes contain a connected part of the boundary of the object, and have associated with them the equation of a plane π_C such that $dist(P, \pi_C) \leq \varepsilon$ for every point P of the object surface S in the cube C associated to the face node, and $dist(Q, S) \leq \varepsilon$ for every point $Q \in \pi_C \cap C$, Figure 1.*

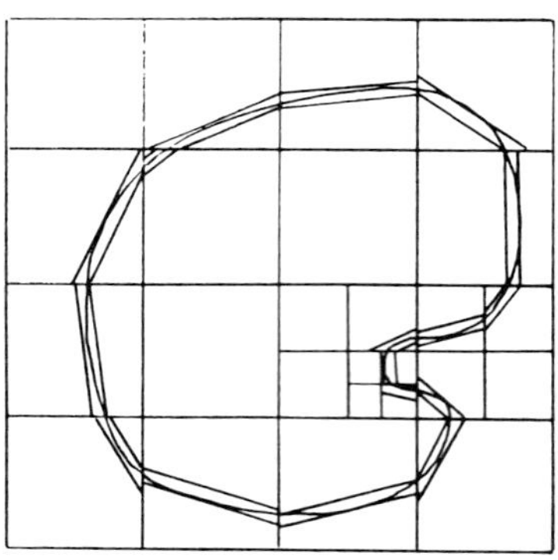

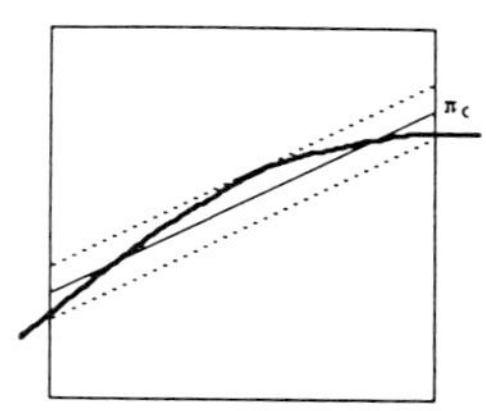

Figure 1. Example of a Face Octree (in 2D) and definition of face nodes.

The part of the surface of the object within a face node must be sufficiently planar. The value ε associated to the face octree controls the degree of approximation of the representation together with the depth of the tree. On the other hand, the space subdivision depends on the local curvature of the surface S while, in G^1 surfaces, it is independent of the patch boundaries existing in the initial boundary representation of S.

Every face node in the face octree representation of a closed free form surface can be interpreted as a band which spans a distance equal to the tolerance at both sides of the plane associated to the node, and which is limited by two planes parallel to it. Because of the definition of a face node, the exact surface must lie in the band. Therefore, the whole boundary of the object is contained in the region defined by the set of bands of the face nodes in the octree [4], Figure 1. Bands of neighbor nodes must overlap at the common boundary, although not completely. The union of all face node bands defines a 'thick surface' with a width of twice the tolerance, which contains the true surface of the object. The thick surface of S is a connected region that can be used as a bound for S in geometric tests, and supply robust region bounds for set operations, [4]. On the other hand, face octrees can be considered as an auxiliary data structure, and stored together with the surface model. The composite model $\{S, FO_\varepsilon(S)\}$ is specially well suited for fast geometric interrogations and volume computations, being that the face octree contains a hierarchical, explicit volume representation of the inside region enclosed by S. They can be refined through further clipping of the surface information in face nodes.

§7. Discussion of These Modeling Schemes

All the methods presented have different strengths and weaknesses. In each case, by present day technology, one has to choose the "right" scheme according to the relevance of its strengths and weaknesses to the application at hand. In this section we would like to present briefly these differences, and explore their interactions, defining in the passing the space that we feel the new model presented in Section 6 can fill. In the next three subsections, we do this by focusing on three central issues: corner compatibility problems, geometric interrogations and set operations.

7.1. Corner Compatibility

Transfinite interpolation with polynomial patches hits an intrinsic problem when one wants to model closed surfaces, namely that one must eventually face corner compatibility conditions. This problem is discussed, for instance, in [34]. It reduces to the fact that, when walking around a vertex, the *twists* of k confluent patches have to be solutions of a linear system

$$\mathbf{A}\Big[D_{12}S_i\Big] = V,$$

where V depends on the data, and $\det(\mathbf{A}) = 1 - (-1)^k$. Thus, the system is singular whenever k is even. If it happens to be compatible, the solution will not be unique. Even more, the solution actually may not exist. Peters [25] has shown that except in a special, easier, case (for n=4), it is sufficient that the boundary curves be compatible with a second fundamental form at the corner for the system to be compatible. This is a strong and somewhat surprising result. The compatibility condition in order to achieve first order visual continuity involves second order information!

Further, even for discrete interpolation, other restrictions are encountered if one attempts to compute the patches one at a time. Suppose we are given a patch (determined in a previous step of the algorithm) and want to connect it with GC^1 continuity to a new neighboring polynomial patch (to be determined) that meets with it along a common boundary curve. For example, consider the Bézier patches depicted in Figure 2. If we think of the left-hand patch as given, and we are trying to paste a new patch on the right hand with geometric continuity, the extreme points R_0 and R_n should be on the planes determined by P_0, Q_0 and P_1 and by P_{n-1}, P_n and Q_n respectively. However, once one of them is fixed, the other is constrained to a line on the corresponding plane, and this limitation is intrinsic to the fact that the patches have the same degree. This can be seen by a simple degrees of freedom-counting argument, as in [5]. There, an algorithm to uncouple these endpoints is presented. As more degrees of freedom are needed, the algorithm proposed operates by perturbing the midpoint Q_k, $k = \lfloor (n/2) \rfloor$ of the original patch. The approach taken starts by considering quadratic coefficient functions

$$\begin{aligned} \alpha(v) &= \alpha_0 B_0^1(v) + \alpha_1 B_1^1(v) \\ \beta(v) &= \beta_0 B_0^1(v) + \beta_1 B_1^1(v) \\ \gamma(v) &= \gamma_0 B_0^2(v) + \gamma_1 B_1^2(v) + \gamma_2 B_2^2(v). \end{aligned} \tag{1}$$

If these functions denote the coefficient functions in the linear dependence of the tangential and cross-boundary derivatives in both patches (using a GC^1 condition like that in [16]), grouping terms of the same degree yields the following continuity conditions:

$$\begin{aligned} &\alpha_0 \frac{n+1-j}{n+1} R_j + \beta_0 \frac{n+1-j}{n+1} Q_j + \alpha_1 \frac{j}{n+1} R_{j-1} + \beta_1 \frac{j}{n+1} Q_{j-1} - \\ &\quad - \left(\alpha_0 \frac{n+1-j}{n+1} + \beta_0 \frac{n+1-j}{n+1} + \gamma_0 \frac{(n+1-j)(n-j)}{n(n+1)} - \right. \\ &\quad\quad\quad \left. -2\gamma_1 \frac{(n+1-j)j}{n(n+1)} \right) P_j - \\ &\quad - \left(\alpha_1 \frac{j}{n+1} + \beta_1 \frac{j}{n+1} + 2\gamma_1 \frac{(n+1-j)j}{n(n+1)} - \gamma_2 \frac{(j-1)j}{n(n+1)} \right) P_{j-1} + \\ &\quad + \gamma_0 \frac{(n+1-j)(n-j)}{n(n+1)} P_{j+1} - \gamma_2 \frac{(j-1)j}{n(n+1)} P_{j-2} = 0 \end{aligned} \tag{2}$$

where $j = 0 \dots n+1$. Let us assume that the degree is $n \geq 4$. Then, equation (2) can first be used to obtain, for $j = 0$, the values α_0, β_0, γ_0 in terms of R_0, P_0, Q_0, P_1:

$$\alpha_0 R_0 + \beta_0 Q_0 - (\alpha_0 + \beta_0 + \gamma_0) P_0 + \gamma_0 P_1 = 0 \tag{3}$$

and, in the same way, α_1, β_1, γ_2 can be obtained in terms of R_n, P_n, Q_n, P_{n-1} by using (2) in the case $j = n+1$:

$$\alpha_1 R_n + \beta_1 Q_n - (\alpha_1 + \beta_1 - \gamma_2) P_n - \gamma_2 P_{n-1} = 0. \tag{4}$$

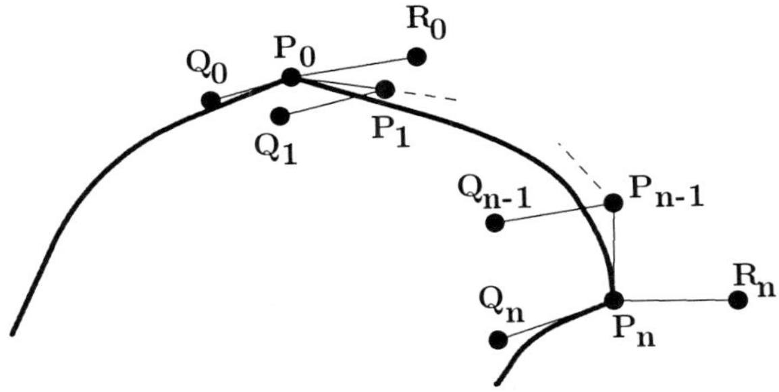

Figure 2. A GC^1 joint.

Setting $j = 1$ and $j = n$ in (2), one obtains expressions for R_1 and R_{n-1} in terms of the data points and the remaining unknown γ_1,

$$\begin{aligned}\alpha_0 \frac{n}{n+1} R_1 = & -\beta_0 \frac{n}{n+1} Q_1 - \alpha_1 \frac{1}{n+1} R_0 - \beta_1 \frac{1}{n+1} Q_0 - \gamma_0 \frac{n(n-1)}{n(n+1)} P_2 + \\ & + \left(\alpha_0 \frac{n}{n+1} + \beta_0 \frac{n}{n+1} + \gamma_0 \frac{n(n-1)}{n(n+1)} - 2\gamma_1 \frac{n}{n(n+1)}\right) P_1 + \\ & + \left(\alpha_1 \frac{1}{n+1} + \beta_1 \frac{1}{n+1} + 2\gamma_1 \frac{n}{n(n+1)}\right) P_0\end{aligned} \tag{5}$$

$$\begin{aligned}\alpha_1 \frac{n}{n+1} R_{n-1} = & -\alpha_0 \frac{1}{n+1} R_n - \beta_0 \frac{1}{n+1} Q_n - \beta_1 \frac{n}{n+1} Q_{n-1} + \\ & + \gamma_2 \frac{(n-1)n}{n(n+1)} P_{n-2} + \\ & + \left(\alpha_0 \frac{1}{n+1} + \beta_0 \frac{1}{n+1} - 2\gamma_1 \frac{1}{(n+1)}\right) P_n + \\ & + \left(\alpha_1 \frac{n}{n+1} + \beta_1 \frac{n}{n+1} + 2\gamma_1 \frac{(n+1-j)j}{n(n+1)} - \right. \\ & \left. - \gamma_2 \frac{(n-1)n}{n(n+1)}\right) P_{n-1}.\end{aligned} \tag{6}$$

In this way, one can proceed from both ends generating two sets of equations. The first one, which corresponds to equation (2) for $j = 2...k$, expresses R_j in terms of R_{j-1}, the data points and γ_1. The second one particularizes equation (2) for $j = n-1...k+1$, and expresses R_{j-1} in terms of R_j, the data points and also γ_1. Both are linear, and start from the values of R_1 and R_{n-1} computed from equations (5), (6). In their last step, however, both recurrences compute the value of the point R_k. As these two values must obviously coincide, both expressions produce a compatibility condition, that must be fulfilled:

$$\omega_k^1 \beta_0 R_k + \omega_k^1 \alpha_0 Q_k = S_0 + \gamma_1 S_1$$

$$\omega_k^0 \beta_1 R_k + \omega_k^0 \alpha_1 Q_k = S_2 + \gamma_1 S_3, \tag{7}$$

where, $\omega_k^0 = \frac{k}{n+1}$ and $\omega_k^1 = \frac{n+1-k}{n+1}$, and the points S_i, $i = 0, 1, 2, 3$ depend only on the data. All the needed coefficients are obtained from equation (3) and (4).

Observe that (7) is a set of 6 linear equations in 7 scalar unknowns (the value of γ_1, and the points R_k, Q_k). As a consequence, the set of feasible locations of Q_k that produce compatibility is a straight line in $\mathbb{R}^3$. It must be observed that the use of cubic coefficient functions instead of quadratic ones in equation (1) does not improve the proposed algorithm; it still yields only one extra degree of freedom as in the scheme proposed here.

Equation (7) can be solved for R_k and Q_k if a value for γ_1 is assumed. In this case, the system is singular if and only if,

$$det(A) = 0 = \omega_k^{0^3} \omega_k^{1^3} (\beta_0 \alpha_1 - \beta_1 \alpha_0)^3. \tag{8}$$

Two different cases arise. If $\frac{\alpha_0}{\beta_0} \neq \frac{\alpha_1}{\beta_1}$, a unique solution exists for every value of γ_1 and both patches can be connected without coupling the endpoints with the proposed algorithm. On the other hand, it is interesting to observe that the other case, $\frac{\alpha_0}{\beta_0} = \frac{\alpha_1}{\beta_1}$ corresponds to the case that Farin solves in his original paper [16] using linear coefficients.

These computations, together with a measure of the importance of the perturbation of Q_k for the quality of the surface, yields an algorithm that can be used to stitch patches of the same degree without the end transversal derivatives being coupled. The price that is paid is that the perturbation of Q_k is bound to produce oscillations in the resulting surface.

These compatibility problems, as posed, are not relevant in the context of algebraic surfaces, where only methods for the interpolation of isolated points have been proposed. Neither are they relevant to recursive subdivision methods, that do not attempt to interpolate a fixed boundary either. Both classical and face octrees model an approximation to the object, where the concept of visual continuity is not meaningful (face octrees are not even C^0!). In connection with this discussion, see also the paper by De Rose, in this same volume, where the concept of *approximate geometric continuity* is proposed to circumvent these problems, and in passing improve the global quality of the surface.

Another solution to the compatibility problem is to use rational patches although they must have a singularity of the derivative. In this connection, one should see [17] where the Gregory square is introduced, and [11] where a modification of it is discussed. The latter is actually at the heart of a commercial solid modeler. The same idea is exploited in the paper by Foley, in this same volume.

7.2 Geometric Interrogations

The whole purpose of setting up a model for an object is to to be able to compute with it. Depending on the application, the interrogation may simply amount to rendering the object on a graphics display or may involve other operations, like boolean operations, interference checking, volumetric or mechanical evaluations, simulation of dynamics, heat or wave propagation, and a very long etcetera. We will make some remarks on boolean operations in the next subsection. Here we concentrate on two very basic and fundamental geometric interrogations that one would like to be able to perform efficiently and robustly on any solid model: the determination of whether a point is inside or outside the solid (or on its boundary), and the intersection of a ray with the object. The two are actually related. One may determine if a point is inside or outside a solid by counting the number of intersections (with multiplicity!) that a ray emanating from the point has with the boundary of the solid. Conversely, to find the intersection of a ray with the boundary of the solid means to find a point on the ray and on the boundary.

In general, this questions are hardest in the parametric models. There, the intersection or the classification of a point amount to solving a system of nonlinear equations. This can be done numerically or symbolically, depending on what further processing needs to be carried out with the outcome of the computation.

In contrast, the implicit or algebraic representations shine here the most. By just substituting the coordinates of a point into the function that describes a surface implicitly, one can easily decide on which side it lies. The intersection with a line can also be computed easily by parametrizing the line and substituting in the defining function. This again produces a nonlinear equation to be solved, but this time it is an equation in just one variable, the parameter along the line.

Somewhere in between lie the octree representations. They both answer these questions easily, but only approximately. For classical octrees, the approximation is given by the maximum subdivision. For face octrees, by the tolerance explicitly chosen by the user. The advantage of the face octrees is the compactness of the representation, if the object is relatively smooth. See [4] for explicit bounds on their rate of growth. For both models the point classification problem is solved by first finding the terminal node that contains the given point —which amounts to a tree traversal— and then checking the nature of the node. For a face octree, a further computation is needed for the face nodes, to classify the point with respect to the face stored in the node. The line intersection problem entails a process of marching along the ray and visiting several nodes, which in turn may require several traversals of the tree, because of backtracking to compute neighboring nodes (see for example [28] for the necessary algorithms).

The case of recursive subdivision reduces to the parametric case if we have a priori knowledge of what is the limiting surface for the subdivision scheme used. Otherwise, an arbitrary number of subdivisions may be necessary to classify a point with respect to the limit surface, if the answer has to be exact.

7.3 Set Operations

Regularized boolean operations (union, intersection, difference) play a central role in solid modelers (see [27]), as they are the main tool to build objects from more simple solids. This entails the computation of the intersection of the boundaries of the operands. In the case of solids with free-form boundaries, this computation is nontrivial because the intersection curve always steps out of the model. For instance the intersection of two polynomial parametric surfaces is a polynomial parametric curve, but of very high degree. This means that the trimmed patches are no longer represented in the model. It also means that delicate computations need to be done, with inherent robustness problems.

Recursive subdivision methods simply do not lend themselves well to this problem, just as they did not in the case of the closely connected point classification problem. In general one needs to have an explicit description of the result of the recursive subdivision or one must do with an approximate answer.

Although intrinsically approximate, adaptive space subdivision methods perform very well in this area. They are inherently robust, as critical comparisons can be made in integer arithmetic (actually all of them in the case of classical octrees). They are also inexpensive computationally, as a single traversal of the operands' octrees is needed. In the case of the Face Octrees, a single traversal allows us to compute a solid "tube" that is guaranteed to contain the intersection curve. The tube's width depends on the tolerance specified by the user. Locally it can be used to formulate starting guesses for numerical algorithms to refine this precision if needed.

References

1. Ayala, D., P. Brunet, R. Juan, and I. Navazo, Object representation by means of non-minimal division quadtrees and octrees, ACM Trans. Graph. **4** (1985), 41–59.
2. Brunet, P., Including shape handles in recursive subdivision surfaces, Computer Aided Geom. Design **5** (1988), 41–50.
3. Brunet, P., and I. Navazo, Geometric modeling using exact octree representation of polyhedral objects, Proc. Eurographics'85, North-Holland, Nice 1985, 159–169.
4. Brunet, P., I. Navazo, and A. Vinacua, A Modeling scheme for the approximate representation of closed surfaces, Geometric Modeling Seminar, Schloss Dagstuhl, July 1991.
5. Brunet, P., and A. Vinacua, Increasing the flexibility of VC^1 connections of Bézier patches, Technical Report 89-12, Polytechnic University of Catalonia
6. Brunet, P., and A. Vinacua, Surfaces in solid modeling, in *Geometric Modeling, Methods and Applications*, H. Hagen and D. Roller (eds.), Springer Verlag, Berlin, 1991, 17–34

7. Carlbom, I., I. Chakravarty, and D. A. Vandersschel, A hierarchical data structure for representing the spatial decomposition of 3D objects, IEEE Comp. Graph. Appl. **5** (1985) 24–31.
8. Casale, M. S., and J. E. Bobrow, A set operation algorithm for sculptured solids modeled with trimmed patches, Computer Aided Geom. Design **6** (1989), 235–247.
9. Catmull, E. E., and J. H. Clark, Recursively generated B-spline surfaces on arbitrary topological meshes, Comput. Aided Design **10** (1978), 350–355.
10. Chaikin, G. M., An algorithm for high speed curve generation, Comp. Graphics and Image Proc. **3** (1974) 346–349.
11. Chiyokura, H., and F. Kimura, A new surface interpolation method for irregular curve models, Comp. Graph. Forum **3** (1984) 209–218.
12. Coquillart, S., Extended free-form deformation: a sculpturing tool for 3D geometric modeling, SIGGRAPH'90 **24** (1990), 187–196.
13. Dahmen, W., Smooth piecewise quadric surfaces, in *Mathematical Methods in Computer Aided Geometric Design*, Tom Lyche and L. L. Schumaker (eds.), Academic Press, New York, 1989, 181–194.
14. Doo, D. V. H., and M. Sabin, Behaviour of recursive subdivision surfaces near extraordinary points, Comput. Aided Design **10** (1978), 356–360.
15. Dyn, N., D. Levin, and J. A. Gregory, A butterfly subdivision scheme for surface interpolation with tension control, ACM Trans. Graph. **9** (1990), 160–169.
16. Farin, G. E., A construction for visual C^1 continuity of polynomial surface patches, Comp. Graphics and Image Proc. **20** (1982), 272–282.
17. Gregory, J. A., Smooth interpolation without twist constraints, in *Computer Aided Geometric Design*, R. Barnhill and R. Riesenfeld (eds.), Academic Press (1974), 71–87.
18. Jones, A. K., Nonrectangular surface patches with curvature continuity, Comput. Aided Design **20** (1988), 325–335.
19. Loop, C., and T. DeRose, Generalized B-spline surfaces of arbitrary topology, SIGGRAPH'90 **24**, ACM 1990, 347–356.
20. Meagher, D., Efficient synthetic image generation of arbitrary 3D objects, Proc. IEEE Computer Society, Conference on pattern recognition and image processing (1982), 473–478.
21. Micchelli, C. A., and H. Prautzsch, Computing surfaces invariant under subdivision, Computer Aided Geom. Design **4** (1987), 321–328.
22. Nasri, A., Polyhedral subdivision methods for free-form surfaces, ACM Trans. Graph. **6** (1987), 29–73.
23. Navazo, I., Geometric modeling of octree encoded polyhedral objects, Doctoral thesis, Univ. Politècnica de Catalunya, Barcelona 1986.
24. Nielson, G. M., A transfinite, visually continuous, triangular interpolant, in *Geometric Modeling: Algorithms and New Trends*, G. E. Farin (ed.), SIAM, Philadelphia, 1987, 235–246.
25. Peters, J., Smooth mesh interpolation with cubic patches, Comput. Aided Design **22** (1990), 109–120.

26. Piper, B. R., Visually Smooth Interpolation with Triangular Bézier Patches, in *Geometric Modeling: algorithms and new trends*, G. E. Farin (ed.), SIAM, Philadelphia, 1987, 221–233.
27. Requicha, A., Representations for rigid solids: theory, methods and systems, Computing Surveys of the ACM, **12** (1980), 437–464.
28. Samet, H., *The Design and Analysis of Spatial Data Structures*, Addison-Wesley, 1989.
29. Sarraga, R., G1 interpolation of generally unrestricted cubic Bézier curves, Report GMR–5424, Computer Science Dept, General Motors Research Laboratories, Warren, Michigan (1986). Also in Computer Aided Geom. Design **4** (1987), 23–39. See also R. Sarraga, Errata: G1 interpolation of generally unrestricted cubic Bézier curves, Computer Aided Geom. Design **6** (1989), 167–172.
30. Satoh, T., and H. Chiyokura, Boolean operations on sets using surface data, *Proceedings of the Symposium on Solid Modeling Foundations and CAD/CAM Applications, Texas, 1991*, J. Rossignac and J. Turner (eds.), ACM Press, 1991, 119–127.
31. Sederberg, T. W., Piecewise algebraic surface patches, Computer Aided Geom. Design **2** (1985), 53–59.
32. Sederberg, T. W., and S. R. Parry, Free-form deformation of solid models, SIGGRAPH'86 **20** ACM, 1986, 151–160.
33. Van Wijk, J. J., Bicubic patches for approximating non-rectangular control-point meshes, Computer Aided Geom. Design **3** (1986), 1–13.
34. Watkins, M. A., Problems in geometric continuity, Comput. Aided Design **20** (1988), 499–502.

P. Brunet and A. Vinacua
Polytechnic University of Catalonia
Diagonal 647, 8^{ena} Planta
08028 Barcelona, Spain
EALVAR@EBRUPC51.BITNET

A New Characterization of Plane Elastica

Guido Brunnett

Abstract. Plane elastic curves are considered in dependence of a tension parameter σ and the squared global curvature maximum κ_m^2. A characterization of these curves by their curvature function is given. A formula is established that gives the maximum turning angle of an elastica as a function depending on κ_m^2 and σ. It is shown that an elastic curve can be represented as a linear combination of its curvature, arc length and energy function and that any curve with this property is an elastica.

§1. Introduction

The search for Smoothing Algorithms in CAGD has induced research on curves and surfaces which minimize functionals with geometrical or physical meaning. The classical examples are elastica which describe the shape of elastic materials. Plane elastica can be defined as the extremals (critical points) of the variational problem

$$\int_0^L \left[\kappa^2(s) + \sigma\right] \, ds \rightarrow \quad \min$$

where κ denotes the curvature, s the arc length of a plane curve x, and σ is constant. The total length L of x is considered to be variable. The set of comparison curves for the problem is the set of all C^∞ curves in the plane with fixed endpoints and fixed tangent directions at these points.

Lee and Forsyth found (see [6]) that the curvature function of an elastica satisfies the differential equation

$$\kappa''(s) + \frac{1}{2}\kappa^3(s) + K\kappa(s) = 0, \quad K \in \mathbb{R}. \tag{1}$$

This equation however is not sufficient for a curve to be an elastica. In Section 2 we will show that for any elastica, the square of the curvature function has a

Mathematical Methods in Computer Aided Geometric Design II

Tom Lyche and Larry L. Schumaker (eds.), pp. 43–56.

ISBN 0-12-460510-9.

global maximum κ_m^2 which is bigger than the tension parameter σ. A solution of (1) which satisfies this condition is the curvature function of an elastic curve. This provides a characterization of the elastica by their curvature function.

In Section 3 various formulas expressing properties of elastica in terms of κ_m^2 and σ are given. This includes a formula that gives the maximum turning angle of an elastica as a function depending on κ_m^2 and σ. The explicit formula for the maximum turning angle allows the determination of an upper bound for κ_m^2 in the case that an inflectional elastica has a turning angle bigger than π. As the main result of this section, we show that a plane elastica has a representation as a linear combination of its curvature, arc length and energy function $E(s) = \int_0^s \kappa^2(\bar{s})\, d\bar{s}$. This property characterizes the class of plane elastica and can be used for speeding up the computation of interpolating elastica.

§2. A Characterization of Elastica by their Curvature Function

Given two points $P, Q \in \mathbb{R}^2$ and two unit vectors $V \in T_P\mathbb{R}^2, W \in T_Q\mathbb{R}^2$ then M denotes the set

$$\begin{aligned} M := \{x : [0,L] \to \mathbb{R}^2 \ &: L \in \mathbb{R}^+, x \in C^\infty[0,L], |x'(s)| = 1 \text{ for } s \in [0,L], \\ &x(0) = P,\ x(L) = Q,\ x'(0) = V,\ x'(L) = W\}. \end{aligned}$$

We consider the problem of minimizing the functional

$$E + \sigma L := \int_0^L \left[\kappa^2(s) + \sigma\right] ds$$

among all curves of M, where κ denotes the curvature and s the arc length parameter of a plane curve x. The integral $E = \int_0^L \kappa^2(s)\, ds$ is called the *energy* of x on $[0, L]$. σ denotes a constant which we call the *tension parameter*. The total length L is considered to be variable. As the functional considered is translational and rotational invariant, we assume in this section that $P = (0,0)$ and $V = (1,0)$.

For $x \in M$ the tangent vector T of x is given by

$$T(s) = (\cos(\Psi(s)),\ \sin(\Psi(s))),$$

where the function Ψ with $\Psi(s) := \int_0^s \kappa(\bar{s})\, d\bar{s}$ gives the turning angle of x. Using this notation, the variational problem can be written as

$$\min_{\Psi \in C^\infty[0,L]} \int_0^L \left([\Psi']^2(s) + \sigma\right) ds$$

under the constraint

$$\int_0^L \begin{pmatrix} \cos\Psi(s) \\ \sin\Psi(s) \end{pmatrix} ds = Q,$$

where the admissible functions $\Psi \in C^\infty[0, L]$ are subject to the constraints:

$$\Psi(0) = 0, \quad \Psi(L) = \psi \quad \text{with} \quad (\cos\psi,\ \sin\psi) = W.$$

According to Lagrange's multiplier rule, the differential equation of this problem is the Euler equation for the integrand

$$F(\Psi, \Psi', s) = \Psi'^2(s) + \sigma + \lambda \cos \Psi(s) + \mu \sin \Psi(s),$$

for some constants $\lambda, \mu \in \mathbb{R}$, i.e.

$$\Psi''(s) = -\frac{\lambda}{2} \sin \Psi(s) + \frac{\mu}{2} \cos \Psi(s). \tag{2}$$

This equation can be found e.g. in [3,6].

In the variational problem the upper limit L of the integral is variable. Therefore the extremals of the problem have to satisfy the following boundary condition (see e.g. [2], p. 571)

$$F(\Psi(L), \Psi'(L), L) - \frac{\partial}{\partial \Psi'} F(\Psi(s), \Psi'(s), s)_{|L} \Psi'(L) = 0$$

which yields

$$\Psi'^2(L) = \kappa^2(L) = \lambda \cos \Psi(L) + \mu \sin \Psi(L) + \sigma. \tag{3}$$

Definition 1. *An arc length parametrized plane curve $x \in C^\infty[0, L]$:*

$$x(s) = \int_0^s \begin{pmatrix} \cos(\Psi(\bar{s}) + \vartheta) \\ \sin(\Psi(\bar{s}) + \vartheta) \end{pmatrix} d\bar{s} + x(0) \tag{4}$$

is called an elastica (or elastic curve) with tension parameter σ, if there are λ, $\mu \in \mathbb{R}$ such that the function Ψ satisfies (2) *for all $s \in [0, L]$ and (3).*

Our first objective is to characterize the class of plane elastica by their curvature function. Lee and Forsyth showed in [6] that the Euler equation (2) implies the differential equation (1) for the curvature function $\kappa(s) = \Psi'(s)$.

Multiplying (1) by $2\kappa'$ and integrating yields the first order differential equation

$$(\kappa'(s))^2 = C - \frac{1}{4}\kappa^4(s) - K\kappa^2(s). \tag{5}$$

A solution κ of (5) for any C is obviously a solution of (1). If C is chosen to be smaller than $-K^2$, the extension of κ on $\mathbb{R}$ is monotone throughout. We will show now that these solutions of (1) can not be considered as curvature functions of plane elastica because their integral does not satisfy the Euler equation of the problem.

Theorem 1. *If $\Psi \in C^\infty(\mathbb{R})$ is a solution of (2) for all $s \in \mathbb{R}$ and σ is given by (3), then $\kappa := \Psi'$ is a solution of $\kappa'' + \frac{1}{2}\kappa^3 - \frac{1}{2}\sigma\kappa = 0$ and κ^2 has a global maximum on $\mathbb{R}$.*

Proof: For the derivation of the differential equation for κ, we rewrite the Euler equation in the form

$$\Psi'' = \kappa' = -a\sin(\Psi - \varphi) \tag{6}$$

by introducing the constants

$$\lambda = 2a\cos\varphi \quad \mu = 2a\sin\varphi.$$

Multiplying (6) by $2\Psi'$ and integrating yields

$$\kappa^2 = 2a\cos(\Psi - \varphi) + A, \tag{7}$$

where $A \in \mathbb{R}$ denotes an integration constant. Comparing (7) with (3) shows that

$$A = \sigma.$$

Differentiating (6) gives

$$\kappa'' = -a\cos(\Psi - \varphi)\kappa. \tag{8}$$

Combining (7) and (8) yields (1) with $K = -\sigma/2$.

To show that κ^2 has a global maximum we observe first that (6) implies the existence of a local extremum of κ. This is because the assumption $\kappa'(s) = \Psi''(s) \neq 0$ for all $s \in \mathbb{R}$ means that Ψ is convex or concave and therefore unbounded while according to (6), κ' has no zeros only if Ψ is bounded. Furthermore, from (6) and (7) it is obvious that any local extremum of κ is a global extremum of κ^2.

We assume now that κ^2 has no global maximum on $\mathbb{R}$. In this situation κ' has exactly one zero s_{min} on $\mathbb{R}$ and κ^2 takes its global minimum in this point. Note that κ is non-zero for any $s \neq s_{min}$ because a zero of κ at a point $\bar{s} \neq s_{min}$ would imply that $\kappa(s_{min}) = 0$ and $\kappa'(\tilde{s}) = 0$ for some point $\tilde{s}$ between $\bar{s}$ and s_{min}. Therefore $(\kappa^2)'$ has no zero besides s_{min}, and κ^2 is strictly monotone increasing on the right of s_{min}. Hence κ is monotone increasing, (resp. decreasing), on the right of s_{min} if κ has positive, (resp. negative), values on the right of s_{min}. The formula

$$\Psi(s) = \int_{s_{min}}^{s} \kappa(\bar{s})\,d\bar{s} - \Psi(s_{min})$$

yields that Ψ is in any case unbounded. This is a contradiction to the assumption because (7) implies the existence of global maxima for κ^2 if Ψ is unbounded. ■

We now give the relation between the global maximum κ_m^2 of κ^2 and the tension parameter σ, and express κ in terms of elliptic functions.

Theorem 2. *Let $\kappa \in C^\infty(I\!R)$ be a solution of the differential equation*

$$\kappa''(s) + \frac{1}{2}\kappa^3(s) - \frac{1}{2}\sigma\kappa(s) = 0 \quad \text{for } s \in I\!R$$

with a global maximum $\kappa_m^2 \neq 0$ of κ^2 on $I\!R$. Then the following statements hold:

(i) $\kappa_m^2 \geq \sigma$

(ii) κ has a zero if and only $\kappa_m^2 > 2\sigma$. In this case κ is given by

$$\kappa(s) = \kappa_m \operatorname{cn}(\sqrt{(\kappa_m^2 - \sigma)/2}(s - s_m) \,|\, k^2) \tag{9}$$

with the parameter

$$k^2 = \frac{\kappa_m^2}{2(\kappa_m^2 - \sigma)}.$$

(iii) $2\sigma - \kappa_m^2 \leq \kappa^2 \leq \kappa_m^2$ for $\kappa_m^2 \leq 2\sigma$. In this case κ is given by

$$\kappa(s) = \kappa_m \operatorname{dn}(\kappa_m(s - s_m)/2 \,|\, \frac{1}{k^2}). \tag{10}$$

Proof: As a global maximum point s_m of κ^2 is also a zero of κ' (5) with $K = -\sigma/2$ implies

$$C = \frac{1}{4}\kappa_m^4 - \frac{1}{2}\sigma\kappa_m^2.$$

Hence (5) takes the form

$$(\kappa')^2 = \frac{1}{4}(\kappa_m^2 - \kappa^2)(\kappa^2 + \kappa_m^2 - 2\sigma). \tag{11}$$

As all quantities in (11) are real, (i) follows from the fact that the term $(\kappa_m^2 - \kappa^2)$ is always non-negative, and therefore the term $(\kappa^2 + \kappa_m^2 - 2\sigma)$ also has to be non-negative for any κ^2.

Using the same argument, we observe that $\kappa^2(s) = 0$ for some s implies that $\kappa_m^2 \geq 2\sigma$ while for $\kappa_m^2 \leq 2\sigma$ κ^2 has to be greater or equal to $2\sigma - \kappa_m^2$.

To express κ in terms of elliptic functions, we proceed as follows. In the case that $\kappa_m^2 \geq 2\sigma$, we substitute $z^2 = (\kappa_m^2 - \kappa^2)/\kappa_m^2$ in (11) and obtain for z the differential equation

$$(z')^2 = \frac{1}{2}(\kappa_m^2 - \sigma)(1 - z^2)(1 - k^2 z^2), \tag{12}$$

where

$$k^2 = \frac{\kappa_m^2}{2(\kappa_m^2 - \sigma)}.$$

(12) is the differential equation of the Jacobi function sn=sn(u) for the argument $u = \sqrt{\frac{1}{2}(\kappa_m^2 - \sigma)}s$ (see [1],p. 114). We therefore obtain

$$z^2(s) = \operatorname{sn}^2(\sqrt{(\kappa_m^2 - \sigma)/2}(s - s_m) \,|\, k^2).$$

The relation $\kappa^2 = \kappa_m^2(1-z^2)$ then gives

$$\kappa^2(s) = \kappa_m^2 \mathrm{cn}^2(\sqrt{(\kappa_m^2-\sigma)/2}(s-s_m)\,|\,k^2) \tag{13}$$

(see [8], p.16). Equation (13) shows that κ indeed has zeros in the case $\kappa_m^2 > 2\sigma$. For $\kappa_m^2 = 2\sigma$ the parameter $k^2 = 1$ and the elliptic function cn degenerates to the hyperbolic function sech=1/cosh which has no zeros.

A solution of (1) that has a zero and a zero of its derivative is a periodic function with the symmetry properties of the sine function (see [9]). In particular, κ takes both values κ_m and $-\kappa_m$. Hence (13) implies (9).

In the case that $\kappa_m^2 \le 2\sigma$ we substitute $z^2 = (\kappa_m^2 - \kappa^2)/2(\kappa_m^2-\sigma)$ in (12) and obtain for z the differential equation

$$(z')^2 = \frac{1}{4}\kappa_m^2(1-z^2)(1-l^2z^2),$$

where

$$l^2 = \frac{1}{k^2} = \frac{2(\kappa_m^2-\sigma)}{\kappa_m^2}.$$

Hence

$$z^2(s) = \mathrm{sn}^2(\kappa_m(s-s_m)/2\,|\,l^2).$$

The relation

$$\kappa^2 = \kappa_m^2(1-l^2z^2)$$

yields

$$\kappa^2(s) = \kappa_m^2 \mathrm{dn}^2(\kappa_m(s-s_m)/2\,|\,l^2)$$

(see [8], p.16). As dn is positive throughout we obtain (10). ■

Theorem 2 implies that the curvature function of an elastic curve extends to a periodic function on $\mathbb{R}$. In the case that $\kappa_m^2 > 2\sigma$, this periodic function is symmetric with respect to any zero of its derivative and antisymmetric with respect to any zero. According to Love [7], we will call this situation **inflectional** because the extension of x has turning points. In the case that $\kappa_m^2 \le 2\sigma$ the extension of κ has no zeros but is still symmetric with respect to any zero of its derivative. The local extrema in this case are κ_m and $\pm(2\sigma - \kappa_m^2)$. The situation is illustrated in Figures 1 –8, where for $\kappa_m = 1$ various curvature functions and the corresponding elastica are shown. If only non-negative tension values are considered, σ is an element of $[0,1]$. With σ increasing from 0 to 1, the curvature function changes continuously from a lemniscate function ($\sigma = 0$) to a constant ($\sigma = 1$). A classification of the different forms of elastica has been given by Euler (see [5]) and also Love (see [7]).

We are now able to characterize the set of elastic curves in the plane by their curvature function.

Theorem 3. *An arc length parametrized plane curve $x \in C^\infty[0, L]$ is an elastica with tension parameter σ if and only if the curvature function κ of x satisfies the differential equation*

$$\kappa''(s) + \frac{1}{2}\kappa^3(s) - \frac{1}{2}\sigma\kappa(s) = 0 \quad \text{for } s \in [0, L] \tag{14}$$

and the extension of κ^2 on $\mathbb{R}$ has a global maximum.

Proof: According to Theorem 1 the curvature function of a plane elastica with tension σ satisfies (14) and also the maximum condition. If $\kappa = 0$, x is the trivial elastic curve. According to Theorem 2, a solution κ of (14) with a global maximum $\kappa_m^2 \neq 0$ of κ^2 is explicitly determined by (9) or (10). It remains to show that for any $\kappa : [0, L] \to \mathbb{R}$ according to (9) or (10) the function $\Psi(s) = \int_0^s \kappa(\bar{s})\, d\bar{s}$ satisfies the Euler equation (6) on $[0, L]$ for some $a, \varphi \in \mathbb{R}$. (6) together with (14) implies (3) which is also required for the elastica.

Let $\bar{\kappa}$ denote the extension of κ on $\mathbb{R}$ and $\bar{\Psi}(s) = \int_0^s \bar{\kappa}(\bar{s})\, d\bar{s}$. Furthermore define $\varphi := \bar{\Psi}(s_m)$, where s_m is a global minimum point of κ^2. Then we achieve in the case that

$$\kappa(s) = \kappa_m \operatorname{cn}(\sqrt{(\kappa_m^2 - \sigma)/2}(s - s_m) \mid k^2)$$

the relation

$$\bar{\Psi}(s) - \varphi = \int_{s_m}^s \bar{\kappa}(\bar{s})\, d\bar{s} = \frac{\sqrt{2}\kappa_m}{\sqrt{\kappa_m^2 - \sigma}} \int_0^u \operatorname{cn}(\bar{u} \mid k^2)\, d\bar{u},$$

where $u = \sqrt{(\kappa_m^2 - \sigma)/2}(s - s_m)$. If $4nK \leq u \leq (4n+2)K$ where K denotes the complete elliptic integral of the first kind and n a non-negative integer, the integral yields

$$\bar{\Psi}(s) - \varphi = 2\operatorname{sign}(\kappa_m) \arccos\big(\operatorname{dn}(u - 4nK)\big).$$

For the integral of cn; see e.g. [8], p. 25. With $r = u - 4nK$ it follows that

$$\begin{aligned}
\sin(\bar{\Psi}(s) - \varphi) &= \operatorname{sign}(\kappa_m) 2 \sin\Big(\arccos\big(\operatorname{dn}(r)\big)\Big)\operatorname{dn}(r) \\
&= \operatorname{sign}(\kappa_m) 2\sqrt{1 - \operatorname{dn}^2(r)}\operatorname{dn}(r) \\
&= \frac{\sqrt{2}\kappa_m}{\sqrt{\kappa_m^2 - \sigma}} \operatorname{sn}(r)\operatorname{dn}(r).
\end{aligned}$$

In the above equations we used that $\operatorname{dn}^2 + k^2\operatorname{sn}^2 = 1$ and that $\operatorname{sn}(r)$ is positive.

For $(4n+2)K < u < 4(n+1)K$ the integral of cn gives

$$\bar{\Psi}(s) - \varphi = -2\operatorname{sign}(\kappa_m) \arccos\big(\operatorname{dn}(u - (4n+2)K)\big)$$

and it follows in the same way as above

$$\sin(\bar{\Psi}(s)-\varphi) = -\frac{\sqrt{2}\kappa_m}{\sqrt{\kappa_m^2-\sigma}}\mathrm{sn}(u-(4n+2)K)\mathrm{dn}(u-(4n+2)K).$$

With

$$\bar{\Psi}''(s) = \bar{\kappa}'(s) = -\frac{1}{\sqrt{2}}\kappa_m\sqrt{\kappa_m^2-\sigma}\,\mathrm{sn}(u)\mathrm{dn}(u)$$

and $\mathrm{sn}(u-4nK) = \mathrm{sn}(u)$, $\mathrm{dn}(u-4nK) = \mathrm{dn}(u)$, $\mathrm{sn}(u-(4n+2)K) = -\mathrm{sn}(u)$, $\mathrm{dn}(u-(4n+2)K) = \mathrm{dn}(u)$, we achieve in both cases

$$\bar{\Psi}'' = -\frac{1}{2}(\kappa_m^2-\sigma)\sin(\bar{\Psi}(s)-\varphi). \tag{15}$$

As $\bar{\Psi}(s) = \Psi(s)$ for $s \in [0, L]$ we have proved the claim in the case that $\kappa_m^2 > 2\sigma$.

In the case that $\kappa_m^2 \le 2\sigma$, κ is given by

$$\kappa(s) = \kappa_m \mathrm{dn}(\kappa_m(s-s_m)/2 \,|\, l^2).$$

Here we get

$$\bar{\Psi}(s)-\varphi = \int_{s_m}^{s}\bar{\kappa}(\bar{s})\,d\bar{s} = 2\int_0^u \mathrm{dn}(\bar{u}\,|\,l^2)\,d\bar{u} \tag{16}$$

with $u = \kappa_m(s-s_m)/2$. We consider the partition $u = pK + r$, where p denotes the biggest integer smaller than u/K and distinguish the cases that p is even or odd. In the first case we get

$$\bar{\Psi}(s)-\varphi = 2p\pi + 2\arcsin\big(\mathrm{sn}(r)\big).$$

Hence

$$\sin(\bar{\Psi}(s)-\varphi) = 2\sqrt{1-\mathrm{sn}^2(r)}\mathrm{sn}(r) = 2\mathrm{sn}(r)\mathrm{cn}(r),$$

where we used that $\mathrm{sn}^2+\mathrm{cn}^2 = 1$ and that $\mathrm{cn}(r)$ is positive. Differentiating κ and using the properties $f(s+4K) = f(s)$, $f(s+2K) = -f(s)$ which hold for sn and cn, one obtains

$$\bar{\Psi}''(s) = \bar{\kappa}'(s) = -(\kappa_m^2-\sigma)\mathrm{sn}(r)\mathrm{cn}(r).$$

Combining the last two equations yields (15). In the case that p is odd, we get

$$\bar{\Psi}(s)-\varphi = 2(p+1)\pi - 2\arcsin\big(\mathrm{sn}(K-r)\big),$$

which leads to

$$\sin(\bar{\Psi}(s)-\varphi) = -2\mathrm{sn}(K-r)\mathrm{cn}(K-r).$$

In this case we have

$$\bar{\kappa}'(s) = -(\kappa_m^2-\sigma)\mathrm{sn}(r+K)\mathrm{cn}(r+K).$$

As $\mathrm{sn}(r+K) = \mathrm{sn}(K-r)$ but $\mathrm{cn}(r+K) = -\mathrm{cn}(K-r)$, combining the last two equations also yields (15) in this case. ■

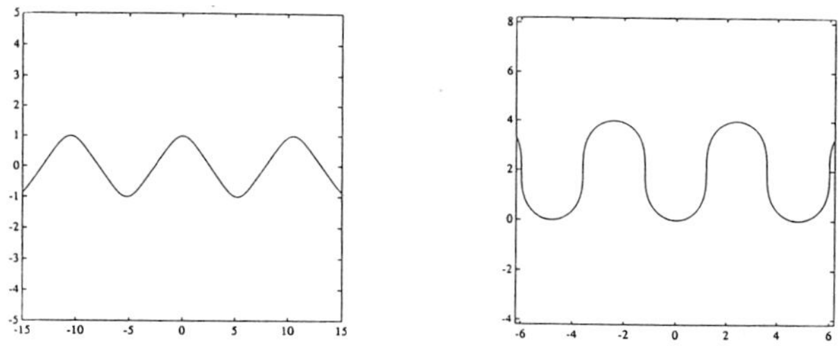

Figure 1. $\sigma = 0$; $\kappa(s) = \kappa_m \operatorname{coslemn}(\kappa_m(s - s_m)/2)$.

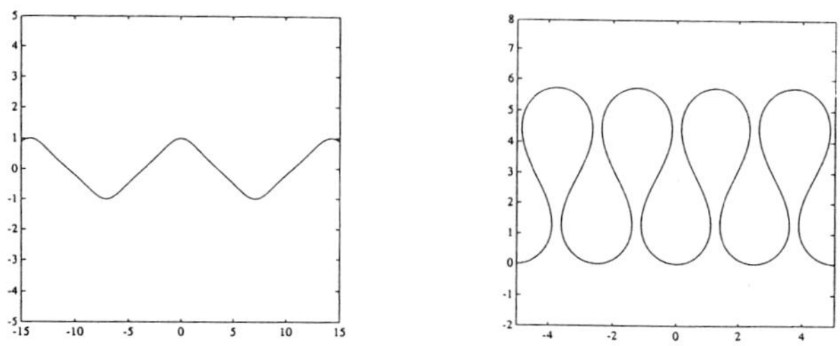

Figure 2. $\sigma = 0.3$; $\kappa(s) = \kappa_m \operatorname{cn}(\sqrt{(\kappa_m^2 - \sigma)/2}(s - s_m) \mid k^2)$.

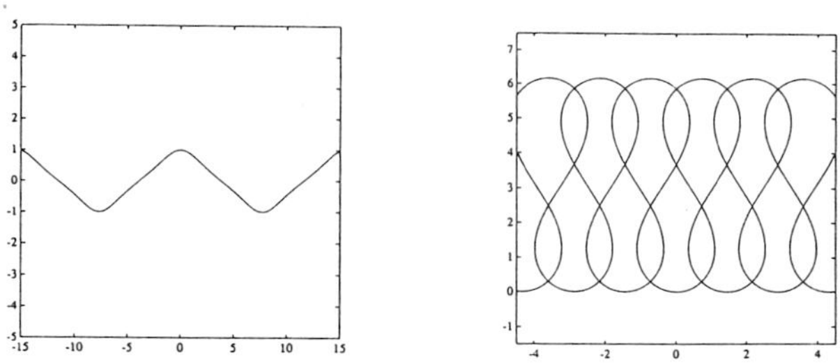

Figure 3. $\sigma = 0.35$; $\kappa(s)$ as in Figure 2.

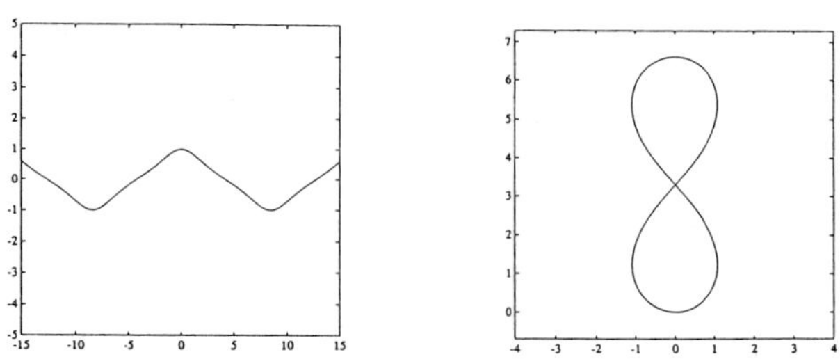

Figure 4. $\sigma = 0.394757217$; $\kappa(s)$ as in Figure 2.

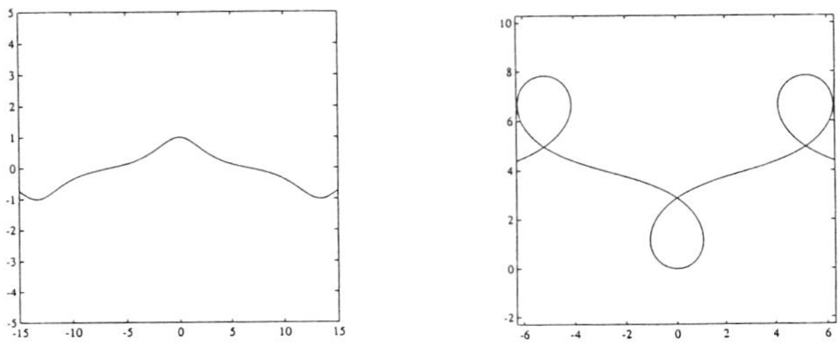

Figure 5. $\sigma = 0.49$; $\kappa(s)$ as in Figure 2.

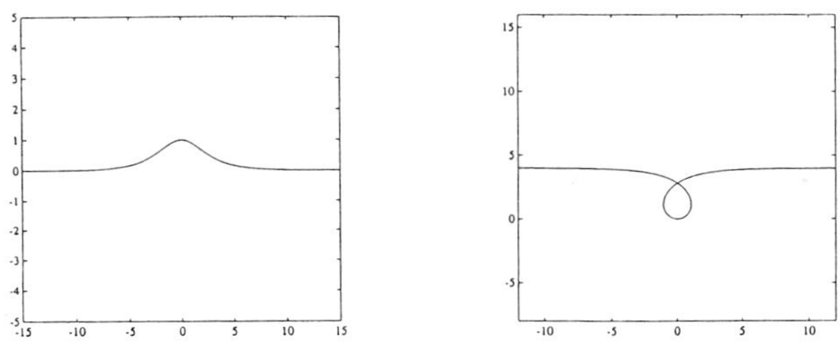

Figure 6. $\sigma = 0.5$; $\kappa(s) = \operatorname{sech}(\kappa_m(s - s_m)/2)$.

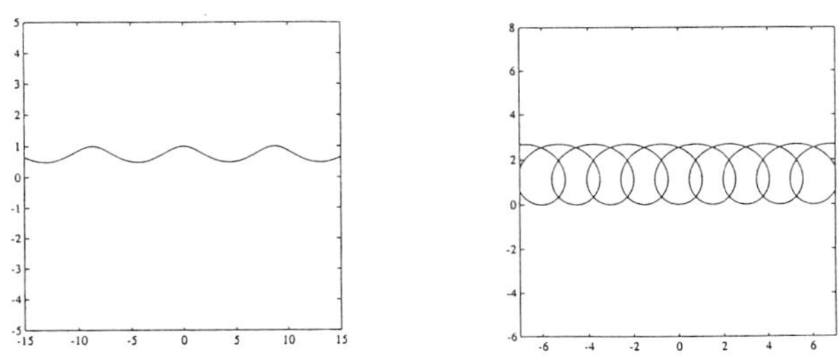

Figure 7. $\sigma = 0.62$; $\kappa(s) = \kappa_m \operatorname{dn}(\kappa_m(s - s_m)/2 \mid \frac{1}{k^2})$.

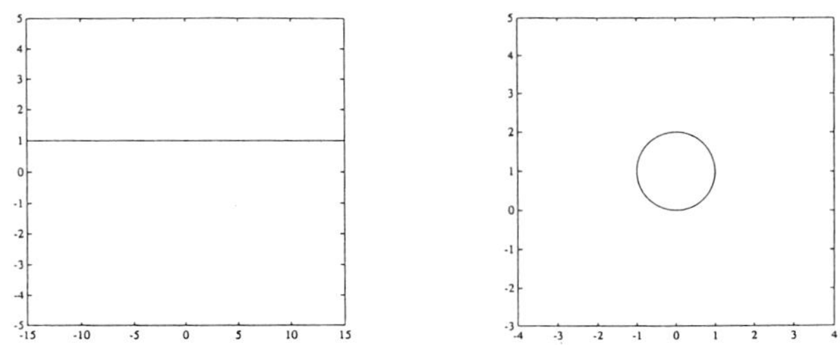

Figure 8. $\sigma = 1$; $\kappa(s) = \kappa_m$.

§3. A Characterizing Representation Theorem

First we state some important formulas for elastic curves. In order to avoid repetitions we use the following convention: if a curve $x : [0, L] \to \mathbb{R}^2$ is an elastica , then κ denotes the (analytic) curvature function of x, $\bar{\kappa}$ denotes the extension of κ on $\mathbb{R}$, κ_m^2 the global maximum of $\bar{\kappa}^2$ and s_m is a number with $\bar{\kappa}^2(s_m) = \kappa_m^2$. Furthermore, $\Psi : [0, L] \to \mathbb{R}$ is the function $\Psi(s) := \int_0^s \kappa(\bar{s})\, d\bar{s}$, $\bar{\Psi}$ denotes the extension of Ψ on $\mathbb{R}$, and $\varphi := \bar{\Psi}(s_m)$. The energy E of x is the function $E(s) = \int_0^s \kappa^2(\bar{s})\, d\bar{s}$. For convenience we also use the notation $C(\psi) = (\cos\psi, \sin\psi)$. ϑ is the angle in $[0, 2\pi[$ with $x'(0) = C(\vartheta)$.

Theorem 4. *For a plane elastic curve x with tension parameter σ the following relations hold:*

(i) $\kappa^2(s) = (\kappa_m^2 - \sigma)\cos(\Psi(s) - \varphi) + \sigma$,
(ii) $E(s) = (\kappa_m^2 - \sigma)\langle C(\varphi + \vartheta), x(s) - x(0)\rangle + \sigma s$,
(iii) $\kappa(s) - \kappa(0) = -\frac{1}{2}(\kappa_m^2 - \sigma)\langle C'(\varphi + \vartheta), x(s) - x(0)\rangle$.

Proof: (i) was established in the proof of Theorem 1 for $\bar{\kappa}$ in form of (7). As $\bar{\kappa}^2$ has the global maximum κ_m^2 we verify that $2a = \kappa_m^2 - \sigma$. In the proof of Theorem 3 it was shown that the constant $\varphi = \int_0^{s_m} \bar{\kappa}(\bar{s})\, d\bar{s}$.

To obtain (ii) we rewrite (i) as

$$\kappa^2(s) = (\kappa_m^2 - \sigma)(\cos(\Psi + \vartheta)\cos(\varphi + \vartheta) + \sin(\Psi + \vartheta)\sin(\varphi + \vartheta)) + \sigma$$

and integrate using (4).

To verify (iii) we write the Euler equation (6) in the form

$$\kappa'(s) = -\frac{1}{2}(\kappa_m^2 - \sigma)(\sin(\Psi + \vartheta)\cos(\varphi + \vartheta) - \cos(\Psi + \vartheta)\sin(\varphi + \vartheta))$$

and integrate. ■

Theorem 2 implies that the turning angle of an elastic curve is bounded in the inflectional case $\sigma \leq \frac{1}{2}\kappa_m^2$ and unbounded otherwise. Formula (i) of Theorem 4 can be used to determine the maximum turning angle of an inflectional elastic curve.

Corollary 5. *The maximum turning angle $\Psi_{max} := \max_{s\in[t,t+T]} |\bar{\Psi}(s)|$ of an elastica x with period T and tension parameter $\sigma \leq \frac{1}{2}\kappa_m^2$ is given by*

$$\Psi_{max} = 2\arccos\left(\frac{\sigma}{\sigma - \kappa_m^2}\right). \tag{17}$$

If $\sigma > \frac{1}{2}\kappa_m^2 \neq 0$, then

$$|\bar{\Psi}(t + T) - \bar{\Psi}(t)| = 2\pi$$

for any $t \in \mathbb{R}$.

Proof: Let a be a zero of $\bar{\kappa}$. Then it follows from the symmetry properties of $\bar{\kappa}$ in the inflectional case that

$$\Psi_{max} = 2\left|\int_a^{a+T/4} \bar{\kappa}(s)\, ds\right| = 2|\bar{\Psi}(a + T/4) - \bar{\Psi}(a)|.$$

With (i) we get

$$\Psi_{max} = 2|\arccos(\frac{\kappa_m^2 - \sigma}{\kappa_m^2 - \sigma}) + \varphi - \arccos(\frac{-\sigma}{\kappa_m^2 - \sigma}) - \varphi|,$$

and therefore (17) holds. In the non-inflectional case, κ is periodic with period $T = 4K/|\kappa_m|$, where K denotes the complete elliptic integral of the first kind. To apply (16), we choose $t = s_m$ and obtain

$$|\bar{\Psi}(s_m + T) - \bar{\Psi}(s_m)| = 4\int_0^K \mathrm{dn}(u|l^2)\,du = 4\arcsin(\mathrm{sn}(K)) = 2\pi. \quad \blacksquare$$

For an inflectional elastica formula x, (17) implies an upper bound for κ_m^2 if the absolute value of the oriented angle between $x'(0)$ and $x'(L)$ is bigger than π.

Corollary 6. *For an inflectional elastica x with tension parameter σ and $\psi := |\Psi(L)| > \pi$:*

$$2 \le \frac{\kappa_m^2}{\sigma} \le 1 - \frac{1}{\cos(\psi/2)}.$$

Proof: It follows from (17) that a turning angle bigger than π can only happen for positive σ. The inflectional nature of x implies therefore the left inequality. The right inequality follows from the fact that ψ has to be less or equal Ψ_{max} which is given by (17). $\blacksquare$

As the main result of this section we show that an elastica can be represented as a linear combination of its curvature, energy and arc length.

Theorem 7. *If x is an elastica with tension parameter $\sigma \neq \kappa_m^2$, then*

$$x(s) = \frac{1}{\kappa_m^2 - \sigma}\begin{pmatrix} \sin(\varphi+\vartheta) & \cos(\varphi+\vartheta) \\ -\cos(\varphi+\vartheta) & \sin(\varphi+\vartheta) \end{pmatrix}\begin{pmatrix} 2(\kappa(s) - \kappa(0)) \\ E(s) - \sigma s \end{pmatrix} + x(0), \quad (18)$$

where $\cos\varphi = (\kappa_0^2 - \sigma)/(\kappa_m^2 - \sigma)$ and $\sin\varphi = 2\kappa_0'/(\kappa_m^2 - \sigma)$.

Proof: A plane curve $x \in C^\infty[0, L]$ has a representation

$$\begin{aligned} x(s) - x(0) &= \int_0^s C(\Psi(\bar{s}) - \varphi + \varphi + \vartheta)\,d\bar{s} \\ &= C(\varphi+\vartheta)\int_0^s \cos(\Psi(\bar{s}) - \varphi)\,d\bar{s} \\ &+ C'(\varphi+\vartheta)\int_0^s \sin(\Psi(\bar{s}) - \varphi)\,d\bar{s}\,. \end{aligned}$$

Applying (ii) and (iii) of Theorem 4 yields

$$x(s) - x(0) = \frac{1}{\kappa_m^2 - \sigma}\Big(C(\varphi+\vartheta)\big(E(s) - \sigma s\big) - C'(\varphi+\vartheta)2\big(\kappa(s) - \kappa(0)\big)\Big),$$

which is equivalent to (18). The formulas for φ follow from (6) and (7) for $s = 0$. ■

Formula (18) provides an explicit representation of an elastica in terms of its curvature function if $\kappa_m^2 \neq \sigma$. Note that in the case $\kappa_m^2 = \sigma$, the elastic curve is a circle of radius $1/|\kappa_m|$ as shown in Figure 8.

Finally we show that elastica are essentially the only curves in the plane that have a representation of the form (18).

Theorem 8. *Let $\kappa : I\!R \to I\!R$ be an arbitrary analytic function with a global maximum κ_m^2 of κ^2, σ a real number smaller than κ_m^2, and $E(s) = \int_0^s \kappa^2(\bar{s})\, d\bar{s}$. An arc length parametrized curve x, where $x(s)$ is given by (18) is an elastica with curvature function κ and tension parameter σ.*

Proof: We assume the case $\varphi + \vartheta = \pi/2$ which can always be achieved by applying a rotation to x. Then

$$x' = \frac{1}{\kappa_m^2 - \sigma}(2\kappa', \kappa^2 - \sigma), \quad x'' = \frac{1}{\kappa_m^2 - \sigma}(2\kappa'', 2\kappa\kappa').$$

x is arc length parametrized if and only if $x' = 1$, i.e.

$$(2\kappa')^2 = (\kappa_m^2 - \sigma)^2 - (\kappa^2 - \sigma)^2. \tag{19}$$

Differentiating (19) yields

$$2\kappa'\kappa'' + \kappa'\kappa(\kappa^2 - \sigma) = 0.$$

Therefore either κ satisfies (14) or κ is constant. For an arc length parametrized curve the determinant $[x', x'']$ is curvature. Here we have

$$[x', x''] = \frac{4}{(\kappa_m^2 - \sigma)^2}\big(2(\kappa')^2\kappa - \kappa''(\kappa^2 - \sigma)\big). \tag{20}$$

If κ is constant the curvature of x is zero, hence x is a trivial elastic curve. If κ satisfies (14) we substitute in (20) κ'' according to (14) and κ'^2 according to (19). This yields

$$[x', x''] = \kappa.$$

As κ satisfies (14) on $[0, L]$ and κ^2 has a global maximum on $\mathbb{R}$, x is an elastic curve according to Theorem 3. ■

The representation formula (18) of an elastic curve is extremely useful for the computation of interpolating elastica. One reason for this is that (18) involves no trigonometric functions and less integrations than the standard representation (4). Therefore (18) is less expensive to evaluate. Furthermore, (18) can be used to find piecewise polynomial approximations of elastic curves based on a spline approximation of the curvature function. The author has established polynomial splines which approximate the curvature functions of plane elastica with high accuracy. These approximations have been used to compute interpolating elastica in representation (4). Using these piecewise polynomial curvature functions together with the new representation (18) one obtains polynomial spline approximations of the elastica itself. These approximations will be discussed in [4].

Acknowledgments. The author thanks Professor Carl de Boor for helpful discussions and enlightening remarks.

References

1. Boehm, K., *Elliptische Funktionen*, Teil 2, Verlagshandlung Goeschen, Leipzig, 1910.
2. Bolza, O., *Vorlesungen ueber Variationsrechnung*, Koehler und Amelang, Leipzig, 1949.
3. Born, M., Untersuchungen über die Stabilität der elastischen Linie in Ebene und Raum, Dissertation, Universität Göttingen, 1906.
4. Brunnett, G., Spline approximation of interpolating elastica, in preparation.
5. Euler, L., Additamentum: De curvis elasticis, in *Methodus inveniendi lineas curvas maximi minimive proprietate gaudentes*, Ser. 1, Vol. 24, Lausanne, 1744.
6. Lee, E. H., and G. E. Forsyth, Variational study of nonlinear spline curves, SIAM Review **15** (1975), 120–133.
7. Love A. E., *A Treatise on the Mathematical Theory of Elasticity*, Dover Publications, New York, 1944.
8. Milne-Thompson, L. M., *Jacobian Elliptic Function Tables*, Dover Publications, New York, 1950.
9. Walter W., *Gewoehnliche Differentialgleichungen*, Springer Verlag, 1976.

Guido Brunnett
Department of Computer Science
Arizona State University
Tempe, AZ 85281
brunnett@asuvax.eas.asu.edu

POLynomials, POLar Forms, and InterPOLation

Paul de Faget de Casteljau

§1. Introduction

Invented in 1958 at CITROËN, the name *Polar Form* recalls the triple occurence of *POL* in the title. The visualization of forms by using pilot-points allows one to avoid numerous singularities, which ordinarily are the delight of traditional mathematicians, but are feared in industrial practice.

It is appropriate to consider two completely independent steps during a calculation: The first step consists in the definition of local elements, like points, normals, etc. which are necessary for the execution of a drawing or a milling, while the algebraic nature of the form, defined by its poles, is kept constant. The second step, the inverse operation, tries to specify the position of the poles according to the given requirements in order to define the form.

Whenever possible, this paper tries to free itself from any superfluos notions coming from calculus, may it be integral calculus or extrema, as soon as the pure algebraic point of view is enough to provide a complete solution. Presented by means of multilinear polar forms, the exclusive use of polynomials of fixed degree n allows to guarantee against all intrusion of the imaginary. In fact, one requires a unique real solution, and consequently no verbal confusion with Cauchy's theory of complex variables could be feared.

§2. Algebraic Definition of Polar Curves

We assume throughout that E_k is a vector space of arbitrary dimension k. Taking the sum of all products of p variables among n given variables $t_1, \ldots, t_n$ we obtain the symmetric functions

$$S_n^p(t_1, \ldots, t_n) = \sum_{\substack{I \subseteq \{1,\ldots,n\} \\ |I|=p}} \prod_{i \in I} t_i, \tag{1}$$

which can also be extracted from a product expansion of $\prod_{i=1}^{n}(1+t_i)$, i.e.

$$\prod_{i=1}^{n}(1+t_i) = \sum_{p=0}^{n} S_n^p(t_1, \ldots, t_n). \tag{2}$$

Mathematical Methods in Computer Aided Geometric Design II
Tom Lyche and Larry L. Schumaker (eds.), pp. 57–68.

ISBN 0-12-460510-9.

Setting the last factor in this product equal to $(1+t)$, one obtains the recurrence relation

$$S_n^p(t_1,\ldots,t_{n-1},t) = S_{n-1}^p(t_1,\ldots,t_{n-1}) + t \cdot S_{n-1}^{p-1}(t_1,\ldots,t_{n-1}), \qquad (3)$$

which uses the symmetric functions in $n-1$ variables $t_1,\ldots,t_{n-1}$, different from t.

Given vectors $A_0,\ldots,A_n \in E_k$, one can then define a *pole* with respect to the n variables $t_1,\ldots,t_n$ as

$$P_{12\ldots n} = P(t_1,\ldots,t_n) = \sum_{p=0}^{n} S_n^p(t_1,\ldots,t_n)A_p. \qquad (4)$$

If all variables take the same value $t_i = t$, one obtains a univariate polynomial curve

$$P(t) = \sum_{p=0}^{n} \binom{n}{p} t^p A_p, \qquad (5)$$

and using the recurrence (3) with $t=1$, we obtain the well-known recurrence relation for the binomial coefficients $\binom{n}{p}$

$$\binom{n}{p} = \binom{n-1}{p} + \binom{n-1}{p-1}. \qquad (6)$$

From now on we assign to these variables fixed parameters, taken from a non-decreasing *knot sequence*, or *chain*,

$$t_0 \le t_1 \le t_2 \le \cdots \le t_i < t_f \le \cdots \le t_m, \quad m \ge n-1, \qquad (7)$$

from which we consider n consecutive elements. Suppose that $\Theta = (\theta_1,\ldots,\theta_{n-1})$ is a substring of $n-1$ elements for which the symmetric functions S_{n-1}^p are known. One can then add a variable t and calculate the corresponding pole by inserting the result of formula (3) into equation (4) to obtain

$$P_\theta(t) = \sum_{p=0}^{n-1} S_{n-1}^p(\theta_1,\ldots,\theta_{n-1})A_p + t\sum_{p=0}^{n-1} S_{n-1}^{p-1}(\theta_1,\ldots,\theta_{n-1})A_p. \qquad (8)$$

Two of these poles, for the values $t=\alpha$, which precedes the substring Θ, and for $t=\lambda$, which follows it, are known, since $(\alpha\Theta)$ and $(\Theta\lambda)$ form a sequence of n consecutive knots, for which the corresponding poles $P_{\alpha\Theta}$ and $P_{\Theta\lambda}$ are given as input to the form. By using a homogeneous variable u in the sense of Euler in the linear equation (8) we can then write

$$P_\theta(t) = (\frac{\partial P}{\partial u})_\theta + t(\frac{\partial P}{\partial t})_\theta. \qquad (9)$$

This defines a derivative $\partial P/\partial t$ (or $\partial P/\partial u$), a polar form of degree $n-1$, from which one knows the values

$$P_{\alpha\theta} = (\frac{\partial P}{\partial u})_\theta + \alpha(\frac{\partial P}{\partial t})_\theta \quad \text{and} \quad P_{\theta\lambda} = (\frac{\partial P}{\partial u})_\theta + \lambda(\frac{\partial P}{\partial t})_\theta. \qquad (10)$$

An elementary calculation shows that these derivatives are obtained from the given poles as

$$(\frac{\partial P}{\partial t})_\theta = \frac{P_{\theta\lambda} - P_{\alpha\Theta}}{\lambda - \alpha} \quad \text{and} \quad (\frac{\partial P}{\partial u})_\theta = \frac{\lambda P_{\alpha\theta} - \alpha P_{\theta\lambda}}{\lambda - \alpha}. \tag{11}$$

When inserted in equation (9), these values determine $P_\theta(t)$. Thus, one has succeeded in inserting into the chain (7) an additional link t that necessarily lies inside the segment (α, λ) (Extrapolation would be dangerous).

By regrouping the coefficients around the poles $P_{\alpha\theta}$ and $P_{\theta\lambda}$ one thus recovers the algorithm of Cox and de Boor for B-splines

$$P_\theta(t) = \frac{\lambda - t}{\lambda - \alpha} P_{\alpha\Theta} + \frac{t - \alpha}{\lambda - \alpha} P_{\Theta\lambda} \tag{12}$$

with the poles as control points.

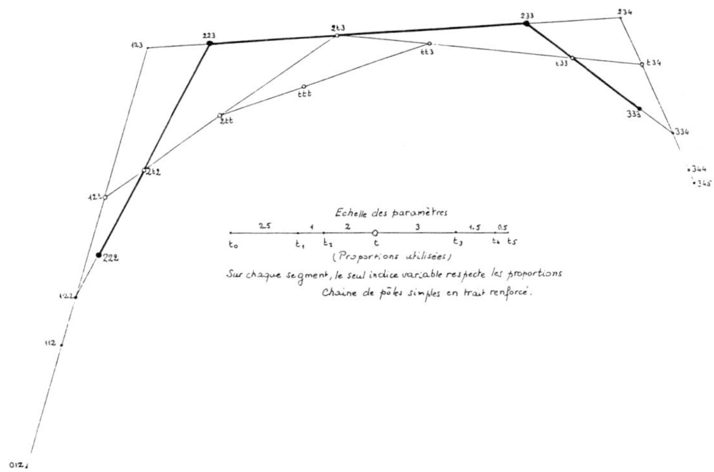

Figure 1. Construction of a point on the curve from its progressive poles. The ratios are given by the numbers over each segment. The corresponding simple poles lie on the bold lines.

In order to define the segment (t_i, t_f), one must thus consider all the poles which allow the insertion of a new parameter t into the segment (t_i, t_f), from the first pole, $P_{t_{i-n+1},\ldots,t_i}$, ending with the knot t_i, up to the last pole, $P_{t_f,\ldots,t_{f+n-1}}$, starting with the knot t_f. The $n-1$ other poles simultaneously include both t_i and t_f and will be replaced by n new poles upon insertion of t.

This operation of knot insertion can be repeated. Repetition of a parameter with maximal multiplicity n then leads to a point on the curve, no matter if this parameter may be t, t_i, or t_f.

The notion of polar derivative can be extended to higher orders. Given a nucleus $\bar{\omega}$ of $p = n - q$ consecutive knots in the knot sequence (7), there are $q + 1$ ways to augment this nucleus to a sequence of n parameters, by adding q_1 elements before and $q_2 = q - q_1$ elements after $\bar{\omega}$. If S_q^k are the symmetric functions in q variables one can thus write

$$P_\omega(t_1, \ldots, t_q) = \sum_{k=0}^{q} S_q^k(t_1, \ldots, t_q)\ (\frac{\partial^q P}{\partial u^{q-k} \partial t^k})_{\bar{\omega}}. \tag{13}$$

The solution of this system of $q+1$ equations gives the values of the homogenuous partial derivatives up to order q. In particular, if all parameters in the nucleus $\bar{\omega}$ are equal to t_i, where t_i is a knot of multiplicity p, the equality of these derivatives for the two segments joining at t_i shows that these two segments join with continuity up to order $q = n - p$.

At a construction parameter t, where $p = 0$, the continuity is maximal and equal to n. If all knots are different ($p = 1$), the continuity still has order $n - 1$. In this situation one obtains the *progressive poles*, the Riesenfeld points of B-splines. Contrary to common belief there exist cases where the break in continuity is intentional: Light lines, connection bolts, etc.

An optimum, to be defined in the following section, occurs when n is close to p^2. Similar to focusing a picture given by an optical instrument one goes from the most blurred picture - the progressive poles - to a focused picture of the points, the focus increasing with the knot multiplicity p.

If the multiplicity p reaches the value n, hence is maximal, only the knots t_i and t_f appear in the definition of the segment (t_i, t_f). In this situation one obtains the *simple poles* of the segment. The simple poles have the privilege to be unique, and to completely characterize the segment (t_i, t_f), without any outside reference.

It was through them that everything started.

We have avoided to actually write down the homogeneous variable u_i in the preceding equations in order not to overburdon these formulas unnecessarily. The indeterminacy caused by homogenization is removed by setting $u_i = 1$, in the manner of Euler.

But one could also have removed this indeterminacy by setting $t_i + u_i = 1$, in the manner of barycentric coordinates. This last alternative has the advantage to simplify generalizations to triangular, tedrahedral, etc. coordinate systems.

The fundamental relation with respect to a normalized base triangle would be $t_i + u_i + v_i = 1$. The surface is then written in terms of simple poles as

$$P(t, u, v) = \sum_{\alpha+\beta+\gamma=n} \binom{n}{\alpha\beta\gamma} t^\alpha u^\beta v^\gamma A_{\alpha\beta\gamma}, \quad t + u + v = 1, \tag{14}$$

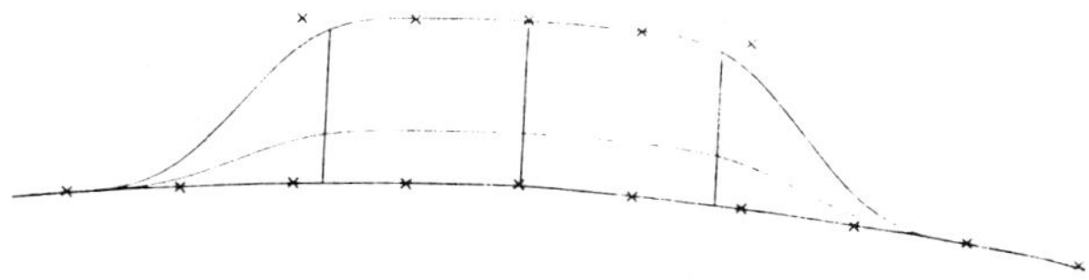

Figure 2. Deformation of a form. The deformation of the poles gives the poles of the deformation. This property is sometimes referred to as "affine invariance".

where only the poles $A_{\alpha\beta\gamma}$ have been added to the development of $(t+u+v)^n$.

In order to extend this to the case of a, necessarily regular, paving of triangles, one establishes two types of triangles. By dividing between the neighbouring pieces of the second type of triangles, one then arrives at a hexagonal paving which better conforms to the spirit of continuity.

Industrial practice teaches us that users object to this kind of paving, preferring instead the use of rectangular tiles; perhaps a reminiscence to the instant where the child wakes up to life and receives all kinds of games with a glaring domination of the right angle, as, e.g., cubes, dominos, etc...

The rectangular tile corresponds to the orderly development of a double product $\prod_{i=1}^{n}(u_i + t_i)\prod_{j=1}^{N}(U_j + T_j)$ (with the desired removal of indeterminacy) and is written as

$$P(t,T) = \sum_{p=0}^{n} \binom{n}{p} t^p \Big(\sum_{P=0}^{N} \binom{N}{P} T^P A_{p,P}\Big). \qquad (15).$$

All the operations we have shown for curves are applicable to each group of knots since the results are insensitive to the order in which these operations are performed. We can thus insert knots into the two independent chains, or take partial derivatives.

However, we may absolutely not mix up two vertices of a tile without seriously infringing the rules of logic and creating a forbidden singularity.

Let us note here that the first singularity of surfaces is manifested by the absence of a tangent plane in one point. For implicit cartesian forms the singularity will be a conic point, for explicit parametric forms we will have two collinear tangents. While from an algebraic point of view parametric surfaces are equivalent to cartesian surfaces, the same does not hold for analytic surfaces: While the result is still true for complete power series expansions, it no longer holds for limited expansions. In other words, in order to know the n-th derivative of one type, we must know derivatives up to order n^2 of the other type.

Nothing stops us from using polar volumes and beyond. We can thus find intermediates in the evolution of forms, progressively between two or several extremes, similar to how cartoonists do for celebrities. This can enrich ideas of form for creation and styling.

A particular case of this is vector addition of poles for the creation of bumps (head lights, interior elements) in a basic form. Better yet, we can thicken a surface sheet by addition and subtraction of poles, e.g. those of the form defined by $z = x^4 + y^4$ for which we take an area defined by $0 \leq x \leq 1$ and $-1 \leq y \leq 1$. In this way we can join a square blade to a hub with rounded corners and rounded edges.

Curves in higher dimensions can be used by means of their projection to E_3. This would be the case for NURBS, but careful! We lose the fundamental convex hull property: If all the pilot points are inside a convex sheet of paper, all construction points will be contained therein. If we recall the time when designers scratched each other's eyes out when it came to drawing large radius circles whose centres were unaccessible, we can appreciate the progres that we have achieved. For computation this allows us to use fixed point math as long as divisions do not intervene.

We must mention another possibility, namely rewriting the poles at a higher degree, an elementary problem. All one has to do is take the polar form of an algebraic equation and simultaneously increment n and p. It is this operation of degree raising which allows us to adjust poles locally. We can also use improper parameters, e.g. $t = w^2$, and then thicken the shape (e.g. a turbin blade) or separate even and odd terms of the form.

There exist polar surfaces that are ruled surfaces, as well as surfaces of revolution (e.g. if $r^2 = t^2 + u^2$ and $z^2 = x^2 + y^2$ one would obtain $x = tf(r^2)$, $y = uf(r^2)$, $z = rf(r^2)$.)

All modifications of forms, and the determination of poles according to given requirements will be discussed in the following section.

§3. Interpolation

It should be obvious that local expansion into a power series can advantageously be replaced by a polar form which realizes the desired continuity of the segments and of the derivatives, thus not allowing the tentative of Tchebycheff. For the usual functions e^x, $\log x$, or for the trigonometric functions, the solution is immediate: If one knows the algebraic nature of a form, it suffices to resort to the polar form.

If not, there is a problem: while it is easy to lengthen a chain of knots, the same is not true for shortening it, since this requires that certain conditions are fullfilled, unless we are content with an approximate solution, in one word with smoothing.

However, and this is of fundamental importance: If the polar curve or surface is given by points calculated from a polynomial of some reasonable degree, it would be wrong if the result of the calculation would not reproduce the algebraic nature of this polynomial exactly.

This introduces the degree of reproduction.

It would be unacceptable if going back and forth between calculating sample points and reconstructing the poles would cause an accumulation of errors. To my knowledge, the competition has hardly been concerned about this "detail". One has to respect the algebraic nature of a form up to the degree r of reproduction before some smoothing may take place beyond r.

Hence, the solution to the interpolation problem poses itself naturally: Given the degree n, the desired order of continuity c, and the degree of reproduction r, each knot is inserted with multiplicity $p = n - c$. The poles of the degree n quasi-interpolating spline are then obtained in groups of p as n-polar values of the Lagrange polynomial of degree r which is effected by $r + 1$ consecutive input points. This guarantees polynomial reproduction up to degree r.

Successive shifts in the input points, by dropping the first point in favour of the next one, successively provide all poles of the interpolating spline in groups of p, thereby respecting the given characteristics (n, c, r) where n is the degree of the spline, c is the order of continuity, and r is the degree of reproduction.

The following example should help to clarify these concepts:

Example: Suppose we wish to construct a quasi-interpolant P to a given curve $E(t)$. For simplicity we assume that the curve samples $E_i = E(t_i)$ are taken at the integers, i.e. $t_i = i$. We wish to construct a quasi-interpolant that satisfies the (n, c, r)-characteristics $(n, c, r) = (8, 5, 7)$. This means that we are looking for a degree 8 spline P with knots at the integers, that is C^5-continuous everywhere and guarantees polynomial reproduction for polynomials of degree ≤ 7.

We start by setting up the knot vector in such a way that every knot appears with multiplicity $p = n - r = 8 - 5 = 3$. A typical substring of the knot sequence is then given, e.g., by

$$\dots 2\ 2\ 2\ 3\ 3\ 3\ 4\ 4\ 4\ 5\ 5\ 5\ 6\ 6\ 6\ 7\ 7\ 7 \dots$$

The generalized poles of the quasi-interpolant P can be grouped in portions of p (according to p essentially different knot patterns), the group associated with the interval $[4, 5]$ being, e.g.,

$$\begin{aligned} P_{33444555} &= P(3,3,4,4,4,5,5,5) \\ P_{34445556} &= P(3,4,4,4,5,5,5,6) \\ P_{44455566} &= P(4,4,4,5,5,5,6,6). \end{aligned}$$

In order to quarantee polynomial reproduction up to order $r = 7$, we wish to determine these generalized poles from a Lagrange polynomial of degree 7, based on $r + 1 = 8$ of the sample values E_i. The most natural choice for the above poles is to determine these poles from the sample values

$$E_1\ E_2\ E_3\ E_4\ E_5\ E_6\ E_7\ E_8.$$

We thus set up the degree 7 Lagrange polynomial

$$L(u) = \sum_{i=1}^{8} L_i^7(u)\ E_i$$

that interpolates the sample values $E_1, \ldots, E_8$. We then raise the degree of this polynomial to $n = 8$, take its n-polar form $L(t_1, \ldots, t_n)$ and evaluate this polar form on the above substrings of the knot vector to obtain $P_{33[\ldots]} = L_{33[\ldots]}$, $P_{3[\ldots]6} = L_{3[\ldots]6}$, and $P_{[\ldots]66} = L_{[\ldots]66}$.

The relationship between the poles $P_{33[\ldots]}$, $P_{3[\ldots]6}$, $P_{[\ldots]66}$ and the sample points $E_1, \ldots, E_8$ is then given by a fixed 3×8-matrix $A_{P,E}$. The matrix $A_{P,E}$ in the example is given as

$$\frac{1}{40320}\begin{pmatrix} -85 & 908 & -5409 & 42644 & 3689 & -1836 & 461 & -52 \\ -154 & 1490 & -7146 & 25970 & 25970 & -7146 & 1490 & -154 \\ -52 & 461 & -1836 & 3689 & 42644 & -5409 & 908 & -85 \end{pmatrix}.$$

After carrying out the analogous steps for the generalized poles $P_{22[3\ldots4]}$, $P_{2[3\ldots4]5}$, $P_{[3\ldots4]55}$ and $P_{44[5\ldots6]}$, $P_{4[5\ldots6]7}$, $P_{[5\ldots6]77}$ we thus obtain a linear relationship

$$\begin{pmatrix} P_{22[333444]} \\ \vdots \\ P_{[555666]77} \end{pmatrix} = M_{P,E} \cdot \begin{pmatrix} E_0 \\ \vdots \\ E_9 \end{pmatrix}$$

that expresses all the poles influencing the quasi-interpolant P on the interval $[4, 5]$ in terms of the sample points $E_0, \ldots, E_9$.

It should be obvious from the above discussion that this method actually has reproductive power $r = 7$: If $E(t)$ is a polynomial of degree ≤ 7 then the Lagrange polynomial $L(t)$ agrees with $E(t)$, and we obtain the representation of $E(t)$ as a B-spline of degree $n = 8$ with generalized poles $\ldots, P_{22[333444]}, \ldots, P_{[555666]77}, \ldots$.

As mentioned above, it is sometimes convenient to represent every segment of a curve in terms of its simple poles. All one has to do to obtain the simple poles is to insert every knot up to multiplicity n. The matrix $M_{P_s,E}$ that expresses the simple poles $P_{44444444}$, $P_{44444445}$, $\ldots$, $P_{55555555}$ of the segment $[4, 5]$ in terms of the sample values $E_0, \ldots, E_9$ is then given as $\frac{1}{80640}$ times the following matrix:

$$\begin{pmatrix} -77 & 616 & -2156 & 4312 & 75250 & 4312 & -2156 & 616 & -77 & 0 \\ -41 & 232 & -140 & -3752 & 75250 & 12376 & -4172 & 1000 & -113 & 0 \\ -13 & -72 & 1436 & -9208 & 70770 & 23048 & -6628 & 1464 & -157 & 0 \\ 0 & -224 & 2234 & -11568 & 61810 & 35840 & -9186 & 1936 & -202 & 0 \\ 0 & -232 & 2264 & -11160 & 49448 & 49448 & -11160 & 2264 & -232 & 0 \\ 0 & -202 & 1936 & -9186 & 35840 & 61810 & -11568 & 2234 & -224 & 0 \\ 0 & -157 & 1464 & -6628 & 23048 & 70770 & -9208 & 1436 & -72 & -13 \\ 0 & -113 & 1000 & -4172 & 12376 & 75250 & -3752 & -140 & 232 & -41 \\ 0 & -77 & 616 & -2156 & 4312 & 75250 & 4312 & -2156 & 616 & -77 \end{pmatrix}$$

An application of the method for smoothing a unit pulse is shown in Fig.3. ■

Asides from their relative length, these calculations present no major difficulty whatsoever! The best compromise between c and r is obtained when the number s of different knots in a pole is close to the multiplicity p ($n = p \cdot s \neq p^2$).

One discovers then that the number of algebraic conditions respected by virtue of continuity and reproduction exceeds the number of free parameters for approximation by a factor of two.

This is very spectacular.

Furthermore, it becomes evident what could be called a particularly intense "mathematical absorption", inspired by the curve $e^{-x^2} \cdot \frac{\sin x}{x}$, although our quasi-interpolating curve stays always piecewise polynomial, and it would be difficult to talk about limits (this hasn't very much bothered Gauss when treating the bell curve e^{-x^2} as limit of the binomial coefficients $\binom{n}{p}$ for $n \to \infty$).

Whether it may be interpolation or smoothing, we are always facing a summation of terms, each of which can be considered as a unit function in one point and zero for all the others. This defines the response to a unit impulse which isn't without reminder to the theory of "wavelets", exceedingly fashionable right now. While the standard wavelet is $e^{-x^2} \cdot \cos x$, I personally prefer $e^{-x^2} \cdot \frac{\sin x}{x}$ which widens the central lobe, as is the case in physics (spherical waves or antenna radiation diagrams). This study also extends itself to the response to unevenness, as, e.g., steps of a stairway, gutters, etc.

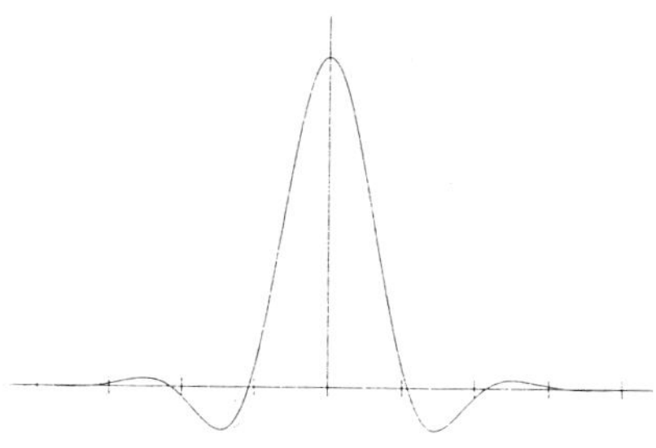

Figure 3. Interpolation or smoothing lead to a summation of unit-pulse responses. The form of this response tells us more than the whole verbal discussion.

All least square fitting, all recourse to calculus has vanished. The result is obtained through the clean path of algebra whose efficiency extends simultaneously to the smoothing of the function and of its first derivative. Nobody could have even dreamed of this otherwise.

A function such as $\frac{1}{x}$, or even better $\frac{1}{|x|}$, hardly seems to feel the effects of its journey to infinity, around $x = 0$, if the smoothing is done by sampling at the odd integers: The error is below 10% in the worst case for $\frac{1}{x}$, and below 1.5% for $\frac{1}{|x|}$.

If one was used to the 57% increase due to Gibbs' phenomenon, observed for moderate spacing, one believes to dream.

A scheme which in addition to interpolation of points also permits interpolation of poles can be set up with the help of the usual determinant. The rows for the points in the corresponding determinant contain y and all the coefficients $\binom{n}{p}t^p$. The rows for the poles contain a pole $P_{t_1,\ldots,t_n}$ and the symmetric functions $S_n^p(t_1,\ldots,t_n)$ with respect to the parameters $(t_1,\ldots,t_n)$ of this pole (Instead of starting with one, it would be more rational to start indexing the rows and columns of this matrix with zero, what the author also has done for the quaternions. The p-th column then deals with the degree p). This is good for moving up towards the progressive poles, i.e. for increasing the fuzziness for one segment.

Figure 4. Application of smoothing (1983) to the turbin blade of a ship.

§4. Conclusion and a Few Historical Remarks

Known in the U.S.A. under the name of 'Blossoming', the theory of polar forms raises the enthusiasm of the young, whereas the veterans of CAD dread the progress for fear to jeopardize the commercial success of software currently in use.

Initially it was unthinkable to use notions with such good performance. The first polar surface ever imagined was defined by 400 points taken from cubics on a parallel surface, calculated with an electomechanical machine fairly archaic by now (Tétractys Olivetti). The entire roll of printout was filled in one afternoon from hell. "This will make a nice cubic bonfire" somebody told me that evening while I was still seeing the numbers dancing in front of my eyes!

The electronic machine that calculated the 'Flying Saucer' (E 101 Burroughs) was still functioning with thyratons of gas (2 D 21) and mechanical contacts! And such were calculated the normal, the motor accelerations, the tolerances, and all the rest...!! One was satisfied with simple poles, being more suggestive, augmented by trial and error fix-ups. The idea to extend the polar form outside the $0, 1$ parameters came to my mind, and I have to thank fate that I forgot it right away.

The first simple pole entered daylight in 1958 at Citroën, who since 1960 taught about it in its school of design. In 1962 one sees the computation and manufacture of a workpiece with rectangular and triangular grids superimposed, also providing controlled joints and spiral sweeping for spherical milling, with the three tolerances calculated up to the third degree: the flying saucer which was the final achievement of about ten attempts including the DS dashboard in the automobile industry.

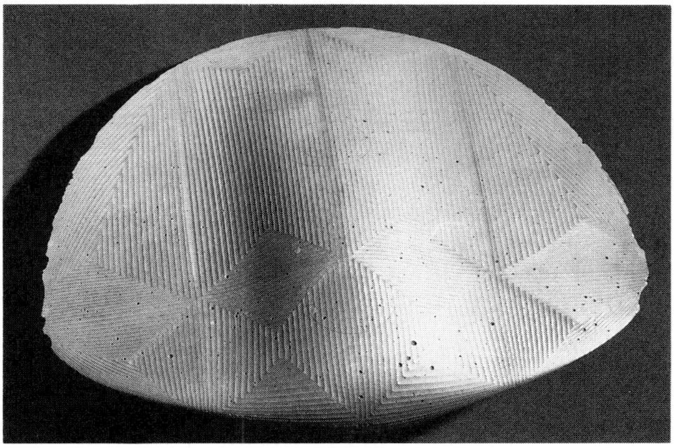

Figure 5. The 'Flying Saucer': Superimposed rectangular and triangular grids, spiral sweeping, tangential joints, the three types of tolerances calculated, manufactured by spherical milling.

The complete mathematisation of a car body was possible, first to be done for the Citroën GS. While up to then this method had been extremely controversial in the automobile industry, it now became fully functional to design forms in this way. However, although very promising, further results were not encouraged, the management probably fearing the "scientist's" desire for perfection. So one then entered an industrial phase. In order to describe the research spirit that motivated our small team let me say that we did consider manufacturing by twisted milling: This would have allowed a bi-contact based upon two nearby curves so that even the manufacturing grooves would have disappeared.

A milling machine with orientable head was at this point being built by Citroën , although not only for this purpose.

My only regret is that no film will transmit to future generations a record of our activities. One day we manufactured a cam with evolutive profile, with the cylindrical mill working on its entire length. Put on a circular platform which in turn was on a divider pivot, the cam became a success the first time through. It didn't have the slightest hint of manufacturing grooves, much to the amazement of the assistants. They had feared a bad choice in the direction of rotation of this truly 5-axis machine, and consequently they had delegated the entire responsibility for this decision on my fragile shoulders. In those moments one is always scared of the worst and of the fear to trigger a giant burst of laughter from which one may never recover.

There was no scarcity of adversity of all kinds, which has only made our stand stronger. It would be unfair here if I would not mention my supervisor, M. de la Boixière, without whom none of this would have existed. He used to say: "When an opponent is loyal, he strengthens you; flattery weakens; but nothing is worse than indifference!"

Acknowledgements. I would like to thank H.-P. Seidel for kindly undertaking the task of translating my paper from French to English.

Paul de Faget de Casteljau
Société des Automobiles Citroën
Centre Technique de Velizy
Route de Gizy chemin vicinal 2
78140 Velizy Villacoublay
France

Pyramid Patches Provide Potential Polynomial Paradigms

A. S. Cavaretta and C. A. Micchelli

Abstract. Recursive schemes are central to curve and surface representation by piecewise polynomials. Many algorithms for the evaluation, conversion, subdivision, differentiation, degree raising of piecewise polynomial curves and surfaces are given in recursive form. This has the advantage of providing computational simplicity and stability while at the same time revealing the beautiful geometric and algebraic structure of these representation forms. This paper is an exploratory presentation into the varied world of pyramidal schemes for the generation of multivariate polynomial surfaces. We focus primarily on lineal polynomials, that is, polynomials formed by products of linear forms. Through the multinomial theorem we are led to dual systems of polynomials whose properties are reflections of those possessed by lineal polynomials. Connections to multilinear symmetric forms, the multiparameter multinomial distribution and the theory of permanents is made.

§1. Introduction.

B-patches were introduced by Seidel in [6] and shown in [4] to agree with the multivariate B-spline on a certain region. Our purpose here is to establish the fundamental properties of B-patches, including dual functionals, subdivision, degree raising, and recursion formulas. All of these issues and others are discussed via a careful systematic study of the generating function for this class of polynomials. This generating function formula we view as a generalized form of the multinomial theorem. The dual polynomials appearing in this formula are explicit, simple products of linear forms. We call them lineal polynomials. Because of the simple structure of lineal polynomials, we readily read off their properties and compute with them. We are led to associated properties of the B-patches as well as other issues concerning the span of lineal polynomials.

Mathematical Methods in Computer Aided Geometric Design II
Tom Lyche and Larry L. Schumaker (eds.), pp. 69–100.

ISBN 0-12-460510-9.

The form of the generating function for the B-patches has suggested viewing it in terms of homogeneous variables. Having adopted this point of view whole-heartedly, we find that all the important properties of lineal polynomials and of B-patches have homogeneous analogues which are easily established. In Section 3 we will consider the pyramid recursion, the de Casteljau algorithm, differentiation formulas, and the dual functionals for homogeneous B-patches. In Section 5, these results are specialized to the non-homogeneous affine setting which includes the Bernstein-Bézier polynomials. Section 6 concerns other matters such as degree raising formulas, subdivision, and the Bernstein approximation operator in the B-patch setting.

B-patches are a special class of polynomials which satisfy a pyramid recursion, as defined in [4,6]. In Section 4 we shall observe that a single lineal polynomial determines a family of polynomials which also satisfies a pyramid recursion. In special cases, these polynomials correspond to the discrete multinomial multiparameter probability distribution. These polynomials, called M-patches, also generalize the Bernstein-Bézier polynomials and we study them from a similar point of view. Many questions of this section remain unanswered, and in particular linear independence of the M-patches has not been established. Nonetheless, we provide some sufficient conditions for their independence.

We will work in $H_n(\mathbb{R}^s)$, the space of homogeneous polynomials of degree n in s variables, where n and s are both positive integers. Therefore we will use the index set $\Gamma = \Gamma_{s,n} \subseteq Z_+^s$ defined by

$$\Gamma_{s,n} := \{\alpha \in Z_+^s : |\alpha| = n\},$$

where

$$\alpha = (\alpha_1, \cdots, \alpha_s)$$

and

$$|\alpha| = \alpha_1 + \cdots + \alpha_s.$$

Fundamental to our study is the notion of a *lineal polynomial* given by

$$a(x) := a(x|X) := \prod_{\ell=1}^{n} (x, x^\ell), \qquad x \in \mathbb{R}^s, \tag{1.1}$$

where

$$X = \{x^\ell : \ell = 1, \cdots, n\} \subseteq \mathbb{R}^s,$$

and (x, y) denotes the standard inner product of x and y in $\mathbb{R}^s$. While the polynomial a depends only on the set X and not on the ordering of the set, we will nevertheless also let X denote the $s \times n$ matrix whose j^{th} column is x^j. Clearly $a(\cdot|X) \in H_n(\mathbb{R}^s)$. Such polynomials have occurred in the literature before, and in particular in Dahmen and Micchelli [3], it is shown that $q \in H_n(\mathbb{R}^s)$ is lineal if and only if for each pair $x, y \in \mathbb{R}^s, q(x + ty)$ has only real roots in the variable $t \in \mathbb{C}$, or it vanishes identically. These polynomials and the entire functions approximated by them are also important in the classification of distributions which are limits of multivariate B-splines [1] and have occurred even earlier in work of Motzkin and Schoenberg [5].

Given the matrix X and any $\alpha \in \Gamma_{s,n}$ we define an $n \times n$ matrix $X(\alpha)$ by repeating the ith row of X exactly α_i times. Then the following identity, given in terms of permanent, is known and useful for us:

$$a(x|X) = \sum_{\alpha \in \Gamma_{s,n}} \frac{x^\alpha}{\alpha!} \operatorname{per} X(\alpha). \tag{1.2}$$

We will demonstrate its validity shortly. Here we have used the standard multinomial notion $x^\alpha = x_1^{\alpha_1} \cdots x_s^{\alpha_s}, \alpha! = \alpha_1! \cdots \alpha_s!$, and per $X(\alpha)$ denotes the permanent of the matrix $X(\alpha)$. Recall that the permanent of an $n \times n$ matrix $B = (b_{ij}), i, j = 1, \cdots, n$ is defined as

$$\operatorname{per} B = \sum_\pi b_{1,\pi(1)} \cdots b_{n,\pi(n)},$$

where the sum is over all permutations π of $\{1, \cdots, n\}$.

Some very special cases of lineal polynomials and the above formula are the monomials x^α when the set X contains the ith unit vector e^i exactly α_i times. Also when the set X is simply a fixed vector $y \in \mathbb{R}^s$ repeated n times, then $a(x) = (x, y)^n$ and the above formula is an instance of the multinomial theorem. This says, by the way, that for $Y = [y, \cdots, y]$,

$$\operatorname{per} Y(\alpha) = n! y^\alpha, \qquad \alpha \in \Gamma_{s,n} \tag{1.3}$$

which is also obvious from the definition above.

Lineal polynomials also occur as the generating function of a certain distribution. If we require that the matrix X be column stochastic, then the jth factor (x, x^j) in the product $a(x|X)$ is the generating function of a single trial of an experiment whose s possible outcomes occur with probability $x_i^j, i = 1, \cdots, s$. Suppose now there are n such independent trials each of whose probability distribution may be different. Let X_i denote the number of times the outcome i occurs in n trials, $i = 1, \cdots, s$ and let $X = (X_1, \cdots, X_s)$. Then we say that $X = (X_1, \cdots, X_s)$ has the (multiparameter) multinomial distribution. If $\alpha \in \Gamma_{s,n}$, then the probability that $X = \alpha$ is given by the coefficient of x^α in (1.2).

§2. Linear Independence of Families of Lineal Polynomials

As remarked above, $a(\cdot|X) \in H_n(\mathbb{R}^s)$ for every $X = \{x^1, \cdots, x^n\}$. Since the dimension of $H_n(\mathbb{R}^s)$ is $|\Gamma_{s,n}| = \binom{n+s-1}{n}$, it is natural to ask when a family $\{a_\alpha : \alpha \in \Gamma_{s,n}\}$ of lineal polynomials is linearly independent and hence a basis for $H_n(\mathbb{R}^s)$. This question is basic for much of what follows.

We have already encountered this linear independence question before in the study of the multivariate B-spline. For the convenience of the reader, we briefly review this connection. Here our discussion takes place in affine space. The multivariate B-spline can be defined as follows: given the $s \times n$ matrix $X = [x^1, \cdots, x^n]$, $x^i \in \mathbb{R}^s, i = 1, \cdots, n$, $M(\cdot|X)$ is the multivariate distribution

$$M(\cdot|X)f := \int_{S_n} f(Xt)dt,$$

where $f \in C_0^\infty(\mathbb{R}^s)$ and

$$S_n = \left\{ t = (t_1, \cdots, t_n) : \sum_{i=1}^n t_i = 1, t_i \geq 0, i = 1, \cdots, n \right\}.$$

When X has rank s, M corresponds to a function which has been intensively studied. We note only the formula

$$\int_{\mathbb{R}^s} \frac{M(t|X)}{(1+i(t,x))^n} dt = \frac{1}{(n-1)! \prod_{k=1}^n (1+i(x,x^k))}, \qquad i = \sqrt{-1}, x \in \mathbb{R}^s$$

from [1]. This formula follows from the definition above by noting that

$$\int_{R^s} \frac{M(t|X)}{(1+i(t,x))^n} dt = \int_{S_n} \frac{1}{(1+i\sum_{k=1}^n t_k(x,x^k))^n} dt = \int_{S_n} q^{(n-1)} \left(\sum_{k=1}^n t_k y_k \right) dt,$$

where $q(t) := (-1)^{n-1}(1+t)^{-1}/(n-1)!$ and $y_k := i(x,x^k)$. This last integral can be evaluated as the $n-1$ divided difference of q at $y_1, \cdots, y_n$ which can be shown to be equal to

$$\frac{1}{(n-1)!(1+y_1)\cdots(1+y_n)} = \frac{1}{(n-1)! \prod_{k=1}^n (1+i(x,x^k))}$$

as claimed above.

To apply the above formula we note the fact that whenever $f \in C_0(\mathbb{R}^s)$ and

$$\int_{R^s} \frac{f(t)}{(1+i(t,x))^n} dt = 0, \quad x \in \mathbb{R}^s$$

then $f = 0$. Now, let $X_\alpha := [x_1^\alpha, \cdots, x_n^\alpha]$ be a family of vectors in $\mathbb{R}^s$ indexed by α in some set Δ. Then we see that the family of multivariate B-splines $\{M(\cdot|X^\alpha) : \alpha \in \Delta\}$ are linearly independent on $\mathbb{R}^s$ if and only if the family of reciprocal polynomials $\{a^{-1}(\cdot|iX^\alpha) : \alpha \in \Delta\}$ are linearly independent. Equivalently, if we let

$$X = \bigcup_{\alpha \in \Delta} X^\alpha = \{x_j^\alpha : \alpha \in \Delta, 1 \leq j \leq s\},$$

the condition is the linear independence of the lineal polynomials $\{a(\cdot|Y^\alpha) : \alpha \in \Delta\}$ where $Y^\alpha := X \backslash X^\alpha$. Results about independence of families of multivariate B-splines appear in [3].

A general criterion for linear independence of lineal polynomials can be given in terms of multilinear symmetric forms in n vector variables $y^1, \cdots, y^n$ where $y^i \in \mathbb{R}^s$, $i = 1, \cdots, n$. These functions are called *polar forms* or *blossoms* and we denote this class by $\mathcal{P}_{s,n}$. It is easily seen using the polar identity that for $p \in \mathcal{P}_{s,n}$, $(\text{diag } p)(y) := p(y, \cdots, y) = 0$ for all $y \in \mathbb{R}^s$ only if p vanishes identically.

We further claim that

$$\dim \mathcal{P}_{s,n} = \dim H_n(\mathbb{R}^s). \tag{2.1}$$

To see this, define the functions

$$\text{per}_\alpha(y^1, \cdots, y^n) = \text{ per } Y(\alpha) \tag{2.2}$$

where $Y = [y^1, \cdots, y^n]$. Clearly $\text{per}_\alpha \in \mathcal{P}_{s,n}$, and

$$\text{per}_\alpha(y, \cdots, y) = n! y^\alpha, \qquad y \in \mathbb{R}^s \tag{2.3}$$

which we observed in (1.3) follows from (1.2) and the multinomial theorem. By our previous remark, it is clear that for any coefficients $\{c_\alpha : \alpha \in \Gamma_{s,n}\}$, $\sum_{\alpha \in \Gamma_{s,n}} c_\alpha \text{ per}_\alpha(y^1, \cdots, y^n)$ vanishes identically if and only if $\sum_{\alpha \in \Gamma} c_\alpha$ $(\text{diag per}_\alpha)(y)$ does.

Since the set of monomials $\{m_\alpha : \alpha \in \Gamma_{s,n}\}$, $m_\alpha(y) := y^\alpha$, are linearly independent in $H_n(\mathbb{R}^s)$, it follows from (2.3) that $\{\text{per}_\alpha : \alpha \in \Gamma_{s,n}\}$ is a linearly independent set in $\mathcal{P}_{s,n}$. Therefore we conclude that

$$\dim \mathcal{P}_{s,n} \geq |\Gamma_{s,n}| = \dim H_n(\mathbb{R}^s). \tag{2.4}$$

To obtain the reverse inequality, let $\{p_\alpha : \alpha \in \Delta\}$ be any other basis for $\mathcal{P}_{s,n}$, where Δ is an appropriate index set. Since p_α is multilinear with n factors, diag $p_\alpha \in H_n(\mathbb{R}^s)$. Since $\{p_\alpha : \alpha \in \Delta\}$ is a basis, for any scalars $\{c_\alpha : \alpha \in \Delta\}$, not all $c_\alpha = 0$,

$$\sum_{\alpha \in \Delta} c_\alpha p_\alpha \neq 0,$$

and therefore by our initial remark

$$\sum_{\alpha \in \Delta} c_\alpha \text{ diag } p_\alpha \neq 0.$$

Thus $\{\text{diag } p_\alpha : d \in \Delta\}$ is a linearly independent set in $H_n(\mathbb{R}^s)$ and so $|\Delta| \leq \dim H_n(\mathbb{R}^s)$, which together with (2.4) establishes the claim (2.1).

Another way of putting these observations is that the map diag $: \mathcal{P}_{s,n} \to H_{s,n}$ is a bijection where in addition we are pointing out that $\{\text{per}_\alpha : \alpha \in \Gamma_{s,n}\}$ spans $\mathcal{P}_{s,n}$.

As an application of the above remark we return to equation (1.2). For each fixed $x \in \mathbb{R}^s$ both sides of (1.2) are symmetric linear forms in $x^1, \cdots, x^n$ and therefore are in $\mathcal{P}_{s,n}$. Using formula (1.3) which readily follows from the definition of permanent, we see by the multinomial theorem that both sides of (1.2) agree when $x^1 = \cdots = x^n = y$ for any $y \in \mathbb{R}^s$. Thus, since the diagonal map is a bijection (1.2) persists for any $x^1, \cdots, x^n$.

These observations lead us to our first result on linear independence of lineal functions.

Proposition 2.1. *Let $X_\alpha = [x_{\alpha,1}, \cdots, x_{\alpha,n}]$ be a collection of $s \times n$ matrices indexed by $\alpha \in \Gamma_{s,n}$. Then $\{a(\cdot|X_\alpha) : \alpha \in \Gamma_{s,n}\}$ is a linearly independent set of lineal polynomials if and only if whenever $p(X_\alpha) = 0, \alpha \in \Gamma_{s,n}$ for $p \in \mathcal{P}_{s,n}$, then $p = 0$ identically.*

Proof: From (1.2) for $\alpha \in \Gamma_{s,n}$

$$a(x|X_\alpha) = \sum_{\beta \in \Gamma_{s,n}} \frac{x^\beta}{\beta!} \text{ per } X_\alpha(\beta).$$

So considering a linear combination with coefficients $\{c_\alpha : \alpha \in \Gamma_{s,n}\}$ we find

$$\sum_{\alpha \in \Gamma_{s,n}} c_\alpha a(x|X_\alpha) = \sum_{\beta \in \Gamma_{s,n}} \frac{x^\beta}{\beta!} \sum_{\alpha \in \Gamma_{s,n}} c_\alpha \text{ per } X_\alpha(\beta).$$

By the linear independence of monomials, this sum vanishes if and only if

$$\sum_{\alpha \in \Gamma_{s,n}} c_\alpha \text{ per } X_\alpha(\beta) = 0, \qquad \beta \in \Gamma_{s,n}. \tag{2.5}$$

But from linear algebra, (2.5) holds non-trivially if and only if there exist scalars d_β such that

$$\sum_{\beta \in \Gamma_{s,n}} d_\beta \text{ per } X_\alpha(\beta) = 0, \qquad \alpha \in \Gamma_{s,n}, \tag{2.6}$$

where $d_\beta \neq 0$ for some $\beta \in \Gamma_{s,n}$. So equivalently

$$p(x_1, \cdots, x_n) := \sum_{\beta \in \Gamma_{s,n}} d_\beta \text{ per}_\beta(x_1, \cdots, x_n)$$

is a non trivial member of $\mathcal{P}_{s,n}$ (recall that the functions per_β are a basis for $\mathcal{P}_{s,n}$) and clearly from (2.6)

$$p(x_{\alpha,1}, \cdots, x_{\alpha,n}) = 0.$$

This establishes the equivalence of our two conditions. ■

Remark 2.1. The conditions on X_α in Proposition 2.1 mean that for any data $\{d_\alpha : \alpha \in \Gamma_{s,n}\}$ the interpolation problem $p(x_{\alpha,1}, \cdots, x_{\alpha,n}) = d_\alpha, \alpha \in \Gamma_{s,n}$ is solvable over $\mathcal{P}_{s,n}$.

As a special case of Proposition (2.1), the so called ridge functions $a_\alpha(x) = (x, y_\alpha)^n$, $\alpha \in \Gamma_{s,n}$, are linearly independent if and only if no non-trivial $p \in H_n(\mathbb{R}^s)$ vanishes at each point y_α since in this case $X_\alpha = [y_\alpha, \cdots, y_\alpha]$.

We turn now to a construction which always yields for $H_n(\mathbb{R}^s)$ a basis consisting of lineal polynomials. The basis to be constructed is determined by a fixed $s \times n$ array W whose $(i,j)^{\text{th}}$ entry $x^{i,j}$ is a vector in $\mathbb{R}^s$. For

$\alpha \in \Gamma_{s,n}$ we choose a set X_α of n vectors according to the following recipe. If $\alpha = (\alpha_1, \cdots, \alpha_s)$ then we choose $x^{i,1}, \cdots, x^{i,\alpha_i}$ from the array for $i = 1, \cdots, s$. If $\alpha_i = 0$, we do not select any vectors from the ith column. The set X_α consists of all of these and since $|\alpha| = n$, X_α has exactly n members, with multiplicities allowed. Put succinctly, we have

$$X_\alpha = \{x^{i,j} : 1 \le j \le \alpha_i, i = 1, \cdots, s\}. \tag{2.7}$$

Then we define

$$a_\alpha(x) := a(x|X_\alpha), \qquad \alpha \in \Gamma_{s,n}. \tag{2.8}$$

Schematically, we think of the vectors of W stacked in s vertical columns where the vertical axis represents the levels $\ell = 1, \cdots, n$. Thus our selection process of X_α from W corresponds to the graph of a histogram with n lattice points below its graph. Our linear independence condition in Theorem 2.1 below means that any choice of one vector from each column must result in a linearly independent set of vectors if the sum of the chosen levels does not exceed $s + n - 1$. In the terminology of [4], the ith vertical column of vectors is thought of as a "cloud" in the neighborhood of the ith vertex of a base simplex in $\mathbb{R}^s$.

Theorem 2.1. *For any integer $s \ge 2$, the lineal polynomials $\{a_\alpha : \alpha \in \Gamma_{s,n}\}$, chosen according to (2.7) and (2.8), form a basis for $H_n(\mathbb{R}^s)$ provided for all $\beta \in \mathbb{Z}^s_+, |\beta| \le n - 1$, the vectors $\{x^{i,\beta_i+1} : i = 1, \cdots, s\}$ are linearly independent in $\mathbb{R}^s$.*

We prove this theorem by a simultaneous induction on s and n. Such a proof relies on the obvious fact

$$X_{\alpha - e^i} = X_\alpha - \{x^{i,\alpha_i}\}.$$

Proof: For $n = 1$, the result is elementary. The linear forms $a_{e^i}(x) = (x, x^{i,1})$, $i = 1, \cdots, s$, are linearly independent if and only if $\{x^{i,1} : i = 1, \cdots, s\}$ is a linearly independent set in $\mathbb{R}^s$.

This trivial case aside, we now fix $s = 2$ and $n \ge 2$. In this case the construction yields $n + 1$ polynomials explicitly given by

$$a_\alpha(x) = \prod_{i=1}^{\alpha_1}(x, x^{1,i}) \prod_{j=1}^{\alpha_2}(x, x^{2,j}), \qquad \alpha \in \Gamma_{2,n}.$$

By assumption, $x^{1,1}$ and $x^{2,j}$ are linearly independent for $j = 1, \cdots, n-1$, and so there exists y for which $(y, x^{1,1}) = 0$ but $(y, x^{2,j}) \ne 0$ for $j = 1, \cdots, n-1$. Therefore $a_\alpha(y) = 0$ for all $\alpha \in \Gamma_{2,n}$ except $\alpha = (0, n)$, and $a_{0,n}(y) \ne 0$. Now suppose

$$\sum_{\alpha \in \Gamma_{2,n}} c_\alpha a_\alpha = 0 \tag{2.9}$$

for certain scalars c_α. Then

$$0 = \sum_{\alpha \in \Gamma_{2,n}} c_\alpha a_\alpha(y) = c_{0,n} a_{0,n}(y)$$

from which we conclude that $c_{0,n} = 0$. For the remaining $c_\alpha a_\alpha, \alpha \neq (0, n)$, which appear in (2.9), the factor $(x, x^{1,1})$ appears in each a_α. This factor we eliminate from (2.9), obtaining

$$\sum_{\alpha \in \Gamma_{2,n-1}} c_{\alpha+e^1} b_\alpha(x) = 0,$$

where

$$b_\alpha(x) = a(x|Y_\alpha), \quad \alpha \in \Gamma_{2,n-1}, \quad x \in \mathbb{R}^2,$$

and the Y_α are selected by our recipe from the array consisting of

$$x^{1,r}, r = 2, \cdots, n$$
$$x^{2,r}, r = 1, \cdots, n-1.$$

Applying the induction hypothesis for $n-1$ (and $s = 2$), we conclude $c_{\alpha+e^1} = 0, \alpha \in \Gamma_{2,n-1}$, and so (2.9) can hold only trivially. So the theorem is proved when $s = 2$.

Let $s \geq 3$ and $n > 1$, and consider the equation

$$\sum_{\alpha \in \Gamma_{s,n}} c_\alpha a_\alpha = 0$$

which we write as

$$\sum_{\substack{\alpha \in \Gamma_{s,n} \\ \alpha_1 = 0}} c_\alpha a_\alpha(x) + \sum_{\substack{\alpha \in \Gamma_{s,n} \\ \alpha_1 \neq 0}} c_\alpha a_\alpha(x) = 0, x \in \mathbb{R}^s. \tag{2.10}$$

Pick an orthogonal transformation $U : \mathbb{R}^s \to \mathbb{R}^s$ satisfying $Ux^{1,1} = (1, 0, \cdots, 0)$ and put $x = U^T[0, y]$ where $[0, y]$ is any vector whose first coordinate is zero and whose remaining $s-1$ coordinates are $y = (y_2, \cdots, y_s) \in \mathbb{R}^{s-1}$. Note that

$$a_\alpha(x) = \prod_{\ell=1}^{s} \prod_{r=1}^{\alpha_\ell} ([0, y], Ux^{\ell,r}). \tag{2.11}$$

Since $Ux^{1,1} = (1, 0, \cdots, 0), a_\alpha(x) = 0$, if $\alpha_1 \neq 0$ and $x = U^T[0, y]$. With x so specialized, (2.10) becomes

$$\sum_{\substack{\alpha \in \Gamma_{s,n} \\ \alpha_1 = 0}} c_\alpha a_\alpha(U^T[0, y]) = 0. \tag{2.12}$$

So the vectors $x^{1,r}, r = 1, \cdots, n$ are not present in equation (2.12). By (2.11) we see that (2.12) is a linear combination of lineal polynomials of degree n in

$s-1$ variables $y = (y_2, \cdots, y_s)$. These lineal polynomials are determined by the array

$$w^{\ell,r} := PUx^{\ell,r}, \quad \ell = 2, \cdots, s, \quad r = 1, \cdots, n,$$

where P is the projection $(x_1, \cdots, x_s) \to (0, x_2, \cdots, x_s)$. Since $Ux^{1,1} = (1, 0, \cdots, 0)$ and U is an orthogonality transformation the vectors $w^{\ell,r}$, $\ell = 2, \cdots, s$, $r = 1, \cdots, n$ inherit the linear independence condition

$$\det |w^{\ell,\alpha_\ell}|_{\ell=2}^{s} \neq 0$$

whenever $(\alpha_2, \cdots, \alpha_s)$ satisfies $\sum_{\ell=2}^{s} \alpha_\ell \leq n-1$. So the induction hypothesis can be applied to (2.12) with s replaced by $s-1$ and we conclude that $a_\alpha = 0$ provided $|\alpha| = n$ and $\alpha_1 = 0$. Then (2.10) reduces to

$$\sum_{\substack{\alpha \in \Gamma_{s,n} \\ \alpha_1 \neq 0}} c_\alpha a_\alpha(x) = 0, \qquad x \in \mathbb{R}^s. \tag{2.13}$$

Clearly the sum in (2.13) contains the common factor $(x, x^{1,1})$. Moreover, no lineal polynomial appearing in (2.13) contains the factors $(x, x^{\ell,n}), \ell = 2, \cdots, s$. So the lineal polynomials

$$h_\beta(x) = \frac{a_\alpha(x)}{(x, x^{11})}, \quad \beta = \alpha - e^1, \quad \alpha_1 \neq 0, \quad x \in \mathbb{R}^s$$

are obtained by our special construction from the array

$$\begin{aligned} &x^{1,r}, r = 2, \cdots, n \\ &x^{\ell,r}, r = 1, \cdots, n-1, \qquad \ell = 2, \cdots, s. \end{aligned} \tag{2.14}$$

So we obtain from (2.13)

$$\sum_{\beta \in \Gamma_{s,n-1}} c_{\beta+e^1} b_\beta(x) = 0, \qquad x \in \mathbb{R}^s. \tag{2.15}$$

The array (2.14) clearly satisfies the linear independence condition necessary to apply the induction hypothesis to (2.15) in which the lineal polynomials have degree $n-1$. Therefore $c_{\beta+e^1} = 0$ for all $\beta \in \Gamma_{s,n-1}$. So (2.13) is a trivial sum and the induction proof of Theorem 2.1 is complete. ∎

We let $\prod_n(\mathbb{R}^s)$ be the space of all polynomials of total degree at most n in s variables and state below an affine (nonhomogeneous) form of the above result.

Corollary 2.1. *Let $\Sigma := \{\tau^{\ell,r} : \ell = 1, \cdots, s+1, r = 1, \cdots, n\}$ be an array of vectors in $\mathbb{R}^s$. For $\alpha \in \Gamma_{s+1,n}$, define $\psi_\alpha(\mathbb{R}^s)$ by*

$$\psi_\alpha(\omega) = \prod_{\ell=1}^{s+1} \prod_{r=1}^{\alpha_\ell} (1 + (\omega, \tau^{\ell,r})), \qquad \omega \in \mathbb{R}^s. \tag{2.16}$$

Then $\{\psi_\alpha : \alpha \in \Gamma_{s+1,n}\}$ is a basis for $\prod_n(\mathbb{R}^s)$ provided for all $\alpha \in \mathbb{Z}_+^s$ satisfying $|\alpha| \leq n-1$, the set $\{\tau^{\ell,\alpha_\ell+1} : \ell = 1, \ldots, s+1\}$ is affinely independent.

Proof: Put $x = (1, \omega)$ and $x^{\ell,r} = (1, \tau^{\ell,r})$, $\ell = 1, \cdots, s+1, r = 1, \cdots, n$ and apply Theorem 2.1. ∎

§3. B-patches for $\mathbf{H_n(\mathbb{R}^s)}$.

Throughout this section we assume that a given $s \times m$ array W of vectors in $\mathbb{R}^s$ satisfies the hypothesis of Theorem 2.1 and that a basis of lineal polynomials $\{a_\alpha^n : \alpha \in \Gamma_{s,n}\}$ for $H_n(\mathbb{R}^s)$ has been determined according to (2.7) and (2.8) for each n, $1 \le n \le m$.

Clearly for fixed y in $\mathbb{R}^s$ the polynomial $(x,y)^n$ is in $H_n(\mathbb{R}^s)$ and so

$$(x,y)^n = \sum_{\alpha \in \Gamma_{s,n}} a_\alpha^n(x) b_\alpha^n(y), \tag{3.1}$$

where the coefficients $b_\alpha^n(y), |\alpha| = n$, are uniquely determined. Viewing these coefficients as functions of y, we call each b_α^n a *B-patch*. As a very special simple case of (3.1), suppose W_b is the array $x^{i,j} = e^i, j = 1, \cdots, n$ and $i = 1, \cdots, s$, where $e^i \in \mathbb{R}^s$ is the ith coordinate vector. Then $a_\alpha^n(x) = x^\alpha$ and the multinomial theorem applied to $(x,y)^n$ in (3.1) shows that $b_\alpha^n(y) = \binom{n}{\alpha} y^\alpha$ which are the Bernstein-Bézier polynomials.

Our first goal is to establish a pyramid recursion for the B-patches directly from (3.1), from which it will follow that $b_\alpha^n \in H_n(\mathbb{R}^s)$. To this end, we note that for any $\alpha \in \Gamma_{s,n}$, the vectors $\{x^{i,\alpha_i+1} : i = 1, \cdots, s\}$ are linearly independent, according to our assumption in Theorem 2.1, and so they determine linear functionals $u_{i,\alpha}$ on $\mathbb{R}^s$ satisfying

$$y = \sum_{i=1}^s u_{i,\alpha}(y) x^{i,\alpha_i+1}, \qquad y \in \mathbb{R}^s. \tag{3.2}$$

We adopt the convention $u_{i,\alpha} = 0, i = 1, \cdots, s$, if α has a negative component. It follows from (2.7) and (2.8)

$$(x,y) a_\alpha^n(x) = \sum_{i=0}^s u_{i,\alpha}(y) a_{\alpha+e^i}^{n+1}(x). \tag{3.3}$$

Multiplying (3.1) as it stands by (x,y) and using (3.3), we obtain

$$\begin{aligned}(x,y)^{n+1} &= \sum_{\alpha \in \Gamma_{s,n}} \sum_{i=1}^s u_{i,\alpha}(y) a_{\alpha+e^i}^{n+1}(x) b_\alpha^n(y) \\ &= \sum_{i=1}^s \sum_{\substack{\alpha \in \Gamma_{s,n+1} \\ \alpha_i \ge 1}} u_{i,\alpha-e^i}(y) a_\alpha^{n+1}(x) b_{\alpha-e^i}^n(y) \\ &= \sum_{\alpha \in \Gamma_{s,n+1}} a_\alpha^{n+1}(x) \sum_{i=1}^s u_{i,\alpha-e^i}(y) b_{\alpha-e^i}^n(y),\end{aligned}$$

where we use our convention concerning $u_{i,\alpha}$ when α has a negative component. Comparing this to (3.1) with n replaced by $n+1$, we conclude that

$$b_\alpha^{n+1}(y) = \sum_{i=1}^s u_{i,\alpha-e^i}(y) b_{\alpha-e^i}^n(y), \qquad \alpha \in \Gamma_{s,n+1}. \tag{3.4}$$

Since the $u_{i,\alpha}$ are all linear forms in y, a straightforward induction based on (3.4) shows that b_α^n is a homogeneous polynomial of degree n in y for all n.

Moreover, $\{b_\alpha^n : \alpha \in \Gamma_{s,n}\}$ forms a basis for $H_n(\mathbb{R}^s)$. To check this, we first observe that for appropriate choices $x_\alpha, \alpha \in \Gamma_{s,n}, \{(x_\alpha, y)^n : \alpha \in \Gamma_{s,n}\}$ itself is a basis for $H_n(\mathbb{R}^s)$. One convenient choice is $x_\alpha = \alpha, \alpha \in \Gamma_{s,n}$, for in this case the linear independence of $(\alpha, y)^n$ follows from the nonvanishing determinant

$$[(\alpha, \beta)^n]_{\alpha,\beta\in\Gamma_{s,n}}$$

whose value is calculated using the multinomial theorem. So since the span of $\{(x_\alpha, y)^n : \alpha \in \Gamma_{s,n}\}$ is all of $H_n(\mathbb{R}^s)$, it follows directly from (3.1) that likewise $\{b_\alpha^n : \alpha \in \Gamma_{s,n}\}$ spans $H_n(\mathbb{R}^s)$. Since $\dim H_n(\mathbb{R}^s) = |\Gamma_{s,n}|$ we conclude that $\{b_\alpha^n : \alpha \in \Gamma_{s,n}\}$ is in fact a basis for $H_n(\mathbb{R}^s)$, hence linearly independent.

Equation (3.1) can be interpreted in terms of the blossom or polar form of $(x, y)^n$, viewed as a function of y. Fixing x, the polar form is

$$a_b^n(y_1, \cdots, y_n) = (x, y_1) \cdots (x, y_n).$$

So evaluating the blossom on each set X_α, we can express (3.1) as

$$(x, y)^n = \sum_{\alpha\in\Gamma_{s,n}} a_b^n(X_\alpha) b_\alpha(y).$$

It follows that any polynomial in $H_n(\mathbb{R}^s)$ can be expressed as a linear combination of the B-patches b_α by evaluating its blossom on the sets X_α to obtain the coefficients. More precisely,

$$p(y) = \sum_{\alpha\in\Gamma_{s,n}} c_\alpha b_\alpha(y),$$

where $c_\alpha = p_b(X_\alpha)$ and p_b is the nth order polar form of p. As an example, we have

$$y^\beta = \sum_{\alpha\in\Gamma_{s,n}} \frac{1}{n!} \operatorname{per}_\beta(X_\alpha) b_\alpha(y), \qquad \beta \in \Gamma_{s,n}.$$

The recursion given by (3.4) could be called the "up" pyramid recursion in that the B-patches of degree $n+1$ are computed from the "neighboring" B-patches of degree n. Dual to this is a "down" recursion which operates on the coefficients in the expression

$$p(y) = \sum_{\alpha\in\Gamma_{s,n}} c_\alpha b_\alpha^n(y). \tag{3.5}$$

By inserting (3.4) into (3.5) we obtain for $\ell = 1, \cdots, n$

$$p(y) = \sum_{\alpha\in\Gamma_{s,n-\ell}} c_\alpha^\ell(y) b_\alpha^{n-\ell}(y), \tag{3.6}$$

where

$$\begin{aligned} c^0_\alpha(y) =& c_\alpha, \quad \alpha \in \Gamma_{s,n} \\ c^\ell_\alpha(y) =& \sum_{i=1}^{s} u_{i,\alpha}(y) c^{\ell-1}_{\alpha+e^i}(y), \qquad \alpha \in \Gamma_{s,n-\ell}. \end{aligned} \tag{3.7}$$

Since $b^0_0 = 1$, we conclude that

$$p(y) = c^n_0(y)$$

so that (3.6) and (3.7) provides the B-patch pyramid scheme which generalizes the de Casteljau algorithm corresponding to the array W_b to homogeneous B-patches.

To obtain a derivative recursion for the B-patches we apply the directional derivative operator D_z to (3.1) in the variable y to obtain

$$n(x,y)^{n-1}(x,z) = \sum_{\alpha \in \Gamma_{s,n}} a^n_\alpha(x) D_z b^n_\alpha(y). \tag{3.8}$$

Using (3.1) with n replaced by $n-1$ we can express the left hand side of (3.8) as

$$n(x,y)^{n-1}(x,z) = n \sum_{j=1}^{s} \sum_{\alpha \in \Gamma_{s,n-1}} u_{j,\alpha}(z)(x, x^{j,\alpha_j+1}) a^{n-1}_\alpha(x) b^{n-1}_\alpha(y),$$

where for each $\alpha \in \Gamma_{s,n-1}$ we have used (3.2) with z in place of y. Putting $\beta = \alpha + e^j$ in the inner sum and using (2.7), (2.8), and our convention on $u_{j,\alpha}(z)$, we obtain

$$\begin{aligned} n(x,y)^{n-1}(x,z) &= n \sum_{j=1}^{s} \sum_{\beta \in \Gamma_{s,n}} u_{j,\beta-e^j}(z) a^n_\beta(x) b^{n-1}_{\beta-e^j}(y) \\ &= n \sum_{\beta \in \Gamma_{s,n}} a^n_\beta(x) \sum_{j=1}^{s} u_{j,\beta-e^j}(z) b^{n-1}_{\beta-e^j}(y). \end{aligned}$$

Comparing this last equation with (3.8) we obtain the differential recursion for B-patches:

Proposition 3.1.

$$D_z b^n_\alpha(y) = n \sum_{j=1}^{s} u_{j,\alpha-e^j}(z) b^{n-1}_{\alpha-e^j}(y), \qquad \alpha \in \Gamma_{s,n}. \tag{3.9}$$

To conclude this section on homogeneous B-patches, we observe that the dual linear functionals $\{\ell_\alpha : \alpha \in \Gamma_{s,n}\}$ satisfying

$$\ell_\alpha(b_\beta) = \begin{cases} 1, & \alpha = \beta \\ 0, & \alpha \neq \beta \end{cases} \tag{3.10}$$

are quite easily identified. For the argument $x = (x_1, \cdots, x_s)$ we write the differential operator $D = \left(\frac{\partial}{\partial x_1}, \cdots, \frac{\partial}{\partial x_s}\right)$ so that for $x^\alpha, \alpha \in \Gamma_{s,n}$, we obtain the nth order partial derivative D^α. Using this notation, we define the homogeneous differential operators

$$\ell_\alpha := \frac{1}{n!} a_\alpha(D), \quad \alpha \in \Gamma_{s,n}. \tag{3.11}$$

Proposition 3.2. *The homogeneous differential operators $\{\ell_\alpha : \alpha \in \Gamma_{s,n}\}$ of (3.11) provide a dual basis for the B-patches, that is, they satisfy (3.10).*

Proof: Applying ℓ_α to $(x, y)^n$ in the variable y, we obtain

$$\ell_\alpha(a_n) = a_\alpha(x).$$

So applying ℓ_α to both sides of (3.1) we obtain

$$a_\alpha(x) = \sum_{\beta \in \Gamma_{s,n}} a^n_\beta(x) \ell_\alpha(b^n_\beta)$$

from which we conclude (3.10) as desired. ■

More properties of B-patches will follow later. Specifically we describe in Section 5 the affine version of Proposition 3.2 in detail. Section 6 contains results about degree raising, conversion, and subdivision for B-patches. But first we consider B-patches in the wider context of pyramid schemes.

§4. Other Pyramid Schemes.

The B-patch pyramid scheme given by (3.7) is but one of a general family of pyramid schemes that can be used to generate homogeneous polynomials. The general case has the form

$$\begin{aligned} c^\ell_\alpha(x) &:= \sum_{i=1}^{s} u^\ell_{i,\alpha}(x) c^{\ell-1}_{\alpha+e^i}(x), \quad \alpha \in \Gamma_{s,n-\ell} \\ c^0_\alpha(\alpha) &:= c_\alpha, \quad \alpha \in \Gamma_{s,n}, \end{aligned} \tag{4.1}$$

where $u^\ell_{i,\alpha}(x)$ are linear functionals in x. Clearly $c^n_0 \in H_\ell(\mathbb{R}^s)$, and c^ℓ_α depends only on the "adjacent" $c^{\ell-1}_\alpha$ of the previous level. Associated with (4.1), is a multilinear version in n vector variables $x^i \in \mathbb{R}^s, i = 1, \cdots, n$. For this, (4.1) is replaced by

$$\begin{aligned} c^\ell_\alpha(x^1, \cdots, x^\ell) &:= \sum_{i=0}^{s} u^\ell_{i,\alpha}(x^\ell) c^{\ell-1}_{\alpha+e^i}(x^1, \cdots, x^{\ell-1}), \quad \alpha \in \Gamma_{s,n-\ell} \\ c^0_\alpha(x) &:= c_\alpha, \quad \alpha \in \Gamma_n. \end{aligned} \tag{4.2}$$

One useful definition of the linear functionals $u_{i,\alpha}^{\ell}$ is given in terms of subsets

$$X^{\ell,\alpha} = \{x^{i,\alpha,\ell} : i = 1, \cdots, s\} \subseteq \mathbb{R}^s, \ell = 0, \cdots, n \quad \alpha \in \Gamma_{s,n-\ell}.$$

Assuming the sets $X^{\ell,\alpha}$ are linearly independent spanning sets of $\mathbb{R}^s$, then we can define $\{u_{i,\alpha}^{\ell}(x) : i = 1, \cdots, s\}$ by

$$x = \sum_{i=1}^{s} u_{i,\alpha}^{\ell}(x) x^{i,\alpha,\ell}. \tag{4.3}$$

We obtain the B-patches via (4.3) and (3.2) provided $x^{i,\alpha,\ell} = x^{i,\alpha_i}$, where

$$W = \{x^{i,j} : 1 \leq i \leq s, 1 \leq j \leq n\} \tag{4.4}$$

is a fixed array as in Section 2. In fact, it was shown in [6], see also [4], that the B-patch scheme is characterized among pyramid schemes (4.1) with (4.3) by the fact that the multilinear functions appearing in (4.2) are symmetric, and hence in $\mathcal{P}_{s,\ell}$.

The study of arbitrary pyramid schemes (4.1) seems to be at present a formidable task. A basic but unresolved question about such schemes concerns the linear independence of the polynomials $\{b_{\alpha}^{\ell} : \alpha \in \Gamma_{s,\ell}\}$ generated by the dual scheme

$$\begin{aligned} b_0^0(x) &= 1 \\ b_{\alpha}^{\ell}(x) &= \sum_{i=1}^{s} u_{i,\alpha-e^i}^{\ell-1}(x) b_{\alpha-e^i}^{\ell-1}(x), \quad \alpha \in \Gamma_{s,\ell}. \end{aligned} \tag{4.5}$$

Even in the following important special case, the linear independence problem has not been completely solved.

Let us define the polynomials $\{M_{\alpha}^{\ell} : \alpha \in \Gamma_{s,\ell}, \ell = 0, \cdots, n\}$ determined by the array (4.4) by the recursion

$$\begin{aligned} M_0^0(y) &= 1 \\ M_{\alpha}^{\ell}(y) &= \sum_{i=1}^{s} u_i^{\ell}(y) M_{\alpha-e^i}^{\ell-1}(y), \qquad \alpha \in \Gamma_{s,\ell}, \qquad \ell = 1, \cdots, n, \end{aligned} \tag{4.6}$$

where

$$y = \sum_{i=1}^{s} u_i^{\ell}(y) x^{i,\ell},$$

so clearly we have a special case of (4.5). In (4.6) we adopt the convention that $M_{\alpha}^{\ell} = 0$ if any component of α is negative. We call the polynomials $M_{\alpha} := M_{\alpha}^{n}, \alpha \in \Gamma_{s,n}$ so generated *M-patches*. Adopting matrix notation we can set

$$u_i^{\ell}(y) = (R_{\ell} y)_i \quad 1 \leq i \leq s, \quad \ell = 1, \cdots, n,$$

where R_ℓ is the $s \times s$ matrix defined by

$$R_\ell = [x^{1,\ell} \cdots x^{s,\ell}]^{-1}.$$

We denote this set of matrices collectively by $\mathcal{R}$. Then a straightforward calculation using (4.6) repeatedly shows that the generating function for $\{M_\alpha = M_\alpha(\cdot|\mathcal{R}) : \alpha \in \Gamma_{s,n}\}$ is

$$(x, R_1 y) \cdots (x, R_n y) = \sum_{\alpha \in \Gamma_{s,n}} x^\alpha M_\alpha(y|\mathcal{R}). \tag{4.7}$$

As the left hand side is a product of quadratic forms, we can symmeterize and replace $R_1, \cdots, R_n$ by $\frac{1}{2}(R_1 + R_1^T), \cdots, \frac{1}{2}(R_n + R_n^T)$. So without loss of generality, we assume that the matrices $R_\ell, \ell = 1, \cdots, n$, are symmetric.

In the special case that $R_1 = \cdots = R_n = I$, the identity matrix, (4.7) and the multinomial theorem identifies the M-patches as the Bernstein-Bézier polynomials, $M_\alpha(x) = \binom{n}{\alpha} x^\alpha$.

One necessary condition for the linear independence of $\{M_\alpha : \alpha \in \Gamma_{s,n}\}$ is that R_ℓ is non-singular, $\ell = 1, \cdots, n$. Indeed, if R_k is singular for some k, then $R_k x^0 = 0$ for some $x^0 \in \mathbb{R}^s \backslash \{0\}$. Hence by (4.7)

$$\sum_{\alpha \in \Gamma_{s,n}} (x^0)^\alpha M_\alpha(y|\mathcal{R}) = 0, \quad y \in \mathbb{R}^s.$$

Proposition 4.1 reveals the essential issue concerning the linear independence of $\{M_\alpha : \alpha \in \Gamma_{s,n}\}$, which we state as

Proposition 4.1. *$\{M_\alpha : \alpha \in \Gamma_{s,n}\}$ is linearly independent if and only if whenever $p \in \mathcal{P}_{s,n}$ then*

$$p(R_1 x, \cdots, R_n x) = 0, \quad x \in \mathbb{R}^s \text{ implies } p = 0. \tag{4.8}$$

Proof: Choose $\{x_\alpha : \alpha \in \Gamma_{s,n}\} \subseteq \mathbb{R}^s$ such that if $q \in H_n(\mathbb{R}^s)$ and $q(x_\alpha) = 0$ for all $\alpha \in \Gamma_{s,n}$ then $q = 0$. In particular then $\det(x_\alpha^\beta) \neq 0$, since this is the matrix which determines the interpolation at the nodes x_α in the standard basis x^β.

We claim condition (4.8) is equivalent to the linear independence of $a(\cdot|X_\alpha)$ where $X_\alpha = [R_1 x_\alpha, \cdots, R_n x_\alpha]$. To see this we choose any $p \in \mathcal{P}_{s,n}$. By our choice of $\{X_\alpha : \alpha \in \Gamma_{s,n}\}$ $p(R_1 x_\alpha, \cdots, R_n x_\alpha) = 0$ for all α if and only if $p = 0$. But by Proposition 2.1 this is equivalent to the linear independence of $\{a(\cdot|X_\alpha) : \alpha \in \Gamma_{s,n}\}$.

However since $R_\ell = R_\ell^T, \ell = 1, \cdots, n$, (4.7) can be rewritten for $x = x_\alpha$ as

$$(y, R_1 x_\alpha) \cdots (y, R_n x_\alpha) = \sum_{\beta \in \Gamma_{s,n}} x_\alpha^\beta M_\beta(y|\mathcal{R}).$$

So for any scalars $\{d_\alpha : \alpha \in \Gamma_{s,n}\}$

$$\sum_{\alpha \in \Gamma_{s,n}} d_\alpha a(y|X_\alpha) = \sum_{\beta \in \Gamma_{s,n}} \Big(\sum_{\alpha \in \Gamma_{s,n}} d_\alpha x_\alpha^\beta \Big) M_\beta(y|\mathcal{R}).$$

Since $\det(x_\alpha^\beta) \neq 0$, the linear independence of $\{M_\beta : \beta \in \Gamma_{s,n}\}$ is equivalent to the linear independence of $\{a(\cdot|X_\alpha) : \alpha \in \Gamma_{s,n}\}$. ■

As an example of the above we consider the case $n = 2$ and $s = 2$. The map which sends an element $p \in \mathcal{P}_{s,n}$ into the polar form of the quadratic polynomial

$$p(R_1(R_2)^{-1}x, x), \qquad x \in \mathbb{R}^s$$

has a matrix representation

$$Y = \begin{pmatrix} x_{11} & \frac{1}{2}x_{21} & 0 \\ x_{12} & \frac{1}{2}(x_{11} + x_{22}) & x_{21} \\ 0 & \frac{1}{2}x_{12} & x_{22} \end{pmatrix}, \qquad X := R^1(R^2)^{-1} := \begin{pmatrix} x_{11} & x_{12} \\ x_{21} & x_{22} \end{pmatrix}$$

in the basis $p_1(x^1, x^2) := x_1^1 x_1^2$, $p_2(x^1, x^2) := \frac{1}{2}(x_1^1 x_2^2 + x_2^1 x_1^2)$ and $p_3(x^1, x^2) := x_2^1 x_2^2$, where $x^i = (x_1^i, x_2^i)$, for $\mathcal{P}_{2,2}$. The determinant of Y is $\frac{1}{2}$ trace $X \cdot$ det X. Hence, according to Proposition 4.1 the corresponding M-patches are linearly independent if and only if

$$\lambda_i + \lambda_j \neq 0, \qquad i, j = 1, 2,$$

where λ_1, λ_2 are the eigenvalues of X.

This observation leads us to a general sufficient condition for the linear independence of the M-patches in terms of computable properties of $R_1, \cdots, R_n$. We state the result as

Theorem 4.1. *Assume the generating function (4.7) is given with R_j nonsingular for $j = 1, \cdots, n$. Assume that there exists two invertible matrices L and K so that $Q_j = LR_jK$ is lower triangular for each j, and let W be the $s \times n$ matrix whose j^{th} column consists of the s diagonal elements (eigenvalues) of Q_j. Then the M patches $\{M_\alpha : \alpha \in \Gamma_{s,n}\}$ given by (4.7) are linearly independent if and only if*

$$\text{per } W(\alpha) \neq 0 \tag{4.9}$$

for all $\alpha \in \Gamma_{s,n}$.

Remark 4.1. Worthy of further comment is the special case $n = 2$ when the hypothesis on reduction to lower triangular form is always satisfied. Then condition (4.8) is equivalent to $p(x, R_2R_1^{-1}x) = 0, x \in \mathbb{R}^s$ implies $p = 0$, and we let W be the $s \times 2$ matrix whose first column is all ones (the eigenvalues of the identity matrix) and whose second column consists of the eigenvalues $\lambda_1, \cdots, \lambda_s$ of $Q = R_2R_1^{-1}$. Then condition (4.9) is simply

$$\lambda_i + \lambda_j \neq 0, \qquad i, j = 1, \cdots, s. \tag{4.10}$$

Interestingly enough, this condition is related to Lyapunov stability for the matrix Q and occurs for example in O. Taussky's generalization of Lyapunov's theorem [8].

To elaborate on this connection, we observe that every $p \in \mathcal{P}_{s,n}$ as a bilinear form corresponds to an $s \times s$ symmetric matrix:

$$p(x, y) = (x, Ty), \quad x, y \in \mathbb{R}^s.$$

Our mapping in Proposition 4.1 takes p into the quadratic form

$$p(x, Wx) = (x, TWx),$$

and obviously the symmetric matrix which corresponds to its polar form is

$$TW + W^t T.$$

Hence, condition (4.10) determines when the map from T into $TW + W^t T$ is onto. The Taussky result says more, namely, there is a unique symmetric matrix S such that

$$SW + W^t S = I$$

and the number of positive eigenvalues of S is the same as the number of eigenvalues of W with positive real parts.

Proof: Using Proposition 4.1 we must characterize condition (4.8). By changing variables, this condition is equivalent to

$$p(Q_1 x, \ldots, Q_n x) = 0, \quad x \in \mathbb{R}^s \text{ implies } p = 0 \tag{4.11}$$

for all $p \in \mathcal{P}_{s,n}$.

Every $p \in \mathcal{P}_{s,n}$ can be written in the form

$$p(x^1, \ldots, x^n) = \sum_{\substack{i \in I \\ i=(i_1,\ldots,i_n)}} x^1_{i_1} \cdots x^n_{i_n} t_{i_1,\cdots,i_n},$$

where $I = \{i = (i_1, \cdots, i_n) : 1 \le i_\ell \le s, \quad \ell = 1, \cdots, n\}$ and $\{t_i\} = \{t_{i_1,\cdots,i_n}\}$ is a symmetric tensor so that

$$t_{\pi(i)} = t_i, \quad \pi(i) = (i_{\pi(1)}, i_{\pi(2)}, \cdots, i_{\pi(n)})$$

for any permutation π of $\{1, \cdots, n\}$. In fact any $p \in \mathcal{P}_{s,n}$ can be identified with a symmetric tensor t via the identification

$$t_{i_1,\cdots,i_n} = p(e^{i_1}, \cdots, e^{i_n}),$$

where e^k is the standard k^{th} unit vector in $\mathbb{R}^s$.

Let $J := \{i \in I : i_1 \le i_2 \le \cdots \le i_n\}$ and $\alpha(i) := (\alpha_1(i), \cdots, \alpha_s(i))$, where $\alpha_j(i)$ = cardinality of $\{\ell : i_\ell = j, \ell = 1, \cdots, n\}, j = 1, \ldots, s, i \in J$. Note that for each $i \in J$ the vector $a(i)$ is in $\Gamma_{s,n}$ and moreover the mapping

$J : i \to \alpha(i)$ is a bijection from J to $\Gamma_{s,n}$. Using these notations, we have the general formula

$$\sum_{i\in I} f(i) = \sum_{i\in J}\sum_{\pi} \frac{1}{\alpha(i)!} f(\pi(i))$$

which holds for any function defined on I. So in particular, for any $p \in \mathcal{P}_{s,n}$ we have

$$\begin{aligned} p(x^1, \cdots, x^n) &= \sum_{i\in J} \frac{1}{\alpha(i)!} \sum_{\pi} p(e^{i_{\pi(1)}}, \cdots, e^{i_{\pi(n)}}) x^1_{i_{\pi(1)}} \cdots x^n_{i_{\pi(n)}} \\ &= \sum_{i\in J} \frac{1}{\alpha(i)!} p(e^{i_1}, \ldots, e^{i_n}) \sum_{\pi} x^1_{i_{\pi(1)}} \cdots x^n_{i_{\pi(n)}} \\ &= \sum_{i\in J} \frac{1}{\alpha(i)!} p(e^{i_1}, \ldots, e^{i_n}) \text{ per } X(\alpha(i)). \end{aligned} \tag{4.12}$$

Now, we use this form to interpret condition (4.11). Let Q be the $s \times n$ matrix $[Q_1 x, \ldots, Q_n x]$. Since by hypothesis each $Q_j, j = 1, \ldots, n$, is lower triangular we find that for any $\alpha \in \Gamma_{s,n}$

$$\text{per} Q(\alpha) = \text{ per } W(\alpha) x^{\alpha} + \text{ lower terms},$$

where the lower terms in the variable $x = (x_1, \ldots, x_s)$ are relative to lexiographic ordering of the monomials $x^{\beta}, \beta \in \Gamma_{s,n}$. Thus we see that the polynomials per $Q(\alpha)$, $\alpha \in \Gamma_{s,n}$, are linearly independent if and only if (4.9) holds.

Since

$$p(Q_1 x, \cdots, Q_n x) = \sum_{i\in J} \frac{1}{\alpha(i)!} p(e^{i_1}, \cdots, e^{i_n}) \text{ per } Q(\alpha(i)),$$

we see that when (4.9) holds $p(Q_1 x, \cdots, Q_n x)$ vanishes identically in x only if $p(e^{i_1}, \cdots, e^{i_n}) = 0$ for all $i \in J$ or in other words $p = 0$. Conversely, if there is an $\alpha \in J$ such that per $W(\beta) = 0$ then by our above remarks there are constants $c_{\alpha}, \alpha \in \Gamma_{s,n}$, such that

$$\sum_{\alpha\in\Gamma_{s,n}} c_{\alpha} \text{ per } Q(\alpha) = 0$$

for all $x \in \mathbb{R}^s$. Thus $p(x^1, \cdots, x^n) := \sum_{\alpha\in\Gamma_{s,n}} c_{\alpha}$ per $X(\alpha)$ is a nontrivial element of $\mathcal{P}_{s,n}$ such that $p(Q_1 x, \cdots, Q_n x) = 0, \quad x \in \mathbb{R}^s$. This completes the proof of Theorem 4.1. ■

As another example of the above result we consider a set of univariate polynomials $c_0(y), \cdots, c_n(y), y \in \mathbb{R}$ defined by the generating function

$$\prod_{i=1}^{n} (\alpha_i + \beta_{iy} y + (1 - \alpha_i - \beta_i y) x) = \sum_{i=0}^{n} c_i(y) x^i. \tag{4.13}$$

When $\alpha_i = 0$ and $\beta_i = 1$ we see that $c_i(y)$ becomes the i-th Bernstein-Bézier polynomial, that is,

$$c_i(y) = \binom{n}{i} y^i (1-y)^{n-i}.$$

In general, we convert (4.13) into homogeneous coordinates in $\mathbb{R}^2$ by setting $y = y_1/y_2$ and $x = x_1/x_2$ to get

$$\prod_{i=1}^{n}(\alpha_i x_2 y_2 + \beta_i y_1 x_2 + (1-\alpha_i)x_1 y_2 - \beta_i x_1 y_1) = \sum_{\alpha_1+\alpha_2=n} M_{\alpha_1\alpha_2}(y_1, y_2) x_1^{\alpha_1} x_2^{\alpha_2},$$

where

$$M_{i,n-i}(y_1, y_2) = y_2^n c_i(\frac{y_1}{y_2}).$$

Thus $\{M_\alpha(y_1, y_2) : \alpha \in \Gamma_{2,n}\}$ are M-patches corresponding to the matrices

$$R_i = \begin{pmatrix} -\beta_i & 1-\alpha_i \\ \beta_i & \alpha_i \end{pmatrix}, \qquad i = 1, \cdots, n.$$

This set of matrices can be simultaneously converted to lower triangular form. In fact

$$\begin{pmatrix} 1 & 1 \\ 0 & 1 \end{pmatrix} \begin{pmatrix} -\beta_i & 1-\alpha_i \\ & \beta_i\alpha_i \end{pmatrix} \begin{pmatrix} 0 & 1 \\ 1 & 0 \end{pmatrix} = \begin{pmatrix} 1 & 0 \\ \alpha_i & \beta_i \end{pmatrix}$$

So the matrix W of Theorem 4.1 in this case is

$$W = \begin{pmatrix} 1 & 1 & \cdots & 1 \\ \beta_1 & \beta_2 & \cdots & \beta_n \end{pmatrix}.$$

For $\alpha = (n-k, k), 0 \le k \le n$ we have

$$\text{per } W(\alpha) = \sigma_k(\beta_1, \cdots, \beta_n), \tag{4.14}$$

where σ_k is the k-th symmetric function in n-variables. This follows either directly from the definitions or from formula (1.2) specialized to the vectors $x^i := (1, \beta_i)^t, i = 1, \cdots, n$. In fact, (1.2) gives the formula

$$\prod_{i=1}^{n}(x_1 + \beta_i x_2) = \sum_{k=0}^{n} \frac{x_1^{n-k}}{(n-k)!} \frac{x_2^k}{k!} \text{ per } W(n-k, k). \tag{4.15}$$

By definition the left hand side is equal to

$$\sum_{k=0}^{n} x_1^{n-k} x_2^k \sigma_k(\beta_1, \cdots, \beta_n).$$

This establishes (4.14) and therefore by Theorem 4.1 the polynomials $c_0(y), \cdots, c_n(y)$ are linearly independent if and only if $\sigma_k(\beta_1, \cdots, \beta_n) \neq 0 \quad k = 0, 1, \cdots, n$. This special case of Theorem 4.1 is due to Stefanus and Goldman [7].

§5. B-patches for $\prod_n(\mathbb{R}^s)$.

In this section we interpret the results of Section 3 in terms of affine coordinates. This requires setting one of the homogeneous variables equal to a constant. In order to obtain s affine variables, we will use the results of Section 3 for $s+1$ homogeneous variables $x = (x_0, \cdots, x_s)$, and subsequently set $x_0 = 1$.

The first result we have for affine coordinates has already been presented as Corollary 2.1. Recall that we began with a fixed array $\Sigma := \{\tau^{\ell,r} : \ell = 1, \cdots, s+1, r = 1, \cdots, n\}$ of vectors in $\mathbb{R}^s$. In Corollary 2.1 we observed that $\{\psi_\alpha : \alpha \in \Gamma_{s+1,n}\}$ is a basis for $\prod_n(\mathbb{R}^s)$, where

$$\psi_\alpha(\omega) = \prod_{\ell=1}^{s+1} \prod_{r=1}^{\alpha_\ell} (1 + (\omega, \tau^{\ell,r})), \omega \in \mathbb{R}^s \tag{5.1}$$

provided $\{\tau^{\ell,\alpha_\ell+1} : \ell = 1, \cdots, s+1\}$ is affinely independent for all $\alpha = (\alpha_1, \cdots, \alpha_{s+1}) \in \mathbb{Z}_+^{s+1}$ with $|\alpha| \le n-1$. This fact is obtained from Theorem 2.1 by the simple device of setting

$$\begin{aligned} x^{\ell,r} &:= (1, \tau^{\ell,r}) \in \mathbb{R}^{s+1}, \quad \ell = 1, \cdots, s+1, \quad r = 1, \cdots, n, \\ x &:= (\omega_0, \omega), \quad \omega \in \mathbb{R}^s, \quad \omega_0 \in \mathbb{R}. \end{aligned} \tag{5.2}$$

Then given any $p \in \prod_n(\mathbb{R}^s)$

$$h(x) := \omega_0^n p(\omega_0^{-1}\omega)$$

is in $H_n(\mathbb{R}^{s+1})$ and

$$p(\omega) = h(1, \omega).$$

These two relations establish an algebraic isomorphism between $H_n(\mathbb{R}^{s+1})$ and $\prod_n(\mathbb{R}^s)$. So the basis $\{a_\alpha : \alpha \in \Gamma_{s+1,n}\}$ of lineal polynomials obtained in Section 2 for $H_n(\mathbb{R}^{s+1})$ becomes the basis $\{\psi_\alpha : \alpha \in \Gamma_{s+1,n}\}$ for $\prod_n(\mathbb{R}^s)$ when we set $\omega_0 = 1$.

Let $\{b_\alpha : \alpha \in \Gamma_{s+1,n}\}$ be the homogeneous B-patches defined by (3.1) and determined by the array (5.2). For $\alpha \in \Gamma_{s+1,n}$ we set

$$\varphi_\alpha(\rho) := b_\alpha(1, \rho), \quad \rho \in \mathbb{R}^s. \tag{5.3}$$

Then it follows just as for $\{\psi_\alpha : \alpha \in \Gamma_{s+1,n}\}$ that $\{\varphi_\alpha : \alpha \in \Gamma_{s+1,n}\}$ is also a basis for $\prod_n(\mathbb{R}^s)$. Moreover from (3.1) we obtain the generating function for the φ_α

$$(1 + (\omega, \rho))^n = \sum_{\alpha \in \Gamma_{s+1,n}} \psi_\alpha(\omega)\varphi_\alpha(\rho). \tag{5.4}$$

This formula, which is our starting point, is identical with an observation made in [6]; it is established there by a recursive geometric argument. Obviously the polar form of $(1 + (\omega, \rho))^n, \omega$ fixed, is

$$q_\omega(\rho^1, \cdots, \rho^n) = (1 + (\omega, \rho^1)) \cdots (1 + (\omega, \rho^n)),$$

so that in (5.4)

$$\psi_\alpha(\omega) = q_\omega(X_\alpha),$$

where X_α is determined from the array (5.2) by our standard recipe. Thus we obtain

Proposition 5.1. *For $p \in \prod_n(\mathbb{R}^s)$, let p_b denote its polar form. Then*

$$p(\rho) = \sum_{\alpha \in \Gamma_{s+1,n}} p_b(X_\alpha)\varphi_\alpha(\rho), \rho \in \mathbb{R}^s.$$

Proof: For appropriate choices of $\omega = \omega_\alpha, \alpha \in \Gamma_{s+1,n}$, the functions $\{(1 + (\omega_\alpha, \rho))^n : \alpha \in \Gamma_{s+1,n}\}$ span $\prod_n(\mathbb{R}^s)$. So the stated result follows from the previous remarks by the linearity of the association of p and its polar form p_b. This proposition appears earlier in [6], see also [4]. ■

The recursions given in Section 3 all depend on (3.2) which given the array (5.2) becomes for $\omega \in \mathbb{R}^s, \alpha \in \mathbb{Z}_+^{s+1}, |\alpha| \le n-1$

$$\begin{aligned} \omega &= \sum_{i=1}^{s+1} \lambda_{i,\alpha}(\omega)\tau^{i,\alpha_i+1} \\ 1 &= \sum_{i=1}^{s+1} \lambda_{i,\alpha}(\omega). \end{aligned} \tag{5.5}$$

So the coefficients $\{\lambda_{i,\alpha}(\omega)\}_{i=1}^{s+1}$ are the barycentric coordinates of τ relative to the simplex with vertices $\{\tau^{i,\alpha_i+1} : i = 1, \cdots, s+1\}$. Then for $\alpha \in \mathbb{Z}_+^{s+1}, 0 < |\alpha| \le n-1$ the "up" recursion (3.4) and "down" recursion become respectively

$$\varphi_\alpha(\omega) = \sum_{i=1}^{s+1} \lambda_{i,\alpha-e^i}(\omega)\varphi_{\alpha-e^i}(\omega), \tag{5.6}$$

and for $\ell = 1, \cdots, n$

$$\begin{aligned} c_\alpha^0(\omega) &= c_\alpha, \quad \alpha \in \Gamma_{s+1,n} \\ c_\alpha^\ell(\omega) &= \sum_{i=1}^{s+1} \lambda_{i,\alpha}(\omega)c_{\alpha+e^i}^{\ell-1}(\omega), \quad \alpha \in \Gamma_{s+1,n-\ell}, \end{aligned} \tag{5.7}$$

which recursively computes $p(\omega) = c_0^n(\omega)$ where

$$p(\omega) = \sum_{\alpha \in \Gamma_{s+1,n}} c_\alpha \varphi_\alpha(\omega).$$

To formulate the differential recursion (3.9) in the affine setting, note that for fixed $\tau \in \mathbb{R}^s$, properties of the directional derivative and (5.3) give

$$D_\tau \varphi_\alpha(\rho) = D_\tau b_\alpha(1, \rho) = D_{(0,\tau)} b_\alpha(y) \tag{5.8}$$

where $y = (\rho_0, \rho), \rho_0 \in \mathbb{R}, \rho \in \mathbb{R}^s$. Therefore (3.9) can applied with $z = (0, \tau)$. Applying D_τ to (5.5) and setting

$$\sigma_{i,\alpha}(\tau) := D_\tau \lambda_{i,\alpha}(\cdot), \quad i = 1, \cdots, s+1, \quad \alpha \in \Gamma_{s+1,n}$$

we obtain

$$\begin{aligned} \tau &= \sum_{i=1}^{s+1} \sigma_{i,\alpha}(\tau)\tau^{i,\alpha_i+1} \\ 0 &= \sum_{i=1}^{s+1} \sigma_{i,\alpha}(\tau). \end{aligned} \tag{5.9}$$

So from (3.9), (5.8), and (5.9) we conclude

$$D_\tau \varphi_\alpha(\rho) = n \sum_{i=1}^{s+1} \sigma_{i,\alpha-e^i}(\tau)\varphi_{\alpha-e^i}(\rho), \quad \alpha \in \Gamma_{s+1,n}. \tag{5.10}$$

As might be expected, obtaining a formula for the dual functionals given by (3.11) suitable for the affine setting requires some effort. Given any function f on $\mathbb{R}^s$,

$$h(t,\rho) := t^n f(t^{-1}\rho) \tag{5.11}$$

is homogeneous of order n and the dual functionals (3.11) can be applied to h provided that f is sufficiently smooth. In the context of the array given by (5.2), each of the polynomials a_α in the variable $x = (t,\rho)$ has the form

$$a(x) = \prod_{i=1}^{n}(t + (\rho, \tau^i)), \quad \tau^i \in \mathbb{R}^s.$$

Writing for x the differentiation $\left(\frac{d}{dt}, D\right)$, where $D = \left(\frac{\partial}{\partial\rho_1}, \cdots, \frac{\partial}{\partial\rho_s}\right)$, we see that dual functionals should have the form

$$(\lambda f)(\rho) := \frac{1}{n!}(a(\frac{\partial}{\partial t}, D)h)(1,\rho), \tag{5.12}$$

where the differentiations are to be evaluated at $(1,\rho)$, $a(\cdot)$ is a homogeneous polynomial, and ρ is *arbitrary*. Carrying out these calculations, we obtain

Theorem 5.1. *The dual functionals λ_α for the B-patches $\{\varphi_\alpha : \alpha \in \Gamma_{s+1,n}\}$ defined by (5.3) (or recursively, by (5.6)) are given by*

$$(\lambda_\alpha f)(\rho) = \frac{1}{n!} \sum_{|\beta| \le n} q_{\alpha,\beta}(\rho)(D^\beta f)(\rho), \tag{5.13}$$

where for $\beta \in \Gamma_{s,n}$ the $q_{\alpha,\beta}$ are defined by the generating function

$$\sum_{\beta \in \Gamma_{s,n}} q_{\alpha,\beta}(\rho)\sigma^\beta = \prod_{\ell=1}^{s+1} \prod_{r=1}^{\alpha_\ell} (\sigma, \tau^{\ell,r} - \rho) \tag{5.14}$$

and the remaining $q_{\alpha,\beta}$, $|\beta| < n$, are found by successive differentiation using

$$\frac{\partial q_{\alpha,\beta}}{\partial \rho_i} = -q_{\alpha,\beta-e^i}, \alpha \in \Gamma_{s+1,n}. \tag{5.15}$$

Proof: By our definitions (5.11) and (5.3)

$$\begin{aligned} h(t,\rho) = t^n\varphi_\alpha(t^{-1}\rho) &= t^n b_\alpha(1,t^{-1}\rho) \\ &= b_\alpha(t,\rho), \end{aligned}$$

and so from (5.12) with a replaced by $a_\gamma, \gamma \in \Gamma_{s+1,n}$

$$\begin{aligned} (\lambda_\gamma\varphi_\alpha)(\rho) &= \frac{1}{n!}(a_\gamma(\frac{\partial}{\partial t}, D)b_\alpha)(1,\rho) \\ &= \delta_{\gamma\alpha} \end{aligned}$$

by (3.10) and (3.11).

With h given by (5.11) it is clear that (5.12) assumes the form (5.13). To identify the polynomials $q_{\alpha,\beta}$, we apply λ_α to the ridge function

$$f(y) = (\sigma, y-\rho)^n,$$

where σ and ρ are fixed in $\mathbb{R}^s$. Let $x_0 := (-(\sigma,\rho),\sigma) \in \mathbb{R}^{s+1}$. Then from (5.11)

$$h(t,y) = (\sigma, y - t\rho)^n = (x_0, x)^n$$

where $x = (t,y)$. So

$$\begin{aligned} \lambda_\alpha f &= \frac{1}{n!}(a_\alpha(\frac{\partial}{\partial t}, D)h)(1,\rho) \\ &= \prod_{\ell=1}^{s+1}\prod_{r=1}^{\alpha_i}(-(\sigma,\rho)+(\sigma,x^{\ell,n})) = \prod_{\ell=1}^{s+1}\prod_{r=1}^{\alpha_i}(\sigma, x^{\ell r}-\rho). \end{aligned}$$

Alternatively if we compute $\lambda_\alpha f$ with the right hand side of (5.13) we obtain

$$\lambda_\alpha f = \sum_{\beta\in\Gamma_{s,n}} q_{\alpha,\beta}(\rho)\sigma^\beta.$$

Equating these two equations yields (5.14).

To obtain (5.15) we make the simple observation that if f is any polynomial of degree n in s variables then by (5.12) $\lambda_\alpha f$ is a constant independent of the evaluation at ρ. Therefore $\frac{\partial}{\partial\rho_i}(\lambda_\alpha f) = 0, i = 1,\cdots,s$. So calculating with the right hand side of (5.13) we conclude

$$\sum_{|\beta|\le n}\left(\frac{\partial q_{\alpha,\beta}(\rho)}{\partial\rho_i} + q_{\alpha-e^i}(\rho)\right) D^\beta f = 0$$

for all $f \in \prod_n(\mathbb{R}^s)$. From this (5.15) follows by an elementary argument. ■

We conclude this section with a short digression into approximation theory based on B-patches of arbitrarily high degree. We begin with a fixed infinite array

$$\Sigma := \{\tau^{\ell,r} : \ell = 1,\ldots,s+1, \quad r = 1,2,\ldots\}$$

which satisfies our standard hypothesis of affine independence. Referring to the system of barycentric coordinates in (5.5), let

$$\Omega := \{\rho \in \mathbb{R}^s : \lambda_{i,\alpha}(\rho) \geq 0, \quad i = 1, \ldots, s+1, \quad \alpha \in \mathbb{Z}_+^{s+1}\}.$$

We assume that the array Σ is so situated that Ω is a nonempty compact domain in $\mathbb{R}^s$. On $C(\Omega)$ we defined the sequence $\{\Phi_n\}_{n=1}^{\infty}$ of approximation operators

$$(\Phi_n f)(\rho) := \sum_{\alpha \in \Gamma_{s+1,n}} f\left(\frac{1}{n}\sum_{\ell=1}^{s+1}\sum_{r=1}^{\alpha_\ell} \tau^{\ell,r}\right)\varphi_\alpha(\rho), \qquad \rho \in \Omega, \quad f \in C(\Omega). \quad (5.16)$$

By the recursion (5.6) it is clear that all the B-patches φ_α are nonnegative on Ω, and hence (5.16) defines a positive linear operator on $C(\Omega)$. By putting $\omega = 0$ in (5.4) we see that Φ_n reproduces constants. Furthermore, identifying the linear term in ω on both sides of (5.4) we readily conclude that Φ_n reproduces all linear functions in $C(\Omega)$. This property in fact motivates the particular functional evaluations which appear in the definition (5.16). For the quadratic functions

$$f_\omega(\rho) = (\omega, \rho)^2, \quad \omega, \rho \in \mathbb{R}^s$$

a straightforward calculation shows that

$$f_\omega(\rho) = \frac{n}{n-1}\left(\Phi_n f_\omega\right)(\rho) - \frac{1}{n(n-1)} \sum_{\alpha \in \Gamma_{s+1,n}} \left(\sum_{\ell=1}^{s+1}\sum_{r=1}^{\alpha_\ell}(\omega, \tau^{\ell,r})^2\right)\varphi_\alpha(\rho). \quad (5.17)$$

Clearly from (5.17), $\Phi_n f_\omega$ converges uniformly to f_ω provided there exist fixed positive constants M and ε such that

$$\sum_{\ell=1}^{s+1}\sum_{r=1}^{\alpha_\ell} \|\tau^{\ell,r}\| \leq M n^{2-\varepsilon} \quad (5.18)$$

for all $\alpha \in \Gamma_{s+1,n}$ with n arbitrary. Therefore the Bohman-Korovkin Theorem yields

Proposition 5.2. *For all $f \in C(\Omega)$, $\Phi_n f$ converges uniformly to f as n tends to infinity provided the array σ satisfies (5.18).*

§6. Degree Raising, Conversion and Subdivision for B-patches.

In this section we provide some information pertaining to the questions of degree raising, conversion and subdivision for B-patches. We begin by examining the algebraic formulation of degree raising. Given a fixed array

$$\Sigma = \{\tau^{\ell,r} : \ell = 1, \cdots, s+1, r = 1, \cdots, n+1\}.$$

Associated with Σ is the array W generated by the vectors $x^{\ell,r} := (1, \tau^{\ell,r})$, $\ell = 1, \cdots, s+1$, $r = 1, \cdots, n+1$. Suppose Σ satisfies the hypothesis of Corollary 2.1. Then there is a matrix $\mathcal{D}^n = (\mathcal{D}^n_{\alpha,\beta})$, $\quad \alpha \in \Gamma_{s+1,n}$, $\beta \in \Gamma_{s+1,n+1}$ such that

$$\varphi^n_\alpha(\rho) = \sum_{\beta \in \Gamma_{s+1,n+1}} \mathcal{D}^n_{\alpha,\beta} \varphi^{n+1}_\beta(\rho). \tag{6.1}$$

This formula follows from the obvious fact that $\Pi_n(\mathbb{R}^s) \subseteq \Pi_{n+1}(\mathbb{R}^s)$. This nesting is essential to the degree raising point of view. To study this matrix $\mathcal{D}^n$ we will convert (6.1) into homogeneous coordinates. For this purpose recall that $b_\alpha(x) = \tau^n \varphi_\alpha(\rho/\tau)$ where $x = (\tau, \rho)$. Hence (6.1) becomes

$$x_1 b^n_\alpha(x) = \sum_{\beta \in \Gamma_{s+1,n+1}} \mathcal{D}^n_{\alpha,\beta} b^{n+1}_\beta(x). \tag{6.2}$$

Using the dual functionals of Proposition 3.2 we get

$$\mathcal{D}^n_{\alpha,\beta} = \frac{1}{(n+1)!} a^{n+1}_\beta(D)(\hat{b}^n_\alpha) \tag{6.3}$$

$\alpha \in \Gamma_{s+1,n}, \beta \in \Gamma_{s+1,n+1}$ and $\hat{b}^n_\alpha(x) := x_1 b^n_\alpha(x)$. To make use of this formula we need a degree lowering formula for our lineal polynomials. This is readily available. For each $\beta = (\beta_1, \cdots, \beta_{s+1}) \in \Gamma_{s+1,n+1}$ choose the least integer $\ell \in \{1, \cdots, s+1\}$ such that $\beta_\ell = \max\{\beta_j : j = 1, \cdots, s+1\}$. Then

$$a^{n+1}_\beta(x) = (x, x^{\ell,\beta_\ell}) a^n_{\beta - e^\ell}(x), \qquad \beta \in \Gamma_{s+1,n+1}. \tag{6.4}$$

This formula is only one possible degree lowering formula, which all have the form

$$a^{n+1}_\beta(x) = \sum_{\gamma \in \Gamma_{s+1,n}} (x, y^{\gamma,\beta}) a^n_\gamma(x), \qquad \beta \in \Gamma_{s+1,n+1}, \tag{6.5}$$

where $Y = \{y^{\alpha,\beta} : \beta \in \Gamma_{s+1,n+1}, \alpha \in \Gamma_{s+1,n}\}$ is some array of vectors in $\mathbb{R}^{s+1}$. Other examples will be given later.

We will now combine (6.5) with (6.3) and the derivative recursion for the B-patches to obtain a recursion for the matrices $\mathcal{D}^1, \cdots, \mathcal{D}^n$. Specifically, from (6.5) and (6.3) we get

$$\mathcal{D}^n_{\alpha,\beta} = \frac{1}{(n+1)!} \sum_{\gamma \in \Gamma_{s+1,n}} a^n_\gamma(D)(D_{y^{\gamma,\beta}}(\hat{b}^n_\alpha)). \tag{6.6}$$

According to the derivative recursion of Proposition 3.1, we have

$$D_{y^{\gamma,\beta}}(\hat{b}^n_\alpha) = n \sum_{i=1}^{s+1} u_{i,\alpha-e^i}(y^{\gamma,\beta}) \hat{b}^{n-1}_{\alpha-e^i} + b^n_\alpha y^{\gamma,\beta}_1.$$

Substituting this formula into (6.6) gives

$$\mathcal{D}^n_{\alpha,\beta} = \frac{1}{(n+1)!} \sum_{\gamma\in\Gamma_{s+1,n}} a^n_\gamma(D) \left\{ y_1^{\gamma,\beta} b^n_\alpha + n \sum_{i=1}^{s+1} u_{i,\alpha-e^i}(y^{\gamma,\beta}) \hat{b}^{n-1}_{\alpha-e^i} \right\}.$$

Thus, using Proposition 3.2 and (6.3) with n replaced by $n-1$, we finally get

$$\mathcal{D}^n_{\alpha,\beta} = \frac{1}{n+1} y_1^{\alpha,\beta} + \frac{n}{n+1} \sum_{i=1}^{s+1} \sum_{\gamma\in\Gamma_{s+1,n}} u_{i,\alpha-e^i}(y^{\gamma,\beta}) \mathcal{D}^{n-1}_{\alpha-e^i,\gamma}. \tag{6.7}$$

We specialize this formula in two ways. First, using (6.4) we identify

$$y^{\gamma,\beta} = \begin{cases} 0, \text{if } \gamma \neq \beta - e^\ell \\ x^{\ell,\beta_\ell}, \text{if } \gamma = \beta - e^\ell \end{cases},$$

and so we obtain

Proposition 6.1. *For $\alpha \in \Gamma_{s+1,n}$ and $\beta \in \Gamma_{s+1,n+1}$ we have*

$$\mathcal{D}^n_{\alpha,\beta} = \frac{1}{n+1} \left(\delta_{\alpha,\beta-e^\ell} + n \sum_{i=1}^{s+1} u_{i,\alpha-e^i}(x^{\ell,\beta_\ell}) \mathcal{D}^{n-1}_{\alpha-e^i,\beta-e^\ell} \right). \tag{6.8}$$

This provides a recursion for the degree raising matrices.

The second identity which we wish to point out requires some further discussion. Note in our initial degree raising formula (6.2) in homogeneous coordinates the bias toward the first coordinate of x. Now, clearly there are a family of relations, s in number, of the type (6.2) which can be written collectively in the form

$$(x,y) b^n_\alpha(x) = \sum_{\beta\in\Gamma_{s+1,n}} (d^n_{\alpha,\beta}, y)\, b^{n+1}_\beta(x), \qquad x, y \in \mathbb{R}^s, \tag{6.9}$$

where $d^n_{\alpha,\beta}$ are vectors in $\mathbb{R}^s$. Comparing with (6.2) we see that $\mathcal{D}^n_{\alpha,\beta} = (d^n_{\alpha,\beta})_1$. We take (6.9) and multiply both sides by $a^n_\alpha(y)$ and then sum over $\alpha \in \Gamma_{s,n}$. Using our basic B-patch formula (3.1) we get

$$(x,y)^{n+1} = \sum_{\beta\in\Gamma_{s+1,n}} \left(\sum_{\alpha\in\Gamma_{s,n}} (y, d^n_{\alpha,\beta}) a^n_\alpha(y) \right) b^{n+1}_\beta(x).$$

Again, appealing to (3.1), now with n replaced by $n+1$, and using the linear independence of the B-patches $\{b^{n+1}_\beta : \beta \in \Gamma_{s+1,n}\}$ we obtain the relation

$$a^{n+1}_\beta(x) = \sum_{\gamma\in\Gamma_{s+1,n}} (x, d^n_{\gamma,\beta}) a^n_\gamma(x). \tag{6.10}$$

Thus, we see that we can choose $y_{\alpha,\beta} = d^n_{\alpha,\beta}$ in our degree lowering formula (6.5). This gives us

Proposition 6.2. *For $\alpha \in \Gamma_{s+1,n}$ and $\beta \in \Gamma_{s+1,n}$ we have*

$$\mathcal{D}^n_{\alpha,\beta} = \sum_{i=1}^{s+1} \sum_{\gamma \in \Gamma_{s+1,n}} u_{i,\alpha-e^i}(d^n_{\gamma,\beta}) \mathcal{D}^{n-1}_{\alpha-e^i,\gamma}. \tag{6.11}$$

We turn now to the algebra associated with changes from one B-patch basis for $H_n(\mathbb{R}^s)$ to another. This algebra is most conveniently expressed in terms of matrix operations on $\mathbb{R}^N$ where $N = \operatorname{card} \Gamma_{s,n}$. As in Section 2 we let W be any $s \times n$ array of vectors in $\mathbb{R}^s$ and we always suppose this array satisfies the linear independence hypothesis of Theorem 2.1. Expanding our notation to indicate the dependence on the array W, we set $a_\alpha(\cdot|W) := a(\cdot|X_\alpha)$, where $X_\alpha \subset W$ is defined by (2.7). The B-patches determined via the generating function (3.1) from these lineal polynomials we similarly denote by $b_\alpha(\cdot|W)$. In addition, we set

$$m_\alpha(x) := \frac{x^\alpha}{\alpha!}.$$

Thus in $\mathbb{R}^N$ there are defined three s-dimensional manifolds

$$\begin{aligned} a(\cdot|W) &:= (a_\alpha(\cdot|W) : \alpha \in \Gamma_{s,n}) \\ b(\cdot|W) &:= (b_\alpha(\cdot|W) : \alpha \in \Gamma_{s,n}) \\ m(\cdot) &:= (m_\alpha(\cdot) : \alpha \in \Gamma_{s,n}) \end{aligned}$$

which we call respectively the lineal manifold, the B-patch manifold, and the monomial manifold. Each of these manifolds extends over all of $\mathbb{R}^N$ in the sense that no one of them lies in any hyperplane of $\mathbb{R}^N$. The components of each of these manifolds collectively give a basis for $H_n(\mathbb{R}^s)$ and so clearly there exist $N \times N$ matrix transformations $A(W)$ and $B(W)$ such that

$$a(\cdot|W) = A(W)m(\cdot) \tag{6.12}$$

and

$$b(\cdot|W) = B(W)m(\cdot). \tag{6.13}$$

Using the natural inner products on $\mathbb{R}^N$ and $\mathbb{R}^s$, equation (3.1) becomes

$$(a(x|W), b(y|W)) = (x, y)^n, x, y \in \mathbb{R}^s.$$

Expanding the right hand side by the multinomial theorem and using (6.12) and (6.13), we obtain

$$(A(W)m(x), B(W)m(y)) = n!(m(x), \mathcal{F}m(y)),$$

where $\mathcal{F} = (\mathcal{F}_{\alpha,\beta})_{\alpha,\beta \in \Gamma_{s,n}}$ is the diagonal matrix

$$\mathcal{F}_{\alpha,\beta} = \alpha! \delta_{\alpha,\beta}.$$

So we obtain the relation

$$A^t(W)B(W) = n!\mathcal{F}. \tag{6.14}$$

Now suppose we have two different arrays W^0 and W^1 and suppose that for $p \in H_n(\mathbb{R}^s)$

$$p(x) = \sum_{\alpha \in \Gamma_{s,n}} c_\alpha^0 b_\alpha(x|W^0) = \sum_{\alpha \in \Gamma_{s,n}} c_\alpha^1 b_\alpha(x|W^1).$$

Using (6.13) on both sides of the above equality, we find that

$$B^t(W^0)c^0 = B^t(W^1)c^1$$

and so

$$c^1 = Sc^0, \tag{6.15}$$

where S the linear transformation given by

$$S^t := B(W^0)B^{-1}(W^1) \tag{6.16}$$

To make effective use of (6.15) and (6.16) it is possible to specialize these formulas when

$$W^1 = TW^0 \tag{6.17}$$

for a given non-singular transformation $T : \mathbb{R}^s \to \mathbb{R}^s$. We first need the relation

$$m(Tx) = \mathcal{E}(T)m(x) \tag{6.18}$$

valid for a particular $N \times N$ matrix $\mathcal{E}(T)$ which we now describe. The key is to recognize $m_\alpha(Tx), \alpha \in \Gamma_{s,n}$, as the particular lineal polynomial

$$\prod_{i=1}^{s} (t_i, x)^{\alpha_i},$$

where t_i is the ith row of T, and we then can apply (1.2) to express this polynomial in terms of the monomial basis. So we find that

$$\mathcal{E}(T) = (\mathcal{E}_{\alpha,\beta})_{\alpha,\beta \in \Gamma_{s,n}},$$

where

$$\mathcal{E}_{\alpha,\beta} = \frac{1}{\alpha!} \text{ per } T_\alpha(\beta). \tag{6.19}$$

and T_α is the $s \times n$ matrix whose columns are the vectors t_i repeated α_i times. As in Section 1 we obtain $T_\alpha(\beta)$ from T_α by repeating its jth row β_j times.

We will find the following two simple properties of the mapping $\mathcal{E}$ useful. If $T = T_2T_1$, then $\mathcal{E}(T) = \mathcal{E}(T_2)\mathcal{E}(T_1)$ as follows immediately from (6.18). Thus, in particular,

$$\mathcal{E}(T^k) = \mathcal{E}(T)^k, \quad k = 0, 1, \cdots . \tag{6.20}$$

Moreover,

$$\mathcal{E}(T^t) = \mathcal{F}^{-1}\mathcal{E}^t(T)\mathcal{F}, \tag{6.21}$$

which follows from (6.19). Alternatively, note that by using the multinomial theorem twice

$$\begin{aligned} n!(\mathcal{F}m(y), m(Tx)) = (y, Tx)^n &= ((T^t y, x))^n \\ &= n!(\mathcal{F}m(T^t y), m(x)). \end{aligned}$$

So

$$(\mathcal{F}m(y), \mathcal{E}(T)m(x)) = (\mathcal{F}\mathcal{E}(T^t)m(y), m(x)),$$

and from this (6.21) follows immediately.

Lemma 6.1. *The lineal and B-patch manifolds transform according to*

$$\begin{aligned} A(TW) &= A(W)\mathcal{E}(T^t) \\ B(TW) &= B(W)\mathcal{E}(T)^{-1} \end{aligned} \tag{6.22}$$

Proof: As the lineal polynomials are products of inner products of vectors

$$a(T^t x|W) = a(x|TW) = A(TW)m(x).$$

On the other hand, by (6.12)

$$a(T^t x|W) = A(W)m(T^t x) = A(W)\mathcal{E}(T^t)m(x).$$

The first equation in (6.22) follows from these two lines. To obtain the second equation from the first, observe from (6.14) and (6.21) that

$$\begin{aligned} B(TW) &= n!A^{-t}(TW)\mathcal{F} \\ &= n!A^{-t}(W)\mathcal{E}^{-t}(T^t)\mathcal{F} \\ &= B(W)\mathcal{F}^{-1}(\mathcal{F}^{-1}\mathcal{E}(T)^t\mathcal{F})^{-t}\mathcal{F} \\ &= B(W)\mathcal{E}(T)^{-1} \end{aligned}$$

as asserted in (6.22). ■

We can now state our result for the composite subdivision operator which results from iterating the map T in (6.17).

Proposition 6.3. *Let* $W^1 = TW$ *and* $W^k = TW^{k-1}$ *for* $k \geq 2$. *Then the composite subdivision operator* S *satisfying*

$$\begin{aligned} c^k &= S_k c \\ B^t(W)c &= B^t(W^k)c^k \end{aligned}$$

is given by

$$S_k = B^{-t}(W)(\mathcal{E}^t(T))^k B^t(W) \tag{6.23}$$

Proof: Using (6.16) repeatedly, we obtain

$$S_k^t = B(W)B^{-1}(W^k).$$

Then from (6.20) and (6.22)

$$\begin{aligned} B(W^k) &= B(T^k W) \\ &= B(W)\mathcal{E}(T^k)^{-1} \\ &= B(W)\mathcal{E}(T)^{-k}. \end{aligned}$$

So

$$S^t = B(W)(\mathcal{E}(T))^k B^{-1}(W)$$

from which the result follows. ■

We conclude this discussion of the subdivision operation by establishing a convergence theorem for the affine B-patches. As in Corollary 2.1 let Σ be a fixed array satisfying the affine independence property given there. We assume that some $\psi \in \Pi_n(\mathbb{R}^s)$ is variously written, via the subdivision operation, as

$$\begin{aligned} \psi(\rho) &= \sum_{\alpha\in\Gamma_{s+1,n}} c_\alpha \varphi_\alpha^n(\rho|\Sigma) \\ &= \cdots = \sum_{\alpha\in\Gamma_{s+1,n}} c_\alpha^k \varphi_\alpha^n(\rho|\Sigma^k), \qquad \rho \in \mathbb{R}^s, \end{aligned} \tag{6.24}$$

where

$$\Sigma^k = \Theta^k \Sigma \tag{6.25}$$

for a given non-singular affine map $\Theta : \mathbb{R}^s \to \mathbb{R}^s$. Under these special circumstances, the convergence result can be stated as

Theorem 6.1. *Let Θ be a contractive, non-singular affine map of $\mathbb{R}^s$ into itself with fixed point ρ_0. Then*

$$\lim_{k\to\infty} c_\alpha^k = \psi(\rho_0), \qquad \alpha \in \Gamma_{s+1,n}.$$

For the proof we require two lemmas which are both easily proved using the Jordan normal form.

Lemma 6.2. *Let R be a $n \times n$ matrix whose highest eigenvalue is one and is simple. Let $Rx = x$ and $R^t y = y$ normalized so that $(x, y) = 1$. Then*

$$\lim_{k\to\infty} (R^k)_{ij} = x_i y_j, \qquad i, j = 1, \ldots, n$$

Lemma 6.3. *Let R be an $s \times s$ matrix with eigenvalues $\lambda_1, \ldots, \lambda_s$. Then $\mathcal{E}(R)$ has eigenvalues $\{m_\alpha(\lambda) : \alpha \in \Gamma_{s,n+1}\}$ where $\lambda = (\lambda_1, \ldots, \lambda_s)$. Consequently,* *if R satisfies the hypothesis of Lemma 6.2, then so does $\mathcal{E}(R)$, and moreover we have*

$$\mathcal{E}(R)m(x) = m(x),$$

i.e., $m(x)$ is the unique eigenvalue corresponding to eigenvalue one for $\mathcal{E}(R)$.

Proof of Theorem 6.1. We write (6.24) in homogeneous variables by associating with Σ the array W obtained by adjoining a first component equal to one to every vector in Σ. Then

$$W^k = T^k W,$$

where

$$T((\tau,\rho)) := (\tau, \tau\Theta(\tau^{-1}\rho), \qquad \tau \in \mathbb{R}, \rho \in \mathbb{R}^s$$

and (6.23) becomes

$$\sum_{\alpha\in\Gamma_{s+1,n}} c_\alpha b_\alpha^n(x|W) = \sum_{\alpha\in\Gamma_{s+1,n}} c_\alpha^k b_\alpha^n(x|W^k).$$

The highest eigenvalue of T is one and its corresponding eigenvector is $x_0 = (1, \rho_0)$. We note that for all k and $e := (1, 1, \ldots, 1)^t$,

$$S_k e = e \tag{6.26}$$

and from (6.22)

$$(S_k)_{\alpha\beta} = \sum_{\gamma\in\Gamma_{s+1,n}} \sum_{\delta\in\Gamma_{s+1,n}} B_{\gamma\alpha}^{-1}(\mathcal{E}^t(T))_{\gamma\delta}^k B_{\beta\delta}.$$

By Lemmas 6.2 and 6.3 applied to the matrix $\mathcal{E}^t(T)$

$$\lim_{k\to\infty} (\mathcal{E}^t(T))_{\alpha\beta}^k = \mu_\alpha m_\beta(x_0), \qquad \alpha, \beta \in \Gamma_{s+1,n},$$

where $\mathcal{E}^t(T)\mu = \mu$ for some $\mu \neq 0$. So

$$\begin{aligned}\lim_{k\to\infty} (S_k)_{\alpha\beta} &= \sum_{\gamma\in\Gamma_{s+1,n}} \sum_{\delta\in\Gamma_{s+1,n}} B_{\gamma\alpha}^{-1}\mu_\gamma m_\delta(x_0) B_{\beta\delta} \\ &= \Big(\sum_{\gamma\in\Gamma_{s+1,n}} B_{\gamma\alpha}^{-1}\mu_\gamma\Big) b_\beta(x_0).\end{aligned} \tag{6.27}$$

By summing over β and using (6.24) we find that

$$\sum_{\gamma\in\Gamma_{s+1,n}} B_{\gamma\alpha}^{-1} u_\gamma = 1, \qquad \alpha \in \Gamma_{s+1,n}$$

or equivalently

$$u = B^t e.$$

So from (6.27) we obtain

$$\lim_{k\to\infty} \sum_{\beta\in\Gamma_{s+1,n}} (S_k)_{\alpha\beta} c_\beta = \sum_{\beta\in\gamma_{s+1,n}} c_\beta b_\beta(x_0) = \psi(\rho_0)$$

which proves the theorem since $c^k = S_k c$. ■

Remark 6.1. We conjecture that Theorem 6.1 remains valid for any sequence of sets Σ^k (not necessarily satisfying equation (6.25)) such that $\lim_{k\to\infty} \Sigma^k = \rho_0$.

References

1. Dahmen, W., and C. A. Micchelli, On limits of multivariate B-splines, Journal d'Analyse Mathématiques **39** (1981), 256–278.
2. Dahmen, W., and C. A. Micchelli, On entire functions of affine lineage, Proc. Amer. Math. Soc. **84** (1982), 344–346.
3. Dahmen, W., and C. A. Micchelli, On the linear independence of multivariate B-splines II: Complete configuration, Mathematics of Computation **41** (1983), 143–163.
4. Dahmen, W., C. A. Micchelli, and H.P. Seidel, Blossoming begets B-spline bases built better by B-patches, Mathematics of Computation, to appear.
5. Motzkin, T. S., and I. J. Schoenberg, On lineal entire functions of n complex variables, Proc. Amer. Math. Soc. **3** (1952), 517–526.
6. Seidel, H.-P., Symmetric recursive algorithms for surfaces: B-patches and the de Boor algorithm for polynomials over triangles, Constructive Approximation **7** (1991), 257–279.
7. Stefanus, Y., and R. N. Goldman, Blossoming Marsden's identity, preprint, 1991.
8. Taussky, O., A generalization of a theorem of Lyapunov, J. Soc. Indust. Appl. Math. **9** (1961) 640–643.

Alfred S. Cavaretta
Department of Mathematics and Computer Science
Kent State University
Kent, Ohio 44242 USA
cavarett@mcs.kent.edu

Charles A. Micchelli
IBM Research Division
T.J. Watson Research Center
Mathematical Sciences Department
Yorktown Heights, NY 10598
cam@yktvmx.bitnet

Implicitizing Rational Surfaces with Base Points by Applying Perturbations and the Factors of Zero Theorem

Eng-Wee Chionh and Ronald N. Goldman

Abstract. Rational surfaces can be implicitized algorithmically by using an explicit resultant expression such as the Macaulay quotient. But this method fails when the parametrization has base points. Fortunately, base points can always be removed by perturbations because a generic rational surface parametrization has no base points. By carefully undoing the perturbations, we can salvage the original implicit representation from the perturbed one. Our main goal is to demonstrate how to undo the perturbations by proving what we call the factors of zero theorem. We also show that the Dixon resultant is a better alternative than true resultant expressions for obtaining the perturbed implicit representation.

§1. Introduction

1.1. Motivation and Goal

Surfaces described parametrically by rational functions are called *rational surfaces*. Rational surface patches are popular in CAGD [1]. They are well suited for rendering, modeling, and analyzing shapes.

Surfaces determined implicitly by polynomial equations are called *algebraic surfaces*. Algebraic surfaces have two particularly nice properties. First, half spaces are defined naturally; distinguishing the inside, boundary, and outside reduces to a trivial sign test. Second, the implicit representation offers surfaces of desired smoothness with the lowest possible degree. For example, implicit quadric patches suffice for tangent plane continuity [5]. But if parametric quadratic patches are used, quartic surfaces usually arise.

Clearly, it is advantageous to have both representations for a given shape. Hybrid representations also promise better algorithms, see [12] for an example.

Mathematical Methods in Computer Aided Geometric Design II

Tom Lyche and Larry L. Schumaker (eds.), pp. 101–110.

ISBN 0-12-460510-9.

Since every rational surface is part of an irreducible algebraic surface [2], we are interested in *implicitization*, that is, in finding the implicit representation of rational surfaces.

There are many ways to implicitize rational surfaces [4]: successive elimination, multivariate resultants, undetermined coefficients, Gaussian elimination, and the theory of field extensions. The method of multivariate resultants is most elegant. Using explicit resultant expressions such as the Macaulay quotient [8,9], we can extend implicitization by resultants from rational plane curves [11] to rational surfaces in a straightforward manner. But there is a pitfall: the method fails when a rational surface parametrization has base points—parameter values at which the quotients for X, Y, Z are $0/0$. Unlike base points of plane curves, surface base points cannot be removed. But they can always be annihilated by perturbations because a generic rational surface parametrization has no base points. The problem is the recovery of the original implicit representation from the perturbed one. We provide a solution by proving what we call the *factors of zero theorem*. A consequence of this theorem is that true resultant expressions are not essential for implicitizing rational surface parametrizations having base points. We shall show that the Dixon resultant [6], which is a customized rather than a true resultant, is generally a better alternative.

§2. Mathematical Preliminaries

2.1. Polynomials

The set of polynomials in the indeterminates $x_1, \cdots, x_k$ with coefficients in an integral domain K is an integral domain denoted by $K[x_1, \cdots, x_k]$. In our discussion, K is always a subfield of $\mathbb{C}$, the field of complex numbers.

A polynomial $p \in K[x_1, \cdots, x_k]$ is *homogeneous* if all the terms of p have the same total degree.

A bivariate polynomial $p \in K[x, y]$ is *bi-degree* (m, n) in (x, y) if the highest degree in x, y is m, n respectively. A general degree n bivariate polynomial is treated as a bi-degree (n, n) polynomial whose total degree is n.

In this paper, a polynomial $p \in K[x_1, \cdots, x_k]$ is said to be *irreducible* if and only if it cannot be written as the product of two non-constant polynomials in $\overline{K}[x_1, \cdots, x_k]$ where $\overline{K}$ is the algebraic closure of K.

The following theorem is very useful for determining whether one polynomial divides another [14]:

Theorem 1. *Let p, $q \in F[x_1, \cdots, x_k]$ where F is an algebraically closed field. If p is irreducible, then p divides q if and only if q vanishes whenever p does.*

2.2 Resultants

For a system of k homogeneous polynomial equations in k variables, there is a polynomial function of the coefficients, known as the *resultant*, that vanishes if and only if the homogeneous system has non-trivial (not all zero) solutions [7]. See [3] for a tutorial on resultants.

The case $k = 1$ is trivial. Bézout and Sylvester resultants are well-known resultant expressions for $k = 2$. When $k \geq 2$, a concise but explicit resultant expression as the quotient of two determinants was provided by Macaulay [8,9]. Since a resultant eliminates k variables from k homogeneous equations, we use the term "multivariate resultant" for $k > 2$.

For three bi-degree (m, n) polynomials whose total degrees are $m + n$, there is a determinant expression for solvability called the Dixon resultant [6,10]. Strictly speaking, the Dixon resultant is a customized rather than a true resultant because certain solutions at infinity are assumed and its vanishing is only a necessary but not a sufficient condition for solvability [3].

2.3. Rational Surfaces, Number of Correspondence, Base Points

A rational surface is a transformation of a plane into 3-space by rational functions (quotients of polynomials). Using projective coordinates, $(r : s : t)$ for the plane and $(w : x : y : z)$ for the 3-space, we can describe a rational surface as

$$w \;:\; x \;:\; y \;:\; z \;=\; \omega(r,s,t) \;:\; \xi(r,s,t) \;:\; \eta(r,s,t) \;:\; \zeta(r,s,t), \qquad (1)$$

where the *parametric polynomials* ω, ξ, η, ζ are homogeneous polynomials of the *same* degree n. The common degree n is known as the *parametric degree* of the parametrization. When $(1 : 0 : 0 : 0)$ is the origin of the 3-space, the affine portion of the rational surface is defined wherever $\omega(r, s, t) \neq 0$ as

$$X = \frac{\xi(r,s,t)}{\omega(r,s,t)}, \qquad Y = \frac{\eta(r,s,t)}{\omega(r,s,t)}, \qquad Z = \frac{\zeta(r,s,t)}{\omega(r,s,t)}. \qquad (2)$$

For a rational surface parametrization, there exists an integer $\mu \geq 1$, known as the *number of correspondence*, such that a general surface point corresponds to μ parameter points [15]. When $\mu = 1$, a rational surface parametrization is *faithful*; otherwise it is *unfaithful*.

For example, the parametrization $w : x : y : z = r^2 + s^2 + t^2 : r^2 : s^2 : t^2$ is unfaithful with $\mu = 4$: a general surface point corresponds to four parameter points $(r : \pm s : \pm t)$.

The parametric polynomials of a surface parametrization define four curves in the projective plane:

$$\omega(r,s,t) = 0, \qquad \xi(r,s,t) = 0, \qquad \eta(r,s,t) = 0, \qquad \zeta(r,s,t) = 0.$$

Common intersections of the four curves are called the *base points* of the parametrization. Clearly a base point corresponds to the surface point $(0 : 0 : 0 : 0)$ which is undefined.

For example, the plane $X + Y + Z - 1 = 0$ can be parametrized as

$$X = \frac{st}{rs + rt + st}, \qquad Y = \frac{rt}{rs + rt + st}, \qquad Z = \frac{rs}{rs + rt + st}.$$

The three parameter points $(1 : 0 : 0)$, $(0 : 1 : 0)$, $(0 : 0 : 1)$ are base points.

2.4. Every Rational Surface is Algebraic

An *algebraic surface* in the projective 3-space $(w : x : y : z)$ is represented implicitly by a homogeneous polynomial equation $g(w, x, y, z) = 0$. When the origin of the 3-space is $(1 : 0 : 0 : 0)$, the affine portion of the algebraic surface is given by $G(X, Y, Z) = 0$ where $G(X, Y, Z) \equiv g(1, X, Y, Z)$ or $g(w, x, y, z) \equiv w^m G(\frac{x}{w}, \frac{y}{w}, \frac{z}{w})$ where m is the degree of g. Note that g and G have the same degree unless g has the plane at infinity, $w = 0$, as a component.

The degree and reducibility of an algebraic surface is the degree and reducibility of its defining polynomial.

The relationship between rational and algebraic surfaces is given by the following theorem [2]:

Theorem 2. *Every rational surface is a subset of a unique irreducible algebraic surface. Furthermore, the coefficients of the implicit polynomial can be taken to be elements of the coefficient field of the parametric polynomials.*

The *implicit degree*, or simply the degree, of a rational surface is the degree of the corresponding irreducible algebraic surface.

The converse of Theorem 2 is not true; there are irreducible algebraic surfaces which admit no rational parametrizations.

2.5. Auxiliary Equations and Implicitization by Resultants

The *auxiliary equations* of parametrization (2) are defined to be

$$\begin{aligned} \xi(r,s,t) - X\omega(r,s,t) &= 0 \\ \eta(r,s,t) - Y\omega(r,s,t) &= 0 \\ \zeta(r,s,t) - Z\omega(r,s,t) &= 0. \end{aligned} \tag{3}$$

The precise relationships between the implicit equation, the resultant of the auxiliary equations, the number of correspondence, and the base points of a rational surface parametrization are summarized by the following theorem [4].

Theorem 3. *For a rational surface parametrization, the resultant of the auxiliary equations is either identically zero if there are base points or the implicit polynomial raised to the power μ, the number of correspondence, if there are no base points.*

§3. The Factors of Zero Theorem

The following result, called the *factors of zero theorem*, is an important property of polynomials. In conjunction with perturbations and elimination, it can be used for implicitizing rational surface parametrizations with base points.

The statement of the theorem may appear complicated, but it asserts an intuitively simple result: if some polynomials with indeterminate coefficients satisfy a non-zero polynomial relationship with coefficients involving these indeterminates, then after specializing some of these indeterminate coefficients, the specialized polynomials will again satisfy a *non-zero* polynomial relationship.

3.1. Theorem and Proof

Theorem 4. *Let K be a field and let $a_1, \cdots, a_k$, $s_1, \cdots, s_n$, $x_1, \cdots, x_m$ be indeterminates. Let D be the integral domain $K[a_1, \cdots, a_k]$ and let $p_1, \cdots, p_m \in D[s_1, \cdots, s_n]$. Suppose that there is a non-zero $f \in D[x_1, \cdots, x_m]$ such that $f(p_1, \cdots, p_m) \equiv 0$. Suppose further that f vanishes when any one of $a_1, \cdots, a_l$ is specialized to $a_1^*, \cdots, a_l^* \in K$, respectively, where $1 \leq l \leq k$. Let $D^* = K[a_{l+1}, \cdots, a_k]$ and let $p_1, \cdots, p_m$ become $p_1^*, \cdots, p_m^*$ when D becomes D^*. Then there is a non-zero $f^* \in D^*[x_1, \cdots, x_m]$ such that $f^*(p_1^*, \cdots, p_m^*) \equiv 0$.*

Proof: Consider $a_1 - a_1^*$, f as polynomials of $\overline{K}[a_1, \cdots, a_k; x_1, \cdots, x_m]$, where $\overline{K}$ is the algebraic closure of K. Since f vanishes whenever $a_1 - a_1^*$ does, by repeated use of Theorem 1, there is an integer $d_1 > 0$ such that

$$\begin{aligned} f &\equiv (a_1 - a_1^*)^{d_1} f_1 \\ f_1^* &\equiv f_1|_{a_1 \leftarrow a_1^*} \not\equiv 0. \end{aligned} \tag{4}$$

Clearly $f_1(p_1, \cdots, p_m) \equiv 0$ and f_1 vanishes whenever f does except for the specialization $a_1 \leftarrow a_1^*$.

These vanishing properties of f_1 are inherited by f_1^*:

1. $f_1^*(p_1^{(1)}, \cdots, p_m^{(1)}) \equiv 0$ where $p_i^{(1)} \equiv p_i|_{a_1 \leftarrow a_1^*}$, $1 \leq i \leq m$.
2. f_1^* vanishes when any a_i is specialized to a_i^*, $2 \leq i \leq l$.

The equalities $f_1^*(p_1^{(1)}, \cdots, p_m^{(1)}) \equiv f_1(p_1, \cdots, p_m)|_{a_1 \leftarrow a_1^*} \equiv 0|_{a_1 \leftarrow a_1^*} \equiv 0$ establish the first vanishing property. To verify the other vanishing property, we see that for $2 \leq i \leq l$:

$$f_1^*|_{a_i \leftarrow a_i^*} \equiv f_1|_{a_1 \leftarrow a_1^*}|_{a_i \leftarrow a_i^*} \equiv f_1|_{a_i \leftarrow a_i^*}|_{a_1 \leftarrow a_1^*} \equiv 0|_{a_1 \leftarrow a_1^*} \equiv 0.$$

The theorem is thus proved when $l = 1$. By applying this special version of the theorem l times, each time specializing some a_i and obtaining a non-zero f_i^*, we prove the general version by producing f_l^* as the f^* required in the theorem. ■

The proof demonstrates more than the existence of f^*; it also suggests a method for finding f^* by properly discarding factors of zero.

3.2. An Example

To illustrate this theorem, consider $D = \mathbb{R}[\alpha, \beta, \gamma]$, where $\mathbb{R}$ is the field of real numbers and α, β, γ are indeterminates. Let $p_1, p_2, p_3, p_4 \in D[s_1, s_2, s_3]$ and $f \in D[x_1, x_2, x_3, x_4]$ be

$$p_1 = s_1 + s_2 + s_3, \qquad p_2 = p_1 + \beta\gamma s_1, \qquad p_3 = p_1 + \alpha\gamma s_2, \qquad p_4 = p_1 + \alpha\beta s_3,$$
$$f = \alpha\beta\gamma[\alpha\beta\gamma x_1 + \alpha(x_1 - x_2) + \beta(x_1 - x_3) + \gamma(x_1 - x_4)].$$

Clearly f is non-zero, vanishes when any of α, β, γ is specialized to zero, and $f(p_1, p_2, p_3, p_4) = \alpha\beta\gamma[\alpha\beta\gamma(s_1 + s_2 + s_3) - \alpha\beta\gamma s_1 - \beta\alpha\gamma s_2 - \gamma\alpha\beta s_3] = 0$.

When α, β, γ are all set to zero, we have $p_1^* = p_2^* = p_3^* = p_4^* = s_1 + s_2 + s_3$. The f^* of the theorem is $x_1 - x_2$, $x_1 - x_3$, $x_1 - x_4$ if the last specialized indeterminate is α, β, γ respectively. Whatever the order of specialization, we always obtain a non-zero f^* such that $f^*(p_1^*, p_2^*, p_3^*, p_4^*) \equiv 0$ as predicted by the theorem.

§4. Implicitization with Base Points Using the Dixon Resultant

4.1. A General Implicitization Strategy

By Theorem 3, if a perturbed parametrization $w : x : y : z = \hat{\omega}(r,s,t) : \hat{\xi}(r,s,t) : \hat{\eta}(r,s,t) : \hat{\zeta}(r,s,t)$. has no base points, then the resultant $\hat{R}$ of the perturbed auxiliary equations is non-zero, and we have

$$\hat{R}(\frac{\hat{\xi}(r,s,t)}{\hat{\omega}(r,s,t)}, \frac{\hat{\eta}(r,s,t)}{\hat{\omega}(r,s,t)}, \frac{\hat{\zeta}(r,s,t)}{\hat{\omega}(r,s,t)}) \equiv 0.$$

It follows from the factors of zero theorem that there is a non-zero R^* such that

$$R^*(\frac{\xi(r,s,t)}{\omega(r,s,t)}, \frac{\eta(r,s,t)}{\omega(r,s,t)}, \frac{\zeta(r,s,t)}{\omega(r,s,t)}) \equiv 0.$$

Thus by Theorem 2, R^* has the implicit polynomial f of the original parametrization $\omega(r,s,t) : \xi(r,s,t) : \eta(r,s,t) : \zeta(r,s,t)$ as a factor. The implicit polynomial f can be found by factorization and back substitution; empirically, we have observed that it can also be found as the GCD of the R^*'s of two different perturbations.

4.2. The Dixon Resultant is a Better Alternative

From the general strategy, we see that what we really need is a non-zero expression $\hat{R}$ that is satisfied identically by the perturbed parametrization. The resultant of the perturbed auxiliary equations, which can be computed as a Macaulay quotient, is certainly such an expression, but another better alternative is the Dixon resultant.

The Dixon resultant is computationally more efficient than the Macaulay quotient because it uses smaller determinants. For a degree n parametrization, the dimension of the Macaulay numerator determinant for three equal degree equations is $4n^2 - n$, but the dimension of the Dixon determinant for a bi-degree (n,n) parametrization of total degree $2n$ is only $2n^2$ [3]. In CAGD, surface parametrizations are usually bi-degree (m,n) of total degree $m+n$. In this situation, the determinant dimension reduces from $4(m+n)^2 - (m+n)$ for Macaulay to $2mn$ for Dixon. Since the cost of evaluating an $N \times N$ determinant is $O(N^3)$, the efficiency gain is substantial.

4.3. The Dixon Resultant

Dixon [6] elegantly extended Cayley's statement of Bézout's method to three bi-degree polynomial equations in two variables

$$\begin{array}{rcccl} f(x,y) & \equiv & \sum_{ij} a_{ij} x^i y^j & = & 0 \\ g(x,y) & \equiv & \sum_{ij} b_{ij} x^i y^j & = & 0 \\ h(x,y) & \equiv & \sum_{ij} c_{ij} x^i y^j & = & 0, \end{array} \tag{9}$$

where $m \geq i \geq 0$, $n \geq j \geq 0$ by considering the determinant

$$\Delta(x,y,\alpha,\beta) \equiv \begin{vmatrix} f(x,y) & g(x,y) & h(x,y) \\ f(\alpha,y) & g(\alpha,y) & h(\alpha,y) \\ f(\alpha,\beta) & g(\alpha,\beta) & h(\alpha,\beta) \end{vmatrix}.$$

Since $\Delta = 0$ when $x = \alpha$ or $y = \beta$, $(x-\alpha)(y-\beta)$ divides Δ and gives the *Dixon expression* $\delta(x,y,\alpha,\beta) = \sum d_{ijkl} x^i y^j \alpha^k \beta^l$, where $m-1 \geq i \geq 0$, $2n-1 \geq j \geq 0$, $2m-1 \geq k \geq 0$, $n-1 \geq l \geq 0$. Thus

$$\delta(x,y,\alpha,\beta) \equiv \begin{pmatrix} x^{m-1}y^{2n-1} \\ x^{m-1}y^{2n-2} \\ \vdots \\ y \\ 1 \end{pmatrix}^T D \begin{pmatrix} \alpha^{2m-1}\beta^{n-1} \\ \alpha^{2m-1}\beta^{n-2} \\ \vdots \\ \beta \\ 1 \end{pmatrix},$$

where D is the dimension $2mn$ coefficient matrix (d_{ijkl}). Each entry d_{ijlk} is of the form $\sum |a_\mu b_\nu c_\lambda|$, where

$$|a_\mu b_\nu c_\lambda| \equiv \begin{vmatrix} a_\mu & a_\nu & a_\lambda \\ b_\mu & b_\nu & b_\lambda \\ c_\mu & c_\nu & c_\lambda \end{vmatrix},$$

so $|D|$ is of degree $2mn$ in the coefficients of each of f, g, h. An algorithm to compute the d_{ijlk}'s from the a's, b's, c's is given by Sederberg [10], but it is equally easy to compute the d_{ijlk}'s symbolically from the expression for $\delta(x,y,\alpha,\beta)$.

The three equations $f = 0$, $g = 0$, $h = 0$ define three curves in the projective plane $(w : x : y)$, where w is a homogenizing variable. If the degree of the curves is $m+n$, in general each curve has $(0:1:0)$, $(0:0:1)$ as a point of multiplicity n and m respectively. Thus two generic curves $af+bg+ch=0$, $a'f+b'g+c'h=0$ must intersect at least n^2 and m^2 times at $(0:1:0)$ and $(0:0:1)$ respectively [14]. When the intersection counts are exactly n^2 and m^2, the determinant $|D|$, known as the *Dixon resultant*, is a resultant in the sense that f, g, h have a common solution $(1:x:y)$ if and only if $|D|$ vanishes [3].

When the three polynomials are the auxiliary equations of a rational parametrization, these intersection conditions are equivalent to saying that the bi-degree (m,n) parametrization has respectively n^2 and m^2 base points at the two vertices of the coordinates system at infinity. This is the canonical base point configuration for tensor product surfaces. Hence in CAGD the Dixon resultant was first applied to implicitize tensor product patches [11].

4.4. Perturbations Ensuring a Non-Zero Dixon Resultant

To obtain a non-zero Dixon resultant, the base point configuration of the parametrization must be canonical. An extreme measure is to force the parametrization to a general bi-degree (m,n) parametrization of total degree $m+n$ by making all the coefficients indeterminate. But actually much less perturbation will suffice.

To see this, we shall set one of the parameters to unity, say $r = 1$, because affine parameters are more convenient than projective parameters for computing the Dixon resultant. A bi-degree (m,n) parametrization, of whatever total degree, can always be perturbed to have total degree $m + n$ by adding λ_w, λ_x, λ_y, λ_z to the parametric polynomials for w, x, y, z respectively; where λ_w, λ_x, λ_y, λ_z is some permutation of the terms $\lambda s^m t^n$, λs^m, λt^n, λ. The Dixon resultant of the auxiliary equations of the perturbed parametrization is non-zero because it has the term $\lambda^{6mn}(x_0 x_1 - x_2 x_3)^{mn}$. The variables x_0, x_1, x_2, x_3 is a permutation of the factors 1, X, Y, Z; such that x_0, x_1, x_2, x_3 represent the affine version of the projective coordinates whose parametric polynomials are perturbed by $\lambda s^m t^n$, λ, λs^m, λt^n respectively. For example, the Dixon resultant for the auxiliary equations of

$$w : x : y : z = \cdots + \lambda s^m : \cdots + \lambda s^m t^n : \cdots + \lambda t^n : \cdots + \lambda$$

has the term $\lambda^{6mn}(XZ - Y)^{mn}$; we have x_0, x_1, x_2, x_3 correspond to X, Z, 1, Y respectively.

§5. An Implicitization Example

Consider the parametrization

$$w : x : y : z = (r^2+s^2)(r^2+t^2) : (r^2+s^2)(r^2-t^2) : 4r^2 st : 2rt(r^2-s^2)$$

of the sphere $X^2+Y^2+Z^2-1=0$. This parametrization has four base points $(r : s : t) = (0 : 1 : 0)$, $(0 : 0 : 1)$, $(1 : i : 0)$, $(1 : -i : 0)$ [4].

Setting $r = 1$ and perturbing the parametrization by adding the terms $\lambda s^2 t^2$, λs^2, λt^2, λ to w, x, y, z respectively, we obtain a bi-degree $(2, 2)$ parametrization of total degree 4 having the canonical base point configuration. The factor λ can be taken out from the 1st, 4th, 5th, and 8th rows of the Dixon determinant of the perturbed auxiliary equations. After discarding the factor λ^4 and then setting $\lambda = 0$, the Dixon resultant is found to be

$$\begin{gathered} 1048576(X^2 + Y^2 + Z^2 + 2XY + 2X + 2Y + 1) \\ (X^2 + Y^2 + Z^2 + 2XZ - 2X - 2Z + 1)(X^2 + Y^2 + Z^2 - 1)^2. \end{gathered}$$

Clearly the square of the implicit polynomial is a factor of the resultant.

The implicit polynomial could also have been found by computing a GCD instead of by factorization. A different perturbation can be done by adding λ, λs^2, λt^2, $\lambda s^2 t^2$ to w, x, y, z respectively. Again the factor λ can be taken out from the 1st, 4th, 5th, and 8th rows of the Dixon determinant of the perturbed auxiliary equations. After discarding λ^4 and then setting $\lambda = 0$, the Dixon resultant is computed to be

$$8388608(X + 1)^2(Y^2 + Z^2)(X^2 + Y^2 + Z^2 - 1)^2.$$

We see that the GCD of the two final non-zero Dixon resultants is the square of the implicit polynomial.

Notice that the given sphere parametrization is unfaithful with the number of correspondence $\mu = 2$. In general, a point on the sphere corresponds to two of the four parameter points

$$(r : s : t) = (1 \ : \ -\frac{z}{y} \pm \sqrt{1 + (\frac{z}{y})^2} \ : \ \pm\sqrt{\frac{w - x}{w + x}}.$$

Thus, in our example the implicit polynomial appears in the salvaged Dixon resultant as a factor to the power μ, just as one might expect from Theorem 3.

§6. Conclusion and Open Problems

We have shown that we can indeed salvage the implicit polynomial from an otherwise zero resultant by using perturbations and the factors of zero theorem. We have also illustrated that the Dixon resultant is more efficient than the Macaulay quotient for obtaining perturbed implicit polynomials.

Empirically we have observed that there are always two perturbations such that the GCD of the final non-zero resultants gives the implicit polynomial to the power μ, the number of correspondence, of the given parametrization. But we have not established these facts rigorously. Our observations deserve further study because it is known that when there are no base points or the base point configuration is canonical for bi-degree, the true resultant, say by the Macaulay quotient or the customized Dixon resultant, gives the implicit polynomial to the power μ exactly.

Acknowledgments. Eng-Wee Chionh is supported by a scholarship from the National University of Singapore. Professor Goldman's participation in this research is supported by NSF grant DMC-8813688.

References

1. Böhm, W., G. Farin, and J. Kahmann, A survey of curve and surface methods in CAGD, Computer Aided Geom. Design **1** (1984), 1–60.
2. Chionh, E. W., and R. N. Goldman, On the existence and the coefficients of the implicit equation for a rational surface, preprint, 1989.
3. Chionh, E. W., and R. N. Goldman, A tutorial on resultants and elimination theory, manuscript, 1989.
4. Chionh, E. W., Base points, resultants, and the implicit representation of rational surfaces, dissertation, University of Waterloo, 1990.
5. Dahmen, W., Smooth piecewise quadric surfaces, in *Mathematical Methods in Computer Aided Geometric Design*, Tom Lyche and L. L. Schumaker (eds.), Academic Press, New York, 1989, 181–193.
6. Dixon, A. L., The eliminant of three quartics in two independent variables, Proc. London Math. Soc. **6** (1908), 49–69 and 473–492.
7. Lefschetz, S., *Algebraic Geometry*, Princeton University Press, 1953.

8. Macaulay, F. S., On some formula in elimination, Proc. London Math. Soc., (May 1902), 3–27.
9. Macaulay, F. S., Notes on the resultant of a number of polynomials of the same degree, Proc. London Math. Soc., (June 1921), 14–21.
10. Sederberg, T. W., Implicit and parametric curves and surfaces, dissertation, Purdue University, 1983.
11. Sederberg, T. W., D. C. Anderson, and R. N. Goldman, Implicit representation of parametric curves and surfaces, Computer Vision, Graphics, and Image Processing **28** (1984), 72–84.
12. Sederberg, T. W., and S. R. Parry, Comparison of three curve intersection algorithms Comput. Aided Design **18:1** (1986).
13. Waerden, B. L. van der, *Modern Algebra*, Frederick Ungar Publishing Co., New York, 2nd edition, 1950.
14. Walker, R. J., *Algebraic Curves*, Princeton Press, 1950.
15. Zariski, O., *Algebraic Surfaces*, Springer-Verlag, 2nd supplemented edition, 1971.

Eng-Wee Chionh
Department of Information Systems and Computer Science
National University of Singapore
Keng Ridge, Singapore 0511
isccew@nusvm.bitnet

R. N. Goldman
Department of Computer Science
Rice University
P. O. Box 1892,
Houston, Texas, 77251
rng@rice.edu

Wavelets and Multiscale Interpolation

Charles K. Chui and Xianliang Shi

Abstract. The most powerful technique for constructing wavelets is the formulation of certain multiresolution analyses (MRA), and the essential ingredient in any MRA is the two-scale relation of a scaling function ϕ that generates this MRA. Once ϕ and its two-scale relation are established, there are various methods available for constructing the companion wavelets ψ and their corresponding dual wavelets $\tilde{\psi}$. It so happens that a two-scale relation with a finite two-scale sequence also describes a subdivision scheme in CAGD. The objective of this paper is to survey the recent development in wavelets and subdivision schemes in this direction, including a discussion of our own most recent work. In particular, results on regularity and characterization of orthonormal and fundamental scaling functions will be given.

§1. Introduction

The wavelet series representation of a square-integrable function defined on the real line $\mathbb{R}$, with respect to a given wavelet ψ, may be viewed as an analogue of the Fourier series expansion of a 2π-periodic square-integrable function, since each of these representations, is generated by a single function, $\psi(x)$ or e^{ix}, respectively. There are, however, at least three essential differences. Firstly, since the operation of dilation on ψ alone is obviously insufficient, the operation of translation is also needed in a wavelet series representation. Secondly, the coefficients in a wavelet series expansion of f contain very important local spectral information of f in terms of the integral wavelet transform (IWT) of f relative to the dual wavelet $\tilde{\psi}$ of ψ. Thirdly, there is freedom in the choice of the wavelet ψ to meet the specifications of the mathematical model of interest. In general, the construction of a wavelet is based on that of its companion function, called a scaling function. For these and other reasons, there is a need for the study of scaling functions and the multiresolution analyses (MRA) they generate. The objective of this paper is to discuss the recent development on wavelets, scaling functions, MRA, and

Mathematical Methods in Computer Aided Geometric Design II
Tom Lyche and Larry L. Schumaker (eds.), pp. 111–133.

ISBN 0-12-460510-9.

the intimate relation between the two-scale relations of scaling functions and subdivision schemes in computer aided geometric design (CAGD).

In the following section, we will state some new results on wavelets and give a complete characterization of orthonormal scaling functions. In Sections 3 and 4, we will give two criteria of scaling functions ϕ that are fundamental, in the sense that $\phi(\mathbf{k}) = \delta_{\mathbf{0},\mathbf{k}}$, $\mathbf{k} \in \mathbb{Z}^s$, and describe a fairly general formulation of symmetric fundamental scaling functions with compact support. In Section 5, some new results on subdivision schemes will be discussed. The final section will be devoted to a study of the regularity of scaling functions and wavelets.

Throughout this paper, the usual multivariate notations will be used, namely: $\mathbf{x} = (x_1, \ldots, x_s)$ and $\omega = (\omega_1, \ldots, \omega_s) \in \mathbb{R}^s$, and

$$\|f\|_p = \begin{cases} \left(\int\limits_{\mathbb{R}^s} |f(\mathbf{x})|^p d\mathbf{x} \right)^{1/p}, & \text{for} \quad 1 \le p < \infty, \\ \operatorname*{ess\,sup}_{\mathbf{x} \in \mathbb{R}^s} |f(\mathbf{x})|, & \text{for} \quad p = \infty. \end{cases}$$

Also, $L^p = L^p(\mathbb{R}^s)$ denotes the Banach space of measurable functions with $\|f\|_p < \infty$, and $C(\mathbb{R}^s)$, the subspace of L^∞ consisting of all continuous functions. If $f \in C(\mathbb{R}^s)$, we denote by

$$\omega_\infty(f, t) = \sup_{0 \le |\mathbf{h}| \le t} \|f(\cdot + \mathbf{h}) - f(\cdot)\|_\infty$$

its modulus of continuity. In addition, $L^p(2\pi)^s$ denotes the space of 2π-periodic functions satisfying $\|f\|_{L^p(2\pi)^s} < \infty$, where the norm $\|f\|_{L^p(2\pi)^s}$ is defined by

$$\|f\|_{L^p(2\pi)^s} = \begin{cases} \left(\frac{1}{(2\pi)^s} \int_{[-\pi,\pi]^s} |f(\mathbf{x})|^p d\mathbf{x} \right)^{1/p}, & \text{if} \quad 1 \le p < \infty, \\ \operatorname{ess\,sup} |f(\mathbf{x})|, & \text{if} \quad p = \infty. \end{cases}$$

Also, $C(2\pi)^s$ denotes the space of 2π-periodic continuous functions with norm $\|f\|_c = \|f\|_\infty$. For any function f defined on $\mathbb{R}^s$, we will consider

$$f_{j,\mathbf{k}}(\mathbf{x}) = 2^{js/2} f(2^j \mathbf{x} - \mathbf{k}), \tag{1.1}$$

where $j \in \mathbb{Z}, \mathbf{k} \in \mathbb{Z}^s$. If $f \in L^1 \cap L^2$ then its Fourier transformation is defined by

$$\hat{f}(\omega) = \int_{\mathbb{R}^s} e^{-i\omega\mathbf{x}} f(\mathbf{x}) d\mathbf{x}.$$

§2. Wavelets and Multiresolution Analysis

Let

$$\Psi := \{\psi_1, \ldots, \psi_{2^s-1}\}$$

be a given family of $2^s - 1$ functions in L^2. We will call Ψ an $\mathcal{R}$-family (Riesz family) and each ψ_ℓ an $\mathcal{R}$-function, if

$$\mathcal{B} := \{\psi_{\ell;j,\mathbf{k}}\colon\ 1 \le \ell \le 2^s - 1, \quad j \in \mathbb{Z}, \mathbf{k} \in \mathbb{Z}^s\} \tag{2.1}$$

forms a Riesz basis of L^2, namely: it is complete and there exist two positive constants A and B such that the inequalities

$$A \sum_{\ell=1}^{2^s-1} \sum_{j,\mathbf{k}} |a_{j,\mathbf{k}}^{(\ell)}|^2 \le \left\| \sum_{\ell=1}^{2^s-1} \sum_{j,\mathbf{k}} a_{j,\mathbf{k}}^{(\ell)} \psi_{\ell;j,\mathbf{k}} \right\|_2^2 \le B \sum_{\ell=1}^{2^s-1} \sum_{j,\mathbf{k}} |a_{j,\mathbf{k}}^{(\ell)}|^2$$

hold for all $\{a_{j,\mathbf{k}}^{(\ell)}\} \in \ell^2$. The constants A and B will be called Riesz bounds of the family Ψ. It is well known that corresponding to every Riesz basis $\mathcal{B}$ as given in (2.1), there is another (unique) Riesz basis

$$\widetilde{\mathcal{B}} := \{\psi_\ell^{j,\mathbf{k}}\colon\ 1 \le \ell \le 2^s - 1, \quad j \in \mathbb{Z}, \mathbf{k} \in \mathbb{Z}^s\}$$

of L^2 with Riesz bounds B^{-1} and A^{-1} such that

$$\langle \psi_{\ell;j,\mathbf{k}}, \psi_m^{p,\mathbf{q}} \rangle = \delta_{\ell,m}\delta_{j,p}\delta_{\mathbf{k},\mathbf{q}},$$

for all $\ell, m = 1, \ldots, 2^s - 1$, $j, p \in \mathbb{Z}$ and $\mathbf{k}, \mathbf{q} \in \mathbb{Z}^s$. $\widetilde{\mathcal{B}}$ is called the basis dual to $\mathcal{B}$, or the dual basis relative to $\mathcal{B}$. However, although $\mathcal{B}$ is generated by an $\mathcal{R}$-family Ψ, it is also known (cf. [16,26,29]) that $\widetilde{\mathcal{B}}$ is not necessarily generated by dilations and translations of an $\mathcal{R}$-family. A simple example is as follows.

Example 2.1 (see [7]). Let $s = 1$ and ψ be an orthonormal wavelet, namely: $\{\psi_{j,k}\}_{j,k\in\mathbb{Z}}$ forms an orthonormal basis of L^2. Set

$$\eta(x) := \psi(x) - \bar{z}\sqrt{2}\,\psi(2x)$$

as suggested by Daubechies, Lemarié, and Meyer (cf. [16]). Then the dual $\{\eta^{j,k}\}_{j,k\in\mathbb{Z}}$ of $\{\eta_{j,k}\}_{j,k\in\mathbb{Z}}$ satisfies

$$\begin{cases} \eta^{0,0}(x) = \displaystyle\sum_{j=0}^{\infty} \psi_{-j,0}(x) z^j, \\ \eta^{0,1}(x) = \psi_{0,1}(x). \end{cases}$$

So, if $\tilde{\eta}$ exists, then

$$\psi_{0,0}(x) = \tilde{\eta}_{0,1}(x+1) = \tilde{\eta}_{0,0}(x) = \sum_{j=0}^{\infty} \psi_{-j,0}(x) z^j$$

or

$$\sum_{j=1}^{\infty} \psi_{-j,0}(x) z^j \equiv 0,$$

which is absurd. ■

If the dual basis $\widetilde{\mathcal{B}}$ is also generated by some $\mathcal{R}$-family

$$\widetilde{\Psi} := \{\widetilde{\psi}_1, \ldots, \widetilde{\psi}_{2^s-1}\}$$

in L^2 in the sense that

$$\psi_\ell^{j,\mathbf{k}} = \widetilde{\psi}_{\ell;j,\mathbf{k}}, \qquad 1 \le \ell \le 2^s - 1, j \in \mathbb{Z}, \mathbf{k} \in \mathbb{Z}^s,$$

where the notation in (1.1) is used, we will then call Ψ a W-family (wavelet family), and each $\psi_\ell \in \Psi$ a wavelet. There are three classes of such wavelets. In this paper we are only interested in orthogonal and semi-orthogonal wavelets.

If $\Psi := \{\psi_1, \ldots, \psi_{2^s-1}\}$ is a W-family such that $\{\psi_{\ell;j,\mathbf{k}}\}$ forms an orthonormal basis of L^2, then we call Ψ an o.n. family (orthonormal family) and each ψ_ℓ an o.n. wavelet (orthogonal wavelet). In this case, we have $\Psi = \widetilde{\Psi}$; that is, an o.n. family is a self-dual family. If for all $\ell \neq \ell'$ or $j \neq j'$,

$$\langle \psi_{\ell,j,\mathbf{k}}, \psi_{\ell',j',\mathbf{k}'} \rangle = 0, \qquad \mathbf{k}, \mathbf{k}' \in \mathbb{Z}^s,$$

we will call Ψ an s.o. family (semi-orthogonal family) and each ψ_ℓ an s.o. wavelet. In this case, if we define

$$W_{\ell,j} = \operatorname{clos}_{L^2}\{\psi_{\ell;j,\mathbf{k}} \colon \mathbf{k} \in \mathbb{Z}^s\}, \quad \ell = 1, \ldots, 2^s - 1; j \in \mathbb{Z},$$

then $\{W_{\ell j}\}$ is mutually orthogonal. If Ψ is an s.o. family, then the dual family $\widetilde{\Psi} = \{\widetilde{\psi}_1, \ldots, \widetilde{\psi}_{2^s-1}\}$ can be determined by

$$\widehat{\widetilde{\psi}}_\ell(\omega) = \frac{\widehat{\psi}_\ell(\omega)}{\sum_{\mathbf{k}\in\mathbb{Z}^s} |\widehat{\psi}(\omega + 2\mathbf{k}\pi)|^2}, \quad 1 \le \ell \le 2^s - 1.$$

Let us list some examples in the one-dimensional setting:

(a) Haar wavelet (Haar, 1911).

$$\psi_H(x) = \chi_{[0,1)}(x) \operatorname{sgn}\left(\frac{1}{2} - x\right);$$

(b) Strömberg wavelets (Strömberg, 1981, cf. [28]);

(c) Meyer wavelets (with compactly supported Fourier transform) (Meyer 1986, cf. [26,27]);

(d) Battle and Lamarié wavelets (exponentially decaying splines) (Battle-Lamarié, 1987 cf. [1,2,22]);

(e) Daubechies wavelets (with compact support) (Daubechies, 1988 cf. [15]);

(f) Chui-Wang wavelets (s.o. wavelet with compact support) (Chui and Wang, 1990; cf. [6,10,11]).

The Daubechies wavelets can be formulated as follows. For each positive integer N, let

$$S_N(e^{-i\omega}) = \sum_{k=0}^{N-1} s_k^N e^{-ik\omega}, \quad s_0^N \neq 0,$$

be defined by

$$|S_N(e^{-i\omega})|^2 = \sum_{k=0}^{N-1} \binom{N-1+k}{k} \sin^{2k} \frac{1}{2}\omega,$$

and denote

$$P_N^D(e^{-i\omega}) = \left(\frac{1+e^{-i\omega}}{2}\right)^N \sum_{k=0}^{N-1} s_k^N e^{-ik\omega} = \frac{1}{2} \sum_{k=0}^{2N-1} p_k^N e^{-ik\omega}. \tag{2.2}$$

Consider

$$\hat{\phi}_N^D(\omega) = \prod_{j=1}^{\infty} P_N^D(e^{-i2^{-j}\omega}).$$

Then the Daubechies wavelets are

$$\psi_N^D(x) = \sum_{k=2-2N}^{1} (-1)^k p_{-k+1}^N \phi_N^D(2x-k).$$

In [12], Cohen studied wavelets which are not semi-orthogonal, and called them "biorthogonal wavelets". Recently, Cohen, Daubechies and Feauveau [14] have considered biorthogonal wavelets with compact support. In this paper, we will call them nonorthogonal wavelets.

All the examples (a)-(f) are based on the formulation of a multiresolution analysis, namely: a ladder of closed subspaces which approximate L^2. More precisely, let $\phi \in L^2$ and denote

$$V_j := \operatorname{clos}_{L^2}\{\phi_{j,\mathbf{k}}\colon \mathbf{k} \in \mathbb{Z}^s\}, \quad j \in \mathbb{Z}.$$

The spaces V_j, $j \in \mathbb{Z}$, are said to constitute a multiresolution analysis (or MRA) of L^2 if the following conditions are satisfied:

$$\cdots \subset V_{-1} \subset V_0 \subset V_1 \subset \cdots; \tag{2.3}$$

$$\operatorname{clos}_{L^2}\left(\bigcup_j V_j\right) = L^2, \qquad \bigcap_{j\in\mathbb{Z}} V_j = \{0\}; \tag{2.4}$$

$$f(\cdot) \in V_j \Leftrightarrow f\left(\cdot - \frac{\mathbf{k}}{2^j}\right) \in V_j \Leftrightarrow f(2\cdot) \in V_{j+1}; \quad \text{and} \tag{2.5}$$

$$\{\phi_{0,\mathbf{k}}\} \text{ forms a Riesz basis of the subspace } V_0. \tag{2.6}$$

We will also call the function ϕ a generator of the MRA $\{V_j\}_{j\in\mathbb{Z}}$. The condition (2.5) implies that there exists a sequence $\{p_\mathbf{k}\} \in \ell^2$ such that

$$\phi(\mathbf{x}) = \sum_{\mathbf{k}\in\mathbb{Z}^s} p_\mathbf{k}\phi(2\mathbf{x} - \mathbf{k}). \tag{2.7}$$

So, the generator ϕ is also called a scaling function and the identity (2.7) the two-scale relation of ϕ. By taking Fourier transformation, we see that (2.7) implies that

$$\hat{\phi}(2\omega) = P(e^{-i\omega})\hat{\phi}(\omega),$$

where

$$P(\mathbf{z}) = \frac{1}{2^s}\sum_{\mathbf{k}\in\mathbb{Z}^s} p_\mathbf{k}\mathbf{z}^\mathbf{k}, \quad \mathbf{z} = e^{-i\omega}. \tag{2.8}$$

We will call P the two-scale symbol of the scaling function ϕ or of the MRA $\{V_j\}_{j\in\mathbb{Z}}$. If $\{\phi(\cdot - \mathbf{k})\}_{\mathbf{k}\in\mathbb{Z}^s}$ forms an orthonormal basis of V_0, then we will call the generator ϕ an orthonormal scaling function (o.n. scaling function). Mallat [25], Daubechies [15], and Cohen [12] have established certain criteria for o.n. scaling functions. Recently we have obtained the following characterization.

Theorem 2.2 *(see [9]). Let $P(e^{-i\omega})$ be in $C(2\pi)^s$ satisfying*

$$\int_0^\pi \frac{\omega_\infty(P,t)}{t}dt < \infty. \tag{2.9}$$

Then the following two statements are equivalent:

(I) $g(\omega) = \prod_{j=1}^\infty P(e^{-i2^{-j}\omega}) \in L^2$ and ϕ, defined by $\hat{\phi}(\omega) = g(\omega)$, is an o.n. scaling function.

(II) $P(e^{-i\omega})$ satisfies the following conditions:

(i) $P(1) = 1$;

(ii) $\sum_{j=0}^{2^s-1} |P(e^{-i(\omega+\pi h^j)})|^2 = 1$, *where $h^0, \ldots, h^{2^s-1}$ are the 2^s vertices of the cube*

$$Q = \{(x_1, \ldots, x_s)\colon\ 0 \le x_j \le 1, j = 1, \ldots, s\}; \quad \text{and}$$

(iii) $\lim_{N\to\infty} \overline{\lim_{j\to\infty}}\, 2^{sj} \int_\Omega |P(e^{-i\omega})P(e^{-i2\omega})\ldots P(e^{-i2^{j-1}\omega})|^2 d\omega = 0,$

where $\Omega = [-\pi, \pi] \backslash [-2^{-j}N, 2^jN]^s$.

The assumption (2.9) is sharp in the sense that it cannot be replaced by

$$\int_0^\pi \frac{\omega_\infty(P,t)}{t}\delta(t)dt < \infty,$$

for some $\delta(t) \downarrow 0, t \to 0^+$.

Of course wavelets can always be constructed from the scaling functions ϕ. Suppose that $\{V_j\}_{j\in\mathbb{Z}}$ is a given multiresolution analysis of L^2 with generator ϕ. Then we can decompose every V_j as $V_j = V_{j-1} \oplus^\perp W_{j-1}$, where W_{j-1} is the orthogonal complement of V_{j-1} in V_j.

Consider the simplest case $s = 1$ and $\{V_j\}_{j\in\mathbb{Z}}$ is a MRA associated with an o.n. scaling function ϕ. Suppose that

$$P(e^{-i\omega}) = \frac{1}{2}\sum_{k\in\mathbb{Z}} p_k e^{-ik\omega}$$

is the two-scale symbol of ϕ. If we define

$$\psi(x) = \sum_{k\in\mathbb{Z}} (-1)^k \bar{p}_{1-k}\phi(2x-k),$$

then for every $j \in \mathbb{Z}$, $\{\psi_{j,k}(x)\}_{k\in\mathbb{Z}}$ forms an orthonormal basis of W_j. Thus $\{\psi_{j,k}(x)\}_{j,k\in\mathbb{Z}}$ is a wavelet basis of L^2.

If ϕ is not an o.n. scaling function, then we can orthogonalize it by considering

$$\hat{\phi}^*(\omega) = \frac{\hat{\phi}(\omega)}{\left(\sum_{k\in\mathbb{Z}} |\hat{\phi}(\omega + 2k\pi)|^2\right)^{1/2}}.$$

On the other hand, we may also construct a wavelet ψ by using the following formula. If $\theta(t) \in L^2_{2\pi}$ and $0 <$ ess inf $\theta(t) \le$ ess sup $\theta(t) < \infty$, then any function $\psi(x)$, defined by

$$\widehat{\psi}(\omega) = \frac{\theta(\omega)\lambda(\frac{\omega}{2}+\pi)}{\lambda(\frac{\omega}{2})}\overline{P(-e^{-i\frac{\omega}{2}})}\hat{\phi}\left(\frac{\omega}{2}\right),$$

is also a wavelet.

§3. Fundamental Scaling Functions

Let ϕ be a scaling function that generates an MRA $\{V_j\}_{j\in\mathbb{Z}}$. We will call ϕ a fundamental scaling function (f. scaling function), if ϕ satisfies the following cardinal interpolatory condition

$$\phi(\mathbf{k}) = \delta_{\mathbf{k},0} \qquad \mathbf{k} \in \mathbb{Z}^s.$$

If, in addition, $\{\phi(\cdot - \mathbf{k})\}_{\mathbf{k}\in\mathbb{Z}^s}$ forms an orthonormal basis of V_0, then we will call ϕ an orthonormal fundamental scaling function (o.f. scaling function).

Lemarié [23] has considered the so-called self-similar interpolating functions, namely: continuous functions ϕ that satisfy

$$\left\|\mathbf{x}^\beta D^\alpha \phi\right\|_\infty < \infty, \qquad \beta \in \mathbb{N}^s, \quad |\alpha| \le N; \tag{3.1}$$

$$\phi(\mathbf{k}) = \delta_{\mathbf{k},0}, \qquad \mathbf{k} \in \mathbb{Z}^s; \quad \text{and} \tag{3.2}$$

$$\phi\left(\frac{\mathbf{x}}{2}\right) = \phi(\mathbf{x}) + \sum_{\mathbf{k}\in\mathbb{Z}^s\backslash 2\mathbb{Z}^s} \phi\left(\frac{\mathbf{k}}{2}\right)\phi(\mathbf{x}-\mathbf{k}). \tag{3.3}$$

He pointed out that if ϕ satisfies the conditions (3.1)-(3.3) for some $N \ge 0$, then ϕ is a fundamental scaling function. Lewis [24] has also established a criterion of o.f. scaling functions. His criterion is parallel to Mallat's criterion for o.n. scaling function (cf. Mallat [25]). Precisely, he obtained the following result. Let

$$P(e^{-i\omega}) = \frac{1}{2}\sum_{k\in\mathbb{Z}} p_k e^{ik\omega}$$

be a 2π-periodic function of ω satisfying $p_k = O((1+k^2)^{-1})$, $k \to \infty$, $P(1) = 1$, and $P(e^{-i\omega}) \ne 0$ for $|\omega| \le \frac{\pi}{2}$. If P satisfies

$$P(e^{-i\omega}) + P(-e^{-i\omega}) = 1, \quad \text{and} \tag{3.4}$$

$$|P(e^{-i\omega})|^2 + |P(-e^{-i\omega})|^2 = 1 \tag{3.5}$$

and ϕ, defined by $\hat{\phi}(\omega) = \prod_{j=1}^{\infty} P(e^{-i2^{-j}\omega})$, is regular, namely: $\phi \in C^1(\mathbb{R})$ and

$$|\phi(x)| + |\phi'(x)| \le C(1+x^2)^{-1},$$

then ϕ is an o.f. scaling function.

In the following, we introduce a method for constructing an o.f. scaling function from a known f. scaling function (cf. [24]). Suppose that P satisfies the conditions (3.4) and (3.5). Substituting $P(-e^{-i\omega}) = 1 - P(e^{-i\omega})$ into (3.5) yields

$$|P(e^{-i\omega})|^2 - Re\, P(e^{-i\omega}) = 0. \tag{3.6}$$

Let

$$P(e^{-i\omega}) = a(e^{-i\omega}) + ib(e^{-i\omega}).$$

Then (3.6) leads to the relation

$$b^2(e^{-i\omega}) = a(e^{-i\omega}) - a^2(e^{-i\omega}). \tag{3.7}$$

By combining this with (3.5), we see that a and b must also satisfy

$$a(e^{-i\omega}) + a(-e^{-i\omega}) = 1,$$
$$b(e^{-i\omega}) + b(-e^{-i\omega}) = 0.$$

Now suppose that ϕ_1 is a given o.n. scaling function with two-scale symbol $P_1(e^{-i\omega})$. Then we have

$$|P_1(e^{-i\omega})|^2 + |P_1(-e^{-i\omega})|^2 = 1.$$

Thus we may select, as a candidate for $a(e^{-i\omega})$, the function $|P_1(e^{-i\omega})|^2$. Such a choice of $a(e^{-i\omega})$ has the following attractive feature. We see that for an o.f. scaling function ϕ defined by $\hat{\phi}(\omega) = \prod_{j\geq 1} P(e^{-i2^{-j}\omega})$, since $|P(e^{-i\omega})| = a^{\frac{1}{2}}(e^{-i\omega}) = |P_1(e^{-i\omega})|$, we have

$$|\hat{\phi}(\omega)| = |\hat{\phi}_1(\omega)|.$$

Thus ϕ is as smooth as ϕ_1.

Suppose that $P_1(e^{-i\omega})$ is a function of the form

$$|P_1(e^{-i\omega})|^2 = \left|\frac{1+e^{-i\omega}}{2}\right|^2 m^2(e^{-i\omega}),$$

where $m(e^{-i\omega})$ is a 2π-periodic even function of ω such that $m(e^{-i\omega}) \neq 0$ for all $\omega \in \mathbb{R}$. This is the case for the spline scaling function of Lemarié and Battle and the compactly supported o.n. scaling function of Daubechies. Now construct $a(e^{-i\omega})$ as described above, namely:

$$a(e^{-i\omega}) = |P_1(e^{-i\omega})|^2 = \cos^{2N}\frac{\omega}{2} m^2(e^{-i\omega}).$$

Thus,

$$\begin{aligned} b^2(e^{-i\omega}) &= a(e^{-i\omega})(1 - a(e^{-i\omega})) = a(e^{-i\omega})a(-e^{-i\omega}) \\ &= \cos^{2N}\frac{\omega}{2}\sin^{2N}\frac{\omega}{2} m^2(e^{-i\omega})m^2(-e^{-i\omega}). \end{aligned}$$

We wish to define a continuous function $b(e^{-i\omega})$ that satisfies $b(e^{-i\omega}) + b(-e^{-i\omega}) = 0$. This is accomplished by setting

$$b(e^{-i\omega}) = \begin{cases} \left(-\frac{i}{4}\right)^N (e^{i\omega} - e^{-i\omega})^N m(e^{-i\omega})m(-e^{-i\omega}), & \text{if } N \text{ is odd} \\ \sigma(e^{-i\omega})\left(-\frac{i}{4}\right)^N (e^{i\omega} - e^{-i\omega})^N m(e^{-i\omega})m(-e^{i\omega}), & \text{if } N \text{ is even,} \end{cases}$$

where σ is defined by

$$\sigma(e^{-i\omega}) = \begin{cases} 1 & \text{if } \ 0 \leq \omega < \pi; \\ -1 & \text{if } \ \pi \leq \omega < 2\pi. \end{cases}$$

By using the criterion obtained by Lewis [24], we see that $P(e^{-i\omega}) = a(e^{-i\omega}) + ib(e^{-i\omega})$ is an o.f. scaling function. Recently we have also established the following two criteria for f. scaling functions and o.f. scaling functions, respectively.

Theorem 3.1. *(see [9]). Let*

$$P(e^{-i\omega}) = \frac{1}{2^s} \sum_{\mathbf{k}\in\mathbb{Z}^s} p_{\mathbf{k}} e^{-i\mathbf{k}\omega},$$

satisfying $P(1) = 1$, (2.9), and

$$\theta_j(P) := \sup_{1\le|\omega|\le 2} |P(e^{-i2\omega})\dots P(e^{-i2^j\omega})| = O(\Phi(2^j)),$$

where $\Phi(t)$ is a monotonic non-increasing positive function on $[0,\infty)$ that satisfies

$$\int_0^\infty \Phi(t)t^{s-1}dt < \infty.$$

Then $g(\omega) = \prod_{j=1}^{\infty} P(e^{-i2^{-j}\omega}) \in L^1 \cap L^2$ and $\phi(t)$, defined by $\hat{\phi}(\omega) = g(\omega)$, belongs to $C \cap L^2$. If, in addition, ϕ satisfies

$$\phi_1(\mathbf{x}) := \sup_{|\mathbf{x}|\le|\mathbf{y}|} |\phi(\mathbf{y})| \in L^1,$$

then the following statements are equivalent:
(I) ϕ is a fundamental scaling function; and
(II) the two-scale sequence of ϕ satisfies

$$p_{2\mathbf{k}} = \delta_{\mathbf{k},0}, \qquad \mathbf{k} \in \mathbb{Z}^s$$

and

$$\sum_{\mathbf{k}\in\Gamma_j} p_{\mathbf{k}} = 1, \qquad j = 1,\dots,2^s-1,$$

where

$$\Gamma_j = \{h^j + 2\pi\ell\colon\ \ell \in \mathbb{Z}^s\}, \qquad j = 0,1,\dots,2^s-1,$$

and the h^j's are the vertices of the cube $[0,1]^s$, with $h^0 = (0,\dots,0)$.

Theorem 3.2. *(see [9]). Under the assumptions of P and ϕ in Theorem 3.1, the following two statements are equivalent:*
(I) ϕ is an orthonormal fundamental scaling function;
(II) the symbol P satisfies

$$p_{2k} = \delta_{\mathbf{k},0}, \qquad \mathbf{k} \in \mathbb{Z}^s,$$

$$\sum_{\mathbf{k}\in\Gamma_j} p_{\mathbf{k}} = 1, \qquad j = 1,\dots,2^s-1,$$

and

$$\sum_{j=0}^{2^s-1} |P(e^{-i(\omega+\pi h^j)})|^2 = 1.$$

§4. Symmetric and Compactly Supported Scaling Functions

In this section we consider the one-dimensional setting and assume throughout that $\phi \in C(\mathbb{R})$ is an f. scaling function with two-scale symbol

$$P(z) = \frac{1}{2} \sum_{k \in \mathbb{Z}} p_k z^k, \qquad z = e^{-i\omega},$$

and symmetric finite two-scale sequences, i.e. $p_k = p_{-k}$, and only finitely many of the p_k's are non-zero. So, ϕ satisfies the following two-scale relation

$$\phi(x) = \sum_{k=-K}^{K} p_k \phi(2x - k),$$

with $p_{_{-K}} = p_{_K} \neq 0$. It is easy to show that

$$\text{supp}\ \phi \subset [-K, K].$$

By Theorem 3.1, we see that $P(1) = 1$ and $P(-1) = P(e^{-i\pi}) = 0$. So in general, we assume that $P(e^{-i\omega})$ is of the form

$$P(e^{-i\omega}) = \left(\frac{1 + \cos\omega}{2} \right)^N S(e^{-i\omega}),$$

where

$$S(e^{-i\omega}) = \sum_{k=-M}^{M} s_k e^{-ik\omega},$$

with $s_k = s_{-k}$ and $S(-1) \neq 0$. Then $P(e^{-i\omega})$ can be written as

$$P(e^{-i\omega}) = \cos^{2N} \frac{\omega}{2} \sum_{k=0}^{K} b_k \cos^k \omega = \cos^{2N} \frac{\omega}{2} \sum_{k=0}^{K} b'_k \left(\sin \frac{\omega}{2} \right)^{2k}.$$

Now, consider $y = \cos^2\left(\frac{\omega}{2}\right)$ and

$$T(y) = \sum_{k=0}^{K} b'_k y^k,$$

so that

$$P(e^{-i\omega}) = y^N T(1 - y).$$

Since $P(e^{-i\omega})$ must satisfy

$$P(e^{-i\omega}) + P(-e^{-i\omega}) = 1, \tag{4.1}$$

we obtain

$$y^N T(1 - y) + (1 - y)^N T(y) = 1. \tag{4.2}$$

According to Daubechies [15], the polynomial

$$T_N(y) = \sum_{j=0}^{N-1} \binom{N-1+j}{j} y^j$$

is the only polynomial of order less than N which solves (4.2) (see also Chui [6]). Hence, we see that the Laurent polynomial

$$P_N^I(z) = z^{-N} \left(\frac{1+z}{2}\right)^{2N} \sum_{j=0}^{N-1} \binom{N-1+j}{j} (-z)^{-j} \left(\frac{1-z}{2}\right)^{2j}, \quad z = e^{-i\omega},$$

or

$$P_N^I(e^{-i\omega}) = \left(\frac{1+\cos\omega}{2}\right)^N \sum_{j=0}^{N-1} \binom{N-1+j}{j} \left(\frac{1-\cos\omega}{2}\right)^j$$
$$:= \frac{1}{2} \sum_{k=-2N+1}^{2N-1} p_k^N e^{ik\omega}$$

solves (4.1).

Since $|T_N(y)| < 2^{2(N-1)}, 0 \le y \le 1$, we have

$$\|P_N^I(e^{-i\omega})\|_\infty < 2^{2(N-1)}.$$

It is easy to check that $P_N^I(e^{-i\omega})$ satisfies all the conditions in Theorem 3.1. Hence, ϕ_N^I, defined by

$$\hat{\phi}_N^I(\omega) = \prod_{j=1}^{\infty} P_N^I(e^{-i2^{-j}\omega}),$$

belongs to $C(\mathbb{R}) \cap L^2$ with support $[-2N+1, 2N-1]$ and is an f. scaling function.

If $T(y)$ is an another solution of (4.2), then $T(y)$ is of the form

$$T(y) = T_N(y) + y^N R(y);$$

and from (4.2) we have

$$R(1-y) + R(y) = 0.$$

The polynomial R is therefore antisymmetric with respect to $y = \frac{1}{2}$, or $R(y) = \widetilde{R}(\frac{1}{2} - y)$, where $\widetilde{R}$ is an odd polynomial. Thus we obtain the following result.

Theorem 4.1. *Suppose that $\phi \in C(\mathbb{R})$ is a compactly supported f. scaling function with symmetric two-scale sequence such that the two-scale symbol P is of the form*

$$P(e^{-i\omega}) = \left(\frac{1+\cos\omega}{2}\right)^N S(e^{-i\omega})$$

with $S(-1) \neq 0$. *Then*

$$S(e^{-i\omega}) = T_N\left(\frac{1-\cos\omega}{2}\right) + \left(\frac{1-\cos\omega}{2}\right)^N R\left(\frac{1}{2}\cos\omega\right),$$

where R *is an odd polynomial.*

Now we discuss the computational aspect of ϕ_N^I. Since $\phi_N^I \in C(\mathbb{R})$, it is uniquely determined by its values on the set of all dyadic rational numbers. Denote

$$\begin{aligned} B_0 &= \mathbb{Z} = \{\ldots, -1, 0, 1, \ldots\} \\ B_1 &= \left\{\frac{k}{2}\colon\ k \in B_0\right\} \\ &\cdots \\ B_m &= \left\{\frac{k}{2}\colon\ k \in B_{m-1}\right\}. \end{aligned}$$

First, $\phi_N^I(k) = \delta_{k,0}$, $k \in \mathbb{Z}$, so that ϕ_N^I is well defined on B_0. According to the two-scale relation, we may compute the values of ϕ_N^I on B_{m+1} from its values on B_m by the following scheme:

$$\phi_N^I\left(\frac{k}{2^{m+1}}\right) = \sum_{j\in\mathbb{Z}} p_j \phi_N^I\left(\frac{k}{2^m} - j\right), \quad k \in \mathbb{Z}. \tag{4.3}$$

However, this scheme is not convenient for applications. Usually, we replace it by the following scheme:

$$\phi_N^I\left(\frac{k}{2^{m+1}}\right) = \sum_{j\in\mathbb{Z}} p_{k-2j} \phi_N^I\left(\frac{j}{2^m}\right), \quad k \in \mathbb{Z}. \tag{4.4}$$

The schemes (4.3) and (4.4) are equivalent, since we have

$$\sum_j p_{k-2j}\phi_N^I\left(\frac{j}{2^m}\right) = \sum_j p_j \phi_N^I\left(\frac{k}{2^m} - j\right).$$

It is perhaps more interesting to note that we can use Lagrange iterative interpolation to compute ϕ_N^I. If we know all the values of ϕ on $B_m = \{\frac{k}{2^m}\colon\ k \in \mathbb{Z}\}$, then we can compute every new point of B_{m+1} as follows. Let

$$y_k = \frac{k}{2^m} + \frac{1}{2^{m+1}}$$

be a new point in $B_{m+1}\backslash B_m$. Suppose that $L_{k,N}$ is the Lagrange interpolation polynomial of degree $2N-1$ satisfying interpolating conditions

$$L_{k,N}\left(\frac{j}{2^m}\right) = \phi\left(\frac{j}{2^m}\right), \quad j = k-N+1, k-N+2, \ldots, k+N.$$

Then we may define

$$\phi\left(\frac{k}{2^m}+\frac{1}{2^{m+1}}\right)=L_{k,N}\left(\frac{k}{2^m}+\frac{1}{2^{m+1}}\right).$$

This iterative interpolation process was first considered by Deslauriers and Dubuc [17]. Here, we introduce it in a somewhat different way. They proved that if we take $\phi(k)=\delta_{k,0}$ as the initial data, then this process produces a continuous function ϕ which satisfies the two-scale relation

$$\phi(x)=\sum_{j\in\mathbb{Z}} p_j\phi(2x-j),$$

where $\{p_j\}$ is defined by

$$p_j=\begin{cases}1, & \text{for} \quad j=0,\\ L_{k,N}^0\left(\frac{1}{2}\right) & \text{for} \quad j=2k-1 \text{ or } 1-2k, \quad k=1,\ldots,N,\\ 0 & \text{for other} \quad j,\end{cases}$$

and where $L_{k,N}^0(x)$ denotes the polynomial of degree $2N-1$ that satisfies the interpolation condition

$$L_{k,N}^0(\ell)=\delta_{k,\ell} \quad (\ell=-N+1,-N+2,\ldots,N-1,N);$$

Deslauriers and Dubuc [17] also showed that the two-scale symbol P of ϕ is of the form

$$P(e^{-i\omega})=\left(\cos\frac{\omega}{2}\right)^{2N} S(e^{-i\omega}),$$

where $S(e^{i\omega})$ is a symmetric trigonometric polynomial of degree $N-1$ of ω. Since such a symbol is unique, we have

$$P(e^{-i\omega})\equiv P_N^I(e^{-i\omega}), \text{ or } \phi(x)\equiv\phi_N^I(x).$$

According to Theorem 4.1, we obtain the following result.

Corollary 4.2. *Let N be any positive integer. Among all the symmetric fundamental scaling functions that generate the same MRA, ϕ_N^I has minimum support.*

§5. Subdivision Schemes

We have introduced the scheme (4.4) in order to compute ϕ_N^I. This scheme can be viewed as a subdivision scheme for the generation of curves and surfaces in computer aided geometric design (CAGD). In general, a subdivision scheme is defined in terms of a mask consisting of a finite set of non-zero coefficients

$$P=\{p_{\mathbf{k}}\colon \mathbf{k}\in\mathbb{Z}^s\},$$

where $s = 1$ for curves, and $s = 2$ for surfaces (defined by some control nets with the topology of a regular grid, where each point has two indices: $y_{i,j}$, $i = 1, \dots, N_1$; $j = 1, \dots, N_2$).

In general, this scheme is given by

$$y_{\mathbf{k}}^{m+1} = \sum_{\mathbf{j} \in \mathbb{Z}^s} p_{\mathbf{k}-2\mathbf{j}} y_{\mathbf{j}}^m, \quad \mathbf{k} \in \mathbb{Z}^s, \quad m \in \mathbb{N}. \tag{5.1}$$

Usually, in practice, the set of initial control points $\{y_{\mathbf{k}}^0 \colon \mathbf{k} \in J_0\}$ is finite, and the set of control points at level $m+1$, denoted by $\{y_{\mathbf{k}}^{m+1} \colon \mathbf{k} \in J_{m+1}\}$, is the maximal set for which the rule

$$y_{\mathbf{k}}^{m+1} = \sum_{\mathbf{j} \in J_m} p_{\mathbf{k}-2\mathbf{j}} y_{\mathbf{j}}^m,$$

coincides with (5.1). Thus, it is easy to see that

$$\begin{aligned} J_{m+1} &= 2J_m + \operatorname{supp}\{P\} \\ &= \{\mathbf{k} \in \mathbb{Z}^s \colon \mathbf{k} = 2\mathbf{j} + \mathbf{j}', \mathbf{j} \in J_m, \mathbf{j}' \in \operatorname{supp} P\}. \end{aligned}$$

If $s = 1$, then there are two rules for defining the new control points

$$\begin{cases} y_{2i}^{k+1} = \sum_{j \in \mathbb{Z}} p_{2i-2j} y_j^k = \sum_{j \in \mathbb{Z}} p_{2j} y_{i-j}^k, & i \in \mathbb{Z}; \text{ and} \\ y_{2i+1}^{k+1} = \sum_{j \in \mathbb{Z}} p_{2i+1-2j} y_j^k = \sum_{j \in \mathbb{Z}} p_{2j+1} y_{i-j}^k, & i \in \mathbb{Z}. \end{cases} \tag{5.2}$$

One of the simplest examples is Chaikin's algorithm [5] given by

$$\begin{cases} y_{2i}^{k+1} = \frac{3}{4} y_i^k + \frac{1}{4} y_{i+1}^k \\ y_{2i+1}^{k+1} = \frac{1}{4} y_i^k + \frac{3}{4} y_{i+1}^k. \end{cases}$$

This algorithm generates a quadratic B-spline curve.

When $p_{2i} = \delta_{i,0}$, the scheme (5.2) becomes

$$\begin{cases} y_{2i}^{k+1} = y_i^k \\ y_{2i+1}^{k+1} = \sum_j p_{1+2j} y_{i-j}^k. \end{cases} \tag{5.3}$$

The scheme (5.3), introduced in the computation of ϕ_N^I, is an interesting example of this case. We will call (5.3) an interpolatory subdivision scheme.

A subdivision scheme (5.1), denoted by S, is said to be convergent, if for every set of initial control point $f^0 = \{f_{\mathbf{j}}^0 \colon \mathbf{j} \in \mathbb{Z}^s\}$, there exists a nontrivial continuous function $f \in C(\mathbb{R}^s)$ such that

$$\lim_{\ell < m \to \infty} \left| (S^m f^0)_{2^{m-\ell}\mathbf{j}} - f\left(\frac{\mathbf{j}}{2^\ell}\right) \right| = 0,$$

for all $\mathbf{j} \in \mathbb{Z}^s$, $\ell \in \mathbb{Z}_+$. The function f is denoted by $S^\infty f^0$. S is called a uniformly convergent subdivision scheme, if for any bounded domain $D \subset \mathbb{R}^s$ and $\varepsilon > 0$, there exists a K such that

$$\left|(Smf^0)_{\mathbf{j}} - f\left(\frac{\mathbf{j}}{2^m}\right)\right| < \varepsilon$$

for all $m > K$, $\mathbf{j} \in \mathbb{Z}^s \cap 2^m D$.

For further study of convergence, computation, and regularity of limit functions of subdivision schemes, the reader is referred to Dyn and Levin [20], Dyn [19], Cavaretta, Dahmen and Micchelli [4], Deslauriers and Dubuc [17], Deslauriers, Dubois and Dubuc [18] and Berger and Wang [3], etc. In particular, Dyn [19] showed that if S is a uniformly convergent subdivision scheme, then

(i) $\sum_{\mathbf{k}\in\Gamma_j} p_{\mathbf{k}} = 1$, where $j = 0, 1, \ldots, 2^s - 1$ and the Γ_j's are defined as in (3.8); and

(ii) its mask $P = \{p_{\mathbf{k}}\colon \mathbf{k} \in \mathbb{Z}^s\}$ determines a unique compactly supported continuous function ϕ with the following properties:

$$\phi(\mathbf{x}) = \sum_{\mathbf{k}\in\mathbb{Z}^s} p_{\mathbf{k}}\phi(2\mathbf{x} - \mathbf{k}),$$

i.e. ϕ is a scaling function, and

$$\sum_{\mathbf{k}\in\mathbb{Z}^s} \phi(\mathbf{x} - \mathbf{k}) \equiv 1.$$

Furthermore, for any f^0, we have

$$S^\infty f^0 = \sum_{\mathbf{k}\in\mathbb{Z}^s} f^0_{\mathbf{k}}\phi(\mathbf{x} - \mathbf{k}).$$

Thus we see that if $\delta = \{\delta_{\mathbf{k},0}\}_{\mathbf{k}\in\mathbb{Z}^s}$, then

$$S^\infty\delta = \phi.$$

We will call the function

$$P(\mathbf{z}) = \frac{1}{2^s}\sum_{\mathbf{k}\in\mathbb{Z}^s} p_{\mathbf{k}}\mathbf{z}^{\mathbf{k}}$$

the symbol of the scheme S defined by (5.1).

Consider the one-dimensional setting. If S is a uniformly convergent subdivision scheme, then

$$\sum_{j\in\mathbb{Z}} p_{2j} = 1 \quad \text{and} \quad \sum_{j\in\mathbb{Z}} p_{2j+1} = 1. \tag{5.4}$$

Let f^0 be an initial data and denote

$$f^k = S^k f^0.$$

Define df^k by

$$(df^k)_j = 2^j (f^k_{j+1} - f^k_j).$$

If S is a subdivision scheme satisfying (5.4), then there exists a subdivision scheme S_1 such that

$$df^k = S_1 df^{k-1}.$$

In fact, if $P(z)$, $z = e^{-i\omega}$, is the symbol of S, then the scheme determined by the symbol

$$P_1(z) = \frac{2z}{1+z} P(z), \quad z = e^{-i\omega},$$

defines a subdivision scheme S_1 with the required property.

Dyn [19] showed that S is a uniformly convergent subdivision scheme if and only if $\frac{1}{2}S_1$ converges uniformly to zero for all initial data f^0.

§6. Regularity

We only restrict our discussion to the one-dimensional setting. First we introduce the following function spaces. Denote by C^α the Hölder space defined as follows. For $\alpha = n + \beta$, n an integer and $\beta \in [0, 1)$, $f \in C^\alpha$, if and only if f is n times continuously differentiable and

$$C(f) := \sup_{x \neq y} \frac{|f^{(n)}(x) - f^{(n)}(y)|}{|x - y|^\beta} < \infty.$$

The Zygmund space Λ^α is defined by $\Lambda^\alpha = C^\alpha$ if $\alpha \notin \mathbb{Z}$, and

$$\Lambda^\alpha = \{f\colon\ |f^{(\alpha-1)}(x+h) - 2f^{(\alpha-1)}(x) + f^{(\alpha-1)}(x-h)| \le c|h|\},$$

if α is an integer. Denote

$$H^\alpha_p = \{f\colon\ \|\hat{f}(\omega)(1 + |\omega|^\alpha)\|_p < \infty\}.$$

It is easy to check that

$$H^{\alpha+1+\varepsilon}_\infty \subset H^\alpha_1 \subset C^\alpha \qquad (\varepsilon > 0).$$

If f has compact support, then

$$f \in C^\alpha \Rightarrow f \in H^\alpha_\infty.$$

In [19], Dyn discussed the regularity of the limit functions of subdivision schemes S, defined by (5.1) and (5.2). She showed that if the symbol corresponding to δ is

$$P(z) = \left(\frac{1+z}{2z}\right)^N R(z), \quad z = e^{-i\omega},$$

and the subdivision scheme with the symbol $R(z)$ converges uniformly, then $S^\infty\delta =$
$\phi \in C^N$.

Suppose that S is an interpolating subdivision scheme. Then, Dyn [19] proved that if S converges to a C^ν-limit, then it reproduces $\mathcal{P}_\nu$ (the set of all polynomials of degree $\le \nu$), namely:

$$\left(k+\frac{1}{2}\right)^\ell = \sum_{j\in\mathbb{Z}} p_{1+2j}(k-j)^\ell, \quad \ell = 0,1,\ldots,\nu; \quad k \in \mathbb{Z}.$$

Furthermore, if S is an interpolatory subdivision scheme which reproduces $\mathcal{P}_\nu$, then the following conditions are equivalent:

(i) S converges uniformly to C^ν-limit functions;

(ii) S_n converges uniformly to $C^{n-\nu}$-limit functions; and

(iii) $\frac{1}{2}S_{\nu+1}$ converges uniformly to zero for any initial control points.

The definition of S_1 was introduced in §5, and the definition S_ν for a general ν is similar.

Consider the symbol

$$P(e^{-i\omega}) = \left(\frac{1+e^{-i\omega}}{2}\right)^N S(e^{-i\omega}),$$

with $P(1) = 1$, $S(-1) \ne 0$. Denote

$$\begin{cases} B_j = \sup_\omega |S(e^{-i\omega})S(e^{-i2\omega})\ldots S(e^{-i2^j\omega})|; \\ b_j = \dfrac{\log B_j}{j\log 2}; \quad \text{and} \\ b = \inf_{j>0} b_j. \end{cases} \tag{6.1}$$

The constant b plays an important role in the consideration of regularity of the scaling function ϕ, defined by

$$\hat\phi(\omega) = \prod_{j=1}^\infty P(e^{-i2^{-j}\omega}).$$

We will call b the critical exponent of P. According to Cohen [12], if $b - N < -1$, then $\phi \in H_1^\alpha \subset C^\alpha$, where $\alpha = H - b - 1 - \varepsilon$ $(0 < \varepsilon < N - b - 1)$. Cohen [12] proved that if $|S(-1)| > |S(1)| = 1$, then $\hat\phi(\omega)$ cannot have a better decay at infinity than $|\omega|^{b-N}$.

How can we estimate the critical exponent?

Cohen and Daubechies [13] pointed out that in several cases, it is possible to use a method based on the transformation

$$\tau\colon\ \omega \to 2\omega \quad \text{modulo} \quad 2\pi$$

and the fixed points of its power τ^n, $n > 0$. Indeed, let ω_0 be a fixed point of τ^n $(n > 0)$ and define its orbit

$$\omega_j = \tau^j \omega_0, \quad j = 0, \ldots, n-1.$$

Since $S(e^{-i\omega})$ has period 2π on ω, we have

$$S(e^{-i2^{nk}\omega_j}) = S(e^{-i\omega_j}), \quad k > 0.$$

This implies that

$$b \geq \frac{1}{n \log 2} \log \left| \prod_{j=0}^{n-1} S(e^{-i\omega_j}) \right|.$$

Now consider the smallest orbit of τ different from $\{0\}$, namely: the pair $\{-\frac{2\pi}{2}, \frac{2\pi}{3}\}$. For many filters

$$|S(e^{-i\frac{2\pi}{3}})| = |S(e^{i\frac{2\pi}{3}})|,$$

so we may obtain critical exponents b by using the following criterion.

Theorem 6.1. *(see Cohen and Daubechies [13]). Suppose that $S(e^{-i\omega})$ satisfies*

$$|S(e^{-i\omega})| \leq |S(e^{-i\frac{2\pi}{3}})| \quad \textit{if} \quad |\omega| \leq \frac{2\pi}{3}$$

and

$$|S(e^{-i\omega})S(e^{-i2\omega})| \leq |S(e^{-i\frac{2\pi}{3}})|^2, \quad \textit{if} \quad \frac{2\pi}{3} \leq |\omega| \leq \pi.$$

Then

$$b = \frac{1}{\log 2} \log |(S(e^{-i\frac{2\pi}{3}})|.$$

By applying Theorem 6.1 to the Daubechies wavelets we obtain the following (see [13]).

Corollary 6.2. *Let b_N^D be the critical exponent associated with the symbol of the Daubechies wavelet ψ_N^D, and α_N^D the Hölder exponent of the corresponding scaling function ϕ_N^D. Then*

$$\lim_{N\to\infty} \frac{b_N^D}{N} = \frac{\log 3}{2 \log 2}$$

and

$$\lim_{N\to\infty} \frac{\alpha_N^D}{N} = \lim_{N\to\infty} \frac{N - b_D^N}{N} = 1 - \frac{\log 3}{2\log 2} \simeq 0.2075.$$

By combining Theorems 4.1 and 6.1, we obtain the following.

Corollary 6.3. *Let b_N^I be the critical exponent associated with the symbol of a symmetric fundamental scaling function ϕ_N^I with minimum support, and α_N^I the Hölder exponent of ϕ_N^I. Then*

$$\lim_{N\to\infty} \frac{b_N^I}{N} = \frac{\log 3}{\log 2}$$

and

$$\lim_{N\to\infty} \frac{\alpha_N^I}{N} = 2 - \frac{\log 3}{\log 2} \simeq 0.415.$$

We have a characterization for the Zygmund space Λ^α by means of the dyadic block operators. Let γ be a real-valued and symmetric Schwartz function satisfying

$$\begin{cases} \hat{\gamma}(\omega) = 0 & \text{if} \quad |\omega| \le \frac{1}{2} \text{ or } |\omega| \ge \frac{5}{2}; \\ \hat{\gamma}(\omega) > 0 & \text{if} \quad \frac{1}{2} < |\omega| < \frac{5}{2}, \end{cases}$$

so that the frequency axis is completely covered by the dyadic dilations of γ. Indeed, we have

$$0 < C_1 \le \sum_{j\in\mathbb{Z}} \hat{\gamma}(2^j\omega) \le C_2, \qquad \omega \ne 0.$$

Define $\Delta_j(f)(x)$ by

$$\hat{\Delta}_j(f) = \hat{\gamma}(2^{-j}\cdot)\hat{f}.$$

The Littlewood-Paley theory tells us that $f \in \Lambda^\alpha$ if and only if $\|f\|_\infty < \infty$ and

$$\|\Delta_j(f)\|_\infty \le C2^{-\alpha j} \qquad \forall j \ge 0.$$

By using this characterization, Cohen and Daubechies [13] determined the Zygmund exponent of ϕ. For any $g \in C_{2\pi}$, we define the transition operator T_g: $L_{2\pi}^p \to L_{2\pi}^p$, $1 \le p \le \infty$, by

$$T_g f(\omega) = g\left(\frac{\omega}{2}\right) f\left(\frac{\omega}{2}\right) + g\left(\frac{\omega}{2}+\pi\right) f\left(\frac{\omega}{2}+\pi\right), \tag{6.2}$$

and define

$$E(N_1, N_2) = \left\{ \sum_{k=N_1}^{N_2} h_k e^{-ik\omega}\colon\ h_j \in \mathbb{C}, N_1 \le j \le N_2 \right\}.$$

Thus we have

$$(f, g) \in [E(N_1, N_2)]^2 \Rightarrow T_g f \in E(N_1, N_2).$$

We now assume that $P(e^{-i\omega})$ is a positive trigonometric polynomial of ω in $E_M = E(-M, M)$, with $P(1) = 1$, $P(-1) = 0$ and

$$P(e^{-i\omega}) = \cos^{2N}\frac{\omega}{2} S(e^{-i\omega}), \tag{6.3}$$

where $S(e^{-i\omega})$ is a trigonometric polynomial of ω and $S(-1) \neq 0$. Note that $N \leq M$. Denote

$$F_N = \left\{ \sum_{k=-M}^{M} h_k e^{-ik\omega} \colon \sum_{k=-M}^{M} k^j h_k = 0, \quad j = 0, \ldots, 2N-1 \right\}.$$

If p_k is the k-th Fourier coefficient of P, the factorization in (6.3) is equivalent to the cancellation conditions

$$\sum_{k=-M}^{M} (-1)^k k^j p_k = 0, \qquad j = 0, \ldots, 2N-1.$$

Then it follows that $1, \frac{1}{2}, \ldots, 2^{-2N+1}$ are the eigenvalues of T_P. The row vectors $t^j = (k^j)_{k=-M,\ldots,M}$ $(0 \leq j \leq 2N-1)$ generate a subspace which is left invariant with respect to T_P and contains an eigenvector for each of these $2N$ eigenvalues. Consequently, the orthogonal subspace F_N is right invariant by T_P.

Cohen and Daubechies [13] have shown that if λ is the largest eigenvalue of the restriction of T_P to F_N and $|\lambda| < 1$, then defining $\alpha = -\frac{1}{\log 2} \log |\lambda|$ and

$$\hat{\phi}(\omega) = \prod_{j=1}^{\infty} P(e^{-i2^{-j}\omega}),$$

we have

(i) $\phi \in \Lambda^{\alpha-\varepsilon}, \varepsilon > 0$; and

(ii) $\phi \in \Lambda^{\alpha}$, when the restriction of T_P to the invariant subspace F_λ corresponding to the eigenvalue λ is purely diagonal.

They also pointed out that if $S(e^{-i\omega}) \neq 0$ for $\omega \in \left[-\frac{\pi}{2}, \frac{\pi}{2}\right]$, then these two estimates are optimal.

Remark. Eirola [27] has recently obtained some results on the regularity of ϕ_J^D and ϕ_N^I in terms of certain Sobolev properties. If we denote

$$\tilde{\alpha}_N^D = \sup\{s\colon\ \phi_N^D \in H_2^s\}$$

and

$$\tilde{\alpha}_N^I = \sup\{s\colon\ \phi_N^I \in H_2^s\},$$

then we have

$$\lim_{N\to\infty} \frac{\tilde{\alpha}_N^D}{N} = 1 - \frac{\log 3}{2\log 2};$$

$$\lim_{N\to\infty} \frac{\tilde{\alpha}_N^I}{N} = 2 - \frac{\log 3}{\log 2}.$$

Acknowledgements. Research supported by NSF Grant DMS 89-01345, ARO Contract DAAL 03-90-G-0091, and NATO Grant CRG 90-0158.

References

1. Battle, G., A block spin construction of ondelettes, Part I: Lemarié functions, Comm. Math. Phys. **110** (1987), 601-615.
2. Battle, G., A block spin construction of ondelettes, Part II: The QFT connection, Comm. Math. Phys. **114** (1988), 93-102.
3. Berger, M. A. and Y. Wang, Multidimensional two-scale dilation equation, in *Wavelets: A Tutorial of Theory and Applications*, C.K. Chui, ed., Academic Press, Boston, 1992, pp. 295-324.
4. Cavaretta, A. S., Dahmen, W. and Micchelli, C. A., *Stationary Subdivision*, Memoirs of the Amer. Math. Soc., to appear.
5. Chaikin, G. M., An algorithm for high speed curve generation, Comp. Graphics and Image Proc. **3** (1974), 346-349.
6. Chui, C. K., *An Introduction to Wavelets*, Academic Press, Boston, 1992.
7. Chui, C. K., and X. L. Shi, Inequalities of Littlewood-Paley type for frames and wavelets, CAT Report #249, Texas A&M University, 1991.
8. Chui, C. K., and X. L. Shi, On a Littlewood-Paley identity and characterization of wavelets, CAT Report #250, Texas A&M University, 1991.
9. Chui, C. K., and X. L. Shi, Characterizations of scaling functions and wavelets, CAT Report #259, Texas A&M University, 1991.
10. Chui, C. K., and J. Z. Wang, On compactly supported spline wavelets and a duality principle, Trans. Amer. Math. Soc. , to appear.
11. Chui, C. K., and J. Z. Wang, A general framework of compactly supported splines and wavelets, J. Approx. Th., to appear.
12. Cohen, A., Doctoral Thesis, Univ. Paris-Dauphine, 1990.
13. Cohen, A., and I. Daubechies, Non-separable bidimensional wavelet bases, preprint.
14. Cohen, A., Daubechies, I. and Feauveau, J.C., Biorthogonal bases of compactly supported wavelets, Comm. Pure and Appl. Math., to appear.
15. Daubechies, I., Orthonormal bases of compactly supported wavelets, Comm. Pure and Applied Math., Vol. XLI (1988), 909–996.
16. Daubechies, I., The wavelet transform, time-frequency localization and signal analysis, IEEE Trans. Inf. Theory, Vol. 36, No. 5 (1990), 961–1005.
17. Deslauriers, G. and S. Dubuc, Symmetric Iterative Interpolation Processes, Constr. Approx. **5** (1989), 49–68.
18. Deslauriers, G., J. Dubois, and S. Dubuc, Multidimensional iterative interpolation, preprint.
19. Dyn, N., Subdivision schemes in CAGD, preprint.
20. Dyn, N. and D. Levin, Interpolating subdivision schemes for the generation of curves and surfaces, in *Multivariate Interpolation and Approximation*, W. Haussmann and K. Jetter, eds., Birkhauser Verlag, Basel, 1990, 91–106.
21. Eirola, T., Sobolev characterization of solutions of dilation equations, Helsinki Univ. of Technology, Inst. Math., Research Report #A292, 1991.

22. Lemarié, P. G., Ondelettes à localisation exponentielle, Journal de Math. Pures et Appl., **67** (1988), 227–236.
23. Lemarié, P. G., Some remarks on wavelet theory and interpolation, preprint.
24. Lewis, R. M., Cardinal interpolating multiresolutions, preprint.
25. Mallat, S., Multiresolution approximations and wavelet orthonormal bases of $L^2(\mathbb{R})$, Trans. Amer. Math. Soc., **315** (1989), 69–87.
26. Meyer, Y., Principe d'incestitude, bases hilbertiennes et algèbres d'opérateurs, Séminaire Bourbaki, No. 662, 1988–1986.
27. Meyer, Y., *Ondelettes et Opérateurs*, (two volumes), Hermann, Paris, 1990.
28. Strömberg, J.O., A modified Franklin system and higher-order spline systems of $\mathbb{R}^n$ as unconditional bases for Hardy spaces, in Conference in *Harmonic Analysis in honor of A. Zygmund, Vol. II*, W. Beckner et al. (eds.), Wadsworth Math. Series, 1983, 475–493.
29. Tchamitchian, Ph., Cacul symbolique sur les opérateurs de Caldèron-Zygmund et bases inconditionnells de $L^2(\mathbb{R}^n)$, C.R. Acad. Sc. Paris, **303** (1986), 215–218.

Charles K. Chui
Center for Approximation Theory
Texas A&M University
College Station, TX 77843-3368
e443cc@tamvm1.bitnet

Xian-Liang Shi
Center for Approximation Theory
Texas A&M University
College Station, TX 77843-3368
and
Hangzhou University
Hangzhou, Zhejiang, 310028
P.R.C.

Decomposition of Splines

M. Dæhlen and T. Lyche

Abstract. In this paper we discuss decomposition techniques and data reduction strategies with emphasis on piecewise polynomial curves and surfaces. The main idea behind the methods is to transform the object (curve or surface) into a sequence of components in order to analyze the original object in a multiresolutional setting. Data reduction is obtained by only keeping coefficients from each component which give "significant" contributions to the reconstruction of the original. A new method combining decomposition and knot removal is given. We present several examples which show data reduction and the special effect of decomposition.

§1. Introduction

As more and more data are being processed by computers, transmission and storage become serious tasks. Surfaces and higher dimensional objects arising in areas like CAGD, computer graphics, image processing, and seismology will in general require huge amounts of data space, and from an applicational point of view the need for new methods which can handle such problems is evident. In this paper we discuss curve and surface decomposition methods, which is a way to transform curves and surfaces so that they more easily can be analyzed and presented in different resolutions.

Recently, numerous papers have been written on the subject of multiresolution analysis and decomposition through wavelet transforms, see [5,6,8,9,18, 19] and references therein. These works primarily discuss decomposition on uniform partitions. We aim here at a more general setting where we investigate decomposition of piecewise polynomials or splines, not only on a uniform partition, but also for non-uniform spline curves and tensor product spline surfaces. However, we will also discuss decomposition on uniform grids defined from box splines.

In this paper we will focus on spline curves and surfaces, and the decomposition schemes presented here can to a certain extent be regarded as a

Mathematical Methods in Computer Aided Geometric Design II

Tom Lyche and Larry L. Schumaker (eds.), pp. 135–160.

ISBN 0-12-460510-9.

generalization of the knot removal strategy for splines which was introduced in [16]. Briefly, this method works as follows: given a spline function the goal is to remove as many knots as possible without perturbing the spline more than a prescribed tolerance. This algorithm was generalized to parametric curves and surfaces in [17], and general shape constraints were incorporated in [3]. We use this technique to adaptively determine the subspaces in the decomposition.

The outline of the paper is as follows: in the next section we give a general introduction to decomposition. Decomposition using nonuniform B-spline curves and surfaces is the topic in Section 3, and in Section 4 we give a short introduction to decomposition using translates and dilates of box splines. In Section 5 we discuss how these methods can be used as data reduction techniques.

§2. Decomposition

The purpose of this section is to give a general description of the decomposition strategies and discuss the principal components needed to construct computational methods. Given a function $f \in S = S_N$, where S_N is a finite dimensional linear space, we say that $f_0, g_1, \ldots, g_N$ is a *decomposition of* f if

$$f = f_0 + g_1 + \ldots + g_N, \quad \text{where} \quad g_i = f_i - f_{i-1}, \tag{2.1}$$

$f_i \in S_i$, and $S_0, \ldots, S_N$ is a nested sequence of subspaces of S_N,

$$S_0 \subset S_1 \subset \cdots \subset S_N. \tag{2.2}$$

Normally, the f_i's are increasingly better approximations to f. This will make the size of each g_i smaller for increasing i. Thus, the relation (2.1) is reminiscent of a Fourier expansion, where f_0 and the first g_i's are the principal components of f. The use of splines instead of trigonometric polynomials in (2.1) allows more local control of each individual component g_i.

In our setting, the spaces S_i thus will be spaces of piecewise polynomials or splines having a basis $\Phi_i = (\phi_{i,\ell})$ of local support functions forming a partition of unity. Thus, each f_i can be written as a linear combination in the form

$$f_i = \sum_{\ell=1}^{m_i} v_{i,\ell}\phi_{i,\ell}. \tag{2.3}$$

Note that we do not require the basis to be orthogonal.

In order to make it clear that $\mathbf{v}_i = (v_{i,\ell})$ is the coefficient vector with respect to a basis Φ_i of some function f_i, we use the notation $\mathbf{v}_i = \mathbf{c}(f_i, \Phi_i)$ or $\mathbf{v}_i = \mathbf{c}(f_i, S_i)$ whenever the basis is clear from the context.

The decomposition algorithms we will discuss only work with the basis coefficients $v_{i,\ell}$. We therefore find it desirable to describe the decomposition only in terms of Euclidian spaces $\mathbb{R}^{m_i}$. In order to do this we consider a sequence of spaces

$$V_i = \mathbb{R}^{m_i}, \quad i = 0, \ldots, N,$$

with $1 \leq m_0 < m_1 < \ldots < m_N$. We define a family of approximation operators $\{\mathcal{P}_j^i\}_{i \leq j}$,

$$\mathcal{P}_j^i : V_j \to V_i, \quad 0 \leq i \leq j \leq N, \tag{2.4}$$

which to a given vector $\mathbf{v}_j$ in V_j finds an approximation $\mathbf{v}_i = \mathcal{P}_j^i \mathbf{v}_j$ in V_i. Moreover, we define a family of refinement operators $\{\mathcal{R}_i^j\}_{i \leq j}$,

$$\mathcal{R}_i^j : V_i \to V_j, \quad 0 \leq i \leq j \leq N \tag{2.5}$$

which "lifts" a given vector $\mathbf{v}_i$ from V_i to $\mathbf{v}_j = \mathcal{R}_i^j \mathbf{v}_i$ in V_j. We assume that the refinement operators $\mathcal{R}_i^j$ satisfy

$$\mathcal{R}_i^j = \mathcal{R}_k^j \mathcal{R}_i^k, \tag{2.6}$$

for any integer k with $i \leq k \leq j$, and that $\mathcal{R}_i^i$ is the identity operator on V_i. When $\mathbf{v}_i = \mathbf{c}(f_i, S_i)$ is the coefficient vector in S_i of some underlying function f_i, the operator $\mathcal{R}_i^j$ produces a new vector $\mathbf{v}_j$ which could be the coefficient vector of the same function in a refined space S_j, hence, we say that $\mathbf{v}_j = \mathcal{R}_i^j \mathbf{v}_i$ is a refined version of $\mathbf{v}_i$ in this case.

We are now in a position to define decomposition in terms of vectors in the spaces $V_0, \ldots, V_N$. We say that the vector $\mathbf{v}_0 \in V_0$ and the sequence of vectors $\mathbf{w}_i \in V_i, i = 1, \ldots, N$ is a *decomposition* of $\mathbf{v}_N \in V_N$ if

$$\mathbf{v}_N = \mathcal{R}_0^N \mathbf{v}_0 + \sum_{i=1}^{N} \mathcal{R}_i^N \mathbf{w}_i, \tag{2.7}$$

where

$$\mathbf{w}_i = \mathbf{v}_i - \mathcal{R}_{i-1}^i \mathbf{v}_{i-1},$$

and

$$\mathbf{v}_i = \mathcal{P}_N^i \mathbf{v}.$$

Using (2.6) it is easy to see that (2.7) is a formal identity. Indeed,

$$\begin{aligned} \mathcal{R}_0^N \mathbf{v}_0 + \sum_{i=1}^{N} \mathcal{R}_i^N \mathbf{w}_i &= \mathcal{R}_0^N \mathbf{v}_0 + \sum_{i=1}^{N} \mathcal{R}_i^N (\mathbf{v}_i - \mathcal{R}_{i-1}^i \mathbf{v}_{i-1}) \\ &= \mathcal{R}_0^N \mathbf{v}_0 + \sum_{i=1}^{N} (\mathcal{R}_i^N \mathbf{v}_i - \mathcal{R}_{i-1}^N \mathbf{v}_{i-1}) \\ &= \mathcal{R}_N^N \mathbf{v}_N = \mathbf{v}_N. \end{aligned}$$

Note that

$$\mathcal{R}_0^N V_0 \subset \mathcal{R}_1^N V_1 \subset \ldots \subset \mathcal{R}_N^N V_N \tag{2.8}$$

defines a nested sequence of subspaces of V_N.

Example 2.1. *Suppose we take* $V_i = \mathbb{R}^{2^i+1}, i = 0, 1, 2, \ldots$, *and define*

$$\begin{aligned}\mathcal{P}_j^i(v_0, v_1, \ldots, v_{2^j}) &= (y_0, y_1, \ldots, y_{2^i}), \quad y_k = v_{k2^{j-i}},\\ \mathcal{R}_{i-1}^i(v_0, v_1, \ldots, v_{2^{i-1}}) &= (y_0, y_1, \ldots, y_{2^i}) \qquad \text{where,}\\ y_{2k} &= v_k, \quad y_{2k+1} = (v_k + v_{k+1})/2,\\ \mathcal{R}_i^j &= \mathcal{R}_{j-1}^j \mathcal{R}_{j-2}^{j-1} \cdots \mathcal{R}_i^{i+1}, \qquad j \geq i.\end{aligned}$$

Thus, $\mathcal{P}_i^j$ *picks out every* 2^{j-i}*th component of a vector of length* $2^j + 1$, *while the* R_i^j *inserts averages between components. For* $N = 2$ *and* $\mathbf{v} = (v_0, v_1, v_2, v_3, v_4)$ *we obtain the decomposition*

$$\begin{aligned}\mathbf{v}_0 &= (v_0, v_4) \in \mathbb{R}^2,\\ \mathbf{w}_1 &= (0, -(v_0 - 2v_2 + v_4)/2, 0) \in \mathbb{R}^3,\\ \mathbf{w}_2 &= (0, -(v_0 - 2v_1 + v_2)/2, 0, -(v_2 - 2v_3 + v_4)/2, 0) \in \mathbb{R}^5.\end{aligned}$$

This corresponds to a decomposition of piecewise linear functions. To see this, for any set A, *let* $\chi(a)$ *be the indicator function for the set* A, *and set*

$$\phi(x) = (1 + x)\chi_{[-1,0)}(x) + (1 - x)\chi_{[0,1)}(x).$$

This is the hat function of height one centered at the origin. For ϕ *we have the refinement equation*

$$\phi(x) = \phi(2x + 1)/2 + \phi(2x) + \phi(2x - 1)/2. \tag{2.9}$$

Given $\mathbf{v} = (v_0, v_1, v_2, v_3, v_4)$, *let*

$$f_2(x) = \sum_{j=0}^{4} v_j \phi(2^2 x - j), \qquad x \in [0, 1]$$

be the piecewise linear function connecting the points $(j/4, v_j), j = 0, 1, \ldots, 4$. *Now, let*

$$\begin{aligned}f_0(x) &= v_0\phi(x) + v_4\phi(x - 1),\\ f_1(x) &= v_0\phi(2x) + v_2\phi(2x - 1) + v_4\phi(2x - 2),\\ g_1(x) &= -(v_0 - 2v_2 + v_4)\phi(2x - 1)/2,\\ g_2(x) &= -(v_0 - 2v_1 + v_2)\phi(2^2 x - 1)/2 - (v_2 - 2v_3 + v_4)\phi(2^2 x - 3)/2.\end{aligned}$$

Using (2.9) it can be seen that $g_1 = f_1 - f_0$ *and* $g_2 = f_2 - f_1$, *so we have the decomposition*

$$f_2 = f_0 + g_1 + g_2$$

corresponding to the spaces $S_i = \text{span}\{\phi(2^i x - j) : j = 0, 1, \ldots, 2^i\}, i = 0, 1, 2.$

■

The calculation of the vectors $\mathbf{w}_i$ in (2.7) can be done in different order, and we now present 3 algorithms to this effect. First, given $\mathcal{P}_j^i$ and $\mathcal{R}_i^j$ we have the following simple algorithm, which finds the decomposition (2.7) of the vector $\mathbf{v}_N \in V_N$.

> **Algorithm 1**
> 1. $\mathbf{v}_0 = \mathcal{P}_N^0 \mathbf{v}_N$
> 2. For $i = 1, \ldots, N$
>   2.1. $\mathbf{v}_i = \mathcal{P}_N^i \mathbf{v}_N$
>   2.2. $\mathbf{w}_i = \mathbf{v}_i - \mathcal{R}_{i-1}^i \mathbf{v}_{i-1}$

Instead of starting by finding an approximation in V_0 and successively finding better approximations by adding components in $V_1, \ldots, V_N$, we can also find the components by starting with the component in V_N and ending up with the basic approximation in V_0.

> **Algorithm 2**
> 1. $\mathbf{v}_N = \mathbf{v}$
> 2. For $i = N, N-1, \ldots, 1$
>   2.1. $\mathbf{v}_{i-1} = \mathcal{P}_N^{i-1} \mathbf{v}_N$
>   2.2. $\mathbf{w}_i = \mathbf{v}_i - \mathcal{R}_{i-1}^i \mathbf{v}_{i-1}$

Algorithm 1 and Algorithm 2 are mathematically equal as long as the spaces $V_0, \ldots, V_N$ are fixed. However, if we let the number of spaces N and each space V_i, $i = 0, \ldots, N-1$ vary depending on the data the algorithms might differ.

We observe in Algorithm 2 that the approximation is always taken from the larger space V_N to V_i for some i. If the approximation operators satisfy

$$\mathcal{P}_j^i = \mathcal{P}_k^i \mathcal{P}_j^k, \tag{2.10}$$

for all integers k with $i \leq k \leq j$ and if $\mathcal{P}_i^j$ is the identity operator for $i = j$ we obtain the following variant of Algorithm 2.

> **Algorithm 3**
> 1. $\mathbf{v}_N = \mathbf{v}$
> 2. For $i = N, N-1, \ldots, 1$
>   2.1. $\mathbf{v}_{i-1} = \mathcal{P}_i^{i-1} \mathbf{v}_i$
>   2.2. $\mathbf{w}_i = \mathbf{v}_i - \mathcal{R}_{i-1}^i \mathbf{v}_{i-1}$

Algorithm 3 is used for orthogonal wavelet decomposition where the operators P_i^j correspond to L_2 projections.

§3. Decomposing Splines

In the previous section we gave an introduction to decomposition in general and presented algorithms for decomposing vectors in $\mathbb{R}^n$. In this section we shall see how this vector decomposition can be used to construct similar algorithms for splines. We will investigate both spline curves and tensor product spline surfaces. Let us start by recalling some basics about splines.

3.1. Splines

B-splines are locally supported piecewise polynomials, and splines are linear combinations of B-splines. The family of spline functions of order k on a partition (knot sequence) $\mathbf{t} = (t_1, \ldots, t_{n+k})$ of an interval $[a,b] \subset \mathbb{R}$ is given by

$$S_{k,\mathbf{t}} = \{\sum_{i=1}^{n} c_i B_i, \quad c_i \in \mathbb{R}\}, \tag{3.1}$$

where $B_i = B_{i,k} = B_{i,k,\mathbf{t}} : \mathbb{R} \to \mathbb{R}$ are B-splines of order k.

For simplicity we assume that $t_1 = \cdots = t_k = a$ and that $t_{n+1} = \cdots = t_{n+k} = b$. Hence, any spline function in $S_{k,\mathbf{t}}$ can take nonzero values only on the interval $[a,b]$. We recall that $B_i(x) > 0$, whenever $t_i < x < t_{i+k}$, $B_i(x) = 0$ for $x \notin [t_i, t_{i+k}]$, and with a proper normalization we have

$$\sum_i B_i(x) = 1 \tag{3.2}$$

for all $x \in [a,b]$. Also for any function $f \in S_{k,\mathbf{t}}$ with coefficient vector $\mathbf{c} = (c_1, \ldots, c_n)^T$ we have the stability estimate

$$K||\mathbf{c}||_{p,\mathbf{t}} \le ||f||_{L^p} \le ||\mathbf{c}||_{p,\mathbf{t}}, \tag{3.3}$$

for some positive constant K ([4]). Here $||\cdot||_{L^p}$ is the classical L^p-norm on $[a,b]$ and

$$||\mathbf{c}||_{p,\mathbf{t}} = ||f||_{\ell^p,\mathbf{t}} = \begin{cases} \left(\sum_i |c_i|^p (t_{i+k} - t_i)/k\right)^{1/p}, & 1 \le p < \infty \\ \max_i |c_i|, & p = \infty, \end{cases} \tag{3.4}$$

is the weighted discrete ℓ^p-norm for splines. In (3.3) we observe that the function itself is bounded from above by its coefficients. This is particularly important when using the decomposition algorithms as a tool for data removal (data reduction).

We need to consider spline spaces defined over a union of knot vectors. Given two spline spaces $S_{k,\tau}$ and $S_{k,\mathbf{t}}$, we define their sum as

$$S_{k,\tau} + S_{k,\mathbf{t}} = \{f_\tau + f_\mathbf{t} \,:\, f_\tau \in S_{k,\tau}, f_\mathbf{t} \in S_{k,\mathbf{t}}\}. \tag{3.5}$$

This is the smallest vector space containing the two spline spaces. With $m_\tau(x)$ denoting the total number of times the number x occurs in the knot vector τ, we define knot sequences $\tau \vee \mathbf{t}$ and $\tau \wedge \mathbf{t}$ by

$$m_{\tau\vee\mathbf{t}}(x) = m_\tau(x) \vee m_\mathbf{t}(x) = \max\{m_\tau(x), m_\mathbf{t}(x)\},$$
$$m_{\tau\wedge\mathbf{t}}(x) = m_\tau(x) \wedge m_\mathbf{t}(x) = \min\{m_\tau(x), m_\mathbf{t}(x)\}.$$

It was shown in [15] that if τ and $\mathbf{t}$ are two arbitrary knot vectors such that $\tau \wedge \mathbf{t}$ contains at least k elements, then

$$S_{k,\tau} + S_{k,\mathbf{t}} = S_{k,\tau\vee\mathbf{t}}.$$

3.2. Refinement

Refinement algorithms for splines are extensively used both as mathematical tools and as tools for numerous practical applications where construction and manipulation of curves and surfaces are important. The algorithms are numerically stable, they are easily implemented on a computer and the algorithms are mathematically well understood.

In order to give a brief description of refinement algorithms for splines, let us assume that $\mathbf{t}_i, \mathbf{t}_j$ are two knot sequences such that $\mathbf{t}_i \subset \mathbf{t}_j$. It follows that $S_i = S_{k,\mathbf{t}_i} \subset S_j = S_{k,\mathbf{t}_j}$. Refinement is now the process of rewriting or "lifting" a spline f in the space S_i to the space S_j. Hence, the spline function itself is unchanged, but is represented with more degrees of freedom in S_j.

Let S_i and S_j be spline spaces of dimension m_i and m_j, respectively. It follows that the coefficient vectors in S_i and S_j are vectors in $\mathbb{R}^{m_i}$ and $\mathbb{R}^{m_j}$, respectively. Hence, we can define the refinement operator $\mathcal{R}_i^j : \mathbb{R}^{m_i} \to \mathbb{R}^{m_j}$ by

$$\mathbf{c}(f, S_j) = \mathcal{R}_i^j \mathbf{c}(f, S_i) \tag{3.6}$$

which takes the B-spline coefficient vector $\mathbf{c}(f, S_i)$ and produces the coefficient vector $\mathbf{c}(f, S_j)$ of a function f with respect to the B-spline basis for S_j. The operator $\mathcal{R}_i^j$ is given by the relation

$$\mathbf{c}(f, S_j) = A_i^j \mathbf{c}(f, S_i), \tag{3.7}$$

where the matrix A_i^j in (3.7) is known as the knot insertion matrix of order k from $\mathbf{t}_i$ and $\mathbf{t}_j$. There is a simple recurrence relation, the *Oslo Algorithm*, for computing the elements of the matrix A_i^j. To give details, let the knot vectors be $\mathbf{t}_i = (\tau_j)_{j=1}^{n+k}$ and $\mathbf{t}_j = (t_q)_{q=1}^{m+k}$. To find the nonzero elements $a_{p,q}$ in row p of A_i^j, we let $\mu = \mu_p$ be the integer so that $\tau_\mu \le t_p < \tau_{\mu+1}$. Then we set $\alpha_{\mu,1}(p) = 1$ and for $r = 2, 3, \ldots k$

$$\alpha_{q,r}(p) = \frac{t_{p+r-1} - \tau_q}{\tau_{q+r-1} - \tau_q}\alpha_{q,r-1}(p) + \frac{\tau_{q+r} - t_{p+r-1}}{\tau_{q+r} - \tau_{r+1}}\alpha_{q+1,r-1}(p).$$

We then have $a_{p,q} = \alpha_{q,k}(p), q = \mu - k + 1, \ldots, \mu$ and $a_{p,q} = 0$ for other values of q. Hence, there are at most k nonzero elements in each row of A_i^j. Moreover, A_i^j is a totally positive, stochastic matrix of full rank. Finally, we remark that if we set $t_p = t_{p+1} = \cdots = t_{p+k-1} = x$ in this recurrence relation, then we obtain $\alpha_{q,k}(p) = B_{q,k}(x)$, the B-splines in (3.1). For further details on refinement and knot insertion for B-splines, see [14].

3.3 Discrete Least Squares

With knot sequences $\mathbf{t}_i \subset \mathbf{t}_j$ it follows, as pointed out in the previous section, that $S_i = S_{k,\mathbf{t}_i}$ is a subspace of $S_j = S_{k,\mathbf{t}_j}$ and that the refinement operator $\mathcal{R}_i^j$ is well defined for splines through equation (3.7). For the approximation operator $\mathcal{P}_j^i : S_j \to S_i$ we have many choices. One possibility is to define $\mathcal{P}_j^i$ as an operator which produce a least squares approximation in S_i to the function in S_j. So, given a spline $f_j \in S_j$ and using the weighted ℓ^2-norm (cf. equation (3.4)) on S_j we find the B-spline coefficients $\mathbf{c}(f_i, S_i)$ of an approximation $f_i \in S_i$ by solving the least squares problem

$$\min ||\mathcal{R}_i^j \mathbf{c}(f_i, S_i) - \mathbf{c}(f_j, S_j)||_{2,\mathbf{t}_j}. \tag{3.8}$$

Hence, we can define the approximation operator $\mathcal{P}_j^i$ such that

$$\mathbf{c}(f_i, S_i) = \mathcal{P}_j^i \mathbf{c}(f_j, S_j) \tag{3.9}$$

is the solution to (3.8). A detailed discussion of this kind of discrete least squares approximation for splines can be found in [16]. Let us just mention that (2.10) holds for least squares operators. It is of course possible to use other approximation schemes, e.g. interpolating f at a sample of points.

We are now equipped with the necessary tools $\mathcal{R}_i^j$ (refinement) and $\mathcal{P}_j^i$ (approximation), which are the basic elements needed to perform a decomposition of a spline function in terms of the coefficients. Let us have a look at an example where we apply Algorithm 1 given in Section 2.

Example 3.1. *Figure 3.1a shows a cubic (order $k = 4$) spline curve f on the knot sequence $\mathbf{t} = (0,0,0,0,1,2,\ldots,76,77,77,77,77)$. From $\mathbf{t}$ we construct the "near" uniform knot sequences*

$$\begin{aligned}
\mathbf{t}_0 &= (0,0,0,0,16,32,48,64,77,77,77,77),\\
\mathbf{t}_1 &= (0,0,0,0,8,16,\ldots,72,77,77,77,77),\\
\mathbf{t}_2 &= (0,0,0,0,4,8,\ldots,72,76,77,77,77,77),\\
\mathbf{t}_3 &= \mathbf{t}.
\end{aligned}$$

Note that $\mathbf{t}_0 \subset \mathbf{t}_1 \subset \mathbf{t}_2 \subset \mathbf{t}_3$. This implies, that the spline spaces $S_i = S_{k,\mathbf{t}_i}$, $i = 0,1,2,3$ satisfy $S_0 \subset S_1 \subset S_2 \subset S_3$ and have dimensions $m_0 = 8$, $m_1 = 13$, $m_2 = 23$ and $m_3 = 80$, respectively. Applying Algorithm 1 with $\mathcal{R}_i^j$ as the refinement operator given in (3.6) and $\mathcal{P}_j^i$ as the approximation operator in (3.9) we obtain the decomposition

$$\mathbf{c}(f_0, S_0) \in \mathbb{R}^{m_0} \quad \text{and} \quad \mathbf{c}(g_i, S_i) \in \mathbb{R}^{m_i}, \quad i = 1,2,3$$

of the coefficient vector $\mathbf{c}(f, S_3) \in \mathbb{R}^{m_3}$. (Since $\mathcal{P}_j^i$ given by (3.9) satisfy (2.10) we will obtain the same results using Algorithms 2 and 3.) Figure 3.1a shows the original spline curve f with coefficients $\mathbf{c}(f, S_3)$. Figure 3.1b indicates the knot sequences $\mathbf{t}_0$ (top) to $\mathbf{t}_3$ (bottom) with $\mathbf{t}_0 \subset \mathbf{t}_1 \subset \mathbf{t}_2 \subset \mathbf{t}_3 = \mathbf{t}$. Figure 3.1c and (d) show the approximation f_0 and error $f - f_0$. Figures 3.1e and 3.1f show the approximation $f_1 = f_0 + g_1$ and error $f - f_1$. Figures 3.1g and 3.1h show the approximation $f_2 = f_0 + g_1 + g_2$ and error $f - f_2$. (The exact function is $f_3 = f_0 + g_1 + g_2 + g_3 = f$.)

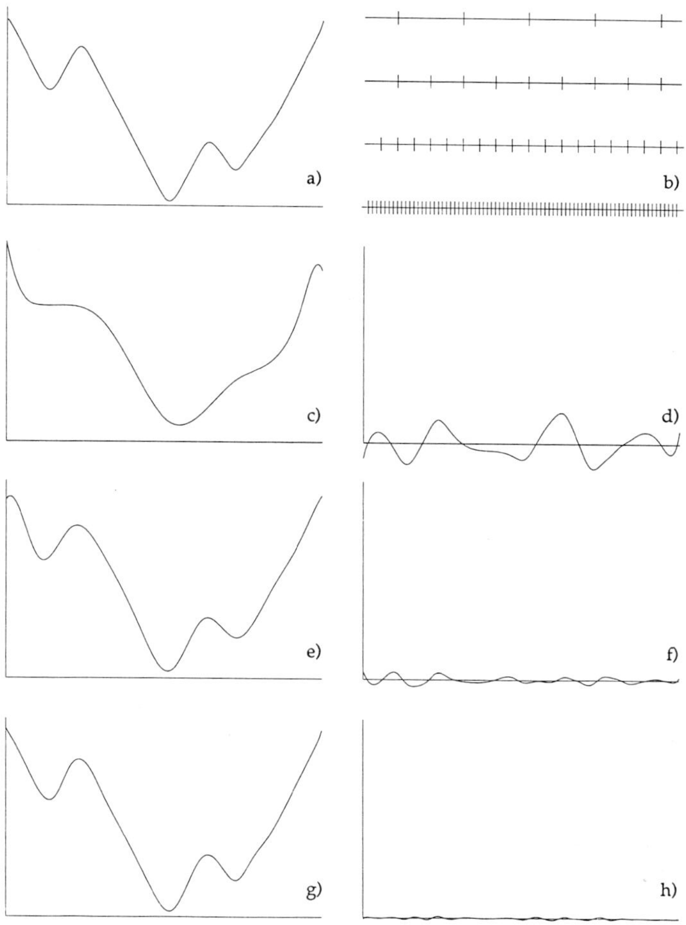

Figure 3.1. Decomposed smooth function, fixed subspaces.

3.5 Adaptive Subspace Selection

In order to perform decomposition as stated in Section 2, we have to select knot sequences and hence define the spline spaces before running the algorithms. In this section we will modify the algorithms by selecting subspaces as a function of results produced at previous steps in the algorithms. Moreover, we will adapt the subspace selection to the given data.

Given $f \in S = S_{k,\mathbf{t}}$ and a real number $\epsilon > 0$, we let $\mathcal{P}^j$ be the discrete least squares operator from S to a subspace S_j given by (3.8). Since the number of spaces is unknown, we omit the subscript on $\mathcal{P}^j$. The refinement operator is as defined in (3.6). However, when the superscript is omitted on $\mathcal{R}_i$, the curve is refined from S_i to the given space S. The third ingredient in this method is a subspace selection operator $\mathcal{F}(S, \mathbf{c}(f, S), \delta)$ which to a given B-spline coefficient vector $\mathbf{c}(f, S)$ of a function f in a spline space S, and a given tolerance δ attempts to find a subspace $T = \mathcal{F}(S, \mathbf{c}(f, S), \delta)$ of S. Ideally, T is a smallest subspace for which an approximation to f within error δ can be found. Below we describe a method which can be used to find a candidate for T.

We now have the following modified version of Algorithm 1 which uses the tolerance ϵ to determine N and the nested subspaces in (2.2). (A similar modification of Algorithms 2 and 3 is also possible.)

Algorithm 4

1. $i = 0;\quad S_{-1} = \emptyset;\quad \mathbf{c}(f_{-1}, S_{-1}) = 0;\quad \mathbf{d} = (d_j) = \mathbf{c}(f, S);$
2. while $(||\mathbf{d}||_\infty > \epsilon)$
   - 2.1. $\delta = (\max_j d_j - \min_j d_j)/2$
   - 2.2. $S_i = S_{i-1} + \mathcal{F}(S, \mathbf{d}, \max\{\delta, \epsilon\})$
   - 2.3. $\mathbf{c}(f_i, S_i) = \mathcal{P}^i \mathbf{c}(f, S)$
   - 2.4. $\mathbf{c}(g_i, S_i) = \mathbf{c}(f_i, S_i) - \mathcal{R}^i_{i-1} \mathbf{c}(f_{i-1}, S_{i-1})$
   - 2.5. $\mathbf{d} = \mathbf{c}(f, S) - \mathcal{R}_i \mathbf{c}(f_i, S_i)$
   - 2.6. $i := i + 1$
3. $N = i;\quad S_N = S;\quad \mathbf{c}(g_N, S_N) = \mathbf{d}$

In addition to the fact that we have to define $\mathcal{F}$, there are two major aspects connected to this algorithm. Firstly, we observe that

$$f = f_N = f_0 + g_1 + \ldots + g_N$$

is an exact decomposition of f as is also the case for the plain algorithms presented in Section 2. Secondly, we observe that the sum

$$f_{N-1} = f_0 + g_1 + \ldots + g_{N-1}$$

is an approximation from $S_{N-1} \subset S = S_N$ to f within ϵ. This follows since the L_∞ norm of a spline is bounded by the ℓ_∞ norm of its coefficients. Hence, we obtain a data reduction when dropping the component g_N. We will discuss data reduction in more detail in Section 5.

The parameter δ is used to select the subspaces S_i in Algorithm 4. We have used a δ which is proportional to the total variation in the coefficients $\mathbf{d}$ of the error $f - f_i$ at the ith level. We divide the total variation by the factor 2. Other choices can be used to give a different number N of components in the decomposition.

Let us turn to a description of the adaptive subspace selection, in the algorithm denoted by $\mathcal{F}$. Given a spline space $S = S_{k,\mathbf{t}}$ of order k on the knot sequence $\mathbf{t}$, and a spline $h \in S$ with B-spline coefficients $\mathbf{c} = \mathbf{c}(h, S)$, we will construct $\mathcal{F}(T, \mathbf{c}, \delta)$ to be a subspace of S. In other words, we want to find a knot sequence $\mathbf{t}' \subset \mathbf{t}$. Moreover, we want $\mathbf{t}'$ to be such that the spline space $\mathcal{F}(S, \mathbf{c}, \delta)$ reflects the behavior of the spline function h for a suitable δ.

For this purpose we propose to use the knot removal strategy for splines introduced in [16]. The knot removal problem can be stated as follows: Given a spline $h \in S_{k,\mathbf{t}}$ and a tolerance $\delta > 0$, find a spline $\hat{h} \in S_{k,\mathbf{t}'}$ with $\mathbf{t}' \subset \mathbf{t}$ as short as possible, but such that

$$||h - \hat{h}||_\infty \leq \delta.$$

For the operator $\mathcal{F}$ we are interested in $\mathbf{t}'$ and not so much the approximation $\hat{h}$. In order to find $\mathbf{t}'$, a weight

$$w_i = ||\mathbf{c}(h, S) - \mathcal{R}'\mathbf{c}(h_i, S')||_\infty$$

is assigned to every interior knot t_i in $\mathbf{t}$. Here $\mathcal{R}'$ is the refinement operator which lifts h_i in $S' = S_{k,\mathbf{t}\setminus\{t_i\}}$ to $S = S_{k,\mathbf{t}}$. Hence, w_i measures the importance of the knot t_i. The spline $h_i \in S'$ can in principle be any approximation to $h \in S$, e.g. a discrete least squares approximation (cf. equation (3.8)). However, a best approximation in the $\ell_{\infty,\mathbf{t}}$-norm (cf. equation (3.4)) has proven favorable.

The sequence of weights w_i are now sorted into groups

$$W(a, b) = \{w_i : a \leq w_i < b\},$$

for

$$(a, b) = (0, \delta/2), (\delta/2, \delta), (\delta, 2\delta), \ldots.$$

The knots in each group are listed in the order in which they occur in the knot sequence.

Let $W_1 = W(0, \delta/2)$, $W_2 = W(\delta/2, \delta)$, and in general for $i \geq 2$, $W_i = (2^{i-3}\delta, 2^{i-2}\delta)$. Suppose we find n_1 knots in W_1, n_2 knots in W_2, n_3 knots in W_3, etc.. Now, if we want to remove n knots where say, $n_1 \leq n \leq n_1 + n_2$, then we remove all n_1 points in W_1 and $n_1 + n_2 - n$ points from W_2. In order to obtain a uniform removal in each group we remove every $\lfloor n_2/(n_1 + n_2 - n) \rfloor^{th}$ knot from W_2, where $\lfloor x \rfloor$ is the largest integer $\leq x$. In this case no knots from W_3 are removed.

The exact number of knots to be removed is found by a binary search. We try to remove half of the knots in $\mathbf{t}$, and compute the best approximation in the $\ell_{2,\mathbf{t}}$-norm (cf. equation (3.8)). Then we measure the distance between the approximation and the original in the $\ell_{\infty,\mathbf{t}}$-norm. If this distance is less than the given tolerance δ, we try to remove 3/4 of the knots, otherwise we try to remove 1/4 of the knots. This process is continued until the exact number of knots that can be removed has been determined.

If the $\mathbf{t}'$ are the knots in $\mathbf{t}$ which are left after this process, then our space is given by $S_{k,\mathbf{t}'} = \mathcal{F}(S, \mathbf{c}(h, S), \delta)$.

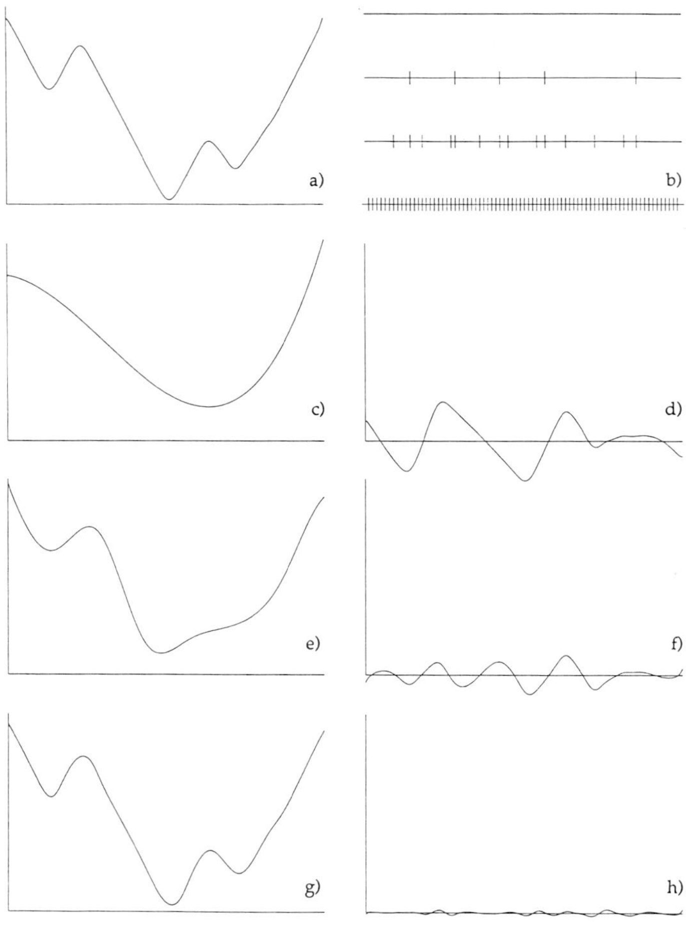

Figure 3.2. Decomposed smooth function, unknown subspaces.

The following two examples use Algorithm 4 for the decomposition of splines.

Example 3.2. *We have given the same curve f and the same knot sequence* $\mathbf{t}$ *as in Example 3.1 (Figure 3.2a) The values $f(x)$ are in the range [0,25], and we have used $\epsilon = 0.5$. Figure 3.2b shows the selected knot sequences* $\mathbf{t}_0 \subset \mathbf{t}_1 \subset \mathbf{t}_2 \subset \mathbf{t}_3 = \mathbf{t}$ *found by the above knot removal strategy. Only interior knots are indicated. Note that the knot sequence* $\mathbf{t}_0$ *has no interior* *knots. Figures 3.2c and 3.2d show the approximation f_0 and error $f - f_0$. Figures 3.2e and 3.2f show the approximation $f_1 = f_0 + g_1$ and error $f - f_1$. Figures 3.2g and 3.2h show the approximation $f_2 = f_0 + g_1 + g_2$ and error $f - f_2$. The function f_2 is an approximation to f within the given tolerance $\epsilon = 0.5$. The function $f_3 = f_0+g_1+g_2+g_3 = f$. For this value of ϵ we obtained three subspaces of $S_{k,\mathbf{t}}$. It is interesting to compare the approximations in Figure 3.1c and Figure 3.2e. Both show an approximation from a space with five interior knots. We observe that the error in Figure 3.2f is smaller than the error shown in Figure 3.1d. Moreover, we observe that the error in Figure 3.2f has a more equioscillating behavior than the error in Figure 3.1d.*

Example 3.3. *The curve in Examples 3.1 and 3.2 are relatively smooth. In this example we will show how a noisy curve are decomposed into smooth components and noisy components. Figure 3.3a shows a curve selected from a solar spectral image. This curve is represented as a cubic spline function $f \in S = S_{4,\mathbf{t}}$, where $\mathbf{t} = 1, 1, 1, 1, 3, 4, \ldots, 253, 254, 256, 256, 256, 256)$ and where the selected numbers from the image are taken as coefficients $\mathbf{c}(f, S) \in \mathbb{R}^{256}$. Figures 3.3c–3.3h show the same sequence as the Figures 3.2c–3.2h. The knot sequences selected by the knot removal strategy are shown in Figure 3.3b. The tolerance used is $\epsilon = 0.3$ corresponding to data in the range $[0, 5]$. The top knot sequence is used for the approximation in Figure 3.3c, while the next two knot sequences are used for the approximation in Figures 3.3e and 3.2g, respectively. The bottom knot sequence in Figure 3.3b corresponds to the exact function in Figure 3.3a.*

3.6 Tensor Product Spline Surfaces

The tensor product B-spline functions

$$\Phi_{i,j}(x, y) = \Phi_{i,j,k,\ell,\mathbf{s},\mathbf{t}}(x, y) = B_{i,k,\mathbf{s}}(x)B_{j,\ell,\mathbf{t}}(y), \quad i = 1, \ldots, m, \quad j = 1, \ldots n$$

are the building blocks of tensor product B-spline surfaces on the knot sequences $\mathbf{s}$ and $\mathbf{t}$. Hence, we have the space of order-(k, ℓ) tensor product spline surfaces given by

$$S^{k,\ell}_{\mathbf{s},\mathbf{t}} = \Big\{F(x, y) = \sum_{i=1}^{m}\sum_{j=1}^{n} c_{i,j}\Phi_{i,j}(x, y)\Big\}. \tag{3.10}$$

We use the notation $\mathbf{c}(F, S^{k,\ell}_{\mathbf{s},\mathbf{t}})$ for the B-spline coefficients $(c_{i,j})^{m,n}_{i,j=1}$.

Given pairs of knot sequences $(\mathbf{s}_i, \mathbf{t}_i)$, $i = 0, \ldots, N$ such that

$$\mathbf{s}_0 \subset \mathbf{s}_1 \subset \ldots \subset \mathbf{s}_N,$$
$$\mathbf{t}_0 \subset \mathbf{t}_1 \subset \ldots \subset \mathbf{t}_N,$$

we define the tensor product spline spaces $S_i = S^{k,\ell}_{\mathbf{s}_i,\mathbf{t}_i}$. It follows that

$$S_0 \subset S_1 \subset \ldots \subset S_N. \tag{3.11}$$

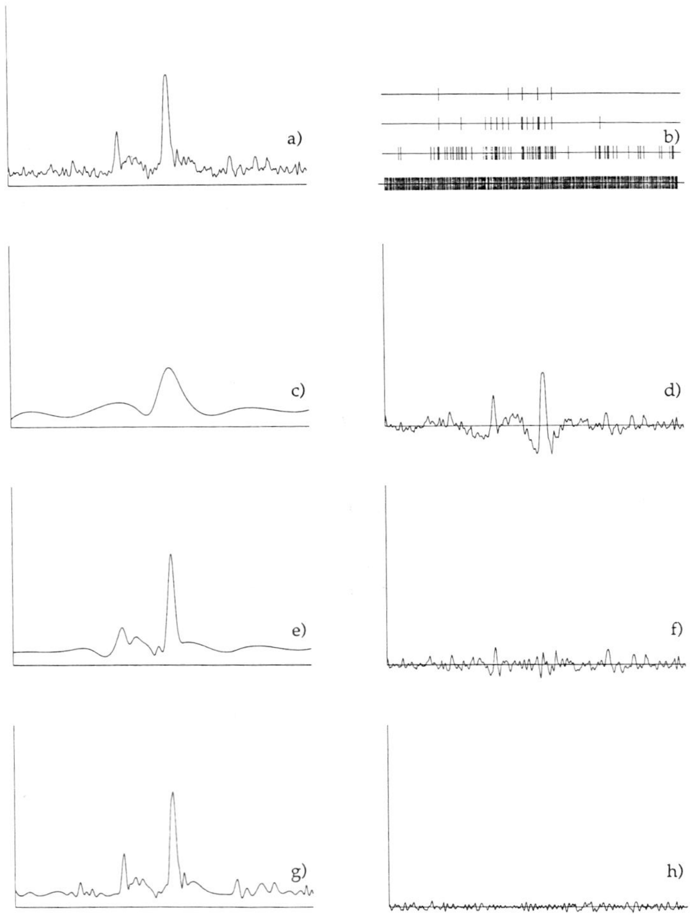

Figure 3.3. Decomposed noisy function, unknown subspaces.

The refinement operator $\mathcal{R}_i^j$ given in (3.6) and approximation operator $\mathcal{P}_j^i$ given in (3.9) are easily extended to tensor product splines, e.g. refinement for tensor products are done by first performing curve refinement (Section 3.2) on each row of coefficients in one direction and then doing the same in the other direction. Details on the particular discrete least squares approximation for tensor products can be found in [17].

Let us apply Algorithm 4 to a surface example. The adaptive subspace selection is done using the knot removal strategy for surfaces described in [17].

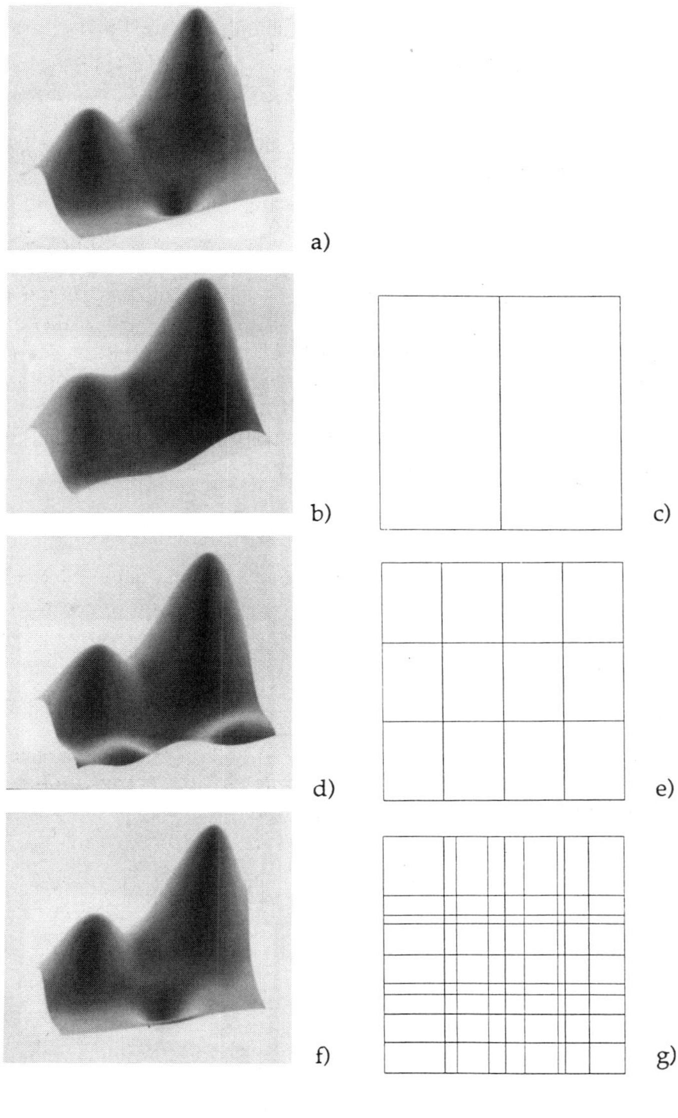

Figure 3.4. Decomposed smooth function, unknown subspaces.

Example 3.4. *The Franke function*

$$\begin{aligned}\Gamma(x,y) = {} & (3/4)\exp\{-((9x-2)^2+(9y-2)^2)/4\} \\ & + (3/4)\exp\{-(9x+1)^2/49-(9y+1)/10\} \\ & + (1/2)\exp\{-((9x-7)^2+(9y-3)^2)/4\} \\ & - (1/5)\exp\{-(9x-4)^2-(9y-7)^2\},\end{aligned}$$

is sampled on a uniform 256×256 *grid over the unit square* $[0,1]\times[0,1]$. *Then the values are taken as coefficients* $\mathbf{c}(F,S)$ *to a bicubic* ($k=\ell=4$) *spline in* $S=S^{k,\ell}_{\mathbf{s},\mathbf{t}}$ *with*

$$\mathbf{s}=\mathbf{t}=(1,1,1,1,3,4,\ldots,253,254,256,256,256,256).$$

The spline F *obtained using Algorithm 4 with* $\epsilon=0.05$, *is shown in Figure 3.4a, and the approximations* F_0, F_0+G_1 *and* $F_0+G_1+G_2$ *are shown in the Figures 3.4b, 3.4d and 3.4f, respectively. The respective knot sequences are illustrated by knot lines on the unit square in the Figures 3.4c, 3.4e and 3.4g. Since we have* $k=\ell=4$ *knots at the end of each knot sequence, and since all interior knots are simple the dimension of the tensor product spaces are* $5\times 4=20$, $7\times 6=42$ *and* $12\times 12=144$, *respectively.*

§4. Box Spline Decomposition

We consider now decomposition on uniform partitions using box splines and one of the first three algorithms. This is essentially the decomposition considered in [8]. We start by giving some facts about box splines. For a more comprehensive introduction, see [10].

4.1 Box Splines

Suppose $\mathbf{X}_\mu=(\mathbf{x}^1,\mathbf{x}^2,\ldots,\mathbf{x}^\mu)$ with $\mathbf{x}^i\in\mathbb{R}^d$ are $\mu\ge d$ vectors in $\mathbb{R}^d$ with $\mathbf{x}^1,\ldots,\mathbf{x}^d$ linearly independent. A box spline with direction vectors $\mathbf{X}_\mu$, is a function $M(\mathbf{u}\,|\,\mathbf{X}_\mu):\mathbb{R}^d\to\mathbb{R}$ defined recursively by

$$M(\mathbf{u}\,|\,\mathbf{X}_\mu)=\int_0^1 M(\mathbf{u}-t\mathbf{x}_\mu\,|\,\mathbf{X}_{\mu-1})\,dt, \tag{4.1}$$

with

$$M(\mathbf{u}\,|\,\mathbf{X}_d)=\begin{cases}1/|\det(\mathbf{X}_d)|, & \text{if } \mathbf{u}\in[\mathbf{X}_d[,\\ 0, & \text{otherwise},\end{cases} \tag{4.2}$$

and where

$$[\mathbf{X}_\mu[=\{t_1\mathbf{x}^1+\ldots+t_\mu\mathbf{x}^\mu\;:\;0\le t_j<1,\quad 1\le j\le\mu\}. \tag{4.3}$$

We consider the situation where the direction vectors of a box spline consist of repetitions of a few nonparallel directions in $\mathbb{R}^2$. Suppose for an integer $r\ge 2$, that

$$\mathbf{X}_\mu=(\overbrace{\mathbf{d}^1,\cdots,\mathbf{d}^1}^{k_1\text{ times}},\overbrace{\mathbf{d}^2,\cdots,\mathbf{d}^2}^{k_2\text{ times}},\ldots,\overbrace{\mathbf{d}^r,\cdots,\mathbf{d}^r}^{k_r\text{ times}}),$$

where $k_1, k_2, \ldots, k_r$ are positive integers with $\mu = \sum_i k_i$. If for $r \leq 4$ we have the standard directions

$$\mathbf{d}^1 = \begin{pmatrix}1\\0\end{pmatrix} = \rightarrow, \quad \mathbf{d}^2 = \begin{pmatrix}0\\1\end{pmatrix} = \uparrow, \quad \mathbf{d}^3 = \begin{pmatrix}1\\1\end{pmatrix} = \nearrow, \quad \mathbf{d}^4 = \begin{pmatrix}1\\-1\end{pmatrix} = \searrow, \tag{4.4}$$

then we denote $M(\mathbf{u}\,|\,\mathbf{X}_\mu)$ by $M_{\mathbf{k}}(\mathbf{u})$, where $\mathbf{k} = (k_1, \ldots, k_r)$. It can be seen that each $M_{\mathbf{k}}$ is nonnegative, zero off $[\mathbf{X}_\mu[$, and that

$$\sum_{\mathbf{i}\in\mathbb{Z}^2} M_{\mathbf{k}}(\mathbf{u}) = 1, \quad \mathbf{u} \in \mathbb{R}^2.$$

Box spline spaces can be formed from one or more box splines by using translations and dilations. We restrict our attention to box splines on the regular grids with two, three or four directions. For any positive integer ν an element of the space

$$S_{\mathbf{k},\nu} = \{p(\mathbf{u}) = \sum_{\mathbf{i}\in\mathbb{Z}^2} c_{\mathbf{i}} M_{\mathbf{k}}(\nu\mathbf{u} - \mathbf{i}) \,:\, c_{\mathbf{i}} \in \mathbb{R}, \mathbf{u} \in \mathbb{R}^2\} \tag{4.5}$$

is called a box spline surface. A box spline surface is a piecewise polynomial of degree $\mu - 2 = \sum_{i=1}^{r} k_i - 2$ with continuity $C^{\mu-2-k_i}$ across a grid line parallel to the direction d_i. The box spline grid consists of the union of all straight lines through each point of the form $\mathbf{i}/\nu$, $\mathbf{i} \in \mathbb{Z}^2$ in the directions $\mathbf{d}_1, \ldots, \mathbf{d}_r$. On any region of the plane not intersected by any grid lines, the nonzero box spline translates $M_{\mathbf{k}}(\nu\mathbf{u} - \mathbf{j})$ are linearly independent for $r \leq 3$, but linearly dependent for $r = 4$.

Corresponding to a region $\Omega \subset \mathbb{R}^2$, we have the box spline space

$$S_{\mathbf{k},\nu}(\Omega) = \{p(\mathbf{u}) = \sum_{\mathbf{i}\in I_{\mathbf{k},\nu}(\Omega)} c_{\mathbf{i}} M_{\mathbf{k}}(\nu\mathbf{u} - \mathbf{i}) \,:\, c_{\mathbf{i}} \in \mathbb{R}, \mathbf{u} \in \Omega\}, \tag{4.6}$$

where

$$I_{\mathbf{k},\nu}(\Omega) = \{\mathbf{i} \,:\, M_{\mathbf{k}}(\nu\mathbf{u} - \mathbf{i}) \neq 0 \quad \text{for some} \quad \mathbf{u} \in \Omega^0\}. \tag{4.7}$$

Here Ω^0 is the interior of the set Ω.

4.2. Refinement

In decomposition algorithms with box splines, we take the spaces S_i in (2.2) to be of the form $S_{\mathbf{k},2^{n_i}}(\Omega)$ for integers $1 \leq n_0 < n_1 < \cdots < n_N$ and a suitable $\Omega \in \mathbb{R}^2$. There are simple, stable, and fast refinement algorithms for such spaces ([7,12]). To describe refinement algorithms for box splines, suppose for simplicity we have a box spline surface p in $S = S_{\mathbf{k},1}$ and that $\Omega = \mathbb{R}^2$. Since $S_{\mathbf{k},1} \subset S_{\mathbf{k},2}$, we can write

$$p(\mathbf{u}) = \sum_{\ell\in\mathbb{Z}^2} c_\ell M_{\mathbf{k}}(\mathbf{u} - \ell) = \sum_{\mathbf{q}\in\mathbb{Z}^2} b_{\mathbf{q}} M_{\mathbf{k}}(2\mathbf{u} - \mathbf{q})$$

for certain box spline coefficients $\mathbf{c} = (c_\ell)$ and $\mathbf{b} = (b_\mathbf{q})$. A typical refinement operator $\mathcal{R}_1^2$ maps $\mathbf{c}$ into $\mathbf{b} = \mathcal{R}_1^2\mathbf{c}$. The recursive computation of $b_\mathbf{q} = b_\mathbf{q}^\mu$ with $\mu = \sum k_i$ is as follows

$$
\begin{aligned}
&b^2_{2\mathbf{q}+\mathbf{r}} = c_\mathbf{q}, \qquad \mathbf{q} \in \mathbb{Z}^2, \quad \mathbf{r} \in \{0,0\},\{1,0\},\{0,1\},\{1,1\}, \\
&\text{For } \ell = 3,\dots,\mu \\
&\quad b^\ell_\mathbf{q} = (b^{\ell-1}_{\mathbf{q}-\mathbf{x}_\ell} + b^{\ell-1}_\mathbf{q})/2, \qquad \mathbf{q} \in \mathbb{Z}^2.
\end{aligned} \tag{4.8}
$$

Here $\mathbf{x}_1 = (1,0), \mathbf{x}_2 = (0,1)$, and $\mathbf{x}_3,\dots,\mathbf{x}_\mu$ are the remaining directions defining $M_\mathbf{k}$. The refinement can also be done for a given $\mathbf{k}$ by computing an explicit *mask*.

In the more general case, where for positive integers $m > n$ we would like to rewrite a function $p \in S_{\mathbf{k},2^n}$ as a function in $S_{\mathbf{k},2^m}$, we simply apply (4.8) $m - n$ times to obtain the refined box spline coefficients.

4.3. Approximations Schemes

In order to use the algorithms presented in Section 2, we also need to define approximation operators $\mathcal{P}$ mapping an element f in some box spline space $S_{\mathbf{k},2^m}(\Omega)$ into an element $\mathcal{P}f \in S_{\mathbf{k},2^n}(\Omega)$ for some $n < m$ and some $\Omega \subset \mathbb{R}^2$. A suitable $\mathcal{P}$ is obtained through interpolation at grid points with possibly a few additional points outside Ω determined by extrapolation. We then obtain a large sparse linear system which has to be solved. Usually iterative methods are advantageous, all the more so since we usually have good starting vectors available for the iteration. It is demonstrated in [1] that the conjugate gradient method is extremely efficient for such problems, and convergence is achieved in a few iterations.

§5. Data Reduction by Decomposition

In this section we will give two modifications of Algorithm 1 where we aim at giving decomposition algorithms which provide data compression. We assume that the vectors $\mathbf{v}_i$ are the basis coefficients in (2.3) corresponding to the decomposition (2.1). Also we assume that the basis in (2.3) satisfies for all i,

$$
\phi_{i,\ell} \ge 0, \qquad \sum_\ell \phi_{i,\ell} = 1. \tag{5.1}
$$

We can then obtain data compression by only storing coefficients in each component, which gives a significant contribution to the reconstruction of the actual object. Similar modifications can also be done with the Algorithms 2, 3 and 4. In order to simplify the outline of the modified algorithms we will introduce the following notation. Given any vector $\mathbf{w} = (w_1,\dots,w_n)$ and an $\epsilon > 0$, let

$$
\mathbf{w}(\epsilon) = (w_1(\epsilon),\dots,w_n(\epsilon)),
$$

where

$$w_i(\epsilon) = \begin{cases} w_i, & \text{if } |w_i| \geq \epsilon, \\ 0, & \text{otherwise.} \end{cases} \tag{5.2}$$

Applying (5.2), we obtain the following simple extension of Algorithm 1. Note that a variable tolerance ϵ_i can be used at each level.

Algorithm 5

1. $\mathbf{v}_0 = \mathcal{P}_N^0 \mathbf{v}_N$
2. For $i = 1, \ldots, N$
   2.1. $\mathbf{v}_i = \mathcal{P}_N^i \mathbf{v}_N$
   2.2. $\mathbf{w}_i = \mathbf{v}_i - \mathcal{R}_{i-1}^i \mathbf{v}_{i-1}$
   2.3. $\mathbf{z}_i = \mathbf{w}_i(\epsilon_i)$

We assume that $\mathbf{v}_0$ is kept unchanged. The truncated decomposition of $\mathbf{v}_N = \mathbf{v}$ is now given by

$$\mathbf{u}_N = \mathcal{R}_0^N \mathbf{v}_0 + \sum_{i=1}^{N} \mathcal{R}_i^N \mathbf{z}_i. \tag{5.3}$$

An efficient storage of the components $\mathbf{z}_i$ is obtained if a majority of the elements in $\mathbf{z}_i$ are equal to zero. There are several possibilities when it comes to efficient bit-packing of objects like $\mathbf{z}_i$, see [13,20,21].

An alternative way of implementing the algorithm is to carry the error made on level i to the next level $i + 1$.

Algorithm 6

1. $\mathbf{y}_0 = \mathbf{v}_0 = \mathcal{P}_N^0 \mathbf{v}_N$
2. For $i = 1, \ldots, N$
   2.1. $\mathbf{v}_i = \mathcal{P}_N^i \mathbf{v}_N$
   2.2. $\mathbf{w}_i = \mathbf{v}_i - \mathcal{R}_{i-1}^i \mathbf{y}_{i-1}$
   2.3. $\mathbf{z}_i = \mathbf{w}_i(\epsilon_i)$
   2.4. $\mathbf{y}_i = \mathcal{R}_{i-1}^i \mathbf{y}_{i-1} + \mathbf{z}_i$

Again we obtain a truncated decomposition of $\mathbf{v}_N$ in the form (5.3). Note that only one refinement operation is necessary for each i in Algorithm 6.

We have the following results which show the error, or more precisely, the distance between the exact vector decomposition $\mathbf{v}_N$ and the truncated vector decomposition $\mathbf{u}_N$:

Lemma 1. *With $\mathbf{u}_N$ as in (5.3) we obtain*

$$\mathbf{u}_N - \mathbf{v}_N = \sum_{i=1}^{N} \mathcal{R}_i^N (\mathbf{z}_i - \mathbf{w}_i), \tag{5.4}$$

using Algorithm 5, and

$$\mathbf{u}_N - \mathbf{v}_N = \mathbf{z}_N - \mathbf{w}_N \tag{5.5}$$

using Algorithm 6.

Proof: The proof of (5.4) is trivial. To prove (5.5) we use statement 2.4 in Algorithm 6,

$$\mathbf{u}_N = \mathcal{R}_0^N \mathbf{v}_0 + \sum_{i=1}^{N} \mathcal{R}_i^N \mathbf{z}_i = \mathcal{R}_0^N \mathbf{v}_0 + \sum_{i=1}^{N} \mathcal{R}_i^N (\mathbf{y}_i - \mathcal{R}_{i-1}^i \mathbf{y}_{i-1}) = \mathbf{y}_N.$$

But from statements 2.2 and 2.4 in Algorithm 6 we see that $\mathbf{y}_i = \mathbf{v}_i + \mathbf{z}_i - \mathbf{w}_i$ for all i. Hence $\mathbf{u}_N = \mathbf{v}_N + \mathbf{z}_N - \mathbf{w}_N$, and the lemma follows. ■

Assuming that $\mathbf{v}_N = \mathbf{c}(f, S_N)$ (the coefficients of the function f in the space S_N), $\mathbf{v}_0 = \mathbf{c}(f_0, S_0)$, $\mathbf{w}_i = \mathbf{c}(g_i, S_i)$, and $\mathbf{z}_i = \mathbf{c}(\hat{g}_i, S_i)$, we have the following lemma.

Lemma 2. *If (5.1) holds and*

$$f_N = f_0 + g_1 + \ldots + g_N$$

is the exact decomposition of a function f and

$$h_N = f_0 + \hat{g}_1 + \ldots + \hat{g}_N$$

is a truncated decomposition corresponding to Algorithm 5 or Algorithm 6, then

$$||h_N - f_N||_{L_\infty} \leq \begin{cases} \sum_i \epsilon_i ||R_i^N||_\infty, & \textit{Alg. 5}, \\ \epsilon_N, & \textit{Alg. 6}. \end{cases}$$

Proof: Using (5.1) and Lemma 1, we have

$$||f_N - h_N||_{L^\infty} \leq ||\mathbf{u}_N - \mathbf{v}_N||_\infty \leq \sum_{i=1}^{N} ||R_i^N(\mathbf{z}_i - \mathbf{w}_i)||_\infty \leq \sum \epsilon_i ||R_i^N||_\infty,$$

for Algorithm 5. The result for Algorithm 6 follows similarly from (5.5). ■

We remark that we have $||R_i^N||_\infty = 1$ for both the B-spline and box spline refinement operators. For the following discussion we assume for simplicity that $||R_i^N||_\infty = 1$.

In both algorithms the choice of the tolerances ϵ_i is important. In Algorithm 5 we observe that the choice of ϵ_i is important in order to obtain a decomposition h_N which does not deviate more than a prescribed tolerance ϵ from the original f. By Lemma 2 we must have $\sum_i \epsilon_i \leq \epsilon$. One straightforward choice is $\epsilon_i = \epsilon/N$. When it comes to Algorithm 6, we observe that all tolerances ϵ_i except ϵ_N can be chosen arbitrary. However, in order to obtain a reasonable decomposition with Algorithm 6 a careful choice of tolerances is necessary.

There are basically two reasons for choosing different tolerances when estimating coefficients $\mathbf{z}_i = \mathbf{c}(\hat{g}_i, S_i)$ of the function $\hat{g}_i$ in S_i:

- Since $S_i \subset S_j$ for $i < j$ and since $S_1, \dots S_N$ are finite dimensional spaces we know that there are fewer coefficients in S_i compared to S_j whenever $i < j$. Hence, from a data reduction point of view it will be natural to use a larger tolerance in S_j than in S_i. Such a strategy will in general reduce the total number of coefficients.
- The basis functions which span S_i have a larger impact on the function h_N than the basis functions which span S_j whenever $i < j$.

One possible choice for ϵ_i, which satisfies $\sum \epsilon_i = \epsilon$ and to some extent fulfill these two requirements is

$$\epsilon_i = \frac{2i\epsilon}{(N)(N+1)}, \quad i = 1, \dots, N. \tag{5.6}$$

Depending on the choice of the operators $\mathcal{P}_j^i$, the spaces $S_0, \dots, S_N$, and the application of interest, the choices of the sequence ϵ_i $i = 1, \dots, N$ can vary.

5.1. Examples

We conclude this paper with two examples. In the first example we will use bilinear tensor product splines on uniform knot sequences, hence the spline spaces $S_i = S_{\mathbf{k},2^i}$, $i = 0, \dots, 4$, of interest are given in (4.6) with $\mathbf{k} = (2,2)$, i.e., translates and dilates of the bilinear function $M_{2,2}$. In the second example we approximate a digital image using a sequence of quartic box spline surfaces in spaces $S_i = S_{\mathbf{k},2^i}$ $i = 0, \dots, 3$, in the form (4.6) with $\mathbf{k} = (2,2,2)$, that is, the spaces S_i are spanned by translates and dilates of the C^2-quartic box spline $M_{2,2,2}$.

Example 5.1. *The data in Color Plate 1 (see also Figure 5.1) shows a contour map of a terrain model given by $m \times n = 357 \times 257$ grid points. The terrain is a geological surface model from an oil reservoir in the North Sea and the black dots in the figure correspond to oil production wells in that area. The white region in the upper left corner is a region where no grid values were given. We performed a simple straightforward constant extrapolation in order to fill the grid. Moreover, the map was extended by a constant extrapolation in the x-direction to $m' = 369$, so that both $m' - 1$ and $n - 1$ got 16 as a common factor. We can then define dyadic subspaces on the grid. We use bilinear tensor product splines (box splines) in the subspaces $S_i = S_{\mathbf{k},2^i}(\Omega)$, $\mathbf{k} = (2,2)$, (see equation (4.6)), and where $\Omega = [0,24] \times [0,16]$. The spline space S_i has dimension $(2^i 24 + 1) \times (2^i 16 + 1)$, $i = 0, \dots, 4$. As approximation operator $\mathcal{P}_N^i$ we used interpolation at the grid points corresponding to S_i i.e.,we take these grid points as coefficients to a bilinear spline. We used a tolerance which varies with the position in the grid, that is, $\epsilon = \epsilon(x, y)$ where*

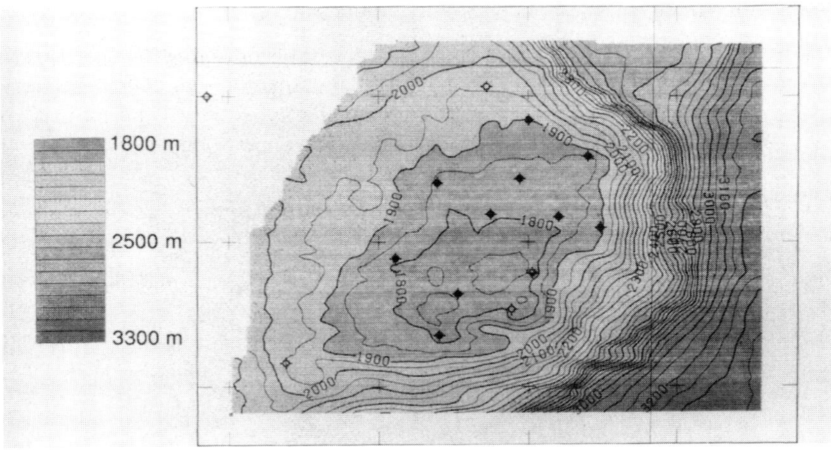

Figure 5.1. Contour map of an oil reservoir.

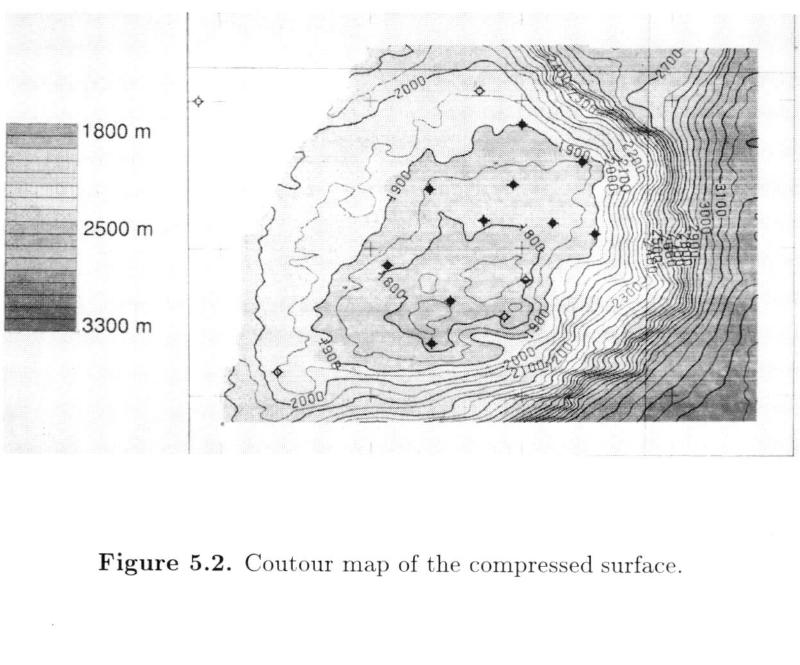

Figure 5.2. Coutour map of the compressed surface.

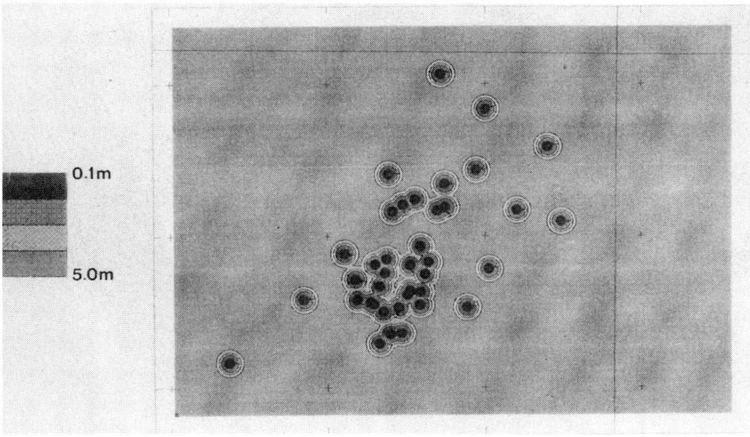

Figure 5.3. The tolerance function.

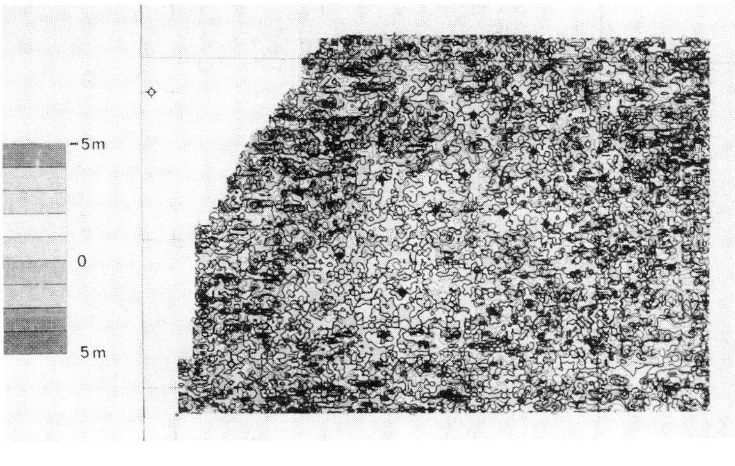

Figure 5.4. The error function.

Figure 5.5. 512×512 image. Original (left) and reconstruction (right).

Figure 5.6. The box spline coefficients..

the overall tolerance is equal to 5 meters decreasing to 0.1 meter at the wells. Using Algorithm 6 and the same tolerance $\epsilon(x,y)$ at all levels we obtain a truncated decomposition (cf. Lemma 2)

$$h_N = f_0 + \hat{g}_1 + \hat{g}_2 + \hat{g}_3 + \hat{g}_4.$$

Color Plate 2 (see also Figure 5.2) shows the function h_N, while the tolerance function $\epsilon(x,y)$ is shown in Color Plate 3 (see also Figure 5.3). Color Plate 4 (see also Figure 5.4) shows the original function f minus the reconstruction h_N. The coefficients and their positions are stored using a standard quad-tree coding scheme, see [13]. Assuming that every point in the original grid is stored with 4 bytes (single precision) we obtain a total of 367Kb (Kilobytes), while the reconstruction (cf. Color Plate 2) is stored with 10.64Kb (2.9%).

Example 5.2. *In this example we apply Algorithm 6 coupled with a noice elimination scheme to a* 512×512 *image. We use with dyadic subspaces* $S_i = S_{\mathbf{k},2^i}(\Omega)$ $i = 0,\dots,N$ *given in (4.6), where* $\mathbf{k} = (2,2,2), N = 3$, *and* $\Omega = [0,63]^2$. *The original image is shown in Figure 5.5a. Each element in the image is stored with eight bits, corresponding to 256 gray levels (0 is black and 255 is white). We have used the approximation scheme* $\mathcal{P}$ *discussed in Section 4.3. As pointed out earlier, Algorithm 6 carries the error (or noise) onto the last component, in this case the function* g_3. *The reconstruction shown in Figure 5.5b is found using* $\epsilon_1 = 8$ *and* $\epsilon_2 = 12$ *in (5.2), while the third component is found by the use of a standard noise elimination scheme (median filter, see [13]) followed by a truncation with* $\epsilon_3 = 10$. *The original image is stored with approximately 264Kb, while the compressed image is stored using 15.5Kb. The coefficients and their position in the image is again coded using a standard quad-tree coding scheme ([13]). Figure 5.6 indicates the position of the nonzero coefficients in the truncated image. The coefficients of* f_0 *are not shown.*

We must point out that the image compression scheme based on a box spline decomposition is not necessarily optimal. However, the general decomposition strategy presented in this paper has interesting applications in connection with image compression and general signal processing. In the papers [2] and [11] these data compression strategies are adapted to digital images and seismic data, respectively.

References

1. Arge, E., M. Dæhlen, and A. Tveito, Box spline interpolation; a computational study, J. Comp. and Appl. Math., to appear.
2. Arge, E. and M. Dæhlen, Data dependent subdivision and image approximation, preprint, 1991.
3. Arge, E., M. Dæhlen, T. Lyche, and K. Mørken, Constrained spline approximation of functions and data based on constrained knot removal, in *Algorithms for Approximation II*, J. C. Mason and M. G. Cox (eds.), Chapman and Hall, London, 1990, 4–20.
4. de Boor, C., Splines as linear combinations of B-splines: A survey, in *Approximation Theory II*, (G. G. Lorentz, C. K. Chui, and L. L. Schumaker (eds.), Academic Press, New York, 1976, 1–47.
5. Burt, P. J. and E. H. Adelson, The Laplacian pyramid as a compact image code, IEEE Trans. Comm. **31** (1983), 532–540.
6. Chui, C., *An Introduction to Wavelets*, Academic Press, Boston, 1992.

7. Cohen, E., T. Lyche, and R. Riesenfeld, Discrete box-splines and refinement algorithms, Computer Aided Geom. Design **1** (1984), 131–148.
8. DeVore, R.A., Jawerth, B., and Lucier, B., Surface compression, preprint 1991.
9. DeVore, R.A., Jawerth, B., and Lucier, B., Image compression through wavelet transform coding, preprint 1991.
10. Dæhlen, M. and T. Lyche, Box splines and applications, in *Geometric Modeling: Methods and Applications*, D. Roller and H. Hagen (eds.), Springer, 1991, 35–95.
11. Dæhlen, M. and Ø. Hjelle, Compact representation of seismic sections, preprint 1991.
12. Dahmen, W. and C. A. Micchelli, Subdivision algorithms for the generation of box-spline surfaces, Computer Aided Geom. Design **1** (1984), 115–130.
13. Jain, A. K., *Fundamentals of Digital Image Processing*, Prentice Hall, Englewoods, 1989.
14. Lyche, T., Discrete B-splines and conversion problems, in *Computation of Curves and Surfaces*, W. Dahmen, M. Gasca, and C. A. Micchelli, (eds), Kluwer Academic Publishers, Dordrecht, 1990, 117–134.
15. Lyche, T. , K. Mørken, and K. Strøm, Conversion between B-spline bases using the generalized Oslo Algorithm, Preprint 1990-12, Inst. for informatikk, Univ. of Oslo.
16. Lyche, T. and K. Mørken, A data reduction strategy for splines with applications to the approximation of functions and data, IMA J. of Num. Anal. **8** (1988), 185–208.
17. Lyche, T. and K. Mørken, Knot removal for parametric B-spline curves and surfaces, Computer Aided Geom. Design **4** (1987), 217–230.
18. Mallat, S., Multiresolution approximations and wavelet orthonormal bases of $L^2(\mathbb{R})$, Trans. Amer. Math. Soc. **315** (1989), 69–87.
19. Meyer, Y., *Ondelettes et Opérateurs I*, Hermann, Paris, 1990.
20. Pennbaker, W. B., Mitchell, J. L. and G. G. Landon Jr., An overview of the principles of the Q-coder adaptive binary arithmetic coder, IBM J. Res. Development, **32** (1988), 717–727.
21. Welch, T.A., A technique for high performance data compression, IEEE Computers **17** (1984), 8–19.

Morten Dæhlen
Center for Industrial Research
P.O.Box 124, Blindern
0314 Oslo 3, Norway
EMAIL: mortend@ifi.uio.no

Tom Lyche
Inst. for Informatikk, Univ. Oslo
P.O.Box 1080, Blindern
0316 Oslo 3, Norway
EMAIL: tom@ifi.uio.no

A Curve Intersection Algorithm with Processing of Singular Cases: Introduction of a Clipping Technique

Marc Daniel

Abstract. A curve intersection algorithm is presented. A clipping technique and computation of all the singular cases (tangency, curve of intersection) are provided. The clipping assumes a modeling with a local scheme. Singular cases processing is achieved without any assumption and according to numerical considerations. Examples are given. Extension of the algorithm to surfaces is discussed.

§1. Introduction

The study of intersections is a fundamental problem in CAGD. It is of interest for both surfaces and curves. As a matter of fact, steps in 2D are usually included in design processes and boundaries of surfaces are space curves. Curve intersection means curve-hyperplane, curve-curve intersection in 2D or 3D (and possibly curve-line in 3D). Our work concerns parametric curves defined by a control polygon like Bézier or NURBS curves. General intersection methods have been proposed [3,4,6] and compared in [7]. A different approach (clipping) is introduced in [8].

The result of a curve intersection computation is generally a set of points. But pieces of a curve can sometimes be found which correspond to degeneracies. A limit case of the previous one is tangency, where there is one theorical intersection point, but both intersecting elements are in fact numerically identical on a short interval containing the point. Approaches for processing of tangency exist (for example in [8] for curves, and [1,5] for surfaces).

We develop a method suitable for all singular cases encountered in curve intersection. A clipping technique is also provided for reducing parametric definition domains of curves. A more detailed presentation is available in

Mathematical Methods in Computer Aided Geometric Design II

Tom Lyche and Larry L. Schumaker (eds.), pp. 161–170.

ISBN 0-12-460510-9.

[2]. In particular, the problem of tolerances is not entirely covered in this presentation.

The general algorithm is adapted from the one proposed in [4]. Each entity is recursively subdivided until it is well approximated by a line segment (condition of linearity). The segment considered usually joins the two extreme control points. The test of linearity computes distances of every control point to the line, support of the segment. It is checked if all the distances are less than a given tolerance $\epsilon 1$. This is a strong condition because it is applied to the convex hull. The elementary intersection can then be computed. The associated detection and processing of singular cases are discussed in Section 3.

The process is sped up by eliminating pieces of curves using a test of overlap from minmax boxes. If the minmax boxes do not intersect, the corresponding curves do not intersect. We combine this test with clipping which first reduces the curves as described in the next section. Section 4 begins with a discussion of the different choosen orientations, followed by some examples. Finally, extension of this algorithm to surfaces is discussed in Section 5.

§2. Clipping

Clipping assumes a local scheme modeling with convex hull property. So it is suitable for B-splines and NURBS with positive weights. The basic property applied is the following:

Property. *Let a NURBS curve C of order k be defined by its control points P_i, its weights ω_i, $i = 0, \ldots, n$, and its knot vector $(t_i)_{i=0,\ldots,n+k}$. If the weights are positive and if k successive points $P_{i-k+1}, \ldots, P_i$ lie on the same side of a hyperplane, then the curve does not cross the hyperplane on the parametric interval $[t_i, t_{i+1}]$.*

Proof: This follows from the convex hull property and the well-known formula

$$for\ t \in [t_i, t_{i+1}]\,,\, t_i < t_{i+1}, \qquad C(t) \;=\; \frac{\displaystyle\sum_{j=i-k+1}^{i} \omega_j\, P_j\, N_{j,k}(t)}{\displaystyle\sum_{j=i-k+1}^{i} \omega_j\, N_{j,k}(t)} \tag{1}$$

The property is summarized in Figure 1. ■

Curve-hyperplane intersection is an immediate application of this property. Recall that if $F(P) = 0$ is the hyperplane equation, then the sign of $F(P)$ characterizes respective positions of the point P and the hyperplane D. In the case of curve-curve intersection, hyperplanes are introduced by constructing a minmax box for one of the two curves. Oriented minmax box could have been computed, according to the construction given by Sederberg

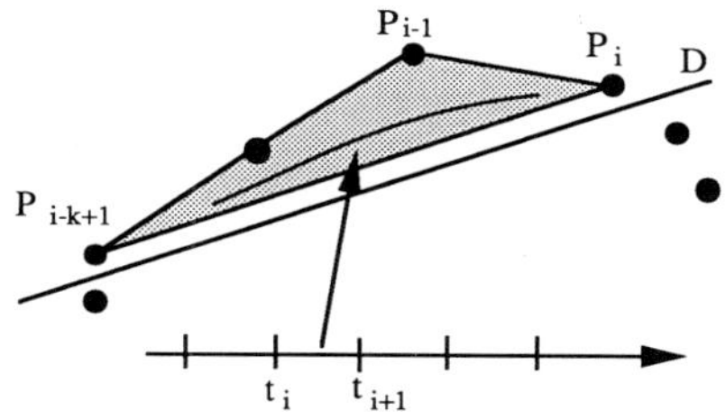

Figure 1. Curve-hyperplane position.

[8] for example. The minmax box is choosen with edges parallel to the reference frame axis. The advantage is that it allows for evident determinations (coordinates comparisons). For plane curves, four lines are defined, separating the plane into nine areas (Figure 2). Clipping succeeds for a parametric interval if the associated piece of curve does not go into the interior of the minmax box (area 5). To ensure this property, it is enough that the corresponding control points never cross one of the lines, which depends on the position of the initial control point studied. For example, if the first point lies in area 1, two possibilities are available: the other points could lie in the areas 1, 2 or 3, or in the areas 1, 4 or 7.

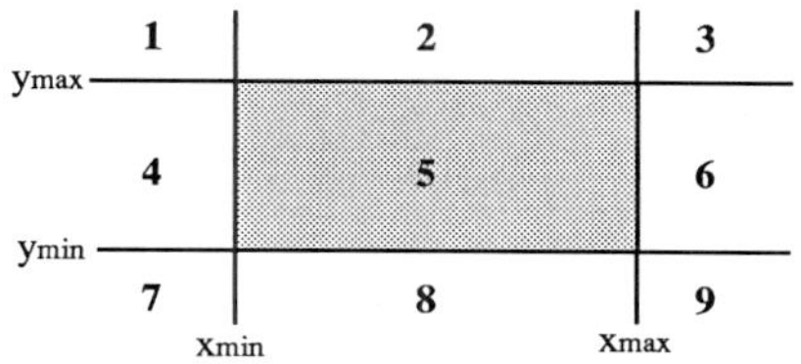

Figure 2. Definition of line and areas using a minmax box.

The method for space curves is the transposition of the previous one: the last box coordinates provides a way for splitting the space into three parts (below the minimum, between the minimum and the maximum, above the maximum). Nine areas are defined on each part. Compared to the plane curves study, there are only some more cases to handle.

Clipping is only acheived at the ends of the curves. The clipping process thus transforms each curve into a single curve, according to the following steps (given for curve-curve intersection):

1.) minmax box 1 determination
2.) curve 2 clipping using box 1

3.) curve 2 reduction
4.) minmax box 2 determination
5.) curve 1 clipping using box 2
6.) curve 1 reduction

Figure 3 illustrates step 2) of the process. Curve 2 (of order 4) could be reduced on $[t_3, t_4]$ and $[t_n, t_{n+1}]$. Curve reduction is obtained with knots insertion algorithms.

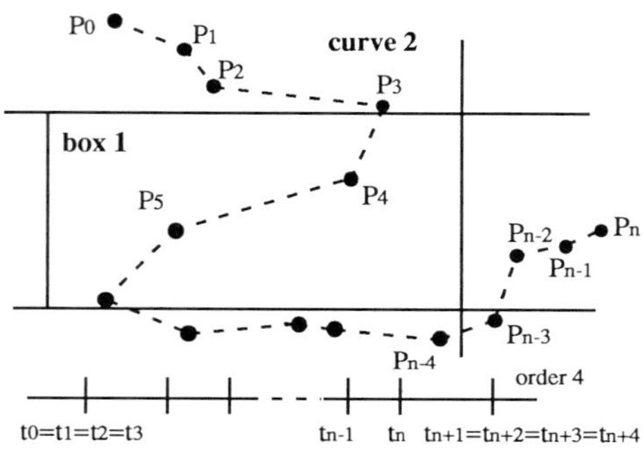

Figure 3. Clipping for curve-curve intersection.

§3. Singular Cases

A numerical approach is used for detection and processing of singular cases. It assumes that each curve verifies the condition of linearity described in Section 1.

For curve-hyperplane intersection, a singular case is obtained if the two extreme control points are at distances to the hyperplane lesser than a given tolerance $\epsilon 2$. The result of this elementary intersection is the entire curve. This corresponds to a line segment in 2D, or to a plane curve in 3D.

The case of curve-curve intersection has to be more detailed: let P_0 and P_n, Q_0 and Q_m, the extreme points of the two curves, $d(P, AB)$ the distance of a point P to the line defined by two points A and B. A singular case is obtained if one of the following two sets of conditions is satisfied:

$$d(P_0, Q_0Q_m) \leq \epsilon 2 \quad \textit{and} \quad d(P_n, Q_0Q_m) \leq \epsilon 2 \tag{2}$$

$$\textit{or} \quad d(Q_0, P_0P_n) \leq \epsilon 2 \quad \textit{and} \quad d(Q_m, P_0P_n) \leq \epsilon 2. \tag{3}$$

Condition (2) holds in the example shown in Figure 4. It is then necessary to compute the cartesian and parametric limits of the intersection curve. The two curves lie on the same line. In fact, three cases can occur, detected with "well-choosen" tolerances.

1.) The result is a curve segment, which corresponds to one of the two curves.
2.) The curves have the same support but do not intersect.
3.) the intersection is reduced to one point.

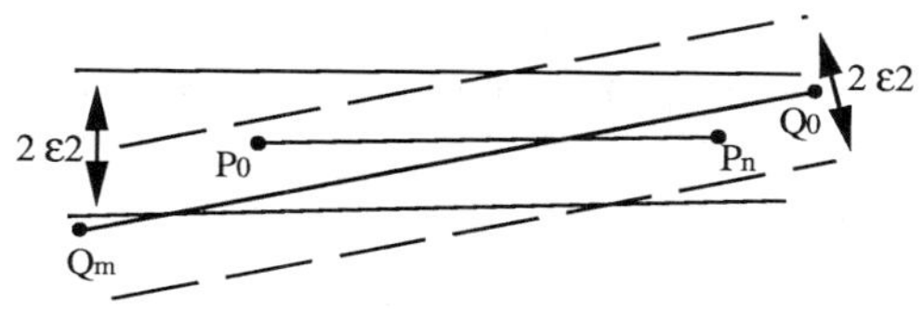

Figure 4. A singular case in curve-curve intersection.

The final step is storage. It is necessary to classify all the elementary solutions and connect them (the extreme common points are identical according to a tolerance) to obtain the complete solution. In addition some undesired points must be removed. It most often occurs in curve-hyperplane intersection: the extreme points of a singular case are found like normal intersection points during the studies of intersection with the neighbouring pieces of curve, as illustrated in Figure 5.

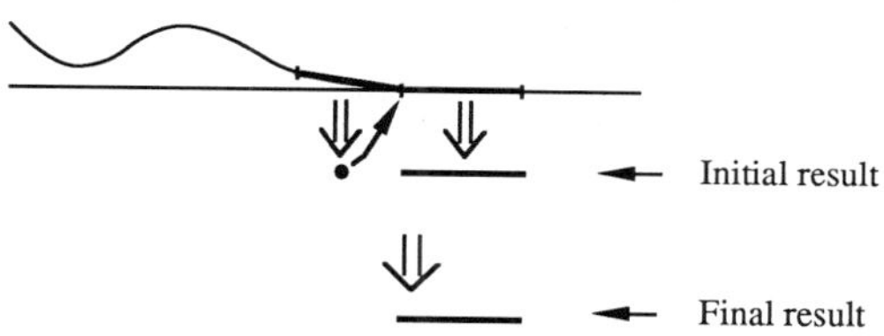

Figure 5. Undesired point removal in curve-line intersection.

The main difficulty, which is included in the general problem of tolerances in intersection problems, is the choice of $\epsilon 2$. It is evidently closely related to the tolerance ϵ of the result, and also to the tolerance $\epsilon 1$ (test of linearity) which is directly dependent on ϵ. In addition, accuracies (number of significant digits) of the numerical quantities manipulated are important. In fact, in curve-curve intersection, distance computations depend on the two control points defining the line. So, accuracy is influenced by the distance between the two points. Below a minimum length, the line cannot be considered as well-defined, which must be taken into account for $\epsilon 2$. This last point is very important for space curves because the test of linearity is verified for very small pieces of curve, due to torsion twist.

Our method detects any degeneracy and thus a contact of any order, but without giving its order. If there is a particular need for it, it has to be found by other means, e.g. by studying the successive hodographs.

§4. Examples

Preliminary discussion. Several points concerning this algorithm may be discussed. A first aspect concerns implementation choices. We decided to use classical minmax boxes which are not always good surrounding areas. In these cases oriented minmax boxes could be more efficient, but involve more computing time for box determination and clipping. In very particular cases, considering the convex hull could *perhaps* save time. For example, it allows for extending the clipping to the interval $[t_{n-1}, t_n]$ in the problem shown in Figure 3. In addition, clipping would have to be applied to all the possible intervals, after modifications of the algorithm. For the the same example, clipping is possible on $[t_{n-4}, t_{n-2}]$. Interest is conditional upon place and number of intersections. In fact, a response to these interrogations needs a very fine and difficult evaluation. *The conclusion certainly depends on cases, and may vary for different classes of problems.*

A second aspect we would like to comment on is the method for processing singular cases. As in every numerical method, a failure is always possible in a test. We encountered this a few times in our numerous trials. A careful study has shown that we were just at the limit value of the tolerance $\epsilon 2$, which demonstrates the sensitivity of the problem. It involves a small hole in the singular case definition, which is not in fact very important. The method could also produce a slight inaccuracy at the ends of singular cases. Elementary curves are sometimes better appoximated by two segments rather that one. Since the process considers a single segment, it follows that a very small piece of curve could be found in addition, or could be lost, as illustrated in Figure 6, which must be considered as a very important zoom. Anyway, the degeneracies are always sufficiently well surrounded.

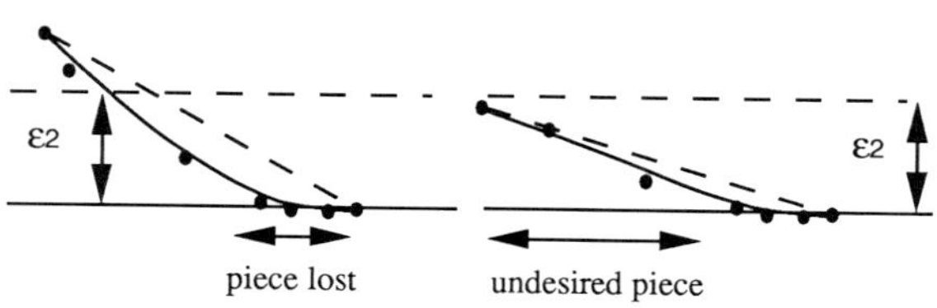

Figure 6. Possible inaccuracy in singular case limit definition.

The implemented version of the algorithm provides good results, as shown in the following three examples. We work with 32 bit floating point arithmetic. So about five significant digits could reasonably be obtained.

Example 1. Intersection between a line and a B-spline curve of order 4 is studied. Intersection points and ends of the singular case (shown with a continuous line) are marked by crosses in Figure 7. A zoom of the degeneracy is also provided in Figure 8 for two different accuracies on the results (10^{-4} and 10^{-5}). It is possible to verify that only small variations in the end points (on very flat areas) exist.

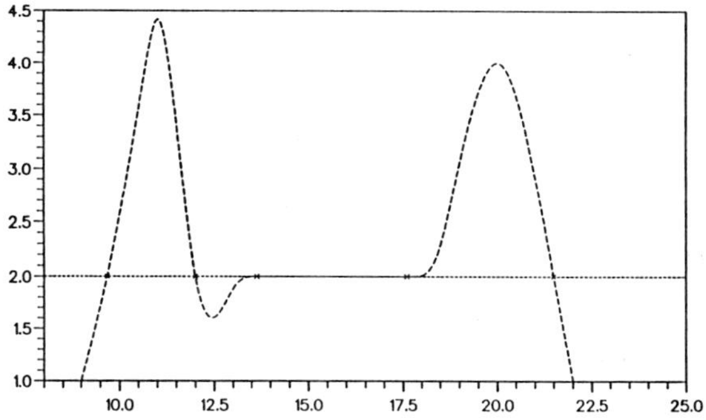

Figure 7. Example 1: curve-line intersection with a singular case.

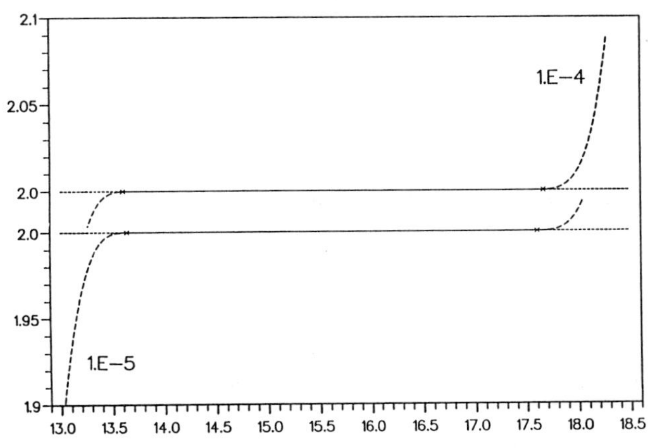

Figure 8. Example 1: zoom of the singular case computed with two different accuracies (10^{-4} and 10^{-5}).

Example 2. The previous curve is again considered. It is defined by 12 control points. The parametric definition domain is thus [0, 9] (uniform parametrization). Self intersection is computed with the previous accuracies. Two tests have been conducted: the curves are described with the same orientation or with the opposite orientations. In all cases, the program finds a single curve segment beginning at parametric value $t = 0$, and ending at $t = 9$. Computing this entire curve of intersection evidently needs more time than computing the results for the other two examples given here.

Example 3. The tangency between two Bézier symetric curves is studied. The exact parametric value for the tangency is $t = 0.5$ for both curves. The abscissa of this point is 2.5. The program finds, with a 10^{-4} accuracy, a singular case which corresponds to the parametric interval [0.4922, 0.5078]. The abscissa variation is $2.477 \leq x \leq 2.523$. No significant change exists on the ordinates. The curve line solution is shown as a continuous line in Figure 9, where a zoom of the tangency area is also plotted. Notice that there is an additional constraint in this example: the tangency lies exactly at the ends of two elementary curves because the first subdivision takes place at this point. This forces the degeneracy to span the two elementary curve segments.

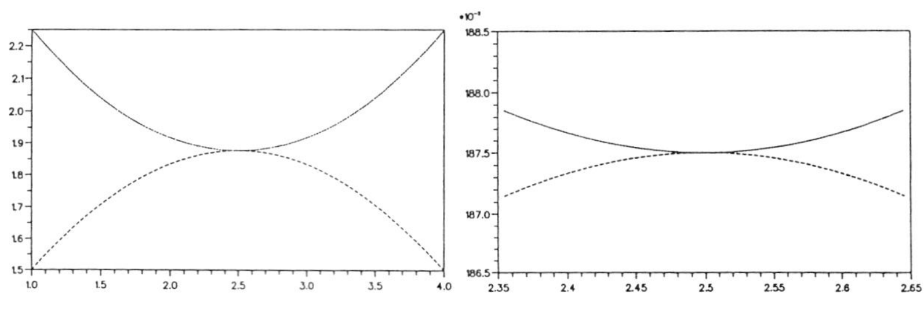

Figure 9. Example 3: tangency between two Bézier curves.

§5. Extension to Surfaces

Clipping and singular cases processing could be introduced as for curves. Note that the general result for surface-intersection is a set of curves and the degeneracies: a set of points and pieces of surfaces.

Elimination of an elementary parametric domain depends on a set of $ku \times kv$ control points, where ku and kv are the two surface orders. The requirement that we get a single surface after clipping is a strong global constraint because it forces us to consider only rows and columns of control points. On the other hand, a local clipping provides several surfaces defined after many expensive (in number of operations) subdivisions, which seems not to be a good solution for this algorithm.

Detection and determination of degeneracies is an extension of the described method for curves. But storage is *the real problem*. In surface-plane intersection, the final result, collation of elementary parametric rectangular domains does not correspond with a rectangular domain. So different links between elementary domains may be considered. In surface-surface intersection each elementary singular case is a polygon. Assembling all these pieces of a puzzle provides the global result. The boundaries of this domain (in space or parametric plane) are curves which have to be determined. Finally, the degeneracies could be detected with such a method (which is very important), but the boundaries, if they are desired, must be obtained differently, using a marching method for example. In this case, starting points are easily found, and contouring is made easier by the approximate knowledge of the interior area.

§6. Conclusion

The general algorithm proposed is reliable for curve-curve intersection. Singular cases are computed with good accuracy. Their processing really increases the algorithm's complexity, but does not slow down the normal case process. Clipping is very efficient for reducing the definition domains of curves if possible. The algorithm could be extended to surfaces, but the storage for singular cases requires particular care.

Acknowledgments. This work has been supported in part by IFREMER (French Institute of Sea Research and Development).

References

1. Barnhill, R. E., and S. N. Kersey, A marching method for parametric surface/surface intersection, Computer Aided Geom. Design **7** (1990), 257–280.
2. Daniel, M., Modélisation de courbes et surfaces par des B-splines. Application à la conception et à la visualisation de formes, Thèse, Université de Nantes, 1989.
3. Koparkar, P. A., and S. P. Mudur, A new class of algorithms for the processing of parametric curves, Comput. Aided Design , **15** (1983), 41–45.
4. Lane, J. M., and R. F. Riesenfeld, A theorical development for the computer generation and display of piecewise polynomial surfaces, IEEE Trans. on P. A. M. I. **2** (1980), 35–46.
5. Markot, R. P., and R. L. Magedson, Solutions of tangential surface and curve intersections, Comput. Aided Design **21** (1989), 421–429.
6. Sederberg, T. W., Implicit and parametric curves and surfaces for computer aided geometric design, PhD, Purdue University, 1983.

7. Sederberg, T. W., and S. R. Patty, Comparison of three curve intersection algorithms, Comput. Aided Design **18** (1986), 58–63.
8. Sederberg, T. W., and T. Nishita, Curve intersection using Bézier clipping, Comput. Aided Design **22** (1990), 538–549.

Marc Daniel
Ecole Centrale de Nantes
1 rue de la Noë
44072 Nantes Cedex 03 FRANCE
daniel @cc01.ensm-nantes.fr

Best Approximations of Parametric Curves by Splines

W. L. F. Degen

Abstract. A normal distance $d_N(C, C')$ for pairs of plane parametric curves C, C' (with given C and some restictions for C') is introduced. For certain classes of Bézier curves $\mathcal{B}'$ we get a differentiable manifold of deviation functions and can apply the theory of non-linear approximation. As our main result, we obtain that the error of the best approximant $C'_0 \in \mathcal{B}'$ has the alternant property as in the case of Chebyshev approximation. An algorithm to calculate C'_0 explicitly and some examples are included.

§1. Introduction

In CAGD the question of how to approximate a given curve within a certain tolerance by a spline of a prescribed type arises frequently, and various error estimates have been developed ([de Boor, Höllig and Sabin '87],[Hoskins and Ponzo '74], [Hanna, Evans and Schweitzer '86],[Sakai and Usmani '90]). In particular, it may be the task to approximately represent the *offset* of a spline which is, in general, not a spline, or to *convert* a given spline of another kind, say of some higher degree or a rational one, into one of a more restricted class at our disposal (see [Dannenberg and Novacki '85], [Hoschek '87]).

In this paper, we concentrate on Bézier curves as approximants; however we neither cover the problems of what number of segments are necessary, nor how the subdivision points have to be distributed. Rather we focus on the subsequent problem: how to approximate a single segment as well as possible.

At that stage the need arises to measure the rate of approximation. It would be tempting to use the well known Hausdorff distance for this purpose. However, being based only on the point sets, it does not reflect the parameter distributions of the two curves C, C'. Therefore, we introduce the *normal distance* $d_N(C, C')$ defined for a certain class $\mathcal{A}(C)$ of "admissible" curves for

Mathematical Methods in Computer Aided Geometric Design II

Tom Lyche and Larry L. Schumaker (eds.), pp. 171–184.

ISBN 0-12-460510-9.

C (Section 2.3). This notion not only overcomes the disadvantages of the Hausdorff distance $d_H(C, C')$, but provides (besides much better convergence properties) simultanously a *deviation function* $\rho : [a, b] \to \mathbb{R}$ for each $C' \in \mathcal{A}(C)$, connected with $d_N(C, C')$ by $d_N(C, C') = \| \rho \|$ (maximum norm). We will show that the normal distance is related to the Hausdorff distance by the inequality $d_H(C, C') \leq d_N(C, C')$ (Proposition 3) and equality holds under an additional condition (Theorem 1).

Denoting by $\mathcal{B}$ a fixed set of Bézier curves with fixed degree n and specified end conditions, and restricting our attention to admissible ones ($\mathcal{B}' := \mathcal{B} \cap \mathcal{A}(C)$), we get a set $\mathcal{M}$ of corresponding deviation functions. We will show (Theorem 2) that $\mathcal{M}$ is a differentiable manifold of dimension m (depending on $\mathcal{B}$) and that each $\rho \in \mathcal{M}$ is differentiable.

By this approach, the approximation problem for parametric curves is reduced to one for functions: to select $\rho \in \mathcal{M}$ such that $\| \rho \|$ is minimal. Now the theory of non-linear approximation, which has made great progress in the last decades (see e.g. [Braess '86]), in particular the theory of [Meinardus and Schwedt '64], can be applied. Thus we obtain the result (Theorem 4) that the deviation function ρ of the best approximation behaves as in the case of classical Chebyshev approximation, expressed by his famous alternant theorem.

So it remains to prove the assumptions of that theory: the global and the local Haar conditions. As to the former, we give the proof for some special classes $\mathcal{B}$. The general case would be implied by a conjecture on Bézier curves. E. Eisele ['91], an assistant of mine, succeeded in proving the general global Haar condition (at least for some neighborhood).The local Haar condition is not so difficult, so a proof is included in this paper.

In the final paragraph we will develop an algorithm to compute explicitly the best approximant. It is based directly on the alternant property written as a non-linear system of m equations which will be solved by a quasi Newton method (replacing the Jacobian by a finite difference approximation). Though this method requires some computational effort, it yields highly accurate results. We give some examples to confirm the theory, and to show that the algorithm is suitable for practical computations.

§2. Distances in the Set of Plane Parametric Curves

2.1. Differentiable Parametric Curves in the Euclidean Plane

We denote by $\mathcal{C}^1$ the set of all compact parametric differentiable curves in the Euclidean plane $\mathbf{E}^2$. As usual in differential geometry (see e.g. [Do Carmo '86]), a *curve* $C \in \mathcal{C}^1$ is an equivalence class of parametric representations $\mathbf{x} : [a, b] \to \mathbb{R}^2$. Here $\mathbf{x}$ denotes a differentiable mapping of some compact real interval $[a, b]$ $(a < b)$ into $\mathbb{R}^2$ (with tangent vector $\mathbf{x}'$ nowhere vanishing); furthermore, two such parametric representations are called *equivalent* (giving rise to the same curve C) if there is a diffeomorphism $\phi : [a, b] \to [c, d]$ between its parameter intervals (with $\phi' > 0$ anywhere), called *parameter transformations*. A *geometric object or property*, described in terms of a special parametric representation, must be invariant under parameter transformations.

2.2. The Hausdorff Distance

In order to supply the family of all compact sets of an Euclidean space $\mathbf{E}^n$ with a metric making this family of point sets into a metric space, Hausdorff ['14] introduced the following definition of a *distance* between two compact sets $A, B \subset \mathbf{E}^n$, which is now called the *Hausdorff distance* :

$$d(A,B) := \max\Big(\max_{p\in A}(\min_{q\in B} d(p,q)), \quad \max_{q\in B}(\min_{p\in A} d(p,q))\Big), \tag{2.1}$$

where $d(p,q)$ denotes the Euclidean distance between the points p and q. This notion has a great importance in the theory of convex sets which is based on Blaschke's famous selection theorem ['14] saying that the family of compact, convex sets in $\mathbf{E}^n$ is sequentially compact.

But it can serve as a metric in $\mathcal{C}^1$ as well, taking for A and B the point sets $\{C\}$, $\{D\}$ of two parametric curves; it may easily be expressed by two parametric representations for C, D in the form

$$\begin{aligned} d_H(C,D) = \max\Big(&\max_{t\in[a,b]} \quad \min_{s\in[c,d]} \parallel \mathbf{x}(t) - \mathbf{y}(s) \parallel, \\ &\max_{s\in[c,d]} \quad \min_{t\in[a,b]} \parallel \mathbf{x}(t) - \mathbf{y}(s) \parallel\Big). \end{aligned} \tag{2.2}$$

Nevertheless, this Hausdorff distance is not an appropriate means to deal with parametric curves because it does not reflect the parameter distribution nor the orientation. Furthermore, there is no convergence property similar to Blaschke's: It is easy to construct a sequence $(C_n)_{n\in\mathbb{N}}$, $C_n \in \mathcal{C}^1$, such that $\lim_{n\to\infty} d_H(C_n, C) = 0$ but, for no reparametrization ϕ, $\mathbf{x}_n(t)$ converges pointwise to $\mathbf{x}(\phi(t))$.

2.3. The Normal Field and Admissible Curves

To avoid the disadvantages of the Hausdorff distance, we suggest another approach which is more appropriate for approximating parametric curves. It is based on two main ideas: First, we start from a *fixed* curve C, given by any of its parametric representations $\mathbf{x} : [a,b] \to \mathbb{R}^2$, to be approximated. Second, we remove from a certain (topological) neighbourhood of C those curves that are "undesired"; calling the remaining ones "admissible" and denoting the set of all admissible curves with respect to C by $\mathcal{A}(C)$, it is sufficient to define the distance $d(C, C')$ for all $C' \in \mathcal{A}(C)$. To give an idea of what we called "undesired", we mention curves $\tilde{C}$, lying in a neighbourhood of C (even very close to C), but going back- and forwards, the case of loops included.

To be more precise, we assume C is at least of class C^2. Then we associate with each point $\mathbf{x}(t) \in C$ that part of the *normal* $\{\mathbf{x}(t) + \rho\mathbf{n}(t) \mid \rho \in \mathbb{R}\}$ for which $\rho\kappa(t) < 1$, (ρ unrestricted if $\kappa(t) = 0$), $\kappa(t)$ being the curvature of C and $\mathbf{n}(t)$ the normal unit vector (with orientation such that $\det(\mathbf{x}'(t), \mathbf{n}(t)) > 0$).

Then the region swept out by these parts of the normals (in general half rays, for inflection points full lines) is called the *normal field* of C and will be denoted by $\mathcal{N}(C)$.

From the Frenet equations one derives immediately that the mapping $(t, \rho) \mapsto \mathbf{x}(t) + \rho\mathbf{n}(t)$ from $\{(t, \rho) \in \mathbb{R}^2 \mid t \in [a, b], \rho\kappa(t) < 1\}$ onto $\mathcal{N}(C)$ is locally one-to-one, i.e., this mapping is a regular parametrization of $\mathcal{N}(C)$.

Now we can explain the precise meaning of "*admissible*":

Definition 1. *A curve $C' \in \mathcal{C}^1$ is called admissible with respect to C if the following conditions hold (C' represented by $\mathbf{y} : [c, d] \to \mathbb{R}^2$:*

1. *For each $s \in [c, d]$ the curve point $\mathbf{y}(s)$ is contained in $\mathcal{N}(C)$; i.e., there exist real values $t \in [a, b]$ and ρ, such that*

$$\mathbf{y}(s) = \mathbf{x}(t) + \rho\mathbf{n}(t) \tag{2.3}$$

$$\rho\kappa(t) < 1. \tag{2.4}$$

2. *For each $t \in [a, b]$, there is exactly one $s \in [c, d]$ such that (2.3) is valid.*
3. *For corresponding values s, t in (2.3), the tangent vector $\dot{\mathbf{y}}(s)$ is not parallel to $\mathbf{n}(t)$.*
4. *The endpoints of C and C' are the same, i.e., $\mathbf{x}(a) = \mathbf{y}(c), \mathbf{x}(b) = \mathbf{y}(d)$.*

Denoting by $\mathcal{B}$ a given set of Bézier curves with fixed degree n and end conditions specified, our approximation problem consists of selecting from $\mathcal{B}' := \mathcal{A}(C) \cap \mathcal{B}$ that one (if it exists), say C_0 , for which $d(C, C_0) = \inf\{d(C, C') \mid C' \in \mathcal{B}'\}$.

The admissibility conditions have further consequences which we will investigate now. Obviously, by (1) and (2), the assignment $t \mapsto s$ defines a *function* $\sigma : [a, b] \to [c, d]$, $s = \sigma(t)$ in (2.3). Since $\mathbf{y}(s)$ is unique and $\mathbf{n}(t) \neq \vec{0}$, ρ is unique too. Thus, there is a second function $\rho : [a, b] \to \mathbb{R}$ (with values $\rho(t)$ such that $\rho(t) \cdot \kappa(t) < 1$)

$$\mathbf{y} \circ \sigma = \mathbf{x} + \rho \cdot \mathbf{n}. \tag{2.5}$$

On the other hand, the values s and ρ can be thought of as the solution of (2.3) as a system of two equations (nonlinear with respect to s) for the unknowns s, ρ. Its functional determinant is $\det(\dot{\mathbf{y}}(s), \mathbf{n}(t))$, (using a dot to indicate derivation with respect to s) hence, by (3), not vanishing. Then the implicit function theorem implies that σ, ρ are *differentiable*. Furthermore, by differentiation of (2.5), we obtain

$$\dot{\mathbf{y}}\sigma' = (1 - \rho\kappa)\mathbf{x}' + \rho'\mathbf{n}. \tag{2.6}$$

By (2.4) the vector at the right hand side of (2.6) does not vanish ($\mathbf{x}', \mathbf{n}$ being linearily independent) and we conclude $\sigma'(t) \neq 0$ for any $t \in [a, b]$. Since by (4), $\sigma(a) = c$, $\sigma(b) = d$, we have $\sigma'(t) > 0$ for all $t \in [a, b]$. Hence σ is a *diffeomorphism*, and therefore (2.5) is an equivalent representation for C' . Thus we have proved

Proposition 1. *For any admissible curve C' with respect to a given curve C (of class C^2), there exist two differentiable functions $\sigma, \rho : [a,b] \to \mathbb{R}$ such that σ is an orientation preserving diffeomorphism and (2.5) is an equivalent representation for C'.*

2.4. The Deviation Function and the Normal Distance

The principal idea of the representation (2.5) can now be seen at once. The function ρ directly measures the distance of corresponding points of C and C':

$$| \rho(t) | = \| \mathbf{y}(\sigma(t)) - \mathbf{x}(t) \|, \; t \in [a,b]. \tag{2.7}$$

So ρ can serve as an error function for the approximation and we can define:

Definition 2. *For any admissible curve C', we call ρ the deviation function of C and*

$$d_N(C, C') := \max_{t \in [a,b]} | \rho(t) | \tag{2.8}$$

the normal distance of C' from C.

These notions of admissible curves and its normal distances do much more than only to exclude some "undesired" aspirants for approximation. From (2.7) one derives immediately:

Proposition 2. *Let $(C_n)_{n \in \mathbb{N}}$ be a sequence of curves, admissible to C, such that $\lim_{n\to\infty} d_N(C, C_n) = 0$. Then C_n converges uniformly to C. Furthermore, this convergence respects the parameter distribution, i.e., for each C_n there is a reparametrization $\sigma_n : [a,b] \to [c_n, d_n]$ such that the point $\mathbf{y}_n(\sigma_n(t))$ converges to $\mathbf{x}(t)$ for each $t \in [a,b]$.*

2.5. Relations Between the Normal and the Hausdorff Distances

Let C, C' be two different curves in $\mathcal{C}^1$, C of class C^2 and C' admissible to C. We want to compare the Hausdorff distance with our normal distance. Let t_0, s_0 be that pair of parameter values where $d_H(C, C')$ is attained. Certainly $d_H(C, C') > 0$, because otherwise to each point $\mathbf{x}(t)$ of C there would exist a point $\mathbf{y}(s)$ coincident with $\mathbf{x}(t)$; this implies $s = \sigma(t)$, $\rho(t) = 0$ because of the uniqueness in (2.5); hence $d_N(C, C') = 0$ and $C = C'$ in contrary to the assumption. Observing (4), we conclude that the extremal distance will be attained at an *interior* point of each curve. Differentiating the distance function $F(s,t) := \| \mathbf{x}(t) - \mathbf{y}(s) \|$, we obtain that $\mathbf{x}'(t_0)$ and $\dot{\mathbf{y}}(s_0)$ both are *orthogonal* to $\mathbf{x}(t_0) - \mathbf{y}(s_0)$. Therefore $\mathbf{y}(s_0)$ lies on the normal of C at t_0, and hence $s_0 = \sigma(t_0)$, which implies $d_H(C, C') = \| \mathbf{x}(t_0) - \mathbf{y}(\sigma(t_0)) \| = | \rho(t_0) | \leq d_N(C, C')$. Thus we have proved

Proposition 3. *For any pair of curves C, C', the first one being of class C^2 and the second admissible to C, the following inequality holds*

$$d_H(C, C') \leq d_N(C, C'). \tag{2.9}$$

Some simple counterexamples show that equality in (2.9) does not hold in general. The reason is that the normals of C may cover some regions of $\mathcal{N}(C)$ more than once. We exclude this possibility by the following

Definition 3. *C is said to have a non-overlapping normal field if each of its points is lying on exactly one normal of C (in this case, the parametrization (t, ρ), described in Section 2.3 is one-to-one).*

With this additional assumption, we can indeed prove the equality of the two distances:

Theorem 1. *For any pair C, C' of curves, C being of class C^2 and having a non-overlapping normal field and C' admissible for C, the normal distance is equal to the Hausdorff distance.*

$$d_N(C, C') = d_H(C, C'). \tag{2.10}$$

Proof: Without loss of generality, we can assume that the parameter t in our representation $t \mapsto x(t)$, $t \in [a, b]$ is the arc length and consequently the Frenet formulas are valid.

We start again with that value $t_1 \in [a, b]$ where $\rho(t_1) = d_N(C, C')$ is attained. As in the previous case, t_1 is in the interior of $[a, b]$, hence $\rho' = 0$, which implies by (2.6) that the tangent vectors of C and C' at $\mathbf{x}(t_1)$ and $\mathbf{y}(s_1)$ (with $s_1 := \sigma(t_1)$) are parallel, thus the joining line being a common normal of C and C'.

Now we consider the circle B around $\mathbf{y}(s_1)$ with radius $r_1 := \rho(t_1) = d_N(C, C')$ and investigate the distance function $r(t) := \| \mathbf{x}(t) - \mathbf{y}(s_1) \|$ of $\mathbf{x}(t)$ (running on C) from the middle point $\mathbf{y}(s_1)$ of B. From the Frenet formulas one derives $r(t_1) = r_1$, $r'(t_1) = 0$, $r''(t_1) = (1 - \kappa(t_1) r_1)/r_1$. In particular, by (2.4), $r''(t_1)$ is positive. Hence, these equations have the geometric meaning that C is tangent to the circle B at the point $\mathbf{x}(t_1)$ and runs outside of B, both for $t < t_1$ and $t > t_1$ in some neighbourhood.

Indeed, if $d_H(C, C')$ were less than $d_N(C, C')$, then by (2.2), there would exist a point $\mathbf{x}(t_2)$ closer to $\mathbf{y}(s_1)$ than $\mathbf{x}(t_1)$, and therefore a point $\mathbf{x}(t_3)$ with t_3 between t_1 and t_2, where C enters into B. But then there would exist t_4 between t_1 and t_3 where $r(t)$ attains its *maximum* $> r_1$. At that point the tangent to C would be parallel to the circle's tangent at the corresponding point (on the line from $\mathbf{y}(s_1)$ to $\mathbf{x}(t_4)$). This implies that the normal of C at $\mathbf{x}(t_4)$ would pass through $\mathbf{y}(s_1)$, which contradicts the assumption contained in Definition 3 because we had two different values t_1, t_4 where the normals of C would pass through the same point $\mathbf{y}(s_1)$ of C'. ■

§3. Non-linear Approximation Theory for Bézier Curves

3.1. The Differential Manifold of Deviation Functions

We start with the set $\tilde{\mathcal{B}}$ of regular Bézier curves $\tilde{\mathcal{B}} \subset \mathcal{C}^1$ of some fixed degree n, for which n is the minimal algebraic degree. Then, for each $C' \in \tilde{\mathcal{B}}$, we have a *unique* parameter representation

$$\mathbf{y}(s) = \sum_{k=0}^{n} B_k^n(s)\mathbf{b}_k, \quad s \in [0,1] \tag{3.1}$$

with the Bernstein polynomials B_k^n of degree n and a sequence of control points $\mathbf{b}_0, ..., \mathbf{b}_n$. Expressing the latter as a vector $\mathbf{b} \in \mathbb{R}^{2n+2}$, and since the degenerate cases are excluded by requiring certain inequalities, we get an open subset $\tilde{D} \subset \mathbb{R}^{2n+2}$ and a bijective mapping $\tilde{\Lambda} : \tilde{D} \to \tilde{\mathcal{B}}$, assigning to each $\mathbf{b} \in \tilde{D}$ the corresponding $C'_{\mathbf{b}} \in \tilde{\mathcal{B}}$ given by (3.1). Thus we have a parametrization which induces a *differentiable structure* on $\tilde{\mathcal{B}}$.

Next we impose some specified end conditions, which must include in each case (4) of Definition 1, i.e.,

$$\mathbf{b}_0 = \mathbf{x}(a), \quad \mathbf{b}_n = \mathbf{x}(b). \tag{3.2}$$

There may be further conditions to be required, for example to guarantee a certain order k of contact with C at the end points. We assume that the end conditions reduce the number of free parameters to m with $1 \le m \le 2n-2$, and that there is an open subset $D \subset \mathbb{R}^m$ together with an injective and differentiable mapping $\psi : D \to \tilde{D}$ such that these end conditions are satisfied if and only if $\mathbf{b} = \psi(\mathbf{p})$ with some $\mathbf{p} \in D$. Composing the two mappings and denoting $\Lambda := \tilde{\Lambda}\Big|_{\psi(D)} \circ \psi$, $\mathcal{B} := \Lambda(D)$ we get a differentiable structure $\Lambda : D \to \mathcal{B}$ on our set of regular Bézier curves of minimal degree n with specified end conditions.

Examples. If no further end conditions beyond (3.2) are given, then, of course, we have $m = 2n-2$ and the $2n-2$ coordinates of $\mathbf{b}_1, ..., \mathbf{b}_{n-1}$ themselves can serve as parameters. The regularity condition can be obtained by $\operatorname{res}(\dot{y}_1(s), \dot{y}_2(s)) \neq 0$ (res being the resultant of the two coordinate polynomials). Thus, for $n = 2$, we get $D = \{\mathbf{b}_1 \in \mathbb{R}^2 \mid \det(\mathbf{b}_1 - \mathbf{b}_0, \mathbf{b}_2 - \mathbf{b}_1) \neq 0\}$. In the case of tangent cubics we have $n = 3, m = 2$; $\mathbf{b}_0, \mathbf{b}_3$ by (3.2), $\mathbf{b}_1 = \mathbf{x}(a) + p_1\mathbf{x}'(a)$, $\mathbf{b}_2 = \mathbf{x}(b) - p_2\mathbf{x}'(b)$; $D = \{\mathbf{p} = (p_1, p_2) \in \mathbb{R}^2 \mid p_1 > 0, p_2 > 0, \operatorname{res}(\dot{y}_1(s), \dot{y}_2(s)) \neq 0\}$.

Now, we take the admissibility conditions (see Definition 1) into account. Since (4) follows from (3.2) and (1), and since (3) is given by $\det(\dot{\mathbf{y}}(s), \mathbf{n}(t)) \neq 0$, the vectors $\mathbf{b}$ satisfying (1), (3) and (3.2) build up an open subset of $\mathbb{R}^{2n+2}$. Thus it remains to show that the same is true for condition (2). If, for an admissible C', in each of its neighbourhoods there would exist C'' not satisfying (2), we would have at least for one $t_1 \in [a,b]$ a branching from exactly one s

for C' to more than one s for C'' in (2.3); but this would imply that C' is tangent to the normal in contradiction to (3). Thus the restriction $\mathcal{B}' = \mathcal{B} \cap \mathcal{A}(C)$ is equivalent to the restriction of $\wedge$ to an open subset $M \subset D$:

$$\wedge' := \wedge \big|_M \quad \wedge' : M \to \mathcal{B}'. \tag{3.4}$$

Thus, the differentiable structure (and the dimension m) are maintained.

Using again the implicit function theorem for the solutions σ, ρ of (2.5) not only for one single C' (as in the proof of proposition 1) but now for the *entire set* $\mathcal{B}'$ *at once*, we get differentiable functions $\hat{\rho}, \hat{\sigma} : M \times [a, b] \to \mathbb{R}$ whose restrictions $\rho_{\mathbf{p}}, \sigma_{\mathbf{p}}$ to $\{\mathbf{p}\} \times [a, b]$ for just one $\mathbf{p} \in M$ are the deviation and reparametrization functions, respectively, for the corresponding Bézier curve $C'_{\mathbf{p}} = \wedge'(\mathbf{p}) \in \mathcal{B}'$. We can summarize these results as follows:

Theorem 2. *Each set $\mathcal{B}'$ of admissible Bézier curves as described above is equipped with a differentiable structure (3.4). Furthermore, there is a global differentiable function $\hat{\rho} : M \times [a, b] \to \mathbb{R}$ such that the mapping*

$$\Gamma : M \to \mathcal{M}, \tag{3.5}$$

assigning to each $\mathbf{p} \in M$ the deviation function $\rho_{\mathbf{p}}$ of the corresponding Bézier curve $C'_{\mathbf{p}}$, is the restriction of $\hat{\rho}$ to $\{\mathbf{p}\} \times [a, b]$; therefore $\mathcal{M}$ is a differentiable manifold (with just one chart (3.5)).

3.2. The Application of the Theory of Meinardus and Schwedt

The theory of Meinardus and Schwedt is concerned with exactly the situation which we found in the previous section: Given $\hat{\rho} : M \times [a, b] \to \mathbb{R}$ as in Theorem 2 and $f \in C[a, b]$; find $\rho_{\mathbf{p_o}} \in \mathcal{M}$ such that $\| \rho_{\mathbf{p_o}} - f \| = inf\{\| \rho_{\mathbf{p}} - f \| \mid \mathbf{p} \in \mathcal{M}\}$. Of course, in our case, f is zero.

Definition 4.

(a) *$\mathcal{M}$ is said to satisfy "the local Haar property at $\mathbf{p}$" if the linear tangent space $T_{\mathbf{p}}\mathcal{M}$ satisfies the classical Haar property, i.e., if each function $\tau \in T_{\mathbf{p}}\mathcal{M}$, $\tau \neq 0$, has at most $m - 1$ zeros.*

(b) *$\mathcal{M}$ is said to satisfy "the global Haar property" if for every pair $\mathbf{p}, \mathbf{q} \in M$, $\rho_{\mathbf{p}} - \rho_{\mathbf{q}}$ has at most $m - 1$ zeros.*

Remark: In our case, the endpoints are *fixed* by (3.2) and the dimension m of $\mathcal{M}$ is already reduced. Therefore we have $\rho_{\mathbf{p}}(a) = \rho_{\mathbf{p}}(b) = 0$ for all $\mathbf{p} \in M$ and *these zeros need not be counted* either in the local Haar property or in the global one.

Theorem 3. *(Meinardus and Schwedt): Let $f \in C[a, b]$ and suppose $\mathcal{M}$ is induced by a global differentiable function $\hat{\rho} : M \times [a, b] \to \mathbb{R}$ (as in Theorem 2) and satisfying the global Haar property. If there is $\mathbf{p} \in M$ such that the local Haar property is satisfied at $\mathbf{p}$ and $\rho_{\mathbf{p}} - f$ is an alternant with $m + 1$ extremal points, then $\rho_{\mathbf{p}}$ is the best approximation to f and $\rho_{\mathbf{p}}$ is unique.*

Proof: (of the local Haar property). The proof will be given for general Bézier curves without additional end conditions (in the other cases only minor modifications have to be made). Denoting by $D_{\mathbf{v}}$ the directional derivative, we get from (2.5)

$$D_{\mathbf{v}}\mathbf{y} \circ \hat{\sigma} + (\dot{\mathbf{y}} \circ \hat{\sigma}) D_{\mathbf{v}}\hat{\sigma} = D_{\mathbf{v}}\hat{\rho}\,\mathbf{n},$$

and hence, at $\mathbf{p}$, observing $\tau = D_{\mathbf{v}}\hat{\rho}(\mathbf{p}, -)$, we obtain

$$det(D_{\mathbf{v}}\mathbf{y} \circ \sigma_{\mathbf{p}}, \dot{\mathbf{y}} \circ \sigma_{\mathbf{p}}) = \tau det(\mathbf{n}, \dot{\mathbf{y}} \circ \sigma_{\mathbf{p}}). \tag{3.6}$$

By the admissibility assumption (3) of Definition 1, $\det(\mathbf{n}, \dot{\mathbf{y}} \circ \sigma_{\mathbf{p}}) \neq 0$ in $[a, b]$; therefore, the zeros of τ are the same as that of the determinant to the left of (3.6). $\sigma_{\mathbf{p}}$ being a diffeomorphism, we can omit it, thus going back to the original parameter s. Now, $\mathbf{y}$ depends linearly on the parameters $p_1, ..., p_m$, which are, in our case, the $2(n-1)$ components of the control points $\mathbf{b}_1, ..., \mathbf{b}_n$. Therefore, splitting $\mathbf{v}$ correspondingly into $n-1$ vectors $\mathbf{v}_k \in \mathbb{R}^2$, $D_{\mathbf{v}}\mathbf{y} = \sum_{k=1}^{n-1} B_k^n(s)\mathbf{v}_k$ and $f(s) := det(D_{\mathbf{v}}\mathbf{y}, \dot{\mathbf{y}}) = \sum_{k=1}^{n-1} \sum_{j=0}^{n-1} B_k^n(s) B_j^n(s) \; det(\mathbf{v}_k, \Delta\mathbf{b}_j)$ is a polynomial of (maximal) degree $2n-1$. Since k runs only from 1 to $n-1$, every B_k^n contains the factor $(1-s)s$. Therefore f has at most $m - 1 = 2n - 3$ zeros in the interior of [0,1]. ■

On the global Haar property. The global Haar property is immediately obtained by convexity arguments for $n = 2$. It becomes much more complicated for $n \geq 3$. For the case of Bézier cubics, tangent to C at $\mathbf{x}(a)$ and $\mathbf{x}(b)$, we found a proof using the special representation of C' in this case. But it is based on tedious technical details, so it will not be reproduced here. A general proof for arbitrary n and all cases of $k\,th$ order contact with C at both the endpoints recently was given by E. Eisele ['91], using topological arguments combined with an analysis for the resolvent of the intersection condition; however, this proof is valid only for some neighborhood of a fixed $C' \in \mathcal{B}'$.

On the other hand, while trying to obtain a general algebraic proof, I came to the following

Conjecture. *Two Bézier curves can have no more intersection points (where they actually cross and are not just tangent) than their control polygons do.*

This would be a very significant generalization of the variation diminishing property. It might be that some restrictions are required; for example, the control polygons, after being closed by a line from the end point $\mathbf{b}_n$ to the beginning point $\mathbf{b}_0$, should be simple, in order to apply Jordan's curve theorem.

In our case, the validity of the conjecture would imply the global Haar property because the intersection points must be *monotone* with respect to the parameters (for both curves) because they are separated by the normals of C. Then, the same is valid for the control polygons and thus the number is less than or equal to $2n - 3$ (the common end points not counted).

Summarizing, we get:

Theorem 4. *Let $\mathcal{B}'$ be as in Theorem 2 (or a submanifold of it) such that the global Haar property holds for it; then, if there is $\mathbf{p} \in M$ such that $\rho_\mathbf{p}$ is an alternant with $m+1$ extremal points, the corresponding Bézier curve $C'_\mathbf{p}$ is the best approximation to C with respect to the normal distance and $C'_\mathbf{p}$ is unique in $\mathcal{B}'$.*

§4. Numerical Methods and Examples

4.1. Calculating the Best Approximation

For the classical Chebyshev approximation, the algorithm of Remez (cf. [Remez '1934]) is a standard method to calculate the best approximant. However, it is based on a *linear* space and the possibility of solving the discrete approximation problem (on a finite set of knots) by a linear system of equations; it cannot be used directly in our non-linear case.

Instead, we consider the conditions for $\rho_\mathbf{p}$ to be an alternant directly as a non-linear system of equations: Assuming that $C'_\mathbf{p}$ already has the maximum number of $m+1$ alternating local extremal values at the knots $t_1, ..., t_{m+1}$ (depending on $\mathbf{p}$), we consider the functions

$$f_j(\mathbf{p}) := \rho_\mathbf{p}(t_{j+i}) + \rho_\mathbf{p}(t_j), \quad (j = 1, ..., m) \tag{4.1}$$

which must have a common zero for the best approximant $C'_{\mathbf{p}_0}$. However, these equations cannot be solved by Newton's method, because the functions (4.1) are not known explicitly. To avoid this problem, we replace the Jacobian matrix $\mathcal{J}f(\mathbf{p})$ in the correction term of Newton's method by the matrix of finite differences $((f_j(\mathbf{p} + \delta_k \mathbf{e}_k) - f_j(\mathbf{p}))/\delta_k)_{j,k\in\{1...m\}}$ (with suitable small values δ_k) and iterate this quasi Newton method until $\mathbf{p}_0$ is attained within numerical precision.

4.2. Examples

The following examples show that the algorithm described in the previous section is robust and converges rapidly, once a good estimate is given to start with.

In the first two examples, we take Bézier parabolas ($n = 2, m = 2$) to approximate given curves. Rather than being practical, the reason for these examples is to show that approximants with exactly three extremal points exist and how the best approximant looks.

The situation in the first example is one of the worst we can imagine, because the given curve has an inflection point which cannot be modeled by a parabola: the normal distance remains rather large (see Figures 1 and 2).

In the second example, the given curve is convex and the approximation is much better (see Figures 3 and 4).

In the third example we calculate the best approximation of a transcendental curve C (part of the exponential function rotated and stretched) by a

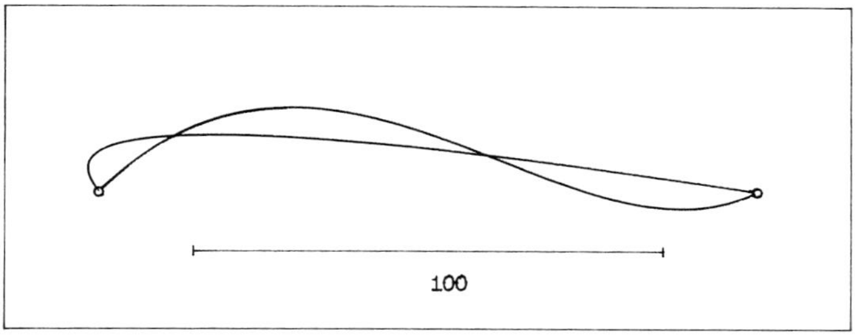

Figure 1. C with inflexion point and its best approximating parabola.

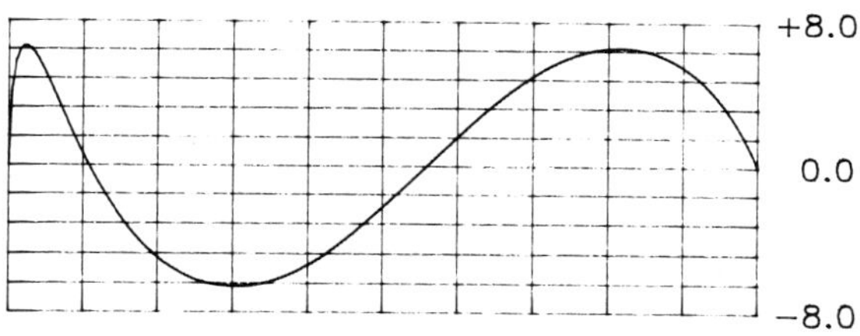

Figure 2. The error function (deviation function) for Example 1.

cubic Bézier curve tangent to C at the end points (see the examples in Section 3.1); we have $n = 3, m = 2$. Now the best approximation can no longer be distinguished from the original curve (see Figure 5).

The plot of the error function (see Figure 6) reveals the alternant property. Furthermore, we note the *high accuracy* of this best approximant: The maximal error (= normal distance) is about 10^{-5} times the length of the curve. We compared this error with some 3rd order polynomial Chebyshev approximants of real valued functions and found that our tangent Bézier cubic yields, in general, better results.

Finally, H. Ruf, a student at our institute investigated in his master's thesis the best approximants of circle arcs (see also [Dokken et al '87]). He compared the best approximants with that ones obtained by Hoschek's conversion method [Hoschek '87], and found that the error varies from 10^{-2} to 10^{-3} times the error of the conversion method with the same degree and a comparable number of intermediate points.

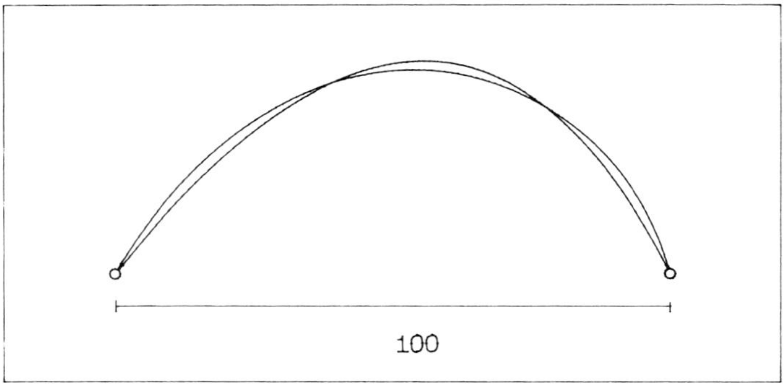

Figure 3. A convex curve and its best approximation by a Bézier parabola.

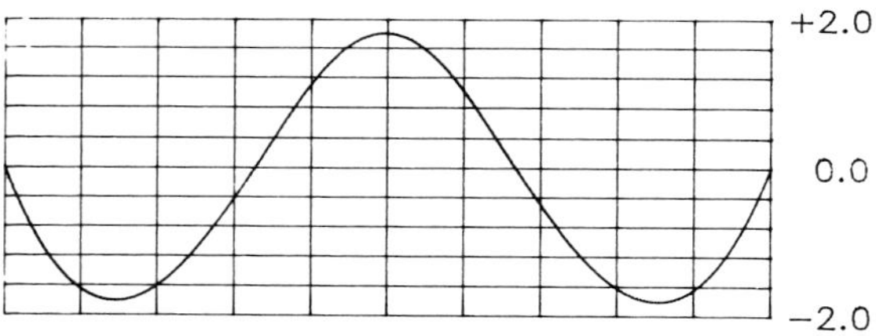

Figure 4. The error function (deviation function) for Example 2.

4.3. Conclusion

Our theoretical results show that the best approximation of a single curve segment by a Bézier curve behaves like the best approximant in Chebyshev approximation of a univariate function, and the algorithm based on the alternant property of its error function *yields best approximants with excellent error bounds* especially in the case of tangent cubic Bézier curves with a reasonable amount of calculation. If a larger curve is to be approximated, one can apply a subdivision procedure in advance, and use known error estimates ([de Boor et al. '87], [Hanna et al. '86], [Sakai and Usmani '90]) to get information on the number of necessary subdivision points and then calculate the best approximation for each segment separately. Fitting together the tangent Bézier cubics, one gets a geometric spline with G^1-continuity; higher order of continuity can be obtained analogously by the corresponding order of contact at the segment's end points.

However, there is one crucial point to be noticed: The region in the parameter domain where the maximal number of $m + 1$ extremal values is

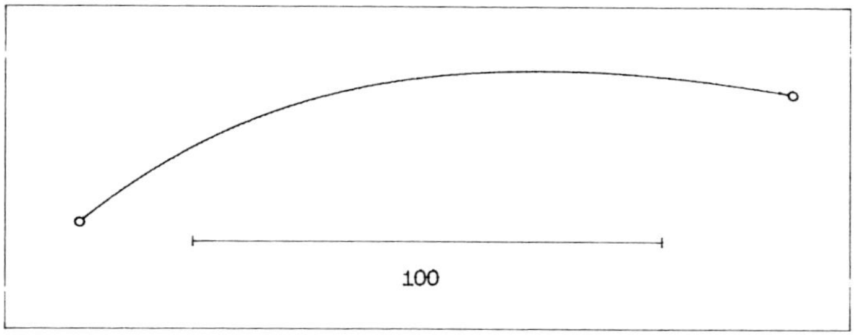

Figure 5. A best approximating tangent cubic to a transcendental curve.

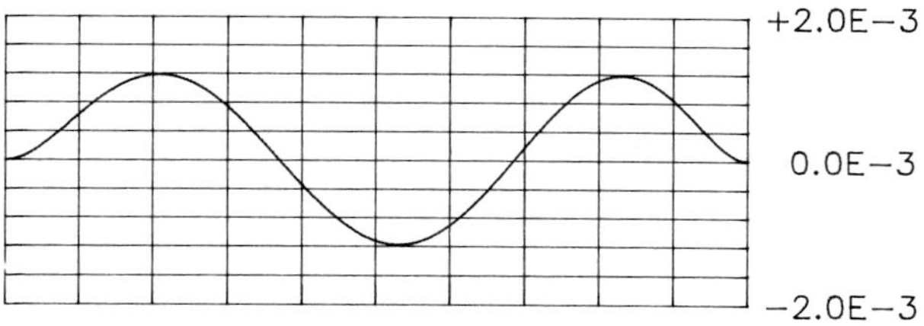

Figure 6. The error function (deviation function) of Example 3.

attained may be very small (in our Example 3 it is about 1% of the entire extension in each direction). Thus one has to be careful in finding a starting parameter vector for the quasi Newton algorithm described in Section 4.1.

References

1. Blaschke, W., Kreis und Kugel (Antrittsrede) Jahresbericht DMV **24** (1915), 195–207.
2. De Boor, C., Höllig, K. and Sabin, M., High accuracy Hermite interpolation, Computer Aided Geometric Design **4** (1987), 269–278.
3. Braess, D., *Nonlinear Approximation Theory*, Springer Verlag, Heidelberg, 1986.
4. Do Carmo, *Differential Geometry of Curves and Surfaces*, Prentice Hall, Englewood Cliffs, NJ, 1976.
5. Chebyshev, P. L., Sur les questions de minima qui se rattachent à la représentation approximative des fonctions, Oeuvres, Tome **1**, St. Petersbourg, 1899, 273–378.
6. Dannenberg, L. and Nowacki, H., Approximate conversion of surface representation with polynomial bases, Computer Aided Geometric Design **2**

(1985), 123–132.

7. Dokken, T., Daehlen, M., Lyche, T. and Mørken, K., Good approximation of circles by curvature continuous Bézier curves, Computer Aided Geometric Design **7** (1990), 33–41.
8. Eisele, E., Chebyshev approxiamtion of planar curves by splines (submitted for publication in Computer Aided Geometric Design).
9. Haar, A., Die Minkowskische Geometrie und die Annäherung an stetige Funktionen, Math. Annalen **78** (1918), 294–311.
10. Hanna, M. S., Evans, D.G. and Schweitzer P.N., On the approximation of plane curves by parametric cubic splines, BIT **26** (1986), 217–232.
11. Hausdorff, F., Grundzüge der Mengenlehre, Verlag Veit, Leipzig 1914, Reprint Chelsea, New York, 1949.
12. Hoschek, J. Approximate conversion of spline curves, Computer Aided Geometric Design **4** (1987), 56–66.
13. Hoschek, J. Spline approximation of offset curves, Computer Aided Geometric Design **5** (1988), 33–40.
14. Hoskins, W. D. and Ponzo, J. P., Some approximation properties of periodic cubic splines, BIT **14** (1974), 152–155.
15. Meinardus, G., Approximation von Funktionen und ihre numerische Behandlung, Springer-Verlag, Heidelberg, 1964.
16. Meinardus, G. and Schwedt, D., Nicht-lineare Approximationen, Arch. nat. Mech. Analysis **17** (1964), 297–326.
17. Mørken, K., Best approximations of circle segments by quadratic Bézier curves, in *Curves and Surfaces*, P.-J. Laurent, A. Le Méhauté and L. L. Schumaker (eds.), Academic Press, New York, 1991, 331–336.
18. Remez, E. J., Sur la détermination des polynômes d'approximation de degreé donnée, Comm. Soc. Math., Kharkov **10** (1934), 41–63.
19. Rice, J. R., The characterization of best nonlinear Chebyshev approximation, Transact. Amer. Math. Soc. **96** (1960), 322–340.
20. Sakai, M. and Usmani, R. A., On orders of approximation of plane curves by parametric cubic splines, BIT **30** (1990), 735–741.

Prof. Dr. W. L. F. Degen
University of Stuttgart
Pfaffenwaldring 57
D-7000 Stuttgart 80, Germany
lbaa@ds0rus1i.bitnet

An Approximately G^1 Cubic Surface Interpolant

Tony DeRose and Stephen Mann

Abstract. In our earlier work on tangent plane continuous surface interpolation, we observed that many interpolation schemes often produce surfaces with drastic curvature variation. The curvature distributions are extreme enough to render the surfaces undesirable for many practicable applications. In an effort to improve the curvature distributions of surface interpolants, we have developed a surfacing method that (empirically) tends to produce surfaces possessing smoothly varying curvature. The surfaces produced by the method are comprised of cubic triangular parametric patches. An interesting feature of the method is that although the surfaces produced are globally only of class C^0, they are "close" to being of class G^1 (tangent plane continuous). Moreover, the extent to which they are not of class G^1 can be made arbitrarily small using a refinement strategy.

§1. Introduction

The problem of constructing a smooth surface to interpolate the vertices of a (usually triangulated) polyhedron has frequently been addressed. For this problem, the interpolating surface is typically considered smooth if it is G^1; that is, if all abutting surface patches possess a common tangent plane along their shared boundary.

Many methods have been devised for solving variants of the problem. Often, however, shaded images of the surfaces they produce do not look "visually smooth," at least in the sense that there is a tendency for the curvature of the surfaces to concentrate along boundary curves, leaving relatively large flat regions on the patch interiors. The purpose of this paper is to introduce two ideas to improve this situation: approximate continuity and the cubic interpolant. Intuitively, our approach is to relax the continuity conditions and to use curvature information at the vertices to construct low degree (cubic) patches.

Mathematical Methods in Computer Aided Geometric Design II

Tom Lyche and Larry L. Schumaker (eds.), pp. 185–196.

ISBN 0-12-460510-9.

The motivation for our work is presented in Section 3. In Section 4, we introduce the idea of approximate continuity. Section 5 reviews the construction of cubic geometric Hermite curves. In Section 6, we present the cubic interpolant patch, which is used in Section 7 to construct a patch network. In Section 8, we present the results of using the method in a number of model situations. We conclude and present topics for further research in Sections 9 and 10.

§2. Notation and Background

We will denote a surface to be approximated as S. A directional derivative of S at p in direction $\vec{u}$ is denoted $S_{\vec{u}}(p)$, where $\vec{u}$ is a vector in the tangent plane of S at p. $S_{\vec{u}\vec{v}}(p)$ will denote two derivatives of S in directions $\vec{u}$ and $\vec{v}$; $\langle\vec{u},\ \vec{v}\rangle$ will be used to denote the dot product of vectors $\vec{u}$ and $\vec{v}$. The normal to S at domain point p is denoted $\hat{N}(p)$; p will be omitted when the point of evaluation is clear.

Our approximation to S will be composed of triangular cubic Bézier patches. A single patch will be referred to as P. The domain of P is given by the triangle rst. The control points of P are denoted with standard multi-index notation as shown in Figure 1.

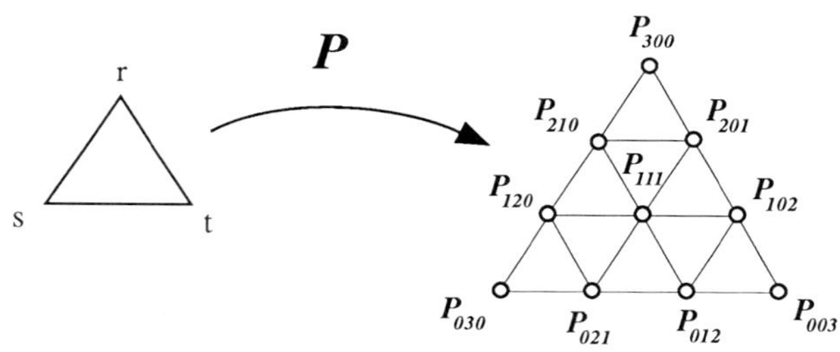

Figure 1. The domain triangle and control points of a cubic Bézier patch P.

It is well known from differential geometry that the curvature properties of a surface S at a point p are characterized by a symmetric bilinear form, called the second fundamental form [13, p. 208]. This form will be denoted $\mathrm{II}(\vec{u},\vec{v})$, where $\vec{u}$ and $\vec{v}$ are vectors in the tangent plane. The second fundamental form is related to the derivatives of the surface according to $\mathrm{II}(\vec{u},\vec{v}) = \langle S_{\vec{u}\vec{v}},\ \hat{N}\rangle$, implying that $\mathrm{II}(\vec{u},\vec{v})$ is the length of the normal component of $S_{\vec{u}\vec{v}}$, a fact that will be used for our surface patch construction in Section 6. We will also use the fact that $\mathrm{II}(\vec{u},\vec{u})$ is proportional to the normal section curvature $k_{\vec{u}}$ corresponding to direction $\vec{u}$; more precisely,

$$k_{\vec{u}} = \frac{\mathrm{II}(\vec{u},\vec{u})}{\langle\vec{u},\ \vec{u}\rangle}.$$

In Section 6, we require that the second fundamental form be specified at each data point. One way to specify a second fundamental form is by fixing three values $\mathrm{II}(\vec{u},\vec{u})$, $\mathrm{II}(\vec{u},\vec{v})$, and $\mathrm{II}(\vec{v},\vec{v})$, where $\vec{u}$ and $\vec{v}$ are any pair of independent tangent vectors [13].

§3. Motivation

There are many variants of the interpolation problem that was informally introduced in Section 1. A more precise statement of a common variant is:

> **Given:** A triangulated set of vertices, where the data stored at each vertex is a position and a normal.
>
> **Find:** A triangular piecewise parametric polynomial G^1 surface that interpolates the points and normals.

This problem occurs when fitting data "scattered" in three dimensions, and when approximating complex surfaces such as offset and algebraic surfaces. Normals are generally easy to obtain when approximating a known surface since they can be sampled directly. Normals are usually harder to obtain when fitting scattered data, often requiring the use of estimation [3] or optimization methods [10, 12]. Although scattered data fitting was the motivating problem for this research, we shall concentrate on function approximation in this paper. Applications of our work to scattered data fitting will be covered in more detail in [9].

Several schemes have been devised to solve the problem posed above. The solutions typically fall into two categories: split domain schemes, including [2, 6, 15, 17]; and convex combination schemes, such as [4, 5, 7, 11] (Hagen [4] additionally interpolates curvature at the vertices). The constructions used by the two classes are quite different. The split domain schemes construct three polynomial patches for each triangular facet, while the convex combination schemes construct a single high degree rational polynomial surface patch for each face. For a comparison of these schemes, see [8].

Despite the differences in the construction techniques, the constructed surfaces exhibit similar shape defects. Curvature plots reveal that the poor shape is due in part to poor curvature distribution over the patches. Our experience indicates that this non-uniform distribution is an indirect consequence of the requirement of producing a G^1 surface. Fairly high degree patches must be used to construct a G^1 join between patches. After satisfying the G^1 conditions, however, these patches have many degrees of freedom remaining which are typically set using heuristics. The poor distribution of curvature is partially a result of the somewhat arbitrary settings of these surplus degrees of freedom.

§4. Approximate Continuity

There are several approaches that may be taken to improve the shapes of the constructed surfaces. One approach is to improve the existing methods by using more principled approaches to setting the remaining degrees of freedom.

In the case of surface approximation, the excess degrees of freedom could be set using additional data sampled from the underlying surface. However, since many of these extra degrees of freedom were introduced in order to construct a G^1 join between surface patches, we chose instead to relax the continuity conditions. Roughly speaking, we consider the following problem:

> **Given:** A scalar ε and a triangulated set of vertices where the data stored at each vertex is a position, a normal and a second fundamental form.
>
> **Find:** A triangular piecewise parametric polynomial C^0 surface that interpolates the positions and normals, and whose discontinuity in surface normal is bounded by ε. It should be noted that the interpolant is not required to interpolate the second fundamental forms; as we describe in Sections 6 and 7, while our method often will interpolate them, the primary use of the second fundamental forms is to set otherwise unconstrained degrees of freedom.

The bound in the discontinuity of the surface normal refers to the maximum angle between two surface normals at any point. We shall refer to a surface satisfying these conditions as being of class ε-G^1. Note that a G^1 surface is ε-G^1 for every $\varepsilon \geq 0$. It has been informally recognized for some time that this weakened form of continuity is adequate for many applications. For example, a milling machine constructs a surface that is only ε-G^1, where ε is governed by the precision of the machine.

§5. Geometric Hermite Curves

Our construction of an ε-G^1 surface is based on a generalization of geometric Hermite curves [1]. In [1] it is shown that it is often possible to construct a planar cubic curve to match position, tangent directions, and curvature information at two points. The only values of a cubic left unspecified by the input data are the lengths of the tangent vectors at the end points (see Figure 2; $\vec{t}_0$ and $\vec{t}_1$ are unit tangent vectors; δ_0 and δ_1 are the unknown lengths). Determining these lengths involves solving a system of two quadratic equations. de Boor, Höllig, and Sabin show that this system does not always have a solution, and when it does, the solution may not be unique.

While most of the curves our construction uses will be planar, there will be times when space curves are needed. The above curve construction technique can be generalized to construct a cubic curve that often matches position, tangents directions, and normal section curvature at two points in 3-space. Peters has recently developed a similar method using quartic space curves [14].

§6. The Cubic Interpolant

The building block for our surface construction technique is a surface patch fit to a triangle of three points. More precisely, given three points, with associated normals and second fundamental forms, $(V_i, \hat{N}_i, \mathrm{II}_i)$, $i \in \{r, s, t\}$, a

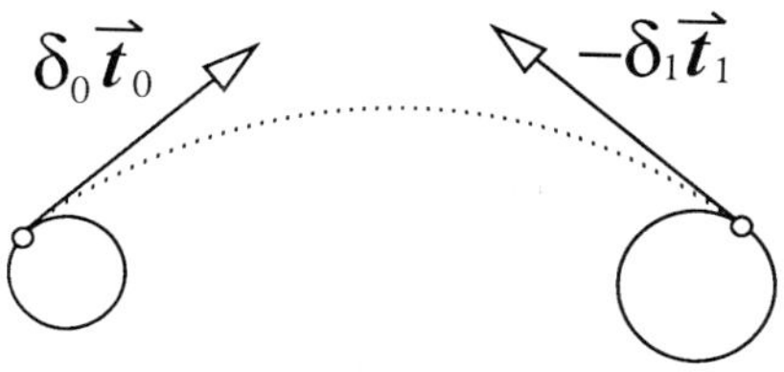

Figure 2. Cubic Geometric Hermite Curves.

cubic polynomial surface patch is constructed that in general interpolates this data. We shall refer to this patch as the cubic interpolant. While the positions and normals are always interpolated, there are cases, shown below, when the second fundamental forms cannot be interpolated with a cubic patch.

The data at each input point imposes eight constraints on the patch: three for position, two for the normal, and three for the second fundamental form. For two linearly independent vectors $\vec{u}$ and $\vec{v}$ in the tangent plane of a patch P at V_r, P must satisfy the following constraints to interpolate the data at the corner V_r :

$$P(r) = V_r, \tag{1}$$

$$\langle P_{\vec{u}}(r),\ \hat{N}_r\rangle = 0, \qquad \langle P_{\vec{v}}(r),\ \hat{N}_r\rangle = 0, \tag{2}$$

$$\langle P_{\vec{u}\vec{u}}(r),\ \hat{N}_r\rangle = -\mathrm{II}_r(\vec{u},\vec{u}), \qquad \langle P_{\vec{v}\vec{v}}(r),\ \hat{N}_r\rangle = -\mathrm{II}_r(\vec{v},\vec{v}), \tag{3}$$

$$\langle P_{\vec{u}\vec{v}}(r),\ \hat{N}_r\rangle = -\mathrm{II}_r(\vec{u},\vec{v}). \tag{4}$$

Similar constraints are imposed by the other two corners. The total number of constraints on the patch is therefore 24. A quadratic polynomial patch only has 18 degrees of freedom and cannot in general be used to solve this problem. A cubic patch is the minimum degree polynomial patch that can be used; it has 30 degrees of freedom in its ten control points.

A cubic polynomial patch generally has $30-24 = 6$ degrees of freedom unconstrained by the input data. These degrees of freedom manifest themselves in the tangents to the boundary curves. Each end tangent has an angular degree of freedom in the tangent plane. We set these degrees of freedom by restricting each boundary of the cubic patch to lie in a plane. The choice of planar boundaries uses three degrees of freedom, while the particular planes selected use the other three degrees of freedom.

There are ten control points in a cubic Bézier patch P (Figure 1). Since the corner control points of the patch are required to be the input data points, there are only seven unknown control points. Interaction of the second order constraints imposes a non-linear system of constraints on these seven points. Fortunately, the system decouples in such a way that each boundary of the patch can be constructed independently. The location of the center control point P_{111} is then given by a 3×3 system of linear equations.

We shall now examine the system of constraints in more detail. Each boundary is constructed by first selecting a plane with normal $\hat{N}$ containing the end points of the boundary. Tangent directions for the boundary curve and curvature in these directions can then be determined, and a planar geometric cubic Hermite is fit to this data. More precisely, for the boundary between V_r and V_s the data to be interpolated is placed in the plane through V_r and V_s perpendicular to the unit vector $\hat{N}$, where $\hat{N}$ is parallel to $(\hat{N}_r + \hat{N}_s) \times (V_s - V_r)$ (Figure 3). The data to be interpolated is given by

$$P_0 = V_r, \qquad P_1 = V_s.$$

$$\vec{t}_0 = \pm(N \times \hat{N}_r)/|N \times \hat{N}_r|, \qquad \vec{t}_1 = \pm(N \times \hat{N}_s)/|N \times \hat{N}_s|,$$

$$k_0 = \mathrm{II}_r(\vec{t}_0, \vec{t}_0)/|N \times \hat{N}_r|, \qquad k_1 = \mathrm{II}_s(\vec{t}_1, \vec{t}_1)/|N \times \hat{N}_s|.$$

The signs of $\vec{t}_0$ and $\vec{t}_1$ are chosen so that $\langle \vec{t}_0,\ P_1 - P_0\rangle$ and $\langle \vec{t}_1,\ P_1 - P_0\rangle$ are positive.

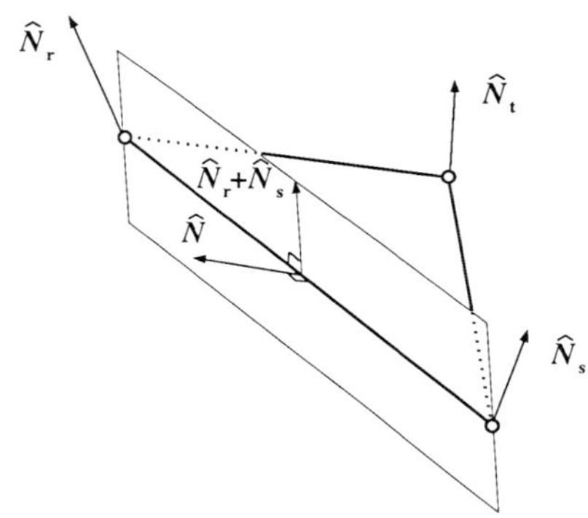

Figure 3. Construction of a planar boundary curve.

After constructing the boundaries, the center control point is the only remaining unknown. Constraints (1) – (3) have been satisfied, leaving only constraint (4) unsatisfied. The center control point influences the mixed partial at the three corners of the patch. The normal component of the mixed partial at the corner V_r is

$$\begin{aligned} \mathrm{II}_r(\vec{u}, \vec{v}) &= -\langle P_{\vec{u}\vec{v}}(r),\ \hat{N}_r\rangle \\ &= -\langle 6(P_{111} + P_{300} - P_{210} - P_{201}),\ \hat{N}_r\rangle \\ &= -6\langle P_{111} - P_{300},\ \hat{N}_r\rangle. \end{aligned}$$

Thus, P will agree with the mixed partial information at V_r if P_{111} is placed anywhere in the plane passing though the point $P_{300} - \frac{\mathrm{II}_r(\vec{u},\vec{v})}{6}\hat{N}_r$ and perpendicular to $\hat{N}_r$. Each corner restricts P_{111} to such a plane. In general, these three planes can be intersected to find a unique position for P_{111} that satisfies all the input constraints.

This construction can fail in two ways. First, the construction of one or more of the boundary curves may fail because no solution exists. In this case, we currently set the lengths of the end tangents to be the distance between the two end points of the curve. It is also possible to have multiple boundary curve solutions, in which case we choose the solution that is closest to an arc length parameterization.

The second way the construction may fail occurs when intersecting the three planes to determine the center point P_{111}. The intersection may not exist, or if it does exist, it may not be unique. A related difficulty is that if the planes become nearly parallel, the intersection calculation becomes numerically unstable. To address these problems, we use singular value decomposition to solve the 3×3 linear system [16]. The principal advantage of singular value decomposition for our purposes is that it automatically handles degeneracies. As the system of equations degenerates, the solution is placed near a user specified point. We choose this point to achieve quadratic precision, as described in [2].

§7. A Piecewise Cubic Surface Scheme

The previous section shows how to fit a single cubic Bézier patch to one triangular face. For a triangular net of data, the above method can be used to construct a cubic patch for each face. A surface constructed in this way is globally C^0, at least G^1 at all vertices, and G^2 at vertices where the construction succeeded on all the surrounding patch boundaries and patch interiors.

The patches of this surface, however, may join with large G^1 discontinuities at interior points of the boundary curves. The maximum discontinuity in surface normals along a boundary can be determined by finding the zeros of a high degree polynomial. An alternate approach would be to bound the discontinuity along each boundary. Currently, however, we simply use the maximum of the deviation between the normals at points sampled along the boundary. If, by any method, the discontinuity along an edge is determined to be small enough, the edge is left unchanged. If the discontinuity is large, then the mesh is refined adaptively in that region as indicated in Figure 4; the actual refinement pattern for a patch depends on how many of its edges are refined.

The refinement process generates new data points. At each new point, the position, normal and surface curvature must be determined. When approximating a known surface, this data is sampled from the surface. The cubic interpolant is then used to construct patches for the new faces, and the process is repeated.

One slight difficulty occurs when the cubic interpolant is being used to approximate a surface with a smooth boundary. The above method restricts the boundaries of the cubic patches to be planar. Since the particular plane used for each patch boundary is chosen independently of the other patches, the

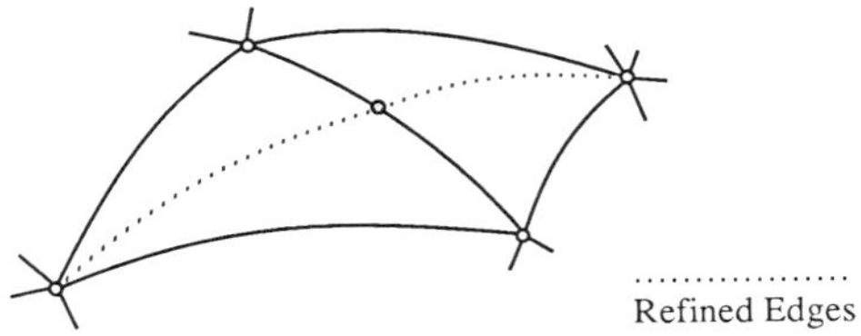

Figure 4. Refinement of the Mesh.

resulting boundary of the surface will not in general be G^1. To achieve a G^1 surface boundary, tangents of the surface along the boundary of the function being approximated are sampled and a non-planar geometric Hermite space curve is fit to the data.

§8. Results

In this section, the cubic interpolant is compared to the method of Shirman-Séquin [17]. The Shirman-Séquin scheme was used as a representative of the G^1 schemes; the other G^1 schemes constructed surfaces with similar shapes. A comparison to the G^1 schemes is somewhat unfair since their developers did not design them for approximating known functions. Additionally, the cubic interpolant makes use of higher order data than the G^1 schemes.

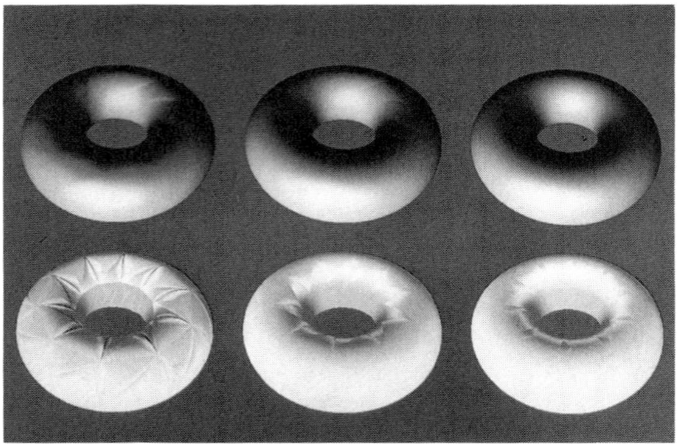

Figure 5. Interpolants to the Torus.

To make a more reasonable comparison, we modified the method of Shirman-Séquin to use the second order data at the vertices. This data was used to construct cubic Hermite boundaries. Thus, in Color Plate 5 (see also

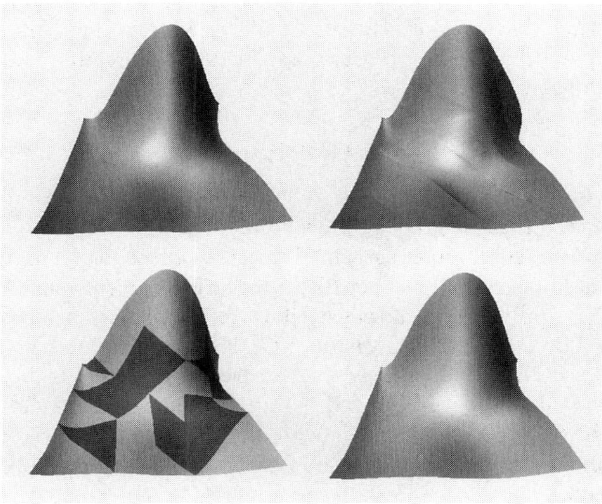

Figure 6. An offset to a bicubic.

Figure 5), the surfaces constructed by Shirman-Séquin and by the cubic interpolant use the same boundary curves; it is only the interior of the patches that are different. Note also that after constructing geometric Hermite boundaries, the G^1 schemes still have degrees of freedom that could be set based on the second order data. However, it is currently unclear how to automatically set these degrees of freedom.

In Color Plate 5 (see also Figure 5), interpolants to a regular sampling of fifty points on the torus are shown. Shaded images of the surfaces constructed appear on the top row. Below each shaded image is a Gaussian curvature plot of the surface. In these plots, the surface has been false colored to indicate the Gaussian curvature of the surface: red areas correspond to positive Gaussian curvature, green regions to zero Gaussian curvature, and blue regions to negative Gaussian curvature.

The surface constructed by the modified Shirman-Séquin scheme appears in the first column. This G^1 surface consists of 300 quartic Bézier patches. The second column shows the cubic interpolant surface for the same data set. Although only C^0, the cubic interpolant surface is ε-G^1 for $\varepsilon = 6.2$ degrees, and is comprised of only 100 cubic Bézier patches. In the third column is the cubic interpolant surface after one level of refinement; this surface is comprised of 160 cubic Bézier patches and is ε-G^1 for $\varepsilon = 1.7$ degrees.

Figure 6 shows the cubic interpolant approximation to a small offset to a bicubic tensor product B-spline surface consisting of nine bicubic patches. In the upper left is the true offset surface; in the upper right is an 18 patch cubic interpolant fit to the corners of the offset to the nine bicubic patches. In the lower right is the cubic interpolant produced after two levels of subdivision. A map showing the level of refinement of the surface appears in the lower left corner; one refinement step was made in the light regions, while two were made in the darker regions.

§9. Conclusions

In order to construct local piecewise G^1 interpolatory schemes, it is necessary to use relatively high degree patches. Once the G^1 conditions have been satisfied, however, many unconstrained degrees of freedom remain. Arbitrary settings of these degrees of freedom lead to surfaces with poor shape.

For many applications, however, it is sufficient for surfaces to be ε-G^1 continuous. The advantages in using ε-G^1 surfaces include the use of lower degree surface patches, which leaves fewer free parameters to be determined. Additionally, because of the relaxed continuity conditions, ε-G^1 schemes are not obliged to solve the "vertex enclosure" problem encountered by G^1 methods. This in turn allows for the use of fewer patches than polynomial split domain schemes.

As a first attempt to exploit approximate continuity, we have presented a piecewise cubic polynomial surface interpolation scheme. This scheme constructs an ε-G^1 surface that is often G^2 at the vertices and always at least G^1 at the vertices. The resulting surfaces show much better distribution of curvature than the published G^1 schemes; moreover, the surfaces are composed of fewer patches each of lower degree.

§10. Future Work

Much work remains to be done in this area. In order to use the cubic interpolant for scattered data, methods need to be devised for estimating normals and curvatures from a set of points. Even if the data initially has such information, new positions, normals, and surface curvature must be generated for the points created by refinement. A reasonable estimation for the position can usually be made by sampling the boundary curve. The estimation of normals and second fundamental forms is more difficult, and we only have preliminary results on the topic. When refining using a simple averaging scheme, the maximum angle between the normals along the boundaries decreased, but the curvature distribution across each patch became less uniform.

The refinement strategy for the cubic interpolant is currently quite primitive; a better strategy should be devised. In particular, an efficient, tight bound of the tangent plane discontinuity along a boundary needs to be developed. Also, the current refinement method often produces "long, skinny triangles." Empirically, the cubic interpolant creates patches with large tangent plane discontinuity for adjacent pairs of such triangles.

An interesting variant of the triangular cubic interpolant would be a tensor product version. Geometric Hermite curves could be constructed for the boundaries of a bicubic tensor product surface, leaving four interior "twist"-points to be set. Unlike the triangular case, each interior control point is influenced by the second order data at only one of the corner vertices, leaving a total of eight degrees of freedom for each patch. When approximating a function, another sample of position, normal, and surface curvature could be made on the interior of the patch. Requiring the interpolant to match this additional data introduces eight more constraints, resulting in a perfect balance between degrees of freedom and constraints.

The G^1 schemes should not be abandoned. Empirically, we have obtained nice shapes with these schemes by manually setting the free parameters. An automatic way of setting them should be devised, perhaps using an optimization method to meet various shape criteria.

Finally, the idea of approximate continuity itself should be explored more fully. The patches of the cubic interpolant have been empirically observed to satisfy properties that lead to relatively smooth variation in curvature. If these properties can be explicitly determined, it may be possible to devise a G^1 scheme with such properties.

Acknowledgments. This work was supported in part by the Xerox Corporation, IBM, Hewlett-Packard, the Digital Equipment Corporation, and the National Science Foundation under grants CCR-8957323 and DMC-8802949.

References

1. de Boor, C., K. Höllig, and M. Sabin, High accuracy geometric Hermite interpolation, Computer Aided Geom. Design **4** (1987), 269–278.
2. Farin, G., Smooth interpolation to scattered 3D data, in *Surfaces in Computed Aided Geometric Design*, R. Barnhill and W. Boehm (eds.), 1983, 43–63.
3. Foley, J., A. van Dam, S. Feiner, and J. Hughes, *Computer Graphics: Principles and Practice*, Addison-Wesley, Second Edition, 1990.
4. Hagen, H., and H. Pottmann, Curvature continuous triangular interpolants, in *Mathematical Methods in Computer Aided Geometric Design*, Tom Lyche and L. L. Schumaker (eds.), Academic Press, New York, 1989, 373–384.
5. Herron, G., Smooth closed surfaces with discrete triangular interpolants, Computer Aided Geom. Design **2** (1985), 297–306.
6. Jensen, T., Assembling triangular and rectangular patches and multivariate splines, in *Geometric Modeling: Algorithms and New Trends*, G. Farin (ed.), SIAM, 1987, 203–220.
7. Longhi, L., Interpolating patches between cubic boundaries, Technical Report T.R. UCB/CSD 87/313, University of California, Berkeley, October 1986.
8. Mann, S., C. Loop, M. Lounsbery, D. Meyers, J. Painter, T. DeRose, and K. Sloan, A survey of parametric scattered data fitting using triangular interpolants, in *Geometric Modeling: Methods and Applications*, H. Hagen (ed.), to appear.

9. Mann, S., Doctoral Dissertation, University of Washington, 1992.
10. Moreton, H. and C. Séquin, Surface design with minimum energy networks, ACM/SIGGRAPH Symposium on Solid Modeling Foundation and CAD/CAM Applications, Austin, Texas, June 5–7, 1991.
11. Nielson, G., A transfinite, visually continuous, triangular interpolant, in *Geometric Modeling: Algorithms and New Trends*, G. Farin (ed.), SIAM, 1987, 235–246.
12. Nielson, G., Interactive surface design using triangular network splines, in *International Conference on Engineering Graphics and Descriptive Geometry*, Volume 2, 1988, 70–77.
13. O'Neill, B, *Elementary Differential Geometry*, Academic Press, 1966.
14. Peters, J., Local generalized Hermite interpolation by quartic C^2 space curves, ACM Trans. Graphics **8** (1989), 235–242.
15. Piper, B., Visually smooth interpolation with triangular Bézier patches, in *Geometric Modeling: Algorithms and New Trends*, G. Farin (ed.), SIAM 1987, 221–233.
16. Press, W., B. Flannery, S. Teukolsky, and W. Vetterling, *Numerical Recipes in C*, Cambridge University Press, Cambridge, 1988.
17. Shirman, L. and C. Séquin, Local surface interpolation with Bézier patches, Computer Aided Geom. Design **4** (1987), 279–295.

Tony DeRose and Stephen Mann
University of Washington FR-35
Seattle, WA USA
derose@cs.washington.edu
smann@cs.washington.edu

On the G^2 Continuity of Piecewise Parametric Surfaces

Wenhui Du and Francis J. M. Schmitt

Abstract. In this paper, we study conditions which guarantee G^2 continuity over piecewise parametric surfaces. We first look at how to control G^2 continuity between any number of patches meeting at a common corner in a general setting; no particular assumptions are made on the mathematical formulation and the number of sides of each patch. Then we study how to ensure G^2 continuity over piecewise surfaces and examine possible approaches for achieving a local definition in a piecewise representation.

§1. Introduction

Controlling geometric continuities over piecewise surfaces has been one of the crucial issues in free-form surface modeling. A significant amount of research work has been devoted to the study of G^1 continuity control between surface patches, and diverse solutions have been developed. G^1 continuity is sufficient in many cases. However, in numerous application fields such as aerodynamics and mechanical fluid analysis, G^2 or higher order continuities are often required.

Previous research efforts on G^2 continuity have mainly concentrated on G^2 smooth connection between parametric curves and on general definition of G^2 continuity. Much less has been done on G^2 continuity control between surface patches. Several authors have considered the G^2 continuity constraints between two isolated adjacent surface patches [2,4,7,14,15,17,18]. Only a few papers have been published on G^2 continuity when an arbitrary number of patches meet at a common corner. Gregory and Hahn [10-12] define the G^2 continuity constraints between adjacent patches in terms of *connecting diffeomorphisms* in the neighbourhood of the boundaries common to adjacent patches. Li [16] defines these constraints by means of reparametrization techniques. Jones [14] interpolates by G^2 continuous Coons patches to a network

Mathematical Methods in Computer Aided Geometric Design II
Tom Lyche and Larry L. Schumaker (eds.), pp. 197–207.

ISBN 0-12-460510-9.

of curves, where the number of incident curves at a node is 4 or an odd number.

Theoretically, we know the constraints which have to be satisfied by surface patches in order to obtain G^2 continuity when they are connected together. But the question remaining to be answered is how to use the theoretical (and somewhat abstract) expression defining these G^2 continuity constraints in a practical construction of G^2 continuous surfaces. The aim of this paper is to propose an approach for developing practical construction methods for piecewise G^2 surfaces with parametric surface patches.

We first recall how G^2 continuity can be characterized (§2), and describe the corresponding constraints between two adjacent surface patches (§3). Then we study how to control G^2 continuity between any number N of patches meeting at a common corner in the general case where no particular assumptions are made on the mathematical formulation and the number of sides of each patch (§4). With these results, we then study how to ensure G^2 continuity over piecewise surfaces, and examine possible approaches for achieving a local definition in a G^2 piecewise representation with parametric surface patches.

§2. Characterization of G^2 Continuity

Different techniques have been developed to characterize G^2 continuity over surfaces: reparametrization [1,3,5,13,16,17], differentiable manifold [5], connecting diffeomorphism [10-12], equality of second order geometric quantities (principal curvatures and directions [17], Dupin indicatrix [15], osculating paraboloid [7]). They reflect the same characteristics of the surfaces, different choices among them leading to similar G^2 continuity constraints. We use here the osculating paraboloid. For G^2 continuity to be satisfied, the osculating paraboloid of two adjacent patches must be identical at every point of their common boundary in addition to the G^0 and G^1 conditions. A complete description of the osculating paraboloid can be found in [6].

§3. G^2 Continuity Between Two Surface Patches

Consider two surface patches $\mathbf{Q}_L$ and $\mathbf{Q}_R$ (cf. Figure 1). They meet with G^2 continuity if: i) they share a common boundary $\mathbf{C}(v)$ (G^0 continuity), ii) their tangent plane is identical along their common boundary (G^1 continuity), and iii) their osculating paraboloid is identical along their common boundary.

Let us clarify the notations used before going into the details of the equations. First, *directional derivatives* will be used to define the tangent plane and the osculating paraboloid instead of the parametric derivatives as is usual. $\mathbf{Q}^{d_1^m d_2^n} = \frac{\partial^{(m+n)}\mathbf{Q}}{\partial^m d_1 \partial^n d_2}$ will denote the mixed directional derivative of order m and n along the two directions d_1 and d_2 respectively, where d_1 is a cross-boundary direction (usually taken as the corresponding parametric directions in the case of a rectangular patch, and the radial direction in a triangular patch), and d_2 the direction along the common boundary between two patches. Secondly, no particular assumptions are made on the mathematical formulation and the number of sides of each patch.

The G^1 continuity condition between $\mathbf{Q}_L$ and $\mathbf{Q}_R$ is usually expressed as follows (cf. [4,8,9,15,17] and Figure 1):

$$\alpha(v)\mathbf{Q}_R^{d_1}(v) + \beta(v)\mathbf{Q}_L^{d_1}(v) + \gamma(v)\mathbf{C}^{d_2}(v) = \mathbf{0}, \quad 0 \le v \le 1, \tag{1}$$

where $\alpha(v)$, $\beta(v)$ and $\gamma(v)$ are functions of v, their form depending on the representation used for $\mathbf{Q}_L$ and $\mathbf{Q}_R$.

The equality of the osculating paraboloid along the common boundary $\mathbf{C}$ gives the following G^2 constraint [2,4,7,14,15,17,18]:

$$\begin{aligned} \alpha^2(v){\mathbf{Q}_R}^{d_1d_1}(v) - \beta^2(v)\mathbf{Q}_L^{d_1d_1}(v) - 2\beta(v)\gamma(v)\mathbf{Q}_L^{d_1d_2}(v) - \gamma^2(v)\mathbf{C}^{d_2d_2}(v) \\ - \delta(v)\mathbf{Q}_L^{d_1}(v) - \tau(v)\mathbf{C}^{d_2}(v) = \mathbf{0}, \qquad 0 \le v \le 1, \end{aligned} \tag{2}$$

where $\delta(v)$ and $\tau(v)$ are functions of v, their form depending on $\mathbf{Q}_L$ and $\mathbf{Q}_R$.

Expressions (1) and (2) together with the G^0 constraint guarantee G^2 continuity between $\mathbf{Q}_L$ and $\mathbf{Q}_R$. With a given surface patch representation and some chosen functions $\alpha(v)$, $\beta(v)$, ..., one can obtain the corresponding constraints on the defining terms of the surface patches (e.g. the control points in the case of Bezier patches) [4,7,14,15,17,18].

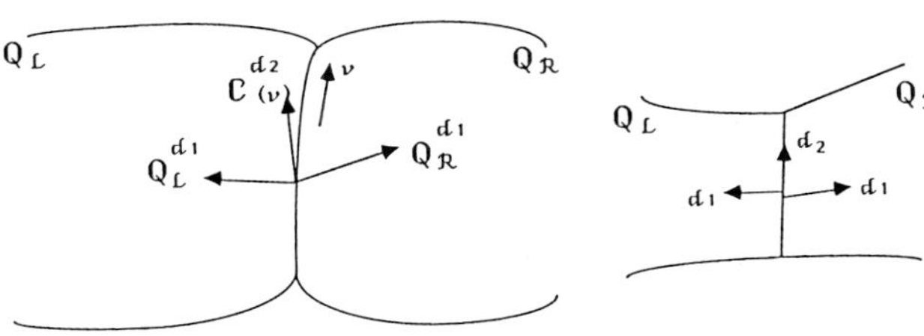

Figure 1. Two adjacent patches meet with G^2 continuity.

§4. G^2 Continuity Around an N-patch Corner

The G^2 smooth connection between two adjacent patches is relatively simple. However, the G^2 smooth connection between any number of patches meeting at a common corner becomes more complex because the G^1 and G^2 constraints along adjacent boundaries intertwine around the common corner. To the knowledge of the authors, up to now this problem has not yet been totally resolved; it makes the construction of G^2 piecewise continuous surfaces very difficult.

In this section, we study how to ensure G^2 continuity between any number N of patches meeting at a common corner; no particular assumptions are made on the mathematical formulation and the number of sides of each patch. We show first how the G^1 and G^2 continuity constraints intertwine around such an N-patch corner, and then how they can be decomposed into subsets and resolved in successive steps. Interesting conclusions are obtained concerning the solution of these intertwined constraints and their propagation towards adjacent patch corners.

4.1. Notations

Let $\mathbf{O}$ denote the corner where N patches $\mathbf{Q}_i$ meet, the index i being taken cyclically (cf. Figure 2). We use the convention that each boundary curve is oriented from the corner $\mathbf{O}$ ($v_i = 0$) towards the corresponding adjacent patch corner (any other choices lead to similar results). With this convention, we can then define unambiguously the functions $\alpha_i(v_i)$, $\beta_i(v_i)$, ... associated to each boundary.

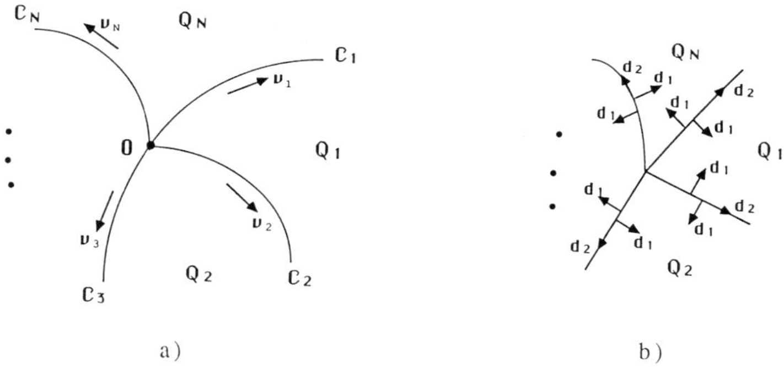

Figure 2. An N-patch corner $\mathbf{O}$.

4.2. Intertwining of G^2 Constraints and Their Decomposition

G^2 continuity between the N patches is obtained if the G^2 constraints are satisfied along each boundary $\mathbf{C}_i$:

$$
\begin{aligned}
G^0: &\quad \mathbf{Q}_{i-1}(v_i) = \mathbf{Q}_i(v_i) = \mathbf{C}_i(v_i),\\
G^1: &\quad \alpha_i(v_i)\mathbf{Q}_i^{d_1}(v_i) + \beta_i(v_i)\mathbf{Q}_{i-1}^{d_1}(v_i) + \gamma_i(v_i)\mathbf{Q}_i^{d_2}(v_i) = \mathbf{0},\\
G^2: &\quad \alpha_i^2(v_i)\mathbf{Q}_i^{d_1d_1}(v_i) - \beta_i^2(v_i)\mathbf{Q}_{i-1}^{d_1d_1}(v_i) - 2\beta_i(v_i)\gamma_i(v)\mathbf{Q}_{i-1}^{d_1d_2}(v_i)\\
&\quad -\gamma_i^2(v_i)\mathbf{Q}_i^{d_2d_2}(v_i) - \delta_i(v_i)\mathbf{Q}_{i-1}^{d_1}(v_i) - \tau_i(v_i)\mathbf{Q}_i^{d_2}(v_i) = \mathbf{0},\\
&\quad 0 \le v_i \le 1, \qquad i = 1, ..., N.
\end{aligned}
$$

The G^0 condition is easy to satisfy and does not cause any difficulty. Let us concentrate on the G^1 and G^2 constraints which we rewrite simply as:

$$\Phi_i(v_i) = \mathbf{0}, \quad \Psi_i(v_i) = \mathbf{0}, \quad i = 1, ..., N. \tag{3}$$

The constraints (3) can share common derivatives along adjacent boundaries only at the corner $\mathbf{O}$, elsewhere they are separated. To study the intertwining of the constraints (3) around the corner $\mathbf{O}$, two equivalent approaches can be used. The first one starts by developing the constraints (3) explicitly with the defining terms of the surface patches and the chosen functions $\alpha_i(v_i)$, ..., and then solves the resulting constraints. While the second one first considers, at the corner $\mathbf{O}$ and in terms of derivatives, the constraints (3) and their directional derivatives of order j along the boundaries $\mathbf{C}_i$ ($\Phi_i^{(j)}(v_i = 0) = 0$, $\Psi_i^{(j)}(v_i = 0) = 0$, $j = 0, 1, ...$), and then transcribes the derivatives into the corresponding defining terms of the surface patches. We take the last one which is more general and allows more intuitive control of the surface shape.

Two situations are to be considered: 1) $\mathbf{Q}_i$ are $C^{(2,2)}$ continuous at the corner $\mathbf{O}$ (e.g. Bézier, B-spline patches) for which the mixed derivative vectors are independent of the differentiation order; 2) $\mathbf{Q}_i$ are not $C^{(2,2)}$ continuous at the corner $\mathbf{O}$ (e.g. Gregory patches) for which the mixed derivative vectors depend on the differentiation order.

In the first case, it is easy to show that the derivative vectors can be shared along adjacent boundaries only in the two following subsets of constraints:

$$\Phi_i(0) = \mathbf{0}, \quad \Phi_i^{(1)}(0) = \mathbf{0}, \quad \Phi_i^{(2)}(0) = \mathbf{0}, \tag{4.1}$$

$$\Psi_i(0) = \mathbf{0}, \quad \Psi_i^{(1)}(0) = \mathbf{0}, \quad \Psi_i^{(2)}(0) = \mathbf{0}, \tag{4.2}$$

in which the derivatives of order up to 4 are involved (cf. Figure 3a where the various derivative vectors are symbolized by signs).

However, in the second case, the dependence on the differentiation order implies that the constraints (3) no longer share any derivatives and remain therefore separate. The resolution of the constraints (3) reduces simply to subsets of independant ones along each boundary $\mathbf{C}_i$.

The fact that the constraints (3) intertwine together in the first situation means that the involved derivative vectors must be defined in a compatible way. The direct resolution of all the intertwined constraints is obviously tedious. The method we propose consists of decomposing the intertwined constraints into subsets and of determining the involved derivative vectors in successive steps.

The constraints sets (4) can be decomposed into four subsets as follows (see Figure 3b):

$$\Phi_i(0) = \mathbf{0}, \qquad i = 1, ..., N. \tag{I}$$

$$\Phi_i^{(1)}(0) = \mathbf{0}, \qquad \Psi_i(0) = \mathbf{0}, \qquad i = 1, ..., N. \tag{II}$$

$$\Phi_i^{(2)}(0) = \mathbf{0}, \qquad \Psi_i^{(1)}(0) = \mathbf{0}, \qquad i = 1, ..., N. \tag{III}$$

$$\Psi_i^{(2)}(0) = \mathbf{0}, \qquad i = 1, ..., N. \tag{IV}$$

Consider now how to determine the derivatives involved in constraints subsets (I) through (IV), the corner $\mathbf{O}$ having already been defined. We do not detail here the whole determination process, but simply describe the resulting constraints and provide general conclusions about the determination of the involved derivatives. By noting that, from the $C^{(2,2)}$ property of $\mathbf{Q}_i$ and the G^0 constraints, $\mathbf{Q}_{i-1}^{d_1}(v_i = 0) = \mathbf{Q}_{i-1}^{d_2}(v_{i-1} = 0)$, $\mathbf{Q}_{i-1}^{d_1 d_1}(v_i = 0) = \mathbf{Q}_{i-1}^{d_2 d_2}(v_{i-1} = 0)$, $\mathbf{Q}_{i-1}^{d_1 d_2}(v_i = 0) = \mathbf{Q}_{i-1}^{d_2 d_1}(v_{i-1} = 0)$, ..., we can further shorten and simplify the notations by discarding the $(v_i = 0)$.

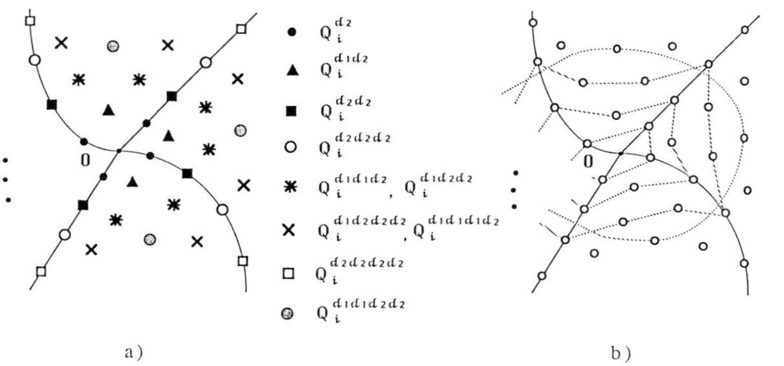

Figure 3. a) Involved derivative vectors in the intertwined G^2 constraints. b) Their decomposition into subsets.

4.3. Resolution of the Intertwined Constraints

4.3.1. First Subset (I)

With the simplified notations, the first subset (I) becomes:

$$\alpha_i \mathbf{Q}_{i+1}^{d_2} + \beta_i \mathbf{Q}_{i-1}^{d_2} + \gamma_i \mathbf{Q}_i^{d_2} = \mathbf{0}, \qquad i = 1, ..., N, \tag{5}$$

in which the N first order derivatives $\mathbf{Q}_i^{d_2}$ are involved (cf. Figure 3b).

Constraints (5) imply in fact the coplanarity of all N of the $\mathbf{Q}_i^{d_2}$. A total of $2(N+1)$ degrees of freedom remains in (5) to define the $\mathbf{Q}_i^{d_2}$ and the α_i, β_i and γ_i. They can be used for local shape control. It is easy to show that the following constraint exists on the α_i and β_i:

$$\prod_{i=1}^{N} \frac{\beta_i}{\alpha_i} = \prod_{i=1}^{N} \frac{\left(\mathbf{Q}_{i+1}^{d_2} \times \mathbf{Q}_i^{d_2}\right) \bullet \mathbf{n}}{\left(\mathbf{Q}_i^{d_2} \times \mathbf{Q}_{i-1}^{d_2}\right) \bullet \mathbf{n}} = 1, \qquad i.e. \qquad \prod_{i=1}^{N} \alpha_i = \prod_{i=1}^{N} \beta_i,$$

where $\mathbf{n}$ is the surface normal vector at the corner $\mathbf{O}$.

4.3.2. Second Subset (II)

Consider now the subset (II). Rewriting these constraints under a matrix product form, we obtain the following constraints in which the $2N$ newly involved derivatives are the $\mathbf{Q}_i^{d_1d_2}$ and $\mathbf{Q}_i^{d_2d_2}$ (cf. Figure 3b):

$$[A_{11}][\mathbf{Q}^{d_1d_2}] + [A_{12}][\mathbf{Q}^{d_2d_2}] = [\mathbf{F}] \tag{6.1}$$

$$[A_{21}][\mathbf{Q}^{d_1d_2}] + [A_{22}][\mathbf{Q}^{d_2d_2}] = [\mathbf{R}] \tag{6.2}$$

$$[A_{11}] = \begin{pmatrix} \alpha_1 & 0 & . & . & 0 & \beta_1 \\ \beta_2 & \alpha_2 & . & . & 0 & 0 \\ . & . & . & . & . & . \\ . & . & . & . & . & . \\ . & . & . & . & . & . \\ 0 & 0 & . & . & \beta_N & \alpha_N \end{pmatrix}, \quad [A_{12}] = \begin{pmatrix} \gamma_1 & . & . & . & 0 \\ . & \gamma_2 & . & . & . \\ . & . & . & . & . \\ . & . & . & . & . \\ 0 & . & . & . & \gamma_N \end{pmatrix},$$

$$[A_{21}] = -2 \begin{pmatrix} 0 & 0 & . & . & 0 & \beta_1\gamma_1 \\ \beta_2\gamma_2 & 0 & . & . & 0 & 0 \\ . & . & . & . & . & . \\ . & . & . & . & . & . \\ . & . & . & . & . & . \\ 0 & 0 & . & . & \beta_N\gamma_N & 0 \end{pmatrix}$$

$$[A_{22}] = \begin{pmatrix} -\gamma_1^2 & \alpha_1^2 & . & . & 0 & -\beta_1^2 \\ -\beta_2^2 & -\gamma_2^2 & . & . & 0 & 0 \\ . & . & . & . & . & . \\ . & . & . & . & . & . \\ . & . & . & . & . & . \\ . & . & . & . & -\gamma_{N-1}^2 & -\alpha_{N-1}^2 \\ \alpha_N^2 & 0 & . & . & -\beta_N^2 & -\gamma_N^2 \end{pmatrix}$$

$$[\mathbf{Q}^{d_1d_2}]^t = [\mathbf{Q}_1^{d_1d_2}\mathbf{Q}_2^{d_1d_2}\ldots\ \mathbf{Q}_N^{d_1d_2}], \quad [\mathbf{Q}^{d_2d_2}]^t = [\mathbf{Q}_1^{d_2d_2}\mathbf{Q}_2^{d_2d_2}\ldots\ \mathbf{Q}_N^{d_2d_2}].$$

The expressions for $[\mathbf{F}]$ and $[\mathbf{R}]$ involve the $\mathbf{Q}_i^{d_2}$ and are easily deduced.

There are $2N$ derivative vectors to be defined under $2N$ constraints. We observe that the solution depends on the values of the α_i, β_i and γ_i. In the general case, constraints (6) give a unique solution for the $\mathbf{Q}_i^{d_1d_2}$ and $\mathbf{Q}_i^{d_2d_2}$. However, in cases where the values of the α_i, β_i and γ_i reduce the rank of the $2N \times 2N$ matrix $[A]$ defined by the four blocks $[A_{ij}]$, $\mathbf{F}$ and $\mathbf{R}$ (and consequently the $\mathbf{Q}_i^{d_2}$) must satisfy some supplementary constraints for the $\mathbf{Q}_i^{d_1d_2}$ and $\mathbf{Q}_i^{d_2d_2}$ to have a solution. These constraints correspond to the linear dependence of the lines of the matrix A. In other words, in degenerate cases, the solutions of the first and second subsets are intertwined, leading to a transfer of the degrees of freedom between the determination of the derivatives involved in these two subsets of constraints.

4.3.3. Third Subset (III)

In subset (III) the newly involved derivatives are $\mathbf{Q}_i^{d_1 d_1 d_2}$, $\mathbf{Q}_i^{d_1 d_2 d_2}$ and $\mathbf{Q}_i^{d_2 d_2 d_2}$ (cf. Figure 3b). $3N$ derivative vectors are now to be defined subject to $2N$ constraints which we rewrite in a matrix product form as follows:

$$[B_{11}][\mathbf{Q}^{d_1 d_1 d_2}] + [B_{12}][\mathbf{Q}^{d_1 d_2 d_2}] + [B_{13}][\mathbf{Q}^{d_2 d_2 d_2}] = [\mathbf{T}] \tag{7.1}$$

$$[B_{21}][\mathbf{Q}^{d_1 d_1 d_2}] + [B_{22}][\mathbf{Q}^{d_1 d_2 d_2}] + [B_{23}][\mathbf{Q}^{d_2 d_2 d_2}] = [\mathbf{W}] \tag{7.2}$$

$$[B_{11}] = \begin{pmatrix} 0 & 0 & . & . & \beta_1 \\ \beta_2 & 0 & . & . & . \\ . & . & . & . & . \\ . & . & . & . & . \\ 0 & . & . & \beta_N & 0 \end{pmatrix}, \quad [B_{12}] = \begin{pmatrix} \alpha_1 & . & . & . & 0 \\ . & \alpha_2 & . & . & . \\ . & . & . & . & . \\ . & . & . & . & . \\ 0 & . & . & . & \alpha_N \end{pmatrix},$$

$$[B_{21}] = \begin{pmatrix} \alpha_1^2 & 0 & . & . & . & -2\beta_1\gamma_1 \\ -2\beta_2\gamma_2 & \alpha_2^2 & . & . & . & 0 \\ . & . & . & . & . & . \\ . & . & . & . & . & . \\ . & . & . & . & . & . \\ 0 & 0 & . & . & -2\beta_N\gamma_N & \alpha_N^2 \end{pmatrix},$$

$$[B_{22}] = -\begin{pmatrix} 0 & 0 & . & . & . & \beta_1^2 \\ \beta_2^2 & 0 & . & . & . & 0 \\ . & . & . & . & . & . \\ . & . & . & . & . & . \\ . & . & . & . & . & . \\ 0 & . & . & . & \beta_N^2 & 0 \end{pmatrix},$$

$$[B_{13}] = [A_{12}], \qquad [B_{23}] = -[A_{12}]^2,$$

$$[\mathbf{Q}^{d_1 d_1 d_2}]^t = [\mathbf{Q}_1^{d_1 d_1 d_2} \ldots \mathbf{Q}_N^{d_1 d_1 d_2}], \quad [\mathbf{Q}^{d_1 d_2 d_2}]^t = [\mathbf{Q}_1^{d_1 d_2 d_2} \ldots \mathbf{Q}_N^{d_1 d_2 d_2}],$$

$$[\mathbf{Q}^{d_2 d_2 d_2}]^t = [\mathbf{Q}_1^{d_2 d_2 d_2} \ldots \mathbf{Q}_N^{d_2 d_2 d_2}].$$

The expressions of $[\mathbf{T}]$ and $[\mathbf{W}]$ now involve the previous derivatives of order 1 and 2 and are easily deduced.

We now have a system of $3N$ derivatives to define subject to $2N$ constraints. Different solutions are possible. It is preferable, in general, that the continuity constraints remain as local as possible. The corresponding solution consists of first defining freely, if possible, all the $\mathbf{Q}_i^{d_2 d_2 d_2}$ which carry the information along the boundaries, then determining $\mathbf{Q}_i^{d_1 d_1 d_2}$ and $\mathbf{Q}_i^{d_1 d_2 d_2}$ which carry the inner information close to the corner $\mathbf{O}$. We observe that this determination depends again on the α_i, β_i and γ_i. In the general case, the matrix B formed by the four blocks B_{11}, B_{12}, B_{21} and B_{22} is of rank $2N$; a unique solution can thus be found for $\mathbf{Q}_i^{d_1 d_1 d_2}$ and $\mathbf{Q}_i^{d_1 d_2 d_2}$. However, in cases where the values of the α_i, β_i and γ_i reduce the rank of the matrix B, in order $\mathbf{Q}_i^{d_1d_1d_2}$ and $\mathbf{Q}_i^{d_1d_2d_2}$ to have a solution, some supplementary constraints (corresponding to the linear dependance of the lines of the matrix B) must be satisfied by $\mathbf{T}$, $\mathbf{W}$ and $\mathbf{Q}^{d_2d_2d_2}$. $\mathbf{T}$ and $\mathbf{W}$ involve derivatives already determined in subsets (I) and (II), the supplementary constraints can therefore merely be imposed on $\mathbf{Q}^{d_2d_2d_2}$. This means that some of the $\mathbf{Q}_i^{d_2d_2d_2}$ can no longer be freely defined, but are determined by these constraints. Conversely, some $\mathbf{Q}_i^{d_1d_1d_2}$ and $\mathbf{Q}_i^{d_1d_2d_2}$ can now be freely defined. This transfer of degrees of freedom causes the propagation of the constraints (7) towards the adjacent patch corners.

4.3.4. Fourth Subset (IV)

It remains to define the N involved derivatives $\mathbf{Q}_2^{d_1d_1d_2d_2}$ (cf. Figure 3b) through the last subset (IV) in which the $2N$ $\mathbf{Q}_i^{d_1d_1d_1d_2}$ and $\mathbf{Q}_i^{d_2d_2d_2d_2}$ are also involved. Rewriting the subset (IV) in matrix product form, we have:

$$[C_{11}][\mathbf{Q}^{d_1d_1d_2d_2}] + [C_{12}][\mathbf{Q}^{d_1d_1d_1d_2}] + [C_{13}][\mathbf{Q}^{d_2d_2d_2d_2}] = [\mathbf{Z}], \qquad (8)$$

$$[C_{11}] = \begin{pmatrix} \alpha_1^2 & 0 & . & . & 0 & -\beta_1^2 \\ -\beta_2^2 & \alpha_2^2 & . & . & 0 & 0 \\ . & . & . & . & . & . \\ . & . & . & . & . & . \\ . & . & . & . & . & . \\ 0 & 0 & . & . & -\beta_N^2 & \alpha_N^2 \end{pmatrix},$$

$$[C_{12}] = [A_{21}], \qquad [C_{13}] = [B_{23}],$$

$$[\mathbf{Q}^{d_1d_1d_2d_2}]^t = [\mathbf{Q}_1^{d_1d_1d_2d_2} \ldots \mathbf{Q}_N^{d_1d_1d_2d_2}],\ [\mathbf{Q}^{d_1d_2d_2d_2}]^t = [\mathbf{Q}_1^{d_1d_2d_2d_2} \ldots \mathbf{Q}_N^{d_1d_2d_2d_2}],$$

$$[\mathbf{Q}^{d_2d_2d_2d_2}]^t = [\mathbf{Q}_1^{d_2d_2d_2d_2} \ldots \mathbf{Q}_N^{d_2d_2d_2d_2}], \qquad [\mathbf{Z}]^t = [\mathbf{Z}_1 \ldots \mathbf{Z}_N].$$

The expressions for Z involve the previous derivatives and are easily deduced.

As mentioned before, it is preferable, in general, that G^2 constraints remain as local as possible. The corresponding solution in this case consists of first defining freely, if possible, all the $\mathbf{Q}_i^{d_1d_1d_1d_2}$ and $\mathbf{Q}_i^{d_2d_2d_2d_2}$ which carry the information along the boundaries, then determining $\mathbf{Q}_i^{d_1d_1d_2d_2}$ which carry the inner information close to the corner $\mathbf{O}$.

It is easy to see:

$$Det[C_{11}] = \prod_{i=1}^{N}(\alpha_i)^2 + (-1)^{2N+1}\prod_{i=1}^{N}(\beta_i)^2 =$$
$$= \prod_{i=1}^{N}(\alpha_i)^2 - \prod_{i=1}^{N}(\beta_i)^2 = 0.$$

Due to the particular structure of C_{11}, it is easy to verify that its rank equals $N-1$. This means that whatever the parity of N and the values of the α_i and β_i, one linear constraint is always imposed on $\mathbf{Z}$, $\mathbf{Q}^{d_1d_1d_1d_2}$ and $\mathbf{Q}^{d_2d_2d_2d_2}$. The derivatives involved in $\mathbf{Z}$ having already been defined through the previous subsets, this constraint can thus merely be imposed on $\mathbf{Q}^{d_1d_1d_1d_2}$ and $\mathbf{Q}^{d_2d_2d_2d_2}$, leading to the propagation of the fourth subset of constraints towards the adjacent patch corners.

§5. Piecewise Representation of G^2 Continuous Surfaces

The results from the last section on the G^2 continuity control around a patch corner forms the basis of the G^2 continuity control over piecewise surfaces. In effect, with a chosen representation for the surface patches, the general process of defining a piecewise G^2 continuous surface can be summarized as follows:

1) determine the derivatives involved in the intertwined G^2 constraints set around each patch corner, and then transcribe them into the corresponding defining terms of the surface patches,

2) determine remaining defining terms at the middle of each common boundary by respecting the remaining G^2 constraints along the boundary,

3) determine any other defining terms not involved in the G^2 continuity constraints along the boundaries of the patches.

In a piecewise representation, it is important to have a local control of the surface shape by maintaining the intertwined G^2 constraints set around each patch corner separate. The solutions for this depend on the representation used for the surface patches, on the choice of the functions α, β, γ, ... (imposed by the data of the application or chosen by the user), and on the connection configuration (number of patches meeting at each patch corner). In the case of a piecewise surface with Bezier patches, the solutions for various commonly used choices for $\alpha(v)$, $\beta(v)$, ..., and for different connection configurations can be found in [7]. For these configurations, the general case without any assumption and several particular cases are considered.

References

1. Barsky, B., The Beta-spline: A local representation based on shape parameters and fundamental geometric measures, Ph.D. Thesis, University of Utah, 1981.
2. Boehm, W., Curvature continuous curves and surfaces, Comput. Aided Design **18** (1986), 105–106.
3. Boehm, W., On the definition of geometric continuity, Comput. Aided Design **20** (1988), 370–372.
4. Degen, W. L. F., Explicit continuity conditions for adjacent Bezier surface patches, Computer Aided Geom. Design **7** (1990), 181–189.
5. DeRose, T., Geometric continuity: a parametrization independent measure of continuity for CAGD, Ph.D. Thesis, UC Berkeley, 1985.

6. DoCarmo, M. P., Differential geometry of curves and surfaces, Prentice-Hall, Inc., Englewood Cliffs, New Jersey, 1976.
7. Du, W., Etude sur la représentation de surfaces complexes : application à la reconstruction de surfaces échantillonnées, Ph.D. Thesis, Télécom Paris, 1988.
8. Du, W. and F. J. M. Schmitt, On the G^1 continuity of piecewise Bezier surfaces : a review with new results, Comput. Aided Design **21** (1990), 556–573.
9. Farin, G., A Construction for the visual C^1 continuity of polynomial surface patches, Comp. Graphics and Image Proc. **20** (1982), 272–282.
10. Gregory, J. A. and H. J. M. Hahn, Geometric continuity and convex combination patches, Computer Aided Geometric Design **4** (1987), 79–89.
11. Gregory, J. A. and H. J. M. Hahn, A C^2 polygonal surface patch, Computer Aided Geometric Design **6** (1989), 69–75.
12. Hahn, H. J. M., Geometric continuous patch complexes, Computer Aided Geom. Design **6** (1989), 55–68.
13. Herron, G., Techniques for visual continuity, in *Geometric Modelling*, G. Farin (ed.), SIAM, Philadelphia, 1987, 163–174. 163–174.
14. Jones A. K., Nonrectangular surface patches with curvature continuity, Comput. Aided Design **20** (1988), 325–335.
15. Kahmann J., Continuity of curvature between adjacent Bezier patches, in *Surfaces in CAGD*, R. E. Barnhill and W. Boehm (eds.), North Holland, Amsterdam, 1983, 65–75.
16. Li H., Shape controllability and geometric continuity in CAGD, Ph.D Thesis (in chinese) of CAD LAB, Institute of Computing Technology, Academic Sinica, Beijing, China.
17. Veron M., G. Ris and J.P. Musse: Continuity of biparametric surface patches, Comput. Aided Design **8** (1976), 276–273.
18. Wassum P., VC^1 und VC^2 ubergangsbedingungen zwischen angrenzenden rechtecks und dreiecks-Bezier-flachen, Preprint No. 1255, Technische Hochschule Darmstadt, 1989.

Wenhui Du
Thomson Digital Image
22, rue Hégésippe Moreau
75018 Paris, FRANCE

Francis J. M. Schmitt
Dept. Telecom Paris, IMA
46, rue Barrault
75013 Paris, FRANCE

Stationary and Non-Stationary Binary Subdivision Schemes

Nira Dyn and David Levin

Abstract. Convergence, smoothness and shape preserving properties of uniform stationary binary subdivision schemes are reviewed. Non-stationary uniform subdivision schemes are presented, including some important examples, such as the up-function and exponential splines. The theory of the stationary case is extended to a certain class of non-stationary schemes.

§1. Introduction

Starting with values $f^0 = \{f_j^0\}_{j=-\infty}^{\infty}$ assigned to the integers, a uniform binary subdivision scheme (BSS) defines recursively values $f^k = \{f_j^k\}_{j=-\infty}^{\infty}$, by the rule

$$\begin{cases} f_{2j}^{k+1} &= \sum\limits_i a_{2i}^{(k)} f_{j-i}^k, \\ f_{2j+1}^{k+1} &= \sum\limits_i a_{2i+1}^{(k)} f_{j-i}^k, \end{cases} \qquad j \in \mathbb{Z}. \tag{1.1}$$

The values $\{f_j^k\}_{j=-\infty}^{\infty}$ are assigned to the binary mesh points $\{2^{-k}j\}_{j=-\infty}^{\infty}$. The rule (1.1) may be rewritten as

$$f_j^{k+1} = \sum_i a_{j-2i}^{(k)} f_i^k, \; j \in \mathbb{Z}, \tag{1.2}$$

and formally as $f^{k+1} = S_k f^k$.

If the coefficients of the rules $\{S_k\}_{k=0}^{\infty}$ are independent of k, the scheme is called stationary, and we denote the rule by S. The fundamental examples for stationary BSS with compactly supported mask $\{a_i\}$, are the schemes

Mathematical Methods in Computer Aided Geometric Design II
Tom Lyche and Larry L. Schumaker (eds.), pp. 209–216.

ISBN 0-12-460510-9.

generating uniform spline functions, while the schemes generating exponential splines serve as the prototype for the class of non-stationary BSS with compactly supported mask $\{a_i\}$, studied in this paper.

The main purpose of BSS analysis is to study the convergence of such processes as $k \to \infty$; to establish the existence of a continuous limit function on $\mathbb{R}$, and to determine its smoothness properties. Other issues important to applications in CAGD are the regularity of curves generated by BSS and their shape preservation [15]. A general treatment of uniform stationary BSS can be found in [1,2,3,5,6,7,8,12,13]. In the present work we review the basic results about stationary BSS and present analogous results for non-stationary BSS with rules $\{S_k\}_{k=0}^{\infty}$ converging to a limit rule as $k \to \infty$. We confine the discussion to univariate BSS for the sake of simplicity and clarity of exposition. The multivariate case will be treated elsewhere [9].

Definition. *A binary subdivision scheme is said to be C^n if for every initial data $f^0 \in \ell^\infty(\mathbb{Z})$ there exists a limit function $f \in C^n(\mathbb{R})$ such that*

$$\lim_{k\to\infty} \sup_{i\in\mathbb{Z}} |f_i^k - f(2^{-k}i)| = 0,$$

and $f \not\equiv 0$ for some initial data f^0.

For a stationary BSS we have the formal relation $f^k = S^k f^0$, and we denote the limit function of a C^0 stationary BSS by $S^\infty f^0$. For non-stationary BSS we have the formal relation

$$f^k = S_{k-1}S_{k-2}...S_0 f^0. \tag{1.3}$$

Let us also denote by *basic limit function of a C^0 BSS* the limit function obtained from the initial data $f^0 = \delta$, i.e. $f_j^0 = \delta_{0,j}$, namely the function

$$\phi_0 = \lim_{k\to\infty} S_k S_{k-1}...S_0\delta. \tag{1.4}$$

It follows that

$$f = \sum_{i\in\mathbb{Z}} f_i^0 \phi_0(\cdot - i), \tag{1.5}$$

and it is enough to analyze the smoothness properties of ϕ_0.

In the stationary case with mask $\{a_i\}$, it follows from (1.5) that $\phi = S^\infty\delta$ is a solution of the functional equation

$$\phi(x) = \sum_{i\in\mathbb{Z}} a_i \phi(2x - i). \tag{1.6}$$

In the non-stationary case we have instead an infinite system of functional equations

$$\phi_m(x) = \sum_{i\in\mathbb{Z}} a_i^{(m)} \phi_{m+1}(2x - i), \; m = 0, 1, 2, ..., \tag{1.7}$$

for the functions

$$\phi_m = \lim_{k\to\infty} S_{m+k}...S_{m+1}S_m\delta, \ m = 0, 1, 2, ..., \tag{1.8}$$

where ϕ_m is the basic limit function of the non-stationary BSS $\{S_k\}_{k=m}^{\infty}$.

Let us define the spaces of functions

$$\Phi_k = \{f \in C(\mathbb{R}) : \ f = \sum_{i\in\mathbb{Z}} g_i\phi_k(2^k. - i), g \in \ell^\infty(\mathbb{Z})\}, k \in \mathbb{Z}_+, \tag{1.9}$$

which by (1.7) satisfy

$$\Phi_0 \subset \cdots \subset \Phi_k \subset \Phi_{k+1} \cdots .$$

Hence, any non-stationary BSS defines a sequence of refined spaces with the inclusion property, i.e., a sequence of spaces which is appropriate for multiresolution analysis [11].

To simplify the presentation of BSS and their analysis, it is convenient to assign to each rule, defined by a compactly supported mask $\{a_i\}$, the polynomial

$$a(z) = \sum_{i\in\mathbb{Z}} a_i z^i.$$

We denote the rule by S_a and the corresponding stationary scheme by $\{S_a\}$. To the rules $S_k, k \geq 0$, defining the non-stationary BSS $\{S_k\}_{k\geq 0}$, we assign the polynomials

$$a_k(z) = \sum_{i\in\mathbb{Z}} a_i^{(k)} z^i.$$

Example 1a. (B-splines). It is well known (see e.g. [5]) that the stationary BSS with the rule defined by the polynomial

$$a(z) = 2^{-m}(1+z)^{m+1}, \tag{1.10}$$

has as a basic limit function the $(m+1)$th order B-spline with integer knots, which is in $C^{m-1}(\mathbb{R})$ and is supported on $(0, m+1)$.

Example 1b. (The up-function). If we construct a non-stationary BSS with

$$a_k(z) = 2^{-k}(1+z)^{k+1}, \ k \geq 0,$$

then the function

$$\phi_0 = \lim_{k\to\infty} S_k S_{k-1}...S_0\delta.$$

is the Rvachev up-function [3,14] which is C^∞ and is supported on (0,2). This is a very nice example of the power of non-stationary BSS. Yet, it does not fall into the class of schemes analyzed in this work.

In Section 2 we review results on stationary schemes. In Section 3 we present extensions of these results to non-stationary schemes with compactly supported rules converging to a limit rule. The proofs of these results will appear in [9]. We also present some examples related to exponential splines, including a BSS which generates circles.

§2. Analysis of Stationary Schemes

The following are basic results about stationary BSS. The first result relates the properties of solutions of the functional equation (1.6) with the convergence of the corresponding stationary BSS.

Theorem 1. (Convergence of BSS and the functional equation [1]). *Let S_a be a rule with a compactly supported mask $\{a_i\}$ satisfying $a(1) = 2$ and $a(-1) = 0$. If $\{S_a\}$ is C^0, then $\phi = S_a^\infty \delta$ is the unique compactly supported continuous solution of the functional equation (1.6) with $\sum_{i\in\mathbb{Z}} \phi(\cdot - i) = 1$. Conversely, if ϕ is a compactly supported continuous solution of the functional equation (1.6), then $\sum_{i\in\mathbb{Z}} \phi(\cdot - i) = 1$ up to normalization, and the BSS $\{S_a\}$ is C^0 whenever*

$$c\|g\|_\infty \leq \| \sum_{i\in\mathbb{Z}} g_i \phi(x-i)\|_\infty, \quad g \in \ell^\infty(\mathbb{Z}). \tag{2.1}$$

The next result is useful for checking if a given stationary BSS is C^0.

Theorem 2. (Convergence of stationary BSS [7].) *Let S_a be a rule with a compactly supported mask $\{a_i\}$ satisfying $a(1) = 2$ and $a(-1) = 0$, and let S_q be the rule corresponding to $q(z)$ where $a(z) = (1+z)q(z)$. The scheme $\{S_a\}$ is C^0 if and only if there exists a positive L such that*

$$\|S_q^L\|_\infty = \mu < 1.$$

In this case [2]

$$|(S_a^\infty f^0)(x) - (S_a^\infty f^0)(y)| \leq C|x-y|^\nu, \tag{2.2}$$

where $\nu = -\frac{1}{L}\log_2 \mu$ and C is a constant depending on $\{a_i\}$ and f^0.

Checking whether a scheme is C^m is reduced by the following theorem to the problem of checking if another scheme is C^{m-1}.

Theorem 3. (Smoothness of stationary BSS [7]). *Consider a stationary BSS $\{S_a\}$ with*

$$a(z) = \frac{1}{2}(1+z)b(z), \tag{2.3}$$

where $\{S_b\}$ is C^0. Then $\{S_a\}$ is C^1, and $(S_a^\infty f^0)' = S_b^\infty(f^0_{\cdot+1} - f^0_\cdot)$.

In Theorem 2 the rule S_q is the difference rule, i.e. the rule transforming the forward differences at level k, $\{\Delta(S_a^k f^0)_j,\ j \in \mathbb{Z}\}$, to the forward differences at level $k+1$. The rule S_b in Theorem 3 is the corresponding divided difference rule. Checking if S_a is C^0 is equivalent to checking if S_q is contractive, while checking if S_a is C^1 is reduced to verifying that S_b is C^0.

Example 1a revisited. That the basic limit function generated by (1.10) is in $C^{m-1}(\mathbb{R})$ follows from Theorems 2,3 and the observation that $\frac{1}{2}(1+z)$ defines a BSS which is contractive (its norm is $\frac{1}{2}$ by (1.1)).

In application of BSS to CAGD, it is desired that the geometric properties of the limit curve be controlled by the geometric properties of the control polygon. The following theorem shows, for a wide class of BSS, that the length of the limit curve, and the variation of its components, are controlled by those of the initial control polygon.

Theorem 4. (Shape preserving stationary BSS [15]). *Consider a stationary BSS* $\{S_a\}$ *with*

$$a(z) = 2^{-m+1}(1+z)^m q(z), \tag{2.4}$$

where $\{S_q\}$ *is* C^0, *and* $q(z) = \sum_{i=0}^{n} q_i z^i$ *with* $q_i \geq 0, 0 \leq i \leq n$. *For* $1 \leq j \leq m$, *and for* $f_i^0 \in \mathbb{R}$, *if*

$$(\Delta^j f^0)_i \geq 0, \forall i \in \mathbb{Z},$$

then, for all $k \geq 1$,

$$(\Delta^j f^k)_i \geq 0, \forall i \in \mathbb{Z}.$$

Furthermore, for $p_i^0 \in \mathbb{R}^d$

$$\sum_{i \in \mathbb{Z}} \|(\Delta^j p^k)_i\| \leq 2^j \sum_{i \in \mathbb{Z}} \|(\Delta^j p^0)_i\|, \tag{2.5}$$

for $1 \leq j \leq m$ *and for all semi-norms* $\|\cdot\|$ *on* $\mathbb{R}^d$.

An important corollary follows if we choose $j = 1$ and $\|\cdot\|_2$ in (2.5). It says that the sequence of lengths of the control polygons created by the scheme is non-increasing, and hence, in the limit, the length of the limit curve is bounded by the length of the initial control polygon. The bound on the variation of the components of the limit curve can be concluded from the choice $j = 1$, $p_i^k \in \mathbb{R}$ in (2.5).

§3. Analysis of Non-stationary Schemes

In this section we consider a limited class of non-stationary BSS $\{S_k\}_{k\geq 0}$ with the following properties:

(i) $\text{supp}\{a_i^{(k)}\} = [0, n_k]$, $n_k \leq n < \infty$,

(ii) $\lim_{k\to\infty} a_i^{(k)} = a_i$ such that

$$\max_{0\leq i\leq n} |a_i^{(k)} - a_i| \leq \theta_k, \ \sum_{k=0}^{\infty} \theta_k < \infty, \tag{3.1}$$

(iii) $\{S_a\}$ is C^0.

Theorem 5. (Convergence of non-stationary BSS). *Under assumptions (i)–(iii), the non-stationary BSS* $\{S_k\}_{k\geq 0}$ *is* C^0.

Example 2. Consider the non-stationary scheme defined by the polynomials

$$a_k(z) = \frac{1}{2}(1 + e^{2^{-k-1}c} z) b(z), \ c \in \mathbb{R},$$

where $\{S_b\}$ is C^0. It follows that

$$\lim_{k\to\infty} a_k(z) = a(z) = \frac{1+z}{2} b(z),$$

and

$$|a_i^{(k)} - a_i| \leq 2^{-k-2} c e^{\frac{c}{2}} \max_{i\in\mathbb{Z}} |b_i|.$$

Hence assumptions (i)–(iii) are satisfied, and the scheme $\{S_k\}_{k\geq 0}$ is C^0. Furthermore, this non-uniform BSS reproduces the exponential function e^{cx}:

$$\sum_{j\in\mathbb{Z}} e^{cj} \phi_0(x-j) = M e^{cx}, \tag{3.2}$$

with $M = \prod_{k=0}^{\infty} \frac{1}{2} b(e^{2^{-k-1}c})$.

Theorem 6. (Convergence of non-stationary BSS and the system of functional equations). *Let $\{a^{(k)}\}_{k\geq 0}$ satisfy assumptions (i)–(iii) and let*

$$\phi_m = \lim_{k\to\infty} S_{m+k} \ldots S_{m+1} S_m \delta.$$

Then the sequence of functions $\{\phi_m(x)\}_{m\geq 0}$, is the unique solution of the infinite system of functional equations

$$\psi_m(x) = \sum_{i\in\mathbb{Z}} a_i^{(m)} \psi_{m+1}(2x-i), \; x \in \mathbb{R}, \; m \geq 0, \tag{3.3}$$

satisfying the conditions

(a) $\lim_{m\to\infty} \sum_{j\in\mathbb{Z}} \psi_m(x-j) = 1$ *uniformly in* $\mathbb{R}$,

(b) $\|\psi_m\|_\infty \leq B$ *for* $m \geq 0$,

(c) $\operatorname{supp} \psi_m \subset (0, n)$.

Conversly, let $\{a^{(k)}\}_{k\geq 0}$ satisfy assumptions (i),(ii), and let $\{\psi_m(x)\}_{m\geq 0}$ be a sequence of continuous functions solving the functional equations (3.3) and satisfying conditions (b) and (c) above. Then, up to normalization, condition (a) is also satisfied. If, moreover, there exists $K \in \mathbb{Z}_+$ such that

$$c\|g\|_\infty \leq \|\sum_{i\in\mathbb{Z}} g_i \psi_k(\cdot - i)\|_\infty, \; g \in \ell^\infty(\mathbb{Z}), \; k \geq K, \tag{3.4}$$

then the non-stationary BSS $\{S_k\}_{k\geq 0}$ determined by $\{a^{(k)}\}_{k\geq 0}$ is C^0.

The following theorem extends Theorem 3 to the non-stationary case.

Theorem 7. (Smoothness of non-stationary BSS). *Consider a non-stationary BSS* $\{S_{a_k}\}_{k\geq 0}$ *with*

$$a_k(z) = \frac{1}{2}(1 + r_k z)b_k(z), \; k > K \in \mathbb{Z}_+, \tag{3.5}$$

where $\{S_{b_k}\}_{k\geq K}$ *is* C^0. *If*

$$\lim_{k\to\infty}(r_k^2 - 1)2^k < \infty, \tag{3.6}$$

and

$$\sum_{k=K}^{\infty} 2^k |r_{k+1}^2 - r_k| < \infty, \tag{3.7}$$

then $\{S_{a_k}\}_{k\geq 0}$ *is* C^1.

Example 3. (Exponential Splines). Consider the non-stationary BSS defined by

$$a_k(z) = \frac{1}{2}(1 + e^{2^{-k-1}c} z)b_k(z), \; k \geq 0, \tag{3.8}$$

where $\{S_{b_k}\}_{k\geq 0}$ is C^0. Here we have $r_{k+1}^2 = r_k$ and $\lim_{k\to\infty}(r_k^2 - 1)2^k = \lim_{k\to\infty} 2^k(e^{2^{-k}c} - 1) = c$. Hence $\{S_{a_k}\}_{k\geq 0}$ is C^1. An important special case of (3.8) is

$$a_k(z) = 2\prod_{\ell=1}^{m} \frac{1}{2}(1 + e^{2^{-k-1}c_\ell} z). \tag{3.9}$$

Theorem 5 and repeated application of Theorem 7 yield that the corresponding non-stationary BSS is C^{m-2}. Furthermore, it turns out that

$$\phi_0 = \lim_{k\to\infty} S_{a_k} S_{a_{k-1}} ... S_{a_0}\delta$$

is the exponential B-spline with integer knots, spanned by $\{e^{c_\ell x} : \ell = 1, ..., m\}$ and supported on $(0, m)$.

Relation (3.8) is the non-stationary analogue of relation (2.3) between the rule S_a and the divided-difference scheme S_b. Here, the rules $\{S_{b_k}\}$ form a non-stationary BSS transforming some non-standard divided differences from level to level.

Example 4 (A non-stationary BSS generating circles). Consider the BSS defined by

$$a_k(z) = \frac{1}{2(1 + \cos\alpha_k)}(1 + z)(1 + e^{i\alpha_k} z)(1 + e^{-i\alpha_k} z),$$

with $\alpha_k = 2^{-k-1}\alpha_0$, $0 \leq \alpha_0 < \pi$. By Theorems 5 and 7 it follows that this scheme is C^1. Up to multiplying constants, the scheme is of the form (3.9) with imaginary exponents. As in (3.2), it reproduces constants, and also $\sin(\alpha_0 x)$ and $\cos(\alpha_0 x)$. This scheme is actually a corner cutting process, and the results in [10] also show that it is C^1. The interesting property of this scheme is that if the initial control polygon is a regular n-gon and we choose $\alpha_0 = \frac{2\pi}{n}$, then the limit curve of the process is the circle circumscribed in that polygon.

Theorem 8. (Shape preserving non-stationary BSS). *All of the results in Theorem 4 hold for a non-stationary BSS $\{S_{a_k}\}_{k\geq 0}$ with*

$$a_k(z) = 2^{-m+1}(1+z)^m q_k(z), \tag{3.10}$$

where $\{S_{q_k}\}_{k\geq 0}$ is C^0, and $q_k(z) = \sum_{i=0}^{n} q_i^{(k)} z^i$ with $q_i^{(k)} \geq 0, 0 \leq i \leq n, \forall k \geq 0$.

References

1. Cavaretta, A. S., W. Dahmen and C. A. Micchelli, *Stationary Subdivision*, Memoirs of the AMS, American Mathematics Society, Providence, 1991.
2. Daubechies, I. and C. Lagarias, Two scale difference equations I and II, SIAM J. Math. Anal., to appear.
3. Derfel, G., N. Dyn and D. Levin, Generalized functional equations and subdivision schemes, in preparation.
4. Deslauriers, G. and S. Dubuc, Symmetric iterative interpolation, Constructive Approximation **5** (1989), 49–68.
5. Dyn, N., Subdivision schemes in computer aided geometric design, in *Advances in Numerical Analysis II, Wavelets, Subdivision Algorithms and Radial Functions*, W.A. Light (ed.), Oxford University Press, 1991, 36–104.
6. Dyn, N., J. A. Gregory and D. Levin, A four-point interpolatory subdivision scheme for curve design, Computer Aided Geom. Design **4** (1987), 257–268.
7. Dyn, N., J. A. Gregory and D. Levin, Analysis of linear binary subdivision schemes for curve design, Constructive Approximation **7** (1991), 127–147.
8. Dyn, N. and D. Levin, Interpolating subdivision schemes for the generation of curves and surfaces, in *Multivariate Interpolation and Approximation*, W. Haussmann and K. Jetter (eds.), Birkhäuser Verlag, Basel, 1990, 91–106.
9. Dyn, N. and D. Levin, Analysis of non-stationary binary subdivision schemes, in preparation.
10. Gregory, J. A. and R. Qu, Non-uniform corner cutting, preprint (1988).
11. Mallat, S. G., Multiresolution approximations and wavelet orthonormal bases of $L_2(\mathbb{R})$, Trans. Amer. Math. Soc. **315** (1989), 69–87.
12. Micchelli, C. A. and H. Prautzsch, Computing curves invariant under halving, Computer Aided Geom. Design **4** (1987), 113–140.
13. Micchelli, C. A. and H. Prautzsch, Uniform refinement of curves, Linear Algebra and Applications, **114/115** (1989), 841–870.
14. Rvachev, V. A., Compactly supported solutions of functional-differential equations and their applications, Russian Math. Surveys **45**:1 (1990), 87–120.
15. Yad-Shalom, I., Monotonicity preserving subdivision schemes, preprint.

MQ–Curves are Curves in Tension

Matthias Eck

Abstract. This paper studies properties of curves with multiquadrics as basis functions (MQ–curves). Firstly, MQ–curves with constant precision are considered. Especially, it is pointed out that the interpolating MQ–curve is a curve in tension. Secondly, a very special MQ–curve with linear precision depending only on four coefficients is used to derive a local basis of C^2–continuous MQ–splines in tension which are an alternative to the well–known exponential C^2–spline in tension.

§1. Introduction

The original multiquadric (MQ) method was discovered in 1971 by Hardy [5] as a global tool to interpolate N data values $\{f_i\ :\ i = 1, \ldots, N\} \subset \mathbb{R}$ according to arbitrarily distributed knots $\mathbf{t}_i \in \mathbb{R}^2$. Then the MQ–interpolant F is of the form

$$F(\mathbf{t}) = \sum_{i=1}^{N} A_i \cdot \sqrt{c^2 + \|\mathbf{t} - \mathbf{t}_i\|^2} \quad , \quad \mathbf{t} \in \mathbb{R}^2$$

where $c \in \mathbb{R}^+$ is a user specified parameter. $\|.\|$ denotes as usual the Euclidean norm on $\mathbb{R}^2$. The coefficients $A_i \in \mathbb{R}$ have to be determined from a symmetric linear $N \times N$ system.

Many successful applications of the MQ–method also in the case of multivariate data $\mathbf{t}_i \in \mathbb{R}^d$ are mentioned in a recently published history that provides an excellent survey of the practical side of the subject [6]. The mathematical breakthrough for the MQ–basis functions and for the more general radial basis functions came with Franke's extensive tests of various methods for interpolation over planar domains [4], where the MQ–surfaces proved to be excellent solutions, and the results of Micchelli [9] concerning the unique solvability of the interpolation problem. A review of this theory is presented by Powell [11].

Mathematical Methods in Computer Aided Geometric Design II
Tom Lyche and Larry L. Schumaker (eds.), pp. 217–228.

ISBN 0-12-460510-9.

In the present paper we investigate the possibility of employing the MQ–method as a free form tool in CAGD. We consider the univariate case with single–valued knots $t_i \in \mathbb{R}$ and vector–valued coefficients $\mathbf{A}_i \in \mathbb{R}^s$ in order to learn more about the case of *parametric MQ–curves* with constant precision. Therefore in Section 2, we turn our attention first to geometrical properties of MQ–curves and their basis functions. Here it will be advantageous to use alternative bell-shaped basis functions. In Section 3, the main application of MQ–curves, the interpolation of given data, is discussed briefly. Here we particularily consider the tension effect of varying the MQ–parameter c on the interpolating curve. Then in Section 4 we derive a new local basis for a very simple MQ–segment with linear precision, where each curve is determined only by four coefficients. This basis, which is comparable to the Bernstein basis in the polynomial case, again points out the tension effect of the MQ–curves. In Section 5 we describe two ways to put such MQ–segments together smoothly using the local basis of Section 4. The first one yields a *C^1–Hermite–spline in tension* by interpolating positions as well as given tangents, whereas the second one uses only the positions to obtain a *C^2–spline in tension*, which is an alternative to the well–known exponential C^2–spline in tension.

§2. Properties of General MQ–Curves

Consider a strictly increasing sequence $\xi = (t_i)_{i=1}^N$ of real knots, and assume that the MQ–parameter $c \in \mathbb{R}^+$ as well as the $N+1$ coefficients $\{\mathbf{A}_1, \ldots, \mathbf{A}_N, \mathbf{K} : \sum_{i=1}^N \mathbf{A}_i = 0\} \subset \mathbb{R}^s$ are given. Then a *MQ–curve segment* $\mathbf{X}$ (with constant precision) over ξ is defined as

$$\mathbf{X}(t) = \sum_{i=1}^{N} \mathbf{A}_i \cdot \Phi_i(t) + \mathbf{K}, \quad t \in [t_1, t_N], \tag{1}$$

with the i–th MQ–basis function $\Phi_i(t) = \sqrt{c^2 + \left(\frac{t-t_i}{L}\right)^2}$ and $L = t_N - t_1$.

The graph of the basis function $\Phi_i(t)$ is simply the positive arc of a hyperbola that possesses the function $\left|\frac{t-t_i}{L}\right|$ as asymptotes; *cf.* [10], where it is also remarked that the MQ–parameter c is therefore obviously something like a tension parameter. Actually, a more local tension control of the very pleasing C^∞ curve $\mathbf{X}(t)$ can be achieved by associating different tension parameters c_i with each $\Phi_i(t)$. In [3] it is pointed out that $c_i = \frac{s}{L} \cdot \min_{j \neq i} |t_j - t_i|$ with a positive constant s is a good choice in dealing with poor data distributions. Other choices are given in [8]. But to avoid misunderstandings, we will always use uniform MQ–parameters c. Note that the representation (1) of the MQ–curve is

(i) *invariant to linear parameter transformations*, because of the unusual division by the interval length L in the basis function $\Phi_i(t)$. Otherwise the coefficients $\mathbf{A}_i$ and the MQ–parameter c have to be changed after such a parameter transformation.

(ii) *non affine invariant*, since the basis functions in (1) do not sum up to 1. Nevertheless, any translation $\mathbf{T}$ can be represented by MQ–curves (1) by setting $\mathbf{A}_i = 0$ for all i and $\mathbf{K} = \mathbf{T}$ which is the reason for the extension of the representation (1) by the constant $\mathbf{K}$. The additional condition $\sum_{i=1}^{N} \mathbf{A}_i = 0$ is not necessary but yields nice properties, as we will see later.

(iii) *not parametrized by the arc length*, since in general, $|\mathbf{X}'(t)| \not\equiv 1$.

The *bell–shaped* basis function of Beatson and Powell [1] is useful to get an affine invariant representation of $\mathbf{X}(t)$. This basis function $\Psi_i(t)$ is introduced for a quasi–interpolation scheme simply as a normalized second divided difference of the MQ–basis $\Phi_i(t)$, whereby the two boundary differences of order one are defined to be of value -1 resp. 1:

$$\begin{aligned} \Psi_i(t) &= \frac{1}{2}\left(\triangle\Phi_i(t) - \triangle\Phi_{i-1}(t)\right), \quad (i = 2, \ldots, N-1) \\ \Psi_1(t) &= \frac{1}{2}\left(\triangle\Phi_1(t) + 1\right), \quad \Psi_N(t) = \frac{1}{2}\left(1 - \triangle\Phi_{N-1}(t)\right), \end{aligned} \tag{2}$$

where $\triangle\Phi_i(t) = L \cdot \frac{\Phi_{i+1}(t) - \Phi_i(t)}{\Delta_i}$ and $\Delta_i = t_{i+1} - t_i$. The functions $\Psi_i(t)$ are suitable for our purposes since the representation (1) can be transformed to

$$\mathbf{X}(t) = \sum_{i=1}^{N} \mathbf{B}_i \cdot \Psi_i(t), \quad \mathbf{B}_i \in \mathbb{R}^s, \quad t \in [t_1, t_N], \tag{3}$$

as follows immediately from $\sum_{i=1}^{N} \mathbf{A}_i = 0$. The associated linear conversion formulas for the coefficients can be easily derived from (2). But more important than these conversion–formulas are the following properties of the functions $\Psi_i(t)$ which hint at the bell–shaped character:

(i) $\Psi_i(t) \geq 0, \quad t \in \mathbb{R}, \quad (i = 1, \ldots, N)$

(ii) $\sum_{i=1}^{N} \Psi_i(t) \equiv 1$

(iii) for $c = 0$: $\Psi_i(t) = H_i(t)$, where $H_i(t)$ is the hat function that has the knots $\{t_{i-1}, t_i, t_{i+1}\}$, that is identical to zero outside the interval $t_{i-1} \leq t \leq t_{i+1}$, and that satisfies the normalization conditions $H_i(t_i) = 1$.

(iv) $\Psi_1(-\infty) = 1$, $\Psi_i(-\infty) = 0, \quad (i = 2, \ldots, N)$

(v) $\Psi_N(+\infty) = 1$, $\Psi_i(+\infty) = 0, \quad (i = 1, \ldots, N-1)$

These properties immediately affect the shape of the curve segment $\mathbf{X}(t)$, $t \in [t_1, t_N]$, as is illustrated in the simple example of four fixed coefficients in Fig. 1 where additionally four different values of c are chosen:

(i) The curve considered on the real line starts in $\mathbf{B}_1$ and ends in $\mathbf{B}_N$.

(ii) The MQ–curve always lies in the *convex hull* of the coefficients $\{\mathbf{B}_i\}$. The coefficients of (3) are *affinely invariant* with respect to $\mathbf{X}(t)$.

(iii) The geometrical meaning of the coefficients is on the other hand very weak, since increasing c causes the MQ–curve segment to shrink together. Here it would be more suitable to have a basis like the Bernstein basis for polynomials where the coefficients (Bézier points) are close to the curve. One way how to get such a basis for MQ–curves is described in Section 4.

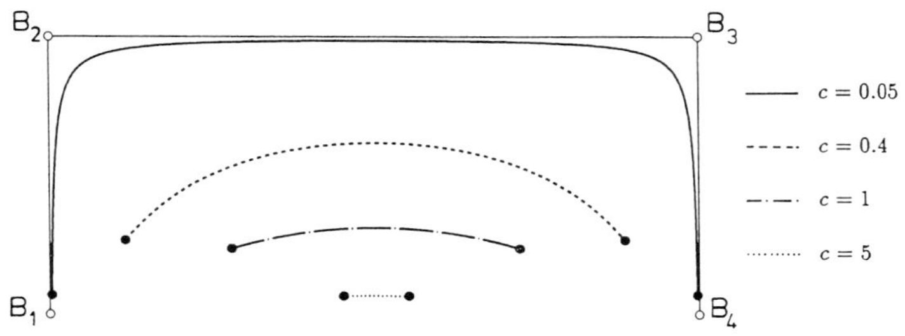

Figure 1. Four MQ–segments with different values of c $(N = 4)$.

§3. Interpolation

This section is adressed to the main application of MQ–curves, the interpolation of a given sequence of N data points $\{\mathbf{P}_i \; : \; i = 1, \ldots, N\} \subset \mathbb{R}^s$ such that $\mathbf{P}_{i-1} \neq \mathbf{P}_i$. The procedure itself contains two steps.

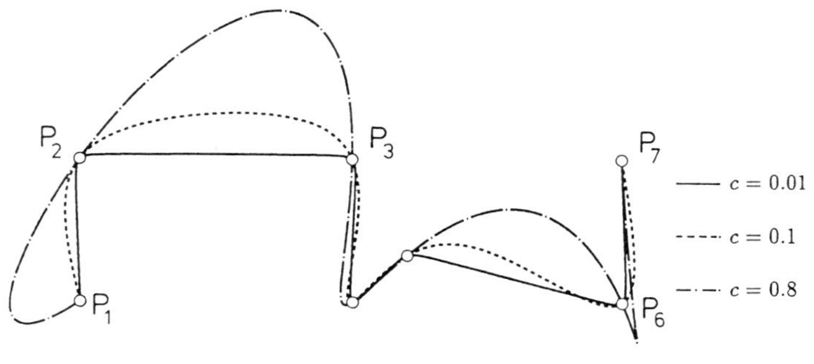

Figure 2. Three interpolating MQ–curves with different values of c.

The first step is to find suitable knots $t_i \in \mathbb{R}$ associated with each point $\mathbf{P}_i$. Instead of dealing with this problem here, we refer to [7]. We will mainly use the *chord length parametrization* recursively defined as $t_1 = 0$ and $t_i = t_{i-1} + \|\mathbf{P}_i - \mathbf{P}_{i-1}\|$, $i \geq 2$. It generates reasonable curves if the given data are smooth.

The second step is the proper interpolation process which is well known. The N interpolation conditions $\mathbf{X}(t_i) = \mathbf{P}_i$ determine a MQ–curve uniquely. Using the representation (1), a symmetric linear $(N + 1) \times (N + 1)$–system of unknown coefficients has to be solved, whereas representation (3) gives a non–symmetric linear $N \times N$–system. The example in Fig. 2 illustrates that this procedure works well; the tension effect of the curve is obvious. As expected, for small c the curve is near the linear spline. Increasing c causes the interpolating curve to become smoother, and it tends finally to an osculating curve.

To learn more about this behaviour, we look at the limit of the interpolating curve if c is increased to infinity. The following statement which was proved explicitly by the author for $l \leq 4$ and later on by M. Buhmann [2] for general l by means of induction is useful:

Theorem 1. *Given l real knots $\{t_j : j = 1, \ldots, l\}$, then there exist $l+1$ real coefficients $\{A_1, \ldots, A_l, K : \sum_{j=1}^{l} A_j = 0\}$ so that*

$$\sum_{j=1}^{l} A_j \cdot \sqrt{c^2 + (t - t_j)^2} + K = p_{l-1}(t) + \mathcal{O}(c^{-2}),$$

where $p_{l-1}(t)$ is a polynomial of exact degree $l-1$ with leading coefficient 1.

Theorem 1 is also valid in more general settings [2], *e.g.*, for basis functions like $(c^2 + (t - t_j)^2)^{\frac{\beta}{2}}$, β odd and ≥ 1, for representations with other precisions than constant, and even for the multivariate case.

Furthermore, since c can be arbitrarily large, it follows as a consequence of the well–known Weierstrass theorem that any continuous function can be approximated uniformly by MQ–functions.

The practical aspect of Theorem 1 is stated in the title of this paper: *the interpolating MQ–curve is a curve in tension* in the sense that the interpolating MQ–curve is near the linear interpolating spline for small c, and is near the interpolating polynomial for large c. In Fig. 3 this behaviour is illustrated where the dotted curve is the interpolating quintic polynomial ($c = \infty$).

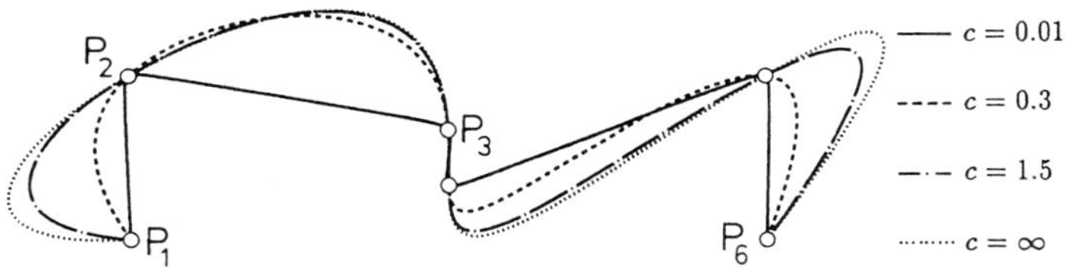

Figure 3. Interpolating MQ–curves illustrating the curve in tension effect.

This property of the MQ–curve may be the reason for the good performance of the MQ–series, also in the multivariate case. The slight tendency to be bulbous at the boundaries of the interpolating curve (surface) can be explained by this 'curve in tension' (or in higher dimensions 'surface in tension') behaviour.

Finally, it should be remarked that of course in practice using (1) or (3), c cannot be chosen arbitrarily high because then the interpolation matrix becomes numerically singular. In general c has to be decreased if the knot distance becomes smaller.

§4. Local Basis Function

In the next two sections, we abandon the consideration of MQ–curves with constant precision for general number N of knots, and will instead investigate a much simpler MQ–segment. This segment $\phi(c,t)$, $c > 0$, $t \in [0,1]$, is defined over only two knots $t_1 = 0$ and $t_2 = 1$ and has linear precision, whereby the coefficients $\{\mathbf{a}_i \,:\, i = 0,\dots,3\} \subset \mathbb{R}^s$ have not to fulfill any condition, *cf.* [11]:

$$\phi(c,t) = \mathbf{a}_0 \cdot \sqrt{c^2+t^2} + \mathbf{a}_1 \cdot \sqrt{c^2+(1-t)^2} + \mathbf{a}_2 \cdot t + \mathbf{a}_3 \cdot (1-t) \quad . \tag{4}$$

We introduce (4) in order to reduce the calculations in determing a local basis of $\phi(c,t)$ with properties similar to the Bernstein basis in the case of polynomials. For this we rewrite (4) as

$$\phi(c,t) = \sum_{i=0}^{3} \mathbf{b}_i \cdot \Omega_i(c,t), \quad t \in [0,1] \tag{5}$$

using the following basis transformation

$$\begin{aligned} \Omega_1(c,t) &= (s(c)+c)^2 \cdot \frac{\beta(c)}{\alpha(c)} \cdot \left(t - s(c)\sqrt{c^2+t^2} + c\sqrt{c^2+(1-t)^2}\right) \\ \Omega_2(c,t) &= \Omega_1(c,1-t) \\ \Omega_3(c,t) &= t - \alpha(c)\cdot\Omega_1(c,t) + (\alpha(c)-1)\cdot\Omega_2(c,t) \\ \Omega_0(c,t) &= \Omega_3(c,1-t), \end{aligned} \tag{6}$$

where the auxiliary functions which appear are defined as

$$s(x) = \sqrt{x^2+1}, \quad \beta(x) = \frac{s(x)}{s(x)+x}, \quad \alpha(x) = \frac{x\cdot s(x)}{2x^2+1+x\cdot s(x)} \quad .$$

Note that the important and frequently used monotone increasing (resp. decreasing) functions $\alpha(x)$ and $\beta(x)$ are always positive for $x > 0$ and have the limits $\alpha(\infty) = \frac{1}{3}$ (resp. $\beta(\infty) = \frac{1}{2}$).

Now, comparing (4) with (5) and (6), we get the 'control'–points $\mathbf{b}_i$ by

$$\begin{pmatrix} \mathbf{b}_0 \\ \mathbf{b}_1 \\ \mathbf{b}_2 \\ \mathbf{b}_3 \end{pmatrix} = \begin{pmatrix} c & s(c) & 0 & 1 \\ c & s(c)-\frac{\alpha(c)}{s(c)} & \alpha(c) & 1-\alpha(c) \\ s(c)-\frac{\alpha(c)}{s(c)} & c & 1-\alpha(c) & \alpha(c) \\ s(c) & c & 1 & 0 \end{pmatrix} \cdot \begin{pmatrix} \mathbf{a}_0 \\ \mathbf{a}_1 \\ \mathbf{a}_2 \\ \mathbf{a}_3 \end{pmatrix} \quad . \tag{7}$$

At a first glance, (5) seems to be more complicated than (4), but actually some nice properties hold. It yields the 'control'–points: ($'$ denotes differentiation of $\phi(c,t)$ regarding to t)

$$
\begin{aligned}
\mathbf{b}_0 &= \phi(c,0)\\
\mathbf{b}_1 &= \phi(c,0)+\alpha(c)\phi'(c,0) = \phi(c,1)+(\alpha(c)-1)\,\phi'(c,1)+\alpha(c)\beta(c)\phi''(c,1)\\
\mathbf{b}_2 &= \phi(c,1)-\alpha(c)\phi'(c,1) = \phi(c,0)+(1-\alpha(c))\,\phi'(c,0)+\alpha(c)\beta(c)\phi''(c,0)\\
\mathbf{b}_3 &= \phi(c,1).
\end{aligned}
\tag{8}
$$

The formulae (8) are very important since, firstly, the points $\mathbf{b}_0$ and $\mathbf{b}_3$ are the two curve end points and, secondly, considering the curve in at least three dimensions, $\mathbf{b}_1$ is in fact the intersection point of the tangent in $\phi(c,0)$ and the osculating plane in $\phi(c,1)$, and vice versa for $\mathbf{b}_2$. This behaviour is also valid for Bézier points in the polynomial case, *cf.* [7].

Now turning our attention to the consideration of the basis functions Ω_i, we are able to confirm immediately from (6) that

(i) $\sum_{i=0}^{3} \Omega_i(c,t) \equiv 1$

(ii) $\Omega_i(c,t) = \Omega_{3-i}(c,1-t), \quad (i=0,\ldots,3) \quad , \quad t \in \mathbb{R}$

Property (i) expresses the affine connection of the 'control'–points with the MQ–curve, since the functions Ω_i can be interpreted as barycentric coordinates. (ii) allows the reordering $\overline{\mathbf{b}}_i = \mathbf{b}_{3-i}$ of the 'control'–points without casually changing the curve. The third property is not obvious because of the differences appearing in (6):

(iii) $\Omega_i(c,t) \geq 0, \quad (i=0,\ldots,3), \quad t \in [0,1], \quad c > 0\ .$

But rearranging the basis functions by introducing the variable $\overline{c} = c^{-1}$, we get, for example, for the second basis function the following identity

$$
\Omega_1(c,t) = \overline{\Omega}_1(\overline{c},t) = \frac{2\,(1+s(\overline{c}))\,\left(2+\overline{c}^2+s(\overline{c})\right)t(1-t)^2}{(1+s(\overline{c}(1-t)))\,\left(\overline{c}^2 t + s(\overline{c}(1-t)) + s(\overline{c})s(\overline{c}t)\right)}, \tag{9}
$$

which ensures positivity for all $t \in [0,1]$ and $c > 0$. Similar arguments are valid for Ω_3, and together with (ii) the positivity of all basis functions is proved. Again (i) and (iii) ensure that the MQ–curve always lies in the *convex–hull* of the 'control'–points.

The evaluation of the Ω_i using (6) for large c (or small $\overline{c}$) is numerically problematic since cancellation occurs. This effect is avoided in (9), which is therefore numerically stable. The remaining basis functions $\Omega_i(c,t) = \overline{\Omega}_i(\overline{c},t)$, $(i=0,2,3)$, can then be determined depending on $\Omega_1(c,t) = \overline{\Omega}_1(\overline{c},t)$ using (6), where the auxiliary function α is computed as

$$
\alpha(c) = \overline{\alpha}(\overline{c}) = \frac{s(\overline{c})}{2+\overline{c}^2+s(\overline{c})} \quad . \tag{10}
$$

Now, considering the limit case $c \to \infty$ (or equivalent $\overline{c} = 0$), we deduce from (9) and (10) together with $s(0) = 1$ and $\overline{\alpha}(0) = \frac{1}{3}$:

(iv) $\lim_{c\to\infty} \Omega_i(c,t) = \overline{\Omega}_i(0,t) = \binom{3}{i} \cdot t^i \cdot (1-t)^{3-i} := B_i^3(t), \quad (i=0,\ldots,3)\ .$

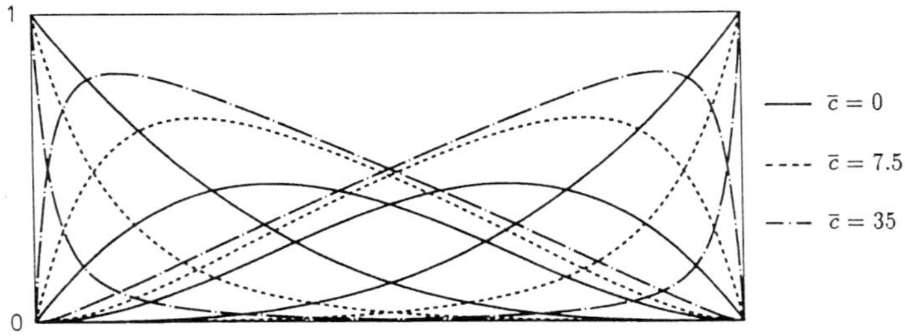

Figure 4. Local basis functions with different values of $\overline{c} = c^{-1}$.

This property corresponds to Theorem 1 since $B_i^3(t)$ denotes the cubic Bernstein polynomials. Fig. 4 shows the four basis functions for $\overline{c} = 7.5$ and $\overline{c} = 35$ as well as for $\overline{c} = 0$.

Furthermore, (iv) allows us to use $\overline{c} = c^{-1}$ like a *tension parameter* to control the shape of the curve segment $\phi(c,t)$ as the resulting curve is pulled to the 'control'–polygon $\{\mathbf{b}_j : j = 0, \ldots, 3\}$ as illustrated in Fig. 5.

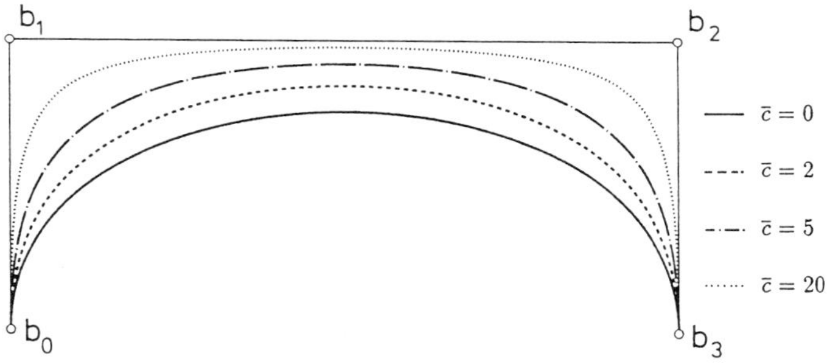

Figure 5. $\phi(c,t)$ with different values of $c = \overline{c}^{-1}$.

§5. Spline Interpolation

In the following we consider piecewise MQ–curve segments as introduced in Section 4 meeting at the knots with a certain continuity class. Firstly, we interpolate given points and tangents and obtain a C^1*–Hermite spline in tension* which is locally determined. Secondly, only points are interpolated by a C^2*–spline in tension*. Here the solution is global.

5.1. C^1–Hermite Spline in Tension

In this section we are interested in MQ–segments $\mathbf{X}_i(t) := \phi\,(c_i, z)$,

$z = \frac{t-t_i}{\Delta_i}$, $t \in [t_i, t_{i+1}]$, $c_i > 0$, $(i = 1, \ldots, N-1)$ such that

$$\mathbf{X}_i(t_i) = \mathbf{P}_i \ , \ \mathbf{X}_i'(t_i) = \mathbf{T}_i \ , \ \mathbf{X}_i(t_{i+1}) = \mathbf{P}_{i+1} \ , \ \mathbf{X}_i'(t_{i+1}) = \mathbf{T}_{i+1} \ . \tag{11}$$

Using the relations (8), the conditions (11) determine the four unknown coefficients $\mathbf{b}_{j,i}$ in an easy manner locally. Here $\mathbf{b}_{j,i}$ denotes the j–th 'control'–point of the i–th segment. Again it is obvious that we can use $\bar{c}_i = c_i^{-1}$ as a *tension parameter* to control the shape of the curve segment $\mathbf{X}_i(t)$. For $\bar{c}_i = 0$ we get the cubic Hermite polynomial whereas increasing $\bar{c}_i$ the curve tends to the straight line from $\mathbf{P}_i$ to $\mathbf{P}_{i+1}$; see Fig. 6.

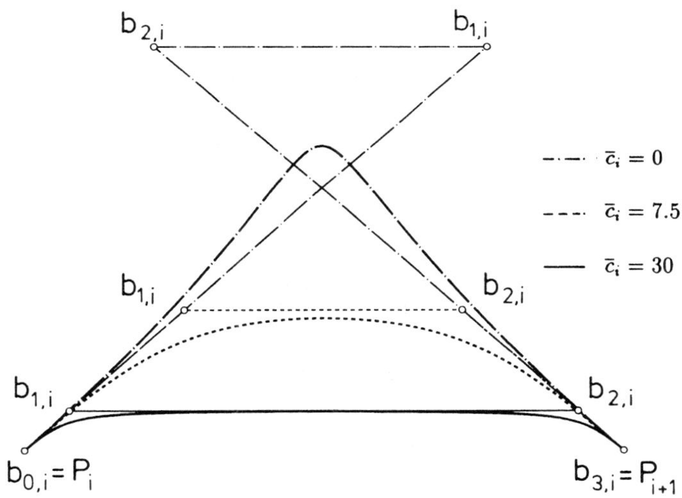

Figure 6. C^1–Hermite spline segment with different values of $\bar{c}_i$.

5.2. C^2–Spline in Tension

The exponential case of C^2–splines in tension is well known and has been investigated extensively. A good overview is given in [13].

In the following we will construct a piecewise C^2–continuous MQ–spline, interpolating N data points $\{\mathbf{P}_i\}$, where each of the $N-1$ segments $\mathbf{X}_i(t) := \phi(c_i, z)$, $c_i > 0$, $t \in [t_i, t_{i+1}]$ is again of the simple form (4).

The first possibility is to use the setting of Pruess [12], who investigated properties of general alternatives to the exponential C^2–continuous spline in tension. Similar to the interpolating cubic polynomial spline, he introduced the unknown second derivatives $\mathbf{P}_i''$ at the knots t_i and an auxiliary function F to rewrite the i–th segment as

$$\mathbf{X}_i(t) = \mathbf{P}_{i+1} \cdot z + \mathbf{P}_i \cdot (1-z) + \mathbf{P}''_{i+1} \cdot \triangle_i^2 \cdot F(c_i, z) + \mathbf{P}''_i \cdot \triangle_i^2 \cdot F(c_i, 1-z) \, .$$

Calculating the function F for the MQ–segment (4), we can show the validity of all the various restrictions which are imposed in [12] on a general function F so that the resulting spline has desirable mathematical and computational properties. The according tension parameter is again $\overline{c}_i = c_i^{-1}$. But for the sake of brevity we skip the details here, and direct our attention to a construction of a local support B–spline type basis of this kind of MQ–spline.

The method consists of the transformation to piecewise defined MQ–curve segments with basis functions Ω_i of the form (6). Then the requirements $\mathbf{X}_{i-1}^{(r)}(t_i) = \mathbf{X}_i^{(r)}(t_i)$ for $r = 0, 1, 2$ of the continuity of two adjacent segments lead to conditions for the 'control'–points $\mathbf{b}_{j,i}$. Using the relations (8), we get the following conditions by introducing an *auxiliary 'control'–point* $\mathbf{d}_{i-2}$:

C^0: $\quad \mathbf{b}_{3,i-1} = \mathbf{b}_{0,i}$

$$C^1: \quad \mathbf{b}_{0,i} = \frac{\alpha_{i-1} \cdot \triangle_{i-1} \cdot \mathbf{b}_{1,i} + \alpha_i \cdot \triangle_i \cdot \mathbf{b}_{2,i-1}}{\alpha_{i-1} \cdot \triangle_{i-1} + \alpha_i \cdot \triangle_i}$$

$$C^2: \quad \begin{aligned} \mathbf{d}_{i-2} &= \frac{(\triangle_{i-1} + \gamma_i \cdot \triangle_i) \cdot \mathbf{b}_{2,i-1} - \gamma_i \cdot \triangle_i \cdot \mathbf{b}_{1,i-1}}{\triangle_{i-1}} \\ &= \frac{(\triangle_i + \overline{\gamma}_i \cdot \triangle_{i-1}) \cdot \mathbf{b}_{1,i} - \overline{\gamma}_i \cdot \triangle_{i-1} \cdot \mathbf{b}_{2,i}}{\triangle_i}, \end{aligned}$$

with

$$\alpha_i = \alpha(c_i), \quad \beta_i = \beta(c_i), \quad \gamma_i = \alpha_i \beta_i \rho_i, \quad \overline{\gamma}_i = \alpha_{i-1} \beta_{i-1} \rho_i$$

and

$$\rho_i = \frac{\alpha_{i-1} \cdot \triangle_{i-1} + \alpha_i \cdot \triangle_i}{\alpha_{i-1} \cdot \beta_{i-1} \cdot (1 - 2\alpha_i) \cdot \triangle_{i-1} + \alpha_i \cdot \beta_i \cdot (1 - 2\alpha_{i-1}) \cdot \triangle_i}.$$

Now, on the other hand, these continuity conditions can be used to construct the 'control'–points $\mathbf{b}_{j,i}$ from a given sequence of auxiliary 'control'–points $\mathbf{d}_i$, given increasing knots t_i and tension values c_i, as indicated in Fig. 7.

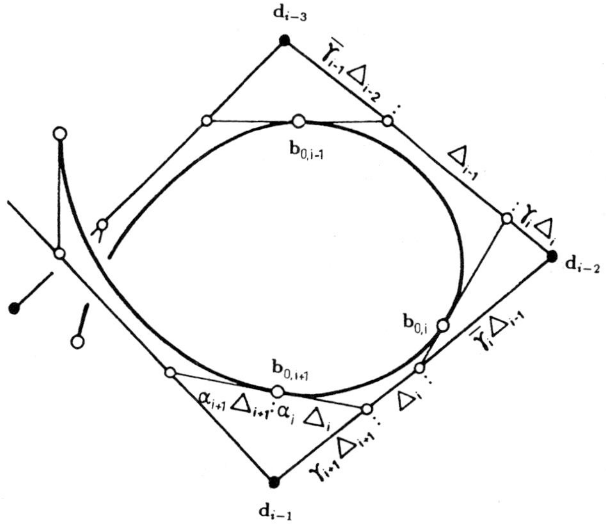

Figure 7. Construction of a C^2–MQ–spline in tension.

We remark here that in the case of $c_i = 0$ and nonvanishing knot differences, the i–th segment degenerates to a part of the straight line from $\mathbf{d}_i$ to $\mathbf{d}_{i+1}$, whereby the tangents are continuous at $\mathbf{X}(t_i)$ and $\mathbf{X}(t_{i+1})$. Moreover, if $t_i = t_{i+1}$ (or equivalent $\triangle_i = 0$), then the MQ–spline is only C^1–continuous at t_i.

Furthermore, the construction gives local support basis functions $G_i(t)$ by simply using the one–dimensional auxiliary 'control'–points $d_j = \delta_{i,j}$, where $\delta_{i,j}$ denotes the usual Kronecker symbol. Thus the C^2–MQ–spline in tension has the B–spline like representation:

$$\mathbf{X}(t) = \sum_i \mathbf{d}_i \cdot G_i(t).$$

Note that the basis functions $G_i(t)$ also form a partition of unity, as immediately follows from the construction. Moreover, the $G_i(t)$ are always positive since $\alpha_i > 0$, $\beta_i > 0$, $\rho_i > 0$ for all i. Also, for increasing c_i, we get the expected uniform limit:

$$\lim_{c \to \infty} G_i(t) = N_{i,3}(t), \quad c = \min_i\{c_i\} \quad .$$

References

1. Beatson, R. K., and M. J. D. Powell, Univariate multiquadric approximation: quasi–interpolation to scattered data, preprint, 1990.

2. Buhmann, M., private communication, 1991.
3. Eck, M., Geometrische Verfahren zur dreidimensionalen Osteotomieplanung, dissertation, TH Darmstadt, 1991.
4. Franke, R., Scattered data interpolation: tests of some methods, Math. Comp. **38** (1982), 181–200.
5. Hardy, R. L., Multiquadric equations of topography and other irregular surfaces, J. Geophys. Res. **76** (1971), 1905–1915.
6. Hardy, R. L., Theory and applications of the multiquadric–biharmonic method, Comput. Math. Applic. **19** (1990), 163–208.
7. Hoschek, J., and D. Lasser, *Grundlagen der geometrischen Datenverarbeitung*, Teubner Verlag, 1989.
8. Kansa, E. J., Multiquadrics – a scattered data approximation scheme with applications to computational fluid dynamics, Comput. Math. Applic. **19** (1990), 127–145.
9. Micchelli, C. A., Interpolation of scattered data: distance matrices and conditionally positive definite functions, Constr. Approx. **2** (1986), 11–22.
10. Pottmann, H., and M. Eck, Modified multiquadric methods for scattered data interpolation over a sphere, Computer–Aided Geom. Design **7** (1990), 313–321.
11. Powell, M. J. D., The theory of radial basis function approximation, preprint, 1990.
12. Pruess, S., Alternatives to the exponential spline in tension, Math. Comp. **33** (1979), 1273–1281.
13. Rentrop, P., and U. Wever, Theory and application of the exponential spline, preprint, 1991.

Matthias Eck
Fachbereich Mathematik
Technische Hochschule Darmstadt
Schlossgartenstr. 7
6100 Darmstadt, GERMANY
db0k@ddathd21.bitnet

Offset Approximation Improvement by Control Point Perturbation

G. Elber and E. Cohen

Abstract. There does not usually exist a closed general NURBs representation of the offset curve (or surface) to a NURBs curve (surface). In a related paper [6] a method was developed to determine a sequence of approximations to the offset curve (surface) of a given curve (surface) with the properties that the global error of each approximation from the true offset can be bounded and the sequence converges to the true offset. In this paper we take the next step and develop a method using the analysis of the offset error function that perturbs the control points of a specified NURBs approximation to an offset of a NURBs curve (surface), so that a better approximation to the offset results.

§1. Introduction

Offset surfaces are very important in modeling and manufacturing, so finding approximations of offset curves and surfaces has undergone extensive research. Creating an offset to a curve is an intuitive operation and has been understood mathematically for more than a hundred years [1, 13, 14]. The offset operator is closed for arcs and lines, i.e. an offset of an arc and a line are an arc and a line, respectively. This is not so for Bézier and NURB curves, in general, so approximations are usually derived.

Two methods for finding approximations to offset curves are commonly used. The first approximates the curve using piecewise lines and arcs and then finds the representation of the exact offset to the approximation. That approach was introduced in [11] and used successfully in [9]. The second method attempts to approximate the offset by directly transforming the curve representation, in particular the control points [2,4,7,8,12]. To improve the accuracy of the approximation in the second method, the original curve is subdivided [7,8,12] or manually refined [2] when the error is above a prespecified

Mathematical Methods in Computer Aided Geometric Design II
Tom Lyche and Larry L. Schumaker (eds.), pp. 229–237.

ISBN 0-12-460510-9.

tolerance level and the same technique is applied to each of the subdivided pieces. The original curve is usually subdivided in the middle of its parametric domain [7,8,12], although that is usually not the optimal location. Curve inflection points have also been considered as splitting points for planar offsets [8].

Neither of these two approaches bounds the offset error globally. In [6] the error of the approximation to the offset is computed as a function, and analyzed to provide a global tolerance bound. Furthermore this error function is also used to automatically identify the regions with larger error. An automatic iterative process is used to refine these regions so that an approximated offset representation is generated whose global error is within a prespecified tolerance.

Section 2 develops the necessary background for this paper. In Section 3 we show how perturbing control points by analyzing the error function results in a better offset approximation. Section 4 provides several results for curves and surfaces.

§2. Background

Let $C(t)$ be a planar regular parameterized curve which without loss of generality, is assumed to be in the $x - y$ plane. An offset curve for $C(t)$ by an amount d is defined mathematically as

$$\hat{\mathcal{C}}_d(t) = C(t) + N(t)d, \tag{1}$$

where $N(t)$ is the unit normal to the curve at t.

Similarly, an offset surface for a surface $S(u, v)$ by an amount d is mathematically defined as

$$\mathcal{S}_d(u, v) = S(u, v) + n(u, v)d, \tag{2}$$

where $n(u, v)$ is the unit normal to the surface at (u, v).

In this paper we will concentrate on characterizing methods for the NURBs representation since the Bézier representation is a subset of it. Given two freeform NURBs curves $C_1(t)$ and $C_2(t)$ (surfaces $S_1(u, v)$ and $S_2(u, v)$), their sum, difference and product are also NURBs curves (NURBs surfaces) [3,5,6,10]. Derivatives of NURBs curves (surfaces) are also NURB curves (surfaces) [3].

Therefore, if $N(t)$ $(n(u, v))$ could be computed and represented as a NURB, so could $\mathcal{C}_d(t)$ $(\mathcal{S}_d(u, v))$, respectively. Unfortunately, however, the general form of a unit normal requires the square root of a function which is usually not representable as either a rational or a piecewise rational. Thus, offsets of freeform curves and surfaces used in computation will, in general, be approximations.

Let $\mathcal{C}_d^a(t)$ be an approximation to the offset curve of $C(t)$ by an amount d (equation 1), and let $\delta(t) = \mathcal{C}_d^a(t) - C(t)$ be the difference curve. Ideally, if $\mathcal{C}_d^a(t) \equiv \mathcal{C}_d(t)$, then $\delta(t) \equiv dN(t)$.

As shown in [6] the magnitude of $\delta(t)$ can be efficiently computed to determine the accuracy of $C_d^a(t)$. Current offset techniques usually test this accuracy by evaluating this magnitude on a prespecified collection of sampled points. Direct representation of $\|\delta(t)\|$ would require the representation of a square root, so instead [6] uses $\psi(t) = \|\delta(t)\|^2$ and compares it to d^2:

$$\psi(t) = \|\delta(t)\|^2 = \delta_x(t)^2 + \delta_y(t)^2 + \delta_z(t)^2, \tag{3}$$

where $\delta_x(t)$, $\delta_y(t)$ and $\delta_z(t)$ are the components of $\delta(t)$.

A NURBs representation for (3) can be determined using multiplicative and additive properties of NURBs. Hereafter, assume that the coefficients of $\psi(t)$ can be computed and $\psi(t)$ is represented as a scalar NURBs curve. If $C_d^a(t)$ were exact, ψ would be a constant curve equal to d^2, so by subtracting d^2 from ψ one could find the error difference curve for a particular approximation:

$$\epsilon(t) = \psi(t) - d^2. \tag{4}$$

The extremal values of the coefficients of ϵ provide a global error measure. It is important to examine the consequences of computing $\epsilon(t)$ instead of $\varepsilon(t) = \|\delta(t)\| - d$, the real error between the exact offset curve and its approximation:

$$\epsilon(t) = \psi(t) - d^2 = \|\delta(t)\|^2 - d^2 = (\varepsilon(t) + d)^2 - d^2 = \varepsilon(t)^2 + 2d\varepsilon(t) \approx 2d\varepsilon(t). \tag{5}$$

In other words, by computing the differences of the squared magnitudes, the resulting error bound is scaled by the magnitude of twice the offset distance, $2d$, which is a constant and therefore easy to control. $\varepsilon(t)^2$ has been ignored since it is much smaller than $2d\varepsilon(t)$, as the error converges to zero.

The problem of finding the global offset error has been reduced to a problem of finding the extrema of a freeform scalar curve. Since the values of a scalar B-spline curve over an interval lie between the maximum and minimum values of the coefficients of the non-zero B-spline functions, a simple and computationally efficient way of locally bounding the curve is immediately available.

An iterative process in which each step uses the direct polygon transformation method [2] to compute offset approximations is used and the error function, $\epsilon(t)$ is computed for it. The curve (or surface) is then refined in the regions with error larger than can be tolerated, and the process repeats itself. The process terminates when the magnitudes of the extrema of $\epsilon(t)$ are within the tolerance.

The above method can be applied equally well to finding errors of offset surfaces. $\delta(t)$, $\psi(t)$ and $\epsilon(t)$ would be simply explicit surfaces, $\delta(u, v)$, $\psi(u, v)$ and $\epsilon(u, v)$, instead of explicit curves.

§3. Getting a Better Approximation of Offsets

In [6] a technique was developed to provide a global bound using an error function and used in conjunction with refinement to reduce maximum error.

Here we show how to reduce the error by using this bound and perturbing control points. For each control point, one could compute the gradient direction that maximizes the local change in the error function and move the control point in that direction. However, such a computation would be extremely expensive and slow. By refining and offsetting in the normal direction, it is known that the sequence of approximations to the offset converges to the exact offset [6]. Therefore, a simple candidate for a preferred direction in which to offset is the normal direction. We also show that perturbing in the normal direction results in an exact representation of offset of circular curves in a rational quadratic representation with no refinement at all. Therefore, the curve or surface normal direction at the node value associated with the respective control point is used as a first order approximation for the gradient to minimize the offset approximation error. The amount that the control point is moved is determined from the amount of error in the offset approximation, as is evaluated from ϵ at that location. The iterative algorithmic process for curves follows:

Algorithm: (Offset improvement using control point perturbation)

> **Input:**
> τ, required offset curve tolerance.
> $C(t)$, input curve.
> $C_d^a(t)$, offset approximation to input curve.
> d, offset distance.
> **Output:**
> $\hat{C}_d^a(t)$, improved curve approximation.
> $C_0^a(t) \Leftarrow C_d^a(t)$
> $i \Leftarrow 0$
> $MaxErr \Leftarrow Infinity$
> Do
>   Compute offset error $\epsilon(t)$ for $C(t)$, $C_i^a(t)$ (equations 3, 4)
> $C_{i+1}^a(t) \Leftarrow C_i^a(t)$ perturbed according to $\epsilon(t)$ at node values.
> $LastMaxErr \Leftarrow MaxErr$
> $MaxErr \Leftarrow \epsilon(t)$ highest error
> $i \Leftarrow i + 1$
> While ($MaxErr > \tau$ and $MaxErr < LastMaxErr$)

The error function is computed in each iteration and each control point is moved in the normal direction by the offset error amount at the node parameter value associated with this control point. This process repeats itself until no improvement is gained in the maximum error (i.e. no convergence) or the required tolerance is achieved. This iterative algorithm is interleaved with refinement to guarantee convergence [6]. In each step Algorithm 1 is invoked. If it improves the approximation to the required tolerance the whole process stops. Otherwise refinement takes place in the regions with high offset error and Algorithm 1 is invoked again.

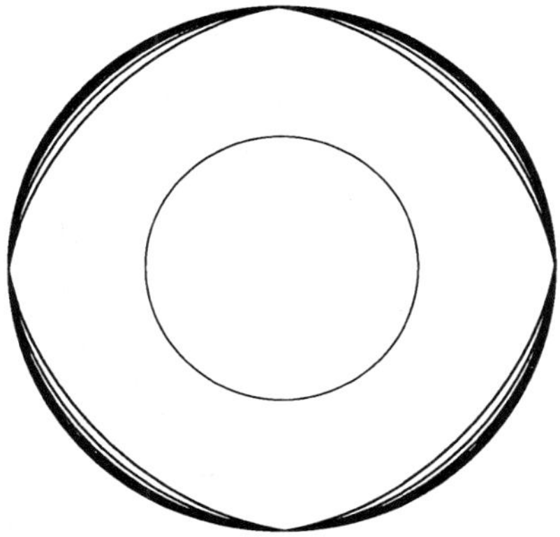

Figure 1. Control points perturbation converges to exact offset circle.

Figure 2. The error function convergence to zero, of the circle in Figure 1.

Algorithm 1 directly extends to surfaces. The process is exactly the same for surfaces where ϵ is a scalar surface, i.e. $\epsilon(u, v)$, and the control mesh is perturbed in the direction of the surface normal.

In Section 4 several examples on the affect of perturbing control points are provided.

§4. Examples

Figure 1 shows a unit circle composed of four 90 degree quadratic arcs. The first offset is obviously underestimated, but it converges quite quickly to the exact offset by moving only the corner points. These points have non-zero error, as can be seen from Figure 2 which also shows the respective error function as the process converges. The points in the error function in Figure 2 in which the error is always zero correspond to the points the circular curve interpolate the control points (the end points of the four quadratic Bézier segments).

Since the normals for the corner control point node values are in the direction (vectors $(\pm a, \pm a)$) pointing to the corner control points of a larger similarly represented circle, this process converges to an exact offset circle, with no refinement.

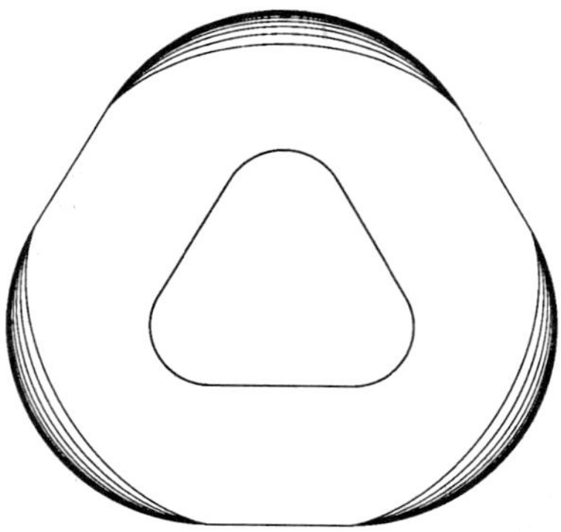

Figure 3. Error convergence to zero, for three 120 degrees arcs in curve.

In Figure 3 a quadratic curve consists of three arcs of 120 degrees and three lines is used. The offset error along the line is zero and no improvement is applied there. The arcs can be improved to the exact representation. The required tolerance of 0.01 terminated this process at that accuracy as can be seen in Table 1.

Step	Error	Comments
1	0.49	
2	0.387	
3	0.291	
4	0.211	
5	0.149	
6	0.104	
7	0.071	
8	0.048	
9	0.033	
10	0.022	
11	0.015	
12	< 0.01	Tolerance is met.

Table 1. Errors in offset of curve in Figure 3 using points perturbation.

Figure 4 is a case in which exact representation of the offset in the NURBs does not exist. Control points perturbation can only improve the result, but refinement is still necessary to meet the required tolerance of 0.04 as can be seen from Table 2.

Figure 4. Error function does not convergence to zero, for general curves.

Step	Error	Comments
1	0.22	
2	0.197	
3	0.187	
4	0.182	
5	0.179	
⋮	⋮	
16	0.178	No improvement - refinement stage
21	0.083	
22	0.040	
23	< 0.04	Tolerance is met.

Table 2. Errors in offset of curve in Figure 4 using control perturbation.

Figure 5 shows the same process applied to a unit sphere. This time the process does not converge to the exact representation since the normals at the node values of the corner points are not in the exact direction (vectors $(\pm a, \pm a, \pm a)$). Even so the improvement gained is quite significant. The right side of Figure 5 is the regular offset while the left side shows the same surface (and same number of control points) after perturbing it. Table 3 provides the convergence steps for this case, up to the prespecified tolerance of 0.1.

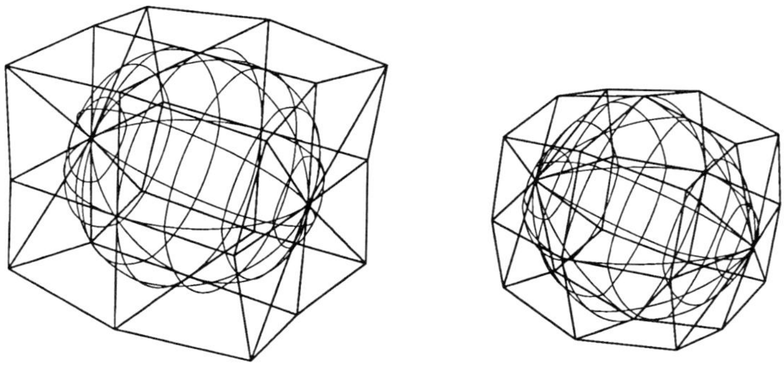

Figure 5. Perturbation can also improve offset surface accuracy.

Step	Error	Comments
1	2.574	
2	1.43	
3	0.964	
4	0.804	
5	0.733	
6	0.691	
⋮	⋮	
20	0.616	No improvement - refinement stage
21	0.296	
22	0.238	
23	0.186	
24	0.146	
25	0.115	
26	< 0.1	Tolerance is met.

Table 3. Errors in offset of sphere in Figure 5 using points perturbation.

Acknowledgments. Supported in part by DARPA (N00014-88-K-0689) and NSF (ASC-89 20219). All opinions, findings, conclusions or recommendations expressed are those of the authors and do not necessarily reflect the views of the sponsoring agencies.

References

1. Bertrand, J., La theorie des courbes a double courbure, Journal de Mathematique Pures et Appliquees **15** (1850), 332–350.
2. Cobb, B., Design of sculptured surfaces using the B-spline representation, Ph.D. thesis, University of Utah, Computer Science Department, June 1984.
3. Farin, G., *Curves and Surfaces for Computer Aided Geometric Design*, Academic Press, Inc., 1990.
4. Farouki, R. T., and C. A. Neff, Some analytic and algebraic properties of plane offset curves, Research Report 14364 (#64329) IBM Research Division, 1989.
5. Farouki, R. T., and V. T. Rajan, Algorithms for polynomials in Bernstein form, Computer Aided Geometric Design **5** (1988), 1–26.
6. Elber, G., and E. Cohen, Error bounded variable distance offset operator for free form curves and surfaces, Technical report No. 91-001, Computer Science, University of Utah, 1991.
7. Hoschek, J. Spline approximation of offset curves, Computer Aided Geometric Design **5** (1988), 33–40.
8. Hoschek, J. and N .Wissel, Optimal approximate conversion of spline curves and spline approximation of offset curves, Computer Aided Design **20** (1988), 475–483.
9. Chou, J. J., Numerical control milling machine toolpath generation for regions bounded by free form curves and surfaces, Ph.D. thesis, University of Utah, Computer Science Department, June 1989.
10. Mørken, K., Some identities for products and degree raising of splines, Constructive Approximation **7** (1991), 195–208.
11. Persson, H., NC machining of arbitrarily shaped pockets, Computer Aided Design **10** (1978), 170–174.
12. Pham, B., Offset Approximation of Uniform B-splines, Computer Aided Design **20** (1988), 471–474.
13. Salkowski, F., Zur Transformation von Raumkurven, Mathematische Annalen **66** (1909), 517–557.
14. Voss, A., Uber Kurvenpaare in Raumen, Sitzungsberichte, Akadamie der Wissenschaften zu Munchen **39** (1909), 106.

Gershon Elber and Elaine Cohen
Computer Science Dept.
University of Utah
Salt lake City
Utah, 84112
cohen@gr.utah.edu
gershon@gr.utah.edu

Curves and Surfaces in Geometrical Optics

Rida T. Farouki and Jean–Claude A. Chastang

Abstract. Elementary optical processes, such as the refraction or reflection of light from a point source by a spherical surface, give rise to wavefronts that may be described precisely by algebraic curves and surfaces of rather high degree. Usually, one is content to "sample" these wavefronts through the expedient technique of ray tracing. By propagating the refracted or reflected wavefronts backward in time, however, a particularly simple "initial" form — the *anticaustic* — may be identified. The refracted or reflected waves can then be regarded as emanating from the anticaustic via Huygens' principle in a homogeneous medium, i.e., they are *offsets* to the anticaustic. The simplest anticaustics of practical interest are (in two dimensions) the ellipse/hyperbola, the limaçon of Pascal, and the Cartesian oval, and (in three dimensions) the corresponding surfaces of revolution — oblate spheroids, hyperboloids of two sheets, and certain cyclides. We use these ideas to derive exact propagating–wavefront equations and to explore some of their interesting geometric properties.

§1. Introduction

Geometrical optics is concerned with describing the propagation of light — and other wave phenomena — in terms of simple geometrical principles. Thus, it may properly be regarded as much a branch of Euclidean geometry (whether algebraic or differential) as of physics. To be sure, such mathematical descriptions are only approximate solutions to the true physical problem, valid when the (unique) wavelength is small compared to characteristic dimensions of the system under consideration, so that "interference" and "diffraction" effects are insignificant. This limited domain is nevertheless of considerable practical and theoretical interest.

With the computer revolution, applied geometrical optics has become virtually synonymous with a (mostly) algorithmic procedure — ray tracing — that is a key ingredient in the design of optical instruments. *Light rays* may be regarded as the trajectories of "particles" of light, parameterized by the elapsed time since their emission. In a homogeneous medium, rays are simply

Mathematical Methods in Computer Aided Geometric Design II

Tom Lyche and Larry L. Schumaker (eds.), pp. 239–260.

ISBN 0-12-460510-9.

uniformly–parameterized straight lines. When they encounter an interface S between disparate homogeneous media $\mathcal{P}$ and $\mathcal{Q}$, however, rays are *refracted* or *reflected*, i.e., they suffer a sudden change in direction and/or speed.

The deflection of a ray impinging on the interface S is determined by the "refractive indices" p and q of the media $\mathcal{P}$ and $\mathcal{Q}$. *Snell's law* states that when a ray incident from $\mathcal{P}$ is refracted into $\mathcal{Q}$, the incident and refracted rays are coplanar with the surface normal at the point of incidence, and the oriented angles α and β that they make with this normal are related by

$$p \sin \alpha \;=\; q \sin \beta \,, \tag{1}$$

as depicted in Figure 1. A ray that is reflected back into the medium $\mathcal{P}$, on the other hand, satisfies the *law of reflection*

$$\beta \;=\; \pi - \alpha \,, \tag{2}$$

α and β being the angles that the incident and reflected rays make with the (coplanar) surface normal. It is understood here that angles are measured in a consistent (e.g., clockwise) sense with respect to the surface normal.

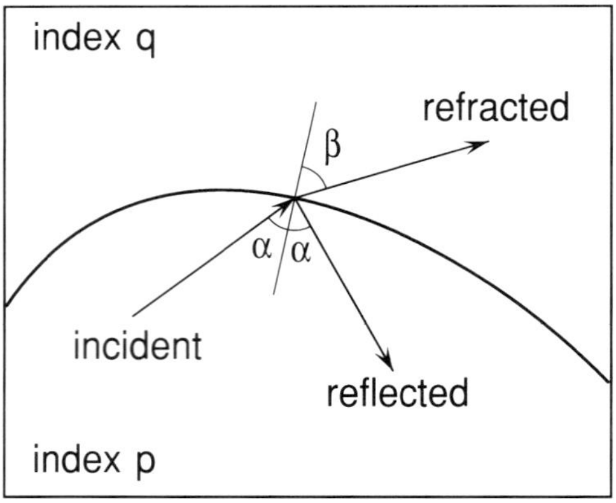

Figure 1. Refraction and reflection of light rays.

The terms *refracted* and *reflected* originate from the Latin words for "broken" and "turned back." The motivation for these terms is evident in the behavior of rays at an interface seen in Figure 1. The law of reflection (2) can be regarded as a special case of Snell's law, with $q = p$. Note also that if $p > q$, equation (1) indicates a *maximum incidence angle* for refraction to occur; rays incident at angles greater than $\alpha_{\max} = \sin^{-1} q/p$ suffer "total internal reflection" — i.e., they *must* be reflected back into the medium $\mathcal{P}$.

The above is an essentially complete characterization of the ray–tracing procedure for piecewise–homogeneous systems. Before we proceed, however, it is worth mentioning a rather more profound formulation that applies in a much broader setting — e.g., inhomogeneous media — and incorporates (1) and (2) as special instances, namely, *Fermat's principle*:

> Among all trajectories connecting an initial point **a** in $\mathcal{P}$ to a final point **b** in $\mathcal{Q}$ (refraction) or in $\mathcal{P}$ (reflection), via an arbitrary point of the interface S, the light ray "chooses" that which will *minimize the total time* in travelling from **a** to **b** (strictly, the time is a "stationary value" rather than a minimum).

Equations (1) and (2) follow directly from this principle if the *speed* of a light ray in $\mathcal{P}$ and $\mathcal{Q}$ is equal to c/p and c/q, respectively, where c is a universal constant. Thus, if L_1 and L_2 are the distances travelled in $\mathcal{P}$ and $\mathcal{Q}$, the quantity minimized is the *optical path*

$$\ell = p\,L_1 + q\,L_2\,. \tag{3}$$

Note also that, in the absence of a refracting/reflecting surface S, the linear nature of light rays in a homogeneous medium is a direct consequence of this "least time" principle.

Notwithstanding the practical utility of ray tracing, from a mathematical viewpoint there is a more complete and satisfying approach to describing light propagation — in terms of *wavefronts*. A wavefront can be regarded as the locus of points corresponding to a fixed time t, or optical path ℓ, on a family of rays (the term "phase" is also used for the parametric variable along rays that identifies successive wavefronts). Thus, tracing rays through an optical system amounts to a discrete (spatial) *sampling* of the wavefronts.

The theory of light propagation in terms of wavefronts is due [17] to the Dutch physicist and mathematician Christiaan Huygens (1629–1695), and is known to explain many optical phenomena that rays — interpreted, according to Newton's classical corpuscular theory, as trajectories of light particles — cannot predict (such effects will not concern us here, however).

In a homogeneous medium, wavefront propagation has a simple, intuitive description that corresponds to the uniform linear motion of rays. *Huygens' principle* states that given a smooth surface W_0 as an initial wavefront, at subsequent times the evolved wavefront W will be an *offset surface* to W_0, at fixed distance d from it — i.e., corresponding points of W_0 and W are distance d apart, measured along their mutual normal directions. The term "offset surface" comes from the field of computer–aided geometric design — classically, such loci were commonly known as "parallel surfaces." Huygens actually formulated the offset surface as the *envelope* of a family of spherical "wavelets" of fixed radius, centered on W_0.

When a wavefront encounters a smooth interface S between homogeneous media $\mathcal{P}$ and $\mathcal{Q}$, however, it suffers a rather complicated change of shape. Our principal goal is to formulate a precise description of this change, when both the original wave W_0 and the interface S are extremely simple (i.e., planar or spherical), within an established mathematical setting — the theory of *plane algebraic curves* [25], which is the study of loci defined by implicit polynomial equations of the form $f(x, y) = 0$. We shall see below that these very simple configurations can incur a surprising algebraic complexity of the refracted or reflected wavefronts.

The mathematical study of wavefronts has a distinguished history. In the XVIIth century it attracted the attention of prominent natural philosophers, such as Descartes, Fermat, Huygens, Tschirnhausen, (Jakob) Bernoulli, and others (see [12] for a brief historical review). Interest in "optical" curves and surfaces was also prevalent among the great Victorian algebraic geometers — e.g., Cayley [8] and Salmon [21]. In modern times, much work has centered on analyzing "topological" characteristics of wavefronts, using the techniques of modern differential geometry [1,6].

Since the configurations that we shall discuss generally possess rotational symmetry, it is not necessary to speak at length of surfaces *per se*. Rather, it will be more economical to restrict our attention to the plane curves that correspond to medial sections of the three–dimensional wavefronts. In the special cases treated below, the axisymmetric wavefront surfaces are offsets to certain *quadrics* and *cyclides* (for more on the latter, see [7,9,19]).

§2. Envelopes, Offset Curves, Evolutes and Involutes

The rays and wavefronts that propagate through an optical system form "conjugate" families — i.e., they cross each other orthogonally. Thus, rays coincide with the *normals* to wavefronts, while wavefronts mark off equal intervals of elapsed time along rays (they are "level curves" for the optical path). In describing the propagation of rays and wavefronts, two special loci — the *caustic* and the *anticaustic* — play a key rôle. In order to understand them, we need to review some basic ideas concerning envelopes, offset curves, and the evolutes and involutes of plane curves.

We begin by noting that a polynomial equation of the form

$$\phi(x, y, \xi) = 0 \tag{4}$$

defines a *family* of plane algebraic curves — each value of the parameter ξ identifying, in general, a distinct member of the family. The *envelope* of (4) is the locus that is tangent to each curve of the family; it may also be regarded as the locus of intersections of successive members ξ and $\xi + \mathrm{d}\xi$ of the family, in the limit $\mathrm{d}\xi \to 0$ (see [3] and [14]).

In principle, the equation $f(x, y) = 0$ of the envelope may be obtained by eliminating the parameter ξ between the equations $\phi = 0$ and $\partial\phi/\partial\xi = 0$, i.e., by taking the resultant with respect to ξ

$$f(x, y) = \text{Resultant}_{\xi}\left(\phi, \frac{\partial\phi}{\partial\xi}\right) \tag{5}$$

of (4) and its derivative. (The vanishing of the resultant of two polynomial equations with respect to a given variable represents a sufficient and necessary condition for the equations to be simultaneously satisfied at some value of that variable [24]. The resultant can be expressed in a number of ways as a determinant in the coefficients of the given equations — see [20].)

In performing this calculation, however, we note that factors extraneous to the desired envelope equation can arise if the family $\phi(x, y, \xi) = 0$ exhibits a locus of singular points, or if there are values of ξ that cause ϕ or $\partial\phi/\partial\xi$ to vanish identically — care must be taken to identify and discard such factors.

A class of envelopes that are of special interest are those that bound a family of circles of fixed radius d whose centers lie along a given curve C. If C is an "initial" wavefront, such an envelope will correspond (in a homogeneous medium) to the "evolved" wavefront at a later time, by virtue of Huygens' principle. We call such envelopes *offset curves*, since each point on them may be regarded as being obtained by a displacement d from the corresponding point of the given curve C, in the direction of its unit normal $\mathbf{n}$.

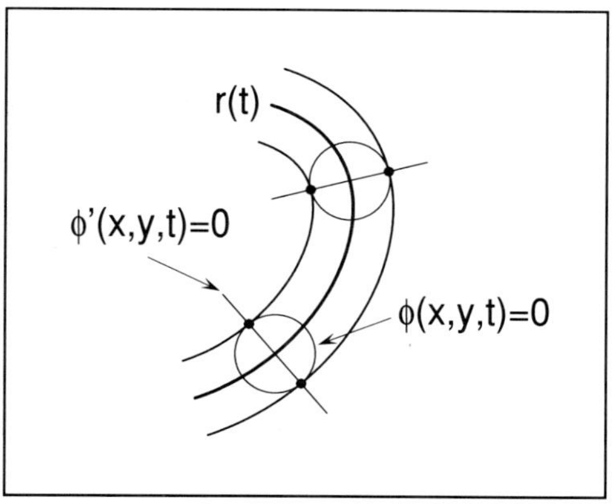

Figure 2. Computing the offset curve equation.

The equation of the offset curve is relatively easy to compute if the given curve C admits a polynomial or rational parameterization $\mathbf{r}(t) = \{x(t), y(t)\}$. The equation of the circle of radius d centered on each point t of C is then

$$\phi(x, y, t) = [x - x(t)]^2 + [y - y(t)]^2 - d^2 = 0, \tag{6}$$

and setting the derivative of (6) to zero defines the *normal line* to C at t

$$-\tfrac{1}{2}\phi'(x, y, t) = x'(t)[x - x(t)] + y'(t)[y - y(t)] = 0, \tag{7}$$

provided $x'(t)$ and $y'(t)$ never vanish simultaneously at any t. Eliminating the parameter t between (6) and (7) then gives the locus of points which, for some t, are the intersections of the normal line at $(x(t), y(t))$ with the circle of radius d centered on $(x(t), y(t))$ — i.e., it gives the offset at distance $|d|$ from C (see Figure 2). A more detailed discussion may be found in [13].

Note that the offset equation obtained in this manner simultaneously defines the "interior" and "exterior" offsets to the given curve C, at distances $-d$ and $+d$, as a single irreducible curve. Taken individually, the interior and exterior offsets are *not*, in general, algebraic curves.

Now the offset to a smooth curve is not necessarily smooth itself. In fact, as we traverse the offset it will exhibit a cusp — or sudden tangent reversal — at each point that happens to be the *center of curvature* for the corresponding point of the given curve C (assuming that the curvature of C is not a local extremum there). We can gain further insight into this behavior by examining the *evolute* of the curve C.

The evolute to a smooth plane curve C is usually defined as the *locus of its centers of curvature*. However, it can also be regarded as the *envelope of the normal lines to* C or the *locus of cusps on the offsets to* C as the offset distance d varies; see Figure 3. The reader may care to verify the equivalence of these definitions — which is, perhaps, not obvious at first sight (and may be subject to certain technical qualifications). Another important property of evolutes is that *all offsets to a given curve* C *share the same evolute with* C. Further discussion of these matters may be found in [5,13].

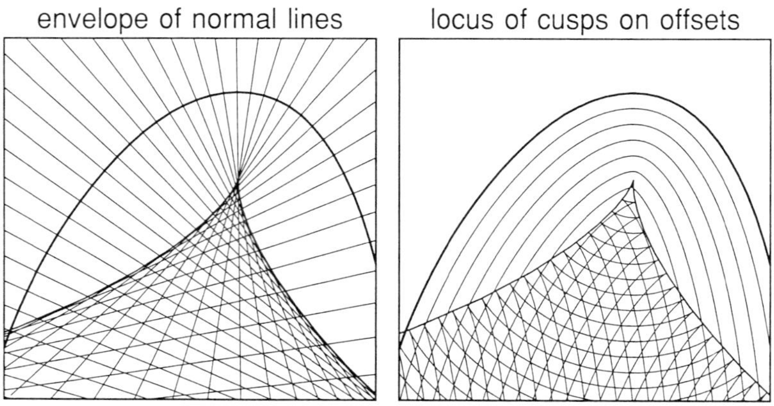

Figure 3. Alternate definitions for the evolute.

A curve that has a given curve C as its evolute is said to be an *involute* of C. Since all members of a family of offset curves share the same evolute, it is clear that any curve has an infinite sequence of involutes — i.e., the involute of a given curve is only defined modulo an arbitrary offset.

Returning to the optical context, suppose we are given a wavefront C and the light rays normal to C. Then the envelope of these rays — i.e., the evolute of C — represents the locus of greatest "concentration" of the rays. This locus was named the *caustic* — from the Greek for burning — by the German mathematician Ehrenfried Walther Tschirnhausen (or Tschirnhaus; 1651–1708), who was interested in igniting fires by focusing sunlight with lenses and mirrors. Hence, in a family of parallel wavefronts propagating in a homogeneous medium, the caustic is the evolute of all members of the family. Conversely, the wavefronts comprise the sequence of involutes to the caustic.

The caustic corresponding to spherical waves emanating from a point source is, of course, simply that point. Of greater interest are the caustics corresponding to spherical waves that have been refracted or reflected at a smooth interface S between two homogeneous media $\mathcal{P}$ and $\mathcal{Q}$ — the envelope of the refracted/reflected rays is then a rather complicated locus. In the case that S is planar or spherical, the caustic surface is rotationally symmetric and it suffices to consider its medial section.

Examples of caustics due to reflection by a spherical mirror are evident in Figure 4, where a sufficiently dense set of rays has been traced so that their envelope after reflection is clearly apparent. The cusped loci seen in Figure 4 are perhaps not unfamiliar to the reader — they may frequently be seen, for example, as luminous reflections from the surface of a cup of coffee; the circular rim of the cup reflects rays from a light bulb in the ceiling (say) onto the coffee, and the reflected rays concentrate along the caustic curve.

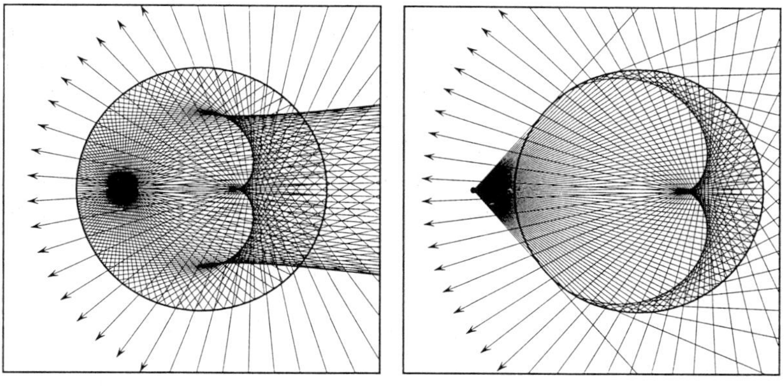

Figure 4. Caustics as envelopes of reflected rays.

The caustic for reflection by a spherical mirror is actually an irreducible curve of degree 6. For a mirror of unit radius centered on the origin, and a source situated at $(a,0)$, the equation of the caustic given by Cayley [8] is:

$$[\,(4a^2-1)(x^2+y^2)-2ax-a^2\,]^3 \;-\; 27a^2(x^2+y^2-a^2)^2y^2 \;=\; 0\,. \qquad (8)$$

Figure 5 illustrates the behavior of this curve as the source location a is varied. For a detailed discussion of how the morphology of (8) depends on the parameter a, see [8] or [16].

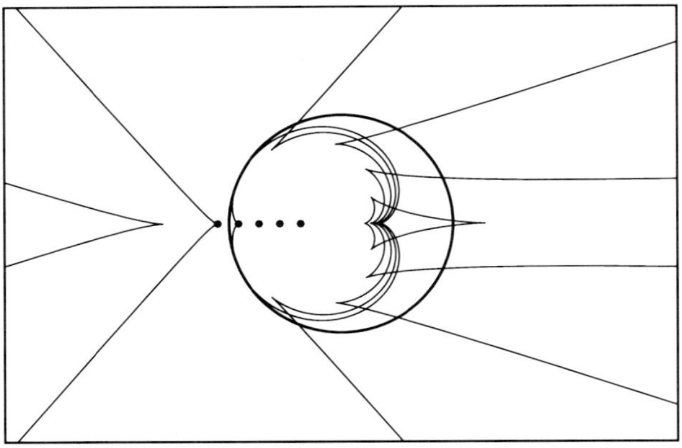

Figure 5. Caustics for reflection by a spherical mirror.

Now if a wave is refracted or reflected at a smooth surface, the wavefronts before and after the refraction/reflection cannot be regarded as members of a *single* family of offset curves — Huygens' principle does not apply directly to this problem. Nevertheless, we may invoke it in the following formal sense:

Suppose a spherical wave issues from a point source at time $t = 0$ and, after refraction or reflection at a smooth interface between two homogeneous media, subsequently assumes some shape W at time t. By propagating W *backward in time in a single homogeneous medium*, we can identify a certain "initial" shape W_0 at time $t = 0$. The significance of W_0 is that its uniform propagation via Huygens' principle — without refraction/reflection — gives rise to the true refracted/reflected wavefront W at the prescribed time t.

The hypothetical "initial" wavefront W_0 is known as the *anticaustic*. Thus, if we are able to identify the anticaustic for any given configuration of source and refracting/reflecting surface, the problem of determining the refracted/reflected wavefronts will reduce to that of computing the offsets to the anticaustic. This idea is illustrated in the simplest non–trivial context — refraction of spherical waves by a plane surface — in Figure 6.

The "anticaustic" concept is due [2] to the Swiss mathematician Jakob Bernoulli (1654–1705) — although it is sometimes erroneously attributed to his brother (and great rival) Johann, as in [4]. We emphasize that it should be regarded only as a formal mathematical construct, i.e., the anticaustic is a "virtual" (rather than a physical) wavefront.

Clearly, the caustic and anticaustic for any system will bear an evolute–involute relationship. Since *all* wavefronts corresponding to a given caustic are involutes of the latter, however, we need further justification for calling W_0 **the** anticaustic. The fact that W_0 represents the "initial" wavefront —

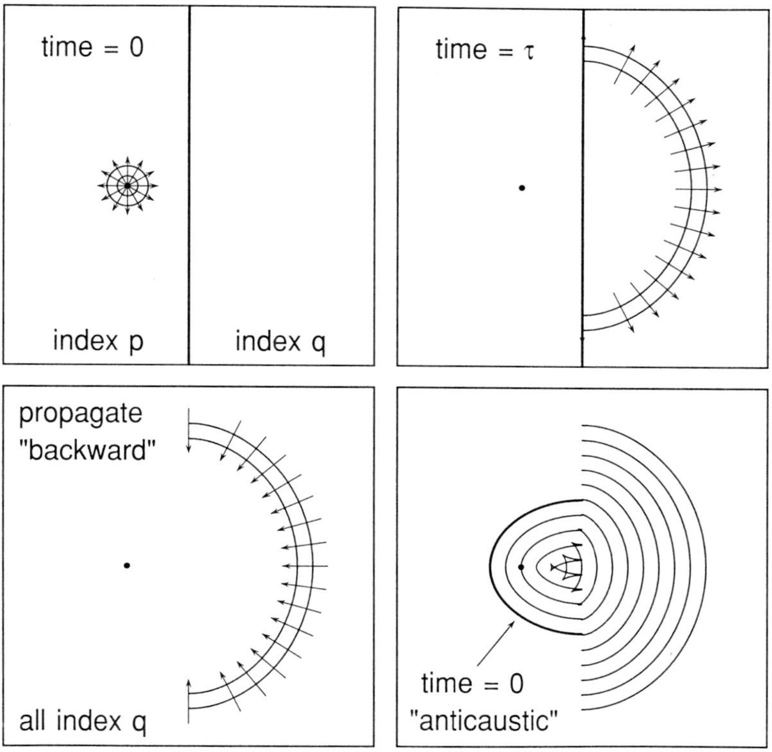

Figure 6. Anticaustic for refraction by a plane surface.

corresponding to zero optical path — means that *its algebraic degree is lower than that of all wavefronts parallel to it.* Thus, it is proper to distinguish W_0 and regard all the refracted/reflected wavefronts as offsets to it.

The anticaustic can be constructed in the following manner. Consider a ray emanating from a point source at (a, b) in the medium $\mathcal{P}$, impinging at a variable point (u, v) on the interface S, and proceeding to the point (x, y) in the medium $\mathcal{Q}$. The optical path from (a, b) to (x, y) via (u, v) is then

$$\ell = p\sqrt{(u-a)^2 + (v-b)^2} + q\sqrt{(x-u)^2 + (y-v)^2}\,, \tag{9}$$

and on setting $\ell = 0$ in (9) and squaring to remove radicals we have

$$\phi(x, y, u, v) = k^2\,[\,(x-u)^2 + (y-v)^2\,] - [\,(u-a)^2 + (v-b)^2\,] = 0\,, \tag{10}$$

where $k = q/p$. Equation (10) defines a family of circles, centered on the points (u, v) of the interface and of variable radius $r = k^{-1}\sqrt{(u-a)^2 + (v-b)^2}$.

The anticaustic is the *envelope* of this family of circles — its equation may be obtained by eliminating the coordinates u and v among the system of three equations $\phi = \phi_u = \phi_v = 0$, taking care to identify extraneous factors that may arise due to any degeneracies in this system of equations. If the interface S admits a rational parameterization $\{u(t), v(t)\}$, we may express (10) as $\phi(x, y, t)$ and then we only need to eliminate the parameter t from the two equations $\phi = \phi_t = 0$ to obtain the anticaustic.

Note that a number of alternate names for the anticaustic can be found in the geometrical optics literature: the *secondary caustic* [8,21], the *orthotomic* [16], the *archetypal wavefront* [23], and the *privileged wavefront* [12].

In summary, the following mutual relationships between rays, wavefronts, caustics, and anticaustics are worth bearing in mind (see Figure 7):

- The refracted/reflected rays are tangents to the caustic, and normals to the anticaustic and the refracted/reflected waves.
- The refracted/reflected waves are offsets to the anticaustic and involutes of the caustic; they cut the refracted/reflected rays orthogonally.
- The caustic is the envelope of the refracted/reflected rays, the locus of cusps on the refracted/reflected waves, and the evolute of the anticaustic.

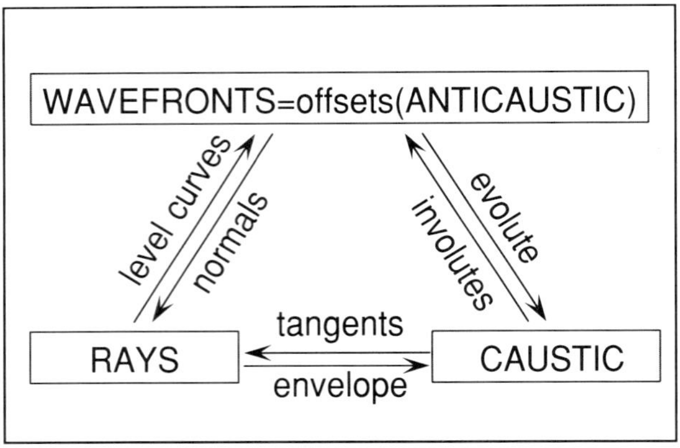

Figure 7. Relationships between rays, wavefronts, and caustics.

§3. Simple Refraction/reflection Problems

We now apply the ideas discussed above to certain configurations that are sufficiently "simple" to admit an explicit derivation of the refracted/reflected wavefront equations. Since we shall be regarding the refracted/reflected waves as offsets to the anticaustic, it is useful at this point to note some generic qualitative features of offset curves.

Figure 8 illustrates the behavior of interior and exterior offsets to an ellipse as the magnitude of the offset distance is increased. While the exterior offsets maintain a "shape" similar to that of the original curve, the interior offsets shrink, "collapse," and then re–emerge with an elongation opposite to that of the original curve and an apparent outward motion.

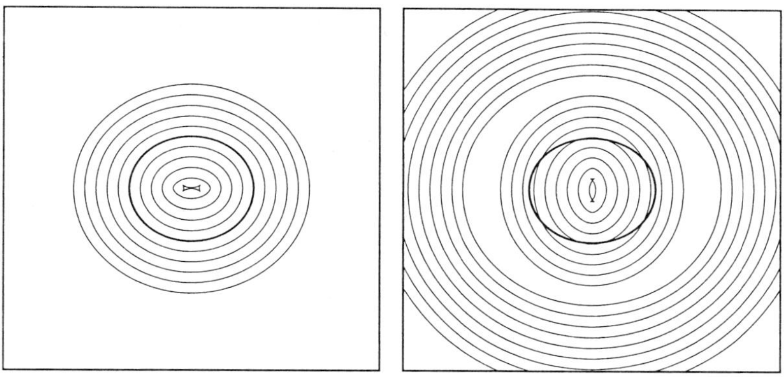

Figure 8. Behavior of "interior" and "exterior" offsets to an ellipse.

In the problems we treat below, the *physical* refracted wavefronts are always *interior* offsets to the anticaustic after emerging from the collapse phase (see Figure 6); the exterior offsets are non–physical. Therefore, we should not expect the "shape" of the refracted/reflected wavefronts to be similar to that of the anticaustic, although they are offsets to it.

3.1. Refraction/reflection of Spherical Waves by a Plane Surface

As is well known, the problem of *reflection* of spherical waves by a plane surface is trivial; the reflected waves are likewise spherical and appear to originate from an "image" source, situated on the opposite side of the mirror and at an equal distance from it.

The problem of *refraction* of spherical waves by a plane surface is not so straightforward (except, of course, if one is content to sample the wavefront by ray tracing). Without loss of generality, we may assume that the source is located at $(-1, 0)$, the refracting plane is $x = 0$, and the ratio of refractive indices q and p for the regions $x > 0$ and $x < 0$ is k $(\neq 1)$. Then it is not difficult to verify that the anticaustic is given by the equation

$$k^2\, x^2 + \frac{k^2}{1-k^2}\, y^2 = 1\,, \tag{11}$$

which defines either an ellipse (as in Figure 6 above) or a hyperbola, according to whether $k < 1$ or $k > 1$. This result was first demonstrated [10] by the French mathematician Joseph Gergonne (1771–1859).

Thus, the refracted wavefronts are (segments of) the "parallel" curves to an ellipse or hyperbola, as suggested already by Figure 6. The offset to (11) at distance $\pm d$ is an irreducible algebraic curve of degree eight, defined by

$$
\begin{aligned}
f_d(x,y) = {} & k^4\,(x^2+y^2)^2\,[\,(1-k^2)x^2+y^2\,]^2 \\
& - 2\,k^2\,[\,k^2(2k^4-3k^2+1)d^2+k^6-k^4-k^2+1\,]\,x^6 \\
& - 2\,k^2\,[\,k^2(3k^4-5k^2+3)d^2-k^6+2k^4-4k^2+3\,]\,x^4y^2 \\
& - 2\,k^2\,[\,k^2(k^4-k^2+3)d^2+3k^4-5k^2+3\,]\,x^2y^4 \\
& - 2\,k^2\,[\,k^2(k^2+1)d^2-2k^2+1\,]\,y^6 \\
& + [\,k^4(6k^4-6k^2+1)d^4+k^2(6k^6-8k^4+4k^2-2)d^2 \\
& \qquad +\,k^8+2k^6-6k^4+2k^2+1\,]\,x^4 \\
& + 2\,[\,k^4(3k^4-k^2+1)d^4-k^2(2k^6-3k^4-3k^2+2)d^2 \\
& \qquad -\,3k^6+4k^4-2k^2+1\,]\,x^2y^2 \\
& + [\,k^4(k^4+4k^2+1)d^4-k^2(6k^4-2k^2+2)d^2 \\
& \qquad +\,6k^4-6k^2+1\,]\,y^4 \\
& - 2\,[\,k^6(2k^2-1)d^6+k^4(3k^4-5k^2+3)d^4 \\
& \qquad +\,k^2(k^6-2k^4+4k^2-3)d^2+k^6-k^4-k^2+1\,]\,x^2 \\
& - 2\,[\,k^6(k^2+1)d^6-k^4(k^4-k^2+3)d^4 \\
& \qquad +\,k^2(3k^4-5k^2+3)d^2-2k^4+3k^2-1\,]\,y^2 \\
& + (k^2d^2-1)^2(k^2d^2+k^2-1)^2\,,
\end{aligned}
\tag{12}
$$

which holds for both $k < 1$ and $k > 1$. A form equivalent to (12) for the offset to an ellipse was obtained by Salmon more than century ago — a remarkable feat without the benefit of computer algebra!

3.2. Refraction/reflection of Spherical Waves by a Spherical Surface

We have seen that wavefronts resulting from the refraction of spherical waves from a point source by a plane surface are algebraic curves of degree 8 — the offsets to an ellipse or a hyperbola. If we replace the plane surface by a spherical one, the "privileged" wavefront (of which the refracted wavefronts are offsets) is a quartic curve, rather than a conic.

Specifically, consider a source located at $(a,0)$, where $|a| < 1$, and let the refractive indices q and p be associated with the regions $x^2+y^2 > 1$ and $x^2+y^2 < 1$. Then, with $k = q/p$, the anticaustic is an irreducible quartic curve known as the *Cartesian oval* and given by the equation

$$
f_0(x,y) \;=\; [\,k^2(x^2+y^2+1)-(a^2+1)\,]^2-4\,[\,(k^2x-a)^2+k^4y^2\,]=0\,. \tag{13}
$$

The identification of Cartesian ovals as the anticaustics corresponding to refraction of spherical waves by a spherical surface is due [10] to the Belgian mathematician Adolphe Quetelet (1796–1874), better known as a statistician. They are named after Descartes since he formulated these curves in a different — but also optical — context some two centuries earlier, in 1637 (considered as a *refracting surface*, rather than an anticaustic, the Cartesian oval has the property that it brings light from a point source to focus at a finite point).

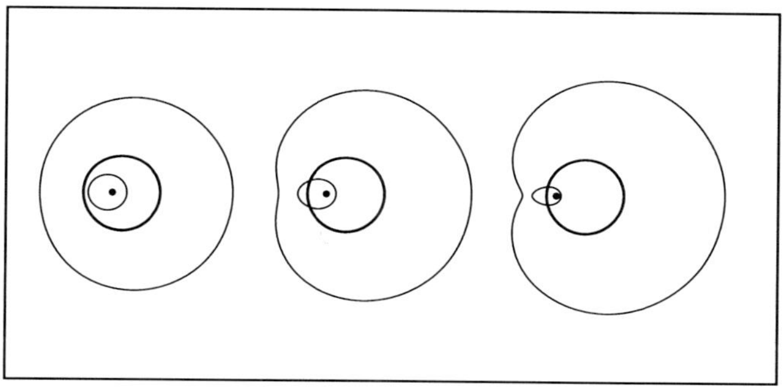

Figure 9. Cartesian ovals with $a = -0.25$, -0.5, and -0.75 (k fixed).

In general, the Cartesian oval is an *elliptic* curve [11] that has no rational parameterization (it is a *bicircular quartic*, having a double point at each of the "circular points at infinity" but no other singular points; a plane quartic should have *three* double points — or one triple point — to be rational). Representative examples are illustrated in Figure 9; note that the two "loops" together form (the real locus of) a single irreducible quartic curve.

The curve (13) has a simple description in terms of "bipolar" coordinates. If r_1 and r_2 are the distances of a variable point $\mathbf{p}$ from the "poles" $\mathbf{o}_1 = (a, 0)$ and $\mathbf{o}_2 = (a/k^2, 0)$, its equation is of the form [18]:

$$\frac{k}{1-k^2}\, r_1 \pm \frac{k^2}{1-k^2}\, r_2 = 1\,. \tag{14}$$

(There are actually three pairs of poles that give an equation of the form (14); see [15] for a more extensive discussion of Cartesian ovals.)

Now for general (i.e., symbolic) values of the source location a, the ratio of indices k, and the distance d, the offset to the Cartesian oval (13) is an algebraic curve of degree 14. Because (13) has no rational parameterization, a "multivariate" resultant technique is required to compute its offset equation. This equation is quite intimidating — it was given in [12] for the special case $k^2 = 2$, and we have subsequently succeeded in computing it for general k. We refrain from reproducing this equation here, relying on illustrations instead.

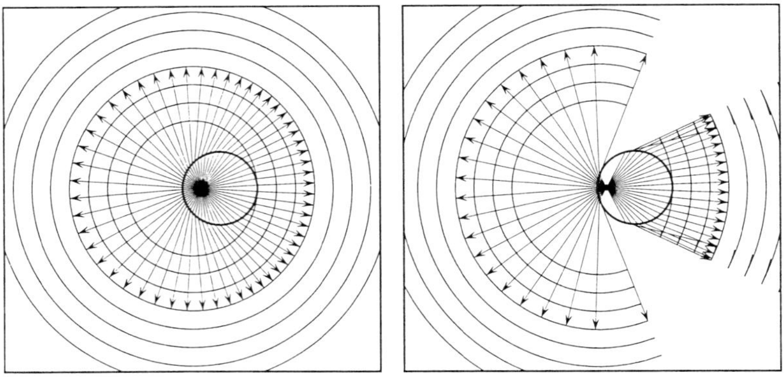

Figure 10. Refraction by a spherical surface.

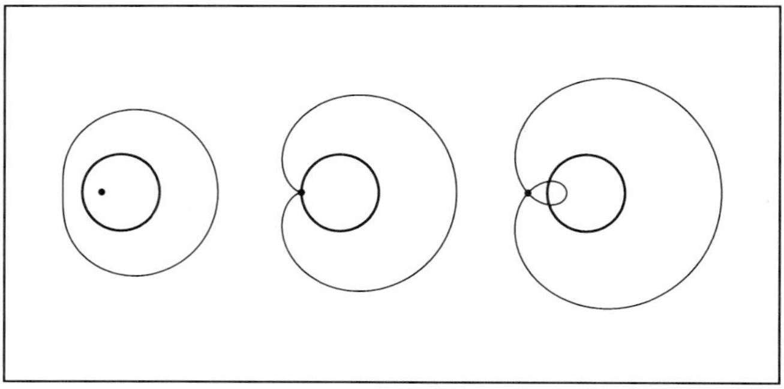

Figure 11. The limaçon with $a = -0.5$, -1.0, and -1.5.

It is interesting to study the behavior of the physical refracted wavefronts in the case $k < 1$ as we increase the magnitude of a from 0 to 1, i.e., as we move the source from the center of the refracting sphere toward its boundary. Initially, the physical refracted wave is a smooth, continuous loop coinciding precisely with the *interior* offset to the *smaller* loop of the Cartesian oval, as seen in Figure 10. A critical juncture is attained when $|a|$ attains the value k — the anticaustic (13) then specializes to a limaçon of Pascal (see below).

When $|a| < k$, all rays from the source arrive at the refracting surface with incidence angles (relative to the local surface normal) less than the maximum value $\alpha_{\max} = \sin^{-1} k$ for refraction into the external medium (see §1 above). When $|a| > k$, however, a certain range of rays will impinge on the spherical surface at angles too great to permit refraction into the external medium. The physical refracted wavefront then "breaks up" into forward– and backward–propagating components that correspond to *parts* of the interior offsets to the smaller and larger loops of the Cartesian oval — see Figure 10.

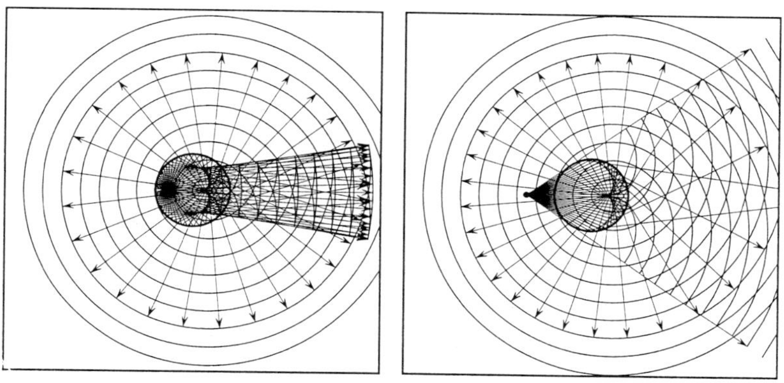

Figure 12. Reflection by a spherical surface.

The *reflection* of spherical waves by a spherical surface may be regarded as the special case $k^2 = 1$ of (13), which is much more tractable. With $k^2 = 1$, the Cartesian oval degenerates into a *limaçon of Pascal* (see Figure 11):

$$f_0(x, y) = (x^2 + y^2 - a^2)^2 - 4(x - a)^2 - 4y^2 = 0 . \tag{15}$$

Fortunately, the limaçon is a *rational* curve — it has the parameterization

$$x(t) = \frac{(2-a)t^4 + 6at^2 - (2+a)}{(1+t^2)^2}, \quad y(t) = \frac{4(1-a)t^3 + 4(1+a)t}{(1+t^2)^2} . \tag{16}$$

Figure 12 illustrates the reflection of spherical waves by a spherical mirror.

The offset to the limaçon (16) is an irreducible algebraic curve of degree 10, whose equation can be obtained by eliminating t between (6) and (7):

$$
\begin{aligned}
f_d(x,y) = {} & a^2\,(x^2+y^2)^5 \;-\; 2\,a\,x\,(x^2+y^2)^4 \\
& -[\,4a^2d^2+4a^4+8a^2-1\,]\,(x^2+y^2)^4 \\
& +2\,a\,[\,3d^2+4(3a^2+2)\,]\,x\,(x^2+y^2)^3 \\
& +2\,[\,3a^2d^4+(2a^4+2a^2-1)d^2 \\
& \qquad +\;3a^6+4a^4-2a^2-4\,]\,(x^2+y^2)^3 \\
& -16a^2\,(x^2+y^2)^2\,(x^2-y^2) \\
& -2\,a\,[\,3d^4-(13a^2+2)d^2 \\
& \qquad +\;2(11a^4+20a^2+4)\,]\,x\,(x^2+y^2)^2 \\
& -[\,4a^2d^6-(4a^4+16a^2+1)d^4 \\
& \qquad -\;2(2a^6-36a^4-27a^2-4)d^2 \\
& \qquad +\;2(2a^8-4a^6-51a^4-36a^2-8)\,]\,(x^2+y^2)^2 \\
& -8\,a^2\,[\,5d^2-8(a^2+1)\,]\,(x^2+y^2)(x^2-y^2) \\
& +2\,a\,[\,d^6-2(13a^2+5)d^4+(13a^4+92a^2+32)d^2 \\
& \qquad +\;4(3a^6-10a^4-20a^2-8)\,]\,x\,(x^2+y^2) \\
& -64\,a^3\,x\,(x^2-y^2) \\
& +a^2\,[\,d^8-4(a^2+3)d^6+2(3a^4+8a^2+26)d^4 \\
& \qquad -\;2(2a^6-2a^4+27a^2+48)d^2 \\
& \qquad +\;a^8-8a^6-4a^4+72a^2+64\,]\,(x^2+y^2) \\
& +2\,a^2\,[\,d^4-4(5a^2+2)d^2-8(a^4-4a^2-2)\,]\,(x^2-y^2) \\
& +2\,a^3\,[\,d^6-(3a^2+10)d^4+(3a^4+2a^2+32)d^2 \\
& \qquad -\;a^6+8a^4-8a^2-32\,]\,x \\
& +a^4\,(d-a-2)(d+a+2)(d-a+2)(d+a-2)\,.
\end{aligned}
\tag{17}
$$

Since the exact forms of refracted/reflected wavefronts are evidently quite complicated, it is often useful in geometrical optics to have *approximations* that are appropriate to limiting regimes — e.g., the "near–field" ($d \to 0$) and "far–field" ($d \to \infty$). It is interesting to see how such approximations can be extracted directly from an exact form such as (17).

We rewrite (17) as a polynomial in the offset distance d

$$
f_d(x,y) \;=\; \sum_{k=0}^{m/2} f_{d,2k}(x,y)\,d^{2k}\,, \tag{18}
$$

where $m = 4$ (i.e., for the limaçon, $f_d(x,y)$ is of degree 8 in d). The particular form (18) arises from the fact that only *even powers* of d occur in the offset equation. The coefficients $f_{d,2k}(x,y)$ in (18) are algebraic curves that depend on the parameter a, namely:

$$
\begin{aligned}
f_{d,8}(x,y) &= a^2(x^2+y^2)\,, \\
f_{d,6}(x,y) &= 2a\,[-2a(x^2+y^2)^2 + x(x^2+y^2) \\
&\quad - 2a(a^2+3)(x^2+y^2) + a^2x\,]\,, \\
f_{d,4}(x,y) &= 6a^2(x^2+y^2)^3 - 6ax(x^2+y^2)^2 \\
&\quad + (4a^4+16a^2+1)(x^2+y^2)^2 \\
&\quad - 4a(13a^2+5)x(x^2+y^2) \\
&\quad + 2a^2(3a^4+8a^2+26)(x^2+y^2) \\
&\quad + 2a^2(x^2-y^2) - 2a^3(3a^2+10)x + a^4\,, \\
f_{d,2}(x,y) &= 2\,[-2a^2(x^2+y^2)^4 + 3ax(x^2+y^2)^3 \\
&\quad + (2a^4+2a^2-1)(x^2+y^2)^3 \\
&\quad + a(13a^2+2)x(x^2+y^2)^2 \\
&\quad + (2a^6-36a^4-27a^2-4)(x^2+y^2)^2 \\
&\quad - 20a^2(x^2+y^2)(x^2-y^2) \\
&\quad + a(13a^4+92a^2+32)x(x^2+y^2) \\
&\quad - a^2(2a^6-2a^4+27a^2+48)(x^2+y^2) \\
&\quad - 4a^2(5a^2+2)(x^2-y^2) \\
&\quad + a^3(3a^4+2a^2+32)x - a^4(a^2+4)\,]\,, \\
f_{d,0}(x,y) &= [\,(x^2+y^2-a^2)^2 - 4(x-a)^2 - 4y^2\,]^2 \\
&\quad \times\,[\,(ax-1)^2 + a^2y^2\,]\,.
\end{aligned}
\tag{19}
$$

Note that $f_{d,2k}(x,y)$ is of degree $n-2k$ in (x,y), where $n=10$ (for the limaçon) is the overall degree of $f_d(x,y)$ in (x,y). Note also that on setting $d=0$ in (18), we obtain the *square* of the original limaçon equation (15), multiplied by the complex lines $a(x \pm \mathrm{i}y) = 1$. The fact that (15) arises with multiplicity 2 is not unexpected, since it will be recalled that $f_d(x,y)=0$ represents both the "interior" and "exterior" offsets to $f_0(x,y)=0$, both of which tend to the latter as $d \to 0$.

From (18) and (19) we thus have the near– and far–field approximations

$$f_{\text{near}}(x,y) \;\approx\; f_{d,0}(x,y) \,+\, d^2 f_{d,2}(x,y) \;=\; 0\,, \tag{20a}$$

$$f_{\text{far}}(x,y) \;\approx\; f_{d,8}(x,y) \,+\, d^{-2} f_{d,6}(x,y) \;=\; 0\,. \tag{20b}$$

In the case of the ellipse/hyperbola anticaustic, the offset equation is also of degree 8 in d. For the general Cartesian oval, however, powers of d up to 12 occur in the offset equation. In both cases, the terms $f_{d,2k}(x,y)$ are of degree $n-2k$ in (x,y), where n is the total degree of the offset $f_d(x,y)$ in (x,y) — 8 for the ellipse/hyperbola, and 14 for the Cartesian oval.

3.3. Refraction/reflection of Spherical Waves by Quadric Surfaces

The refraction and reflection of spherical waves by quadric surfaces other than planes and spheres poses a much more difficult problem. The resulting anticaustics are, in general, no longer axisymmetric and it will be necessary to derive explicit *surface* equations for them. In addition to this breakdown of axisymmetry, it can be expected that the anticaustics — and corresponding offset wavefronts — for refraction/reflection at general quadric surfaces will be of much higher degree than in the examples treated above. An indication of this can be gained from special examples, e.g., quadrics of revolution with sources on the principal axis, in which case the anticaustic is axisymmetric.

Consider, for example, the refraction of spherical waves from a source at $(1, 0)$ by the parabola $x = y^2$. Using the same techniques as above, we find that the medial section of the anticaustic is an irreducible curve of degree *six*,

$$\begin{aligned} f_0(x,y) = {} & 16k^6\,(x^2+y^2)\,[\,x^2+(1-k^2)y^2\,]^2 \\ & -\ 8k^6\,[\,4\,(k^2+1)x^4 \\ & \qquad +(4k^4-5k^2+8)x^2y^2 \\ & \qquad +(5k^4-9k^2+4)y^4\,]\,x \\ & +\ k^4\,[\,8\,(2k^6+5k^4+8k^2-5)x^4 \\ & \qquad +4\,(8k^6-3k^4+33k^2-20)x^2y^2 \\ & \qquad +(k^6-29k^4+68k^2-40)y^4\,] \\ & -\ 2k^4\,[\,4\,(k^6+7k^4-k^2-7)x^2 \\ & \qquad +(k^6+2k^4+45k^2-28)y^2\,]\,x \\ & +\ k^2\,[\,(k^8-4k^6-10k^4-100k^2+33)x^2 \\ & \qquad -(\,19k^6-67k^4+73k^2-33)y^2\,] \\ & +\ 8k^2(k^2+1)(k^4+6k^2-3)x\ -\ (k^4+6k^2-3)^2\,, \end{aligned} \tag{21}$$

where k denotes the ratio of refractive indices. The above expression — which represents only a special case — should be compared with the relatively simple anticaustics (11) and (13) for the general problems of refraction of spherical waves at planar and spherical surfaces. Correspondingly, the degree of the offset to (21) is likely to be much higher than the offset degrees (8 for the ellipse/hyperbola, and 14 for the Cartesian oval) obtained in the latter cases.

Aspheric refracting surfaces are rarely employed in practice. Apart from the difficulty of fabricating them, the ray–tracing analysis they incur is more involved — equation (21) offers a compelling indication of why this is so.

3.4. Refraction of Spherical Waves by a Planar "Slab"

We have seen above how to derive exact descriptions of the "deformed" wavefronts that result when spherical waves are refracted or reflected by a *single* surface. Optical instruments, of course, generally rely upon multiple refracting/reflecting surfaces for their image–forming capabilities. It is therefore of interest to see how the above ideas can be generalized to accommodate a sequence of surfaces. We shall discuss only the simplest case: refraction of spherical waves by a homogeneous "slab" of material of uniform thickness.

We begin by noting that equation (11) for the anticaustic corresponding to the refraction of spherical waves emitted from $(-1,0)$ by the plane $x=0$ can be generalized in a straightforward manner to an arbitrary source location (a,b) and refracting plane $x=c$ (assuming that $a<c$) to give

$$k^2\,(x-c)^2 + \frac{k^2}{1-k^2}\,(y-b)^2 = (c-a)^2\,, \tag{22}$$

which again defines an ellipse or a hyperbola, centered on the point (c,b).

Consider the refraction of spherical waves emitted by a point source at $(-1,0)$ in a medium of index p by a plane parallel slab of index q occupying the region $0 \le x \le h$ (the index p being associated with $x>h$ also).

By taking $(a,b)=(-1,0)$ and $c=0$ in (22), the anticaustic corresponding to the *first* refraction (i.e., at the plane $x=0$) is just that given by equation (11). It may be represented by the rational parameterization

$$X(t)=1-t^2\,,\; Y(t)=2\sqrt{1-k^2}\,t\,,\; W(t)=k(1+t^2) \quad \text{if } k<1\,, \tag{23a}$$

$$X(t)=1+t^2\,,\; Y(t)=2\sqrt{k^2-1}\,t\,,\; W(t)=k(1-t^2) \quad \text{if } k>1\,. \tag{23b}$$

We deal with the second refraction by considering *each point* on the ellipse (23a) or the hyperbola (23b) as a point source. From equation (22) with $a=X(t)/W(t)$, $b=Y(t)/W(t)$, and $c=h$, we find that, for such a source, the anticaustic corresponding to the second refraction is

$$\frac{(x-h)^2W^2(t)}{k^2} + \frac{[\,yW(t)-Y(t)\,]^2}{k^2-1} = [\,hW(t)-X(t)\,]^2\,, \tag{24}$$

provided that $W(t)\neq 0$. In writing (24), it should be noted that the ratio of indices appropriate to the second refraction is $1/k$.

For each t such that $W(t)\neq 0$, (24) defines a family of hyperbolae or ellipses, according to whether k is less than or greater than unity. The overall anticaustic for the complete slab refraction problem is the *envelope* of this family. Now a closer inspection of (24), based on the observation that

$$k^2(1-k^2)X^2(t)+k^2Y^2(t) \equiv (1-k^2)W^2(t) \tag{25}$$

for both (23a) and (23b), reveals that equation (24) actually contains $W(t)$ as a factor. This must be discarded before attempting to compute the envelope of the curves (24); otherwise it will introduce extraneous terms corresponding to the parameter values, $t=\pm\mathrm{i}$ for (23a) or ± 1 for (23b), at which $W(t)=0$.

There remains then the expression

$$\begin{aligned}\phi(x,y,t) = [\,k^2y^2-(1-k^2)(x-h)^2+(1-k^2)(1+k^2h^2)\,]\,W(t)\\ -\;2k^2yY(t)\;-\;2k^2(1-k^2)hX(t)\;=\;0\,,\end{aligned} \tag{26}$$

and on taking the resultant of (26) and its derivative $\partial\phi/\partial t$, we find that the anticaustic is formally the product of *three hyperbolae* or *three ellipses*, according to whether $k < 1$ or $k > 1$:

$$(k^2 - 1)(x - h + 1)^2 + k^2y^2 = k^2(k^2 - 1)h^2 , \tag{27a}$$

$$(k^2 - 1)(x - h - 1)^2 + k^2y^2 = k^2(k^2 - 1)h^2 , \tag{27b}$$

$$(k^2 - 1)(x - h)^2 + k^2y^2 = (k^2 - 1)(1 + kh)^2 . \tag{27c}$$

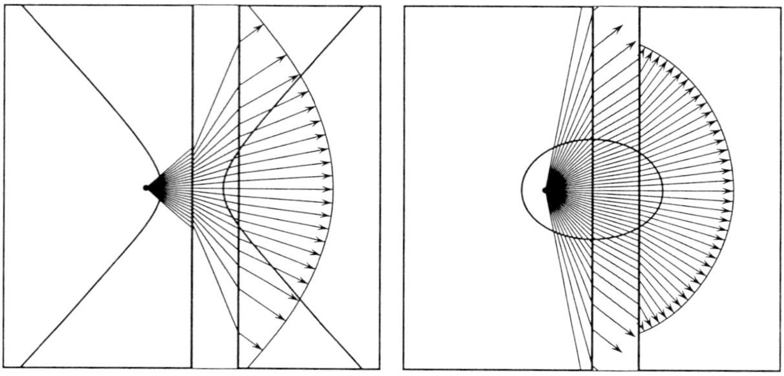

Figure 13. Refraction by a plane parallel "slab".

Figure 13 illustrates the slab refraction problem. For $k < 1$ and $k > 1$, the physical refracted wave corresponds to the (interior) offset of only the *first* of the three conics (27). This simple example shows that "compounding" refractions and reflections does not necessarily lead to an explosive growth in the degree of the overall anticaustic equation. It would be of great interest to repeat such calculations for the case of a planar and a spherical surface (plano–convex or plano–concave lenses), or two spherical surfaces (bi–convex, bi–concave, or meniscus lenses), to see whether the resulting anticaustics are products of relatively low–degree curves — of which perhaps only one gives rise to the physical wavefront. We are currently pursuing this matter.

§4. Concluding Remarks

In elementary optics texts, the precise forms of wavefronts that arise in the refraction of light from a point source by simple (planar or spherical) surfaces are frequently dismissed as mathematical monstrosities. The reader is encouraged to succumb directly to the temptation to simplify the problem at hand by solving it "incompletely" — namely, *sampling* the wavefronts by tracing a discrete ray–set. Quoting from [22], for example, we have (p. 40):

> *The mathematical expression for the shape of the refracted wavefront is so complex as to make it impracticable for purposes of computation*

for refraction by a plane, while for refraction by a sphere (p. 60):

> *... a wavefront that is originally spherical or plane does not have a geometrically simple form after refraction at a spherical surface. This makes it practically impossible to analyze the passage of light through an optical instrument in terms of wave surfaces ...*

Naturally, such statements fail to satisfy the inquisitive mathematical mind.

Drawing upon theory of offset curves developed in the CAGD context [13] and the power of modern computer algebra systems, we offer concrete mathematical characterizations for these "elementary" wavefronts — they are (the surfaces of revolution generated by) algebraic curves of degree 8 and 14. Furthermore, as illustrated in Color Plate 6, the geometry of anticaustics and wavefronts can be of great aesthetic appeal (red = rays, green = anticaustic, and light/dark blue = exterior/interior offsets to anticaustic).

Apart from their theoretical interest, however, it is perhaps premature to be drawing conclusions about the "practical" implications of the equations presented above. In principle, the exact wavefront equations encompass all the information furnished by standard approximations (paraxial rays, near– and far–fields, etc.), but determining the best means for extracting and employing this information is a substantive research project in its own right.

References

1. Arnold, V. I., S. M. Gusein–Zade, and A. N. Varchenko, *Singularities of Differentiable Maps*, Vol. 1, Birkhäuser, Boston, 1985.
2. Bernoulli, J., Lineæ cycloidales, evolutæ, ant–evolutæ, causticæ, anti–causticæ, peri–causticæ, Acta Eruditorum, May 1692.
3. Boltyanskii, V. G., *Envelopes*, MacMillan, New York, 1964.
4. Brocard, H., and T. Lemoyne, *Courbes Géométriques Remarquables*, Vol. I, Albert Blanchard, Paris, 1967.
5. Bruce, J. W., and P. J. Giblin, *Curves and Singularities*, Cambridge Univ. Press, 1984.
6. Bruce, J. W., P. J. Giblin, and C. G. Gibson, On caustics of plane curves, Amer. Math. Monthly **88** (1981), 651–667.
7. Casey, J., On cyclides and sphero–quartics, Phil. Trans. R. Soc. London **171** (1871), 585–721.
8. Cayley, A., A memoir upon caustics, Phil. Trans. R. Soc. London **147** (1857), 273–312; Supplementary memoir upon caustics, Phil. Trans. R. Soc. London **157** (1867), 7–16.
9. Cayley, A., On the cyclide, Quart. J. Pure Appl. Math. **12** (1873), 148–165.
10. Chasles, M., *Aperçu Historique sur l'Origine et le Développement des Méthodes en Géométrie*, Jacques Gabay, Paris, 1989, reprint.
11. Du Val, P., *Elliptic Functions and Elliptic Curves*, Cambridge Univ. Press, 1973.

12. Farouki, R. T., and J.–C. A. Chastang, Evolving wavefronts as algebraic curves, IBM Research Report RC16381, 1990.
13. Farouki, R. T., and C. A. Neff, Analytic properties of plane offset curves, Comput. Aided Geom. Design **7** (1990), 83–99; Algebraic properties of plane offset curves, Comput. Aided Geom. Design **7** (1990), 101–127.
14. Fowler, R. H., *The Elementary Differential Geometry of Plane Curves*, Cambridge Univ. Press, 1929.
15. Gomes Teixeira, F., *Traité des Courbes Spéciales Remarquables Planes et Gauches* (3 volumes), Chelsea, New York, 1907, reprint.
16. Herman, R. A., *A Treatise on Geometrical Optics*, Cambridge Univ. Press, 1900.
17. Huygens, C., *Treatise on Light*, Dover, New York, 1962, reprint.
18. Lockwood, E. H., *A Book of Curves*, Cambridge Univ. Press, 1967.
19. Maxwell, J. C., On the cyclide, Quart. J. Pure Appl. Math. **9** (1868), 111–126.
20. Salmon, G., *Lessons Introductory to the Modern Higher Algebra*, Chelsea, New York, 1960, reprint.
21. Salmon, G., *A Treatise on the Higher Plane Curves*, Chelsea, New York, 1960, reprint.
22. Sears, F. W., *Optics* (3rd edition), Addison–Wesley, Reading, MA, 1974.
23. Stavroudis, O. N., *The optics of rays, wavefronts, and caustics*, Academic Press, New York, 1972.
24. Uspensky, J. V., *Theory of Equations*, McGraw–Hill, New York, 1948.
25. Walker, R. J., *Algebraic Curves*, Springer–Verlag, New York, 1978.

Rida T. Farouki and Jean–Claude A. Chastang
IBM Thomas J. Watson Research Center,
P.O. Box 218, Yorktown Heights, NY 10598, USA.
farouki @ibm.com, chastan @ibm.com

Evaluation and Properties of the Derivative of a NURBS Curve

Michael S. Floater

Abstract. In this paper we describe two natural ways to express the first derivative of a rational B-spline curve in terms of its control points and weights. These expressions generalise the established ones for non-rational curves and offer numerically stable methods for computing the derivative. Some useful properties are derived from one of them.

§1. Introduction

Many applications of B-spline curves require the evaluation of derivatives. Examples include Newton iteration in intersection algorithms, calculating extremal points, blending, and offsetting. It is sometimes desirable also to calculate bounds on the magnitude of at least the first derivative, see Filip, Magedson, & Markot [4] and on the direction of the first derivative (the hodograph), see Sederberg & Meyers [7].

There are two known formulas for the first derivative of a (non-rational) B-spline curve. For simplicity we can refer to these as the *hodograph formula*, equations (5–6), and the *tangent formula*, equation (8), respectively. The aim of this article is to extend these results to NURBS curves and to derive some useful properties from them. A good introduction to the application of NURBS is given by Piegl & Tiller [6].

Corresponding expressions are also well established for Bézier curves and they were recently extended to rational Bézier curves, see Floater [5]. Some of the results in this article use similar techniques to those used for rational Bézier curves. Others warrant a different approach, see Proposition 5 for example.

The reasons for the naming convention for the two formulas is (a) that the hodograph property follows from the first, and (b) the fact that the curve

Mathematical Methods in Computer Aided Geometric Design II
Tom Lyche and Larry L. Schumaker (eds.), pp. 261–274.

ISBN 0-12-460510-9.

is tangential to the vector joining the two penultimate points in the de Boor algorithm follows from the second. This second property is well known in the B-spline community. It can be demonstrated using an alternative approach. One can insert enough knots at the chosen point to subdivide the curve there. Since each subcurve is in Bézier form at that point, one obtains the tangent property. The advantage of the tangent formula itself and indeed of the hodograph formula are that they offer stable methods for computing the first derivative.

Another possible approach to evaluating a NURBS curve and its derivatives is through the well established fact that each of its segments can be converted to rational Bézier form by inserting the appropriate number of knots at either end. This means that we could find the derivative of a NURBS curve at a point by converting the curve locally to Bézier form and then applying formulas for the derivative of a Bézier curve. But this is unnecessary as well as inefficient, and it is clearly superior to use (14–15) or (18) directly.

The usual method of calculating the mth derivative $R^{(m)}(u)$ is to write the quotient R as $R = a/b$. Then $bR = a$ so that, by the Leibnitz formula, $b'R + bR' = a'$, $b''R + 2b'R' + bR'' = a''$, etc. Then R, R', R'' etc. can be computed in sequence. It is difficult, on the other hand, to envisage a general sequence of expressions of the form (14–15) for higher derivatives. In fact by considering a NURBS representation for a conic section (R has just one segment and degree 2) one can see that it is impossible to write R'' in the form

$$R'' = \sum_{i=2}^{L+n-1} \kappa(u)(P_i - 2P_{i-1} + P_{i-2}),$$

unlike non-rational B-splines. Unless R represents a parabola, the direction of R'' cannot be constant. In conclusion, the general Leibnitz method is probably the best in applications. The forms of equations (14–15) and (18) we derive in this paper are rather special to the first derivative.

A strong motivation behind the mathematics was to derive the hodograph property for NURBS curves. This property is well known for non-rational B-splines (it follows immediately from (5–6)) and has been used to great advantage in intersection algorithms by, for example, Sederberg & Meyers [7]. The property has been proved before for rational Bézier curves by Sederberg & Wang [8], and via a different approach by Floater [5].

The following example illustrates how useful the hodograph property is in intersection problems. Suppose that we are given a NURBS curve $R(u)$ in 2-D and we wish to find out whether it intersects itself. To try to obtain information about R we use the hodograph property, Theorem 6. Regarding the vectors $(P_i - P_{i-1})$ as points in 2-D, it is clear that the set $C \subset \mathbb{R}^2$ of all points X of the form

$$X = \sum_{i=1}^{L+n-1} \gamma_i(P_i - P_{i-1}),$$

where $\gamma_i \geq 0$, is an infinite cone whose apex is located at the origin. C is

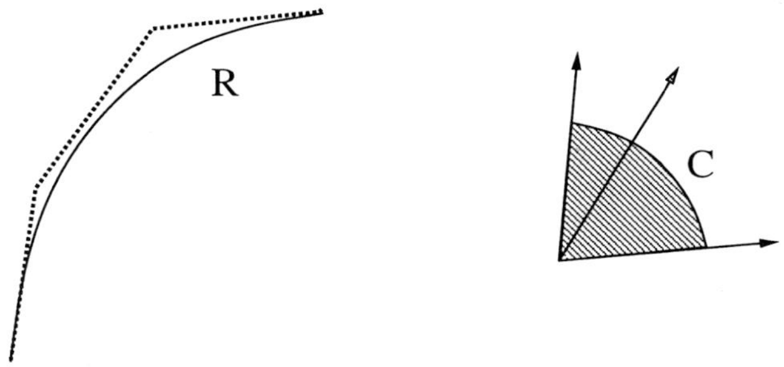

Figure 1. The cone angle is less than 180°.

known as the *direction cone* for R. Equation (14) shows that, for any u, $R'(u)$ lies inside the cone.

Now if the curve is to intersect itself, it must turn through at least 180°. Therefore if the direction cone has an angle of less than 180° the curve cannot intersect itself, see Figure 1. If the angle is any greater, on the other hand, the direction cone does not tell us anything, see Figure 2. In the second case, however, we could subdivide the curve in the hope that each subcurve *does* have a cone angle of less than 180°. This idea can be the starting point for a recursive subdivision algorithm to find all self-intersection points:

Algorithm 1

1. If cone angle $< 180°$, then finish.
2. Else
   2.1. Subdivide the curve into subcurves R_1 and R_2.
   2.2. Find self-intersection points of R_1.
   2.3. Find self-intersection points of R_2.
   2.4. Find intersection points of R_1 and R_2.

This scheme assumes we have another algorithm for intersecting two NURBS curves R_1 and R_2. Once again we can use the direction cones C_1 and C_2 to advantage as part of a recursive subdivision/iteration algorithm. This time we have to consider both tangent directions for each curve. Set

$$\hat{C}_1 = \{X \in \mathbb{R}^2 : X \in C_1 \text{ or } -X \in C_1\},$$

and define $\hat{C}_2$ similarly, see Figure 3. The scheme also makes use of the bounding boxes B_1 and B_2 for R_1 and R_2.

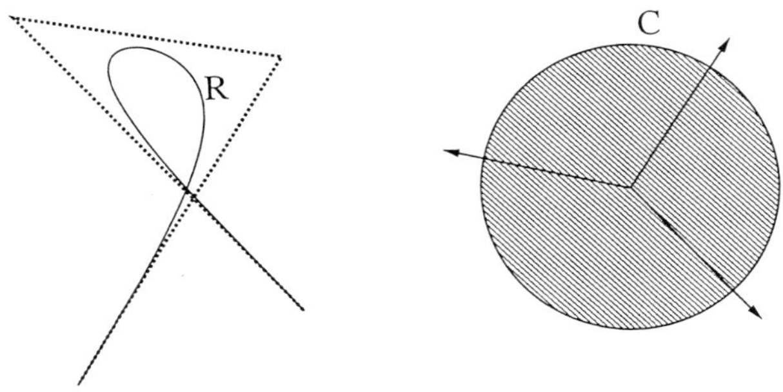

Figure 2. The cone angle is greater than 180°.

Algorithm 2

1. If $B_1 \cap B_2 = \emptyset$, then finish.
2. If $\hat{C}_1 \cap \hat{C}_2 = 0$, then there is at most one intersection. Stop subdividing and use Newton iteration.
3. Else
   3.1. Subdivide R_1 into subcurves R_{11} and R_{12}.
   3.2. Subdivide R_2 into subcurves R_{21} and R_{22}.
   3.3. Find $R_{11} \cap R_{21}$, $R_{12} \cap R_{21}$, $R_{11} \cap R_{22}$, and $R_{12} \cap R_{22}$.

In Section 2 we set out definitions and properties essential to the results in Section 3 where we present generalizations of the hodograph and tangent formulas to NURBS curves. The hodograph equation is then used to derive both the hodograph property in Section 4 and an upper bound on R' in Section 5.

§2. Definitions

To begin, we review some basic properties of non-uniform B-spline curves of both non-rational and rational type. In order to prove rigorous results we have to be specific and consistent about notation. We first of all describe some basic properties of ordinary B-splines and then go on to NURBS.

Following the notation of Farin [3], consider the B-spline curve containing L polynomial segments (some of which may be null). It is specified by the knot sequence $\{u_0 \leq \cdots \leq u_{L+2n-2}\}$ and the control points $P_0, \ldots, P_{L+n-1} \in \mathbb{R}^3$. The curve is defined by

$$S(u) = \sum_{i=0}^{L+n-1} N_{i,n}(u) P_i \tag{1}$$

for $u_{n-1} \leq u \leq u_{L+n-1}$, where $N_{i,n}$ is the ith basis function of degree n. $N_{i,n}$ is only nonzero in the interval $[u_{i-1}, u_{i+n}]$. Note that, strictly speaking, one

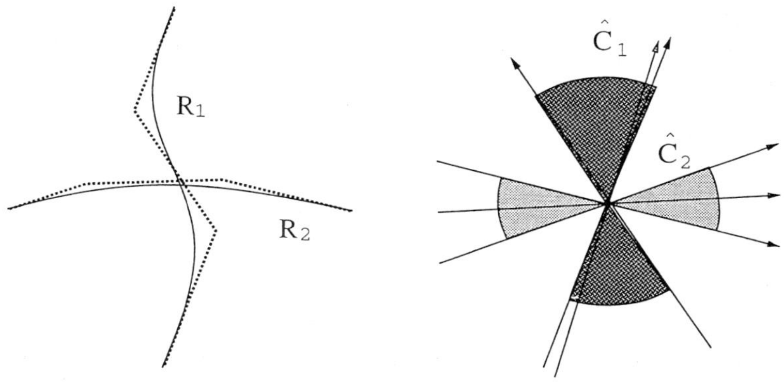

Figure 3. $\hat{C}_1$ and $\hat{C}_2$ intersect only at the origin.

ought to include two more knots u_{-1} and u_{L+2n-1} to justify the last statement, but these knots have no influence over the curve (for $u_{n-1} \leq u \leq u_{L+n-1}$) and are therefore omitted. The basis functions can be defined recursively by the Mansfield, de Boor, Cox recursion, see de Boor [2]:

$$N_{i,k}(u) = \left(\frac{u - u_{i-1}}{u_{i+k-1} - u_{i-1}}\right) N_{i,k-1}(u) + \left(\frac{u_{i+k} - u}{u_{i+k} - u_i}\right) N_{i+1,k-1}(u) \tag{2}$$

for $k \in \{1, \ldots, n\}$ and

$$N_{i,0} = \begin{cases} 1, & \text{if } u_{i-1} \leq u < u_i; \\ 0, & \text{otherwise.} \end{cases}$$

The B-spline basis functions are generalisations of the Bernstein ones; by putting $L = 1$, $u_0 = \cdots = u_{n-1} = 0$, and $u_n = \cdots = u_{2n-1} = 1$ we find $N_{i,n} = B_{i,n}$, the ith Bernstein polynomial of degree n.

The curve $S(u)$ itself can be defined recursively by the de Boor algorithm, due to de Boor [1]. Suppose that $u_I \leq u < u_{I+1}$. By considering the support of the basis functions one can deduce that the only control points which influence the curve in this parameter interval are $P_{I-n+1}, \ldots, P_{I+1}$. Let $P_{i,0}(u) = P_i$ for $i = I - n + 1, \ldots, I + 1$, and for $u_I \leq u < u_{I+1}$. Then define the functions $\alpha_{i,k}(u)$ and $P_{i,k}(u)$ as

$$\alpha_{i,k} = (u - u_{i-1})/(u_{i+n-k} - u_{i-1}) \tag{3}$$

and

$$P_{i,k} = (1 - \alpha_{i,k})P_{i-1,k-1} + \alpha_{i,k}P_{i,k-1} \tag{4}$$

for $k = 1, \ldots, n$ and $i = I - n + k + 1, \ldots, I + 1$. Then it can be shown that $S(u) = P_{I+1,n}$, see Farin [3]. In the special case $u = u_I$ it is only necessary to apply $n-r$ steps of the algorithm, where r is the multiplicity of the knot u_I and then $S(u) = P_{I+1,n-r}$.

In order to make comparisons for the results in Section 3, we state here the hodograph and tangent formulas for the first derivative of a B-spline.

The hodograph formula, given in Farin [3], is the following description of the derivative of $S(u)$:

$$S'(u) = \sum_{i=1}^{L+n-1} \lambda_i(u)(P_i - P_{i-1}), \tag{5}$$

where

$$\lambda_i(u) = nN_{i,n}(u)/(u_{n+i-1} - u_{i-1}). \tag{6}$$

In order to demonstrate this, one needs the following relationship between the derivative of a B-spline basis function and lower degree basis functions, see de Boor [1]:

$$N'_{i,k}(u) = \left(\frac{k}{u_{i+k-1} - u_{i-1}}\right) N_{i,k-1}(u) - \left(\frac{k}{u_{i+k} - u_i}\right) N_{i+1,k-1}(u) \tag{7}$$

for $k \in \{1, \ldots, n\}$. The hodograph formula (5–6) then follows directly from differentiating (1), applying (7), and making a change of subscripts.

The tangent formula for B-splines is as follows. For the sake of completeness we prove this here.

Proposition 1. *Given the B-spline curve of degree n defined by equation (1), the derivative of $S(u)$ is*

$$S'(u) = n\left(\frac{P_{I+1,n-1}(u) - P_{I,n-1}(u)}{u_{I+1} - u_I}\right) \tag{8}$$

where the $P_{i,n-1}$ terms are defined in (3–4).

Proof: We prove, by mathematical induction, the more general statement

$$P'_{i,k} = \frac{k(P_{i,k-1} - P_{i-1,k-1})}{u_{i+n-k} - u_{i-1}} = k\alpha'_{i,k}(P_{i,k-1} - P_{i-1,k-1}). \tag{9}$$

When $k = 1$ the equation can be seen to hold by differentiating equation (4). Now we show that (9) holds for any $k \in \{2, \ldots, n\}$ provided it holds when k is replaced by $k-1$.

Differentiating (4) with respect to u gives

$$P'_{i,k} = -\alpha'_{i,k}P_{i-1,k-1} + (1-\alpha_{i,k})P'_{i-1,k-1} + \alpha'_{i,k}P_{i,k-1} + \alpha_{i,k}P'_{i,k-1}.$$

Now by the induction step we can expand $P'_{i-1,k-1}$ and $P'_{i,k-1}$ so that

$$\begin{aligned} P'_{i,k} =& \alpha'_{i,k}(P_{i,k-1} - P_{i-1,k-1}) \\ &+ (k-1)(1-\alpha_{i,k})\alpha'_{i-1,k-1}(P_{i-1,k-2} - P_{i-2,k-2}) \\ &+ (k-1)\alpha_{i,k}\alpha'_{i,k-1}(P_{i,k-2} - P_{i-1,k-2}). \end{aligned}$$

The second two terms are now rearranged by applying the following identities:

$$\alpha_{i,k}\alpha'_{i,k-1} = \alpha'_{i,k}\alpha_{i,k-1},$$

$$(1-\alpha_{i,k})\alpha'_{i-1,k-1} = \alpha'_{i,k}(1-\alpha_{i,k-1}),$$

$$(1-\alpha_{i,k})\alpha'_{i-1,k-1} - \alpha_{i,k}\alpha'_{i,k-1} = \alpha'_{i,k}(1-\alpha_{i,k-1}) - \alpha'_{i,k}\alpha_{i-1,k-1}.$$

These are verified in a straightforward manner from (3). The rearrangement yields

$$\begin{aligned} P'_{i,k} =& \alpha'_{i,k}(P_{i,k-1} - P_{i-1,k-1}) \\ &+ (k-1)\{\alpha'_{i,k}(1-\alpha_{i,k-1})P_{i-1,k-2} + \alpha'_{i,k}\alpha_{i,k-1}P_{i,k-2} \\ &- \alpha'_{i,k}(1-\alpha_{i-1,k-1})P_{i-2,k-2} - \alpha'_{i,k}\alpha_{i-1,k-1}P_{i-1,k-2}\} \\ =& \alpha'_{i,k}(P_{i,k-1} - P_{i-1,k-1}) + (k-1)\alpha'_{i,k}(P_{i,k-1} - P_{i-1,k-1}) \end{aligned}$$

and we are finished. ■

Now we can define the NURBS curve as a generalisation of the the previously defined B-spline. The NURBS curve is specified from precisely the same data except that there are, in addition, (positive) weights $w_0, \ldots, w_{L+n-1} \in \mathbb{R}$ associated with the control points. The curve is defined to be

$$R(u) = \frac{\sum_{i=0}^{L+n-1} N_{i,n}(u)w_iP_i}{\sum_{i=0}^{L+n-1} N_{i,n}(u)w_i}. \tag{10}$$

Note that by setting all the weights to one we get back to (1).

There is an analogous rational version of the de Boor algorithm, see Farin [3]:

$$\alpha_{i,k} = (u-u_{i-1})/(u_{i+n-k} - u_{i-1}) \tag{11}$$

and

$$w_{i,k} = (1-\alpha_{i,k})w_{i-1,k-1} + \alpha_{i,k}w_{i,k-1} \tag{12}$$

and

$$w_{i,k}P_{i,k} = (1-\alpha_{i,k})w_{i-1,k-1}P_{i-1,k-1} + \alpha_{i,k}w_{i,k-1}P_{i,k-1} \tag{13}$$

with the same subscripts as before. One then finds $R(u) = P_{I+1,n}$. Again, setting all weights to one reproduces the non-rational scheme (3–4).

§3. The Derivative of a Rational B-spline Curve

Throughout this article we will only consider one parameter value u at a time so there is no ambiguity in dropping the u. We will also assume that R is differentiable at u. The only situation where this might not be so is when u is a knot value whose multiplicity is n. First we generalize the hodograph formula to NURBS curves.

Proposition 2. *The derivative of the NURBS curve R defined by equation (10) is*

$$R'(u) = \sum_{i=1}^{L+n-1} \lambda_i(u)(P_i - P_{i-1}), \tag{14}$$

where

$$\lambda_i(u) = \frac{1}{w_{I+1,n}^2(u)} \sum_{j=0}^{i-1} \sum_{k=i}^{L+n-1} (N'_{k,n}(u)N_{j,n}(u) - N_{k,n}(u)N'_{j,n}(u))w_j w_k. \tag{15}$$

Proof: Suppose that our parameter value u lies in the knot interval $[u_I, u_{I+1})$. The only basis functions at u which are nonzero are $N_{I-n+1}, \ldots, N_{I+1}$ so we can rewrite (1) as

$$R(u) = \sum_{i=I-n+1}^{I+1} \mu_i(u) w_i P_i, \qquad \mu_i(u) = \frac{N_{i,n}(u)}{\sum_{k=I-n+1}^{I+1} N_{k,n}(u) w_k} = \frac{N_{i,n}(u)}{w_{I+1,n}(u)}.$$

Then

$$\begin{aligned} \mu'_i =& \frac{1}{w_{I+1,n}^2} \left(\sum_{k=I-n+1}^{I+1} N_{k,n} w_k N'_{i,n} - \sum_{k=I-n+1}^{I+1} N'_{k,n} w_k N_{i,n} \right) \\ =& \frac{1}{w_{I+1,n}^2} \sum_{k=I-n+1}^{I+1} (N'_{i,n} N_{k,n} - N_{i,n} N'_{k,n}) w_k. \end{aligned}$$

Now, at any u, R' is in the form

$$R' = \sum_{i=I-n+1}^{I+1} a_i P_i,$$

where $a_i = \mu'_i(u) w_i$ and we wish to rewrite it in the form

$$R' = \sum_{i=I-n+2}^{I+1} \lambda_i (P_i - P_{i-1}). \tag{16}$$

By successively eliminating the control point with the lowest subscript, we find

$$\begin{aligned} R' =& a_{I-n+1} P_{I-n+1} + a_{I-n+2} P_{I-n+2} + \cdots + a_{I+1} P_{I+1} \\ =& - a_{I-n+1}(P_{I-n+2} - P_{I-n+2}) + (a_{I-n+1} + a_{I-n+2}) P_{I-n+2} \\ & + \cdots + a_{I+1} P_{I+1} \\ =& - a_{I-n+1}(P_{I-n+2} - P_{I-n+2}) - (a_{I-n+1} + a_{I-n+2})(P_{I-n+3} - P_{I-n+2}) \\ & + (a_{I-n+1} + a_{I-n+2} + a_{I-n+3}) P_{I-n+3} + \cdots + a_{I+1} P_{I+1} \\ =& \cdots \\ =& - \sum_{i=I-n+1}^{I+1} \sum_{j=I-n+1}^{i-1} a_j (P_1 - P_{i-1}) + \sum_{j=I-n+1}^{I+1} a_j P_{I+1}. \end{aligned}$$

Now the coefficient of the P_{I+1} term is the sum of all the a_j, and this is equivalent to

$$\frac{1}{w_{I+1,n}^2(u)} \sum_{j=I-n+1}^{I+1} \sum_{k=I-n+1}^{I+1} (N'_{j,n}(u)N_{k,n}(u) - N_{j,n}(u)N'_{k,n}(u))w_j w_k$$

which, by symmetry, is zero. Therefore, setting

$$\lambda_i = - \sum_{j=I-n+1}^{i-1} a_j,$$

we can express R' in the form of (16) and then

$$\begin{aligned}\lambda_i(u) &= \frac{-1}{w_{I+1,n}^2(u)} \sum_{j=I-n+1}^{i-1} \sum_{k=I-n+1}^{I+1} (N'_{j,n}(u)N_{k,n}(u) - N_{j,n}(u)N'_{k,n}(u))w_i w_k \\ &= \frac{1}{w_{I+1,n}^2(u)} \sum_{j=I-n+1}^{i-1} \sum_{k=I-n+1}^{I+1} (N'_{k,n}(u)N_{j,n}(u) - N_{k,n}(u)N'_{j,n}(u))w_i w_k.\end{aligned}$$

Again by symmetry,

$$\sum_{j=I-n+1}^{i-1} \sum_{k=I-n+1}^{i-1} (N'_{k,n}(u)N_{j,n}(u) - N_{k,n}(u)N'_{j,n}(u))w_i w_k = 0,$$

so these terms can be removed from the expression and we arrive at

$$\lambda_i(u) = \frac{1}{w_{I+1,n}^2(u)} \sum_{j=I-n+1}^{i-1} \sum_{k=i}^{I+1} (N'_{k,n}(u)N_{j,n}(u) - N_{k,n}(u)N'_{j,n}(u))w_i w_k. \quad (17)$$

Lastly, due to the fact that all basis functions are zero in $[u_I, u_{I+1})$ except for $N_{I-n+1}, \ldots, N_{I+1}$, we can extend the subscript ranges so that this expression becomes valid for any parameter $u \in [u_{n-1}, u_{L+n-1}]$. Therefore,

$$R'(u) = \sum_{i=1}^{L+n-1} \lambda_i(u)(P_i - P_{i-1}),$$

where

$$\lambda_i(u) = \frac{1}{w_{I+1,n}^2(u)} \sum_{j=0}^{i-1} \sum_{k=i}^{L+n-1} (N'_{k,n}(u)N_{j,n}(u) - N_{k,n}(u)N'_{j,n}(u))w_j w_k.$$

This completes the proof. ■

In the next section, we show that every $\lambda_i(u)$ is nonnegative for all u. Now we generalize the tangent formula to NURBS curves.

Proposition 3. *The derivative of the NURBS curve R defined by equation (10) is*

$$R'(u) = n\frac{w_{I,n-1}(u)w_{I+1,n-1}(u)}{w_{I+1,n}^2(u)}\left(\frac{P_{I+1,n-1}(u) - P_{I,n-1}(u)}{u_{I+1} - u_I}\right), \tag{18}$$

where the $P_{i,k}$ $w_{i,k}$ terms are defined in (11–13).

Proof: Let $R(u) = a(u)/b(u)$. Then

$$R'(u) = (b(u)a'(u) - b'(u)a(u))/b^2(u). \tag{19}$$

Observe that the numerator $a(u)$ and denominator $b(u)$ of $R(u)$ are themselves non-rational B-spline curves of dimensions 3 and 1 respectively. The control points of $b(u)$ are precisely the weights w_i of $R(u)$. The weight functions $w_{i,k}(u)$ computed from the de Boor scheme (3–4) (with $P_{i,k}$ replaced by $w_{i,k}$) are identical to those computed from the rational scheme (11–13). Therefore, applying Proposition 1 means that

$$b' = n(w_{I+1,n-1} - w_{I,n-1})/(u_{I+1} - u_I).$$

Now consider $a(u)$. Its ith control point is the product w_iP_i which we can rename Q_i say. From the de Boor algorithm (3–4), we find

$$Q_{i,k} = (1 - \alpha_{i,k})Q_{i-1,k-1} + \alpha_{i,k}Q_{i,k-1}.$$

Then Proposition 1 implies

$$a' = n(Q_{I+1,n-1} - Q_{I,n-1})/(u_{I+1} - u_I).$$

But by considering (13) one can see that at each step in the algorithm

$$Q_{i,k}(u) = w_{i,k}(u)P_{i,k}(u),$$

and so

$$b' = n(w_{I+1,n-1}P_{I+1,n-1} - w_{I,n-1}P_{I,n-1})/(u_{I+1} - u_I).$$

The substitution of these identities into (19) implies

$$\begin{aligned}
&(u_{I+1} - u_I)w_{I+1,n}^2R' \\
&\quad= n\{((1 - \alpha_{I+1,n})w_{I,n-1} + \alpha_{I+1,n}w_{I+1,n-1}) \\
&\qquad\quad (w_{I+1,n-1}P_{I+1,n-1} - w_{I,n-1}P_{I,n-1}) \\
&\qquad - (w_{I+1,n-1} - w_{I,n-1}) \\
&\qquad\quad ((1 - \alpha_{I+1,n})w_{I,n-1}P_{I,n-1} + \alpha_{I+1,n}w_{I+1,n-1}P_{I+1,n-1})\} \\
&\quad= n\{w_{I+1,n-1}w_{I,n-1}P_{I+1,n-1} - w_{I,n-1}w_{I+1,n-1}P_{I,n-1}\}
\end{aligned}$$

as claimed. ■

We could, in fact, obtain a more general expression. Using exactly the same proof but with different subscripts (see the proof of Proposition 1), it follows that

$$P'_{i,k}(u) = k\frac{w_{i-1,k-1}w_{i,k-1}}{w_{i,k}^2}\left(\frac{P_{i,k-1} - P_{i-1,k-1}}{u_{i-n+k} - u_{i-1}}\right).$$

§4. The Hodograph Property

In this section we apply Proposition 2 (the hodograph formula) to show that NURBS curves have the hodograph property. A Bézier or B-spline curve $K(u)$ with control points $P_0, \ldots, P_m$ is said to have the hodograph property if the derivative of the curve can be expressed in the form

$$K'(u) = \sum_{i=1}^{m} \lambda_i(u)(P_i - P_{i-1}),$$

where $\lambda_i \geq 0$ for any i or u.

Since it has already been shown that R' is in this form it is a matter of showing $\lambda_i(u) \geq 0$. In order to do this we present two intermediate results (Lemma 4 and Proposition 5). Note the similarity of (20) to (2).

Lemma 4. *For any $i \in \{I-n+2, \ldots, I+1\}$ and $k \in \{1, \ldots, n\}$, the derivative of $N_{i,k}$ satisfies the identity*

$$N'_{i,k} = \frac{k}{k-1}\left\{\left(\frac{u - u_{i-1}}{u_{i+k-1} - u_{i-1}}\right) N'_{i,k-1} + \left(\frac{u_{i+k} - u}{u_{i+k} - u_i}\right) N'_{i+1,k-1}\right\}. \tag{20}$$

Proof: First, we write the left-hand side of (20) in terms of $N_{i,k-1}$ and $N_{i+1,k-1}$ using equation (7). We then go a step further by expanding each of these two terms using (2) and the left-hand side of (20) becomes

$$aN_{i,k-2} + bN_{i+1,k-2} + cN_{i+2,k-2},$$

where a, b, c are functions of u and the knots.

The right-hand side of (20) can also be expanded using (7) applied to each of the terms $N'_{i,k-1}$ and $N'_{i+1,k-1}$. The right-hand side of (20) then becomes

$$AN_{i,k-2} + BN_{i+1,k-2} + CN_{i+2,k-2},$$

where A, B, C are also functions of u and the knots.

It is now a straightforward algebraic exercise to show that $a = A$, $b = B$, and $c = C$. ■

By combining the identities (2) and (20) we now obtain the following result; see Figure 4.

Proposition 5. *Suppose that $N_{i,n}$ and $N_{j,n}$ are any two B-spline basis functions of degree n over any knot sequence and $j > i$. Then*

$$D_{i,j,n}(u) = N'_{j,n}(u)N_{i,n}(u) - N_{j,n}(u)N'_{i,n}(u) \geq 0 \tag{21}$$

for every $u \in \mathbb{R}$ for which $N_{i,n}$ and $N_{j,n}$ are differentiable.

Proof: It follows by induction on the degree n. Recall that the support of $N_{i,n}$ is $[u_{i-1}, u_{i+n}]$ and the support of $N_{j,n}$ is $[u_{j-1}, u_{j+n}]$. Suppose $n = 1$.

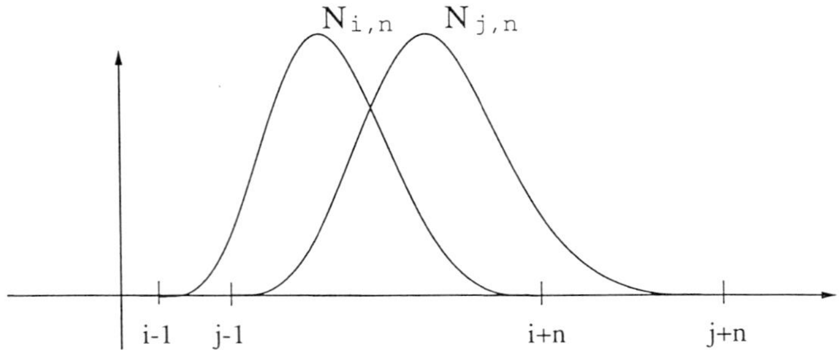

Figure 4. $N'_{j,n}(u)N_{i,n}(u) - N_{j,n}(u)N'_{i,n}(u) \geq 0$ for all u.

Then the only non-trivial case occurs when $j = i+1$ and $u_i < u < u_j$. In that case

$$\begin{aligned} N'_{j,1}(u)N_{i,1}(u)-&N_{j,1}(u)N'_{i,1}(u) \\ &= N'_{i+1,1}(u)N_{i,1}(u) - N_{i+1,1}(u)N'_{i,1}(u) \\ &= \left(\frac{1}{u_{i+1}-u_i}\right)\left(\frac{u_{i+1}-u}{u_{i+1}-u_i}\right) - \left(\frac{-1}{u_{i+1}-u_i}\right)\left(\frac{u-u_i}{u_{i+1}-u_i}\right) \\ &= \frac{1}{u_{i+1}-u_i} \geq 0. \end{aligned}$$

Now we show that (21) holds assuming it holds when n is replaced by $n-1$ and i and j are replaced by any other integers. We may assume that $j-1 < i+n$; otherwise the supports of the two basis functions do not coincide. Now suppose $j-1 < u < i+n$. If one rewrites (2) applied to $N_{i,n}$ and $N_{j,n}$ as

$$N_{i,n} = a_1 N_{i,n-1} + a_2 N_{i+1,n-1}$$

and

$$N_{j,n} = b_1 N_{j,n-1} + b_2 N_{j+1,n-1}$$

then, from (20),

$$N'_{i,n} = \frac{n}{n-1}(a_1 N'_{i,n-1} + a_2 N'_{i+1,n-1})$$

and

$$N'_{j,n} = \frac{n}{n-1}(b_1 N'_{j,n-1} + b_2 N'_{j+1,n-1}),$$

where $a_1, a_2, b_1, b_2 \geq 0$. By substituting these expressions into $D_{i,j,n}$ we find that $D_{i,j,n}$ is equivalent to

$$\frac{n}{n-1}(a_1 b_1 D_{i,j,n-1} + a_1 b_2 D_{i+1,j,n-1} + a_2 b_1 D_{i,j+1,n-1} + a_2 b_2 D_{i+1,j+1,n-1}).$$

But $j > i$ implies $j+1 > i+1$, $j+1 > i$ and $j \geq i+1$. Thus all terms in this expression are non-negative by the induction step except possibly for $D_{i+1,j,n-1}$ when $j = i+1$ but then $D_{i+1,j,n-1} = D_{j,j,n-1} = 0$ anyway. Therefore $D_{i,j,n} \geq 0$ as required. ■

Note that if the two basis functions above happen to be Bézier basis functions, then the proposition can be proved more directly, see Floater [5].

Theorem 6. *NURBS curves have the hodograph property.*

Proof: Proposition 2 demonstrates that R can be written as a linear combination of the direction vectors $P_i - P_{i-1}$, i.e. in the form of equation (14). It remains to show that each coefficient $\lambda_i(u)$ given by equation (15) is non-negative for all u. But this is now elementary because we know the weights are all positive and Proposition 5 shows that whenever $k > j$,

$$N'_{k,n}(u)N_{j,n}(u) - N_{k,n}(u)N'_{j,n}(u) \geq 0$$

for all u. ■

§5. Bounds on the Derivative

As a second application of Proposition 2, we compute an upper bound on the first derivative of a NURBS curve in terms of the control points, knots, and weights from the hodograph formula.

Proposition 7. *The first derivative of the NURBS curves defined by (10) is bounded as follows.*

$$\|R'(u)\| \leq n\frac{W^2}{w^2} \max_{i=1,\ldots,L+n-1} \left\{ \frac{\|P_i - P_{i-1}\|}{u_{i+n-1} - u_{i-1}} \right\}, \tag{22}$$

where

$$w = \min_{i=0,\ldots,L+n-1}\{w_i\} \qquad \textit{and} \qquad W = \max_{i=0,\ldots,L+n-1}\{w_i\}.$$

Proof: First we bound $\lambda_i(u)$ in (15). Suppose $u_I \leq u < u_{I+1}$. Note that because the function $w_{I+1,n}$ is constructed from linear interpolations of the weights, it too is bounded below by w, i.e. $w_{I+1,n}(u) \geq w$ for any u. Then, from (17),

$$\begin{aligned}
|\lambda_i| &\leq \frac{W^2}{w^2} \sum_{j=I-n+1}^{i-1} \sum_{k=i}^{I+1} (N'_{k,n}N_{j,n} - N_{k,n}N'_{j,n}) \\
&= \frac{W^2}{w^2} \sum_{j=I-n+1}^{i-1} \sum_{k=I-n+1}^{I+1} (N'_{k,n}N_{j,n} - N_{k,n}N'_{j,n}).
\end{aligned}$$

Now it is a well known result from spline theory that

$$\sum_{k=I-n+1}^{I+1} N_{k,n} = 1.$$

Further, by using equation (7),

$$\sum_{j=I-n+1}^{i-1} N'_{j,n} = -n \frac{N_{i,n-1}}{u_{i+n-1} - u_{i-1}}$$

and, since $N_{I+2,n-1} = 0$ for $u_I \leq u < u_{I+1}$,

$$\sum_{k=I-n+1}^{I+1} N'_{k,n} = 0.$$

Therefore

$$|\lambda_i| \leq n \frac{W^2}{w^2} \frac{N_{i,n-1}}{u_{i+n-1} - u_{i-1}}.$$

Since this bound on λ_i is independent of the particular parameter range chosen, we substitute it into (14) and obtain

$$|R'(u)| \leq n \frac{W^2}{w^2} \max_{i=1,\ldots,L+n-1} \left\{ \frac{\|P_i - P_{i-1}\|}{u_{i+n-1} - u_{i-1}} \right\}$$

as required. ■

References

1. de Boor, C., On calculating with B-splines, J. Approx. Theory **6** (1972), 50–62.
2. de Boor, C., *A Practical Guide to Splines*, Springer-Verlag, New York, 1978.
3. Farin, G., *Curves and surfaces for computer aided geometric design*, Academic Press, San Diego, 1988.
4. Filip, D., R. Magedson, and R. Markot, Surface algorithms using bounds on derivatives, Computer Aided Geom. Design **3** (1986), 295–311.
5. Floater, M. S., Derivatives of rational Bézier curves, Computer Aided Geom. Design , to appear.
6. Piegl, L., and W. Tiller, Curve and surface constructions using rational B-splines, Computer-Aided Design **19** (1987), 485–498.
7. Sederberg, T. W., and R. J. Meyers, Loop detection in surface patch intersections, Computer Aided Geom. Design **5** (1988), 161–171.
8. Sederberg, T. W., and X. Wang, Rational hodographs, Computer Aided Geom. Design **4** (1988), 333-335.

Michael S. Floater
Centre for Industrial Research
P. O. Box 124, Blindern
0314 Oslo 3, NORWAY
floater@bivrost.cs.si.no

Hybrid Cubic Bézier Triangle Patches

Thomas A. Foley and Karsten Opitz

Abstract. With two additional control points, we modify the cubic Bézier triangle patch to obtain a compact representation for some procedurally defined triangular patches. The hybrid form can be evaluated with the de Casteljau algorithm, and with a slight variation we can easily compute first order derivatives. A C^1 interpolant to scattered data with cubic precision is given that is significantly smoother than when linear cross boundary derivatives are used.

§1. Introduction

A modification of the cubic Bézier triangular patch is presented which can be used to solve the following scattered data interpolation problem. Given N distinct points (x_i, y_i) in the plane and N real values z_i, the problem is to construct a C^1 bivariate function $F(x, y)$ that satisfies $F(x_i, y_i) = z_i$, for $i = 1, \cdots, N$. Several applications and many methods for solving this problem are given in [1,8,9,10,17].

One class of scattered data interpolants involves triangulating the domain vertices and defining $F(x, y)$ piecewise over each triangle. It has been shown that it is not generally possible to solve the scattered data interpolation problem with cubic polynomials defined over the triangulated data. The Clough-Tocher method (see [6]) overcomes the limitation by first splitting each domain triangle into three triangles and then a piecewise cubic interpolant is formed over the finer triangulation.

Another class of triangle patch solutions to this problem involves the discretization of a transfinite interpolant, such as the BBG interpolant in [3] and the side-vertex method in [14]. These methods have a possible advantage over the Clough-Tocher method in that they don't require splitting each domain triangle into three triangles. A possible disadvantage of these methods is that they are generally implemented in a procedural approach by cubic Hermite

Mathematical Methods in Computer Aided Geometric Design II
Tom Lyche and Larry L. Schumaker (eds.), pp. 275–286.

ISBN 0-12-460510-9.

operators and a weighted combination of 3 partial interpolants is formed. This lack of a simple closed form also makes it difficult to compute partial derivatives and surface normals that are needed in most geometry processing applications [2].

The motivation for this research was to obtain a closed form solution for the discrete side-vertex method in a scattered data environment. The hybrid cubic Bézier triangle patch introduced in the following section solves this problem. Instead of 10 control points that are used to define a cubic Bézier patch, the hybrid form uses 12 control points. The hybrid form differs from the standard cubic in that the inner cubic Bézier point is a rational quadratic average of three control points, with the nine boundary control points having the same influence as the standard cubic patch. The hybrid patch can then be evaluated as a standard cubic using the Bernstein or de Casteljau form. With a small perturbation of the de Casteljau algorithm, we can easily compute tangent planes, directional derivatives and surface normals. The hybrid patch also satisfies the convex hull property.

The behavior of the hybrid patch on each triangle edge is identical to a standard cubic and it is easy to obtain a C^1 composite surface. The nice geometric properties of cubic Bézier patches allow us to compute the inner control points for the hybrid patch that yields a method with cubic precision. This approach for computing the cross boundary derivatives produces a significantly smoother interpolant than the commonly used approach which uses linearly varying cross boundary derivatives.

§2. Hybrid Cubic Bézier Triangle

Before describing the hybrid form, we first review some basic properties of the cubic Bézier triangle patch that can be found in [7]. Assume that we are given 10 control points $b_I = b_{ijk}$, where $|I| = i + j + k$ and $0 \leq i, j, k \leq 3$. If $U = (u_0, u_1, u_2)$ are the barycentric coordinates of a point in the domain triangle, then the cubic Bézier patch can be defined in Bernstein form by

$$B(U) = \sum_{|I|=3} b_I B_I(U),$$

$$B_I(U) = B_{ijk}(U) = \frac{3!}{i!j!k!} u_0^i u_1^j u_2^k.$$

An alternative method for evaluating $B(U)$ is the following de Casteljau algorithm. For $r = 1, 2, 3$ and $|I| = 3 - r$, compute

$$b_I^r(U) = u_0 b_{I+e0}^{r-1}(U) + u_1 b_{I+e1}^{r-1}(U) + u_2 b_{I+e2}^{r-1}(U),$$

where $b_I^0(U) = b_I$, $e0 = (1, 0, 0)$, $e1 = (0, 1, 0)$ and $e2 = (0, 0, 1)$. It is shown in [7] that $B(U) = b_{000}^3(U)$ and that the tangent plane at $B(U)$ is the plane through the three points b_{100}^2, b_{010}^2 and b_{001}^2.

The hybrid cubic Bézier triangle patch $H(U)$ is now described. Suppose that we are given 12 control points p_0, p_1, p_2 and $b_I = b_{ijk}$, where $|I| = 3$ and $|I| \neq (1,1,1)$. The hybrid patch is defined by

$$H(U) = \sum_{|I|=3} b_I B_I(U), \tag{1}$$

$$b_{111} = b_{111}(U) = w_0(U)p_0 + w_1(U)p_1 + w_2(U)p_2, \tag{2}$$

$$w_i(U) = \frac{u_j u_k}{u_0 u_1 + u_0 u_2 + u_1 u_2}, \qquad i \neq j \neq k.$$

We will later show that each point p_i can be used to control the cross-boundary derivatives of $H(U)$ along the edge $u_i = 0$. The points p_0, p_1 and p_2 will be referred to as *inner control points* and the remaining nine points b_I, $I \neq (1,1,1)$, will be called *boundary control points* of the patch.

The rational weight functions were used in [14] for convex combinations of triangular elements. These weight functions satisfy $w_i(u_0, u_1, u_2) = 1$ when $u_i = 0$, $w_i(u_0, u_1, u_2) = 0$ when $u_j = 0$ and $i \neq j$, and

$$w_0(U) + w_1(U) + w_2(U) = 1$$

for all U. Although the weight functions have singularities at the domain triangle vertices, the resulting patch $H(U)$ in (1) will be C^1 with removable singularities at the vertices. This result can be proved in a number of ways, and we will verify it as a result of the de Casteljau algorithm.

The term *hybrid* is used because it is used in [19] for curves with one control point that varies in some interval. The hybrid triangle patch is closely related to the approach in [4] which represents a discrete version of Gregory's square with variable control points.

The surface $H(U)$ can be considered as a hybrid cubic Bézier patch in the sense that for each U, we first compute $b_{111} = b_{111}(U)$ by (2) and then $H(U)$ can be evaluated as a standard cubic Bézier patch using either the Bernstein form or the de Casteljau algorithm. The de Casteljau algorithm can also be used to obtain directional derivatives and the tangent plane to $H(U)$ if the following modifications are made. The details are straightforward and tedious, thus only a sketch is given.

Following the notation and the argument used in [7], let U_0 and U_1 be two barycentric points in the domain triangle and let $d = U_1 - U_0$. With the directional derivative denoted by D_d, we have that

$$D_d H(U) = 3 \sum_{|I|=1} b_I^2(U) B_I^1(d) + B_{111}^3(d) D_d(b_{111}(U)), \tag{3}$$

where $b_I^2(U)$ are points computed in the de Casteljau algorithm and B_I^n is the triangular Bernstein basis function of degree n. From (3) it can be shown that the tangent plane at $H(U)$ is through the three points c_0, c_1 and c_2, where

$$\begin{aligned} c_0 &= b_{100}^2 + 2w_0^2(U)[u_0u_2(p_1 - p_0) + u_0u_1(p_2 - p_0)], \\ c_1 &= b_{010}^2 + 2w_1^2(U)[u_1u_2(p_0 - p_1) + u_0u_1(p_2 - p_1)], \\ c_2 &= b_{001}^2 + 2w_2^2(U)[u_1u_2(p_0 - p_2) + u_0u_2(p_1 - p_2)]. \end{aligned} \tag{4}$$

From (3) and (4) it follows that the directional derivative is

$$D_dH(U) = 3(d_0c_0 + d_1c_1 + d_2c_2), \tag{5}$$

where $d = (d_0, d_1, d_2)$.

To summarize the evaluation process, for a given U, we first compute $b_{111}(U)$ by (2) and then apply the de Casteljau algorithm which generates a sequence of points b_I^r. The point on the surface is simply $H(U) = b_{000}^3$, while the tangent plane and directional derivatives in (4) and (5) use perturbations of the points b_{100}^2, b_{010}^2 and b_{001}^2. The ability to compute exact directional derivatives and surface normals permits the hybrid Bézier patch to be used in geometry processing algorithms.

A key property of the hybrid patch is that it behaves like three cubic Bézier triangle patches on the boundary edges. The following theorem will allow us to easily form C^1 composite patches that interpolate scattered data.

Theorem 1. *The hybrid patch $H(U)$ is a C^1 function with removable singularities at the vertices of the domain triangle. Furthermore, the positional and first derivative values of $H(U)$ along the edge $u_i = 0$ are identical with those of the standard cubic Bézier patch $F_i(U)$ whose inner control point is $b_{111} = p_i$ and whose nine boundary control points are $b_I, I \neq (1, 1, 1)$.*

Proof: As the barycentric coordinates approach $U = (1, 0, 0)$ from any direction, all of the perturbation points in (4) approach zero and we have at $U = (1, 0, 0)$ that

$$c_0 = b_{100}^2 = b_{300}, c_1 = b_{010}^2 = b_{210} \quad \text{and} \quad c_2 = b_{001}^2 = b_{201}.$$

As U approaches $(1, 0, 0)$, we have that $H(U)$ approaches b_{300} and the tangent plane approaches the plane spanned by b_{300}, b_{210}, and b_{201}. This is the same behavior as the standard cubic at $U = (1, 0, 0)$ and the same holds at the other two vertices.

Since we have established the appropriate behavior at the vertices, we will consider the edge $u_0 = 0$ and assume that u_1 and u_2 are nonzero. As u_0 approaches zero, we have that $b_{111}(U)$ approaches p_0, thus the positional values of $H(U)$ on this edge is simply the cubic Bézier curve with control points $b_{030}, b_{021}, b_{012}$ and b_{003}. The perturbation terms in (4) also approach zero, thus the tangent plane to $H(U)$ is through the three points b_{100}^2, b_{010}^2 and b_{001}^2, and the directional derivatives in (5) are $D_d(H(U)) = 3(d_0b_{100}^2 + d_1b_{010}^2 + d_2b_{001}^2)$. These are the same values that are generated by the de Casteljau algorithm applied to the standard cubic Bézier patch $F_0(U)$ on the edge $u_0 = 0$. ■

The hybrid patch in (1) can also be written in the form

$$H(U) = \sum_{I \neq (1,1,1)} b_I B_I(U) + \sum_{i=0}^{2} p_i w_i(U) B_{111}(U). \tag{6}$$

Since the three weight functions $w_i(U)$ sum to unity (as do the Bernstein basis functions), we have that

$$\sum_{I \neq (1,1,1)} B_I(U) + \sum_{i=0}^{2} w_i(U) B_{111}(U) = 1$$

for all U. All of these functions are non-negative on the domain triangle, thus (6) is a convex combination. The hybrid patch satisfies the convex hull property on the triangle with respect to its twelve control points.

§3. Non-parametric Hybrid Patches

For our discussion on C^1 continuity of adjacent hybrid patches, we will restrict our attention to the non-parametric or function case. In this setting, assume that the domain triangles are in the xy plane and that the control points b_{ijk} have xy coordinates of the point whose barycentric coordinates are $(i/3, j/3, k/3)$. The inner control points p_0, p_1 and p_2 have xy coordinates whose barycentric coordinates are $(1/3, 1/3, 1/3)$.

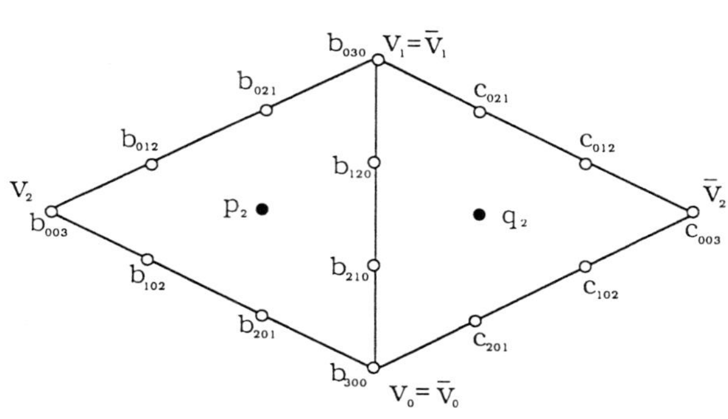

Figure 1. Two adjacent hybrid cubic patches.

Consider the two adjacent triangles in Figure 1 whose vertics are V_i and $\bar{V}_i$, with $V_0 = \bar{V}_0$ and $V_1 = \bar{V}_1$. Let H be the hybrid patch defined on the left triangle with boundary control points b_I and inner control points p_j. Let $\bar{H}$ be the hybrid patch on the right with boundary control points c_I and inner control points q_j. For positional continuity across the edge V_0V_1, we need that the four control points along that edge for H coincide with the similar points for $\bar{H}$. This is the same as with standard cubic patches, thus we simply label them as $b_{300}, b_{210}, b_{120}$ and b_{030} in Figure 1.

For C^1 continuity across the common edge, by Theorem 1 we need that each of the following sets of four points be coplanar: $\{b_{300}, b_{210}, b_{201}, c_{201}\}$, $\{b_{030}, b_{120}, b_{021}, c_{021}\}$, $\{b_{210}, b_{120}, p_2, q_2\}$. The first two sets are identical with standard cubic Bézier patches. The third set indicates that we have a great deal of flexibility on the inner control points to attain C^1 continuity once the boundary control points have been selected. The inner pcint p_2 can take on any value, and then the point q_2 can be computed to be planar because p_2 and q_2 have no influence on the remaining edges. The other inner control points p_0, p_1, q_0 and q_1 are free to be chosen to attain C^1 continuity across the other edges to adjacent patches. Methods for choosing the inner control points are discussed in the following section.

The hybrid patch can also be considered to be a rational convex combination of three cubic Bézier patches $F_i(U)$, described in Theorem 1. Since the weight functions sum to unity, it follows that

$$H(U) = \sum_{i=0}^{2} w_i(U) F_i(U). \tag{7}$$

In this form, we see that the hybrid Bézier patch can be used to represent the non-parametric discrete side-vertex method in [14] that uses cubic Hermite operators. This side-vertex method is generally implemented as a procedurally defined patch which uses radial cubic Hermite operators from one vertex to the opposite side. If the boundary curves are cubic polynomials, then the weight functions following (2) can be used. In the discrete setting, the boundary curves are generally cubics and the cross boundary derivatives are usually assumed to vary linearly from one vertex to the next. If the cross boundary derivatives vary quadratically (or linearly), then it can be shown that the procedurally defined partial interpolants are bivariate cubics, hence they are of the form $F_i(U)$ described in Theorem 1.

It is straightforward to also show that the discrete BBG patch with cubic blending, cubic boundary curves and quadratically varying cross boundary derivatives can be represented with the hybrid Bézier patch (1) using

$$w_i(u_0, u_1, u_2) = u_j u_k((1 - u_j)^{-1} + (1 - u_k)^{-1}),$$

where $i \neq j \neq k$. Following the excellent characterization theorem on related triangular patches in monomial form in [13], any weight function of the following form can be used with (1) and (2)

$$w_i(U) = \frac{u_j u_k f(u_i, u_j, u_k)}{u_0 u_1 f(u_2, u_0, u_1) + u_0 u_2 f(u_1, u_0, u_2) + u_1 u_2 f(u_0, u_1, u_2)},$$

where f is a polynomial symmetric in the second two variables.

Following our presentation at the Biri conference, we became aware of similar work done in [11]. Their C^1 scattered data interpolant used linearly varying cross boundary derivatives and a hybrid patch in the form (7) with weight functions

$$w_i(u_0, u_1, u_2) = \frac{u_j^2 u_k^2}{u_0^2 u_1^2 + u_0^2 u_2^2 + u_1^2 u_2^2}.$$

§4. Cross Boundary Derivatives

For interpolation of scattered data using a triangle based method, the following steps are involved. The points $V_i = (x_i, y_i)$ in the plane are triangulated using some local or global optimal criteria, such as those discussed in [1,5,10,18]. The next step involves estimating partial derivatives of the underlying function at the vertices V_i based upon some or all of the data (x_i, y_i, z_i). Several of these techniques are given in [8,10,11,15,16,20]. With these first order partial derivatives, a piecewise cubic network can be defined over the edges of the triangulation. In Bézier form, this network of cubics provides all of the boundary control points $b_I, I \neq (1,1,1)$, for each triangle patch.

Denote the partial derivative estimates at vertex V_i by $S_x(V_i)$ and $S_y(V_i)$. The boundary control points along the edge $V_1V_2(u_0 = 0)$ for the triangle $V_0V_1V_2$ are computed by $b_{030} = z_1, b_{003} = z_2$,

$$b_{021} = z_1 + \frac{1}{3}((x_2 - x_1)S_x(V_1) + (y_2 - y_1)S_y(V_1)), \tag{8}$$

$$b_{012} = z_2 - \frac{1}{3}((x_2 - x_1)S_x(V_2) + (y_2 - y_1)S_y(V_2)).$$

The other edges are similar.

In order to represent a C^1 scattered data interpolant using hybrid Bézier patches, we need to compute the three inner control points for each triangle. A simple approach used in many methods is to assume that the cross boundary derivatives vary linearly along the edges of all triangles. The top images in Figure 2 (see also Color Plate 7) exhibit the poor behavior of using linearly varying cross boundary derivatives on the $N = 25$ point set from test function $f_1(x, y)$ in [8]. The Delaunay triangulation and exact partial derivatives at the vertices V_i are used in this and subsequent figures.

The bottom images in Figure 2 use the following method for selecting the inner control points so that the interpolant will have cubic precision. It should be noted that both the top and bottom surfaces use the same derivatives at the vertices and the same boundary curves on all edges of the triangles. Only the inner control points (hence the cross boundary derivatives) are different.

Suppose that the boundary control points for each patch have been computed by (8), thus the control points adjacent to each vertex will be coplanar. Referring to Figure 1, if the boundary control points of the two adjacent triangles are from a single cubic polynomial, then we compute p_2 and q_2 so that the cubic will be reproduced. Let (r, s, t) be the barycentric coordinates of $\bar{V}_2$

with respect to the triangle $V_0V_1V_2$. If both patches form a single cubic, then from subdividing Bézier cubics it can be shown that

$$c_{102} = r^2 b_{300} + 2rsb_{210} + 2rtb_{201} + s^2 b_{120} + 2stp_2 + t^2 b_{102}, \tag{9}$$

$$c_{012} = r^2 b_{210} + 2rsb_{120} + 2stb_{021} + s^2 b_{030} + 2rtp_2 + t^2 b_{012}. \tag{10}$$

Reversing the triangles in Figure 1, if (u, v, w) are the barycentric coordinates of V_2 with respect to the triangle $\bar{V}_0\bar{V}_1\bar{V}_2$, we obtain

$$b_{102} = u^2 c_{300} + 2uvc_{210} + 2uwc_{201} + v^2 c_{120} + 2vwq_2 + w^2 c_{102}, \tag{11}$$

$$b_{012} = u^2 c_{210} + 2uvc_{120} + 2vwc_{021} + v^2 c_{030} + 2uwq_2 + w^2 c_{012}. \tag{12}$$

The overdetermined system for p_2 in (9) and (10) will have a solution if the boundary control points of the two patches are from a single cubic. The following observations and computations do not assume that they are from one cubic.

Note that (9) can't be solved for p_2 if $st = 0$. Since (r, s, t) are the barycentric coordinates of $\bar{V}_2$ with respect to $V_0V_1V_2$, we have that $t < 0$ and thus $r + s > 0$ because $r + s + t = 1$. If $s = 0$ and (9) has no solution, then $r > 0$ and (10) will have a solution for p_2. Similarly, at least one of (11) and (12) can be solved for q_2.

For the moment, suppose that r, s, u and v are nonzero and let $p_2^{(1)}$, $p_2^{(2)}$, $q_2^{(1)}$ and $q_2^{(2)}$ be the solutions of equations (9) through (12), respectively. It is straightforward to show that the sets $\{b_{210}, b_{120}, p_2^{(1)}, q_2^{(1)}\}$ and $\{b_{210}, b_{120}, p_2^{(2)}, q_2^{(2)}\}$ are each coplanar. It can also be shown that the quadratic Bézier patch with control points $\{b_{300}, b_{210}, b_{120}, b_{201}, b_{102}, p_2^{(1)}\}$ is the same quadratic as the one defined with control points $\{b_{300}, b_{210}, b_{120}, c_{201}, c_{102}, q_2^{(1)}\}$. A similar statement can be made about the control points listed in (10) and (12).

We compute p_2 by adding equations (9) and (10), and then we solve for p_2. Similarly, equations (11) and (12) are added before solving for q_2. The coefficient of p_2 in the sum of (9) and (10) is $2(r + s)t$, which is never zero because $t < 0$ and $r + s > 0$. Solving for p_2 in this manner is the same as setting $p_2 = \alpha p_2^{(1)} + \beta p_2^{(2)}$, where $\alpha = s/(r + s)$ and $\beta = r/(r + s)$. Although $p_2^{(1)}$ is undefined when $s = 0$, we have that $\alpha = 0$ and $\beta = 1$, thus $p_2 = p_2^{(2)}$ in this case. This case arises when V_2, V_0 and $\bar{V}_2$ are collinear.

If the positional and first derivative values at all vertices are from a single cubic polynomial, then the cubic will be reproduced using this method. In this case, we have that $p_2 = p_2^{(1)} = p_2^{(2)}$ (assuming that $rs \neq 0$) and that $p_0 = p_1 = p_2$ for each patch. Cubic precision is also obtained locally on one triangle patch if the function and derivative values at the vertices of that triangle and the adjacent triangles come from a single cubic polynomial. It should be noted that you can't obtain local cubic precision without considering

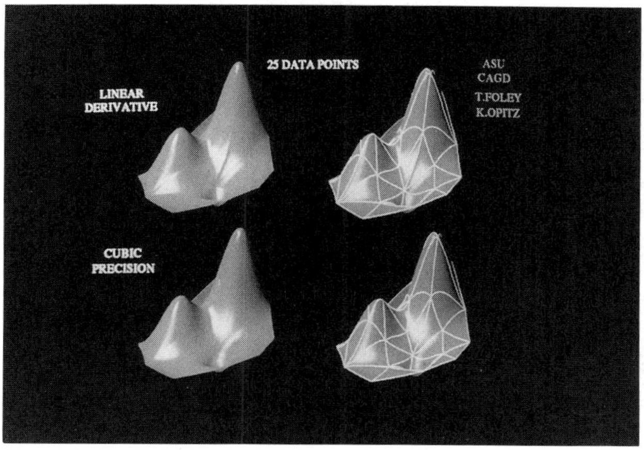

Figure 2. C^1 hybrid patch interpolants.

adjacent triangles because any multiple of $u_0 u_1 u_2$ will have zero position and derivative values at the vertices of that triangle.

A minor modification of the cubic precision method is required for boundary triangles because adjacent triangles are used. If the edge $u_i = 0$ is on the boundary and edges $u_j = 0$ and $u_k = 0$ are interior, we first compute p_j and p_k with the cubic precision method and then define p_i as the average. If both $u_i = 0$ and $u_j = 0$ are boundary edges and $u_k = 0$ is an interior edge, we first compute p_k with the cubic precision method and then set $p_i = p_j = p_k$. In this case, the hybrid patch will be a standard cubic patch.

To obtain local or global cubic precision from the given scattered data, the partial derivatives need to be approximated with a method that also has cubic precision. One approach would be to use local least squares approximation to the data with a cubic polynomial. If the data and the derivatives do not come from a cubic, the cubic precision method can still be used to select the inner control points.

The bottom images in Figure 2 (see also Color Plate 7) use the cubic precision method and there is a clear improvement in the visual smoothness over the top images which used linearly varying cross boundary derivatives. The observed maximum and RMS errors for these methods do not differ much because the errors are influenced more by the boundary curves, which are identical for the two methods. The ridges in the linearly varying method in Figure 2 were not apparent in most earlier publications because the plots were rendered as a wire frame mesh over a uniform rectangular mesh. For rendering the surfaces, we decompose each domain triangle into a fine triangular mesh and we use the exact surface normals computed from (4).

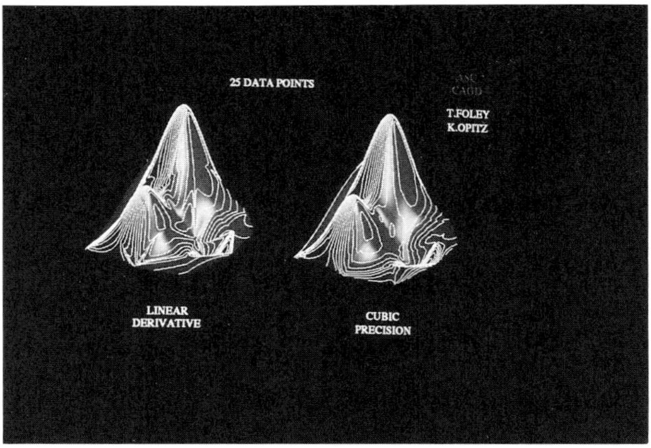

Figure 3. Isophotes on the same surfaces in Figure 2.

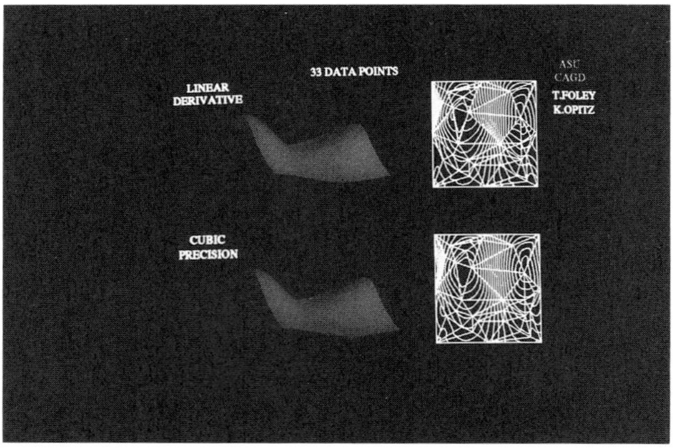

Figure 4. Surface and isophote plots on the triangulated domain.

A surface interrogation technique used in [12,16] that is useful for observing the smoothness of a C^1 function is to display isophotes. Isophotes are points on the surface that have equal diffuse light intensity and they can be computed by contouring the function whose values are the angle between the surface normal and the light direction. Isophotes are similar to the reflection lines used in [6], except that the strip lights are concentric circles. If the surface is G^k, then the isophotes are G^{k-1}.

Figure 3 displays isophote curves on the same surfaces shown in Figure 2 and it is observed that the isophotes for the cubic precision method are generally much smoother than those for the linearly varying cross boundary derivatives. The discontinuities of the tangents of the isophote curves occur on the edges of the triangle patches.

Figure 4 (see also Color Plate 8) contains shaded surface and isophote plots of the interpolants to the $N = 33$ point data set generated by the "saddle" test function $f_3(x, y)$ in [8]. The visual difference between the two shaded images in Figure 4 is not significant because the data has less variation and because the lighting model uses a low specular component. However, the isophotes on the right illustrate that the surface normals of the two interpolants are significantly different. The display of the domain triangles together with the isophotes enables the viewer to easily visualize where the linearly varying cross boundary derivatives are not effective.

§5. Concluding Remarks

The scattered data interpolants of the previous section were applied to other data in [8] and the results are consistent with those shown here, even on the $N = 100$ point sets. We also implemented a method used in [6] for selecting the inner control points so that the patches were approximately C^2 across the triangle edges. The results were effective and nearly identical to those of the cubic precision method. Another approach that we implemented for selecting the inner control points involved making each patch approximately equal to some fixed target cubic. Referring to Figure 1, if b_{111} is the inner control point of a target cubic on the left and c_{111} is one for a target cubic on the right, we choose p_2 and q_2 to approximate b_{111} and c_{111} subject to the C^1 coplanarity condition.

The hybrid cubic Bézier triangle patch method for scattered data interpolation is simple to program, computationally faster than procedural methods, does not require any triangles to be split, and only requires storage of one value per edge in addition to the partial derivatives at the vertices. The evaluation of the position and tangent plane of the hybrid patch is only slightly more costly than evaluating a standard cubic. The hybrid patch can be considered as a rational Bézier patch of degree 5 over degree 2, but is faster to evaluate and requires less storage if implemented in the hybrid cubic form.

Acknowledgments. A portion of this work was supported by the NSF and AFOSR Grant DMS-9116930 awarded to Arizona State University, and by an Associated Western Universities sabbatical fellowship at Lawrence Livermore Lab under the Department of Energy contract W-7405-Eng-48.

References

1. Barnhill, R. E., Representation and approximation of surfaces, in *Mathematical Software III*, J. R. Rice (ed.), Academic Press, New York, 1977, 69–120.
2. Barnhill, R. E. (ed.), *Geometry Processing for Design and Manufacturing*, SIAM, Philadelphia, 1991.

3. Barnhill, R. E., G. Birkhoff, and W. J. Gordon, Smooth interpolation in triangles, J. Approx. Th. **8** (1973), 114–128.
4. Chiyokura, H., and F. Kimura, Design of solids with free-form surfaces, Computer Graphics **17** (1983), 289–298.
5. Dyn, N., D. Levin, and S. Rippa, Algorithms for the construction of data dependent triangulations, in *Algorithms for Approximation II*, J. C. Mason and M. G. Cox (eds.), Chapman and Hall, London, 1990, 185–192.
6. Farin,G., A modified Clough-Tocher interpolant, Computer Aided Geom. Design **2** (1985), 19–27.
7. Farin, G., *Curves and Surfaces for Computer Aided Geometric Design*, Academic Press, New York, 1990.
8. Franke, R., Scattered data interpolation: tests of some methods, Math. Comp. **38** (1982), 181–200.
9. Franke, R., Recent advances in the approximation of surfaces from scattered data, in *Topics in Multivariate Approximation*, C. K. Chui, L. L. Schumaker and F. Utreras (eds.), Academic Press, New York, 1987, 175–184.
10. Franke, R., and G. M. Nielson, Scattered data interpolation and applications: a tutorial and survey, in *Geometric Modeling: Methods and Their Applications*, H. Hagen and D. Roller (eds.), Springer, Heidelberg, 1991, 131–160.
11. Goodman, T. N. T., and H. Said, A C^1 triangular interpolant suitable for scattered data interpolation, Tech. Report, Univ. of Dundee, 1990.
12. Hagen, H., T. Schreiber, and E. Gschwind, Methods for surface interrogation, in *Visualization '90*, A. Kaufman (ed.), IEEE Press, Los Alamitos, CA, 1990, 187–193.
13. Herron, G. F., A characterization of certain C^1 discrete triangular interpolants, SIAM J. Numer. Anal. **22** (1985), 811–819.
14. Nielson, G. M., The side-vertex method for interpolation in triangles, J. Approx. Th. **25** (1979), 318–336.
15. Nielson, G. M., A method for interpolation of scattered data based upon a minimum norm network, Math. Comp. **40** (1983), 253–271.
16. Pottmann, H., Scattered data interpolation based upon generalized minimum norm networks, Constr. Approx. **7** (1991), 247–256.
17. Schumaker, L. L., Fitting surfaces to scattered data, in *Approximation Theory II*, G. G. Lorentz, C. K. Chui and L. L. Schumaker (eds.), Academic Press, New York, 1976, 203–268.
18. Schumaker, L. L., Triangulation methods, in *Topics in Multivariate Approximation*, C. K. Chui, L. L. Schumaker and F. Utreras (eds.), Academic Press, New York, 1987, 219–232.
19. Sederberg, T. W., and M. Kakimoto, Approximating rational curves using polynomial curves, in *NURBS for Curve and Surface Design*, G. Farin (ed.), SIAM, Philadelphia, 1991, 149–158.
20. Stead, S. E., Estimation of gradients from scattered data, Rocky Mt. J. Math. **14** (1984), 265–279.

Modelling Geological Structures Using Splines

Lars A. Frøyland, Arne Lakså and Jan Pajchel

Abstract. Our aim is to build a three dimensional geological model from interpreted seismic data using a spline representation of the geometry and other geological properties. The seismic interpretation is usually done in the time domain, and after a model is build in this domain, we perform a so called depth conversion, which transforms the model to depth domain. The input seismic interpretation yields no unique three dimensional geological model, which implies that we need a close interaction with the user of the system.

§1. Introduction

Our objective is to make a geological modelling tool, and produce a resulting model accessible for simulation programs, e.g. to perform a seismic simulation in order to convert the model from time domain to depth domain.

A *geology model* attempts to describe the physical properties of a part of the earth. The earth is composed of different geological layers, each of different age. These layers may be divided into volume blocks, each block representing some property, e.g density and sonic speed. These properties are again represented as continuous and smooth property functions, hence, all discontinuities in the properties are described by interfaces (surfaces) between blocks.

A *geological model* is a set of property functions which can be evaluated anywhere in the model. In addition, it must contain geometric information about the blocks and the interfaces between them.

The geological data used for building the model are well data and seismic data. Physical properties are measured in the wells, and seismic data are used in order to determine the geometry of the geological layers. Seismic data are echoes from geological structures, and the typical frequency content of seismic data is 5-80 Hz, which implies that the interpreted seismic data are the low frequency components of the real interfaces in the model. The selected low

Mathematical Methods in Computer Aided Geometric Design II
Tom Lyche and Larry L. Schumaker (eds.), pp. 287–296.

ISBN 0-12-460510-9.

frequency information in the given data is the main reason why we want to build models containing smooth interfaces. Moreover, smooth interfaces require less amount of data, which enables us to display huge numbers of interfaces simultaneously. This is important in order to obtain a realistic picture of the models.

§2. Building Geological Models

Triangular representations ([7]) and bi-cubic interpolating splines ([2]) have been used earlier for surface representation in three-dimensional geological modelling. In this paper we use approximation schemes and tensor product spline surfaces. One reason for using approximation schemes, and not interpolation, is that there are uncertainties in the data and we want to construct surfaces with no high frequencies (eg. if the data are low frequency seismic). Moreover, we want the surfaces to be relatively smooth in order to obtain more stable and fast algorithms for seismic simulation [3]. The main reason for using tensor product splines is that we are equipped with algorithms surface/surface intersection and data reduction [6]. Hence, we obtain surfaces which consist of a minimum of redundant information. In our case we could have used other surface representation schemes, e.g. box splines [1], however, some essential modelling tools are still lacking.

The input data are interpreted as seismic data or horizon maps. Horizon maps are depth values in a grid produced by some mapping system. Interpreted seismic data are two-way travel time values either in a grid from an interpretation station or along random lines digitized from seismic sections.

To generate a model from these data sets we have to go through the following four step process:

1.) Preparing data for surface approximation.
2.) Surface approximation (scattered data approximation).
3.) Volume construction (Boolean operations).
4.) Approximation of property functions.

If the input data are two-way travel time data the model must be converted to a depth model. Depth conversion by ray tracing is a discrete transformation, giving data points to be used in a new model building process, see [4]. The volume modelling starts with an editing process.

2.1. Data Editing

The purpose of data editing is to prepare the data for surface approximation:

1.) The interpreted points must be separated in groups containing no sharp edges; hence each group describes some smooth interface. Sharp edges are modelled by a surface/surface intersection.
2.) Different groups from different data sets describing one common smooth geological interface must be connected to avoid tangential intersection and geological inconsistency. This is due to some hueristic decision made by the user.

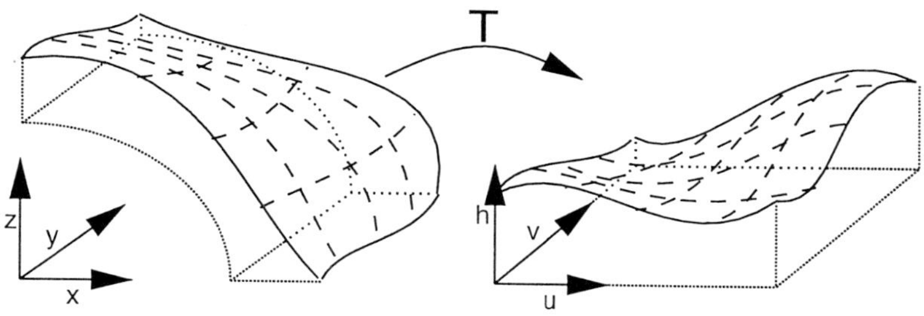

Figure 1. Transformation example.

2.2. Surface Approximation

Each surface in the model is described by a transformation of a scalar function defined on some basis domain, e.g. a plane, a cylinder, a sphere or different combinations of those. In the present version of the system we use affine transformations of ordinary explicit functions. When the overall surface is defined through a combination of different affine transformations, the internal interfaces must be treated so that the overall transformation becomes one to one (cf. standard offset problems). Hence, the scalar function is an offset surface which gives the height of the surface above some underlying basis domain.

Given points $(x_i, y_i, z_i) \in \mathbb{R}^3$, we define a transformation $T : \mathbb{R}^3 \to \mathbb{R}^3$ by $T : (x_i, y_i, z_i) \to (u_i, v_i, h_i)$. The approximation is now given by $T^{-1} f(u, v)$, where $f(u, v)$ is some function which satisfies the relation

$$|f(u_i, v_i) - h_i| < \epsilon \tag{1}$$

for any given tolerance $\epsilon > 0$. This is illustrated in Figure 1. The steps in the surface approximation are:

1.) Estimate the transformation T and transform the point set.
2.) Extrapolate in order to obtain a point set on a rectangular domain.
3.) Interpolate the point set using a triangulation and a piecewise polynomial functions on these triangles [9] or using radial basis functions [8].
4.) Sample the triangulation on a fine mesh.
5.) Estimate knot sequences and take the mesh found in 4) as coefficients of a tensor product spline surface.
6.) Perform data reduction [6].

2.2.1. Transformation

Let us assume that we use a planar basis domain. As soon as we have found the basis planar domain, we have the transformation T. From the data editing we get a point set $\mathcal{P} = \{p_i\}_{i=1}^n$, $p_i \in \mathbb{R}^3$. A so called *main plane* to this point set is given by a mean point

$$m = \frac{\sum_{i=1}^n p_i}{n}$$

and a normalized normal N minimizing

$$E = \sum_{i=1}^n \langle p_i - m, N\rangle^2,$$

where $\langle \cdot, \cdot \rangle$ is the standard dot product in $\mathbb{R}^3$. In order to simplify the notation, we introduce a new point set $\mathcal{Q} = \{q_i\}_{i=1}^n$, where $q_i = p_i - m$, and solve the following minimization problem:

Problem 1. *Find N, $\|N\| = 1$, so that $\lambda = \sum_{i=1}^n \langle q_i, N\rangle^2$ is minimized. With $q_i = (q_{ix}, q_{iy}, q_{iz})^T \in \mathbb{R}^3$, this can be written*

$$\lambda = N^T A N, \tag{2}$$

where A is the 3×3 matrix

$$A = \begin{pmatrix} \sum_{i=1}^n q_{ix}^2 & \sum_{i=1}^n q_{ix}q_{iy} & \sum_{i=1}^n q_{ix}q_{iz} \\ \sum_{i=1}^n q_{iy}q_{ix} & \sum_{i=1}^n q_{iy}^2 & \sum_{i=1}^n q_{iy}q_{iz} \\ \sum_{i=1}^n q_{iz}q_{ix} & \sum_{i=1}^n q_{iz}q_{iy} & \sum_{i=1}^n q_{iz}^2 \end{pmatrix}. \tag{3}$$

The matrix A is positive definite if the point set $\mathcal{Q}$ spans $\mathbb{R}^3$, and the solution of the minimization problem is the eigenvector associated with the smallest eigenvalue of A.

Definition 1. *The least rectangle is the rectangle in the main plane surrounding the projections of all points onto the main, which has the smallest area.*

To compute the least rectangle, we first find the convex hull of the given point set. We then find the least rectangle among all rectangles where one edge in the convex hull is a part of one of the edges of the rectangle. This follows from the following theorem.

Theorem 1. *Given any point set $\mathcal{P}$, the least rectangle surrounding this point set will always have at least two points on at least one edge.*

This theorem follows from the following lemma:

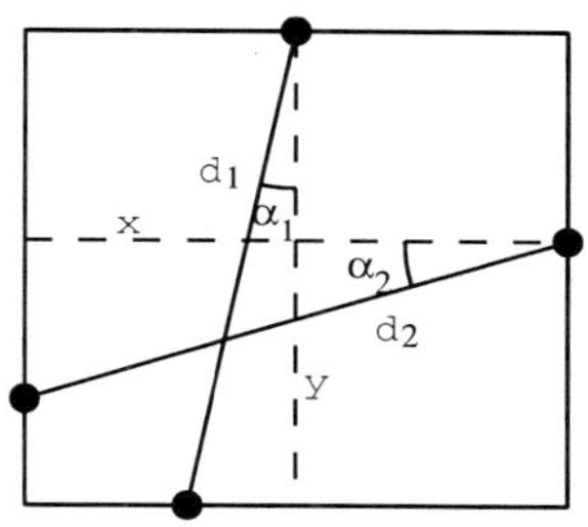

Figure 2. A surrounding rectrangle with one point on each edge.

Lemma 1. *Suppose there is only one point on each of the edges of the surrounding rectangle. Then it is always possible to reduce the area of the rectangle by a rotation.*

Proof: Suppose there is only one point on each of the edges of the surrounding rectangle. With the notation shown in Figure 2, we observe that $x = d_1 \cos\alpha_1$ and $y = d_2 \cos\alpha_2$. Hence, the area of this rectangle is

$$a = xy = d_1 d_2 \cos\alpha_1 \cos\alpha_2. \tag{4}$$

Without loss of generality we assume that $|\alpha_1| + |\alpha_2| \leq \frac{\pi}{2}$. Moreover, we observe that $\delta = \alpha_2 - \alpha_1$ is constant.

Letting $\alpha_2 = \alpha_1 + \delta$, we differentiate the area a with respect to the variable α_1. We obtain

$$\begin{aligned} a &= d_1 d_2 \cos\alpha_1 \cos(\alpha_1 + \delta) \\ \frac{da}{d\alpha_1} &= -d_1 d_2 (\sin\alpha_1 \cos(\alpha_1 + \delta) + \cos\alpha_1 \sin(\alpha_1 + \delta)) \\ \frac{d^2 a}{d\alpha_1^2} &= -2 d_1 d_2 (\cos\alpha_1 \cos(\alpha_1 + \delta) - \sin\alpha_1 \sin(\alpha_1 + \delta)) \\ &= -2 d_1 d_2 \cos(2\alpha_1 + \delta) \\ &= -2 d_1 d_2 \cos(\alpha_1 + \alpha_2). \end{aligned} \tag{5}$$

Combining (3) and the fact that $|\alpha_1| + |\alpha_2| \leq \frac{\pi}{2}$, we obtain

$$\cos(\alpha_1 + \alpha_2) \geq 0 \Rightarrow \frac{d^2 a}{d\alpha_1^2} \leq 0, \tag{6}$$

which implies that a rotation in at least one direction will reduce the area of the surrounding rectangle. ■

All points in the set $\mathcal{P}$ are transformed to a space with origin in the mean point m and the axes parallel to the edges of the surrounding rectangle and the normal N; hence we obtain

$$\mathcal{P}' = T\mathcal{P}. \tag{6}$$

2.2.2. Extrapolation

By construction, we have that all points are internal in the surrounding rectangle. In order to ensure that an actual surface has unique intersections with neighbouring surfaces, we must extrapolate the surface. This is done by extending the surrounding rectangle with a given confidence margin and estimating the function values at these for corners, a constant extrapolation or a linear extrapolation.

1.) Constant extrapolation: The height is the mean of the heights of the n closest points to the corner.
2.) Linear extrapolation: The intersection between the local main plane of the n closest points and the basis plane normal at the corner.

2.2.3. Triangulation, Sampling and Smoothing

To be able to resample the data, we need a function interpolating the data set $\mathcal{P}'$ in order to evaluate at arbitrary positions. We have used the standard Delaunay triangulation and the interpolants given in [9]. The coefficients to a tensor product spline surface is found by sampling this interpolant.

Before this sampling process is performed, we construct a sufficiently dense sampling vectors $\tau = (\tau_1, \ldots, \tau_n)$ in both directions in order to obtain a tensor product spline surface within some given tolerance to the data set $\mathcal{P}'$. The knot sequence $\mathbf{t}$ is given by $t_1 = \ldots = t_k = \tau_1$, $t_{n+1} = \ldots = t_{n+k} = \tau_n$ and

$$t_i = (\tau_{i-k+1} + \ldots \tau_{i-1})/(k-1).$$

This is indicated in Figure 3 for $k = 4$.

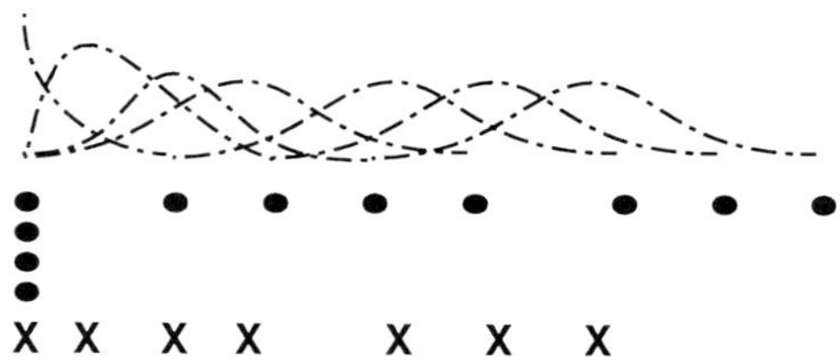

Figure 3. The knot vector, dots = knots and X = sample vector.

The algorithms for knot removal presented in [6] has two important purposes in our system.

1.) When knots are removed, we obtain smoother surfaces, which is of crucial importance in seismic simulation.
2.) Knot removal provides data reduction. This is very important because of the very dense data set obtained from the sampling process.

2.3. Volume Generation

Each volume is described by a set of faces which are trimmed tensor product surfaces. To generate the volumes, we start with a box of surfaces surrounding the total model. Then we try to divide this volume by finding a set of consistent combined faces, whose outer edges completely lies in the volume faces. This procedure is repeated recursively until no new volume division is possible.

2.3.1. Trim Curves

A trim curve is represented by an approximation of the parts of the intersection curves in the parametric plane of the surface. We are using a B-spline curve within some absolute tolerance from the real trim curve in three-space. The only restriction on the tolerance of the trim curve is that it must be useful for display purposes. There must be a one–to–one mapping from the B-spline curve to the real trim curve.

Trim curves are used for displaying and to decide if a point is on the face or not. There are two possible situations:

1.) First, it is easy to decide whether a point is inside or outside the trim curve if the distance to the trim curve is greater than the tolerance.
2.) If the distance is less than the tolerance, we iterate normally to the trim curve to evaluate the distance so we can decide on which side the point lies.

2.3.2. Intersection

We are using the SI Spline Library (SISL) [5] in order to compute all intersections. This library distinguishes between finding the topology of the intersection, which means finding guide points on intersection curves, and approximating the intersection curve by a B-spline curve to a tolerance. The detection is done by a combination of Newton iterations and subdivision. The approximation is done by a marching algorithm. We find the intersections in two different domains, in $\mathbb{R}^3$ and in the parameter planes. In $\mathbb{R}^3$ we use different methods for the detection of intersections:

(i) First we find a point on an intersection curve between two surfaces.
(ii) Then we make an approximation of the intersection curve by marching with a guide point as start point.
(iii) Then we intersect this intersection curve with all surfaces. When we find an intersection point we recursively repeat (ii).

In the parameter space we intersect curve by curve to make trim segments. We then traverse trim segments to find closed loops.

2.4. Properties

The property functions are smooth functions inside the volume it belongs to. It can be a constant, it can depend on the distance from the faces, or it can be a general scalar function of three parameters (x, y, z). In all cases it is an approximation of discrete input data.

§3. Program Structures

We are using three separate programs attached through interprocess comunication and shared memory. The three programs are:

1.) the geology model,
2.) the user dialogue, and
3.) the graphic display.

As we see in Figure 4, these three separate programs interact via the queues, and the different toolboxes are linked to the model. It is also possible to include new programs into the structure, e.g. a seismic simulator.

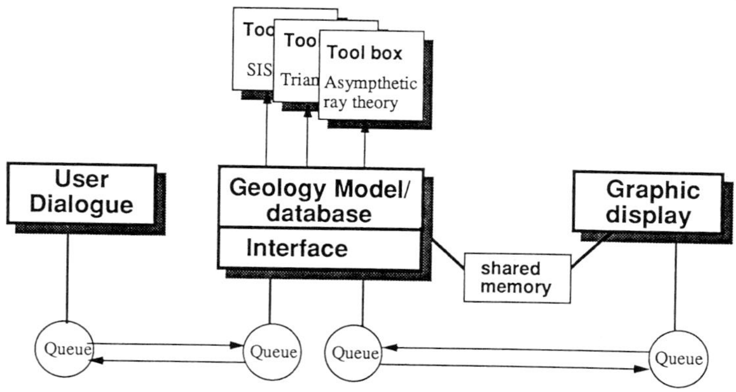

Figure 4. The program structures.

3.1. Model

The program is a combined builder and a data base. The data structure has three parts, the data part, the surface part, and the volume part (Figure 5). The reason for more than one model is that we often need at least two models, one in the time domain and one in the depth domain.

3.2. Dialogue User-interface

The user dialogue is based on X-windows and MOTIF and has two parts:

1.) The server part which can be a message/warning/error window, different prompt windows, or an editor window.
2.) The work windows holding icons representing model objects and the relation between the model objects.

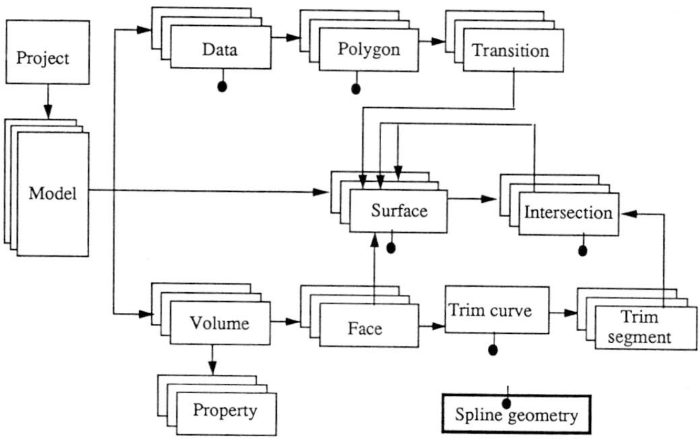

Figure 5. The data structure of the model.

3.3. Graphic User-interface

The graphic display is three dimensional and consists of scenes, where objects, lights, and cameras can be included, removed, and in other ways manipulated. A part of the data editing is done on the graphic display where we can pick out or exclude points by bounding them.

References

1. Dæhlen, M. and T. Lyche, Box splines and applications, in *Geometric Modeling*, H. Hagen and D. Roller (eds.), Springer-Verlag, Berlin, 1991, 35–93.
2. Gjøystdal, H., J. E. Reinhardsen, and K. Åstebøl, Computer representation of complex 3-D geological structures using a new "solid modeling" technique, Geophysical Prospecting **33** (1985), 1195–1211.
3. Hanyga, A., Implicit curve tracing and point-to-curve ray tracing, Seismo-series **56**, Seismological Observatory, University of Bergen, 1991.
4. Hubral, P. and T. Krey, *Interval Velocities from Seismic Reflection Time Measurements*, K. L. Larner (ed.), Society of Exploration Geophysicists, Tulsa, 1980.
5. Laksâ, A., and M. Floater, Reference Manual SI Spline Library version 2.1, Senter for industriforskning, Oslo, Norway, 1990.
6. Lyche, T., and K. Mørken, A data-reduction strategy for splines with applications to the approximation of functions and data, IMA J. Numer. Anal. **8** (1987), 185–208.

7. Mallet, J. L., Three-dimensional graphic display of disconnected bodies, Mathematical Geology **20** (1988), 977–990.
8. Powell, M. J. D., The theory of radial basis function approximation in 1990, DAMTP 1990/NA11 Desember 1990, University of Cambridge.
9. Renka, R. J.and A. K. Cline, A triangle-based C^1 interpolation method, Rocky Mt. J. Math. **14**, (1984), 223–237.

Arne Laksâ
Senter for Industriforskning
Forskningsveien 1
N-0314 Oslo Norway
laksaa@bivrost.cs.si.no

Lars A. Frøyland, Jan Pajchel
Norsk Hydro E&P Research Centre
Lars Hillesgt. 30
N-5008 Bergen Norway
blaf@bg.hbg.hydro.no, bray@bg.hbg.hydro.no

Wonderful Triangle: A Simple, Unified, Algorithmic Approach to Change of Basis Procedures in Computer Aided Geometric Design

Ronald N. Goldman and Phillip J. Barry

Abstract. Many important algorithms in computer aided geometric design can be viewed as change of basis procedures. These include evaluation, subdivision, knot insertion, knot deletion, interpolation, degree elevation, and differentiation. We provide a simple, unified, algorithmic approach to such change of basis procedures for a wide class of polynomial and spline bases. Along the way we shall observe that many well-known algorithms and formulas for polynomials and splines, including synthetic division, forward differencing, Horner's method, the de Casteljau algorithm, the de Boor algorithm, Boehm's derivative algorithm, Boehm's knot insertion algorithm, the Oslo algorithm, Marsden's identity, and the binomial theorem are simply variations on a common theme.

The beauty of nature lies in detail, the message in generality.

—Stephen Jay Gould, *Wonderful Life*

§1. Introduction

Change of representation is a rich theme with many variations in computer graphics and geometric modeling. Shape representation may be polynomial or spline, implicit or parametric, procedural or geometric, boundary file or constructive solid geometry, approximate or exact.

Various conversion algorithms are associated with these different representations. Polygonal approximations are used in computer graphics to speed up rendering algorithms; degree reduction [18] and knot removal procedures [23] are applied in computer aided geometric design to reduce the amount of data in the representation. In algebraic geometry, implicitization algorithms

Mathematical Methods in Computer Aided Geometric Design II

Tom Lyche and Larry L. Schumaker (eds.), pp. 297–320.

ISBN 0-12-460510-9.

use resultants to convert from a rational parametric to an implicit representation [33], and uniformization procedures apply geometric techniques to convert from the implicit to the rational parametric form [1]. In solid modeling, evaluation algorithms have commonly been used to convert from a constructive solid geometry representation to a boundary file representation [29], and more recently algorithms have also been developed that convert a boundary file into constructive solid geometry [35].

All these conversion algorithms are important because different representations provide different information about the model. Furthermore, certain analysis procedures are better adapted to specific representations; thus it is often crucial to change to a particular representation in order to perform the required analysis on the model.

Here we shall restrict our attention to exact change of basis algorithms for polynomial or piecewise polynomial representations. We shall consider, in particular, two general types of polynomial and spline bases: progressive [20,26] and Pólya [4,5]. These two classes are important because, as we shall see, they include many commonly used schemes in computer aided geometric design, including the monomial, Newton, Lagrange, Bernstein, and B-spline bases. We shall provide a simple, unified, algorithmic framework for constructing new, or justifying known, change of basis procedures between any two bases in these two classes. These change of basis procedures include algorithms for evaluation, subdivision, knot insertion, knot deletion, interpolation, degree elevation, and differentiation — algorithms which are fundamental to geometric modeling.

§2. Progressive and Pólya Bases

We begin by reviewing the fundamentals of progressive and Pólya bases.

A progressive basis is akin to a B-spline basis over a single knot interval. Progressive bases of degree n are defined by $2n$ knots $t_1, \ldots, t_n, t_{n+1}, \ldots, t_{2n}$ where $t_{i+n} \neq t_j$ for $1 \leq i < j \leq n$. Given $n+1$ control points $P_0, \ldots, P_n$, we can construct a progressive curve $p(t)$ recursively using the following algorithm due to de Boor [11,12,18].

de Boor algorithm: Let

$$p_j^0(t) = P_j \qquad\qquad j = 0, \ldots, n,$$

$$p_j^r(t) = \frac{t_{j+n+1-r} - t}{t_{j+n+1-r} - t_j} p_{j-1}^{r-1}(t) + \frac{t - t_j}{t_{j+n+1-r} - t_j} p_j^{r-1}(t), \tag{1}$$

where $r = 1, \ldots, n$ and $j = r, \ldots, n$. Then $p(t) = p_n^n(t)$.

This recursive algorithm is conveniently represented as a triangular array of computations, as illustrated in Figure 1a. The functions along the edges of this graph are the coefficients on the right hand side of the recursion, simplified by omitting the normalizing constants in their denominators. To retrieve these constants, notice that if two edges point into the same node, their labels sum to one; therefore their common denominator is given by the sum of their numerators. Notice also that whatever label enters a node also exits the node in the same direction. Thus to fully label the diagram, we need only the labels along the bottom level. These labels are simple to remember: $t - t_k$ labels the arrow exiting P_k to the left, and $t_{k+n+1} - t$ the arrow exiting P_k to the right. One of the purposes of this paper is to gain some insight into other algorithms arising from, or related to, this labeling scheme.

It follows by induction on r that $p_j^r(t)$ is a polynomial of degree r that depends linearly on the control points $P_{j-r}, \ldots, P_j$. Thus there are $n+1$ degree n polynomials $b_0(t), \ldots, b_n(t)$ depending only on the knots $t_1, \ldots, t_{2n}$ such that

$$p(t) = p_n^n(t) = \sum_k b_k(t) P_k.$$

We shall see in Section 3 that the polynomials $b_0(t), \ldots, b_n(t)$ always form a basis, called a *progressive basis*, for the space of polynomials of degree $\leq n$.

The basis function $b_k(t)$ represents all paths from the control point P_k at the base of the triangle to the curve $p_n^n(t)$ at the apex of the triangle (see Figure 1a). Thus we can compute the polynomial $b_k(t)$ directly from the de Boor algorithm by setting $P_j = \delta_{jk}$. Alternatively, we can reverse the arrows in the de Boor algorithm, place a 1 at the apex of the triangle, run the computations downward, and read the basis functions off the base of the triangle (see Figure 1b). This procedure works because paths from the apex to the base are equivalent to paths from the base to the apex. Reversing the arrows in the de Boor algorithm yields the Cox-de Boor-Mansfield recurrence for progressive basis functions [11,12]. This trick of reversing the arrows is an important device that we shall employ again when studying change of basis algorithms for Pólya curves.

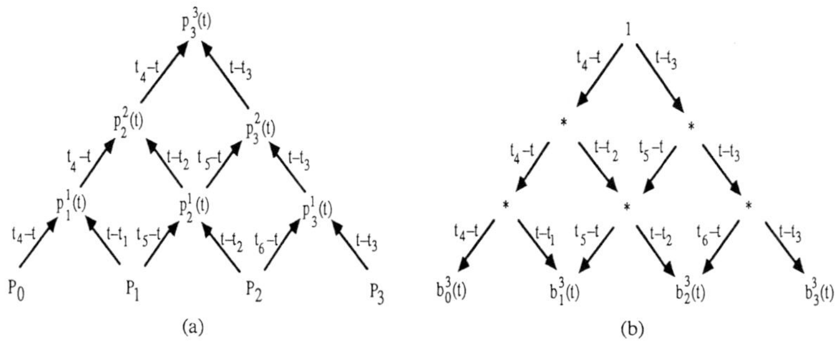

Figure 1. (a) The de Boor algorithm for progressive cubic curves, and (b) the the Cox-de Boor-Mansfield recurrence for cubic progressive basis functions.

We can extend our notion of a progressive basis by homogenizing the de Boor algorithm — *i.e.*, by homogenizing the numerator and denominator of the coefficient functions on the right hand side of (1). If we replace the affine variable t by the homogeneous variable (t, w) and the affine knots t_k by the homogeneous knots (t_k, w_k), then these functions become

$$\frac{t_{j+n+1-r} - t}{t_{j+n+1-r} - t_j} \longrightarrow \frac{t_{j+n+1-r}w - tw_{j+n+1-r}}{t_{j+n+1-r}w_j - t_jw_{j+n+1-r}}$$

$$\frac{t - t_j}{t_{j+n+1-r} - t_j} \longrightarrow \frac{tw_j - t_jw}{t_{j+n+1-r}w_j - t_jw_{j+n+1-r}}$$

and we obtain Figure 2a. This diagram is more complicated than Figure 1a — note particularly that the functions entering a node no longer sum to one — but now certain knots lead to much simpler schemes, as in Figure 2b.

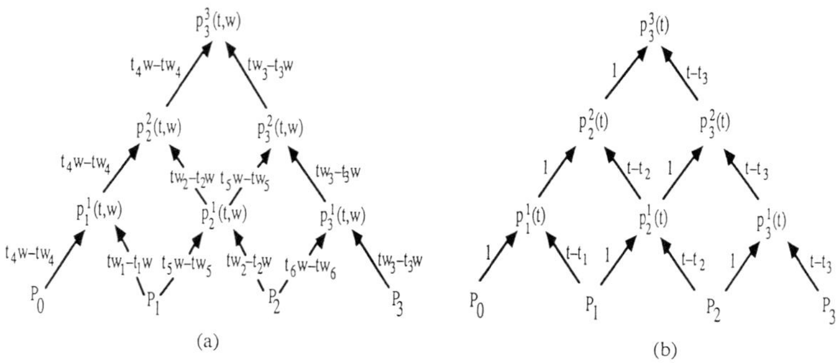

Figure 2. (a) the homogeneous de Boor algorithm for cubic curves, and (b) the homogeneous de Boor algorithm for cubic curves with $(t, w) = (t, 1)$ and knots at $(t_1, 1), (t_2, 1), (t_3, 1), (1, 0), (1, 0), (1, 0)$.

Pólya bases of degree n, like progressive bases, depend on $2n$ parameters $t_1, \ldots, t_n, t_{n+1}, \ldots, t_{2n}$, with $t_{i+n} \neq t_j$ for $1 \leq i < j \leq n$. Here, however, the parameters represent the roots of the Pólya basis functions $\psi_0(t), \ldots, \psi_n(t)$. Explicitly, these basis functions are defined by setting

$$\psi_k(t) = (t_{k+1} - t) \cdots (t_{k+n} - t) \qquad 0 \leq k \leq n.$$

For example, the cubic Pólya basis functions are given by

$$\psi_0(t) = (t_1 - t)(t_2 - t)(t_3 - t)$$

$$\psi_1(t) = (t_2 - t)(t_3 - t)(t_4 - t)$$

$$\psi_2(t) = (t_3 - t)(t_4 - t)(t_5 - t)$$

$$\psi_3(t) = (t_4 - t)(t_5 - t)(t_6 - t).$$

As with progressive bases, we can homogenize the parameters in a Pólya basis. This amounts to replacing each factor $(t_k - t)$ by the factor $(t_kw - tw_k)$. Notice, that if $(t, w) = (t, 1)$ and $(t_k, w_k) = \delta := (1, 0)$, then $(t_k w - t w_k) \longrightarrow 1$; *i.e.*, the factor is simply omitted. We shall see in Section 3 that the progressive and Pólya bases are related more intimately than simply sharing the same parameters $t_1, \ldots, t_{2n}$ initially suggests.

There are many interesting special cases of progressive and Pólya bases. We list the most important of these bases along with their defining parameters and explicit formulas in Table 1. These explicit formulas for the Pólya bases are easy to derive; the explicit formulas for the corresponding progressive bases are derived in [20]. Note that for certain bases, the normalizing constants differ from their usual values; however, since these are easily adjusted we will not concern ourselves with them here. The symbol "—" in the table means that no simple explicit formula exists.

Table 1: Parameters and explicit formulas for selected progressive and Pólya bases of degree n.

Bases	Parameters	Formulas
progressive/	$t_1, \ldots, t_n, t_{n+1}, \ldots, t_{2n}$	—
Pólya	$t_{i+n} \neq t_j \quad 1 \leq i < j \leq n$	$\psi_k(t) = (t_{k+1} - t) \cdots (t_{k+n} - t)$
Bernstein/	$a, \ldots, a, b \ldots, b$	$b_k(t) = \binom{n}{k} \frac{(t-a)^k (b-t)^{n-k}}{(b-a)^n}$
Bernstein	$a \neq b$	$\psi_k(t) = (-1)^{n-k} (t-a)^{n-k} (b-t)^k$
power/	$u_1, \ldots, u_n, u_0, \ldots, u_{n-1}$	$b_k(t) = (t - u_k)^n / \Pi_{j \neq k}(u_j - u_k)$
Lagrange	$u_j \neq u_i$ when $j \neq i$	$\psi_k(t) = \Pi_{j \neq k}(u_j - t)$
semi-Bernstein/	$t_1, \ldots, t_n, b, \ldots, b$	—
half-Bernstein	$t_i \neq b$	$\psi_k(t) = [\Pi_{j=k+1}^{n}(t_j - t)](b - t)^k$
monomial/	$a, \ldots, a, \delta, \ldots, \delta$	$b_k(t) = \binom{n}{k}(t-a)^k$
monomial	$a = (a, 1), \ \delta = (1, 0)$	$\psi_k(t) = (-1)^{n-k}(t-a)^{n-k}$
Newton dual/	$t_1, \ldots, t_n, \delta, \ldots, \delta$	$b_k(t) = (-1)^{n-k}[t_k, \ldots, t_n](t - \cdot)^n$
Newton	$t_i = (t_i, 1), \ \delta = (1, 0)$	$\psi_k(t) = (t_{k+1} - t) \cdots (t_n - t)$

Many of these bases, such as the B-splines, the Bernstein basis, the power basis, the Lagrange basis, and the monomial basis are well-known; others such as the semi-Bernstein, half-Bernstein, and Newton dual are more obscure, but will play important roles later in this paper. The Bernstein and monomial bases are both progressive and Pólya bases (though the sequential order and normalizing coefficients differ); we shall take advantage of this observation in Section 8.

The coefficients of a curve with respect to different bases provide us with important information about the curve (see Table 2).

Table 2: Selected bases and the meaning of their coefficients.

Basis	Coefficients
Bernstein	Bézier points
B-spline	de Boor points
monomial	derivatives
Newton	divided differences
Lagrange	interpolation points

The purpose of many algorithms is to compute certain particular coefficients in Table 2. To obtain this information, one must convert the curve to the desired representation. This observation then defines our goal: *to transform any progressive or Pólya representation to any other progressive or Pólya representation.*

§3. Dual Functionals

There are many ways to construct change of basis procedures for progressive and Pólya curves. One can, *e.g.*, set up and solve a system of linear equations. This is a very general technique which will serve for any two bases. However, in general, this approach is $O(n^3)$. We would like to do better for progressive and Pólya bases. An alternative technique is to use dual functionals. In this section we will discuss dual functionals for progressive bases. While the dual functionals are important throughout the rest of the paper, computationally they are not as simple as one would like. In subsequent sections we show how to organize the dual functionals to avoid messy calculations.

To introduce the dual functionals, we begin with an operator that takes as arguments two functions f and g, and is defined by

$$[f, g] = \sum_{r=0}^{n} [(-1)^{n-r}/n!][f(\tau)]^{(n-r)}[g(\tau)]^{(r)}. \tag{2}$$

This operator has three important properties:

(i) $[f, g]$ is linear in each parameter f, g;
(ii) $[f, g]$ is a constant independent of τ when f, g are polynomials of degree at most n;
(iii) $[(u - \cdot)^n, g] = g(u)$ if g is a polynomial of degree at most n.

The first property follows easily from the linearity of differentiation. The second property can be proved by differentiating the right hand side of (2) by the product rule and then observing that all the terms cancel because of the alternating sign. The last property follows by substituting $f(t) = (u - t)^n$ on the right hand side, evaluating at $\tau = u$, and observing that all terms vanish except the $r = 0$ term, which reduces to $g(u)$.

Now let $\{b_j(t)\}$ be a progressive basis and let $\{\psi_k(t)\}$ be the Pólya basis with the same defining parameters $t_1, \ldots, t_{2n}$. We have the following theorem due to de Boor.

Theorem. $[\psi_i, b_j] = \delta_{ij}$.

Proof: See [2,3,12,13,14]. ■

Corollary. $S(t) = \sum_j c_j b_j(t) \Leftrightarrow [\psi_i, S] = c_i$.

Proof: This follows immediately from the theorem and the linearity of the operator. ■

By the theorem, the operators $\{\lambda_i(S) := [\psi_i, S]\}$ are the dual functionals for the progressive polynomials $\{b_j(t)\}$. Thus, the progressive polynomials do indeed form a basis by the corollary. We could use these dual functionals to compute the coefficients of any polynomial S with respect to the basis $\{b_j(t)\}$. However, we shall see there is often a better way to proceed that avoids the computation of the derivatives in (2). This is considered further in the next section.

§4. Wonderful Triangle

Consider the triangle in Figure 3a. What is so special about this triangle? To begin, the labels $\alpha\beta\gamma$ at the nodes have four possible interpretations:

1. real numbers: $\alpha\beta\gamma =$ an unordered triple of three real numbers
2. Pólya polynomials: $\alpha\beta\gamma = (\alpha - t)(\beta - t)(\gamma - t)$
3. dual functionals: $\alpha\beta\gamma() = [(\alpha - t)(\beta - t)(\gamma - t),\ ]$
4. polynomial coefficients: $\alpha\beta\gamma(S) = [(\alpha - t)(\beta - t)(\gamma - t), S]$.

The main point here, and what makes this triangle so noteworthy, is that for each of these interpretations the labels along the edges of the diagram are the same. As we shall see, the first interpretation makes the diagram easy to compute; the last interpretation makes it useful.

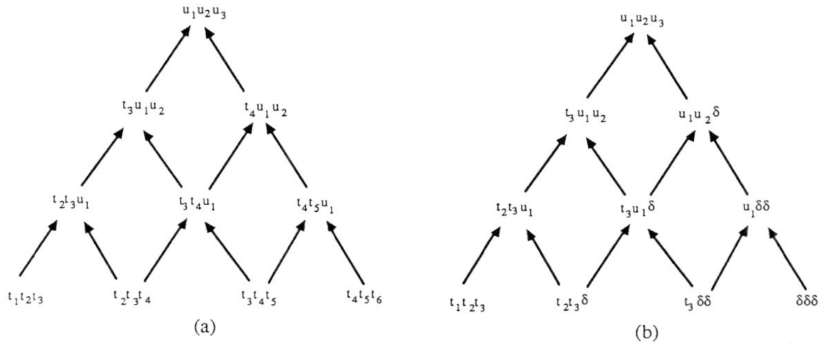

Figure 3. (a) The wonderful triangle, and (b) a homogeneous variant.

How do we label the edges in the diagram? If we interpret the nodes as triples of real numbers, then the labeling is simply barycentric coordinates. Indeed, notice that in any three node subtriangle, all but one of the factors are the same at all three nodes. The three unequal factors u_i, t_j, t_k can be related by barycentric coordinates:

$$u_i = \frac{t_k - u_i}{t_k - t_j} t_j + \frac{u_i - t_j}{t_k - t_j} t_k. \tag{3}$$

These barycentric coefficients then provide the labels for computing the upper node in terms of the two lower nodes whose arrows point into it. The entire diagram is just a way of computing the node at the apex from the nodes at the base of the triangle, just like the de Boor algorithm (Figure 1a). In fact, if $u_i = t$ for all i, then the labels along the edges are exactly the same as the labels in the de Boor algorithm for a progressive curve with knots $t_1, \ldots, t_{2n}$! This observation is not a coincidence; we shall explain the reason shortly.

Next consider the interpretation of this triangle as a collection of Pólya polynomials. Since barycentric coordinates sum to 1, we can write

$$t = \frac{t_k - u_i}{t_k - t_j} t + \frac{u_i - t_j}{t_k - t_j} t. \tag{4}$$

Subtracting (4) from (3) yields

$$(u_i - t) = \frac{t_k - u_i}{t_k - t_j}(t_j - t) + \frac{u_i - t_j}{t_k - t_j}(t_k - t).$$

Multiplying this equation by the common factors at the three nodes, we again see that these barycentric coefficients can be used to compute the upper node in terms of the two lower nodes whose arrows point into it. Letting ψ_L, ψ_R, and ψ_U denote the Pólya polynomials at the lower left, lower right, and upper nodes of the subtriangle, we get

$$\psi_U(t) = \frac{t_k - u_i}{t_k - t_j}\psi_L(t) + \frac{u_i - t_j}{t_k - t_j}\psi_R(t). \tag{5}$$

Thus we can interpret the nodes as Pólya polynomials and the labels along the edges will be exactly the same as in the interpretation with real numbers.

What about the dual functionals? Recall that $[\ ,\ ]$ is a linear operator. Apply it to both sides of (5) to obtain

$$[\psi_U(t),\ ] = \frac{t_k - u_i}{t_k - t_j}[\psi_L(t),\ ] + \frac{u_i - t_j}{t_k - t_j}[\psi_R(t),\ ]. \tag{6}$$

Again, the labels are the same. Finally, if we apply this entire triangle of dual functionals to some polynomial S, then

$$[\psi_U(t), S] = \frac{t_k - u_i}{t_k - t_j}[\psi_L(t), S] + \frac{u_i - t_j}{t_k - t_j}[\psi_R(t), S]. \tag{7}$$

Thus, in all four interpretations, the labels along the edges are identical.

The last interpretation says that if we have the coefficients for a polynomial S in terms of a progressive basis with knots $t_1, \ldots, t_{2n}$, then by running the triangle with appropriate values of $u_1, \ldots, u_n$ we can compute any coefficient of S with respect to any other progressive basis. *The wonderful triangle is just an elegant organization of the dual functionals.* It avoids computations other than barycentric coordinates — in particular, it avoids the derivatives in the dual functional formula. Moreover, it leads to efficient change of basis algorithms — in the next section we show that finding *all* the coefficients with respect to any other progressive basis requires no more than two triangles, and is thus an $O(n^2)$ algorithm.

Note that although we always draw our diagrams only for cubic curves, these remarks apply generally for all degrees; the proofs are much the same.

The original knots must satisfy $t_{i+n} \neq t_j$ for $1 \leq i < j \leq n$. A certain few choices of the new knots may result in either a final new knot vector or an intermediate knot vector, occurring during the computation of the triangle, that does not satisfy an analogous condition. For some of the algorithms discussed below this will not present a problem; for other algorithms it will. We are concerned with the general structures of these algorithms, and will therefore assume, throughout the remainder of the paper, that the knot values chosen will not cause any problems.

The triangle also has another, related, interpretation. Let

$$b(u_1, u_2, u_3) = [(u_1 - t)(u_2 - t)(u_3 - t), S].$$

Then b is what is now commonly called the *blossom* [26,27,31] or *polar form* [15,16,28] of the polynomial S. The blossom is defined by three properties: a multiaffine property, a symmetry property, and a diagonal property. These properties correspond to properties of the triangle. For example, the symmetry of the blossom corresponds to the fact that the triples are unordered, or to the fact that a Pólya basis function is symmetric in its roots. The multiaffine property corresponds to (3) above. Finally, the diagonal property, $b(t, t, t) = S(t)$, follows from Property (iii) of the operator [,].

Like progressive bases, the triangle can be homogenized. The rule for labeling the edges of the homogenized triangle is given by

$$(u_i, v_i) = \frac{t_k v_i - u_i w_k}{t_k w_j - t_j w_k}(t_j, w_j) + \frac{u_i w_j - t_j v_i}{t_k w_j - t_j w_k}(t_k, w_k).$$

Notice that this combination is no longer affine.

While homogenizing generally complicates the triangle, setting some of the knots to $\delta = (1, 0)$ actually simplifies it greatly. Consider how δ may appear in any three node subtriangle. There are three possible cases, shown in Figure 4. Notice that the scheme with δ on the base is the simplest computationally since there is no division and only one multiplication. Several of the algorithms described in the next section are based upon this simple observation.

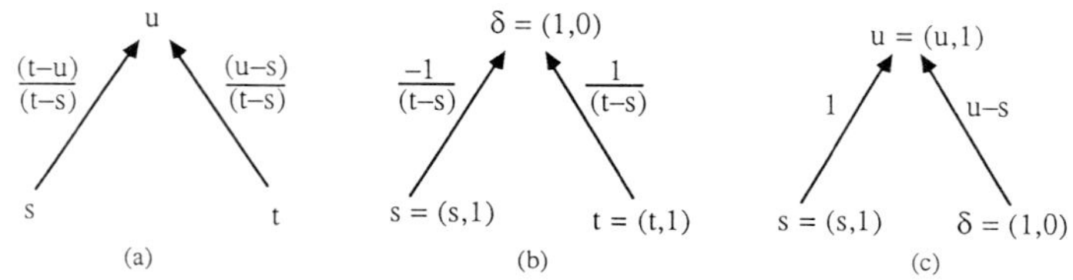

Figure 4. The effect of δ in a three node subtriangle.

§5. Algorithms — Knot Insertion

Wonderful triangles generate many wonderful algorithms. In this section we present some of these algorithms, beginning with an algorithm for change of basis from any progressive representation to any other progressive representation. Then we discuss knot insertion, evaluation, and differentiation. As we shall see below, all these algorithms can be interpreted in terms of knot insertion; thus we shall call algorithms which arise from applying the triangle to progressive curves *knot insertion algorithms*.

At heart, the triangle is a change of basis procedure. At its base lie the coefficients of a polynomial S with respect to the progressive basis for the knot sequence $t_1, \ldots, t_n, t_{n+1}, \ldots, t_{2n}$, and along the two lateral edges lie the coefficients of S with respect to the progressive bases for the knot sequences $t_1, \ldots, t_n, u_1, \ldots, u_n$ and $u_n, \ldots, u_1, t_{n+1}, \ldots, t_{2n}$. Because of this interpretation, from now on we shall diagram the triangle schematically as shown in Figure 5b. This diagram is simply a convenient shorthand notation for Figure 5a. Note the reversal of the knots $u_n, \ldots, u_1$ along the lateral right edge of the triangle.

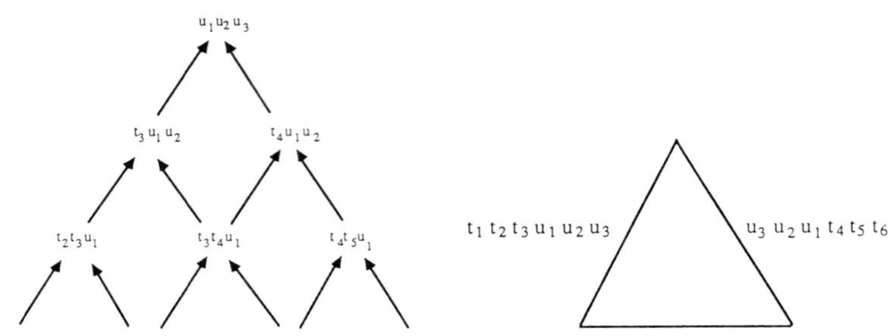

Figure 5. (a) The triangle for cubic curves, and (b) a schematic equivalent.

We can convert from any progressive representation to any other progressive representation in two steps. First we introduce knots $u_{n+1}, \ldots, u_{2n}$, and then knots $u_n, \ldots, u_1$. This algorithm is illustrated in Figure 6. The remainder of the algorithms in this section are special cases of this procedure.

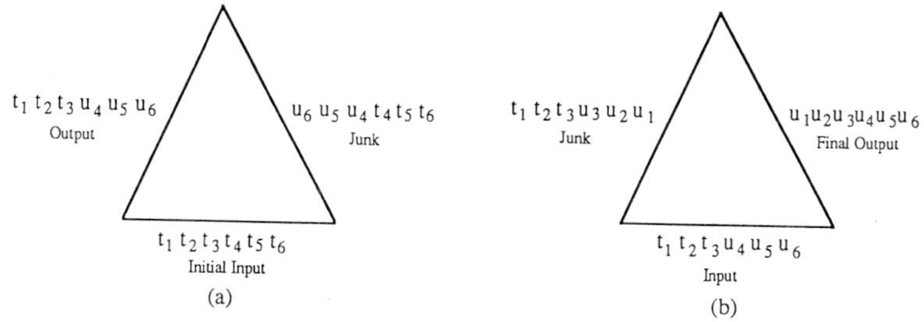

Figure 6. Two stage algorithm for converting from any cubic progressive form to any other cubic progressive form: (a) inserting the knots u_4, u_5, u_6, and (b) inserting the knots u_3, u_2, u_1.

We can think of the triangle as a knot insertion procedure. Begin with the coefficients of S with respect to the progressive basis for $t_1, \ldots, t_{2n}$. The triangle (see Figure 5a) inserts knots $u_1, \ldots, u_n$ between the knots t_n and t_{n+1}, and produces the coefficients of S with respect to the progressive bases for the new knot sequences $t_1, \ldots, t_n, u_1, \ldots, u_n$ and $u_n, \ldots, u_1, t_{n+1}, \ldots, t_{2n}$. If, instead of using the entire triangle, we use only the bottom level of computations in Figure 5a, then we obtain Boehm's knot insertion algorithm [9] for inserting one knot between t_n and t_{n+1}. *In fact, each level of the triangle corresponds to the insertion of one of the* u_i. Note, however, that rather than each level computing n new points, as Boehm's algorithm would do, only the values needed for the next level are found. Another difference to keep in mind is that here knot insertion is being applied to progressive polynomial curves, which do not have their knots ordered by value (*i.e.*, the knots need not be nondecreasing), but rather by their position in the knot vector.

Another commonly used knot insertion algorithm is the Oslo algorithm. One version of this algorithm (the "Oslo II" algorithm of [17]) is generated simply by running the triangle with the appropriate u_i and noting that the point at the apex of the triangle is a new control point for the curve represented over the refined knot vector. Other versions of the Oslo algorithm given in [17,22] also follow easily from the triangle [7].

Because of the close ties between knot insertion and the triangle, all the algorithms in this section can be be interpreted as special knot insertion procedures. This justifies calling them all knot insertion algorithms.

There are also many other knot insertion techniques which can be explained by the triangle [7]. We mention two more here. The first is a two stage procedure that provides another variant of the Oslo algorithm [19]. This procedure inserts $n+1$ knots into a degree n B-spline curve between t_n and t_{n+1} by first inserting the knots $u_1, \ldots, u_n$, and then inserting the knots $u_{n+1}, \ldots, u_2$. The output, read off the lateral edges of two triangles, are the control points for the B-spline curve with the extended knot sequence $t_1, \ldots, t_n, u_1, \ldots, u_{n+1}, t_{n+1}, \ldots, t_{2n}$. This procedure is illustrated in Figure

7. Note that the right lateral edge of Figure 7b can be used as input if further knots are to be inserted between u_{n+1} and t_{n+1}.

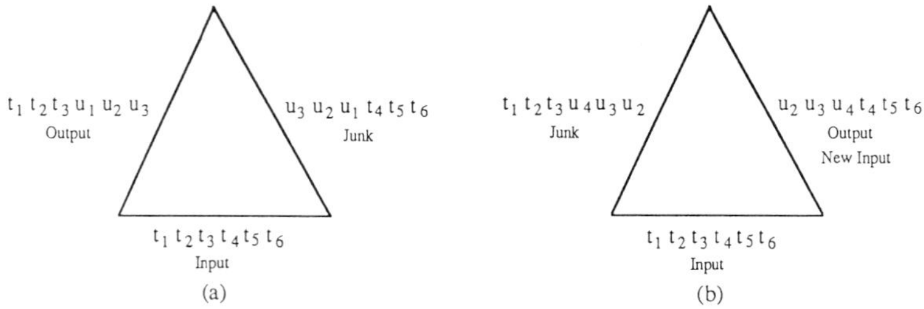

Figure 7. A variant of the Oslo algorithm: (a) inserting the knots u_1, u_2, u_3, and (b) inserting the knots u_4, u_3, u_2.

Barry and Goldman [6] and Sankar, Silbermann, and Ferrari [34] have observed that knot insertion can sometimes be sped up by factoring through the Newton dual basis — *i.e.*, the progressive basis with knots $t_1, \ldots, t_n, \delta, \ldots, \delta$. To begin, transform from progressive to Newton dual form by inserting an n-fold knot at δ. One variant then proceeds in two stages, as in the algorithm directly above, inserting knots $u_1, \ldots, u_n$ and $u_{n+1}, \ldots, u_2$ between t_n and δ. The output, illustrated in Figure 8, consists of control points over the new knot sequence $t_1, \ldots, t_n, u_1, \ldots, u_n$, together with coefficients for a new Newton dual startup over the knot sequence $u_2, \ldots, u_{n+1}, \delta, \ldots, \delta$. The advantage of this approach is that knot insertion using the Newton dual form requires no division and only a single multiplication per node (see Figure 4).

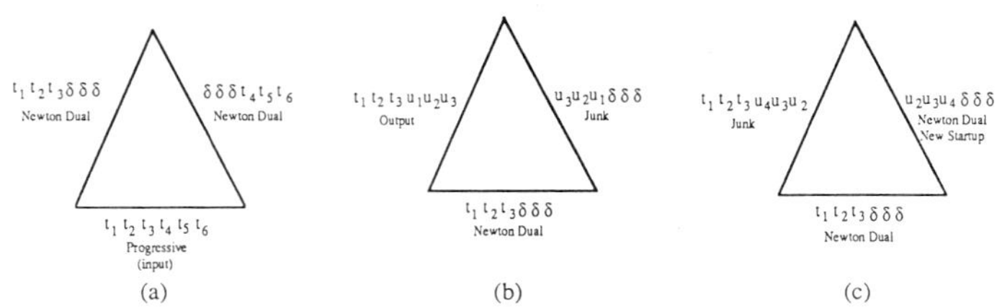

Figure 8. Factored knot insertion: (a) initializing by converting to Newton dual form; (b) inserting u_1, u_2, u_3; and (c) reinitializing by inserting u_4, u_3, u_2.

Consider next evaluation algorithms. We know from the properties of the operator $[\ ,\ ]$ that $(u-t)^n(S) = [(u-\cdot)^n, S] = S(u)$. Thus, to evaluate S at $t = u$, we need only insert an n-fold knot at u. Doing so, we obtain Figure 9a. Important special cases of this evaluation procedure are shown in Figures 9b and 9c.

Notice that Figure 9a can be interpreted as a change of basis procedure for converting from arbitrary progressive to semi-Bézier form. Indeed, the semi-Bézier form is often used in geometric modeling for the first and last

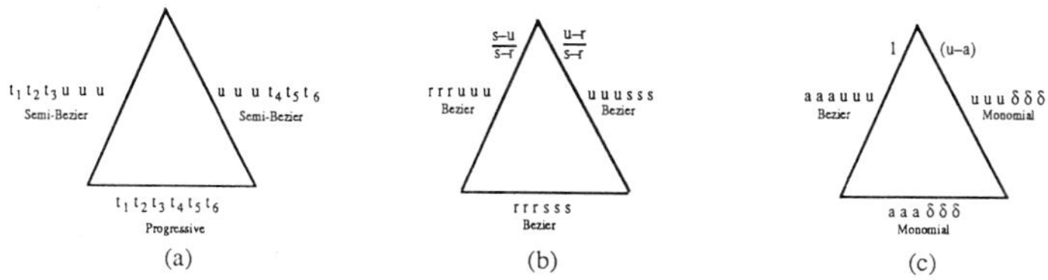

Figure 9. Evaluation algorithms for progressive curve schemes: (a) arbitrary progressive curves, (b) Bézier curves, and (c) monomial form. The labels near the apex indicate the edge labels.

segments of a B-spline curve to ensure interpolation to the end points of the control polygon. The labels along the arrows of this triangle are precisely the labels in the de Boor algorithm. Thus we could have started with the dual functionals and used them to derive the de Boor algorithm. This observation is important not only because it gives us insight into the labeling scheme of the de Boor algorithm, but also because it reverses the standard development where the de Boor algorithm is used to derive the dual functionals [14]. In fact, it is precisely this dual functional approach that permits us to derive a similar recursive evaluation algorithm for geometrically continuous splines [8].

The most famous algorithm where an n-fold knot is inserted into a progressive curve is the de Casteljau evaluation algorithm for Bézier curves [15,18]. From Figure 9b it is clear why this algorithm also generates a subdivision procedure for Bézier curves.

Figure 9c illustrates another special case — evaluating, at $t = u$, curves written in terms of the monomial basis at a. We will return to the properties of this conversion again shortly.

Finally, consider derivatives. Boehm's derivative algorithm [10] is closely related to factored knot insertion. Remember that the coefficients of the monomial basis at a are, up to constant multiples, the derivatives of the curve at $t = a$. Thus to find the derivatives of a progressive curve, we need only transform it into monomial form — *i.e.*, to the knot sequence $a, \ldots, a, \delta, \ldots, \delta$. This can again be done in two stages: first transform to Newton dual form by inserting an n-fold knot at δ, and then convert from Newton dual to monomial form by inserting an n-fold knot at a, as shown in Figure 10.

Notice that once we have converted to Newton dual form, we can find the derivatives at any point by inserting an n-fold knot at that value. But recall that n-fold knot insertion is equivalent to conversion to semi-Bézier form, which, in turn, is equivalent to evaluation. Thus the second stage of Boehm's derivative algorithm is simply the de Boor evaluation algorithm for the Newton dual form. One reason for the importance of the Newton dual basis is that the derivatives are a byproduct of the evaluation procedure (see Figure 10b (or 11c below)).

We can differentiate a Bézier curve at either of its endpoints, or equivalently, convert it to monomial form at either endpoint, in one stage, without transforming to Newton dual form. All we need to do is to insert an n-fold

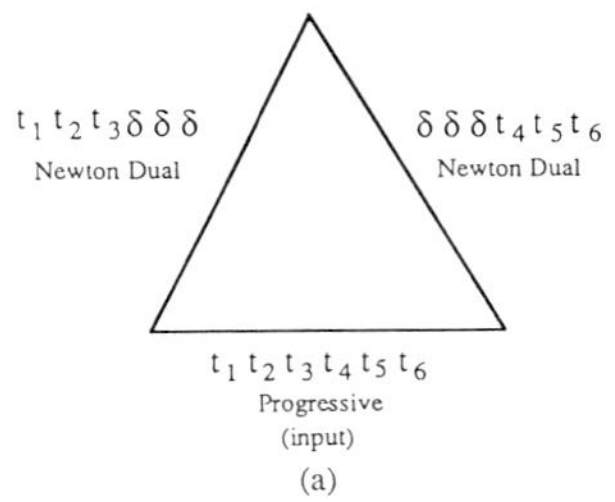

Figure 10. Boehm's derivative algorithm: (a) Converting from progressive to Newton dual form, and (b) converting from Newton dual to monomial form.

knot at δ and then read the results off the lateral edges of the triangle as illustrated in Figure 11a. Notice that except for the normalizing constant $(b-a)$ in the denominators, this algorithm is just differencing.

Finally, we can convert from monomial form at a to monomial form at b by inserting an n-fold knot b. This procedure is useful if we want to find the derivatives at $t = b$ given the derivatives at $t = a$. Classically this algorithm is known as synthetic division [30]. Notice, however, that not only do we get the monomial coefficients at $t = b$, but also as a bonus the Bézier coefficients emerge along the left lateral edge of the diagram in Figure 11b. Thus the standard synthetic division algorithm can be used to convert from monomial to Bézier form.

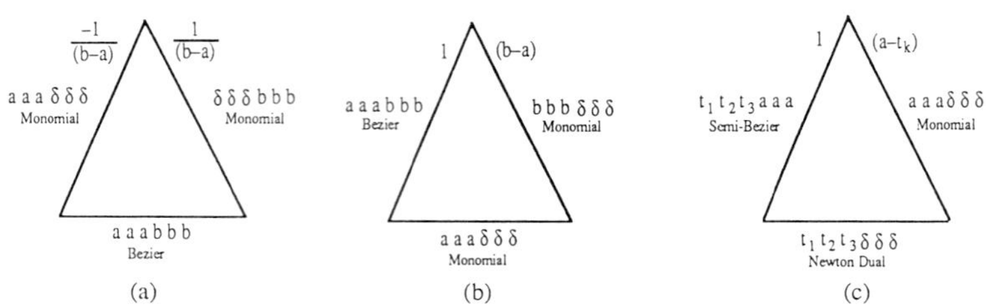

Figure 11. Special derivative algorithms: (a) Bézier $\leftrightarrow$ differencing, (b) monomial $\leftrightarrow$ synthetic division, and (c) Newton dual $\leftrightarrow$ evaluation.

Many other algorithms of this sort can be easily constructed. For example, the de Boor differentiation algorithm [12] for B-spline curves, which represents the first derivative at $t = a$ in terms of B-splines of lower degree, is simply knot insertion at $\delta, a, \ldots, a$ [20]. Or we could use the triangle to convert from piecewise monomial to B-spline form by factoring through the Newton dual basis [21]. This algorithm is basically the inverse of Boehm's derivative algorithm. Once we have the fundamental idea of knot insertion combined with the technique of the wonderful triangle together with some useful examples of progressive bases, these algorithms are very straightforward to design.

§6. Principles of Duality

Our goal is to transform any progressive or Pólya representation to any other progressive or Pólya representation. So far we have only shown how to convert between any two progressive representations. In this section we show how to develop change of basis procedures for Pólya curves.

Again we shall apply a variant of the wonderful triangle. The main insight here is that transformations between Pólya bases are closely related to transformations between progressive bases with corresponding knot sequences. This is because the Pólya bases represent the dual functionals for the progressive bases.

To begin, consider two progressive bases $b(t) = (b_0(t), \ldots, b_n(t))$ and $b^*(t) = (b_0^*(t), \ldots, b_n^*(t))$. Let $P = (P_0, \ldots, P_n)$ be the coefficients for a polynomial curve $S(t)$ relative to the basis $b(t)$, and let $Q = (Q_0, \ldots, Q_n)$ be the coefficients of $S(t)$ relative to the basis $b^*(t)$. Then

$$S(t) = b(t) \cdot P^t = b^*(t) \cdot Q^t$$

where $\cdot$ denotes matrix multiplication and P^t and Q^t denote the transpose of P and Q, respectively. Suppose now that M is the change of basis matrix for these progressive basis. Then

$$b(t) = b^*(t) \cdot M \Leftrightarrow Q^t = M \cdot P^t.$$

Thus the change of basis matrix is equivalent to the matrix that transforms the coefficients. Similar observations hold for any two bases, in particular for two Pólya bases.

The triangle is simply a way of representing a progressive change of basis matrix M. The coefficients P are placed along the base of the triangle and the coefficients Q emerge along one of the lateral edges of the triangle. The transformation matrix M for a single triangle is triangular (recall that for a general progressive to progressive transformation we need to use two triangles), and each level in the triangle corresponds to a 1-banded factor of M.

But remember the triangle also has a dual Pólya interpretation. Here one set of Pólya basis functions is placed along the base of the triangle, and a new set of Pólya basis functions emerges along one of the lateral edges of the triangle. If $\psi(t) = (\psi_0(t), \ldots, \psi_n(t))$ and $\psi^*(t) = (\psi_0^*(t), \ldots, \psi_n^*(t))$ are the Pólya bases corresponding to the progressive bases $b(t)$ and $b^*(t)$, then by the dual Pólya interpretation of the triangle, we can replace P by $\psi(t)$ and Q by $\psi^*(t)$, and get

$$\psi^*(t)^t = M \cdot \psi(t)^t \Leftrightarrow \psi^*(t) = \psi(t) \cdot M^t.$$

Thus M^t is the change of basis matrix for the Pólya bases, but in the opposite direction from the corresponding progressive bases. This observation is our

First Principle of Duality: *Transformations M between progressive bases are equivalent to transformations M^t between corresponding Pólya bases in the other direction.*

For progressive curves not only do we know the transformation matrix, but we also have an algorithm which takes as input the coefficients for one basis and produces as output the coefficients for another basis. We would like to have a similar algorithm for Pólya curves. Our second principle of duality leads to just such an algorithm.

Second Principle of Duality: *Reversing the arrows in the wonderful triangle converts M to M^t.*

Before we prove the second principle is valid, let us explain exactly what it means. For progressive curves, the triangle algorithms all have the structure illustrated in Figure 12a with coefficients relative to one progressive basis placed at the base of the triangle and the coefficients relative to another progressive basis emerging along one of the lateral edges of the triangle. The arrows in the diagram point upwards, and the computations proceed up from the base of the triangle row by row until the apex is reached.

For Pólya transformations we reverse the direction of each arrow, but we keep the same labels along the arrows. Now the algorithm proceeds from the apex to the base of the triangle, as illustrated in Figure 12b. The coefficients relative to one Pólya basis are placed along a lateral edge of the triangle and the coefficients relative to another Pólya basis emerge along the base of the triangle. The arrows in the diagram point down, and the computations proceed down from the apex of the triangle row by row until the base is reached. For nodes with asterisks, the value is computed in the usual manner by multiplying the value along each arrow which enters the node by the value of the node from which the arrow emerges and adding the results. For nodes with coefficients Q_k, the same procedure is applied and then Q_k is added to the result.

Why does this technique work? Consider the progressive transformation $M = (M_{ij})$ illustrated in Figure 12a. By construction

$$Q^t = M \cdot P^t \Leftrightarrow Q_i = \sum_j M_{ij} P_j.$$

Thus M_{ij} represents all paths from the coefficient P_j at the base of the triangle to the coefficient Q_i along the lateral edge of the triangle. Now consider the Pólya transformation $N = (N_{ij})$ illustrated in Figure 12b. Again, by construction

$$P^t = N \cdot Q^t \Leftrightarrow P_i = \sum_j N_{ij} Q_j.$$

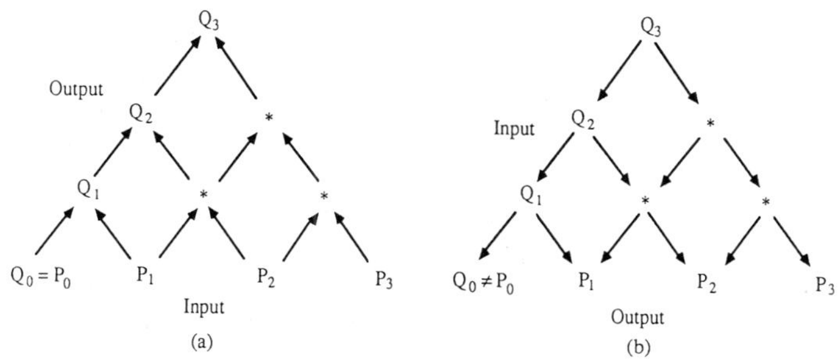

Figure 12. Change of basis algorithms: (a) progressive curves, and (b) Pólya curves.

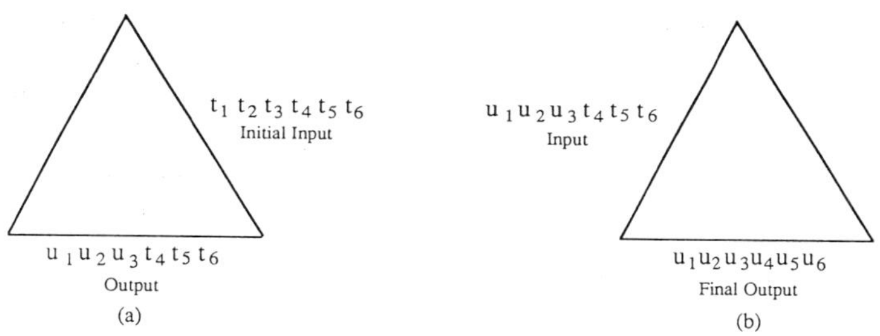

Figure 13. Two stage algorithm for converting from any cubic Pólya basis to any other cubic Pólya basis: (a) deleting the knots t_1, t_2, t_3, and (b) deleting the knots t_6, t_5, t_4.

Therefore,

$$N_{ij} = \text{ all paths from } Q_j \text{ to } P_i = \text{ all paths from } P_i \text{ to } Q_j \; = \; M_{ji}$$

so indeed $N = M^t$.

§7. More Algorithms — Knot Deletion

As a consequence of our two principles of duality, for every progressive algorithm, there is a dual Pólya algorithm. To emphasize this duality between progressive and Pólya algorithms, we shall call these Pólya change of basis procedures *knot deletion algorithms*, since they are dual to the knot insertion algorithms for progressive curves. In this section we will consider a few examples of these knot deletion algorithms.

We begin with a general procedure for converting any Pólya representation to any other Pólya representation. First we delete the knots $t_1, \ldots, t_n$ and introduce $u_n, \ldots, u_1$, and then we delete the knots $t_{2n}, \ldots, t_{n+1}$ and introduce $u_{n+1}, \ldots, u_{2n}$ (see Figure 13).

As was the case with the general change of basis procedure for progressive curves, this change of basis procedure for Pólya curves allows us to access information about curves in Pólya form (see Tables 1 and 2). We can find points on the curve (Lagrange coefficients), the Bézier representation (Bernstein coefficients), divided differences (Newton coefficients), or derivatives (monomial coefficients). We will look at some important special cases of this general algorithm, but first we briefly discuss knot deletion techniques.

There is a simple two term formula for deleting a single knot [5]. This formula is dual to Boehm's knot insertion technique, and is obtained by considering only the bottom level of computations in the triangle.

There are also other possible ways to arrange the dual functionals to generate knot deletion techniques; however, knot deletion appears, in general, to be less elegant than knot insertion. For example, given a progressive curve, the Oslo II algorithm discussed in the previous section allows one to find any coefficient with respect to an arbitrary progressive basis by running a single triangle. We are unaware of an analogous algorithm for Pólya curves.

Next consider evaluation. Note that all the Pólya basis functions, save the nth, vanish at the nth knot (symmetrically, all save the 0th vanish at the $n+1^{st}$ knot). Thus to evaluate a Pólya curve S at $t = a$, it suffices to choose any basis ψ_i with $t_n = a$, find the coefficient of ψ_n in the representation of S in this basis, and divide by $\psi_n(a)$, which is simple to compute. (If the Pólya basis is normalized to sum to 1, then $\psi_n(a) = 1$, and $S(a)$ will be the coefficient of ψ_n.) The coefficient of ψ_n appears at the lower right corner of Figure 13a, so we need only compute the right edge of that triangle; the remainder of the diagram can be ignored. The computation of the right edge has the same general structure as Horner's algorithm for the monomial representation (see Figure 14). Indeed, when the Pólya basis is specialized to the monomial basis, this Pólya evaluation algorithm is essentially Horner's method.

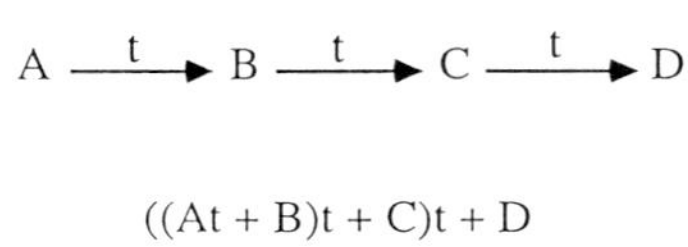

Figure 14. Horner's method.

To differentiate a Pólya curve, recall that the coefficients of the monomial basis at a are, up to constant multiples, the derivatives of the curve at $t = a$. Thus to find the derivatives of a Pólya curve, we can transform it into monomial form — *i.e.*, to the basis with knot sequence $a, \ldots, a, \delta, \ldots, \delta$. Again this can be done in two stages: first transform to Newton form by deleting the knots $t_{2n}, \ldots, t_{n+1}$ and replacing them by an n-fold knot at δ, and then convert from Newton to monomial form by deleting the knots $t_1, \ldots, t_n$ and replacing them by an n-fold knot at a. This two stage algorithm is illustrated in Figure 15.

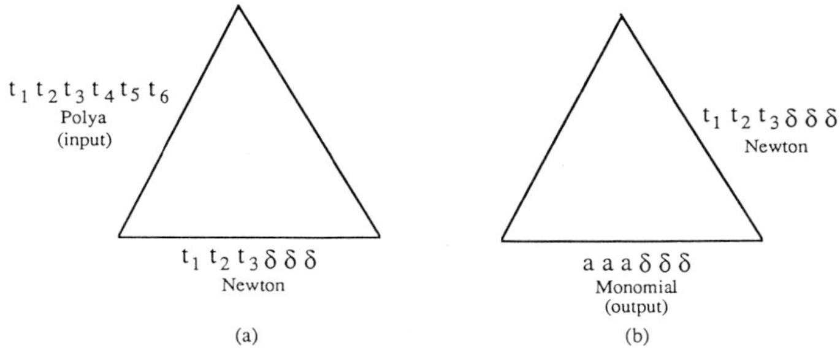

Figure 15. Pólya derivative algorithm: (a) Converting from Pólya to Newton form, and (b) converting from Newton to monomial form.

Notice that this algorithm is complimentary to Boehm's differentiation algorithm for progressive curves. In the progressive differentiation algorithm we factor through the Newton dual basis; in the Pólya differentiation algorithm, we factor through the Newton basis.

Having seen an application of knot deletion involving the Newton form, let us return for a moment to evaluation algorithms. The bottom level of Figure 3b (see also Figure 4) provides an extremely simple conversion from the Newton dual form over $a_1, \ldots, a_n, \delta, \ldots, \delta$ to the Newton dual form over $a_2, \ldots, a_{n+1}, \delta, \ldots, \delta$. The dual algorithm then gives a simple way to convert from Newton form over $a_2, \ldots, a_{n+1}, \delta, \ldots, \delta$ to Newton form over $a_1, \ldots, a_n, \delta, \ldots, \delta$. Because of this simple procedure for changing the nth knot of a curve in Newton form, one can evaluate a Pólya curve at a sequence of values by transforming first to Newton form, and then making repeated use of the Newton to Newton transformation. The resulting technique is essentially the forward differencing technique for evaluating a polynomial.

As with progressive curves, the triangle provides a general framework for developing many Pólya transformation algorithms. Moreover, it is also a useful tool for exposing the deeper connection between progressive and Pólya bases. We conclude this section with an example.

Recall that the de Boor evaluation algorithm for progressive curves at $t = a$ is just conversion to semi-Bézier form. Thus in the dual Pólya interpretation, we are converting from half-Bézier to general Pólya form. In this dual algorithm, the coefficient of the function $(a - t)^n$ is located at the apex of the triangle. Suppose we let this coefficient be 1, and all the other coefficients of the half-Bernstein basis functions be 0. The resulting algorithm is precisely the one given in Figure 1b, only with a new interpretation: this new algorithm represents the function $(a-t)^n$ in terms of an arbitrary Pólya basis $\psi_0(t), \ldots, \psi_n(t)$. We already know the coefficients of $\psi_0(t), \ldots, \psi_n(t)$; they are just the paths from the apex to the base of the triangle, *i.e.*, $b_0(a), \ldots, b_n(a)$. Thus we conclude that Figure 1b is also the Pólya knot deletion diagram for the equation

$$(a-t)^n = \sum_k b_k(a)\psi_k(t).$$

This formula is Marsden's identity [25]. Thus the Cox-de Boor-Mansfield recurrence and Marsden's identity are complimentary. Notice that Marsden's identity reduces to the binomial theorem when the Pólya basis is specialized to the monomial basis.

One can generalize Marsden's identity by replacing $(a-t)^n$ by an arbitrary Pólya basis function with roots at $a_1, \ldots, a_n$. Then the knot deletion algorithm represents the function $(a_1-t)\cdots(a_n-t)$ in terms of an arbitrary Pólya basis $\psi_0(t), \ldots, \psi_n(t)$. Again the coefficients of $\psi_0(t), \ldots, \psi_n(t)$ are the paths from the apex to the base of the triangle. By definition these paths are discrete B-splines [24]; equivalently, they are the blossoms $b_k(a_1, \ldots, a_n)$ of the progressive basis functions $b_k(a)$, since by construction they are symmetric, multiaffine, and reduce to $b_k(a)$ when $a_i = a$ for $i = 1, \ldots, n$. Thus we conclude that

$$(a_1-t)\cdots(a_n-t) = \sum_k b_k(a_1, \ldots, a_n)\psi_k(t).$$

This formula can also be derived by blossoming Marsden's identity [36], or from the properties of discrete B-splines [24].

§8. Still More Algorithms — Factoring the Transformation

We have shown how to convert between any two progressive representations by knot insertion, and between any two Pólya representations by knot deletion. But how do we convert from a progressive to a Pólya representation, or from a Pólya to a progressive representation?

Happily, the Bernstein and monomial bases are both progressive and Pólya bases. Thus, to convert from any progressive to any Pólya basis, or conversely from any Pólya to any progressive basis, we can simply factor the transformation through the Bernstein or monomial basis. For example, to convert from Pólya to progressive, convert from Pólya to Bernstein by knot deletion, and then from Bernstein to progressive by knot insertion. We can use this algorithm, *e.g.*, to perform interpolation by converting from Lagrange to B-spline form, thereby facilitating the de Casteljau construction of Catmull-Rom splines [15,16].

Factoring transformations through an appropriate basis has been a common theme throughout this paper. Factored knot insertion factors through the Newton dual basis; efficient differentiation factors through the Newton or Newton dual basis.

Other important change of basis algorithms can also be constructed by factoring the transformation. Consider degree elevation. For Pólya curves there is a simple two term formula that expresses any degree n Pólya curve as a degree $n+1$ Pólya curve over a knot vector whose knots are closely related to those of the original representation [4]. (It turns out that the dual formula to this degree elevation identity is the two term differentiation formula for the corresponding progressive basis [2].) This procedure, in conjunction with the general knot deletion algorithm, provides a method of representing any degree n Pólya curve as an arbitrary degree $n+1$ Pólya curve. No similar simple technique is known for progressive curves. We can, however, factor degree elevation through the Bernstein basis, where the algorithm is very simple [18], or through the monomial basis, where the algorithm is trivial — simple annex higher order zeros [21]. Similarly, products of progressive basis functions can also be computed by factoring through either the Bernstein or monomial bases.

§9. Summary, Conclusions, and Future Work

We began with a simple goal: to transform from any progressive or Pólya representation to any other progressive or Pólya representation. We achieved this goal in three stages: knot insertion, knot deletion, and factoring. In each stage we applied the wonderful triangle in a different way, reading it bottom up, or top down, or in both directions as the case required.

For progressive curves an important theme emerged: work in the dual space. The Pólya basis represents the dual functionals for the progressive basis, and the Pólya basis functions are much simpler to understand than the progressive basis functions. They are defined by an easy explicit formula, they factor into linear factors, and adjacent basis functions have only one root not in common. Progressive basis functions are defined by a recurrence which, at least initially, is much more difficult to analyze.

The wonderful triangle is just a way of organizing the dual functionals for progressive bases to simplify computations. This organization provides progressive algorithms for evaluation, differentiation, and knot insertion. Since the Pólya basis functions represent the progressive dual functionals, for every progressive algorithm, there is a corresponding dual Pólya algorithm. This observation yields evaluation, differentiation, and knot deletion algorithms for Pólya curves. Thus the triangle provides a simple, unified, algorithmic approach to important procedures for a wide class of polynomial and spline bases. Many well known algorithms for polynomials and splines, such as synthetic division, forward differencing, Horner's method, the de Casteljau algorithm, the de Boor algorithm, Boehm's differentiation algorithm, Boehm's knot insertion algorithm, the Oslo algorithm, Marsden's identity, and the binomial theorem, are variations on the triangle. Furthermore, the triangle reveals complimentary relationships between many seemingly disparate results, *e.g.*, between Marsden's identity and the Cox-de Boor-Mansfield recurrence.

Much of the general framework here carries over to geometrically continuous splines [8,32]. In this setting the Pólya polynomials are replaced by geometrically continuous piecewise polynomials with roots at the specified parameter values, but with this one change the standard properties of the operator [,] and de Boor-Fix formula remain valid [3]. By analyzing relationships between the geometrically continuous Pólya polynomials, we can construct a wonderful triangle for geometrically continuous splines; only the labels on the edges are more complicated. The geometrically continuous triangle again provides us with a framework for constructing recursive evaluation algorithms, blossoming analogues, and knot insertion procedures for geometrically continuous spline bases [8]. Thus not only does the triangle provide a unified approach to many known B-spline curve algorithms, but it also provides a method for extending these algorithms to other important bases.

Another theme that emerged from this discussion is factoring. Fast knot insertion factors through the Newton dual basis; efficient differentiation factors through the Newton or Newton dual basis. To convert from Pólya to progressive or from progressive to Pólya form, we factor through the Bernstein or monomial basis; to degree elevate progressive curves or to compute products of progressive basis functions, we can again factor through the Bernstein or monomial basis.

Though we have developed a general framework for generating change of basis algorithms, many questions still remain open. We have not discussed the numerical stability of these algorithms. While some of the algorithms use only convex combinations, many do not. Thus we expect that at least some of these procedures are numerically unstable. Certainly forward differencing, which in our framework is a knot deletion algorithm, is known to be unstable.

The list in Table 1 of progressive and Pólya bases includes many, but by no means all, of the bases used in computer aided geometric design. We would like to develop change of basis procedures for bases not in this list, such as the Hermite and Chebyshev bases.

Finally, we would like to extend our change of basis techniques to multivariate B-splines. Again, we expect that working in the dual space with the multivariate analogues of the Pólya polynomials may provide the proper framework for developing evaluation, blossoming, subdivision, knot insertion, interpolation, degree elevation, and differentiation algorithms for multivariate splines.

References

1. Abhyankar, S., and C. Bajaj, Automatic parametrization of rational curves and surfaces II: Cubics and cubicoids, Comput. Aided Design **19** (1987), 499–502.
2. Barry, P. J., A. D. DeRose, and R. N. Goldman, B-splines, Pólya curves, and duality, J. Approx. Th. **65** (1991), 3–21.
3. Barry, P. J., N. Dyn, R. N. Goldman, and C. A. Micchelli, Identities for piecewise polynomials determined by connection matrices, Aequationes Mathematicae, to appear.
4. Barry, P. J., and R. N. Goldman, Interpolation and approximation of curves and surfaces using Pólya polynomials, Computer Vision, Graphics, and Image Processing **53** (1991), 137–148.
5. Barry, P. J., and R. N. Goldman, Shape parameter deletion for Pólya curves, Numerical Algorithms **1** (1991), 121–138.

6. Barry, P. J., and R. N. Goldman, Factored knot insertion, *Knot Insertion and Deletion Algorithms for B-spline Modeling*, T. Lyche and R. N. Goldman (eds.), SIAM, to appear.
7. Barry, P. J., and R. N. Goldman, Knot insertion algorithms, submitted for publication.
8. Barry, P. J., R. N. Goldman, and C. A. Micchelli, Knot insertion algorithms for geometrically continuous splines, in preparation.
9. Boehm, W., Inserting new knots into B-spline curves, Comput. Aided Design **12** (1980), 199–201.
10. Boehm, W., G. Farin, and J. Kahmann, A survey of curve and surface methods in CAGD, Computer Aided Geom. Design **1** (1984), 1–60.
11. de Boor, C., On calculating with B-splines, J. Approx. Th. **6** (1972), 50–62.
12. de Boor, C., *A Practical Guide to Splines*, Springer-Verlag, New York, 1972.
13. de Boor, C., and G. Fix, Spline approximation by quasi-interpolants, J. Approx. Th. **8** (1973), 19–45.
14. de Boor, C., and K. Höllig, B-splines without divided differences, in *Geometric Modeling: Algorithms and New Trends*, G. Farin (ed.), SIAM, Philadelphia, 1987, 21–27.
15. de Casteljau, P., *Formes a Poles*, Hermes, Paris, 1985.
16. de Casteljau, P., *Le Lissage*, Hermes, Paris, 1990.
17. Cohen, E., T. Lyche, and R. Riesenfeld, Discrete B-splines and subdivision techniques in computer aided geometric design and computer graphics, Computer Graphics and Image Processing **14** (1980), 87–111.
18. Farin, G., *Curves and Surfaces for Computer Aided Geometric Design: A Practical Guide*, Academic Press, San Diego, 1988.
19. Goldman, R. N., Blossoming and knot insertion algorithms for B-spline curves, Computer Aided Geom. Design **7** (1990), 69–81.
20. Goldman R. N., and P. J. Barry, Algorithms for progressive curves: Extending B-spline and blossoming techniques to the monomial, power, and Newton dual forms of a curve, *Knot Insertion and Deletion Algorithms for B-spline Modeling*, T. Lyche and R. N. Goldman (eds.), SIAM, to appear.
21. Lee, E. T. Y., private communication.
22. Lyche, T., and K. Mørken, Making the Oslo algorithm more efficient, SIAM J. Numer. Anal. **23** (1986), 663–675.
23. Lyche, T., and K. Mørken, A data reduction strategy for splines with applications to the approximation of functions and data, IMA Journal of Numerical Analysis **8** (1988), 185–208.
24. Lyche, T., Discrete B-splines and conversion problems, in *Computation of Curves and Surfaces*, W. Dahmen, M. Gasca, and C. A. Micchelli, (eds.), NATO ASI Series C: Mathematical and Physical Sciences **307** (1990), 117–134.
25. Marsden, M. J., An identity for spline functions with applications to

variation-diminishing spline approximation, J. Approx. Th. **3** (1970), 7–49.
26. Ramshaw, L., *Blossoming: A Connect-the Dots Approach to Splines*, Digital Equipment Research Center Technical Report 19, Palo Alto, California, 1988.
27. Ramshaw, L., Béziers and B-splines as multiaffine maps: in *Theoretical Foundations of Computer Graphics and CAD*, R.A. Earnshaw (ed.), NATO ASI Series F **40**, Springer-Verlag, New York, (1988), 757–776.
28. Ramshaw, L., Blossoms are polar forms, Computer Aided Geom. Design **6** (1989), 323–358.
29. Requicha, A., and H. Voelcker, Boolean operations in solid modeling: Boundary evaluation and merging algorithms, Proceedings of the IEEE, **73** (1985), 30–44.
30. Schumaker, L. L., *Spline Functions: Basic Theory*, Wiley, New York, 1981.
31. Seidel, H.-P., A new multiaffine approach to B-splines, Computer Aided Geom. Design **6** (1989), 23–32.
32. Seidel, H.-P., Geometric constructions and knot insertion for β-splines of arbitrary degree, submitted for publication.
33. Sederberg, T., D. Anderson, and R. N. Goldman, Implicit representation of parametric curves and surfaces, Computer Vision, Graphics, and Image Processing **28** (1984), 72–84.
34. Sankar, P., M. Silbermann, and L. Ferrari, Curve and surface generation and refinement based on a high speed derivative algorithm, submitted for publication.
35. Shapiro, V., and D. Vossler, Construction and optimization of CSG representation, Comput. Aided Design **23** (1991), 4–20.
36. Stefanus, Y., and R. N. Goldman, Blossoming Marsden's identity, submitted for publication.

Ronald N. Goldman
Computer Science Department
Rice University
P.O. Box 1892
Houston, TX, 77251 USA
rng@rice.edu

Phillip J. Barry
Computer Science Department
University of Minnesota
4-192 EE/CSci Bldg., 200 Union St. SE
Minneapolis, MN 55455 USA
barry@cs.umn.edu

An Arbitrary Mesh Network Scheme Using Rational Splines

John A. Gregory and P. K. Yuen

Abstract. A C^1 surface scheme is described which interpolates points defined on an arbitrary mesh network. The scheme involves the blending of 'strip functions' developed from a rational spline method. The rational spline provides interval and point tension weights which can be used to control the shape of the surface scheme.

§1. Introduction

Most 'free-form' or 'sculptured' modelling systems for Computer Aided Design are based on the representation of the surface as a complex of 'rectangular patches'. In this paper, however, we wish to give a progress report on some research into the construction of a C^1 surface which interpolates points defined on an arbitrary 3-D mesh network. The construction is based on the blending of 'strip functions' defined on the mesh, which here are developed from a rational spline method.

The rational spline provides 'interval' and 'point tension' parameters to control the shape of the strip functions and hence of the blended surface. The strip functions contain position and tangent plane information and are blended together to form polygonal patches appropriate to the topological structure of the mesh.

We begin, in Section 2, by considering the topology and data structure of the mesh network, where for the purposes of this paper some simplifying assumptions are made. The rational spline strip functions are developed in Section 3, and the blending of these strip functions is considered in Section 4. The paper is completed in Section 5 with some examples and concluding remarks concerning the future direction of the research.

Mathematical Methods in Computer Aided Geometric Design II

Tom Lyche and Larry L. Schumaker (eds.), pp. 321–329.

ISBN 0-12-460510-9.

§2. The Mesh Network

The mesh network is described by a set of 3-D directed *polygonal lines* whose *vertices* are a given set of interpolation points. The polygonal lines are assumed to intersect at certain of the vertices in such a way as to define a mesh network of *faces*. The faces are topologically equivalent to either triangular, rectangular or, in general, polygonal domains. Each *edge* of a face is defined by a signed subset of one of the polygonal lines, where the sign indicates whether it has the positive or negative direction of that line.

The mesh network is assumed equivalent to an orientable surface (thus having an inside and outside) and the edges of the faces, together with their vertices, are labelled in an anti-clockwise order when viewed from the outside. The data structure thus consists of:

- *Polygonal Lines* E_i, $i = 1, \ldots, n$, each defined by an ordered set of vertices denoted by $\mathbf{P}_{i,j} \in \mathbb{R}^3$, $j = 1, \ldots, n_i$.
- *Faces* M_i, $i = 1, \ldots, m$, defined by their anti-clockwise ordered edges, namely, signed subsets of the polygonal lines.

It can be observed that the vertices divide into the two classes of *face vertices* and *internal edge vertices*. Here, we make the simplifying assumptions that the face and internal edge vertices form mutually exclusive sets and that exactly two of the polygonal lines meet or intersect at any face vertex. (Thus every face vertex is labelled exactly twice with the above notation.) This allows the consistent definition of the strip functions of the following section, with respect to an appropriate local rectangular cartesian coordinate system. In particular, a consistent tangent plane will be defined at each face vertex.

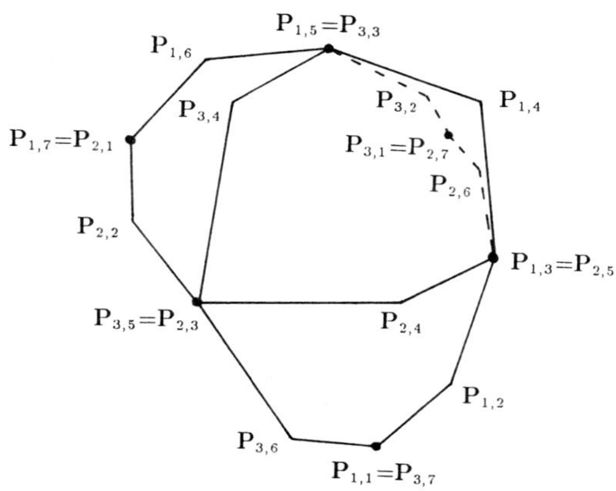

Figure 2.1. Example of a mesh network.

As a concrete example, consider the mesh network shown in Figure 2.1. Here, there are three directed polygonal lines E_i, $i = 1, \ldots, 3$, through a total of fifteen vertices. The vertices decompose into a set of six face vertices and nine internal edge vertices, giving a set of four triangular faces M_i, $i = 1, \ldots, 4$. (The edges of the face hidden by the persective view of the mesh network are shown with broken lines as an aid to understanding the three dimensional figure.)

§3. The Rational Spline Strip Functions

We begin by constructing a set of rational spline curves, each associated with a given polygonal line E_i and interpolating the ordered vertices which define E_i. This will provide the basic design tool for the construction of the arbitrary mesh network surface, since the surface will be constrained to contain the spline network.

The ordered vertices of each polygonal line decompose into face vertices and interior edge vertices. The ordered face vertices are associated with increasing integer knots, and we choose to distribute the knots associated with the interior edge vertices between these integers, in proportion to their number between the subsequent face vertices. This gives a knot vector which may not be appropriate for standard parametric cubic spline interpolation, and we thus wish to provide *shape control parameters* which allow easy manipulation of the spline curve. This is achieved here by the use of a C^2 rational cubic spline method, see [6], which provides interval and point tension control of the curve. (An alternative choice would be the C^1 weighted ν-spline described in Foley [2], which also provides interval and point tension control.)

Given a set of values $\mathbf{P}_i \in \mathbb{R}^3$, $i = 0, \ldots, N$, defined at knots s_i, $i = 0, \ldots, N$, with interval spacing $h_i := s_{i+1} - s_i > 0$, then a parametric C^1 piecewise rational cubic function $\mathbf{p} : [s_0, s_n] \to \mathbb{R}^3$ is defined by

$$\mathbf{p}|_{[s_i,s_{i+1}]}(s) :=$$
$$\frac{(1-\theta)^3\alpha_i\mathbf{P}_i + \theta(1-\theta)^2(\gamma_i+\alpha_i)\mathbf{V}_i + \theta^2(1-\theta)(\gamma_i+\beta_i)\mathbf{W}_i + \theta^3\beta_i\mathbf{P}_{i+1}}{(1-\theta)^2\alpha_i + \theta(1-\theta)\gamma_i + \theta^2\beta_i}. \tag{3.1}$$

Here,

$$\theta|_{[s_i,s_{i+1}]} := (s - s_i)/h_i, \tag{3.2}$$

and

$$\mathbf{V}_i := \mathbf{P}_i + \frac{\alpha_i}{\gamma_i + \alpha_i} h_i \mathbf{D}_i, \quad \mathbf{W}_i := \mathbf{P}_{i+1} - \frac{\beta_i}{\gamma_i + \beta_i} h_i \mathbf{D}_{i+1}, \tag{3.3}$$

where the scalar weights $\alpha_i, \beta_i, \gamma_i$ are assumed positive and $\mathbf{D}_i$, $i = 0, \ldots, N$, denote first derivative values defined at the knots. Thus $\mathbf{p}$ has the Hermite interpolation properties that

$$\mathbf{p}(s_i) = \mathbf{P}_i \text{ and } \mathbf{p}^{(1)}(s_i) = \mathbf{D}_i, \quad i = 0, \ldots, N. \tag{3.4}$$

Suppose the 'end conditions' $\mathbf{D}_0$ and $\mathbf{D}_N$ are given. Then the unknowns $\mathbf{D}_i$, $i = 1, \ldots, N$, can be found by imposing the C^2 consistency constraints

$$\mathbf{p}^{(2)}(s_i^+) = \mathbf{p}^{(2)}(s_i^-), \qquad i = 1, \ldots, N-1. \tag{3.5}$$

These constraints lead to a tridiagonal system of equations in the unknowns, which is diagonally dominant if $\gamma_i > \alpha_i$ and $\gamma_i > \beta_i$, $i = 1, \ldots, N-1$. We thus have a sufficient condition for the existence of a unique, easily computed solution to the consistency constraints.

The scalar weights of the piecewise rational form provide shape control parameters for the C^2 curve which results from this interpolatory process, see [6], where a local support basis representation is also discussed. In particular, we choose

$$\beta_{i-1} = \alpha_i := 1/\lambda_i, \tag{3.6}$$

giving weights λ_i, $i = 0, \ldots, N$, and γ_i, $i = 0, \ldots, N-1$, for the rational spline scheme, where the diagonal dominance condition is now $\gamma_i\lambda_i > 1$, $i = 1, \ldots, N-2$.

The weight λ_i is called a *point tension parameter*, associated with the knot s_i, since a tangent discontinuity is introduced into the spline curve at $\mathbf{P}_i$ as $\lambda_i \to \infty$. The weight γ_i is called an *interval tension parameter*, associated with the interval $[s_i, s_{i+1}]$, since the curve is tightened to the line segment joining $\mathbf{P}_i, \mathbf{P}_{i+1}$ as $\gamma_i \to \infty$. These properties are illustrated in Figure 3.1, where the cubic spline interpolant of the first curve is given by the use of the default parameters $\lambda_i := 1$, $i = 0, \ldots, N$, and $\gamma_i := 2$, $i = 0, \ldots, N-1$.

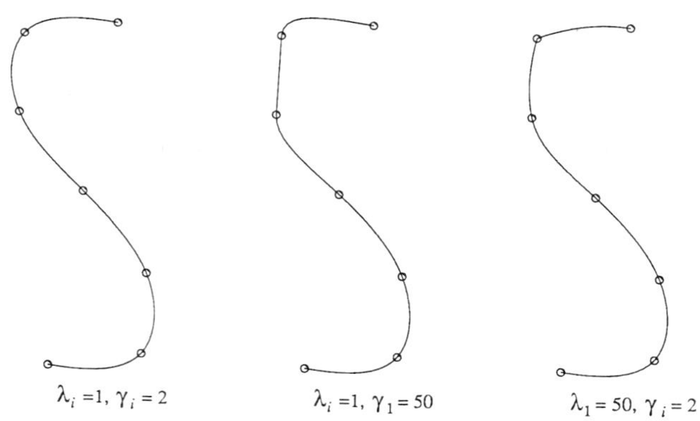

Figure 3.1. Tension behaviour of the rational spline.

Application of the rational spline method to the vertex data along each polygonal line E_i of the arbitrary mesh network, using the previously defined knot distribution, results in a network of spline curves. These curves define a *wire frame model* of a surface which has the same topological structure as the mesh network and the shape of the wire frame model can be manipulated by varying the control parameters which govern the shape of each curve. At each face vertex of the mesh network there is a unique tangent plane for the wire frame surface, defined by the derivative parameters (i.e. tangent vectors) of the two spline curves which meet or intersect at that vertex. In particular, at any integer knot of a rational spline curve, a *'cross boundary'* tangent vector is defined.

We now define a bivariate *'strip function'* extension of the rational spline method on $[s_0, s_N] \times [\varepsilon, \varepsilon]$, for some $\varepsilon > 0$, by the Taylor form

$$\mathbf{p}^{ext}(s,t) := \mathbf{p}(s) + t\mathbf{p}_t(s), \tag{3.6}$$

where $\mathbf{p}_t(s)$ denotes the piecewise cubic Hermite interpolant to a set of cross boundary tangent and *'twist'* vectors, defined on a subset of the rational spline knots. Here, the cross boundary tangent denotes a derivative along the t direction and the twist vector denotes a $(1,1)$ derivative parameter along the s and t directions. The twist vectors can be considered as derivative unknowns to be determined by imposing C^2 constraints across the knot subset. In this case $\mathbf{p}_t(s)$ is the cubic spline interpolant to the cross boundary tangent vectors, where we assume that appropriate end conditions have been imposed. It can be observed that

$$\mathbf{p}^{ext}(s,0) = \mathbf{p}(s) \quad \text{and} \quad \partial_t \mathbf{p}^{ext}(s,0) = \mathbf{p}_t(s), \tag{3.7}$$

and hence the strip function extension contains position and tangent plane information about the rational spline curve.

The rational spline extension method is considered along each curve of the wire frame model, where the cross boundary tangents are given on integer subsets of the spline knots. Since these cross boundary tangents are defined with respect to the local parameterizations of their corresponding spline curves, it is necessary to change certain of their signs for them to be consistent with the t parameterization on the strip function domain. This sign information can be obtained from the face table data structure. The twist vectors for the rational spline extensions are initially determined by imposing C^2 spline constraints on the piecewise cubic Hermite cross boundary tangent function, see above. This will give incompatible twist vector values at each face vertex, corresponding to the two rational spline extensions which meet or intersect at that vertex. This incompatibility must be allowed for in the construction of the surface scheme of the following section or, alternatively, the strip functions can be redefined by taking the average of the two twist vector parameters at each face vertex. In either case, the surface model is now one of rational spline strip functions which can be blended to form a C^1 surface.

§4. Blending the Strip Functions on Polygonal Domains

Consider a face of the arbitrary mesh network. At a face vertex, two edges meet and along each edge a piece of strip function is defined, consistent with a C^1 surface. More precisely, our definition and parameterization of the strip functions implies that the two strip function pieces can be written in terms of a (u,v) rectangular cartesian coordinate system about each vertex as

$$\mathbf{p}(0,v) + u\mathbf{p}_u(0,v),\ 0 \le u \le 1, \quad \text{and} \quad \mathbf{p}(u,0) + v\mathbf{p}_v(u,0),\ 0 \le v \le 1, \tag{4.1}$$

cf (3.6). Here, after appropriate simple affine transformations, $\mathbf{p}(0,v)$ and $\mathbf{p}(u,0)$ represent the pieces of the rational spline curves, and $\mathbf{p}_u(0,v)$ and $\mathbf{p}_v(u,0)$ represent the cross boundary tangent functions, between the face vertices of each edge. Moreover, these Taylor functions are consistent with a C^1 surface $\mathbf{p}(u,v)$. In particular, a C^1 extension of the two strip functions is defined for $(u,v) \in [0,1]^2$ by the 'rationally corrected' Boolean sum Taylor interpolant, see Gregory [3],

$$\mathbf{p}(u,v) := \mathbf{p}(0,v) + u\mathbf{p}_u(0,v) + \mathbf{p}(u,0) + v\mathbf{p}_v(u,0) - [1 \quad u] \begin{bmatrix} \mathbf{p}(0,0) & \mathbf{p}_v(0,0) \\ \mathbf{p}_u(0,0) & \mathbf{t}(u,v) \end{bmatrix} \begin{bmatrix} 1 \\ v \end{bmatrix}, \tag{4.2}$$

where

$$\mathbf{t}(u,v) := \frac{u\mathbf{p}_{v,u}(0,0) + v\mathbf{p}_{u,v}(0,0)}{u+v}, \tag{4.3}$$

and $\mathbf{p}_{u,v}(0,0) := \partial_u\partial_v\mathbf{p}(0,0)$, $\mathbf{p}_{v,u}(0,0) := \partial_v\partial_u\mathbf{p}(0,0)$ denote the twist vectors at the vertex. If $\mathbf{p}_{u,v}(0,0) \neq \mathbf{p}_{v,u}(0,0)$, the rational term introduces singularities which are removable to first order at the vertex. If $\mathbf{p}_{u,v}(0,0) = \mathbf{p}_{v,u}(0,0)$, then

$$\mathbf{t}(u,v) := \mathbf{p}_{u,v}(0,0), \tag{4.3}$$

giving a standard Boolean sum form.

Now each face of the mesh network is topologically equivalent to a polygonal domain, and the final surface is constructed by blending the Boolean sum Taylor interpolants on each of these polygons in the following manner. This follows the approach of Charrot and Gregory [1], see also [3] and [5].

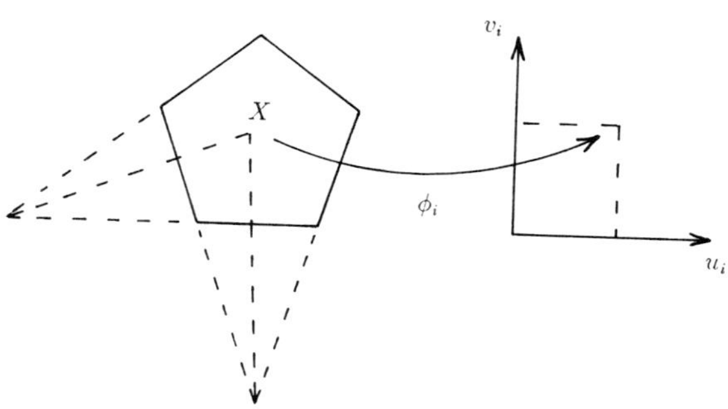

Figure 4.1. The central projection coordinate chart.

Let Ω denote a generic regular polygon with vertices $X_i,\ i = 0,\dots,N-1$. Associated with each vertex X_i there is a Boolean sum Taylor interpolant $\mathbf{p}_i(u,v)$ say, which must be reparameterized onto the polygonal domain. This is achieved by 'central projections'. Thus let

$$(1-u_i)X_i + u_iX_{i+1} \ \text{ and } \ (1-v_i)X_i + v_iX_{i-1} \tag{4.4}$$

define the points of intersection of the two central projections through $X \in \Omega$ with the sides of the polygon as shown in Figure 4.1. Then

$$\phi_i(X) := (u_i, v_i) := \left(\frac{d_{i-1}}{d_{i-1} + d_{i+1}}, \frac{d_i}{d_{i-2} + d_i} \right), \tag{4.5}$$

defines the central projection coordinate chart about X_i, where $d_i(X)$ denotes the perpendicular distance from X to the side $X_i X_{i+1}$.

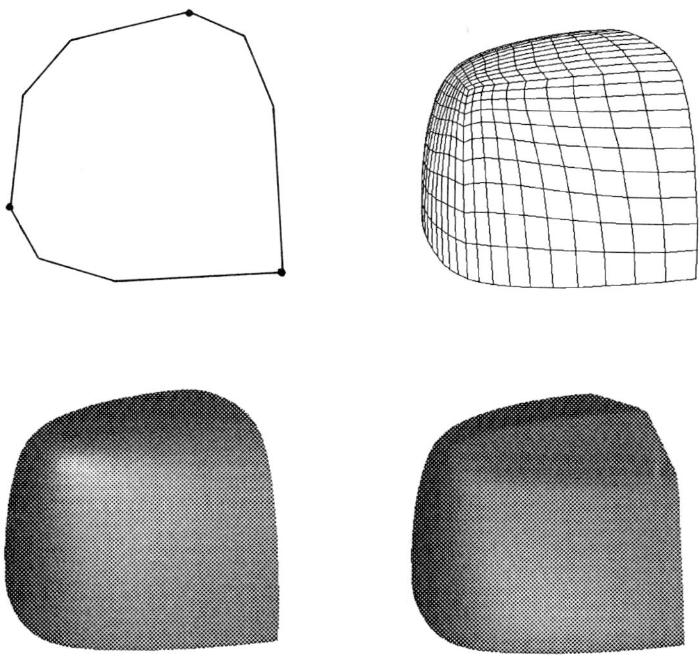

Figure 5.1. A single triangular patch.

The blended polygonal surface patch is now defined by

$$\mathbf{q}(X) := \sum_{i=0}^{N-1} w_i(X)\, \mathbf{p}_i(\phi_i(X)), \tag{4.6}$$

where

$$w_i(X) := \frac{\prod_{j \neq i-1,i} {d_j}^2}{\sum_{k=0}^{N-1} \prod_{j \neq k-1,k} {d_j}^2}, \tag{4.7}$$

defines a weight function which is constructed such that $\mathbf{p}_i$ contributes, in position and tangent plane, only on those edges where it interpolates a strip function.

§5. Examples and Concluding Remarks

We conclude by giving two simple examples of the arbitrary mesh network scheme. The first case, shown in Figure 5.1, is that of a single triangular patch, defined with zero twists, and illustrates the effect of increasing the interval tension parameter between two internal edge vertices.

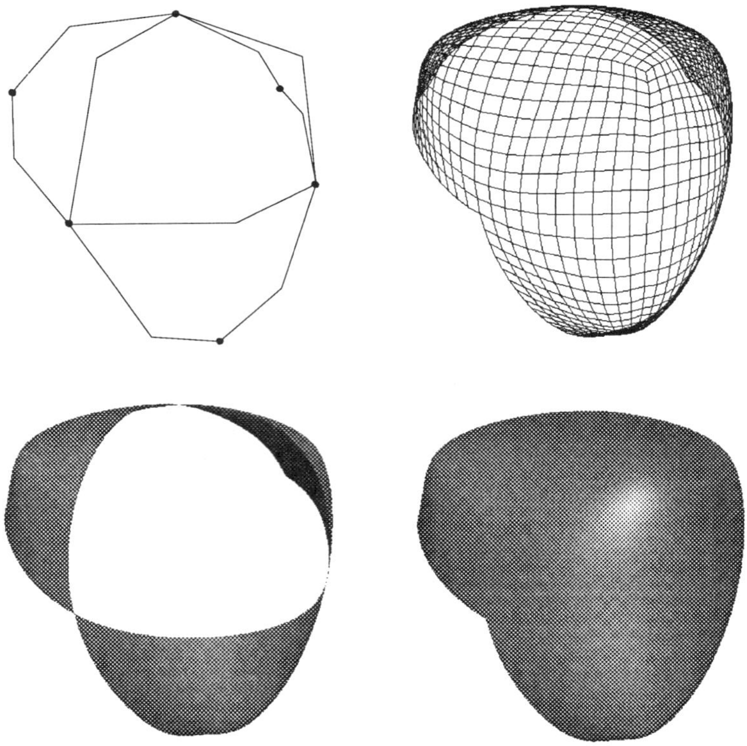

Figure 5.2. A mesh network of triangular patches.

The second case, shown in Figure 5.2, is of the mesh network of four triangular patches, given as the concrete example of Section 2. In the figure, the rational spline network is that defined by the special case of cubic spline interpolation. The figure shows that the blending approach proposed here gives a smooth, tangent plane continuous surface.

Our current research is directed towards allowing more than two curves to meet or intersect at any face vertex, thus removing a major restriction on the topology of the network proposed here. This involves modification of the derivative parameters of the rational spline network at such face vertices, in order that a consistent tangent plane is defined there. It is also proposed to investigate shape control of the interior of a patch, other than through the control of its boundary. Finally, the method can be extended to the construction of higher order continous surfaces, although more care is now needed in the construction of the blends of the strip functions, as is discussed in [4] and [5].

Acknowledgments. This work was supported by the SERC ACME Grant GR/F80838, in collaboration with Siemens Nixdorf Information Systems Ltd.

References

1. Charrot, P., and J. A. Gregory, A pentagonal surface patch for computer aided geometric design, Computer Aided Geom. Design **1** (1984), 87–94.
2. Foley, T. A., Interpolation with interval and point tension controls using weighted ν-splines, ACM Trans. Math. Software **13** (1987), 68–96.
3. Gregory, J. A., C^1 rectangular and non-rectangular surface patches, in *Surfaces in Computer Aided Geometric Design*, R. E. Barnhill and W. Boehm (eds.), North Holland, Amsterdam, 1983, 25–33.
4. Gregory, J. A., and J. Hahn, Geometric continuity and convex combination patches, Computer Aided Geom. Design **4** (1987), 79–89.
5. Gregory, J. A., J. Hahn, and V.K.H. Lau, High order continuous polygonal patches, TR/06/91, Mathematics & Statistics Department, Brunel University, 1991.
6. Gregory, J. A., M. Sarfraz, and P. K. Yuen, A rational spline with interval and point tension, TR/01/92, Mathematics Dept., Brunel Univ., 1992.

John A. Gregory and P. K. Yuen
Department of Mathematics and Statistics
Brunel University
Uxbridge, UB8 3PH, U.K.
John.Gregory@brunel.ac.uk
Pak-Kay.Yuen@brunel.ac.uk

Bézier Curves and Surface Patches on Quadrics

Josef Hoschek

Abstract. Bézier curves, Bézier spline curves, Bézier tensor product patches and triangular patches on quadrics (sphere, hyperboloid of one or two sheets, paraboloid) are constructed with the help of quadratic maps ϕ_i. The maps transform integral Bézier curves and patches from a parameter space (u, v, w) to rational Bézier curves and patches on quadrics. Curves of degree n in (u, v, w)-space lead to Bézier curves of degree $2n$. Tensor product patches of degree (n, m) in (u, v, w)-space lead to degree $(2n, 2m)$ while triangular patches of degree n in (u, v, w)-space lead to triangular patches of degree $2n$.

§1. Introduction

Recently a considerable amount of research has been done on the problem of developing rational Bézier and B-spline schemes [16,17,18,20]. Quadratic Bézier curves and surfaces are of special interest because quadratic Bézier curves lead to the well known conic sections, while special biquadratic tensor product surfaces lead to quadrics – conic sections and quadrics are of importance in almost all solid modeling systems because they have been widely used in industry [3,15,19].

Here we proceed in a somewhat different manner from the above references: We introduce a technique for the construction of Bézier curves or Bézier spline curves on quadrics; we construct Bézier tensor product surface patches or Bézier spline surfaces on quadrics, and we develop Bézier triangle patches on quadrics of (nearly) arbitrary degree. Here we extend results on quadric patches on conics developed in [1,2,6,7,21]. The method can be extended also to B-spline representations.

Mathematical Methods in Computer Aided Geometric Design II
Tom Lyche and Larry L. Schumaker (eds.), pp. 331–342.

ISBN 0-12-460510-9.

§2. A Suitable Map

The main idea for our construction comes from number theory [11]. It is well known [5,10,12] that a Pythagorean triple of real numbers

$$A^2 + B^2 = C^2 \tag{1}$$

is of the form

$$A = c(a^2 - b^2), \quad B = 2cab, \quad C = c(a^2 + b^2).$$

We can extend this theorem to $\mathbb{R}^3$ and get as real solutions of

$$A^2 + B^2 + C^2 = D^2, \tag{2a}$$

$$A = d(a^2 - b^2 - c^2),\ B = 2abd,\ C = 2acd,\ D = d(a^2 + b^2 + c^2). \tag{2b}$$

Now we give (2) a new interpretation: If (2a) is the equation of a sphere in homogeneous coordinates x_0, x_1, x_2, x_3,

$$x_1^2 + x_2^2 + x_3^2 = x_0^2, \tag{3a}$$

then a point on the sphere will have the following rational representation (which we denote by ϕ_1):

$$x_1 = d(u^2 - v^2 - w^2), \quad x_2 = 2uvd, \quad x_3 = 2uwd, x_0 = d(u^2 + v^2 + w^2) \tag{3b}$$

with coordinates in $\mathbb{R}^3$

$$x = \frac{x_1}{x_0}, \quad y = \frac{x_2}{x_0}, \quad z = \frac{x_3}{x_0}$$

and (u, v, w) as parameters. Without loss of generality we can choose $d = 1$. (3b) can be interpreted as projective parametrization [8].

We can easily extend our construction to an ellipsoid – instead of (3b) we have to set up

$$x_1 = a(u^2 - v^2 - w^2), \quad x_2 = 2buv, \quad x_3 = 2cuw, x_0 = u^2 + v^2 + w^2 \tag{4}$$

with a, b, c as half-axes of the ellipsoid. (4) is an affine map of the spherical representation (3b) – therefore in the following we will only discuss the affine normal forms of the quadrics.

If we consider a *hyperboloid of one sheet* with the normal equation

$$x_1^2 + x_2^2 - x_3^2 = x_0^2, \tag{5a}$$

similar to (3b), we get as the rational representation ϕ_2:

$$x_1 = u^2 - v^2 + w^2, \quad x_2 = 2uv, \quad x_3 = 2uw, \quad x_0 = u^2 + v^2 - w^2. \tag{5b}$$

The extension to the other nondegenerated quadrics leads to the following: for the *hyperboloid of two sheets*

$$x_1^2 - x_2^2 - x_3^2 = x_0^2, \tag{6a}$$

we have the rational representation ϕ_3:

$$x_1 = u^2 + v^2 + w^2, \quad x_2 = 2uv, \quad x_3 = 2uw, \quad x_0 = u^2 - v^2 - w^2; \tag{6b}$$

for the *elliptic paraboloid*

$$x_1^2 + x_2^2 = x_3 x_0, \tag{7a}$$

we have the rational representation ϕ_4:

$$x_1 = w(u+v), \quad x_2 = w(u-v), \quad x_3 = 2w^2, \quad x_0 = u^2 + v^2; \tag{7b}$$

for the *hyperbolic paraboloid*

$$x_1^2 - x_2^2 = x_3 x_0, \tag{8a}$$

we get the rational representation ϕ_5:

$$x_1 = w(u+v), \quad x_2 = w(u-v), \quad x_3 = w^2, \quad x_0 = 4uv. \tag{8b}$$

If we consider the equations (3) - (8) as maps ϕ_i from a (u, v, w) space to $\mathbb{R}^3$, we have the key idea of our paper:

> *Take a curve/surface out of a parametric space (u, v, w) and map it by the mapping ϕ_i (3b-8b) onto the corresponding quadratic surface (3a-8a). The map of the curve/surface out of the parameter space will be a curve/surface on the corresponding quadratic surface.*

§3. Bézier Curves on Quadrics

In the parametric space, we choose a Bézier curve of degree n

$$\mathbf{U} = \sum_{i=0}^{n} \mathbf{U}_i B_i^n(t), \tag{9}$$

with $\mathbf{U} = (u, v, w)^\top, \mathbf{U}_i = (u_i, v_i, w_i)^\top$, and substitute (9) in one of the mappings ϕ_i. We can establish that for the products of Bernstein polynomials, the following identity holds ([9,14])

$$B_i^n(t) B_k^m(t) = \frac{\binom{n}{i}\binom{m}{k}}{\binom{n+m}{i+k}} B_{i+k}^{n+m}(t). \tag{10}$$

In the *spherical case*, by (9) and (10), the mapping ϕ_1 leads to a Bézier curve $\mathbf{X}$ of degree $2n$ on the unit sphere. The Bézier points $\mathbf{b}_i$ of $\mathbf{X}$ follow from (10), (9), and (3b). By comparison of the coefficients, we get

$$\mathbf{b}_0 = \frac{1}{\beta_0}\begin{pmatrix} u_0^2 - v_0^2 - w_0^2 \\ 2u_0v_0 \\ 2u_0w_0 \end{pmatrix} \quad \text{with} \quad \beta_0 = u_0^2 + v_0^2 + w_0^2,$$

$$\mathbf{b}_1 = \frac{1}{\beta_1}\begin{pmatrix} (u_0u_1)^2 - (v_0v_1)^2 - (w_0w_1)^2 \\ 2(u_0v_1 + u_1v_0) \\ 2(u_0w_1 + u_1w_0) \end{pmatrix}$$
$$\text{with} \quad \beta_1 = (u_0u_1)^2 + (v_0v_1)^2 + (w_0w_1)^2$$

$$\vdots \tag{11a}$$

$$\mathbf{b}_{n-1} = \frac{1}{\beta_{n-1}}\begin{pmatrix} (u_{n-1}u_n)^2 - (v_{n-1}v_n)^2 - (w_{n-1}w_n)^2 \\ 2(u_{n-1}v_n + u_nv_{n-1}) \\ 2(u_{n-1}w_n + u_nw_{n-1}) \end{pmatrix}$$
$$\text{with} \quad \beta_{n-1} = (u_{n-1}u_n)^2 + (v_{n-1}v_n)^2 + (w_{n-1}w_n)^2,$$

$$\mathbf{b}_n = \frac{1}{\beta_n}\begin{pmatrix} u_n^2 - v_n^2 - w_n^2 \\ 2u_nv_n \\ 2u_nw_n \end{pmatrix} \quad \text{with} \quad \beta_n = u_n^2 + v_n^2 + w_n^2.$$

Figure 1a contains a spherical Bézier curve of degree 6; Figure 1b shows a quartic C^1-Bézier spline curve with four segments. The spline curve is constructed with help of the well known Bézier techniques (cf. [10]). For convenience we have chosen as representation of the control points

$$\mathbf{U}_i = \lambda_i(\cos\phi_i, \sin\phi_i\cos\psi_i, \sin\phi_i\sin\psi_i)^\top. \tag{11b}$$

The corresponding values are given in the figures.
If we insert (9) in one of the other mapping equations ϕ_i, we obtain Bézier curves of degree $2n$ on the other quadrics. Figure 2 shows a quartic Bézier curve on a *hyperboloid of one sheet.* As representation of the points in the parameter space we have chosen

$$\mathbf{U}_i = \lambda_i(\cos\phi_i, \sin\phi_i\cosh\psi_i, \sin\phi_i\sinh\psi_i). \tag{11c}$$

We now can establish

Theorem 1. *An integral Bézier-curve of degree n in the parameter space, with corresponding rational representations ϕ_i, leads to rational Bézier-curves of degree $2n$ on the corresponding quadratic surface.*

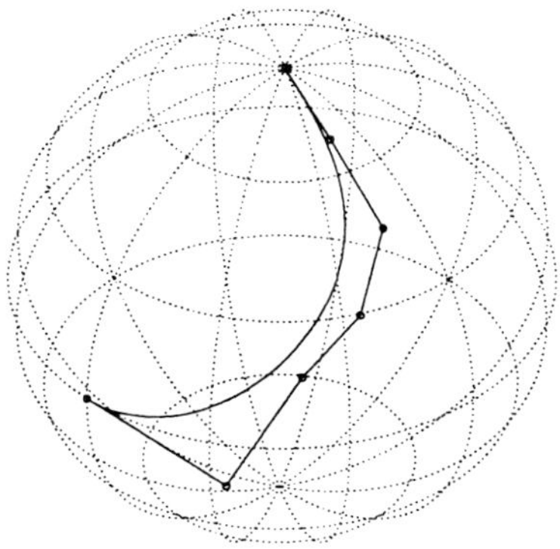

	0	1	2
φ	0	36°	45°
ψ	60°	0	45°
λ	1	0,6	1

Figure 1a. Spherical Bézier curve of degree 6.

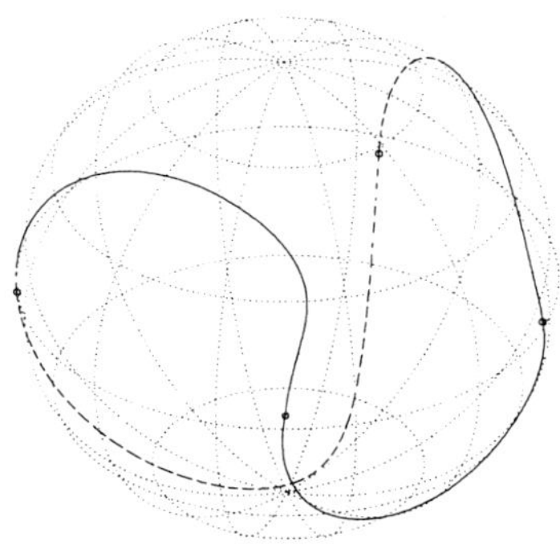

i	1	2	3	4
φ_{0i}	22,5°	67,5°	-77,5°	-32,5°
φ_{1i}	1.485	-1.546	-1.526	-1.407
φ_{2i}	67,5°	-77,5°	-32,5°	22,5°

i	1	2	3	4
ψ_{0i}	-5°	5°	-5°	5°
ψ_{1i}	1.448	1.597	-1.455	1.552
ψ_{2i}	5°	-5°	5°	-5°

Figure 1b. A closed quartic spherical Bézier spline curve.

§4. Bézier Patches on Quadrics

Analogous to the above results, instead of a Bézier curve we now choose a tensor product Bézier patch of degree n, m ($\mu, \nu \in [0, 1]$) in the parameter

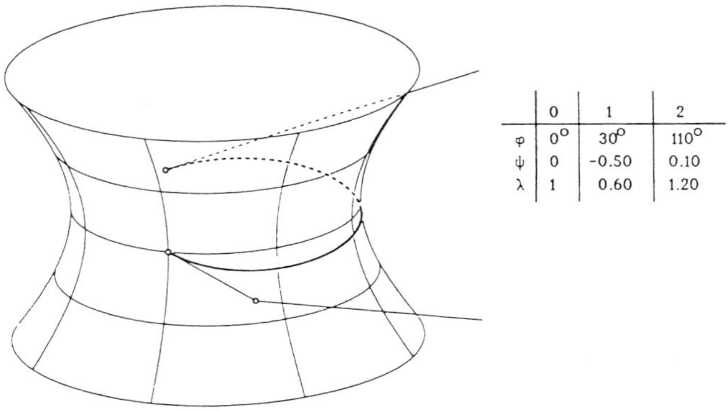

Figure 2. Quartic Bézier curve on a hyperboloid.

space

$$\mathbf{U} = \sum_{i=0}^{n} \sum_{k=0}^{m} \mathbf{U}_{ik} B_i^n(\mu) B_k^m(\nu). \tag{12}$$

With (10) and the substitutions ϕ_i, we get rational tensor product Bézier patches of degree $(2n, 2m)$ on the corresponding quadric. Figure 3 shows quadratic tensor product patches on a sphere and on a hyperboloid.

We can extend our construction to triangular Bézier patches on a quadratic surface if we choose triangular Bézier patches of degree n in the parameter space (cf. [2,4,21])

$$\mathbf{U} = \sum_{i=0}^{n} \mathbf{U}_{ijk} B_{ijk}^n(r, s, t) \tag{13a}$$

with $i + j + k = n, r + s + t = 1, r, s, t \geq 0$ and B_{ijk}^n as generalized Bernstein polynomials. For the generalized Bernstein polynomials we have the product formula

$$B_{ijk}^n B_{lpq}^m = \frac{n!m!}{(n+m)!} \frac{(i+l)!(j+p)!(k+q)!}{i!j!k!l!p!q!} B_{i+l,j+p,k+q}^{n+m} \tag{13b}$$

with $(i + l) + (j + p) + (k + q) = n + m$. The substitution of (13a) with (13b) in the rational representations ϕ_i leads to triangular Bézier patches of degree $2n$ on the corresponding quadric. Thus we have

Theorem 2a. *An integral tensor product Bézier patch of degree n, m in the parametric space leads, with a rational representation ϕ_i, to a rational tensor product Bézier patch of degree $(2n, 2m)$ on the corresponding quadratic surface.*

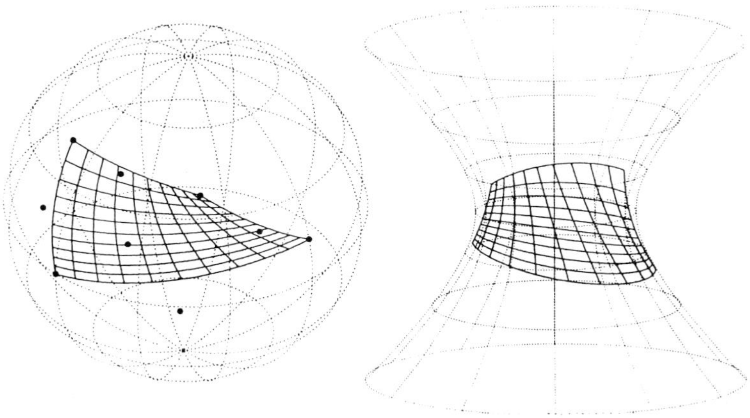

Figure 3. Quadratic tensor product patch on a sphere and on a hyperboloid of one sheet.

Theorem 2b. *An integral triangular Bézier patch of degree n in the parametric space leads, with a rational representation ϕ_i, to a rational triangular Bézier patch of degree $2n$ on the corresponding quadratic surface.*

Figure 4 contains triangular Bézier patches on a sphere and on a hyperboloid.

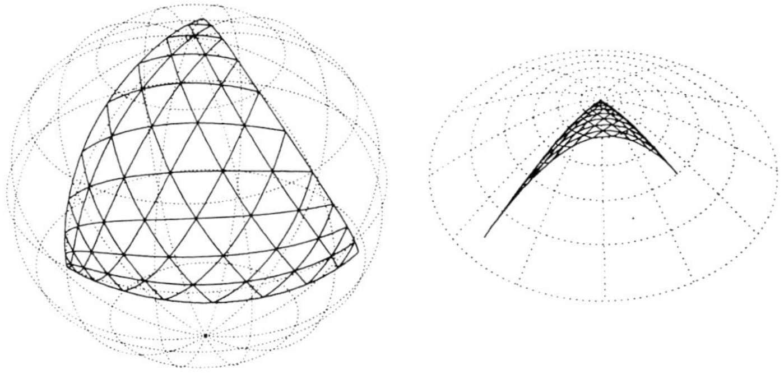

Figure 4. Quartic triangular Bézier patch on a sphere and quartic triangular Bézier patch on a hyperboloid of two sheets.

§5. General Solution

To discuss the magnitude of Bézier curves or Bézier patches which follow from a map ϕ_i, we determine the inverse map ϕ_i^{-1}. If we consider the *spherical case*, we get from (3b)

$$u = \sqrt{\frac{x_0 + x_1}{2}}, \qquad v = \frac{x_2}{2u}, \qquad w = \frac{x_3}{2u}. \tag{14}$$

To give an interpretation of this inverse map, we consider $n = 1$ in (9): The points $\mathbf{U}_0$, $\mathbf{U}_1$ and the origin determine a plane $\overline{\epsilon}$ in the parameter space which may have the equation $au + bv + cw = 0$ (a, b, c const.). The image of $\overline{\epsilon}$ in $\mathbb{R}^3$ is the plane ϵ

$$a(x + 1) + by + cz = 0 \tag{15}$$

with $x = \frac{x_1}{x_0}, y = \frac{x_2}{x_0}, z = \frac{x_3}{x_0}$. If $\overline{\epsilon}$ moves in the parametric space, i.e. a, b, c change, all images of these planes run through a fixed point $\mathbf{F}(-1, 0, 0)$ in $\mathbb{R}^3$.

Remark. Because of this fixed point $\mathbf{F}$, the mapping ϕ_i can be interpreted as a *stereographic projection.*

With $n = 1$, the line segment $(\mathbf{U}_0, \mathbf{U}_1)$ in (u, v, w)-space is mapped to a conic section through the images of $\phi_i(\mathbf{U}_0)$ and $\phi_i(\mathbf{U}_1)$ in $\mathbb{R}^3$. These images and the fixed point $\mathbf{F}$ determine the plane ϵ. Therefore only one conic section through the boundary points $\phi_i(\mathbf{U}_0), \phi_i(\mathbf{U}_1)$ is determined. The osculating plane of this conic section meets the fixed point $\mathbf{F}$. Therefore our construction leads to only one conic section through two points $\phi_i(\mathbf{U}_0), \phi_i(\mathbf{U}_1)$ – obviously there is an infinite set of conic sections through two points on a quadric!

To determine *all* conic sections through two boundary points, we can establish that the maps (3b) – (8b) are true for all *isomorphic maps* of the corresponding quadric into itself. For a quadric with a central point, such an isomorphic map can be set up as

$$\begin{pmatrix} \xi \\ \eta \\ \zeta \end{pmatrix} = A \begin{pmatrix} x \\ y \\ z \end{pmatrix}. \tag{16}$$

With such a transformation the fixed point $\mathbf{F}$ will be transformed in an oblique position on the sphere. The coefficients a_{ik} of the matrix A are determined by the type of the considered quadric: If we have a sphere, the isometric map is defined by the condition

$$\xi^2 + \eta^2 + \zeta^2 = x^2 + y^2 + z^2. \tag{17a}$$

(17) determines a spherical rotation – such motions can be described by orthogonal matrices. Therefore A can have the structure [9]

$$A = \begin{pmatrix} c_2c_3 - c_1s_2s_3 & -c_2s_3 - c_1s_2c_3 & s_1s_2 \\ s_2c_3 + c_1c_2s_3 & -s_2s_3 + c_1c_2c_3 & -s_1c_2 \\ s_1s_3 & s_1c_3 & c_1 \end{pmatrix} \tag{17b}$$

with $c_2 = \cos\omega_2$, $s_1 = \sin\omega_1$, ..., and $\omega_1, \omega_2, \omega_3$ as parameter of the motion.

If we insert our mapping ϕ_1 directly in (16,17), we get that the representations of the Bézier points (11a) are not yet true because these points must be transformed too. Therefore we will change our technique: For instance, for the construction of all conic sections through two points it is very convenient if the boundary points are given as spherical points like

$$\begin{aligned} \mathbf{b}_0 &= (\cos\phi_0 \cos\psi_0, \cos\phi_0 \sin\psi_0, \sin\phi_0)^\top, \\ \mathbf{b}_2 &= (\cos\phi_1 \cos\psi_1, \cos\phi_1 \sin\psi_1, \sin\phi_1)^\top. \end{aligned} \tag{18}$$

This means that we require boundary points as in representation (18) after transformation (16,17). Therefore we first have to transform the boundary points (18) to a suitable position by multiplication with $A^{-1} = A^T$. If we choose as images of $\mathbf{b}_j$ in the parametric space (u, v, w):

$$\begin{aligned} u_i &= \sqrt{[a_{11}\cos\phi_i \cos\psi_i + a_{21}\cos\phi_i \sin\psi_i + a_{31}\sin\phi_i + 1] : 2} \\ v_i &= [a_{12}\cos\phi_i \cos\psi_i + a_{22}\cos\phi_i \sin\psi_i + a_{32}\sin\phi_i] : 2u_i \\ w_i &= [a_{13}\cos\phi_i \cos\psi_i + a_{23}\cos\phi_i \sin\psi_i + a_{33}\sin\phi_i] : 2u_i \\ u &= u_0 B_0^1(t) + u_1 B_1^1(t), \\ v &= v_0 B_0^1(t) + v_1 B_1^1(t), \\ w &= w_0 B_0^1(t) + w_1 B_1^1(t). \end{aligned} \tag{19a}$$

for $i = 0, 1$, and insert in ϕ_1 (equation (3b)) we get for $t = 0$ or $t = 1$

$$\begin{aligned} x &= a_{11}\cos\phi_i \cos\psi_i + a_{21}\cos\phi_i \sin\psi_i + a_{31}\sin\phi_i \\ y &= a_{12}\cos\phi_i \cos\psi_i + a_{22}\cos\phi_i \sin\psi_i + a_{32}\sin\phi_i \\ z &= a_{13}\cos\phi_i \cos\psi_i + a_{23}\cos\phi_i \sin\psi_i + a_{33}\sin\phi_i. \end{aligned} \tag{19b}$$

If we now substitute (19b) in (16) and use (17), we have (18) as boundary points of the required conic section.

To get an oblique conic section through two points $\mathbf{b}_0, \mathbf{b}_2$, we have to proceed in the following way:

- choose $\mathbf{b}_0, \mathbf{b}_2$ with respect to (18), insert ϕ_i, ψ_i in (19);
- choose $\omega_1, \omega_2, \omega_3$ in the transformation (17);
- insert (19) in (3b) to get (x, y, z)-coordinates;
- the use of the transformation (16) for these (x, y, z)-coordinates leads to the corresponding conic section on the (ξ, η, ζ)-sphere (see Figure 5).

If we use the same technique for quadratic *tensor product patches* (or corresponding *triangular patches* in $\mathbb{R}^3$) with given corner points, for $n = 1$ we get all patches whose osculating planes of the boundary conic sections meet in an (oblique) fixed point on the sphere.

Certainly the transformation (17) and the assumptions (18,19) can be used analogously for curves and surface patches of higher degree ($n > 1$).

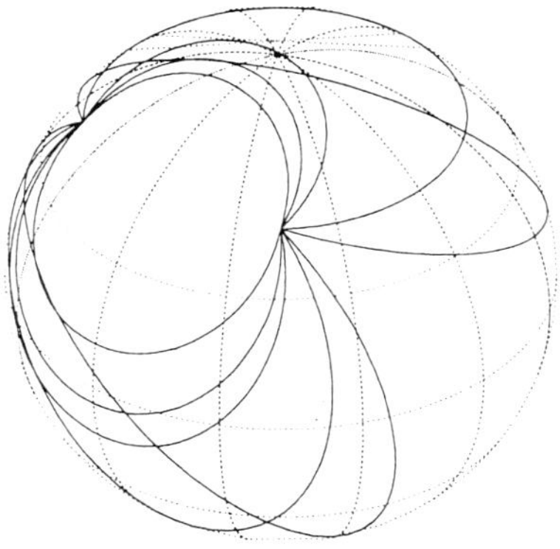

Figure 5. Conic sections (circles) through two points of a quadric (sphere).

In a similar way, we can discuss the isomorphic maps of the other quadratic surfaces: For the *hyperboloid of one sheet* we have the isometric condition

$$\xi^2 + \eta^2 - \zeta^2 = x^2 + y^2 - z^2.$$

The transformation matrix A with respect to (16) has the structure

$$A = \begin{pmatrix} c_2c_3 - ch_1s_2s_3 & -c_2s_3 - ch_1s_2c_3 & sh_1s_2 \\ s_2c_3 + ch_1c_2s_3 & -s_2s_3 + ch_1c_2c_3 & -sh_1c_2 \\ -sh_1s_3 & -sh_1c_3 & ch_1 \end{pmatrix} \tag{20}$$

with $sh_1 = \sinh(\omega_1), ch_1 = \cosh(\omega_1), c_2 = \cos\omega_2, \ldots$. A convenient representation for a point on a hyperboloid of one sheet is

$$\mathbf{b}_i = (\cosh\phi_i \cos\psi_i, \cosh\phi_i \sin\psi_i, \sinh\phi_i)^\top. \tag{21}$$

For a *hyperboloid of two sheets* we have the isomorphic condition

$$\xi^2 - \eta^2 - \zeta^2 = x^2 - y^2 - z^2,$$

which lead to the isomorphic transformation

$$A = \begin{pmatrix} ch_2ch_3 + c_1sh_2sh_3 & -ch_2sh_3 - c_1sh_2ch_3 & s_1sh_2 \\ -sh_2ch_3 - c_1ch_2sh_3 & sh_2sh_3 + c_1ch_2ch_3 & -s_1ch_2 \\ -s_1sh_3 & s_1ch_3 & c_1 \end{pmatrix}. \tag{22}$$

A convenient representation of a point on a hyperboloid of two sheets is

$$\mathbf{b}_i = (\cosh\phi_i \cosh\psi_i, \cosh\phi_i \sinh\psi_i, \sinh\phi_i)^\top. \tag{23}$$

In both cases the transformations analogous to (19a,b) lead to the required representations. Note that the inverse matrix of (20) and (22) is not the transpose of the matrix A!

Remark: The matrices (20), (22) follow from (17b) with help of complex transformations: In the transformation (16) for (20) the z-coordinate must be complex; for (22) the y- and z-coordinates must be complex. Additionally, transformation formulas like $\sin(i\omega) = i\ \sinh(i\omega)$, $\omega \in \mathbb{R}$ are used.

§6. Conclusion

Our construction of Bézier curves and tensor product Bézier surfaces on quadrics can be extended to B-spline curves and tensor product B-spline surfaces on quadrics with the help of results from [13]. Details will be discussed in a further contribution.

For $k = 2$, such transformations are well known: In the parametric space we have a line, in $\mathbb{R}^3$ we get circles on a sphere, for instance. The multiplicity of the new knots of the B-spline representation must be two [9,16,18,20].

References

1. Boehm, W., and D. Hansford, Parametric representation of quadric surfaces, Math. Meth. and Num. Anal, to appear.
2. Boehm, W., and D. Hansford, Bézier patches on quadrics, in G. Farin, editor, Rational Curves and Surfaces, SIAM, Philadelphia, 1991, to appear.
3. Farin, G., *Curves and Surfaces for Computer Aided Geometric Design*, 2nd Edition, Academic Press, 1990.
4. Farin, G., B. Piper, and A. Worsey, The octant of a sphere as a nondegenerate triangular Bézier patch, Computer Aided Geom. Design **4** (1987), 329–332.
5. Farouki, R. T., and T. Sakkalis, Real Rational Curves are not "Unit Speed", Computer Aided Geom. Design **8** (1991), 151–157.
6. Geise, G., and U. Langbecker, Finite quadric segments with four conic boundary curbes, Computer Aided Geom. Design **4** (1990), 141–150.
7. Hansford, D., Boundary curves with quadric precision for a tangent continuous scattered data interpolant, PhD Thesis, Arizona State University, 1991.
8. Hoffmann, C. M., *Geometric & Solid Modeling*, Morgan Kaufmann, 1989.
9. Hoschek, J., and D. Lasser, *Grundlagen der geometrischen Datenverarbeitung*, Teubner, 2nd ed., 1992.
10. Hoschek, J., G. Seemann, Spherical Splines, Math. Methods and Num. Anal., to appear.
11. Hunter, J., *Number Theory*, Oliver & Boyd, Edinburgh, 1964.
12. Kubota, K. K., Pythagorean triples in unique factorization domains, American Mathematical Monthly **79** (1972), 503–505.
13. Morken, K., Some identities for products and degree raising of splines, Constr. Approx. **7** (1991), 195–208.

14. Lasser, D., Composition of tensor product Bézier representations, submitted to Computing.
15. Lee, E., The rational Bézier representation for conics, in *Geometric Modeling: Algorithms and New Trends*, G. Farin (ed.), SIAM, Philadelphia, 1987, 3–19.
16. Piegl, L., On the use of infinite control points in CAGD, Computer Aided Geom. Design **4** (1987), 155–166.
17. Piegl, L., Less data for shapes, IEEE Computer Graphics & Applications **7** (1987), 48–49.
18. Piegl, L., On NURBS, A survey, IEEE Computer Graphics & Applications **11** (1991), 55–71.
19. Teller, S., and C. Sequin, Modeling implicit quadrics and free-form surfaces with trimmed rational quadric Bézier patches, Technical report, Computer Science, Berkeley, 1990.
20. Tiller, W., Rational B-splines for curve and surface representation, IEEE Computer Graphics & Applications **3** (1983), 61–69.
21. Warren, J., and S. Lodha, A Bézier representation for quadric surface patches, submitted to Computer Aided Geom. Design .

Josef Hoschek
Dept. of Mathematics
University of Technology
Schlossgartenstrasse 7
D-6100 Darmstadt
Germany
xbr1d15d@ddathd21.bitnet

Monotonicity Preserving Interpolation Using C^2 Rational Cubic Bézier Curves

M. K. Ismail

Abstract. A scheme for generating interpolating monotone curves is described. The scheme is local, stable and provides parameters for modifying the shape of the curve. A sufficient condition for the curve to be monotone is derived and an explicit representation of the C^2 piecewise rational cubic Bézier function which is used to interpolate the given monotone data is described. A number of examples are given.

§1. Introduction

A variety of curve drawing methods have been discussed in the literature for interpolating given sets of monotone data. Fritsch and Carlson [5] give necessary and sufficient conditions for a C^1 piecewise cubic polynomial curve to be monotone, and also derive simpler sufficient conditions. Butland [1] and Brodlie [2] have proposed formulae for determining first derivative values at the data points which satisfy these sufficient conditions. Results obtained from using these formulae have been shown in [4]. The use of quadratic splines, which produce a C^1 interpolant, has been presented by McAllister and Roulier [7] and Schumaker [8]. Delbourgo and Gregory [3] and Gregory [6] have studied the problem of monotonicity and convexity preserving interpolation using rational piecewise cubic interpolants. In [3] a local scheme which has C^1 continuity is presented, whereas in [6], a global C^2 scheme is discussed.

In this paper, a local method for constructing a C^2 interpolant which preserves monotonicity is described, using piecewise rational cubics. The explicit form of the Bézier representation is discussed in §2. A sufficient condition for a curve to preserve monotonicity, and the corresponding choices of the weights and first derivative values are given in §§3,4. An algorithm for generating a

Mathematical Methods in Computer Aided Geometric Design II
Tom Lyche and Larry L. Schumaker (eds.), pp. 343–350.

ISBN 0-12-460510-9.

curve, with subsequent interactive modification (if required) is presented in §5, and output from an implementation of the algorithm is presented in §6.

Note that the scheme presented in this paper only preserves monotonicity. In particular it will be seen that changes to the values of the first derivatives at the given data points are needed for interpolating a set of comonotone data. An alternative method, which also has C^2 continuity, is being investigated for dealing with such data sets. An outline of that method is included in the final section of the paper.

§2. Rational Cubic Bézier Representation in an Interval

Let (x_i, f_i), $i = 1, ..., n$ denote data points, where $x_1 < x_2 < \cdots < x_n$ and $f_1, f_2, ..., f_n$ are n real numbers. Let $h_i = x_{i+1} - x_i$ be the width of the interval $[x_i, x_{i+1}]$, $i = 1, ..., n-1$ and $\Delta_i = (f_{i+1} - f_i)/h_i$ be the slope of the straight line joining the two data points (x_i, f_i) and (x_{i+1}, f_{i+1}).

A rational cubic Bézier function on an interval $[x_i, x_{i+1}]$, $i = 1, ..., n-1$ is defined as follows. Let $\theta = (x - x_i)/h_i$, $0 \leq \theta \leq 1$. Then the rational cubic Bézier function may be represented in the form

$$\begin{aligned} s(x) &= s_i(\theta) \\ &= \frac{A_i\alpha_i(1-\theta)^3 + 3B_i\beta_i\theta(1-\theta)^2 + 3C_i\gamma_i\theta^2(1-\theta) + D_i\delta_i\theta^3}{\alpha_i(1-\theta)^3 + 3\beta_i\theta(1-\theta)^2 + 3\gamma_i\theta^2(1-\theta) + \delta_i\theta^3}, \end{aligned} \tag{1}$$

where α_i, β_i, γ_i and δ_i are called weights and must be chosen so that the denominator is non-zero, and A_i, B_i, C_i, and D_i are appropriate real constants.

Therefore, the problem is that we want to find values for the scalars A_i, B_i, C_i, D_i and positive values for the weights so that the rational cubic (1) has the following interpolatory properties:

$$s(x_i) = f_i, \qquad s'(x_i) = d_i, \qquad s''(x_i) = c_i, \tag{2}$$

where d_i and c_i denote the given first and second derivative values respectively, at the data points.

Let $S_i = d_i/3$, $T_i = d_{i+1}/3$, $K_i = c_i/6$ and $L_i = c_{i+1}/6$. Substituting $x = x_i$ and $x = x_{i+1}$ into both (1) and its first derivative, it can be shown that the first two interpolatory properties in (2) will be satisfied if the scalars A_i, B_i, C_i and D_i are given by

$$A_i = f_i, \quad B_i = f_i + \frac{\alpha_i}{\beta_i}h_iS_i, \quad C_i = f_{i+1} - \frac{\delta_i}{\gamma_i}h_iT_i, \quad D_i = f_{i+1}. \tag{3}$$

In addition, after differentiating (1) twice, and after some simplifications, it can be shown that the third interpolatory property in (2) will be satisfied if

$$\begin{aligned} K_i &= \frac{1}{h_i}[S_i - \frac{3\beta_i}{\alpha_i}S_i + \frac{\gamma_i}{\alpha_i}\Delta_i - \frac{\delta_i}{\alpha_i}T_i], \\ L_i &= \frac{1}{h_i}[-T_i + \frac{\alpha_i}{\delta_i}S_i - \frac{\beta_i}{\delta_i}\Delta_i + \frac{3\gamma_i}{\delta_i}T_i]. \end{aligned} \tag{4}$$

Clearly the imposition of C^2 continuity at the knots x_i requires that

$$K_i = L_{i-1}, \qquad \text{for all } i = 2, \ldots, n-1. \tag{5}$$

§3. The Monotonicity Conditions

Suppose we are considering a monotone increasing set of data so that $f_1 \leq f_2 \leq \cdots \leq f_n$. Hence $\Delta_i \geq 0, i = 1, \ldots, n-1$. The case of a monotone decreasing set of data can be treated in a similar manner.

We now derive a condition for a function (1) to be monotone in an interval $[x_i, x_{i+1}], i = 1, \ldots, n-1$. $s(x)$ is said to be monotone if $s'(x) \geq 0$ for all $x \in [x_i, x_{i+1}], i = 1, \ldots, n-1$. After some simplifications, the first derivative with respect to θ can be written in the form

$$s'(\theta) = \frac{\sum_{j=0}^{5} \omega_{ji} \theta^j (1-\theta)^{5-j}}{[H_i(\theta)]^2}, \qquad 0 \leq \theta \leq 1, \tag{6}$$

where

$$H_i(\theta) = \alpha_i (1-\theta)^3 + 3\beta_i \theta (1-\theta)^2 + 3\gamma_i \theta^2 (1-\theta) + \delta_i \theta^3$$

and the constants $\omega_{ji}, j = 0, 1, 2, \ldots, 5$ are given by

$$\begin{aligned}
\omega_{0i} &= 3\alpha_i^2 S_i, \\
\omega_{1i} &= 6\alpha_i \gamma_i (\Delta_i - \frac{\delta_i}{\gamma_i} T_i) + 3\alpha_i^2 S_i, \\
\omega_{2i} &= 3\alpha_i \delta_i \Delta_i + 6\alpha_i \gamma_i (\Delta_i - \frac{\delta_i}{\gamma_i} T_i) + 9\beta_i \gamma_i (\Delta_i - \frac{\alpha_i}{\beta_i} S_i - \frac{\delta_i}{\gamma_i} T_i), \\
\omega_{3i} &= 3\alpha_i \delta_i \Delta_i + 6\beta_i \delta_i (\Delta_i - \frac{\alpha_i}{\beta_i} S_i) + 9\beta_i \gamma_i (\Delta_i - \frac{\alpha_i}{\beta_i} S_i - \frac{\delta_i}{\gamma_i} T_i), \\
\omega_{4i} &= 6\beta_i \delta_i (\Delta_i - \frac{\alpha_i}{\beta_i} S_i) + 3\delta_i^2 T_i, \\
\omega_{5i} &= 3\delta_i^2 T_i.
\end{aligned} \tag{7}$$

Note that $H_i(\theta)$ is strictly positive for all $0 \leq \theta \leq 1$. Hence $s'(x) \geq 0$ if each $\omega_{ji} \geq 0, j = 0, \ldots, 5$. If we assume that $S_i, T_i \geq 0, i = 1, \ldots, n$, it follows from (7) that $\omega_{ji} \geq 0, j = 0, \ldots, 5$ if there exists $S_i, T_i, \alpha_i, \beta_i, \gamma_i$ and δ_i such that

$$\Delta_i - \frac{\alpha_i}{\beta_i} S_i - \frac{\delta_i}{\gamma_i} T_i \geq 0. \tag{8}$$

§4. Determination of the Weights and the First Derivatives

In this section, the contruction of a C^2 monotonicity preserving curve interpolation scheme is discussed. For the moment we assume that the data is strictly monotone. We want to find values for the first derivatives, and positive values for the weights, which satisfy the monotonicity condition (8), while also ensuring satisfaction of (5). A simple method is to use the monotonicity condition (8) as a basis for choosing the weights. These are chosen in terms of the ratios of the endpoint derivatives and the slope of the secant line. This also allows the introduction of a local parameter in each interval which can be used to modify the shape of the curve.

In particular, the values of the weights in the interval $[x_i, x_{i+1}], i = 1, \ldots, n-1$, are chosen as follows:

$$\frac{\beta_i}{\alpha_i} = \frac{q_i S_i}{\Delta_i}, \quad \frac{\gamma_i}{\delta_i} = \frac{r_i T_i}{\Delta_i}, \quad \frac{\delta_i}{\alpha_i} = \frac{S_i}{T_i}, \tag{9}$$

where the shape parameters q_i and r_i are nonnegative numbers. Note that we may always assume one of the weights to have the value 1. Substitution of (9) into (8) shows that the monotonicity condition (8) is *always satisfied* if

$$\frac{1}{q_i} + \frac{1}{r_i} \leq 1.$$

In addition, substituting (9) into (4) gives

$$K_i = \frac{1}{h_i}[\frac{-3S_i^2 q_i}{\Delta_i} + r_i S_i],$$

$$L_i = \frac{1}{h_i}[\frac{3T_i^2 r_i}{\Delta_i} - q_i T_i].$$

From (5) it follows that

$$\frac{1}{h_i}[\frac{-3S_i^2 q_i}{\Delta_i} + r_i S_i] = \frac{1}{h_{i-1}}[\frac{3T_{i-1}^2 r_{i-1}}{\Delta_{i-1}} - q_{i-1} T_{i-1}].$$

From the fact that $S_i = T_{i-1}$, it then follows that

$$3S_i = \frac{(q_{i-1}h_i + r_i h_{i-1})\Delta_{i-1}\Delta_i}{q_i h_{i-1}\Delta_{i-1} + r_{i-1} h_i \Delta_i}, \quad , i = 2, \ldots, n-1. \tag{10}$$

This therefore gives an expression for the choice of first derivative, and the effect of the shape parameters q_i and r_i on the rational cubic interpolant is that varying these parameters will clearly change S_i and S_{i+1}. After testing on several sets of data, it was found that the variation of these parameters simultaneously did not give a clear indication of the scale of the impact on the shape of the curve segment of the individual parameters. Hence in subsequent testing, it was decided that this scheme should have just one shape parameter per interval, i. e. $q_i = r_i$ for all i. Note that to guarantee monotonicity we must have $r_i \geq 2$. Thus replacing $3S_i$ by d_i in (10), we obtain

$$d_i = \frac{(r_{i-1}h_i + r_ih_{i-1})\Delta_{i-1}\Delta_i}{r_ih_{i-1}\Delta_{i-1} + r_{i-1}h_i\Delta_i}, \quad , i = 2, \ldots, n-1. \tag{11}$$

It should be noted that formula (11) will reduce to Butland's formula [1] if the data is equally spaced and each $r_i = r$ for all $i = 1, \ldots, n-1$. For the end derivatives, the method proposed by Butland [1] may be used. Note also that for all $r_i \geq 2$, the first derivatives in the form of (11) satisfy the inequality

$$min(\Delta_{i-1}, \Delta_i) \leq d_i \leq max(\Delta_{i-1}, \Delta_i) \tag{12}$$

and increasing r_i, for a particular $1 \leq i \leq n-1$, while keeping all other r_i constant will pull gradients d_i and d_{i+1} towards the gradient of the line from (x_i, f_i) to (x_{i+1}, f_{i+1}) and vice-versa. Thus, there are at most three curve segments which will be affected by a change in r_i, the curve segment i and its two adjacent segments.

Now suppose that the data points are not strictly monotone, such as in the case $f_1 < \cdots < f_{i-1} < f_i = f_{i+1} < f_{i+2} < \cdots < f_n$. In order for $s(x)$ to be monotone in $[x_1, x_n]$ it is necessary that $d_i = d_{i+1} = 0$ and $s(x) = f_i$, for $x \in [x_i, x_{i+1}]$. Since $d_i = 0$, the weights for the curve segment $i-1$ are computed as follows:

$$\frac{\beta_{i-1}}{\alpha_{i-1}} = \frac{r_{i-1}d_{i-1}}{3\Delta_{i-1}}. \tag{13}$$

Substituting (13) into equation (4) involving K_{i-1}, and solving $K_{i-1} = L_{i-2}$, gives values for γ_{i-1} and δ_{i-1}. Note however that at the knot (x_i, f_i), the interpolant is only C^1. Similarly, as $d_{i+1} = 0$, the weights for the curve segment $i+1$ are determined in an analogous manner.

§5. Modification of the Curve

In practice, we choose $r_i = r = 2$ for each i, $i = 1, \ldots, n-1$ to generate an initial default curve. Once this curve has been generated and displayed, then the shape of any of the curve segments can be modified interactively by assigning a new value to a parameter r_i, $1 \leq i \leq n-1$. The behaviour of the rational cubic Bézier curve with respect to the shape parameters r_i as $r_i \to \infty$ can be identified if the Bézier curve (1) is written in terms of r_i. This can be done by substituting (3) and (9) into (1) and after some simplifications, we obtain

$$s(x, r_i) = (1-\theta)f_i + \theta f_{i+1} + e_i(x, r_i), \tag{14}$$

where

$$e_i(x, r_i) = \frac{\theta(1-\theta)h_i[d_i\Delta_i\theta^2 - d_id_{i+1}\theta - d_{i+1}\Delta_i(1-\theta)^2]}{d_i\theta^3 + \Delta_i^{-1}d_id_{i+1}r_i\theta(1-\theta) + d_{i+1}(1-\theta)^3}. \tag{15}$$

Noting that d_i and d_{i+1} satisfy (12) for any $r_i \geq 2$, it follows that the numerator of (15) will remain bounded but the denominator will approach infinity as the parameters $r_i \to \infty$. Thus the rational cubic Bézier curve (14) converges uniformly to the piecewise linear interpolant,

$$(1-\theta)f_i + \theta f_{i+1}$$

as $e_i(x, r_i)$ converges uniformly to zero. Hence, the effect of varying r_i is that increasing r_i will 'tighten' the curve segment i.

An algorithm for generating the interpolating monotone curve is now described. A default curve is first generated by setting $r_i = 2$ for each $i = 1, \ldots, n-1$. The shape of any curve segments can then be altered interactively by varying the values of r_i, $r_i \geq 2$ and noting that from (14)–(15), the curve segment i will tend to a linear segment by allowing $r_i \to \infty$.

Algorithm.

1. Input number of data points, n.
2. Input data $\{x_i\}_1^n$ and $\{f_i\}_1^n$.
3. For $i = 1, \ldots, n-1$
   (a) Set $r_i = 2$ for the generation of a default curve.
   (b) Use equation (11) to compute d_i.
   (c) Set $\alpha_i = 1$. Use equation (9) to compute β_i, γ_i and δ_i.
   (d) Use equation (3) to compute the coefficients A_i, B_i, C_i and D_i.
   (e) Use equation (1) to generate the curve segment i.
4. *Modification of the shape of the curve.* Repeat step 3 except that in step (a), set new values for selected r_i where $r_i \geq 2$.

§6. Numerical Examples

We use two example data sets which were used by Fritsch & Carlson [5]:

$x =$	0	2	3	5	6	8	9	11	12	14	15
$y =$	10	10	10	10	10	10	10.5	15	30	60	85

and

$x =$	7.99	8.09	8.19	8.7	9.2	10
$y =$	0	2.76429e-5	4.37498e-2	0.169183	0.469428	0.943740

$x =$	12	15	20
$y =$	0.998636	0.999919	0.999994

The result of applying the above algorithm to the two given data sets is shown in Figures 1 and 2. For each data set, three different curves are drawn which are given by (a), (b) and (c). A default curve is shown by (a), while (b) shows the effect of a uniform increase in each r_i with $r_i = r = 20$. The effect of varying individual r_i is shown in (c) where $r_9 = 10$ in 1(c) and $r_5 = 4$ in 2(c). We conclude from these figures that this scheme is able to produce a visually pleasing C^2 interpolating curve which can be modified interactively by the user.

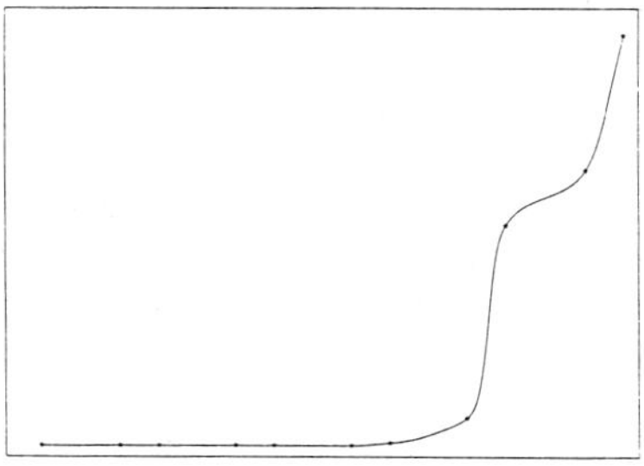

Figure 1a.

Figure 2a.

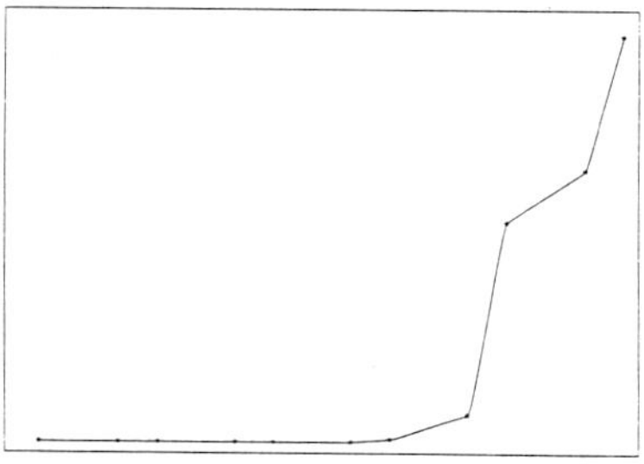

Figure 1b.

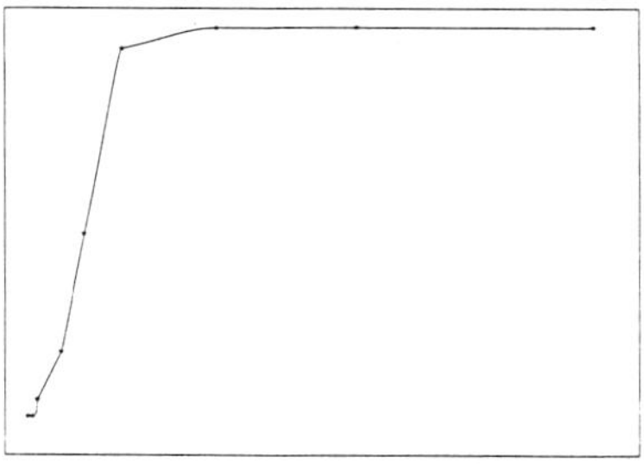

Figure 2b.

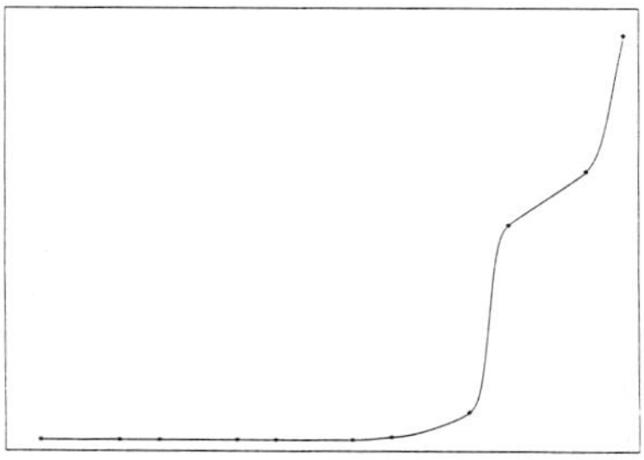

Figure 1c.

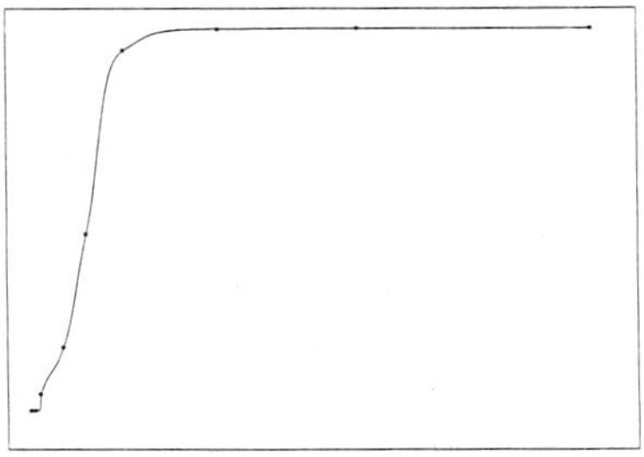

Figure 2c.

§7. Conclusion

In this paper, a simple scheme for generating C^2 interpolating curves has been proposed. The scheme is local and the resulting curve preserves the monotonicity of the data. The formulation of the scheme introduces parameters which provide for interactive modification to the shape of the curve.

Note however that the method described would not necessarily generate satisfactory curves in the case of comonotone data, by virtue of (11), which forces a zero gradient at a local extremum. To avoid this problem, an alternative scheme is currently being investigated, based upon a more general solution of (4) and (8), and requiring the solution of a quadratic equation in order to find the weights. A sufficient condition has been established which guarantees positive weights. This scheme also provides parameters for modifying the shape of the curve. Finally, the preservation of convexity is also being addressed. It is hoped to report on the results of these studies in the near future.

Acknowledgments. The author wishes to thank Dr. K. Unsworth for his comments and valuable suggestions which has helped to improve the paper. Thanks also to the University of Malaya, Kuala Lumpur, Malaysia for their financial support.

References

1. Butland, J., A method of interpolating reasonable-shaped curves through any data, Proc. Computer Graphics **80**, Online publications Ltd, Middlesex, UK, 1980, 409–422.
2. Brodlie, K. W., (eds.), *A review of methods for curve and function drawing, Mathematical methods in Computer Graphics and Design*, Academic Press, New York, 1980.
3. Delbourgo, R. and J. A. Gregory, Shape preserving piecewise rational interpolation, SIAM J. Sci. Statist. Comput. **6** (1985), 967–976.
4. Fritsch, F. N. and J. Butland, A method for constructing local monotone piecewise cubic interpolants, SIAM J. Sci. Statist. Comput. **5** (1984), 300–304.
5. Fritsch, F. N. and R. E. Carlson, Monotone piecewise cubic interpolation, SIAM J. Numer. Anal. **17** (1980), 238–246.
6. Gregory, J. A., Shape preserving spline interpolation, Comput. Aided Design **18** (1986), 53–57.
7. McAllister, D. F. and J. A. Roulier, An algorithm for computing a shape-preserving osculatory quadratic spline, ACM Trans. Math. Software **7** (1981), 331–347.
8. Schumaker, L.L., On shape preserving quadratic spline interpolation, SIAM J. Numer. Anal. **20** (1983), 854–864.

Minimization of Interpolating Spline Curves with Bounded Derivatives

Per Erik Koch

Abstract. Among all curves that pass through given points in space without exceeding specified bounds on the derivatives of the components, we want to find that curve which possesses the smallest parameter-interval. This problem comes from trajectory planning for robot manipulators. The job, for example welding or painting, should be done as fast as possible without velocities, acceleration or jerk being so large that the machinery quickly is worn out. The problem is reformulated as a minimization problem with a fixed parameter interval so that we can use standard minimization algorithms in order to find the best knots and nodes for the polynomial spline curves.

§1. The Problem

We consider the set of spline curves interpolating to given points and at the same time satisfying given constraints on the size of its derivatives. Among these splines we want to pick the one with the smallest parameter-interval. This problem comes from trajectory planning for robot manipulators. The motion is specified by a sequence of positions and orientations of the end effector of a robot manipulator. In applications such as welding and painting, it is necessary for the robot manipulator to follow the shape of the object on which it is working. The job should be done as fast as possible without the velocity, acceleration or jerk being too big.

The application of cubic splines to the generation of smooth robot joint trajectories is described in [1,3,10]. In [10] fourth degree polynomials are used on the first and last interval. Here we will use splines of general degree represented by B-splines, cf. [7]. See also [12].

The problem is the following: given points $y_i = (y_{i,j})_{j=1}^{d}$, $i = 1, 2, \ldots, n$, and bounds $(M_{i,j})_{j=1}^{d}$, $i = 1, 2, \ldots, q$, find the shortest interval $[0, T]$ so that

Mathematical Methods in Computer Aided Geometric Design II
Tom Lyche and Larry L. Schumaker (eds.), pp. 351–358.

ISBN 0-12-460510-9.

there is a vector valued function $s(x) = (s_j(x))_{j=1}^d$, $0 \le x \le T$ and abscissas $0 \le \tau_1 < \tau_2 < \cdots < \tau_n \le T$ such that

$$s_j(\tau_i) = y_{i,j}, \qquad i \le n,\ j \le d \tag{1}$$

and

$$\|s_j^{(i)}\|_p \le M_{i,j}, \qquad i \le q,\ j \le d. \tag{2}$$

Here $p > 1$. We choose to work among the polynomial spline curves so that each component function s_j will be a spline of order k on the interval $[0, T]$. Since we want to have continuous acceleration and even jerk, we only consider $k \ge 5$. In the next section we will reformulate the problem so that we can work on the fixed interval $[0, 1]$ all the time.

§2. Minimization on the Unit Interval

The reformulation will be constructed by linear scaling. If f is any sufficiently smooth scalar function defined on $[0, T]$, set $g(u) = f(u/T)$, $u \in [0, 1]$. We easily find $\|g^{(i)}\|_{L_p[0,T]} = T^{1/p-i}\|f^{(i)}\|_{L_p[0,1]}$. Hence, we must have

$$T \ge \max_{j,i}\left(\frac{\|s_j^{(i)}\|_{L_p[0,1]}}{M_{i,j}}\right)^{1/(i-1/p)}. \tag{3}$$

The least value of T possible in this case is given by the right hand side of (3). We therefore define the function

$$T = \max_{j\le d, i\le q}\left(\frac{\|s_j^{(i)}\|_{L_p[0,1]}}{M_{i,j}}\right)^{1/(i-1/p)} \tag{4}$$

as a function of the knots and the nodes. The nodes are $\tau_1, \ldots, \tau_n$ and the knots are $\{t_i\}_{i=1}^{n+k}$, where the first k knots lie at 0 and the last k at 1. In our case we want to have zero velocity and acceleration at the start and stop. We therefore take $\tau_3 = \tau_2 = \tau_1 = 0$ and $\tau_n = \tau_{n-1} = \tau_{n-2} = 1$ and use Hermite interpolation. The unknown parameters are the $n - k$ interior knots and the $n - 6$ interior nodes.

The minimization problem becomes that of finding the interior knots and nodes so that T takes its minimum. We have tried the quasi-Newton method (DFP-method, see [5]) and the conjugate-gradient method (FRPR-method, see. [4]). Both these methods reduce T substantially in the first few iterations. They slow down as two or more components of T take the maximum simultaneously. The trouble here is that when this happens T is no longer a continuously differentiable function of its parameters. What we do is to change to a slower but more robust minimization teqhnique, that of Powell, see [11]. This method does not use derivatives. We use one iteration of Powell's method, and then we go back to quasi-Newton or conjugate gradient iterations. In this way we iterate and after a while we normally reach a local minimum. It depends upon the initial values which local minimum we reach. Therefore finding good initial values is important.

Let us regard the curve as the result of an automobile driving through the given points with given bounds for the velocities, acceleration and jerks. If the driver uses the chordlength method he will get an awful acceleration at the points. Perhaps it is better to try to drive with uniform acceleration or uniform jerk given by the upper bounds between points. Lee [8] recommends the centripetal method to obtain nice curves interpolating the data. He shows that this generally will give nicer parametrization than the chord-length method. We do not want small centripetal acceleration, we rather want to drive with maximal jerk or acceleration in order to drive through in minimal time. Here we will look at a general exponent method, cf. [8], where the nodes are chosen as

$$\tau_{i+1} - \tau_i = \frac{\mid y_{i+1} - y_i \mid^e}{\sum_{j=1}^{n-1} \mid y_{j+1} - y_j \mid^e} \quad , i = 3, \ldots, n-3, \tag{5}$$

where $| \cdot |$ denotes the Euclidian distance. The choice of the exponent e is important. For $e = 1$ we obtain the chord-length method, so we prefer to use $0 < e < 1$. Choices of e around 0.2 and 0.3 usually give better results. The knots at the start of the minimization are chosen as

$$t_i = \frac{1}{k-1} \sum_{j=i-k+1}^{i-1} \tau_j, \qquad i = k+1, \ldots, n, \tag{6}$$

cf. deBoor[2, p. 219].

Before we look at examples, let us try to see what can happen. Suppose that $s_j^{(i)}$ gives the maximum uniquely in (4). Set $\alpha = 1/(i - 1/p)$. Let θ be any inner node. For simplicity, we drop the index j. Differentiation of (4) gives

$$\frac{\partial T}{\partial \theta} = (\alpha / M_{i,j}^{\alpha}) \|s^{(i)}\|_p^{\alpha - p} \cdot \int_0^1 \mid s^{(i)}(t) \mid^{p-1} sgn(s^{(i)}(t)) \frac{\partial}{\partial \theta}(s^{(i)}(t))\, dt.$$

We see that $\partial T / \partial \theta = 0$ is automatically satisfied if $\partial s^{(i)}(t)/\partial \theta = 0$ identically on $[0, 1]$. The representation of the spline component s is by B-splines, $s = \sum_j c_j B_{j,k}(t)$. The interpolation condition means that $Bc = y$, where the (i, j) element in the B-matrix is equal to $B_{i,k}(\tau_j)$ or derivatives in the case of Hermite interpolation and $c = (c_j)_j$. Differentiation gives $(\partial B / \partial \theta) c + B(\partial c / \partial \theta) = 0$. When the Schoenberg-Whitney condition is satisfied the matrix B is non-singular. Hence,

$$\frac{\partial c}{\partial \theta} = -B^{-1} \frac{\partial B}{\partial \theta} c. \tag{7}$$

Differentiating $s = \sum_j c_j B_{j,k}$, gives

$$\frac{\partial s^{(i)}(t)}{\partial \theta} = \sum_j \frac{\partial c_j}{\partial \theta} B_{j,k}^{(i)}(t).$$

This is identically zero if $\partial c_j/\partial\theta = 0$ for every j. By (7), $(\partial B/\partial\theta)c = 0$. The elements of the matrix $\partial B/\partial\theta$ are zero except for the row where the point θ comes in. Hence, $\sum_j c_j(\partial B/\partial\theta)(\theta) = 0$, i. e.,

$$s'(\theta) = 0. \tag{8}$$

We often see this in practice, that at least some of the optimal nodes are zeroes of the derivative of the leading component spline. To obtain sufficient conditions for $\partial T/\partial t_i$ to be zero is much more complex.

§3. Examples

The algorithm and the graphical interface have been programmed in C by Ola Hendseth [6] using the X-Windows system. The user decides which derivatives of which components he wants to look at. A maximum of 10 components may be viewed at the same time. The interpolation points are plotted as small circles on the zeroth derivative of the components. The user can either input his own nodes and knots or let the nodes and knots be chosen automatically by (5) and (6). Both the initial and the optimized curves are shown at the end of the execution.

Let us first look at a simple scalar example where the ordinates are given by $0, 0, 0, 2, -1, 0, 0, 0$. The three zeroes at the beginning and end indicate that the value, first and second derivative should be zero at the end points. The constraints for the first and second derivative where both equal to 10, in the max-norm. We used cubic splines, the nodes $0, 0, 0, 0.3, 0.7, 1, 1, 1$ and knots $0, 0, 0, 0, 0.2, 0.4, 0.6, 0.8, 1, 1, 1, 1$. Using these nodes and knots we got the graph in Figure 1.

Next we chose nodes and knots according to (5) and (6) with exponent $e = 0.25$. See Figure 2. The minimization program stopped after 9 iterations. The optimized spline is also shown in Figure 2. The value of the function T has gone down from 5.20 in Figure 1, to 3.22. Notice that both the inner nodes of the optimized spline satisfy equation (8).

Our next example is taken from robotics. A PUMA 600 robot manipulator with six joints is considered as a numerical example, see [7]. The points are given in Table 1. Here $n = 14$ and $d = 6$. The 4 columns of zero vectors correspond to zero velocity and zero acceleration at the start and stop. So we use Hermite interpolation at the two boundaries with $\tau_1 = \tau_2 = \tau_3 = 0$ and $\tau_{12} = \tau_{13} = \tau_{14} = 1$.

15	0	0	30	50	90	130	90	45	-10	-30	-50	0	0
10	0	0	25	30	15	-20	-55	-70	-20	0	10	0	0
50	0	0	70	150	200	120	35	-10	50	60	50	0	0
15	0	0	20	40	80	80	40	-60	-100	-60	-30	0	0
10	0	0	30	10	-40	-60	10	50	-40	-20	10	0	0
6	0	0	20	40	80	70	10	-10	15	30	20	0	0

Table 1. The given points.

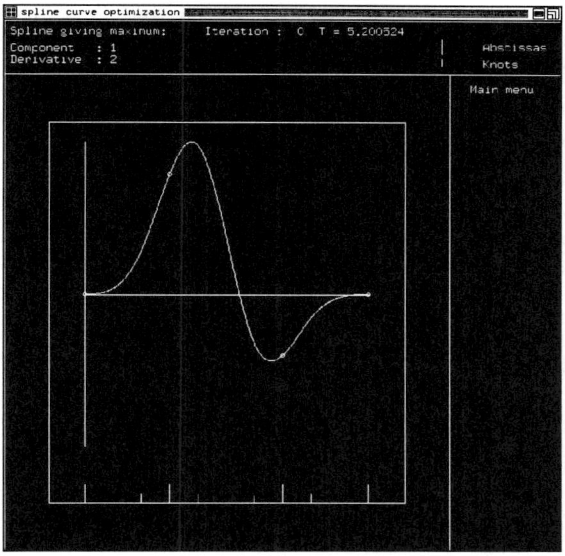

Figure 1. Cubic spline interpolant.

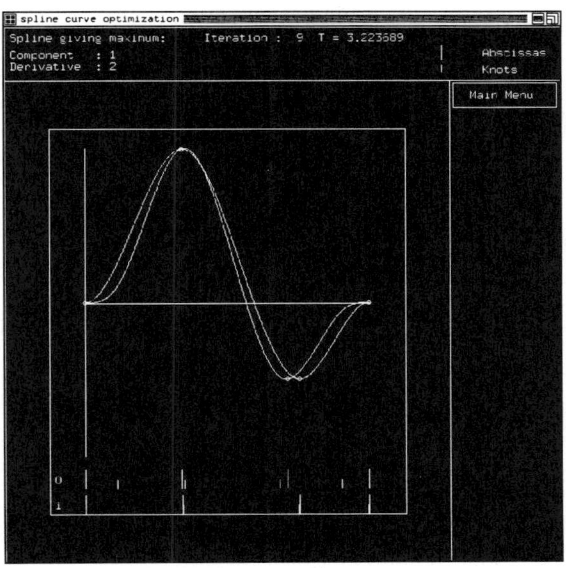

Figure 2. Initial and optimized cubic spline.

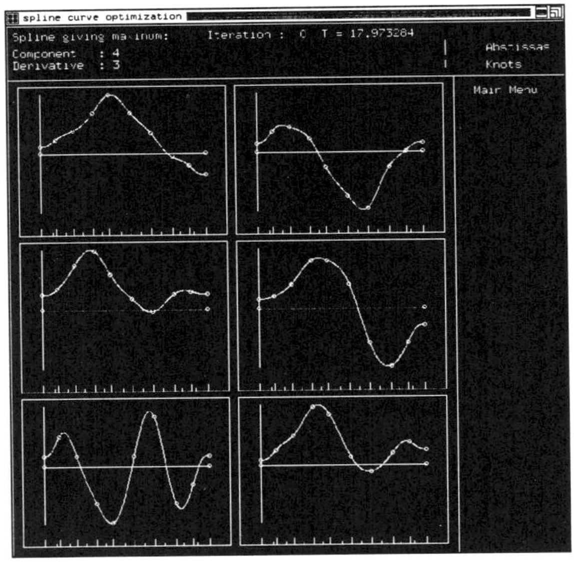

Figure 3. The initial spline components.

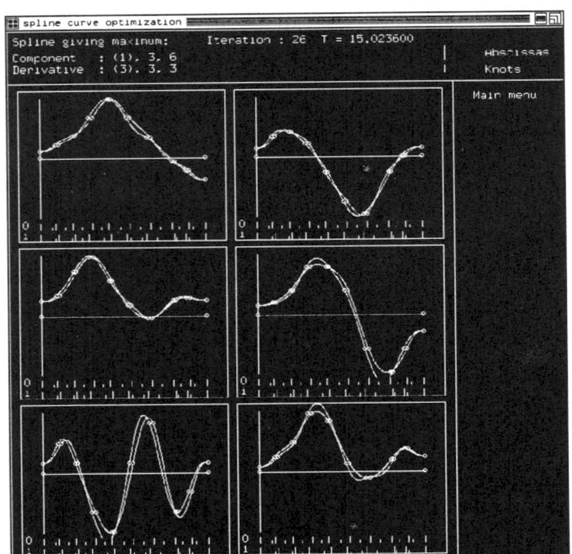

Figure 4. The initial and optimized spline components.

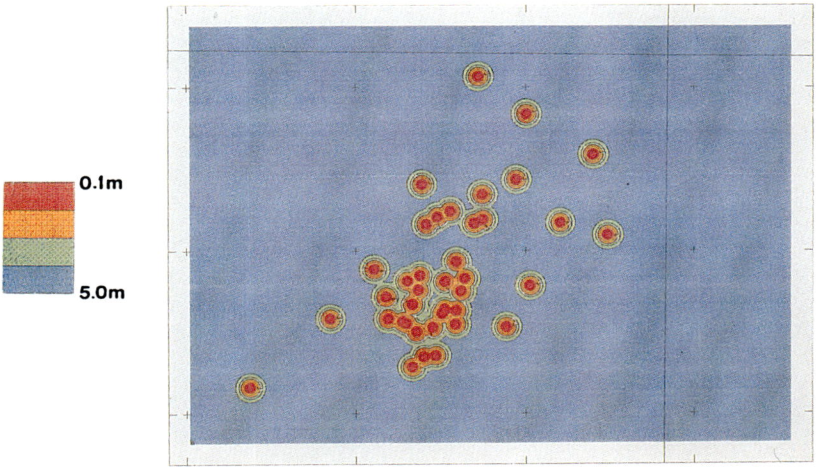

Plate 3. The tolerance function.

Plate 4. The error function.

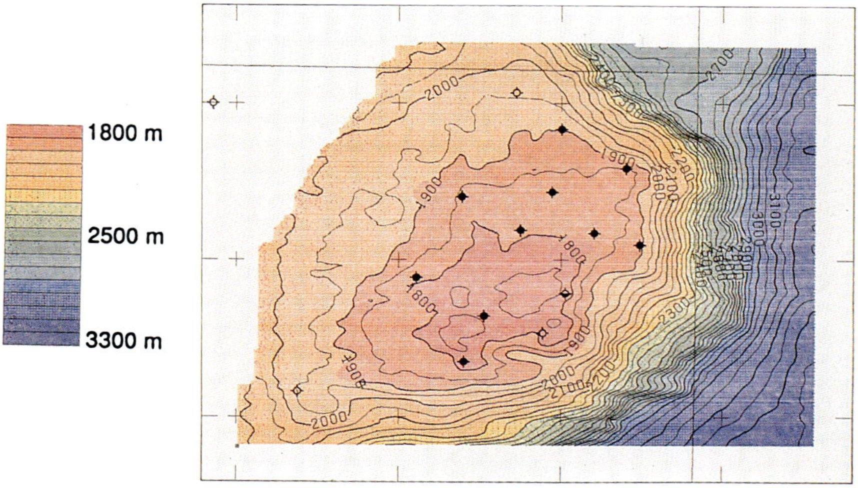

Plate 1. Contour map of an oil reservoir.

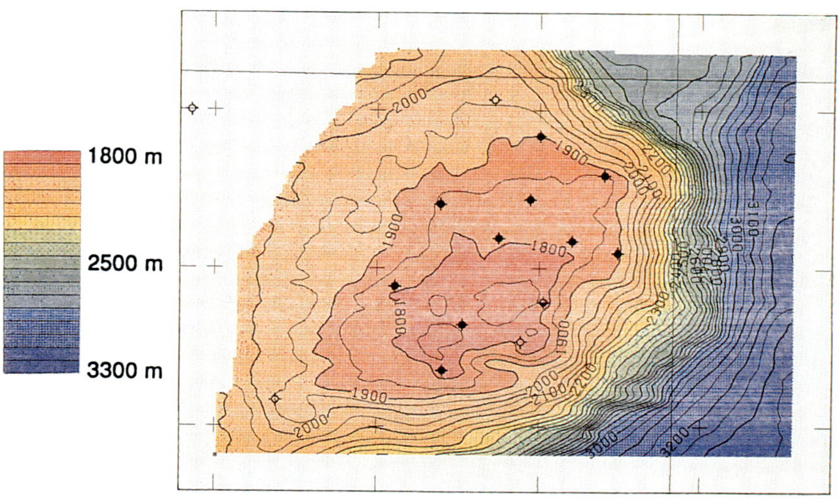

Plate 2. Contour map of the compressed surface.

nodes	1.416	1.502	4.807	6.252	8.226	10.050	12.003	13.543	
knots	1.103	2.305	3.868	5.570	7.334	9.133	10.956	12.655	13.898

Table 3. Optimal nodes and knots.

We tried to reduce T even more by increasing the order k. The CPU time increased considerably, but we did not manage to reduce T below 15.0. It seems that the global minimum lies around 15, and so this method seems to give sufficiently good results.

Acknowledgments. Special thanks go to Ola Hendseth for many interesting discussions during his thesis work and a valuable program system.

References

1. Bollinger, J., and N. Duffie, Computer algorithms for high speed continuous path robot manipulator, Ann. CIRP **28** (1979), 391–395.
2. de Boor, C., *A Practical Guide to Splines*, Springer Verlag, 1978.
3. Cook, C. C., and C. Y. Ho, The application of spline functions to trajectory generation for computer-controlled manipulators, Digital Systems for Industrial Automation **1** (1982), 325–333.
4. Fletcher, R. and C. M. Reeves, Function minimization by conjugate gradients, Computer J. **7** (1964), 149–154.
5. Fletcher, R. and M. J. D. Powell, A rapidly convergent descent method for minimization, Computer J. **6** (1963), 163–168.
6. Hendseth, Ola, masters degree at Norges Tekniske Høgskole (in Norwegian), 1991.
7. Koch, P. E. and K. Wang, The introduction of B-splines to trajectory planning for robot manipulators, Modelling, Identification and Control **9** (1988), 69–80.
8. Lee, E. T. Y., Choosing nodes in parametric curve interpolation, Comput. Aided Design **21** (1989), 363–370.
9. Lee, E. T. Y., Energy, fairness and a counterexample, Comput. Aided Design **22** (1990), 37–40.
10. Luh, J. Y. S., Lin, C. S., and P. R. Chang, Formulation and optimization of cubic polynomial joint trajectory for industrial robots, IEEE Trans. Automatic Control **28** (1983) 1066–1074.
11. Powell, M. J. D., An efficient method of finding the minimum of a function of several variables without calculating derivatives, Computer J. **7** (1964), 155–162.
12. Yamamoto, M., Ozaki, H. and A. Mohri, Planning of a manipulator joint trajectories by an iterative method, Robotica **6** (1988), 101–105.

The constraints for the first, second and third derivative are given in Table 2, corresponding to the max-norm.

Component	1	2	3	4	5	6
1st. deriv.	100	95	100	150	130	110
2nd. deriv.	45	40	75	70	90	80
3rd. deriv.	60	60	55	70	75	70

Table 2. The given constraints.

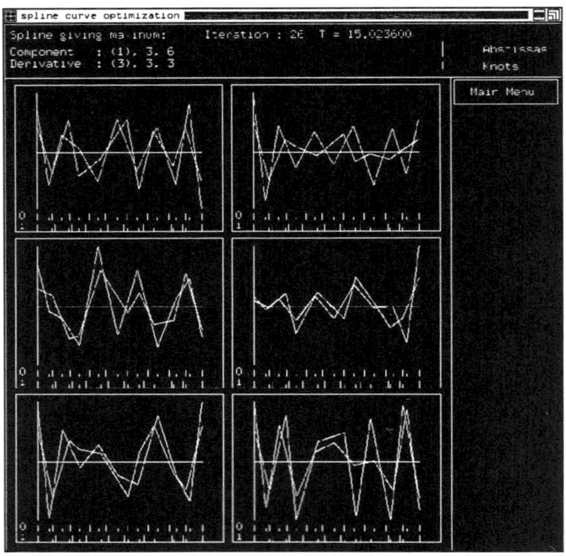

Figure 5. The third derivatives.

We choose to use a spline of order $k = 5$. Using 5 knots at zero and 5 knots at 1, we have 9 internal knots left. Also, there are $14 - 6 = 8$ internal nodes. After a little experimentation we find 0.25 to be a useful exponent in (5). The internal knots were then chosen as in (6). Using this parametrization, the function T became 17.97. See Figure 3. Values below 20 are usually difficult to obtain without minimization. Then we started to minimize. After 26 iterations with a combination of the quasi-Newton and Powell method the value of T was reduced to 15.02. See figure 4. The three component functions 1, 3 and 6 obtained the maximum in (4) simultaneously in the third derivative. The third derivative of the splines in Figure 4 are given in Figure 5. The optimal internal nodes and knots are given in Table 3.

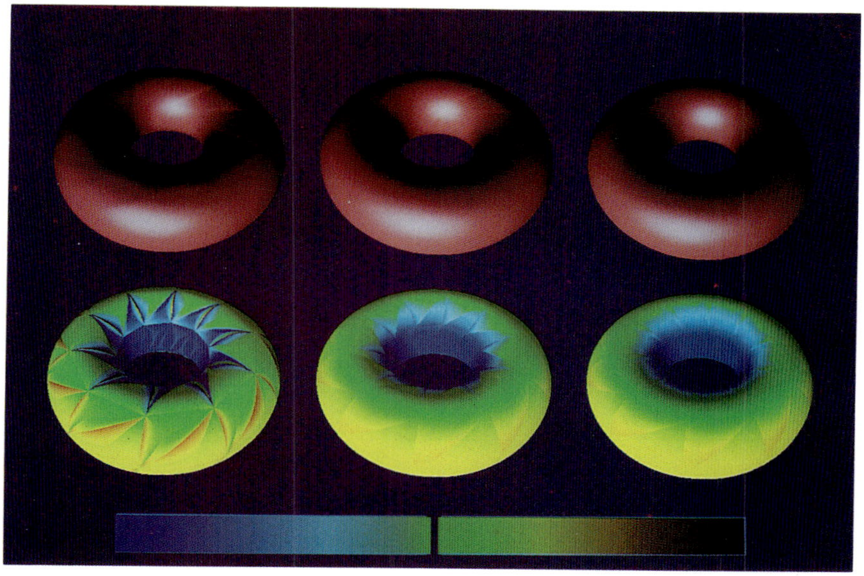

Plate 5. Interpolants to the Torus.

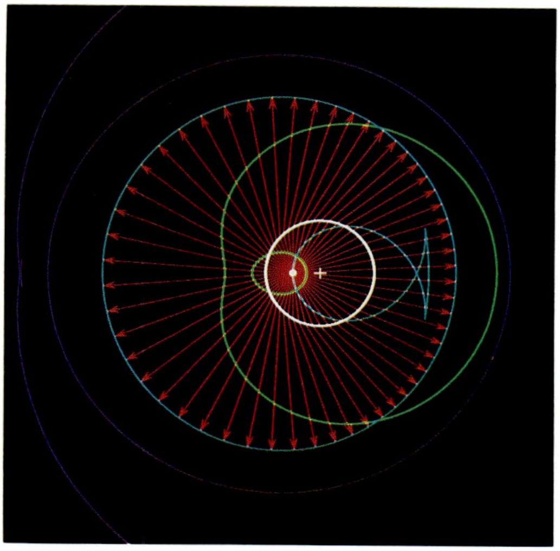

Plate 6. Rays, anticaustics, and offsets.

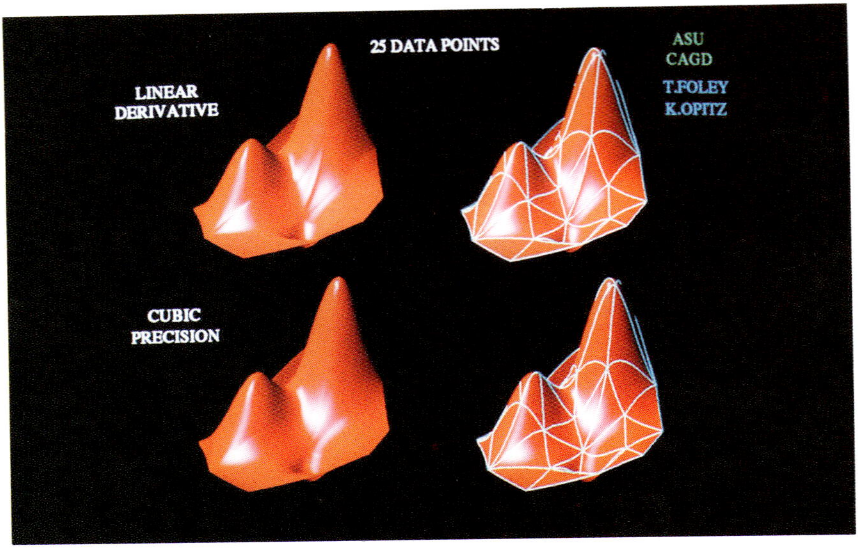

Plate 7. C^1 hybrid patch interpolants.

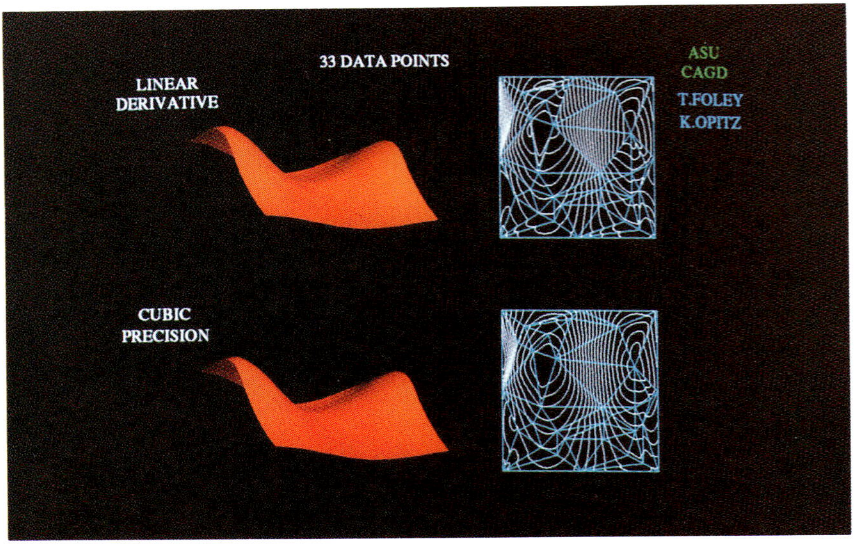

Plate 8. Surface and isophote plots on the triangulated domain.

On Piecewise Quadratic G^2 Approximation and Interpolation

J. Kozak and M. Lokar

Abstract. Let $T_1, T_2, \ldots, T_n$ be a sequence of points in the plane. The problem is to find a parametric curve C such that $C(t_i) \approx T_i$, all i, for some values of the parameter t_i. As a working tool, C is taken to be a piecewise curvature continuous quadratic spline curve. As long as T_i preserve convexity, this produces a satisfactory and efficient constructive algorithm. The key point is the fact that breakpoints in general differ from points T_i.

§1. Introduction

Consider the following problem. Let

$$T_1, T_2, \ldots, T_n \in \mathbb{R}^2$$

be a sequence of points in the plane. We want to construct a curve C that mimics in some way the underlying curve. The curve should be geometrically smooth, i.e. the tangent direction and the curvature should be continuous. In order to keep the curve as simple as possible, piecewise quadratic parametric splines in the Bézier form are taken.

The usual approach is to place the breakpoints of the curve at the given points (for example [2,4,5]). In this way interpolation is straightforwardly achieved. But on the other hand several constraints on the position of given points must be satisfied.

A data set $T_1, T_2, \ldots, T_n$ is called *noninflecting* if all angles

$$\alpha_i := \angle(T_{i+1} - T_i, T_i - T_{i-1}), \quad 1 \leq i \leq n$$

measured in counterclockwise direction, are in $(0°, 180°)$ or in $(180°, 360°)$.

Mathematical Methods in Computer Aided Geometric Design II

Tom Lyche and Larry L. Schumaker (eds.), pp. 359–366.

ISBN 0-12-460510-9.

Since the curvature of a G^2 parametric quadratic spline curve nowhere vanishes (if the curve does not reduce to a straight line), we cannot have inflection points, so this condition has to be satisfied. We assume it holds.

But beside this constraint, as shown in [5], to achieve the solvability and the uniqueness of the interpolant, the above mentioned approach also requires that the data should be *nonreflecting* and *nondegenerate*. This places further, somewhat unnatural restrictions on consecutive direction changes. Consider the following picture of the triangle, with the direction of the curve prescribed at the end points. By placing breakpoints at the interpolation points, it is obvious that not even G^1 continuity can be achieved.

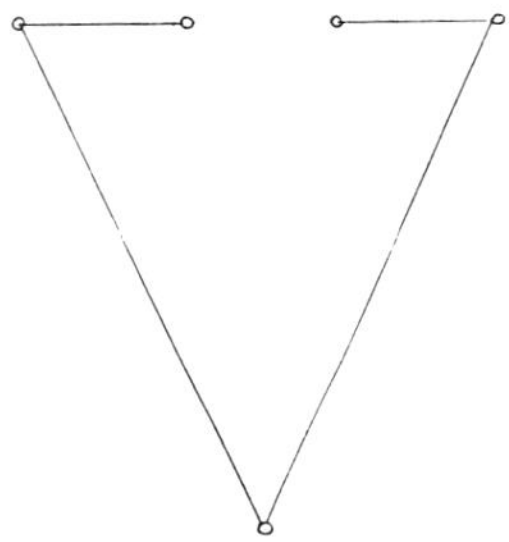

Figure 1. No G^1 piecewise quadratic interpolant.

There is another very practical reason for choosing a different placing of the breakpoints. Suppose one is given a metal or a wooden plate and is requested to cut a smooth shape out of it. In the first step, an ordinary cheap cutting device is applied to produce a control polygon consisting of straight edges. This rough shape is then cut to the final shape with NC machines. Quite clearly breakpoints should be different from the control points to assure that the final shape lies inside the control polygon.

§2. Placing the Breakpoints

The midpoint of the control polygon is fixed at the point T_i given by

$$b_{2i} := T_i.$$

So only the breakpoints have to be determined. The requirement on the tangent continuity forces the breakpoint b_{2i+1} to be put on the line between $T_i = b_{2i}$ and $T_{i+1} = b_{2i+2}$

$$b_{2i+1} = T_i + k_{2i+1}(T_{i+1} - T_i), \quad 0 < k_{2i+1} < 1.$$

The parameter k_{2i+1} has to be chosen in such a way that the resulting curve will be not only G^1 continuous but also G^2 continuous.

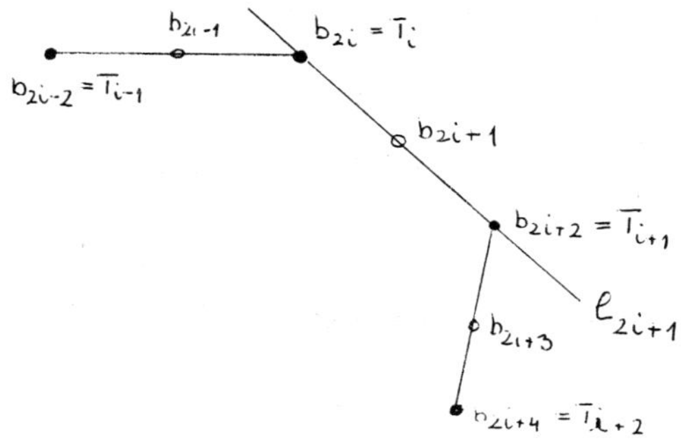

Figure 2. Placing the breakpoints.

As stated in [1], the following condition must be satisfied:

$$\frac{\operatorname{dist}(b_{2i-1}, l_{2i+1})}{||b_{2i} - b_{2i+1}||_2^2} = \frac{\operatorname{dist}(b_{2i+3}, l_{2i+1})}{||b_{2i+2} - b_{2i+1}||_2^2},$$

where l_{2i+1} is the straight line connecting b_{2i} and b_{2i+2}. As assumed above, the given data set is noninflecting, so straightforward calculation yields

$$\begin{aligned}
||b_{2i} - b_{2i+1}||_2^2 &= k_{2i+1}^2 ||T_{i+1} - T_i||_2^2 \\
||b_{2i+2} - b_{2i+1}||_2^2 &= (1 - k_{2i+1})^2 ||T_{i+1} - T_i||_2^2 \\
\operatorname{dist}(b_{2i-1}, l_{2i+1}) &= (1 - k_{2i-1}) A_{2i+1} \\
\operatorname{dist}(b_{2i+3}, l_{2i+1}) &= k_{2i+3} B_{2i+1},
\end{aligned}$$

where

$$A_{2i+1} := \sqrt{||T_{i-1} - T_i||_2^2 - \frac{\langle T_{i+1} - T_i, T_{i-1} - T_i \rangle^2}{||T_{i+1} - T_i||_2^2}}$$

and

$$B_{2i+1} := \sqrt{||T_{i+2} - T_{i+1}||_2^2 - \frac{\langle T_{i+1} - T_i, T_{i+2} - T_{i+1} \rangle^2}{||T_{i+1} - T_i||_2^2}}$$

Here $\langle \bullet, \bullet \rangle$ denotes the dot product. The G^2 continuity conditions now read

$$\frac{1 - k_{2i-1}}{k_{2i+1}^2} A_{2i+1} = \frac{k_{2i+3}}{(1 - k_{2i+1})^2} B_{2i+1}, \quad i = 1, 2, \ldots, n-1. \tag{1}$$

Two boundary conditions have to be added. Perhaps the most natural is to take $b_1 = T_0$ and $b_{2n+1} = T_{n+1}$ which implies

$$k_1 = 0, \quad k_{2n+1} = 1.$$

With $x_i := 2k_{2i+1} - 1, c_i := A_{2i+1} > 0, d_i := B_{2i+1} > 0$, this reduces to the nonlinear system for the unknowns x_i:

$$\begin{aligned} x_0 &= -1, \\ \frac{1 - x_{i-1}}{(1 + x_i)^2} c_i &= \frac{1 + x_{i+1}}{(1 - x_i)^2} d_i \qquad i = 1, 2, \ldots, n-1, \\ x_n &= 1. \end{aligned} \tag{2}$$

Quite clearly one is interested only in solutions that satisfy

$$-1 \le x_i \le 1, \qquad i = 1, 2, \ldots, n-1, \tag{3}$$

and existence as well as uniqueness have to be proved.

Lemma 1. *The system of equations (2) has a solution that satisfies (3).*

Proof: This system can be rewritten as

$$x = f(x)$$

with $x := (x_i)_{i=0}^n, f(x) := (f_i(x))_{i=0}^n$, where

$$\begin{aligned} f_0 &:= -1 \\ f_i(x) &:= f_i(x_{i-1}, x_i, x_{i+1}) = \\ &= \frac{1 - r_i\sqrt{\frac{1+x_{i+1}}{1-x_{i-1}}}}{1 + r_i\sqrt{\frac{1+x_{i+1}}{1-x_{i-1}}}}, \quad r_i := \sqrt{\frac{d_i}{c_i}} > 0, \qquad i := 1, \ldots, n-1 \\ f_n &:= 1. \end{aligned}$$

Clearly, the map $x \to f(x)$ maps a compact convex set $[-1, 1]^{n+1}$ into itself. Thus by Brouwer's fixed point theorem ([3]) it has a fixed point, which is the solution of our system. ■

The following technical observation will establish uniqueness.

Lemma 2. *Let $r \in \mathbb{R}^+$ and $x, y, z \in (-1, 1)$ be such that the following relation is satisfied*

$$y = \frac{1 - r\sqrt{\frac{1+z}{1-x}}}{1 + r\sqrt{\frac{1+z}{1-x}}}.$$

Then

$$\frac{\partial y}{\partial x} = -\frac{r\sqrt{\frac{1-x}{1+z}}}{(r + \sqrt{\frac{1-x}{1+z}})^2(1-x)} < 0,$$

regardless of the choice of z, r and

$$\frac{\partial y}{\partial z} = -\frac{r\sqrt{\frac{1-x}{1+z}}}{(r+\sqrt{\frac{1-x}{1+z}})^2(1+z)} < 0,$$

again independently of the choice of x, r. Also

$$\frac{\partial z}{\partial x} < 0, \quad \frac{\partial z}{\partial y} < 0.$$

Theorem. *The solution of the system (2) satisfying (3) is unique.*

Proof: The following induction claim will be the key point of the proof. The system

$$x_j = f_j(x_{j-1}, x_j, x_{j+1}), \qquad j = i, i+1, \ldots, n, \quad 1 \le i \le n, \tag{4}$$

has a unique solution $(x_j)_{j=i}^n$ with $-1 \le x_j \le 1$ for all j for any $x_{i-1} \in [-1, 1]$ and

$$\frac{\partial x_i}{\partial x_{i-1}} < 0,$$

independently of $x_{i+1}, x_{i+2}, \ldots, x_n$.

Take $i = n$. Then the equation

$$x_n = \frac{1 - r_n\sqrt{\frac{2}{1-x_{n-1}}}}{1 + r_n\sqrt{\frac{2}{1-x_{n-1}}}}$$

obviously has a solution $x_n \in [-1, 1]$ for any $x_{n-1} \in [-1, 1]$. Lemma 2 implies the second relation. Assume now that the claim holds for some i, $1 \le i \le n$, and add the equation

$$x_{i-1} = f_{i-1}(x_{i-2}, x_{i-1}, x_i) \tag{5}$$

to the previous system. If x_{i-2}, x_{i-1}, x_i satisfy (5), Lemma 2 implies

$$\frac{\partial x_i}{\partial x_{i-2}} < 0$$

independently of x_{i-1} as well as $\frac{\partial x_{i-1}}{\partial x_{i-2}} < 0$ independently of x_i. But then, if $x_i, x_{i+1}, \ldots, x_n$ also satisfy (4), then by the induction hypothesis,

$$\frac{\partial x_i}{\partial x_{i-2}} > 0.$$

Thus there can be at most one value of x_i such that both (4) and (5) are satisfied, which brings the induction to step $i-1$. ■

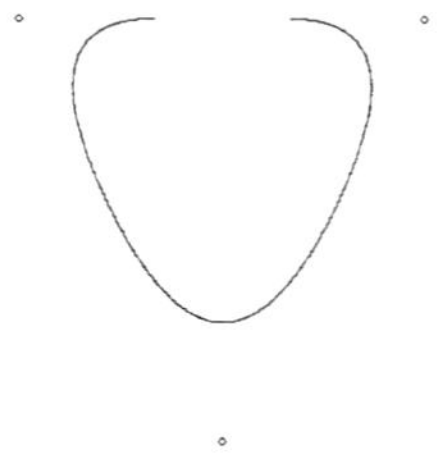

Figure 3.
Approximation of a triangle.

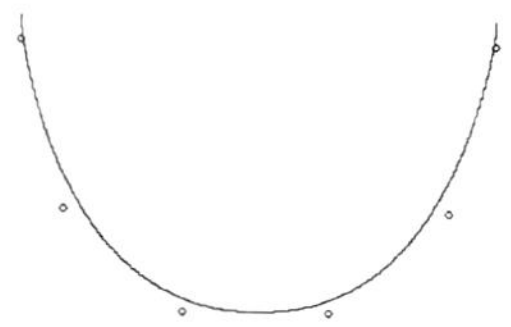

Figure 4.
The first Litostroj example:
Approximation.

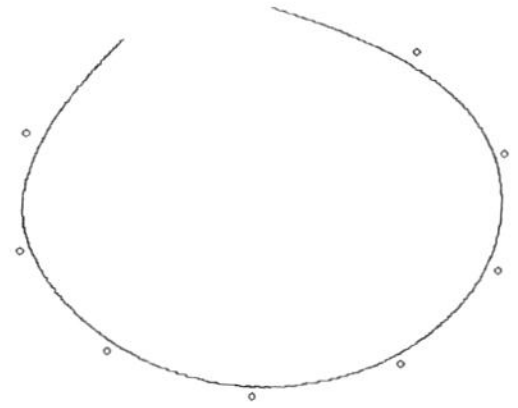

Figure 5.
The second Litostroj example:
Approximation.

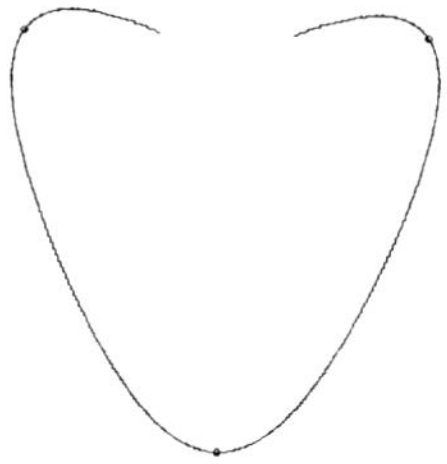

Figure 6.
Interpolation of a triangle.

§3. Examples

The following figures present some curves whose control points were obtained with the described process. The last two examples originate from one of the laboratories of the Litostroj company.

It should also be noted that the numerical solution of the nonlinear system (2) by Newton's method is very efficient. This is not surprising. Jacobian of the system in the equivalent form

$$\begin{aligned} x_0 &= -1 \\ \frac{1-x_{i-1}}{1+x_i^2}\,\frac{1-x_i^2}{1+x_{i+1}} &= \frac{c_i}{d_i}, \qquad i=1,\ldots,n-1 \\ x_n &= 1 \end{aligned}$$

vanishes nowhere,

$$\operatorname{diag}(\frac{1}{f_i})(\mathcal{J}(f(x)) = \operatorname{tridiag}(\frac{1}{x_{i-1}-1}, \frac{4}{x_i^2-1}, \frac{1}{x_{i+1}+1}),$$

and thus local quadratic convergence is assured.

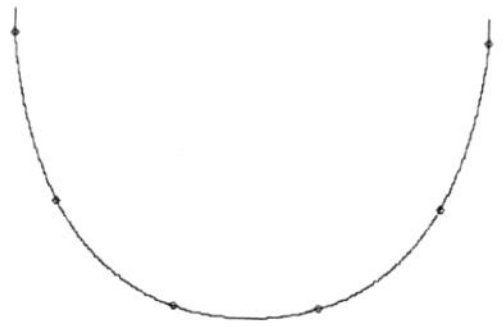

Figure 7.
The first Litostroj example:
Interpolation.

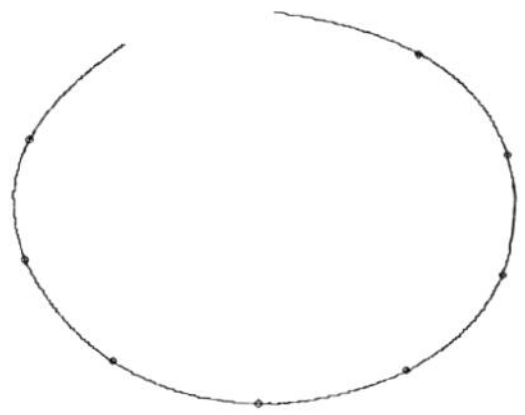

Figure 8.
The second Litostroj example:
Interpolation.

§4. Interpolation

The solution obtained in the previous section is in many practical situations completely satisfactory. However, if one insists on interpolation, the previous procedure should be somewhat extended. The idea is based on the iterative repositioning of the control points, and the algorithm described in the previous sections.

To begin, suppose one is given three points b_1, b_2, b_3 that should be interpolated by a quadratic curve. Quite clearly, b_1 and b_3 should also be the edge points of the control polygon. One is left with the free control point B with the additional requirement

$$Q(t) := b_1(1-t)^2 + 2Bt(1-t) + b_3t^2$$
$$Q(\hat{t}) = b_2$$

for some $\hat{t} \in (0,1)$. This implies that the point B can be placed anywhere on the hyperbola

$$B(t) = (b_2 - b_1(1-t)^2 - b_3t^2)/(2t(1-t)).$$

$B(t)$ is not unique, and an additional requirement has to be added. In particular, one might require that $B(t)$ should be as close to b_2 as possible. In this paper, the value $\hat{t}$

$$\hat{t} = \frac{||b_1 - b_2||}{||b_1 - b_2|| + ||b_3 - b_2||}$$

is suggested.

Now to the general case. Assume the control points $b_{2i+1}, i = 0, 1, \ldots, n$ are given. Then $b_{2i} = T_i$ is replaced by

$$b_{2i} = (T_i - b_{2i-1}(1-\hat{t}_i)^2 - b_{2i+1}\hat{t}_i^2)/(2\hat{t}_i(1-\hat{t}_i))$$

and

$$\hat{t}_i = \frac{||b_{2i-1} - T_i||}{||b_{2i-1} - T_i|| + ||b_{2i+1} - T_i||}$$

But now b_{2i+1} should be recomputed. To summarize the algorithm:

$b_{2i} = T_i, \quad i = 1, 2, \ldots, n$

```
repeat
```

calculate b_{2i+1} as in previous section, $i = 1, \ldots, n-1$,

$$\hat{t}_i = \frac{||b_{2i-1} - T_i||}{||b_{2i-1} - T_i|| + ||b_{2i+1} - T_i||} \quad i = 1, \ldots, n-1,$$

$b_{2i} = b_{2i}(\hat{t}_i), \quad i = 1, \ldots, n-1,$

`until` only small changes in b_{2i}.

Numerical experiments show that convergence is assured, and only a couple of iterations are needed. The proofs and additional comments about the method are in preparation.

§5. More Examples

Figures 7 and 8 show two more examples of interpolation of the data set used in Section 3.

References

1. Boehm, W., Farin, G. and Kahmann, J., A survey of curve and surface methods in CAGD, Computer Aided Geom. Design **1** (1984), 1–60.
2. de Boor, C., Höllig, K. and Sabin, M., High accuracy geometric Hermite interpolation, Computer Aided Geom. Design **4** (1987), 269–278.
3. Ortega, J.M., Rheinboldt, W.C., *Iterative Solution Of Nonlinear Equations In Several Variables*, Academic Press, New York, 1970.
4. Schaback, R., On global GC^2 convexity preserving interpolation of planar curves by piecewise Bezier polynomials, NAM-Bericht 64, 1988.
5. Schaback, R., Interpolation with piecewise quadratic visually C^2 Bezier polynomials, Computer Aided Geom. Design **6** (1989), 219–233.

Jernej Kozak, Matija Lokar
University of Ljubljana
Jadranska 19
Ljubljana, Slovenija
kozak@uni–lj.ac.mail.yu
matija.lokar@uni–lj.ac.mail.yu

Non-affine Blossoms and Subdivision for Q-Splines

Rekha Kulkarni, Pierre-Jean Laurent
and Marie-Laurence Mazure

Abstract. A parametric Q-spline curve is a C^2 piecewise polynomial function of degree 4 for which each polynomial section satisfies an affine relation. For such a curve, a non-affine blossom can be defined and a corresponding non-affine subdivision algorithm can be derived.

§1. Introduction

Given a knot vector

$$T = \{a = t_0, t_1, t_2, \ldots, t_n, t_{n+1} = b\} \tag{1}$$

with $t_i < t_{i+1}$, $i = 0, \ldots, n$ $(n \geq 2)$, a weight function w, measurable on $[a, b]$ and satisfying $0 < m \leq w(t) \leq M$ for all $t \in [a, b]$, and n real numbers y_i, $i = 1, \ldots, n$, let us consider the following minimization problem:

$$\min_{\substack{x \in H^2[a,b] \\ x(t_i) = y_i \\ i=1,\ldots,n}} \int_a^b w(t)(x''(t))^2 \, dt, \tag{2}$$

where $H^2[a, b]$ is the space of real functions defined on $[a, b]$ such that x' is absolutely continuous on $[a, b]$ and $x'' \in L^2[a, b]$.

Using classical results (cf [6] p. 194 and 210, [5]), a function $\sigma \in H^2[a, b]$ satisfying the interpolating conditions $\sigma(t_i) = y_i, i = 1, \ldots, n$, is the unique solution of (1) iff:

$$\sigma''(t) = \frac{1}{w(t)} \sum_{i=1}^{n} \lambda_i (t - t_i)_+, \tag{3}$$

Mathematical Methods in Computer Aided Geometric Design II
Tom Lyche and Larry L. Schumaker (eds.), pp. 367–380.

ISBN 0-12-460510-9.

where the λ_i's satisfy the two following conditions

$$\sum_{i=1}^{n} \lambda_i = 0, \qquad \sum_{i=1}^{n} \lambda_i t_i = 0. \tag{4}$$

In case w is piecewise constant on the knot vector T (i.e. $w(t) \equiv w_i$ for $t \in (t_i, t_{i+1})$), then the solution is a cubic spline, but it is only C^1 (cf [2, 10, 3]); the second derivative satisfies $w(t_i^-)\sigma''(t_i^-) = w(t_i^+)\sigma''(t_i^+)$.

If w is a smoother polynomial spline, for instance a continuous piecewise linear function, then the solution becomes C^2 but it is not piecewise polynomial anymore. In the present study, we will choose a weight function of the following form:

$$w(t) = 1/q(t), \tag{5}$$

where the function q is continuous and piecewise linear on the knot vector T. We will denote by $q_i = q(t_i), i = 0, \ldots, n+1$, the values of q at the knots.

As both the weight function w and the function $p(t) = \sum_{i=1}^{n} \lambda_i (t - t_i)_+$ are continuous, it is clear that $\sigma''(t) = q(t)p(t)$ is continuous. More precisely, σ'' is a piecewise polynomial function of degree 2 vanishing outside of $[t_1, t_n]$. Thus the solution σ itself is a C^2 piecewise polynomial function of degree 4 satisfying the classical end conditions $\sigma''(a) = \sigma^{(3)}(a) = \sigma^{(4)}(a) = 0$ (and similarly at b).

But on each interval (t_i, t_{i+1}), we have to express that σ''/q is a polynomial of degree 1. It can be shown (cf. [4, 5]) that this is accomplished iff

$$(\frac{\sigma''}{q})' = \frac{\sigma^{(3)}}{q} - \frac{\sigma'' q'}{q^2} \tag{6}$$

takes the same value at two distinct points of the interval. This leads to the following definition of the Q-spline space associated with a knot vector T and a continuous function q which is piecewise linear on T. Let us denote by $QS(T, q)$ the set of real functions s defined on $[a, b]$ such that:

1) s is a polynomial of degree 4 in each sub-interval $[t_i, t_{i+1}], i = 0, \ldots, n$,
2) $s \in C^2[a, b]$,
3) In each sub-interval (t_i, t_{i+1}), s satisfies a q-relation:

$$\frac{s^{(3)}(t_i^+)}{q_i} - \frac{a_i}{q_i^2} s^{(2)}(t_i) = \frac{s^{(3)}(t_{i+1}^-)}{q_{i+1}} - \frac{a_i}{q_{i+1}^2} s^{(2)}(t_{i+1}) \quad i = 0, \ldots, n, \tag{7}$$

where $a_i = q'(t) = (q_{i+1} - q_i)/(t_{i+1} - t_i)$, $t \in (t_i, t_{i+1})$.

QS(T, q) is a $(n+4)$-dimensional linear space. There are many other equivalent ways of expressing the q-relation (7); cf [5].

The solution of (2) is the unique funtion σ belonging to QS(T, q), which satisfies the interpolating conditions $\sigma(t_i) = y_i$, $i = 1, \ldots, n$, and the end conditions $\sigma''(a) = \sigma^{(3)}(a) = \sigma''(b) = \sigma^{(3)}(b) = 0$. In [4, 5] it is shown that the solution σ of (2) can be computed by solving a tri-diagonal linear system, and that the corresponding smoothing spline can be computed by solving a block-2×2 tri-diagonal linear system.

A polynomial function of degree 4 satisfying a q-relation of the form (7) can be controlled by 4 pseudo-Bézier points. For such a polynomial we show in Section 2 that a non-affine subdivision algorithm can be defined by 3 steps as for a polynomial of degree 3.

More generally, a Q-spline curve in $\mathbb{R}^d$ can be controlled by $n+4$ poles. The corresponding B-splines are based on at most 4 consecutive sub-intervals of the knot vector (as in the case of classical cubic splines). Using a non-affine blossom for the entire Q-spline, we show in Section 3 that a non-affine subdivision algorithm can also be defined. Throughout this section we follow the method introduced by H. P. Seidel in [8, 9].

§2. Polynomials of Degree 4 Satisfying a q-relation

Let F be a polynomial function of degree 4 defined on $[0,1]$ with values in $\mathbb{R}^d$, and let P_i, $i = 0, \ldots, 4$, be the corresponding Bézier points relative to $[0,1]$. Let us suppose that each component of this polynomial function satisfies the same q-relation, that is

$$\frac{F^{(3)}(0)}{q_0} - \frac{a}{q_0^2} F''(0) = \frac{F^{(3)}(1)}{q_1} - \frac{a}{q_1^2} F''(1) \tag{8}$$

with $q_0 = q(0)$, $q_1 = q(1)$, $a = q'(t) = q(1) - q(0)$, $t \in [0,1]$, where q is a polynomial of degree 1 on [0, 1]. Using the classical expressions for the derivatives in terms of the Bézier points, this relation is equivalent to

$$P_2 = \eta M + (1-\eta) N \tag{9}$$

with

$$\begin{aligned} M &= P_0 + u(P_1 - P_0) \\ N &= P_4 + v(P_3 - P_4), \end{aligned} \tag{10}$$

where

$$\begin{aligned} \eta &= \frac{q_1^2 + 2q_0q_1}{q_0^2 + q_1^2 + 4q_0q_1} \\ u &= \frac{q_1}{q_1 + 2q_0} \\ v &= \frac{q_0}{q_0 + 2q_1}. \end{aligned} \tag{11}$$

The points M and N (together with $A = P_0$ and $B = P_4$) will be called the *pseudo-Bézier points*. They can be chosen arbitrarily and then the 5 Bézier points can be deduced (cf Figure 1).

Now, consider a subdivision of the interval [0,1] into sub-intervals $[0,t]$ and $[t,1]$ where $0 < t < 1$. Using the de Casteljau subdivision algorithm, we can obtain the 5 Bézier points relative to $[0,t]$ (and those relative to $[t,1]$).

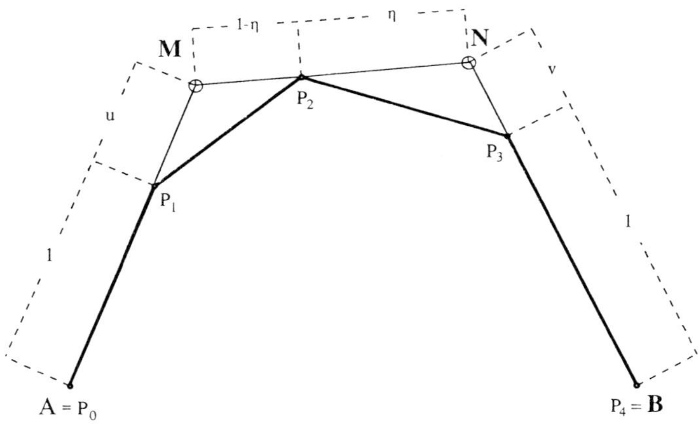

Figure 1. The pseudo-Bézier points.

As the polynomial satisfies a q-relation, these 5 points satisfy a relation of the form (9,10) but with q_1 replaced by $q_t = q(t)$ in the definition (11) of η, u, v. Let us denote by $A_0 = A$, M_0, N_0 and B_0 the corresponding pseudo-Bézier points. In the same way, let us denote by $A_1 = B_0$, M_1, N_1 and $B_1 = B$, the 4 pseudo-Bézier points corresponding to the 5 Bézier points relative to $[t, 1]$ (replacing q_0 by q_t in the definition (11) of η, u, v).

A natural question now arises: Is it possible to compute directly these new pseudo-Bézier points by using a modified 3 step subdivision algorithm (as for a polynomial of degree 3)?

By a direct computation it is possible to prove that the line M_0N_0 meets the line MN at a point Z, and that the line N_1M_1 meets the line MN at the *same* point Z (see Figure 2). This means that we can obtain the new pseudo-Bézier points in the following way:

first step:	put M_0 on $[AM]$
	put Z on $[MN]$
	put N_1 on $[NB]$
second step:	put N_0 on $[M_0Z]$
	put M_1 on $[ZN_1]$
third step:	put $B_0 = A_1$ on $[N_0M_1]$.

This algorithm is exactly the de Casteljau algorithm except that the rates are not just t and $1 - t$. In fact the rates are rational functions of t.

This property can be interpreted in terms of a non-affine blossom. Suppose first, that the points A, M, N and B are affinely independent in $\mathbb{R}^d$.

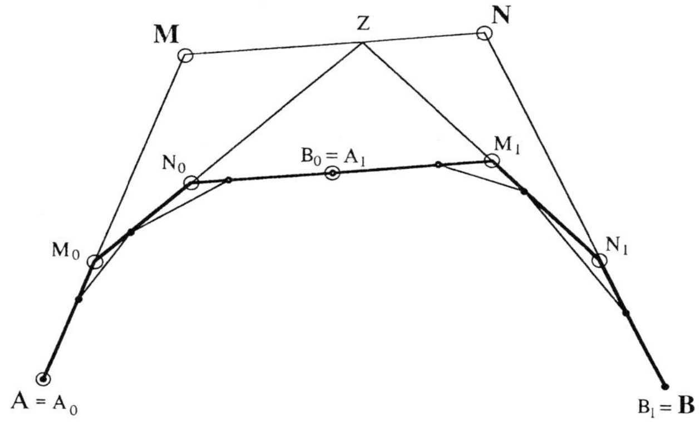

Figure 2. Subdivision for the pseudo-Bézier points.

Let us denote by $\mathcal{T}(u) = \text{osc}_1 F(u)$ the tangent flat of order 1 of F at $F(u)$, i.e. the tangent line

$$\mathcal{T}(u) = \{F(u) + \lambda_1 F'(u), \lambda_1 \in \mathbb{R}\}, \tag{12}$$

and by $\mathcal{P}(u) = \text{osc}_2 F(u)$ the tangent flat of order 2 of F at $F(u)$, i.e. the osculating plane

$$\mathcal{P}(u) = \{F(u) + \lambda_1 F'(u) + \lambda_2 F''(u), \lambda_1 \in \mathbb{R}, \lambda_2 \in \mathbb{R}\}. \tag{13}$$

For three distinct abscissae u, v, w, we define the polar form or blossom $f(u, v, w)$ by the following geometric property (cf [8, 9]):

$$f(u, v, w) := \mathcal{P}(u) \cap \mathcal{P}(v) \cap \mathcal{P}(w). \tag{14}$$

By a direct argument it can be proved that the intersection exists and is reduced to a single point. But this result is also a consequence of a more general property of Q-splines that we will obtain in the next section.

In case two abscissae are coalescent, we define (for $w \neq u$) $f(u, u, w) = f(u, w, u) = f(w, u, u)$ by

$$f(u, u, w) := \mathcal{T}(u) \cap \mathcal{P}(w). \tag{15}$$

Also, we put $f(u, u, u) := F(u)$.

The function f is clearly symmetric. For fixed u, v, $f(u, v, w)$ moves along the line $\mathcal{P}(u) \cap \mathcal{P}(v)$ if $u \neq v$ and $\mathcal{T}(u)$ if $u = v$, but it is non-affine with respect to w. For example, we can write:

$$f(0, 0, t) = (1 - \alpha(t)) f(0, 0, 0) + \alpha(t) f(0, 0, 1), \tag{16}$$

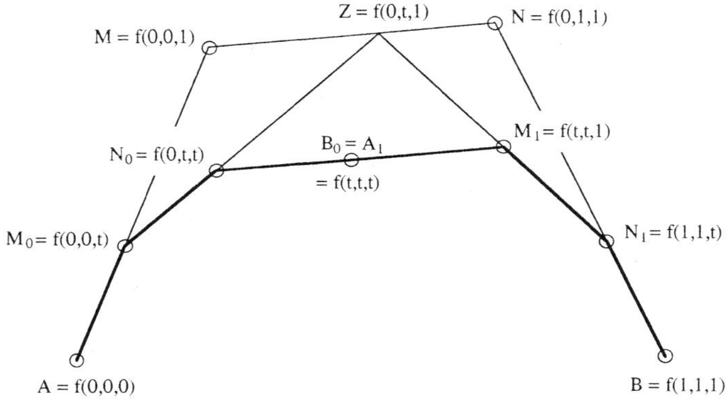

Figure 3. Non affine blossoming for a polynomial satisfying a q-relation.

where α is a rational function of t.

From the classical properties of the Bézier points, we have

$$\mathcal{T}(0) = \text{aff}(P_0, P_1) = \text{aff}(A, M)$$

$$\mathcal{P}(1) = \text{aff}(P_2, P_3, P_4) = \text{aff}(M, N, B).$$

Thus, we see that

$$f(0,0,1) = M \quad \text{and} \quad f(0,1,1) = N, \tag{17}$$

and of course $f(0,0,0) = A$ and $f(1,1,1) = B$.

Using the same argument on $[0,t]$ and on $[t,1]$, we obtain

$$f(0,0,t) = M_0, \quad f(0,t,t) = N_0 \tag{18}$$

and

$$f(t,t,1) = M_1, \quad f(t,1,1) = N_1. \tag{19}$$

Now, using the fact that $f(u,v,w)$ moves along a line when only one argument is varying, we see that the point $f(0,t,1)$ is

- on the line $\text{aff}(f(0,t,0), f(0,t,t)) = \text{aff}(M_0, N_0)$
- on the line $\text{aff}(f(1,t,1), f(t,t,1)) = \text{aff}(N_1, M_1)$
- on the line $\text{aff}(f(0,0,1), f(0,1,1)) = \text{aff}(M, N)$,

i.e., $f(0,t,1)$ is the point Z that we introduced above.

The situation is summarized in Figure 3.

In the case where F reduces to a polynomial function of degree 3 (which occurs when $q_0 = q_1$ so that $u = v = 1/3$ and $\eta = 1/2$), the blossom f is just the classical tri-affine blossom. In the general situation, the blossom becomes a rational function of each argument. Thus the subdivision algorithm for a polynomial of degree 4 satisfying a q-relation has the same structure as the classical de Casteljau algorithm for a polynomial function of degree 3 except that the rates are rational functions of t instead of just $t : 1 - t$.

If the pseudo-Bézier points A, M, N, B are not affinely independent, a non-affine blossom can still be defined with the same properties by introducing 4 affinely independent points $\hat{A}$, $\hat{B}$, $\hat{N}$, $\hat{B}$ (eventually in another space) such that $A = \mathcal{A}(\hat{A})$, $M = \mathcal{A}(\hat{M})$, $N = \mathcal{A}(\hat{N})$, and $B = \mathcal{A}(\hat{B})$, where $\mathcal{A}$ is an affine mapping. Then the blossom is defined by $f(u, v, w) = \mathcal{A}(\hat{f}(u, v, w))$, where $\hat{f}$ is the blossom associated with $\hat{A}$, $\hat{M}$, $\hat{N}$, $\hat{B}$.

§3. Q-spline Curves

Suppose that S is a *parametric Q-spline curve*, that is a mapping $[a, b] \to \mathbb{R}^d$ each component of which is a Q-spline, i.e. an element of $QS(T, q)$. Let us denote by $P_{i,j}$, $i = 0, \ldots, n$, $j = 0, \ldots, 4$ the Bézier points of each polynomial section and by $A_i = P_{i,0}$, M_i, N_i, $B_i = P_{i,4}$ the corresponding pseudo-Bézier points.

For simplicity, we will assume that the t_i are equidistant. The C^2-continuity at t_i implies the classical relations between the Bézier points:

$$
\begin{aligned}
P_{i-1,4} &= P_{i,0} \\
P_{i-1,4} - P_{i-1,3} &= P_{i,1} - P_{i,0} \\
P_{i-1,4} - 2P_{i-1,3} + P_{i-1,2} &= P_{i,2} - 2P_{i,1} + P_{i,0},
\end{aligned} \tag{20}
$$

and these relations imply the following relations between the pseudo-Bézier points:

$$
N_{i-1} + \frac{\eta_{i-1}}{k_i}(N_{i-1} - M_{i-1}) = M_i + (\frac{1-\eta_i}{k_i})(M_i - N_i), \tag{21}
$$

where $k_i = \frac{2-(v_{i-1}+u_i)}{2+(v_{i-1}+u_i)}$. Let us call P_i, $i = 1, \ldots, n$, the point defined by the left hand side (or the right hand side) of the relation (21). The situation is summarized in Figure 4. At both ends we have a special construction: $P_0 = M_0$, and $P_{-1} = A_0 = P_{0,0}$ (and similarly $P_{n+1} = N_n$ and $P_{n+2} = B_n = P_{n,4}$).

The spline curve is well defined by the $n+4$ poles P_i ($i = -1, \ldots, n+2$). From these poles we can determine the pseudo-Bézier points A_i, M_i, N_i, B_i and deduce the Bézier points $P_{i,j}$, $j = 0, \ldots, 4$, and finally compute $S(t)$, for $t \in [t_i, t_{i+1}]$. Note that at each step, we have an affine combination with positive coefficients (convex combination). Thus we have

$$
S(t) \in \mathrm{co}(P_i).
$$

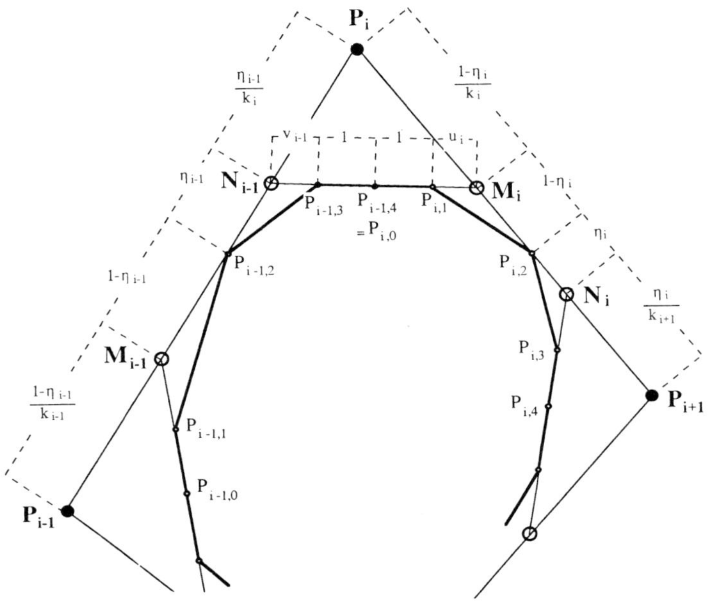

Figure 4. The poles of a Q-spline curve.

More precisely, the pseudo-Bézier points A_i, M_i, N_i, B_i depend only on the 4 poles P_{i-1}, P_i, P_{i+1}, P_{i+2}, and thus $S(t)$ for $t \in [t_i, t_{i+1}]$ depends only on these 4 poles, i.e.

$$S(t) \in \text{co}(P_{i-1}, P_i, P_{i+1}, P_{i+2}), \; i = 0, \ldots, n. \tag{22}$$

Suppose that the P_i's belong to $\mathbb{R}$ $(d = 1)$, $i = -1, \ldots, n+2$, and that they are defined by

$$P_i = \begin{cases} 1, & \text{if } i = k \\ 0, & \text{if } i \neq k. \end{cases}$$

Let us denote by $B_k(t)$, $t \in [a, b]$, the corresponding Q-spline. We will call it the kth q-B-spline. From (22) we have

$$0 \leq B_k(t) \leq 1, \qquad \text{for all } t \in [a, b]. \tag{23}$$

The support of B_k consists of at most 4 consecutive intervals defined by the knot vector (as for classical cubic spline). We have

$$\text{Support}(B_k) = \begin{cases} [a, t_{k+2}], & k = -1, 0, 1 \\ [t_{k-2}, t_{k+2}], & k = 2, \ldots, n-1 \\ [t_{k-2}, b], & k = n, n+1, n+2. \end{cases} \tag{24}$$

Obviously we have

$$S(t) = \sum_{k=-1}^{n+2} B_k(t) P_k, \tag{25}$$

and, in particular, when all P_k are equal to 1,

$$\sum_{k=-1}^{n+2} B_k(t) \equiv 1. \tag{26}$$

The B_k are linearly independent.

In order to give the main result, we need the following two definitions: A triplet (u, v, w) of points in $[a, b]$ will be said to be *admissible* if, defining the new knot vector

$$\begin{aligned} T^* &= T \cup \{u, v, w, \} \\ &= \{t_0^*, t_1^*, \ldots, t_{n^*}^*, t_{n^*+1}^*\} \end{aligned}$$

with $t_i^* < t_{i+1}^*$, $i = 0, \ldots, n^*$ $(n \leq n^* \leq n+3)$, we have

$$\text{co}(\{u, v, w\}) \cap T^* \subset \{u, v, w\}. \tag{27}$$

(This latter inclusion means intuitively that the triplet involves only consecutive points of T^*).

As $S \in C^2[a, b]$, we can define the tangent flats of order 1 or 2 of S at $S(u)$ as in Section 2:

$$\mathcal{T}(u) = \{S(u) + \lambda_1 S'(u), \lambda_1 \in \mathbb{R}\}$$

$$\mathcal{P}(u) = \{S(u) + \lambda_1 S'(u) + \lambda_2 S''(u),\ \lambda_1 \in \mathbb{R},\ \lambda_2 \in \mathbb{R}\}.$$

Then we have the following result:

Theorem. *Suppose*

$$\textit{any 4 consecutive } P_i \textit{ are affinely independent.} \tag{28}$$

Then the function $f : \mathbb{R}^3 \to \mathbb{R}^d$ can be defined for every admissible triplet (u, v, w) in the following way:

- *if u, v, w are distinct:*

$$f(u, v, w) := \mathcal{P}(u) \cap \mathcal{P}(v) \cap \mathcal{P}(w) \tag{29}$$

- *if $u = v \neq w$, $f(u, u, w) = f(u, w, u) = f(w, u, u)$ is defined by:*

$$f(u, u, w) := \mathcal{T}(u) \cap \mathcal{P}(w) \tag{30}$$

- *if $u = v = w$:*

$$f(u, u, u) := S(u) \tag{31}$$

The function f will be called the blossom of S.

Proof: We will need the following property, the proof of which is not given here:

Lemma. *If any 4 consecutive P_i are affinely independent, then the pseudo-Bézier points A_i, M_i, N_i, B_i are also affinely independent, $i = 0, \ldots, n$; and conversely.*

a) *First we define $f(u, v, w)$ when u, v, and w belong to the knot vector T.* As the triplet is admissible, if u, v, w are distinct, we necessarily have $u = t_{i-1}$, $v = t_i$, $w = t_{i+1}$. Now, observing that all the points M_i, $P_{i,2}$, N_i are distinct, strictly between P_i and P_{i+1} and in order (except for $M_0 = P_0$ and $N_n = P_{n+1}$), that $P_{i-1,3}$, $P_{i-1,4} = B_{i-1} = A_i = P_{i,0}$, $P_{i,1}$ are also distinct, strictly between N_{i-1} and M_i and in order, we have (see Fig. 4):

$$\begin{aligned}\mathcal{P}(t_i) &= \operatorname{aff}(P_{i,0}, P_{i,1}, P_{i,2}) = \operatorname{aff}(A_i, M_i, N_i) \\ &= \operatorname{aff}(P_{i-1,2}, P_{i-1,3}, P_{i-1,4}) = \operatorname{aff}(M_{i-1}, N_{i-1}, B_{i-1}) \\ &= \operatorname{aff}(P_{i-1}, P_i, P_{i+1}).\end{aligned} \tag{32}$$

Thus, the intersection (29) is given by

$$\operatorname{aff}(P_{i-2}, P_{i-1}, P_i) \cap \operatorname{aff}(P_{i-1}, P_i, P_{i+1}) \cap \operatorname{aff}(P_i, P_{i+1}, P_{i+2}).$$

As $P_{i-2}, P_{i-1}, P_i, P_{i+1}$ are affinely independent, the intersection of the first two terms is equal to $\operatorname{aff}(P_{i-1}, P_i)$, and as $P_{i-1}, P_i, P_{i+1}, P_{i+2}$ are affinely independent, the intersection of this set with $\operatorname{aff}(P_i, P_{i+1}, P_{i+2})$ is equal to $\{P_i\}$. Thus we have

$$f(t_{i-1}, t_i, t_{i+1}) = \{P_i\}. \tag{33}$$

Similarly, we have

$$\mathcal{T}(t_i) = \operatorname{aff}(N_{i-1}, M_i). \tag{34}$$

Thus, if two parameters are equal, for example $u = v = t_i$, $w = t_{i+1}$, we have

$$\begin{aligned}f(t_i, t_i, t_{i+1}) &= \mathcal{T}(t_i) \cap \mathcal{P}(t_{i+1}) \\ &= \operatorname{aff}(N_{i-1}, M_i) \cap \operatorname{aff}(P_i, P_{i+1}, P_{i+2}) \\ &= \{M_i\}.\end{aligned} \tag{35}$$

In the same way, we have

$$f(t_i, t_{i+1}, t_{i+1}) = \mathcal{P}(t_i) \cap \mathcal{T}(t_{i+1}) = \{N_i\}. \tag{36}$$

b) *Suppose now that the triplet is arbitrary but admissible.*
Consider the new knot vector

$$T^* = T \cup \{u, v, w\} \tag{37}$$

and the space $QS(T^*, q)$. As $T^* \supset T$, we have $QS(T, q) \subset QS(T^*, q)$ and the parametric Q-spline curve

$$S(t) = \sum_{k=-1}^{n+2} B_k(t) P_k \tag{38}$$

can be expressed relatively to this new knot vector T^*:

$$S(t) = \sum_{k=-1}^{n^*+2} B_k^*(t) P_k^* \tag{39}$$

($n \leq n^* \leq n + 3$). Any interval $[t_i^*, t_{i+1}^*]$ is included in some interval defined by the initial knot vector T. As the corresponding pseudo-Bézier points are affinely independent, the pseudo-Bézier points A_i^*, M_i^*, N_i^*, B_i^* are also affinely independent. Note that the dimension of the affine variety spanned by Bézier points (or pseudo-Bézier points) is independent of the interval. Thus using the lemma, we see that any 4 consecutive P_k^* are affinely independent. Thus we can obtain $f(u, v, w)$ as in section a). For example, if $u = t_{i-1}^*$, $v = t_i^*$, $w = t_{i+1}^*$ (distinct) then

$$f(u, v, w) = \mathcal{P}(u) \cap \mathcal{P}(v) \cap \mathcal{P}(w) = P_i^*. \tag{40}$$

If $u = v = t_i^*$ and $w = t_{i+1}^*$ ($\neq u$), then

$$f(u, u, w) = \mathcal{T}(u) \cap \mathcal{P}(w) = M_i^*. \tag{41}$$

The other cases are similar. ■

Properties of f:

The function f is clearly symmetric.
For u, v fixed, $f(u, v, w)$ moves along the line

$$\mathcal{P}(u) \cap \mathcal{P}(v) \quad \text{if} \quad u \neq v, \text{ and } \mathcal{T}(u) \text{ if } u = v \tag{42}$$

but the function $w \to f(u, v, w)$ is non-affine.

For example:

1) If $u = t_i$, $v = t_{i+1}$, the triplet (u, v, w) is admissible for $w \in [t_{i-1}, t_{i+2}]$ and we have

$$f(t_i, t_{i+1}, w) = (1 - \alpha(w)) f(t_i, t_{i+1}, t_{i-1}) + \alpha(w) f(t_i, t_{i+1}, t_{i+2}), \tag{43}$$

where α is piecewise rational and C^1 on $[t_{i-1}, t_{i+2}]$ i.e. is a rational function on each of the 3 sub-intervals and is C^1 at t_i and t_{i+1}.

2) If $u = t_i$ and $t_i < v < t_{i+1}$, then the triplet (u, v, w) is admissible for $w \in [t_{i-1}, t_{i+1}]$, and

$$f(t_i, v, w) = (1 - \beta(w)) f(t_i, v, t_{i-1}) + \beta(w) f(t_i, v, t_{i+1}), \tag{44}$$

where β is piecewise rational and C^1 on $[t_{i-1}, t_{i+1}]$.

Obviously if u, v, w belong to $[t_i, t_{i+1}]$, the function f coincides with the blossom of the polynomial function of degree 4 (satisfying the q-relation) on $[t_i, t_{i+1}]$ as defined in Section 2.

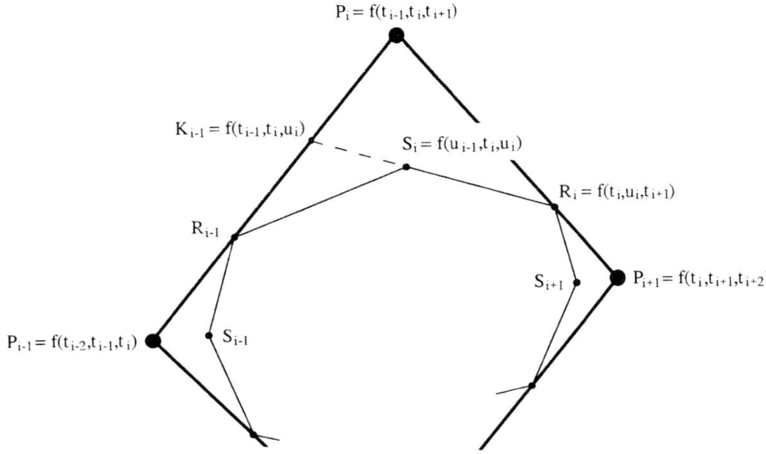

Figure 5. Subdivision algorithm for a Q-spline curve.

If the assumption (28) is not satisfied, it is still possible to define the non-affine blossom with the same properties by considering points $\hat{P}_i$ (eventually in another space) satisfying (28) and such that $P_i = \mathcal{A}(\hat{P}_i)$, where $\mathcal{A}$ is an affine mapping. Then the blossom f is defined by

$$f(u, v, w) = \mathcal{A}(\hat{f}(u, v, w)), \tag{45}$$

where $\hat{f}$ is the blossom associated with the $\hat{P}_i$.

A subdivision algorithm:

This notion of non-affine blossom allows us to define a subdivision algorithm for Q-spline curves.

Suppose that the Q-spline curve S is defined by the poles P_i, $i = -1, \ldots, n+2$. We consider intermediate abscissae $u_i = \frac{t_i + t_{i+1}}{2}$, $i = 0, \ldots, n$ and we denote by U the new knot vector

$$U = \{t_0, u_0, t_1, \ldots, t_n, u_n, t_{n+1}\}. \tag{46}$$

We wish to compute the new poles of S corresponding to U:

$$R_i = f(t_i, u_i, t_{i+1}), \quad i = 0, \ldots, n$$
$$S_i = f(u_{i-1}, t_i, u_i), \quad i = 1, \ldots, n,$$

and also

$$S_0 = f(t_0, t_0, u_0),\ R_{-1} = f(t_0, t_0, t_0) = P_{-1},\ S_{n+1} = f(u_n, t_{n+1}, t_{n+1})$$

and $R_{n+1} = f(t_{n+1}, t_{n+1}, t_{n+1}) = P_{n+2}$.

First, we can compute $R_i = f(t_i, t_{i+1}, u_i)$, as an affine combination of $P_i = f(t_i, t_{i+1}, t_{i-1})$ and of $P_{i+1} = f(t_i, t_{i+1}, t_{i+2})$. In order to compute S_i, we first compute $K_{i-1} = f(t_{i-1}, t_i, u_i)$ as an affine combination of $P_{i-1} = f(t_{i-1}, t_i, t_{i-2})$ and of $P_i = f(t_{i-1}, t_i, t_{i+1})$, and then we can obtain $S_i = f(t_i, u_i, u_{i-1})$ as an affine combination of $K_{i-1} = f(t_i, u_i, t_{i-1})$ and of $R_i = f(t_i, u_i, t_{i+1})$; see Figure 5.

Observe that all triplets for which we computed f are admissible. The algorithm is similar to the classical subdivision algorithm for cubic splines, except that the rates are different.

The notion of Q-spline has been extended to the case where the C^2 continuity at the t_i is replaced by G^2 continuity (cf. [5]). It is clear that a notion of non-affine blossom can also be defined in this case and thus a non-affine subdivision can be developed.

References

1. Barsky, B. A., *Computer Graphics and Geometric Modelling using Beta-splines*, Springer Verlag, 1988.
2. Cinquin, Ph., Splines unidimensionelles sous tension et bidimensionnelles paramétrées : deux applications médicales, Thèse, Université de Saint-Etienne, 28 Octobre, 1981.
3. Foley, T. A., Local control of interval tension using weighted splines, Computer Geometric Aided Design **3** (1986), 281–294.
4. Kulkarni, R., and P. J. Laurent, Pseudo-cubic weighted splines can be C^2 or G^2, in *Curves and Surfaces*, P.-J. Laurent, A. Le Méhauté, and L. L. Schumaker (eds.), Academic Press, New York, 1991, 271–274.
5. Kulkarni, R., and P. J. Laurent, Q-splines, Numerical algorithms **1** (1991), 45–73.
6. Laurent, P. J., *Approximation et Optimisation*, Hermann, Paris, 1972.
7. Ramshaw, L., Blossoming: A connect-the-dots approach to splines, Digital, SRC Techn. Report # 19, June 21, 1987.
8. Seidel, H. P., Universal splines and geometric continuity, in *Curves and Surfaces*, P.-J. Laurent, A. Le Méhauté, and L. L. Schumaker (eds.), Academic Press, New York, 1991, 437–444.
9. Seidel, H. P., Geometric constructions and knot insertion for geometrically continuous spline curves of arbitrary degree, Research Rep. CS, 90, 24, CGL, University of Waterloo; Trans. on Graphics, to appear.
10. Salkauskas, K., C^1 splines for interpolation of rapidly varying data, Rocky Mountain Journal of Mathematics **14** (1) (1984), 239–250.

Rekha Kulkarni
Dept. of Mathematics
Indian Institute of Technology,
Powai, Bombay, 400076, India.
dhruv!marpkia@shakti.ernet.in

Pierre-Jean Laurent
Marie-Laurence Mazure
LMC-IMAG
Université Joseph Fourier,
B. P. 53 X
38041 Grenoble (France)
pjl@imag.fr

On a Class of Data Parametrizations: Variations on a Theme of Epstein, II

E. T. Y. Lee

Abstract. This paper extends the work of Epstein and a previous paper of the author. The main result: In interpolating a planar data sequence by a parametric C^2 cubic spline (with *e.g.*, the *natural* end conditions, or the *periodic* condition for closed data sets), if the knot spacings are proportional to a fixed power of the distances between successive data points, then *corners* cannot appear in the interior of the resulting curve if and only if the exponent lies in $[\frac{1}{2}, 1]$.

§1. Introduction

This paper is a sequel to [4]. We begin by briefly reviewing some of the previous results. Let Q_i, $0 \le i \le n$, be a sequence of points in the plane, consecutive ones being distinct (though the sequence may be closed, *i.e.*, $Q_0 = Q_n$). Suppose one wishes to construct a C^2 cubic spline $P(t) = (x(t), y(t))$ passing through the points. (C^k here always refers to the smoothness with respect to the parameter t.) Usually one interpolates the data at a chosen set of knots $t_0 < \cdots < t_n$ so that $P(t_i) = Q_i$, $i = 0, \ldots, n$. This choice of knots (or *data parametrization*) uniquely determines the spline P when suitable end conditions are prescribed. Two popular parametrizations are the *uniform* ($t_i = i$) and the *chord length* ($t_i - t_{i-1} = |Q_i - Q_{i-1}|$, the usual Euclidean distance between successive points). These are special cases ($s = 0$ and 1) of the simple class

$$\mathcal{P}_s : \quad t_i - t_{i-1} = |Q_i - Q_{i-1}|^s ,$$

where s may be any *real* number. The case $s = \frac{1}{2}$, called the *centripetal* parametrization, has been proposed and studied at some length [3]; generally

Mathematical Methods in Computer Aided Geometric Design II
Tom Lyche and Larry L. Schumaker (eds.), pp. 381–390.

ISBN 0-12-460510-9.

it provides a more satisfactory alternative to either the uniform or the chord length parametrization.

It is of concern what effects the various parametrizations have on the resulting curves; in particular, whether such singularities as corners or cusps might be unwittingly introduced. In [2], Epstein gave an example of a closed data sequence for which the C^2 periodic cubic spline constructed under $\mathcal{P}_0$ contains a *corner* at a data point. (See [4] for constructions, or Proposition 2 below.) The main result of Epstein is that such corners *cannot* occur if the spline is constructed under $\mathcal{P}_1$. A simpler proof of this theorem is given in [4], along with a proof that neither can corners occur under $\mathcal{P}_{1/2}$.

It seems natural to ask what happens with other values of s. (This is not merely an idle exercise, as other values have been experimented with; see [5].) The proofs in [4], however, do not seem extendable. It is the purpose of this paper to give still another proof which does cover all cases (see Theorem 1).

§2. Corner Prohibiting Parametrizations

How can a corner appear in the interior, $t = \tau$, of a parametric C^2 curve? (Interior means $\tau \in (t_0, t_n)$; for a periodic curve, every point is interior.) Clearly we must have $P'(\tau) = 0$. If $P''(\tau) \neq 0$, then the tangent at $P(\tau)$ is parallel to $P''(\tau)$. Thus a corner occurs at $P(\tau)$ only if

$$P'(\tau) = P''(\tau) = 0; \tag{1}$$

and so for a *cubic* spline, it can occur only at a data point where the two adjoining cubic pieces are in fact two line segments meeting at a proper angle. Conversely, if (1) holds *at a data point* and if the two adjoining line segments meet properly, we have a corner. (See [4] for more discussions; if $P'(\tau) = 0$ but $P''(\tau) \neq 0$, one gets a *cusp* or a 'fold'.) We now make an observation which frees us from considering only a specific type of spline.

Proposition 1. *Under any parametrization that depends only on successive chord lengths, the following are equivalent:*

1) *There exists a C^2 cubic interpolating spline P such that $P' = P'' = 0$ at one end, and $P'' = 0$ (or $P' = 0$) at the other end.*
2) *There exists a C^2 cubic interpolating spline P having an interior corner, with $P'' = 0$ (or $P' = 0$) at one of the end points.*
3) *There exists a periodic C^2 cubic interpolating spline having a corner.*

The proof is simple; but to avoid ambiguities we need to first agree on some terminology. Given two curves P_1 on $[0, a]$ and P_2 on $[0, b]$, the curve

$$P(t) = \begin{cases} P_1(t), & 0 \le t \le a \\ P_2(t-a) + P_1(a) - P_2(0), & a \le t \le a+b \end{cases}$$

is said to be obtained by *attaching* P_2 to P_1. (In case $P_2(0) = P_1(a)$, we also say that P is the *concatenation* of P_1 and P_2.) A second elementary construction is *extension by reflection*. Given P on $[0, a]$, and a line ℓ through

$P(a)$, let $\widehat{P}$ be the curve obtained by reflecting P across ℓ. Then the extension $\widetilde{P}$ is obtained by attaching the *reversed* $\widehat{P}$ to P; more precisely,

$$\widetilde{P}(t) = \begin{cases} P(t), & 0 \le t \le a \\ \widehat{P}(2a - t), & a \le t \le 2a. \end{cases}$$

Two special cases are useful. Let $P \in C^2$. (i) Suppose $P''(a) = 0$ but $P'(a) \neq 0$. Let ℓ be the line *normal* to P at $P(a)$. It is clear that $\widehat{P}'(a) = -P'(a)$, so that $\widetilde{P}'(a^+) = -\widehat{P}'(a^-) = P'(a^-) = \widetilde{P}'(a^-)$. Since $P''(a) = 0$, it follows that $\widetilde{P} \in C^2$. (ii) On the other hand, if $P'(a) = 0$ but $P''(a) \neq 0$, then with ℓ the *tangent* line through $P(a)$, we have $\widehat{P}''(a) = P''(a)$, and since $P'(a) = 0$, it again follows that $\widetilde{P} \in C^2$.

Proof of Proposition 1: Let P be defined on $[0, a]$. For 1) $\Rightarrow$ 2), suppose $P'(0) = P''(0) = 0$. Rotate a copy of P about $P(0)$ by a proper angle, reverse it, and attach P to this copy. Then a corner occurs at the juncture. For 2) $\Rightarrow$ 1) or 3) $\Rightarrow$ 1), simply clip at the corner, *i.e.*, consider only the appropriate portion of the curve.

For 1) $\Rightarrow$ 3), suppose $P' = P'' = 0$ at $t = 0$ and $P'(a) \neq 0, P''(a) = 0$. Let $\widetilde{P}$ be the extension of P across the *normal* at $P(a)$. Then $\widetilde{P} \in C^2$. One can then extend $\widetilde{P}$ to $\widetilde{\widetilde{P}}$ by reflecting across the line joining $P(0)$ and $\widetilde{P}(2a)$ to obtain a C^2 periodic curve with two corners at $P(0)$ and $\widetilde{P}(2a)$. In case this should lead to two 'folds' there, one can instead first rotate a copy of $\widetilde{P}$ about $P(0)$, attach it to $\widetilde{P}$ (at $\widetilde{P}(2a)$), then reflect this resulting curve about the line joining its two end points. On the other hand, if $P'(a) = 0, P''(a) \neq 0$ instead, let $\widetilde{P}$ be the extension of P by reflection across the *tangent* at $P(a)$. The rest of the procedure follows as before. (In case $P'(a) = P''(a) = 0$, the construction is even simpler: there is no need for the first extension $\widetilde{P}$.) ■

In the above proof, by an operation on P (*e.g.* a symmetry or a rigid motion) we really mean the same operation on the data sequence of P followed by interpolation. (*Cf.* the construction in Remark 2; more detailed discussions are given in the second paragraph of Remark 3.)

Remark 1. Interior corners may well occur even under $\mathcal{P}_1$, if no restriction at all is placed at either of the end points; see [4] for an example. Note also the emphasis on C^2. Corners in a parametric C^1 curve are easy to come by: If $P'(a) = 0 \neq P''(a)$, just extend P by reflection across any line ℓ through $P(a)$ not in the direction of the tangent or normal. Note that $\widetilde{P}$ is only C^1 at $t = a$ (and the corner is not joined by straight line segments).

Remark 2. As an example of reflections, let a standard Bézier cubic P have control points (0,0), (0,0), (2,0), (3,1). Then $P'(0) = 0 \neq P''(0)$ and as is easily checked, $P'(1) \perp P''(1)$. The latter implies that, after reflecting across the normal at $P(1)$, the extension $\widetilde{P}$ remains C^2 at $t = 1$. Two more reflections across tangents therefore lead to the data sequence

x:	0	3	4	5	8	5	4	3	0
y:	0	1	4	1	0	−1	−4	−1	0

for which the periodic cubic spline interpolant contains four cusps, as shown in Figure 1. Note since the successive data points are equally spaced, any $\mathcal{P}_s$ will lead to the same curve. Thus we have the observation (which, regrettably, I missed in [4]):

Cusps can occur under any $\mathcal{P}_s$.

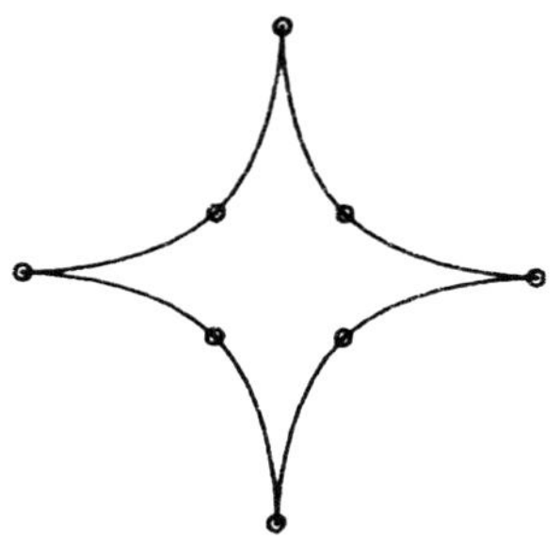

Figure 1. A C^2 periodic cubic spline with cusps.

In view of Proposition 1, we make the following definition, with no explicit reference to corners. The main result of this paper is stated next.

Definition. *$\mathcal{P}_s$ is said to be corner prohibiting provided that for every C^2 cubic interpolating spline P under this parametrization, if $P' = P'' = 0$ at one end point, then neither P' nor P'' can vanish at the other end point.*

Theorem 1. *$\mathcal{P}_s$ is corner prohibiting if and only if $s \in [\frac{1}{2}, 1]$.*

Even though $\mathcal{P}_0$ is not corner prohibiting, in practice corners are rarely seen. This will be made precise in Theorem 2.

§3. Proof of the Theorem

A parametric cubic spline P, interpolating *à la* $\mathcal{P}_s$, can be regarded as a concatenation of cubic Béziers P_i, each restricted to $[0, \delta_i]$:

$$P_i(t) = \sum_{j=0}^{3} B_j^i \binom{3}{j} \left(\frac{t}{\delta_i}\right)^j \left(1 - \frac{t}{\delta_i}\right)^{3-j}, \quad \text{with} \ \ \delta_i = |B_3^i - B_0^i|^s, \quad (i \geq 0).$$

Let $P \in C^2$ with $P_0'(0) = P_0''(0) = 0$. Assume also $P_0(0) = 0$, then

$$B_0^0 = B_1^0 = B_2^0 = 0, \quad B_3^0 = B \neq 0.$$

We will write $P_{i,-}'$ for $P_i'(0)$ and $P_{i,+}'$ for $P_i'(\delta_i)$, *etc.* The proof consists in showing that, because of C^2, whatever extension (by concatenation) of P_0

cannot result in a piece P_i with $P'_{i,+} = 0$ or $P''_{i,+} = 0$, if $s \in [\frac{1}{2}, 1]$; whereas for s outside that range, this can always be done .

For the ith piece, $B_0^i = B_3^{i-1}$, and B_1^i, B_2^i are also determined by the data of the previous piece (see (6) below); thus we are only free to choose B_3^i. It is best to do this and carry out the proof in terms of *complex numbers*. So from now on, all points, vectors, while retaining their previous notations, will be represented as elements in $\mathbb{C}$. Let

$$B_3^i = B_0^i + a_i \xi_i B, \qquad \text{with} \quad a_i > 0, \;\; |\xi_i| = 1 \; (\xi_i \in \mathbb{C}), \tag{2}$$

and $a_0 = \xi_0 = 1$. Thus a_i, ξ_i represent respectively the length and orientation of the ith data segment relative to those of the initial segment. Let

$$\mu_i = \frac{\delta_{i+1}}{\delta_i} = \left(\frac{a_{i+1}}{a_i}\right)^s. \tag{3}$$

Lemma 1. *For all $i \geq 0$, let U_i, V_i be defined by*

$$P'_{i,+} = \frac{3}{\delta_i} U_i B, \qquad P''_{i,+} = \frac{6}{\delta_i^2} V_i B. \tag{4}$$

Then $U_0 = V_0 = 1$ and for $i \geq 1$,

$$\begin{aligned} U_i &= a_i \xi_i - 2\mu_{i-1} U_{i-1} - \mu_{i-1}^2 V_{i-1}\,, \\ V_i &= a_i \xi_i - 3\mu_{i-1} U_{i-1} - 2\mu_{i-1}^2 V_{i-1}\,. \end{aligned} \tag{5}$$

Furthermore, for $i \geq 0$,

$$\begin{aligned} B_1^{i+1} &= B_0^i + (a_i \xi_i + \mu_i U_i) B, \\ B_2^{i+1} &= B_0^i + (a_i \xi_i + 2\mu_i U_i + \mu_i^2 V_i) B. \end{aligned} \tag{6}$$

Proof: Induction. For $i = 0$, (6) is obtained by the C^2 continuity at $P_{0,+}$ (exactly as in the second paragraph below). Now assume the truth for $i - 1$. For (5), use $P'_{i,+} = 3(B_3^i - B_2^i)/\delta_i$ and (2). With B_2^i as given by (6) (induction hypothesis), we get

$$B_3^i - B_2^i = B_0^i + a_i \xi_i B - B_0^{i-1} - (a_{i-1}\xi_{i-1} + 2\mu_{i-1} U_{i-1} + \mu_{i-1}^2 V_{i-1}) B.$$

Replacing B_0^i by $B_3^{i-1} = B_0^{i-1} + a_{i-1}\xi_{i-1} B$ gives (5) for U_i. Similarly for V_i.

For (6), we have $(B_1^{i+1} - B_0^{i+1})/\delta_{i+1} = U_i B/\delta_i$ from $P'_{i+1,-} = P'_{i,+}$; from this we get the first equation of (6) after replacing B_0^{i+1} by $B_3^i = B_0^i + a_i \xi_i B$. Similarly for B_2^{i+1}, using $P''_{i+1,-} = P''_{i,+}$. ■

Proof of Theorem 1: First, we prove that for $s \in [\frac{1}{2}, 1]$,

$$\Re(\,(-1)^i U_i) \geq a_i^s\,, \qquad \Re(\,(-1)^i V_i) \geq a_i^{2s}\,, \tag{7}$$

where $\Re$ denotes the real part. This implies $P'_{i,+} \neq 0$, $P''_{i,+} \neq 0$ by (4) and proves the major half of the theorem.

The proof of (7) is again by induction. From (3) and (5),

$$(-1)^i U_i = (-1)^i a_i \xi_i + 2\Big(\frac{a_i}{a_{i-1}}\Big)^s (-1)^{i-1} U_{i-1} + \Big(\frac{a_i}{a_{i-1}}\Big)^{2s} (-1)^{i-1} V_{i-1}\,;$$

hence by the induction hypothesis,

$$\begin{aligned}\Re(\,(-1)^i U_i) &\geq -a_i + 2\Big(\frac{a_i}{a_{i-1}}\Big)^s a_{i-1}^s + \Big(\frac{a_i}{a_{i-1}}\Big)^{2s} a_{i-1}^{2s} \\ &= a_i^s(2 + a_i^s - a_i^{1-s}).\end{aligned} \tag{8}$$

Similarly

$$\Re(\,(-1)^i V_i) \geq a_i^s(3 + 2a_i^s - a_i^{1-s}). \tag{9}$$

Consider now

$$\phi_1(a) = 2 + a^s - a^{1-s}, \qquad \phi_2(a) = 3 + 2a^s - a^{1-s} \tag{10}$$

defined for $a > 0$. We claim that, for $s \in [\frac{1}{2}, 1]$, $\phi_1(a) > 1$ and $\phi_2(a) > a^s$. This, together with (8) and (9), then proves (7). To see this, note that $0 \leq 1 - s \leq s$, and that therefore if $a \geq 1$, $\phi_1(a) \geq 2$ and if $a < 1$, $\phi_1(a) \geq 1 + a^s$. For ϕ_2, just note that $\phi_2(a) = \phi_1(a) + 1 + a^s$.

It is elementary to show that, on the other hand, for $s \notin [\frac{1}{2}, 1]$, both ϕ_1 and ϕ_2 will have exactly one positive zero. (I leave off the proof, but instead remark that the difference between this and the previous situation is that as $s \searrow 1$ or $s \nearrow \frac{1}{2}$, the location of the zero moves off to infinity.) This fact implies the other half of the theorem: It suffices to consider only two cubic segments P_0, P_1 that are collinear, *i.e.*, $\xi_1 = 1$. Then by (4) and (5),

$$P'_{1,+} = -\frac{3a_1^s}{\delta_1}\phi_1(a_1)B, \qquad P''_{1,+} = -\frac{6a_1^s}{\delta_1^2}\phi_2(a_1)B;$$

and we need only choose a_1 as the positive zero of ϕ_1 to insure $P'_{1,+} = 0$, or as the positive zero of ϕ_2 to make $P''_{1,+}$ vanish. ∎

The last paragraph of the proof can be further explored. With a_1 the positive zero of ϕ_2, we have a line segment through 0, B, $(1 + a_1)B$ as a C^2 cubic spline P with $P' = P'' = 0$ at the origin and $P'' = 0 \neq P'$ at $(1 + a_1)B$. Extending P to $\widetilde{P}$ by reflecting across the normal at $(1 + a_1)B$, we get a line segment with $\widetilde{P}' = \widetilde{P}'' = 0$ at both ends. One can then scale and rotate copies of $\widetilde{P}$ and attach them to fit to any polygon. (Scaling changes derivatives, but does not affect C^2 in the attachments, since each segment has both derivatives zero at both ends.) In particular, we have the somewhat curious result:

Proposition 2. *Any closed n–gon can be represented as a periodic C^2 cubic spline, by interpolating under any $\mathcal{P}_s$ with $s \notin [\frac{1}{2}, 1]$, at $4n+1$ appropriately chosen points.*

As an example, take $s = 0$. From (10), $a_1 = 5$. For a triangle as shown in Figure 2, the appropriate data points for uniform parametrization are:

x :	0	1	6	11	12	12	12	12	12	11	6	1	0
y :	0	0	0	0	0	1	6	11	12	11	6	1	0

In both Figures 1 and 2, the curves are computed from the listed data points by the interpolation routine CUBSPL in [1], suitably modified to allow the periodic end condition. That is, they are not drawn as concatenations of individual cubic pieces already derived (which would be only too easy).

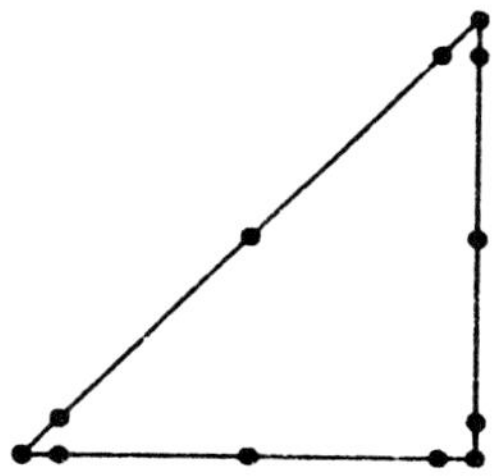

Figure 2. A triangle as a C^2 periodic cubic spline.

Remark 3. In the definition of $\mathcal{P}_s$, the norm used is the ordinary Euclidean (ℓ_2) norm, whereas in Epstein's original result for $\mathcal{P}_1$, it can be *any* norm. It is easy to see that the present proof of Theorem 1 can be extended to any norm that derives from some inner product; but this is not much of an extension. I do not know whether Theorem 1 remains valid (for $s \neq 1$) with any norm.

I should remark that, even if Theorem 1 holds for a general norm, one must be careful as to what it means. In the arguments leading to the equivalences in Proposition 1, the invariance of the interpolation scheme with respect to rigid motions and reflections has been implicitly assumed. This no longer holds with an arbitrary norm. That is, if the entire data sequence is reflected or rotated, the knot spacings determined by $\mathcal{P}_s$ for the new sequence will generally be different from those for the original sequence. Hence the resulting interpolant will not be the same as obtained by reflecting or rotating the original interpolant; and attachments or extensions by reflections (of the data points) will not result in C^2 splines. However, the parts 2) $\Rightarrow$ 1) and 3) $\Rightarrow$ 1) of that proposition clearly remain, hence the validity of Theorem 1 does imply that, for $s \in [\frac{1}{2}, 1]$, no corner of the type described in 2) or 3) can occur. It does *not*, however, guarantee the existence of such a corner if $s \notin [\frac{1}{2}, 1]$.

§4. Uniform Parametrization of Nearly Evenly Spaced Data

For any data sequence $\mathcal{Q} = (Q_i)_0^n$, $Q_i \neq Q_{i+1}$, let

$$\rho_{\mathcal{Q}} = \max_{1\le i\le n} |Q_i - Q_{i-1}| \Big/ \min_{1\le i\le n} |Q_i - Q_{i-1}|$$

be the *data mesh ratio*. This gives a measure of the non–uniformity in the spacings of the data points. For evenly spaced points ($\rho_{\mathcal{Q}} = 1$), all $\mathcal{P}_s$ are identical, hence even under $\mathcal{P}_0$, no corner (of the types described in Proposition 1) can occur. "By continuity", one would expect the same if the data points are nearly evenly spaced (as is generally the case in actual practice).

Theorem 2. *There exists a constant ρ_0 such that, for all $\mathcal{Q}$ with $\rho_{\mathcal{Q}} \le \rho_0$, if P is the C^2 cubic spline interpolating to $\mathcal{Q}$ under $\mathcal{P}_0$, with $P' = P'' = 0$ at one end, then neither P' nor P'' can vanish at the other end.*

The theorem holds with $\rho_0 = 1 + \sqrt{3}$, but to avoid distractions, it will be first proved for a slightly smaller constant $\rho_0 = \frac{5}{2}$.

Proof: With $P_{0,-} = P'_{0,-} = P''_{0,-} = 0$ and $s = 0$, Lemma 1 gives

$$\begin{pmatrix} U_i \\ V_i \end{pmatrix} = a_i\xi_i \begin{pmatrix} 1 \\ 1 \end{pmatrix} - \begin{pmatrix} 2 & 1 \\ 3 & 2 \end{pmatrix} \begin{pmatrix} U_{i-1} \\ V_{i-1} \end{pmatrix} \tag{11}$$

for $i \ge 1$. It follows that

$$\begin{pmatrix} U_i \\ V_i \end{pmatrix} = a_i\xi_i \begin{pmatrix} \alpha_0 \\ \beta_0 \end{pmatrix} - a_{i-1}\xi_{i-1} \begin{pmatrix} \alpha_1 \\ \beta_1 \end{pmatrix} + \cdots + (-1)^i \begin{pmatrix} \alpha_i \\ \beta_i \end{pmatrix}, \tag{12}$$

where $\alpha_0 = \beta_0 = 1$ and for $i \ge 0$

$$\begin{pmatrix} \alpha_{i+1} \\ \beta_{i+1} \end{pmatrix} = \begin{pmatrix} 2 & 1 \\ 3 & 2 \end{pmatrix} \begin{pmatrix} \alpha_i \\ \beta_i \end{pmatrix}. \tag{13}$$

From (12), we have

$$\begin{aligned} |U_i| &\ge \alpha_i - (a_i\alpha_0 + a_{i-1}\alpha_1 + \cdots + a_1\alpha_{i-1}), \\ |V_i| &\ge \beta_i - (a_i\beta_0 + a_{i-1}\beta_1 + \cdots + a_1\beta_{i-1}). \end{aligned} \tag{14}$$

Since $a_i = |B_3^i - B_0^i|/|B| = |Q_i - Q_{i-1}|/|Q_1 - Q_0| \le \rho_{\mathcal{Q}}$, (14) gives

$$|U_i| \ge \alpha_i - \rho_{\mathcal{Q}}\, c_i, \qquad |V_i| \ge \beta_i - \rho_{\mathcal{Q}}\, d_i, \tag{15}$$

with $c_i = \alpha_0 + \cdots + \alpha_{i-1}$ and $d_i = \beta_0 + \cdots + \beta_{i-1}$.

It is clear from (13) that α_i, β_i, c_i, d_i all grow without bound. However, induction will show that, for $i \ge 1$,

$$\frac{\alpha_i}{c_i} > \frac{5}{2}, \qquad \frac{\beta_i}{d_i} > \frac{5}{2}. \tag{16}$$

For we have $\alpha_1 = 3$, $\beta_1 = 5$ by (13) and $c_1 = d_1 = 1$. Also it follows easily from (13) that $\alpha_i/\beta_i \in (\frac{1}{2}, \frac{2}{3})$. Thus (with the induction hypothesis),

$$\frac{\alpha_{i+1}}{c_{i+1}} = \frac{2\alpha_i + \beta_i}{\alpha_i + c_i} = \frac{2 + \frac{\beta_i}{\alpha_i}}{1 + \frac{c_i}{\alpha_i}} > \frac{2 + \frac{3}{2}}{1 + \frac{2}{5}} = \frac{5}{2},$$

and similarly for β_{i+1}/d_{i+1} (using $\alpha_i/\beta_i > \frac{1}{2}$).

Thus, with $\rho_0 = \frac{5}{2}$, (15) shows that $|U_i| > 0$, $|V_i| > 0$ whenever $\rho_Q \le \rho_0$, and this proves the theorem in view of (4) and Lemma 1. ■

That Theorem 2 holds for an improved constant $\rho_0 = 1 + \sqrt{3}$ is seen by replacing (16) by

Lemma 2. *The sequences α_i/c_i and β_i/d_i are strictly decreasing with common limit $\rho_0 = 1 + \sqrt{3}$.*

This follows from a sequence of elementary steps:

Lemma 3. *For all $i \ge 0$, $3\alpha_i^2 - \beta_i^2 = 2$.*

Proof: Use (13). ■

Lemma 4.

$$\begin{aligned} c_i + d_i &= \alpha_i - 1\,, \\ 3c_i + d_i &= \beta_i - 1\,; \end{aligned} \tag{17}$$

hence for $i \ge 1$,

$$c_i = (\beta_i - \alpha_i)/2, \qquad d_i = (3\alpha_i - \beta_i - 2)/2\,. \tag{18}$$

Proof: With M the square matrix in (13), $\binom{c_i}{d_i} = (I + M + \cdots + M^{i-1})\binom{1}{1}$. Hence $(M - I)\binom{c_i}{d_i} = (M^i - I)\binom{1}{1} = \binom{\alpha_i - 1}{\beta_i - 1}$, giving (17). ■

Lemma 5. *α_i/c_i and β_i/d_i are strictly decreasing.*

Proof:

$$\frac{\alpha_i}{c_i} - \frac{\alpha_{i+1}}{c_{i+1}} = \frac{\alpha_i^2 - (\alpha_{i+1} - \alpha_i)c_i}{c_i c_{i+1}},$$

and the numerator is, by (13), (18), and Lemma 3,

$$\alpha_i^2 - (\alpha_i + \beta_i)c_i = \alpha_i^2 - (\alpha_i + \beta_i)(\beta_i - \alpha_i)/2 = (3\alpha_i^2 - \beta_i^2)/2 = 1.$$

Similarly, we find

$$\frac{\beta_i}{d_i} - \frac{\beta_{i+1}}{d_{i+1}} = \frac{(3\alpha_i + \beta_i) - 3}{d_i d_{i+1}} > 0. \quad ■$$

Proof of Lemma 2: Let $\alpha_i/c_i \to \lambda$. Then $d_i/c_i \to \lambda - 1$ by the first equation of (17). Also, Lemma 3 shows that $\beta_i/\alpha_i \to \sqrt{3}$. Hence, again by (17),

$$\frac{1}{\sqrt{3}} \leftarrow \frac{\alpha_i}{\beta_i} = \frac{c_i + d_i + 1}{3c_i + d_i + 1} \to \frac{\lambda}{\lambda + 2},$$

giving $\lambda = 1 + \sqrt{3}$. Furthermore, β_i/d_i converges to the same limit, as

$$\frac{\beta_i}{d_i} = \frac{\beta_i}{\alpha_i}\frac{\alpha_i}{c_i}\frac{c_i}{d_i} \to \frac{\sqrt{3}\lambda}{\lambda - 1} = \lambda. \quad \blacksquare$$

Remark 4. Lemma 2 can also be proved by straightforward computation (though rather tedious in proving the monotone convergence). The matrix M above has eigenvalues $\mu_{1,2} = 2 \pm \sqrt{3}$. By Cayley–Hamilton,

$$M^k = \gamma_k + \eta_k M, \qquad \text{where} \qquad \mu_i^k = \gamma_k + \eta_k \mu_i, \quad i = 1, 2.$$

From these one can compute α_i, β_i, c_i, d_i explicitly.

Remark 5. It is intuitively clear that Theorem 2 should also hold for any $\mathcal{P}_s$, $s \notin [\frac{1}{2}, 1]$. However, the proof here makes use of a significant simplification (11) arising from $s = 0$. For other values of s, I am not able to proceed along similar lines, and so must leave it as an open problem.

References

1. deBoor, C., *A Practical Guide to Splines*, Springer–Verlag, New York, 1978.
2. Epstein, M. P., On the influence of parametrization in parametric interpolation, SIAM J. Numer. Anal. **13** (1976), 261–268.
3. Lee, E. T. Y., Choosing nodes in parametric curve interpolation, Comput. Aided Design **21** (1989), 363–370.
4. Lee, E. T. Y., Corners, cusps, and parametrizations: variations on a theorem of Epstein, SIAM J. Numer. Anal. **29** (1992), to appear.
5. Topfer, H.–J., Models for smooth curve fitting, in *Numerical Methods of Approximation Theory 6*, L. Collatz, G. Meinardus and H. Warner (eds.), ISNM 59, Birkhäuser, Basel, 1982, 209–224.

E. T. Y. Lee
Boeing Commercial Airplane Group
P. O. Box 3707, Mail Stop 6A–WA
Seattle, WA 98124
USA

Wavelets and Image Compression

Bradley J. Lucier

Abstract. In this paper we present certain results about the compression of images using wavelets. We concentrate on the simplest case of the Haar decomposition and compression in L^2. Further results about compression in L^p, $p \neq 2$ are mentioned.

§1. Introduction

Transmitting digital images at video rates requires a tremendous transfer of information. For example:

- If we assume that an image has 512 rows, each of which consists of 512 picture elements, or pixels, then we have $512 \times 512 = 262,144$ pixels.
- If the image is monochrome, and there are 256 different intensity levels with $0 \equiv$ black and $255 \equiv$ white, then each image requires $262,144 \times 8 = 2,097,152$ bits.
- In color, if each of the red, green, and blue components requires a monochrome image, we have $6,291,456$ bits per image.
- Full motion video requires at least 24 frames a second, or $150,994,944$ bits/second.
- High definition television (HDTV) may display four times as much information, requiring $603,979,776$ bits/second.

For comparison, the familiar Ethernet networks commonly used to connect groups of workstations have a peak transfer rate of $10,000,000$ bits/second; common fiber-optic networks have transfer rates of $140,000,000$ bits/second.

Thus, algorithms that compress the information used to represent images are of great interest to engineers. One set of algorithms attempts to recreate at the receiver the precise image sent by the transmitter; these algorithms are known as lossless. Other compression algorithms are willing to lose certain features of the images that will be little noticed by the human observers

Mathematical Methods in Computer Aided Geometric Design II

Tom Lyche and Larry L. Schumaker (eds.), pp. 391–400.

ISBN 0-12-460510-9.

(who, after all, are the consumers of these images) in the hope of gaining higher compression ratios. Wavelet methods of image compression lie within the class of lossy algorithms.

Rather than attempting to give a summary or an outline of the use of wavelet transforms in image compression, we will present almost completely the arguments used to analyze compression in $L^2(\Omega)$ of functions in certain smoothness classes called Besov spaces after applying the well-known Haar transform. This study discusses the basic ideas of wavelet decompositions, compression of wavelet decompositions, error, quantization, and smoothness of images. Perhaps the only thing new here, beyond the presentation, is the observation that the error of compression in $L^2(\Omega)$ does not depend on any unknown constants, mainly due to our choice of the norm used to measure the smoothness of images. Otherwise, the results here are known.

Our particular view of image compression using wavelets is taken from the paper [2]; the survey paper [4] indicates an approach to wavelets that motivates both [2] and this paper. Further information about wavelets can be found in, for example, the book by Meyer [9] and the papers by Mallat [8] and Daubechies [1]. The paper [10] surveys rather broadly the application of wavelets to signal and image processing, including image compression.

§2. Wavelet Decomposition of Images: The Haar Transform

In this section we present a particular way of viewing the well-known Haar transform. We hope that this simple presentation will introduce the reader to the more general wavelet transforms used in image compression.

To be specific, we consider images with 512 rows and 512 columns of pixels, each of which can take an integer grey-scale value ranging from 0 (which represents black) to 255 (which represents white). Such an image requires $2^{18} = 262144$ 8-bit bytes of information to store. With a natural scaling of both the domain and range variables, we consider the image to be a function f that maps the unit square $\Omega := [0,1]^2$ into $[0,1)$. We write

$$f = P_9 f = \sum_{1\times 1\ I} p_I \phi_I, \tag{1}$$

where the sum is over all (1×1 blocks of) pixels I, p_I is the pixel value at I, and ϕ_I is the characteristic function of I, i.e., $\phi_I(x) = 1$ if $x \in I$ and $\phi_I(x) = 0$ otherwise. There are $2^9 \times 2^9$ such pixels; hence the notation P_9.

One can achieve a certain amount of compression simply by averaging the values of the pixels over 2×2 blocks of pixels and storing or transmitting only the block averages rather than the values of the individual pixels. We therefore introduce

$$P_8 f = \sum_{2\times 2\ I} p_I \phi_I,$$

where now p_I is the average of the pixel values on I. (This definition is consistent with what we used in (1).) There are $2^8 \times 2^8$ blocks of 2×2 pixels, which is why we used P_8. This process achieves a compression ratio of 4 to 1, i.e., a mildly distorted version of the image is specified using only one fourth as many numbers as before.

There is no reason to stop this process, and we calculate $P_7 f$, $P_6 f$, all the way down to $P_0 f$, which is specified solely by the average intensity of the pixels over the entire image. We now write the image itself as

$$\begin{aligned} f = P_9 f &= (P_9 f - P_8 f) + (P_8 f - P_7 f) + \cdots + (P_1 f - P_0 f) + P_0 f \\ &= \sum_{1\times 1\, I} d_I \phi_I + \sum_{2\times 2\, I} d_I \phi_I + \cdots + P_0 f \\ &= \sum_I d_I \phi_I + P_0 f, \end{aligned}$$

where the final sum is over all dyadic blocks of pixels. The coefficients d_I are obtained from differences of the approximations at adjacent levels.

We began with 2^{18} pixels p_I, and we now have 4/3 as many coefficients d_I, but this redundancy is easily removed as follows. At the finest level, it is clear that we have subtracted from each pixel value the average over the 2×2 block containing it. Therefore, we organize each set of 4 coefficients d_I into blocks as

$$\boxed{\begin{matrix} a & b \\ c & d \end{matrix}} = \alpha \boxed{\begin{matrix} + & - \\ + & - \end{matrix}} + \beta \boxed{\begin{matrix} - & - \\ + & + \end{matrix}} + \gamma \boxed{\begin{matrix} - & + \\ + & - \end{matrix}}, \tag{2}$$

since $a+b+c+d=0$; here, $+$ denotes $+1$ and $-$ denotes -1. By combining the coefficients d_I and their corresponding characteristic functions ϕ_I into groups of 4, we come up with a new representation using only 3 coefficients and the alternating basis functions indicated by the boxes on the right side of (2). In this way we remove 1/4 of the coefficients, i.e., we have removed the redundancy.

We let the set Ψ consist of the three functions

$$\begin{aligned} \psi^{(1)} &= \phi_{[0,1/2)\times[0,1)} - \phi_{[1/2,1)\times[0,1)}, \\ \psi^{(2)} &= \phi_{[0,1)\times[0,1/2)} - \phi_{[0,1)\times[1/2,1)}, \\ \psi^{(3)} &= \phi_{[0,1/2)\times[0,1/2)} - \phi_{[0,1/2)\times[1/2,1)} - \phi_{[1/2,1)\times[0,1/2)} + \phi_{[1/2,1)\times[1/2,1)}. \end{aligned}$$

For each $\psi \in \Psi$, $k \geq 0$, and $j = (j_1, j_2)$ with $0 \leq j_1 < 2^k$ and $0 \leq j_2 < 2^k$ (we denote this set of j by $\mathbf{Z}^2(k)$), we introduce the function $\psi_{j,k}(x) := 2^k \psi(2^k x - j) = 2^k \psi(2^k(x - j/2^k))$ for $x \in \Omega$. This is the scaled (by 2^k) dyadic dilate (by 2^k) and translate (by $j/2^k$) of ψ, and it has support on the dyadic square $2^{-k}\Omega + j/2^k$. By applying the rewrite rule (2) to the coefficients d_I, we write the image f in terms of the functions $\psi_{j,k}$ as

$$f = \sum_{k\geq 0} \sum_{j\in \mathbf{Z}^2(k)} \sum_{\psi \in \Psi} c_{j,k,\psi} \psi_{j,k} + P_0 f; \tag{3}$$

here $c_{j,k,\psi} = 0$ if $k > 8$. This is the *Haar decomposition* of the image f.

§3. Image Compression in $L^2(\Omega)$

Although we have argued in [2] that one achieves better compressed images when attempting to minimize the error in $L^1(\Omega)$ rather than $L^2(\Omega)$, the L^2 theory is simple enough that a complete presentation can be made in the following paragraphs.

The collection of functions $\{\psi_{j,k} \mid k \geq 0,\ j \in \mathbf{Z}^2(k),\ \psi \in \Psi\}$ is orthogonal. This is easily checked—if the supports of two such functions do not intersect, then their inner product is zero; if the support of one is strictly contained in the support of a second, then the inner product is zero because the integral of each $\psi_{j,k}$ is zero and the second function is constant on the support of the first; and one can check by hand what happens when the supports of two such functions coincide. Because of the scaling by 2^k, each $\psi_{j,k}$ has $L^2(\Omega)$-norm one, i.e., the entire set is orthonormal. Adding the characteristic function ϕ_Ω to this set preserves this orthonormality; the resulting set of functions forms a complete orthonormal basis for $L^2(\Omega)$. Thus, any f for which

$$\|f\|_{L^2(\Omega)} := \left(\int_\Omega |f(x)|^2\,dx\right)^{1/2} < \infty$$

has an expansion of the form (3), with $c_{j,k,\psi} = \int_\Omega f\psi_{j,k}$. Therefore,

$$\begin{aligned}\|f\|^2_{L^2(\Omega)} &= \sum_{k\geq 0}\ \sum_{j\in\mathbf{Z}^2(k)}\ \sum_{\psi\in\Psi} \|c_{j,k,\psi}\psi_{j,k}\|^2_{L^2(\Omega)} + \|p_\Omega\phi_\Omega\|^2_{L^2(\Omega)}\\ &= \sum_{k\geq 0}\ \sum_{j\in\mathbf{Z}^2(k)}\ \sum_{\psi\in\Psi} |c_{j,k,\psi}|^2 + |p_\Omega|^2.\end{aligned}$$

We wish now to find an approximation $\tilde{f}$ to f of the form

$$\tilde{f} = \sum_{(j,k,\psi)\in\Lambda} c_{j,k,\psi}\psi_{j,k} + P_0 f,$$

where Λ is a finite set, say with no more than N elements. We want to do this in a way that minimizes the $L^2(\Omega)$ error,

$$\|f - \tilde{f}\|^2_{L^2(\Omega)} = \sum_{(j,k,\psi)\notin\Lambda} |c_{j,k,\psi}|^2.$$

Because we have an explicit expression for the error, we see immediately that the best choice is to put the triples associated with the N largest values of $|c_{j,k,\psi}|$ into Λ. Thus, we have found an exact, explicit solution to our minimization problem.

This choice of Λ implicitly defines another parameter ϵ, which is

$$\epsilon := \inf_{(j,k,\psi)\in\Lambda} |c_{j,k,\psi}|.$$

One can start with ϵ, and put into Λ only those triples for which $|c_{j,k,\psi}| \geq \epsilon$. This is known as *threshold coding*. The number of terms N in Λ is not known until one is finished, but this method does not require sorting the set $\{c_{j,k,\psi}\}$, which takes more time than to calculate the coefficients themselves! In fact, it is better to calculate *quantized* coefficients

$$\tilde{c}_{j,k,\psi} = \text{round}(\frac{c_{j,k,\psi}}{2\epsilon})\, 2\epsilon,$$

where round(x) is the nearest integer to the real number x. In this way we store or transmit the code round$(c_{j,k,\psi}/(2\epsilon))$ and reconstruct

$$\tilde{f} = \sum_{k\geq 0} \sum_{j\in \mathbf{Z}^2(k)} \sum_{\psi\in\Psi} \tilde{c}_{j,k,\psi}\psi_{j,k} + P_0 f. \tag{4}$$

Two important properties that all these methods of calculating $\tilde{c}_{j,k,\psi}$ satisfy are

$$|c_{j,k,\psi} - \tilde{c}_{j,k,\psi}| \leq \epsilon \tag{5}$$

and

$$\text{if } |c_{j,k,\psi}| < \epsilon, \text{ then } \tilde{c}_{j,k,\psi} = 0. \tag{6}$$

§4. Error, Smoothness, and Quantization

In this section we investigate the connection between the quantization parameter ϵ, the error in our approximation (4), the number of nonzero quantized coefficients $\tilde{c}_{j,k,\psi}$, and the smoothness of the image in suitable function spaces. The results we describe are special cases of results in [5] and [3] about nonlinear approximation using wavelets.

To measure the smoothness of images, we use the Besov spaces $B^\alpha_q(L^p(\Omega))$ for $\alpha > 0$, $0 < p < \infty$, and $0 < q < \infty$. Roughly speaking, these spaces have α derivatives in $L^p(\Omega)$, with the parameter q indicating finer gradations of smoothness. A precise definition of these spaces can be found, for example, in our paper [2]. The scale of spaces $B^\alpha_\tau(L^\tau(\Omega))$, $\alpha > 0$, $1/\tau = \alpha/2 + 1/2$ is important for compression in $L^2(\Omega)$. It is a special case of results in [6] or [7] or ... that one can define the Besov space norm of a function in $B^\alpha_\tau(L^\tau(\Omega))$ in terms of the size of the Haar or wavelet coefficients of that function. In particular, if the $c_{j,k,\psi}$ are given by (3), then one can define

$$\|f\|^\tau_{B^\alpha_\tau(L^\tau(\Omega))} := \sum_{k\geq 0} \sum_{j\in \mathbf{Z}^2(k)} \sum_{\psi\in\Psi} |c_{j,k,\psi}|^\tau + |p_\Omega|^\tau \tag{7}$$

for $0 < \alpha < 1$, i.e., for $1 < \tau < 2$. (The equivalence holds for higher values of α when one uses smoother wavelets, e.g., the wavelets of Daubechies [1].)

Several interesting properties follow immediately from (7). Most importantly, it follows that $B^\alpha_\tau(L^\tau(\Omega))$ is embedded continuously into $L^2(\Omega)$, because

$$\|f\|_{L^2(\Omega)} = \left(\sum_{k\geq 0}\sum_{j\in\mathbf{Z}^2(k)}\sum_{\psi\in\Psi}|c_{j,k,\psi}|^2 + |p_\Omega|^2\right)^{1/2}$$
$$\leq \left(\sum_{k\geq 0}\sum_{j\in\mathbf{Z}^2(k)}\sum_{\psi\in\Psi}|c_{j,k,\psi}|^\tau + |p_\Omega|^\tau\right)^{1/\tau} = \|f\|_{B^\alpha_\tau(L^\tau(\Omega))}.$$

With these preliminaries we can prove the following theorem.

Theorem 1. *Assume that $f \in B^\alpha_\tau(L^\tau(\Omega))$ for some $0 < \alpha < 1$ and $1/\tau = \alpha/2 + 1/2$, and that $c_{j,k,\psi}$ are the coefficients of the Haar decomposition (3). Choose an $\epsilon > 0$, and define $\tilde{c}_{j,k,\psi}$ by any method that satisfies (5) and (6). Let $\tilde{f}$ be given by (4). Then*

(1) *The number N of nonzero coefficients $\tilde{c}_{j,k,\psi}$ satisfies*

$$N \leq \epsilon^{-\tau}\|f\|^\tau_{B^\alpha_\tau(L^\tau(\Omega))}. \tag{8}$$

(2) *The error $f - \tilde{f}$ satisfies*

$$\|f - \tilde{f}\|_{L^2(\Omega)} \leq 2^{1/\tau}N^{-\alpha/2}\|f\|_{B^\alpha_\tau(L^\tau(\Omega))}. \tag{9}$$

Proof: We let $\gamma_{j,k,\psi} := c_{j,k,\psi} - \tilde{c}_{j,k,\psi}$ for $\psi \in \Psi$, $k \geq 0$, and $j \in \mathbf{Z}^2(k)$. Then

$$f - \tilde{f} = \sum_{k\geq 0}\sum_{j\in\mathbf{Z}^2(k)}\sum_{\psi\in\Psi}\gamma_{j,k,\psi}\psi_{j,k},$$

each $|\gamma_{j,k,\psi}| \leq \epsilon$, and either $\tilde{c}_{j,k,\psi} = 0$, in which case $|\gamma_{j,k,\psi}| = |c_{j,k,\psi}|$, or $\tilde{c}_{j,k,\psi} \neq 0$, in which case $|c_{j,k,\psi}| \geq \epsilon \geq |\gamma_{j,k,\psi}|$. Therefore,

$$\sum_{k\geq 0}\sum_{j\in\mathbf{Z}^2(k)}\sum_{\psi\in\Psi}|\gamma_{j,k,\psi}|^\tau \leq \sum_{k\geq 0}\sum_{j\in\mathbf{Z}^2(k)}\sum_{\psi\in\Psi}|c_{j,k,\psi}|^\tau \leq \|f\|^\tau_{B^\alpha_\tau(L^\tau(\Omega))}. \tag{10}$$

We note that for each nonzero $\tilde{c}_{j,k,\psi}$, $|c_{j,k,\psi}| \geq \epsilon$. Therefore, (10) implies that

$$N\epsilon^\tau \leq \|f\|^\tau_{B^\alpha_\tau(L^\tau(\Omega))},$$

and thus (8) follows.

The coefficients $\gamma_{j,k,\psi}$ can be partitioned into sets $\Lambda_1, \ldots, \Lambda_M$, with

$$M \leq \epsilon^{-\tau}\|f\|^\tau_{B^\alpha_\tau(L^\tau(\Omega))},$$

such that

$$\sum_{\gamma_{j,k,\psi}\in\Lambda_n}|\gamma_{j,k,\psi}|^\tau \leq 2\epsilon^\tau, \quad n = 1, \ldots, M. \tag{11}$$

This is accomplished simply by sorting $|\gamma_{j,k,\psi}|$ in decreasing order and adding $\gamma_{j,k,\psi}$ to Λ_1 until the sum in (11) is greater than ϵ^τ. The process is repeated with the remaining coefficients added to Λ_n, $n = 2, \ldots, M$. Because each term is individually less than ϵ^τ, (11) follows. Because each sum is at least ϵ^τ, the bound on M is immediate.

It follows from (11) that if we define

$$f_n := \sum_{\gamma_{j,k,\psi} \in \Lambda_n} \gamma_{j,k,\psi} \psi_{j,k},$$

then

$$\|f_n\|_{L^2(\Omega)} \le \|f_n\|_{B^\alpha_\tau(L^\tau(\Omega))} \le 2^{1/\tau}\epsilon,$$

the set $\{f_n\}$ is orthogonal, and $f - \tilde{f} = \sum_{n=1}^M f_n$. Thus

$$\begin{aligned}\|f - \tilde{f}\|^2_{L^2(\Omega)} &= \sum_{n=1}^{M} \|f_n\|^2_{L^2(\Omega)} \\ &\le \epsilon^{-\tau} \|f\|^\tau_{B^\alpha_\tau(L^\tau(\Omega))} 2^{2/\tau}\epsilon^2 \\ &\le 2^{2/\tau}\epsilon^{\alpha\tau} \|f\|^\tau_{B^\alpha_\tau(L^\tau(\Omega))}\end{aligned}$$

since $2 - \tau = \alpha\tau$. Thus, using our previous bound,

$$\epsilon^{\alpha\tau} \le N^{-\alpha} \|f\|^{\alpha\tau}_{B^\alpha_\tau(L^\tau(\Omega))},$$

and we have

$$\begin{aligned}\|f - \tilde{f}\|^2_{L^2(\Omega)} &\le 2^{2/\tau} N^{-\alpha} \|f\|^{\alpha\tau+\tau}_{B^\alpha_\tau(L^\tau(\Omega))} \\ &= 2^{2/\tau} N^{-\alpha} \|f\|^2_{B^\alpha_\tau(L^\tau(\Omega))}\end{aligned}$$

since $\alpha\tau + \tau = 2$. Taking square roots of both sides yields (9). ■

We remark that the proof given here is a variant of a proof in [2]. Note also that because of the way we expressed the $B^\alpha_\tau(L^\tau(\Omega))$ norm of f, there are no unknown (or unknowable!) constants in (8) and (9).

§5. Compression in $L^p(\Omega)$, $1 < p < \infty$

In this section we outline the results when one tries to minimize the $L^p(\Omega)$ norm of the error instead of the $L^2(\Omega)$ norm of the error.

If $f \in L^p(\Omega)$ for $1 < p < \infty$ we still obtain the Haar decomposition (3) for f. To ensure that an approximation $\tilde{f}$ of the form (4) has small error in $L^p(\Omega)$, one chooses a positive parameter ϵ and coefficients $\tilde{c}_{j,k,\psi}$ that satisfy

$$\|(c_{j,k,\psi} - \tilde{c}_{j,k,\psi})\psi_{j,k}\|_{L^p(\Omega)} \le \epsilon \tag{12}$$

and

$$\text{if } \|c_{j,k,\psi}\psi_{j,k}\|_{L^p(\Omega)} < \epsilon, \text{ then } \tilde{c}_{j,k,\psi} = 0. \tag{13}$$

It is immediately seen that properties (12) and (13) reduce to (5) and (6) when $p = 2$.

When approximating in $L^p(\Omega)$, the scale of Besov spaces $B^\alpha_\tau(L^\tau(\Omega))$, where now $\alpha > 0$ and $1/\tau = \alpha/2 + 1/p$, is most important. Again, one can show that the norm

$$\|f\|^\tau_{B^\alpha_\tau(L^\tau(\Omega))} = \sum_{k\geq 0} \sum_{j\in \mathbf{Z}^2(k)} \sum_{\psi\in\Psi} \|c_{j,k,\psi}\psi_{j,k}\|^\tau_{L^p(\Omega)} + |p_\Omega|^\tau$$

is equivalent to the usual norm for the space $B^\alpha_\tau(L^\tau(\Omega))$ when $0 < \alpha < \min(1, 2/p)$. It is not so obvious that $B^\alpha_\tau(L^\tau(\Omega))$ is embedded in $L^p(\Omega)$, but this is true; see [6]. One can prove the following theorem; see [2] or [3]:

Theorem 2. *Assume that $f \in B^\alpha_\tau(L^\tau(\Omega))$ for some $0 < \alpha < \min(1, 2/p)$ and $1/\tau = \alpha/2+1/p$, and that $c_{j,k,\psi}$ are the coefficients of the Haar decomposition (3). Choose $\epsilon > 0$, and define $\tilde{c}_{j,k,\psi}$ by any method that satisfies (12) and (13). Let $\tilde{f}$ be given by (4). Then there exists a constant C, independent of f, such that*

(1) *The number N of nonzero coefficients $\tilde{c}_{j,k,\psi}$ satisfies*

$$N \leq \epsilon^{-\tau} \|f\|^\tau_{B^\alpha_\tau(L^\tau(\Omega))}.$$

(2) *The error $f - \tilde{f}$ satisfies*

$$\|f - \tilde{f}\|_{L^p(\Omega)} \leq CN^{-\alpha/2} \|f\|_{B^\alpha_\tau(L^\tau(\Omega))}.$$

One can ask whether these methods and estimates are optimal. In an asymptotic sense, the answer is yes. In other words, if for a particular f there is a constant C and best approximations $\tilde{f}_N$, $N > 0$, with at most N nonzero Haar coefficients such that

$$\|f - \tilde{f}_N\|_{L^p(\Omega)} \leq CN^{-\alpha/2}$$

for some $0 < \alpha < \min(1, 2/p)$, then f is in the Besov space $B^\beta_\sigma(L^\sigma(\Omega))$ for all $0 < \beta < \alpha$ and $1/\sigma = \beta/2 + 1/p$. In fact, one has the equivalence

$$\left(\sum_{N=1}^{\infty} [N^{\alpha/2} \|f - \tilde{f}_N\|_{L^p(\Omega)}]^\tau \frac{1}{N} \right)^{1/\tau} < \infty \Longleftrightarrow \|f\|_{B^\alpha_\tau(L^\tau(\Omega))} < \infty \qquad (14)$$

and the left side of (14) is an equivalent norm for $\|f\|_{B^\alpha_\tau(L^\tau(\Omega))}$ (modulo constants); see [3]. Thus, our approximation algorithms are as good as *any* algorithms that use wavelets.

By calculating explicitly $\|\psi_{j,k}\|_{L^p(\Omega)}$, one sees that (12) is equivalent to

$$|c_{j,k,\psi} - \tilde{c}_{j,k,\psi}| 2^{2k(\frac{1}{2} - \frac{1}{p})} < \epsilon.$$

The parameter p determines the relative importance of errors in intensity (given by $2^k|c_{j,k,\psi} - \tilde{c}_{j,k,\psi}|$) and the spatial frequency of the functions $\psi_{j,k}$ (this influence is measured by the term $2^{2k/p}$). Thus, varying p allows one to

Figure 1. The green component of the color image lena.

Figure 2. Lena compressed in $L^1(\Omega)$ at 29.37 to one.

vary the relative importance of *contrast* and *frequency* in the introduction of errors to f to obtain $\tilde{f}$. One can use this flexibility to attempt to match the high-frequency response of the human visual system itself. We claim in [2] that when one follows this approach, one obtains that $p = 1$ roughly matches the high-frequency response of the human visual system.

The attentive reader will note that the above theorems were stated for $1 < p$, and, in fact, the theorems are *false* when using Haar wavelets, or any other orthogonal wavelets, when $p \leq 1$. These algorithms can be modified, however, to use redundant, non-orthogonal representations (using, e.g., the

ϕ_I) to work in $L^p(\Omega)$ for $0 < p \leq 1$; see [2] and [3]. Alternatively, one can work in the Hardy spaces $H^p(\Omega)$.

§6. Examples

Here we present two figures to illustrate the use of the compression algorithms. Figure 1 is the green component of the commonly used color image *lena*. Figure 2 is *lena* compressed using the $L^1(\Omega)$ compression algorithm and a modified Haar transform. The compressed image uses 12,068 nonzero coefficients, which occupy 8,925 bytes; see [2] for complete details. Thus, we have achieved a compression ratio of 29.37 to one.

Acknowledgments. This work was supported in part by National Science Foundation Grant DMS-9006219, Office of Naval Research Contract N00014-91-J-1152, and the Army High Performance Computing Research Center.

References

1. Daubechies, I., Orthonormal bases of compactly supported wavelets, Comm. on Pure and Appl. Math. **XLI** (1988), 909–996.
2. DeVore, R., B. Jawerth, and B. Lucier, Image compression through wavelet transform coding, IEEE Trans. Information Theory, to appear.
3. DeVore, R., B. Jawerth, and V. Popov, Compression of wavelet decompositions, Amer. J. Math., to appear.
4. DeVore, R., and B. Lucier, Wavelets, Acta Numerica, to appear.
5. DeVore, R., and V. Popov, Free multivariate splines, Constr. Approx. **3** (1987), 239–248.
6. DeVore, R., and V. Popov, Interpolation of Besov spaces, Trans. Amer. Math. Soc. **305** (1988), 397–414.
7. Frazier, M., and B. Jawerth, A discrete transform and decompositions of distribution spaces, J. of Functional Analysis **93** (1990), 34–170.
8. Mallat, S., Multiresolution approximations and wavelet orthonormal bases of $L^2(\mathbf{R})$, Trans. Amer. Math. Soc. **315** (1989), 69–87.
9. Meyer, M., *Ondelettes et Opérateurs I: Ondelettes*, Hermann, Paris, 1990.
10. Rioul, O., and M. Vetterli, Wavelets and signal processing, IEEE Signal Proocessing Magazine **8** (1991), 14–38.

Bradley J. Lucier
Department of Mathematics
Purdue University
West Lafayette, IN 47907
lucier@math.purdue.edu

Lower Bounds on the Dimension of Bivariate Spline Spaces and Generic Triangulations

Carla Manni

Abstract. We consider spaces of piecewise polynomials of degree n and smoothness $k < n$, defined over a polygonal partition of a simply connected domain of $\mathbb{R}^2$. We provide a lower bound for the dimension of the space which can exceed Schumaker's lower bound even for generic triangulations if n is "small" compared to k.

§1. Introduction

Let $\Omega \subset \mathbb{R}^2$ be a simply connected polygonal domain and let $\Delta := \{\Omega_i, i = 1, \ldots, \omega\}$ be a partition of Ω. Here and throughout, we shall assume that Δ is a polygonal partition of Ω, i.e. for each i, Ω_i is a polygon.

We are interested in the space of bivariate splines of degree n and smoothness $k, n > k \geq 0$, associated with the given partition:

$$S_n^k(\Omega, \Delta) := \{s : s \in C^k(\Omega), s_{|\Omega_i} \in \mathbb{P}_n, \forall\, \Omega_i \in \Delta\},$$

where $\mathbb{P}_n$ is the $\frac{(n+1)(n+2)}{2}$ dimensional linear space of polynomials of total degree n. This paper deals with the dimension of $S_n^k(\Omega, \Delta)$ (see [2,6,15,16] and references therein).

For general values of n and k and for arbitrary partitions, only lower and upper bounds on the dimension are known: [12,13,14,16]. If Δ is a triangulation, dimension formulae have been established in cases $n \geq 3k + 2$, [8,11], $n = 4, k = 1$, [1], and $n = 3k + 1$ for nondegenerate triangulations, [2].

With regard to partitions which are not necessarily triangulations, formulae for the dimension have been given for quasi-cross-cut partitions [5], and for general rectilinear partitions if $n \geq k - 2 + 2\lceil \frac{k+1}{\eta - 1} \rceil$, [14], or if $n \leq k + \frac{k+1}{D}$, [13], where η, D depend on the structure of Δ. As a general rule, the value

Mathematical Methods in Computer Aided Geometric Design II

Tom Lyche and Larry L. Schumaker (eds.), pp. 401–412.

ISBN 0-12-460510-9.

of the dimension of $S_n^k(\Omega, \Delta)$ not only depends on n, k and the combinatoric properties of the partition, but also on the relative position of the vertices of Δ, [7,10]. In a recent interesting paper, Billera [3] introduced the concept of *generic triangulations* in order to avoid complications due to the geometric dependence; and, via a homological approach, computed the dimension of the spaces $S_n^1(\Omega, \Delta)$ for generic triangulations.

In many cases, [1,2,3,8,11,14], the dimension of $S_n^k(\Omega, \Delta)$ agrees with Schumaker's lower bound [16], so it is reasonable to conjecture as in [12], that this bound gives the exact dimension of $S_n^k(\Omega, \Delta)$ for generic triangulations.

In this paper, by generalizing [13] we present a new lower bound for the dimension and we prove that if n is "small enough" compared to k, its value can exceed the lower bound given in [16], even if Δ is a generic triangulation. This refutes the conjecture in [12].

§2. The Main Result

We introduce some notations. Given a partition Δ, a vertex of Δ is called an *interior vertex* if it is in the interior of Ω. Let $P_i = (x_i, y_i)$, $i = 1, \ldots, V$, be the vertices of Δ ordered so that P_i, $i = 1, \ldots, v < V$, are the interior vertices. For $i = 1, \ldots, v$, let

$$\begin{aligned}
I_i &= \{j : P_j \textit{ is adjacent to } P_i,\ 1 \le j \le V\}, \\
l_{is} &= \textit{oriented edge joining } P_i \textit{ to } P_s, \\
\epsilon_i &= \textit{number of edges of } \Delta \textit{ emanating from } P_i, \\
e_i &= \textit{number of edges of } \Delta \textit{ crossing } P_i \textit{ with different slopes}, \\
E &= \textit{number of interior edges of } \Delta, \\
E_d &= \textit{number of edges joining two interior vertices}, \\
\epsilon &= \textit{number of interior edges of } \Delta \textit{ crossing an interior vertex}, \\
N_c &= \textit{number of cross} - \textit{cuts (i.e. line segments with both endpoints} \\
&\qquad \textit{on } \partial\Omega) \textit{ of } \Delta, \\
N_c^+ &= \textit{number of cross} - \textit{cuts containig at lest one interior vertex}, \\
N_c^i &= \textit{number of cross} - \textit{cuts crossing } P_i, \\
F^i &= \textit{number of rays (i.e. line segments joining an interior vertex} \\
&\qquad \textit{to } \partial\Omega) \textit{ crossing } P_i, \\
\eta_i &= N_c^i + F^i, \\
\alpha &= \frac{(n+1)(n+2)}{2}, \\
\beta &= \frac{(n-k)(n-k+1)}{2}, \\
\phi &= \frac{(k+1)(k+2)}{2},
\end{aligned}$$

We shall prove the following result.

Theorem 1. *Let Δ be a partition of a simply connected polygonal domain $\Omega \subset \mathbb{R}^2$, then*

$$dim\ S_n^k(\Omega, \Delta) \geq \alpha + \beta E - \sigma \tag{1}$$

where

$$\begin{aligned}\sigma &= min(\beta\epsilon - \tau,\ \gamma),\\ \tau &= \sum_{j=1}^{n-k}\sum_{i=1}^{v}(j\eta_i - (k+1+j))_+ + \beta N_c^+,\end{aligned}$$

$$\gamma = \sum_{j=1}^{n-k}\sum_{i=1}^{v} min(k+1+j, je_i) = v(\alpha - \phi) - \sum_{j=1}^{n-k}\sum_{i=1}^{v}(k+1+j-je_i)_+,$$

$$(x)_+ = max(0, x).$$

§3. The Lower Bound

It is well known [4] that an element of $S_n^k(\Omega, \Delta)$ is determined by one polynomial of total degree n and by $2E$ polynomials, $q_{ij} \in \mathbb{P}_{n-k-1}$, which must satisfy the following conformality conditions [5]:

$$\sum_{j \in I_i} [l_{ij}(x, y)]^{k+1} q_{ij}(x, y) \equiv 0, \qquad i = 1, \ldots, v, \tag{2}$$

$$q_{ij} \equiv -q_{ji}, \tag{3}$$

where

$$l_{ij}(x, y) = l_{ji}(x, y) = a_{ij}x + b_{ij}y - (a_{ij}x_i + b_{ij}y_i) = 0,\ a_{ij}^2 + b_{ij}^2 > 0,$$

is the equation of the straight line containing edge l_{ij}.

For $i = 1, \ldots, v$, let us consider the isomorphism:

$$\mathcal{L}_i\ : \mathbb{P}_{n-k-1} \to \mathbb{P}_{n-k-1},$$

$$\mathcal{L}_i := \mathcal{I} + \mathcal{D}_{1,i} + \frac{1}{2!}\mathcal{D}_{2,i} + \ldots + \frac{1}{(n-k-1)!}\mathcal{D}_{n-k-1,i},$$

where $\mathcal{I}$ denotes the identity operator and

$$\begin{aligned}\mathcal{D}_{1,i} &:= x_i\frac{\partial}{\partial x} + y_i\frac{\partial}{\partial y},\\ \mathcal{D}_{j,i} &:= \mathcal{D}_{1,i}\mathcal{D}_{j-1,i}, \qquad j = 2, \ldots .\end{aligned}$$

Putting

$$p_{ij}(x, y) = \mathcal{L}_i q_{ij}(x, y),$$

the linear system (2)-(3) becomes (see [14])

$$\sum_{j\in I_i}[a_{ij}x+b_{ij}y]^{k+1}p_{ij}(x,y)\equiv 0, \qquad i=1,\dots,v, \tag{4}$$

$$\mathcal{L}_j p_{ij}(x,y)+\mathcal{L}_i p_{ji}(x,y)\equiv 0. \tag{5}$$

System (4)-(5) involves $2\beta E$ unknowns, but $\beta(E-E_d)$ of them, associated with the edges emanating from boundary vertices, appear in (5) only, so they are determined by the others explicitly. Then

$$dim\ S_n^k(\Omega,\Delta)=\ \alpha+\beta(E+E_d)-rank\ M, \tag{6}$$

where M is the matrix of the linear system:

$$\sum_{j\in I_i}[a_{ij}x+b_{ij}y]^{k+1}p_{ij}(x,y)\equiv 0, \qquad i=1,\dots,v, \tag{7}$$

$$\mathcal{L}_j p_{ij}(x,y)+\mathcal{L}_i p_{ji}(x,y)\equiv 0, \qquad 1\le i<j\le v. \tag{8}$$

Putting

$$p_{ij}(x,y)=\sum_{r=0}^{n-k-1}p_{ij}^{(r)}(x,y)=\sum_{r=0}^{n-k-1}\sum_{l=0}^{r}a_{rl}x^l y^{r-l},$$

with a suitable arrangement of the equations and unknowns, M has the following structure:

$$M=\begin{pmatrix} M^1 & 0 & & \dots & 0\\ 0 & M^2 & & & 0\\ \vdots & & & & \vdots\\ 0 & \dots & & \dots & M^v\\ & & & & \\ & & L & & \\ & & & & \end{pmatrix}.$$

Each $M^i, i=1,\dots,v$ is the diagonal block matrix of the equations (7) related to the vertex P_i and it contains the influence of the edges emanating from this vertex. To be more precise:

$$M^i=\begin{pmatrix} M^i_{n-k} & 0 & \dots & 0\\ 0 & M^i_{n-k-1} & & 0\\ \vdots & & & \vdots\\ 0 & \dots & \dots & M^i_1 \end{pmatrix},$$

where M^i_r is the $(r+k+1)$ by $r\epsilon_i$ matrix containing the equations (7) involving $p_{ij}^{(r-1)}(x,y), j\in I_i$. To each couple of collinear edges crossing P_i corresponds a couple of blocks of equal columns in M^i_r. Hereafter with the term *columns of M^i_r associated to the edge l_{ij}* we will refer to the columns of M^i_r corresponding to the coefficients of $p_{ij}^{(r-1)}(x,y)$. If $V_{r,t}^i$ is the submatrix of M_r^i consisting of the columns associated to t edges with different slopes crossing P_i, then ([17])

$$rank\ V_{r,t}^i = \ min(k+1+r, rt).$$

The matrix L contains equations (8), so it controls the interaction of the interior vertices. We have:

$$L = \begin{pmatrix} \Lambda_1 \\ \vdots \\ \Lambda_{E_d} \end{pmatrix},$$

where each Λ_s is a matrix with β rows and it contains the equations (8) for two polynomials p_{ir}, p_{ri} associated with an edge joining two interior vertices. If L_r denotes the matrix of $\mathcal{L}_r$ with respect to the basis $\{y^{n-k-1},\ y^{n-k-2}x, \ldots,\ x^{n-k-1}, \ldots,\ y,\ x,\ 1\}$, the only non zero columns in Λ_s are the columns of L_r and L_i and they correspond to the columns of M^i (M^r) related to p_{ir} (p_{ri}). To be more precise the first $(n-k)$ columns of L_r are aligned with the $(n-k)$ columns of M_{n-k}^i associated to $p_{ir}^{(n-k-1)}(x,y)$ and so on.

Lemma 1. *With the notation introduced before:*

$$rank\ M \leq (\epsilon + E_d)\beta - \tau.$$

To prove Lemma 1, it is useful to introduce some additional notations. Once a ray (cross-cut) with endpoints P_t, P_r has been assigned, it will be referred to as the ray (cross-cut) P_tP_r. Let us consider the lexicographical arrangement in $\mathbb{R}^2$ (i.e. $(x_i, y_i) < (x_j, y_j)$, iff $x_i < x_j$ or $x_i = x_j$ and $y_i < y_j$) and let the interior vertices be ordered.

A ray P_tP_r, $P_t \in \overset{\circ}{\Omega}$ is called *progressive (regressive)* if $P_t < P_r$ ($P_t > P_r$). For each edge l_{is}, $i = 1, \ldots, v$, overlying a cross-cut or a ray (see Fig. 1), define

$$\rho_{is} = \begin{cases} 1, \ if\ P_i < P_s\ and\ l_{is}\ overlies\ a\ progressive\ ray \\ 2, \ if\ P_i > P_s\ and\ l_{is}\ overlies\ a\ cross-cut\ or\ a\ regressive\ ray \\ -1, \ if\ P_i > P_s\ and\ l_{is}\ overlies\ a\ progressive\ ray \\ -2, \ if\ P_i < P_s\ and\ l_{is}\ overlies\ a\ cross-cut\ or\ a\ regressive\ ray, \end{cases}$$

$$i = 1, \ldots, v.$$

Proof of Lemma 1: We construct a set C of columns of M on the basis of the following steps ($i = 1, \ldots, v$):

1) select T ($T = min(k+1+j, j\eta_i)$) independent columns in M_j^i, ([17]), $j = 1, \ldots, n-k$, choosing at first all the possible columns associated to the edges l_{is} with $\rho_{is} = 1$, after the ones associated to the edges with $\rho_{is} = 2$, until the amount T is reached.

2) If $\rho_{is} = -1, -2$ and P_s is an interior vertex, choose the β columns associated to l_{is} (j, columns in each M_j^i, $j = 1, \ldots, n-k$).

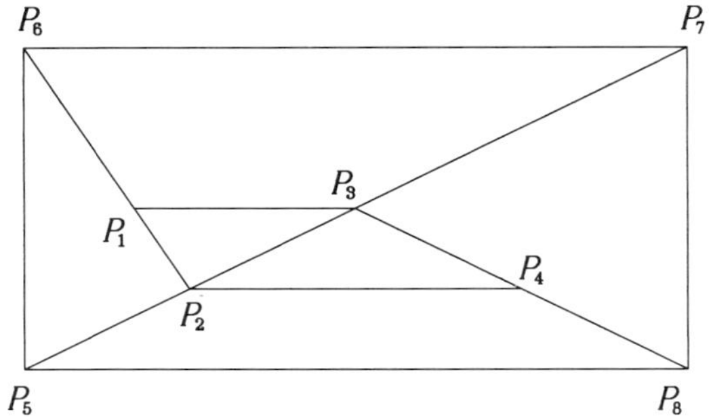

Figure 1. P_3P_8 : progressive rays, P_2P_6 : regressive ray, P_5P_7 : cross-cut, $\rho_{34} = \rho_{48} = 1$, $\rho_{16} = \rho_{21} = \rho_{25} = \rho_{32} = 2, \rho_{43} = -1$, $\rho_{12} = \rho_{23} = \rho_{37} = -2$.

Since for every edge l_{is} with $\rho_{is} = -1$ (-2) there exists one edge l_{ir}, collinear to l_{is}, with $\rho_{ir} = 1$ (2) the sets of columns selected at the steps 1) and 2) are disjoint.

Let $\widehat{M}$ be the submatrix of M consisting of the columns of C and let $\widehat{M}_j^i$, $\widehat{L}_i$, $\widehat{L}$ be respectively the submatrices of M_j^i, L_i, L consisting of the columns which are part of columns in C.

We shall prove that every column in $M \backslash \widehat{M}$ associated to an edge overlying a cross-cut or a ray is linearly dependent on the columns of $\widehat{M}$. Since each column associated to an edge with $\rho_{is} = -1$ is in $\widehat{M}$, we have to investigate three cases.

a) Let us consider a column in $M \backslash \widehat{M}$, α_{is}^t, $1 \le t \le n-k$, associated with an edge l_{is} with $\rho_{is} = 1$. This column has at most two nonzero subcolumns: the first one, $\alpha_{is}^{t,M}$, is a column of M_t^i; the second one, if there is one, is a column of L_s. Because of the ordering chosen in 1), $\alpha_{is}^{t,M}$, is linearly dependent on the columns of $\widehat{M}_t^i$ associated to edges l_{ir} with $\rho_{ir} = 1$. If P_r, P_s are interior vertices, the columns forming this linear combination and α_{is}^t give rise to columns in L which, being associated to edges with $\rho_{ir} = 1$, are linearly dependent on the columns of $\widehat{L}$ associated to the edges l_{ri} with $\rho_{ri} = -1$. These columns have nonzero subcolumns in M_j^r, $P_i < P_r$, $j = 1, \ldots, n-k$. Because of the ordering chosen in 1) each of these columns is linearly dependent on the columns of $\widehat{M}_j^r$ associated the edges l_{rq} with $\rho_{rq} = 1$ so that we can repeat the same construction made before. In this way we can obtain a linear combination consisting of columns of $\widehat{M}$ associated alternatively to edges with $\rho_{rq} = 1$, $\rho_{qr} = -1$, until over each involved ray we reach the vertex adjacent to $\partial\Omega$. At this point the construction of the linear combination ends.

b) Let β_{is}^t, $1 \le t \le n-k$, be a column in $M \setminus \widehat{M}$ associated to an edge l_{is} with $\rho_{is} = 2$. As in a) the subcolumn of β_{is}^t contained in M_t^i is linearly dependent on the columns of $\widehat{M}_t^i$, that is on the columns associated to edges l_{ir} with $\rho_{ir} = 1$ or $\rho_{ir} = 2$. If P_r, P_s are interior vertices, the columns forming this linear combination and β_{is}^t give rise to columns in L.

The columns associated with edges with $\rho_{ir} = 1$ are linearly dependent on the columns of $\widehat{L}$ associated to edges l_{ri} with $\rho_{ri} = -1$, so that for each of them we can construct a linear combination as in a) by using progressive rays only. Conversely, the columns associated to the edges with $\rho_{ir} = 2$ are linearly dependent on the columns of $\widehat{L}$ associated with edges l_{ri} with $\rho_{ri} = -2$. These columns have nonzero subcolumns in M_j^r, $P_i > P_r$, $j = 1, \ldots, n-k$, where we can repeat the same construction previously made. In this way we can obtain a linear combination consisting of columns of $\widehat{M}$ associated to edges with $\rho_{rq} = 1, 2$, $\rho_{qr} = -1, -2$, until over each involved ray or cross-cut we reach the vertex adjacent to $\partial\Omega$.

c) Let γ_{is}^t, $1 \le t \le n-k$, be a column in $M \setminus \widehat{M}$ associated to an edge l_{is} with $\rho_{is} = -2$. Then l_{is} overlies a cross-cut and $P_s \in \partial\Omega$, hence l_{is} is collinear to an edge l_{ib} with $\rho_{ib} = 2$. Therefore in M_t^i there is a column equal to γ_{is}^t and we can repeat the same construction as in b).

Note that in any case the construction of the linear combination proceeds forward over the progressive rays and backward over the regressive rays and cross-cuts so that there are no loops.

Summarizing, the $R = \sum_{j=1}^{n-k} \sum_{i=1}^{v} (j\eta_i - (k+1+j))_+$ columns considered in a) and b), and the βN_c^+ columns considered in c) are linearly dependent on the columns of $\widehat{M}$. Then the lemma is proved by observing that M has $(\epsilon + E_d)\beta$ columns. ■

Since by Schumaker's well-known lower bound ([16])

$$dim\ S_n^k(\Omega, \Delta) \ge \alpha + \beta E - \gamma, \tag{9}$$

Theorem 1 follows from Lemma 1 and (6).

Remark 1. *For every partition if* $v = 1$, $\beta\epsilon - \tau = \gamma$. *Let us now compare the lower bound* $\alpha + \beta E - (\beta\epsilon - \tau)$ *with (9) if* Δ *is a triangulation such that* $\epsilon_i = e_i$, $i = 1, \ldots, v$, $v > 1$. *In* Δ *each cross-cut or ray reduce to only one edge; then* $N_c^+ = 0$. *Cutting* Δ *over each cross-cut we obtain* $N_c + 1$ *subtriangulations,* $\Delta_1, \ldots, \Delta_{N_c+1}$, *such that the values* γ *and* $\beta\epsilon - \tau$ *for* Δ *are the sum of the analogous quantities for* $\Delta_1, \ldots, \Delta_{N_c+1}$. *Hence we assume* $N_c = 0$, *so that* $\epsilon = E$. *For such a triangulation we put*

$$\begin{aligned} V_b &= \{i : i = 1, \ldots, v,\ \eta_i \ge 2\}, \\ V_2 &= \{i : i = 1, \ldots, v,\ \eta_i = 2\}, \\ v_b &= cardinality\ of\ V_b, \\ v_2 &= cardinality\ of\ V_2. \end{aligned}$$

Then

$$\gamma \le v(\alpha - \phi) = \frac{v}{2}(n-k)(n+k+3),$$

$$\tau = \sum_{i \in V_b} \sum_{j=1}^{n-k} (j\eta_i - (k+1+j))_+.$$

By induction on v_b, we obtain

$$V \geq v + \; max(3, 2 + \sum_{i \in V_b} (\eta_i - 2)).$$

Since, $E = 2v + V - 3$, *(cf. [9]),*

$$\beta\epsilon - \tau - \gamma \geq \beta(3v + \; (\sum_{i \in V_b} (\eta_i - 2) - 1)_+) - \tau - \frac{v}{2}(n-k)(n+k+3) =$$

$$= v(n-k)(n-2k) + \beta(\sum_{i \in V_b} (\eta_i - 2) - 1)_+ - \tau.$$

Since $\eta_i \geq 2, \; i \in V_b$, *if* $n \geq 2k$, *we obtain:*

$$\tau \leq \beta \sum_{i \in V_b} (\eta_i - 1) - v_b(k+1)(n-k) + v_2(k+1)\frac{k}{2} + (v_b - v_2)\theta,$$

where

$$\theta = \sum_{j=1}^{n-k} (k + 1 - 2j)_+.$$

Then, since $v \geq v_b$,

$$\beta\epsilon - \tau - \gamma \geq \begin{cases} v_b\beta - v_2(k+1)\frac{k}{2} \; if \; v_b = v_2, \\ (v_b - 1)\beta - v_2(k+1)\frac{k}{2} - (v_b - v_2)\theta \; if \; v_b > v_2, \end{cases}$$

and since $v_b \geq 2$,

$$\beta\epsilon - \tau - \gamma \geq 0 \; if \; \begin{cases} n \geq 2k, \; v_b = v_2, \\ n \geq k - \frac{1}{2} + \sqrt{(k^2 + k)\frac{v_b}{v_b - 1} + \frac{1}{4}}, \; v_b > v_2, \; k \geq 2, \\ n \geq 2, \; k = 1. \end{cases}$$

Remark 2. *If Δ is a quasi-cross-cut partition [5], $\eta_i = e_i, \; i = 1, \ldots, v$ and $\sum_{i=1}^{v} e_i = \epsilon - N_c^+$, then the lower bound (1) agrees with (9), giving the exact dimension of the space (cf. [5]).*

§4. Generic Triangulations

As mentioned in the introduction, the problem of computing the dimension of space $S_n^k(\Omega, \Delta)$ can not be solved for general values of n and k without taking into account the exact position of the vertices of Δ. Therefore in [3] Billera proposed the problem of determining "generically" the dimension of $S_n^k(\Omega, \Delta)$ if Δ is a triangulation To be more precise let us introduce the following definition, [3]:

Definition 1. *Let Δ be a triangulation with V vertices. A property of Δ holds for generic embeddings of Δ in the plane, or holds generically, if there is a nonzero real polynomial p of $2V$ variables such that the property is true for any embedding of Δ in $\mathbb{R}^2$, with vertex locations $P_i = (x_i, y_i)$, $i = 1, \ldots, V$, such that $p(x_1, y_1, \ldots, x_V, y_V) \neq 0$.*

Furthermore Billera proved that $dim\ S^1_n(\Omega, \Delta) = \alpha + \beta E - v(\alpha - \phi)$ for generic triangulations. Observing that this value agrees with the lower bound (9) if there are no collinear edges in each vertex of Δ, in [12] Jia proposed the following conjecture:

Conjecture 1. *If Δ is a generic triangulation then:*

$$dim\ S^k_n(\Omega, \Delta) = \alpha + \beta E - \hat{\gamma}, \tag{10}$$

where

$$\hat{\gamma} = v(\alpha - \phi) - \sum_{j=1}^{n-k} \sum_{i=1}^{v} (k + 1 + j - j\epsilon_i)_+.$$

This conjecture is false if n is "small" compared to k and Δ. We can refute the conjecture observing that if

$$n \leq k + \frac{k+1}{e}, \qquad e + 1 := \min_{1 \leq i \leq v} \epsilon_i,$$

we have

$$\hat{\gamma} = \sum_{j=1}^{n-k} \sum_{i=1}^{v} min(k+1+j, j\epsilon_i) \geq \sum_{j=1}^{n-k} \sum_{i=1}^{v} min(k+1+j, j(e+1)) = \beta(e+1)v.$$

Let us now consider the well-known Morgan-Scott triangulation Δ_{MS} shown in Fig. 2 (cf. [10]). In this case, $e = 3$, $(e+1)v > E$, and hence for $n \leq \frac{4k+1}{3}$, $\hat{\gamma} > \beta E$.

If we consider a sufficiently small perturbation of the position of each vertex of Δ_{MS} (cf. Fig.3), we get an open neighborhood U of $(x_1, y_1, \ldots, x_6, y_6)$ in $\mathbb{R}^{12}$, such that for any embedding, Δ, of Δ_{MS} in $\mathbb{R}^2$ with vertex location in U, the values E, e, e_i, ϵ_i, $i = 1, 2, 3$, do not change. Hence if $n \leq \frac{4k+1}{3}$, from (10) we have the following contradiction

$$dim\ S^k_n(\Omega, \Delta) < \alpha = dim\ \mathbb{P}_n.$$

Since U is an open set of $\mathbb{R}^{12}$ there are no nonzero polynomials p of 12 variables such that $p_{|U} \equiv 0$. Hence (10) does not hold for generic embeddings of Δ_{MS} in the plane.

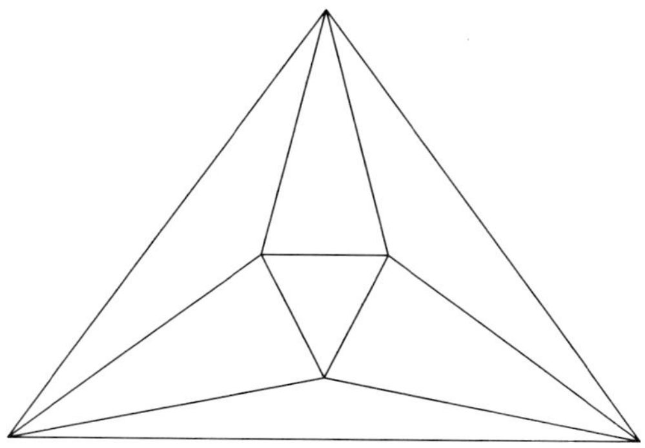

Figure 2. $\Delta_{MS}: \; E = 9, \; e = 3, \; (e+1)v = 12.$

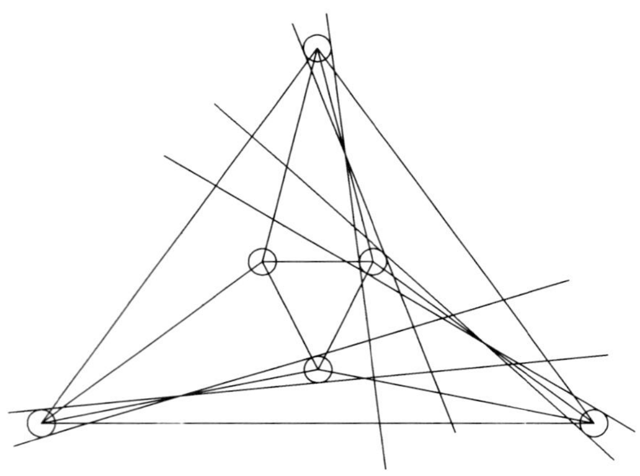

Figure 3. A small perturbation of the position of vertices of Δ_{MS} does not change the values of $E, e, e_i, \epsilon_i, \; i = 1, 2, 3.$

Moreover, Theorem 1 shows that the dimension of $S_n^k(\Omega, \Delta)$ for generic triangulations can exceed the values (10) and (9) even if they are not less than α and each interior edge emanates from an interior vertex.

Let us consider the triangulation Δ_2 in Fig. 4. As in the previous case considering a sufficient small perturbation of the position of each vertex of Δ_2 we obtain embeddings of Δ_2 in the plane having the same values of $E, \; \epsilon, \; v, \; V,$ $\epsilon_i, \; \eta_i, \; i = 1, 2,$ as Δ_2.

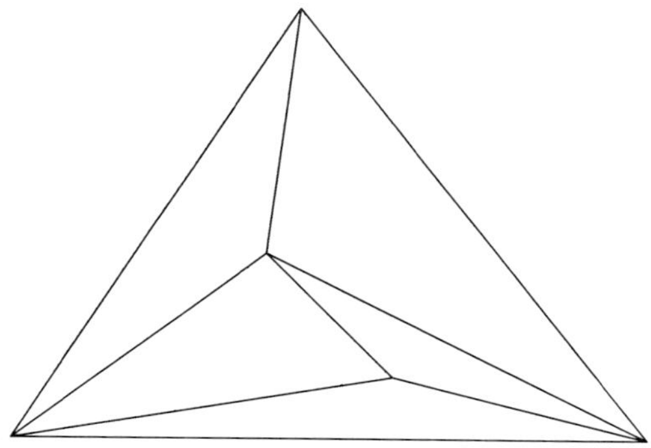

Figure 4. $\Delta_2: \; E=\epsilon=6, \; v=2, \; \epsilon_1=4, \; \epsilon_2=3, \; \eta_1=3, \; \eta_2=2..$

In this case,

$$\hat{\gamma}=\sum_{j=1}^{n-k} min(k+1+j, j4)+\sum_{j=1}^{n-k} min(k+1+j, j3),$$

$$\tau=\sum_{j=1}^{n-k}(3j-(k+1+j))_+ +\sum_{j=1}^{n-k}(2j-(k+1+j))_+.$$

If $\frac{3k+1}{2}<n<2k$ and $k>1$, with a simple manipulation the lower bound (1) exceeds $\alpha+\beta E-\hat{\gamma}$ if

$$n<\frac{3k}{2}+\frac{1}{2}\sqrt{\frac{k^2+2k}{3}}, \; \forall\, k\geq 2,$$

while $\beta E-\hat{\gamma}\geq 0$ if

$$n\geq \frac{3k}{2}+\frac{k+2}{2\sqrt{6}}, \; \forall\, k\geq 2.$$

Example 1. *Consider the space* $S_9^5(\Omega,\Delta_2)$*. Then*

$$\beta E=60, \; \beta\epsilon-\tau=58, \; \gamma=\hat{\gamma}=59,$$

hence there exist generic embeddings, Δ*, of* Δ_2 *in the plane such that:*

$$dim \; S_9^5(\Omega,\Delta)\geq \alpha+2>\alpha+\beta E-\hat{\gamma}.$$

References

1. Alfeld, P., B. Piper, and L. L. Schumaker, An explicit basis for C^1 quartic bivariate splines, SIAM J. Numer. Anal. **24** (1987), 891–911.
2. Alfeld, P., and L. L. Schumaker, On the dimension of bivariate spline spaces of smoothness r and degree $d=3r+1$, Numer. Math. **57** (1990), 651–661.

3. Billera, L. J., Homology of smooth multivariate splines: generic triangulations and a conjecture of Strang, Trans. Amer. Math. Soc. **310** (1988), 325–340
4. Chui, C. K. and R. H. Wang, On smooth multivariate spline functions, Math. Comp. **41** (1983), 131–142.
5. Chui, C. K. and R. H. Wang, Multivariate spline spaces, J. Math. Anal. Appl. **34** (1983), 197–221.
6. Dahmen, W. and C. A. Micchelli, Recent progress on multivariate splines, in *Approximation Theory V*, C. K. Chui, L. L. Schumaker, and J. Ward (eds.), Academic Press, New York, 1983, 27–121.
7. Diener, D., Geometry dependence of the dimension of space of piecewise polynomials on rectilinear partitions, preprint.
8. Dong, H., Spaces of bivariate spline functions over triangulations, Approx. Th. & Its Appl. **7** (1991), 56–75.
9. Ewing, D. J., A. J. Fawkes and J. R. Griffiths, Rules governing the number of nodes and elements in a finite element mesh, Int. J. for Num. Meth. in Eng. **2** (1970), 597–601.
10. Gmelig, R. H. J. and P. R. Pfluger, On the dimension of the space of quadratic C^1 splines in two variables, Approx. Th. & Its Appl. **4** (1988), 37–54.
11. Ibrahim, A. Kh., and L. L. Schumaker, Superspline spaces of of smoothness r and degree $d \geq 3r+2$, Constr. Approx. **7** (1991), 401–423.
12. Jia, R. Q., Lower bounds on the dimension of spaces of bivariate splines, in *Multivariate Approximation and Interpolation*, W. Hausmann and K. Jetter (eds.), Birkauser Verlag, Berlin, 1990, 155–165.
13. Manni, C., On the dimension of bivariate spline spaces over rectilinear partitions, Approx. Th. & its Appl. **7** (1991), 23–34.
14. Manni, C., On the dimension of bivariate spline spaces on generalized quasi-cross-cut partitions, J. Approx. Th. , to appear.
15. Schumaker, L. L., On the dimension of spaces of piecewise polynomials in two variables, in *Multivariate Aproximation Theory*, W. Schempp and K. Zeller (eds.), Birkauser Verlag, Berlin, 1979, 396–412.
16. Schumaker, L. L., Bounds on the dimension of spaces of multivariate piecewise polynomials, Rocky Mt. J. Math. **14** (1984), 251–264.
17. Schumaker, L. L., On spaces of piecewise polynomials in two variables, in *Splines and Aproximation Theory*, Singh et al. (eds.), Dordrecht, Reidel, 1984, 151–197.

Carla Manni
Dip. di Informatica
Universitá di Pisa
Corso Italia 40
56100 Pisa, Italy
bini@icnucevm.cnuce.cnr.it.
font1@ifiidg.bitnet

Geometric Contact of Order p Between Two Surfaces

Marie-Laurence Mazure

Abstract. A notion of geometric contact of order p is defined, leading, as in the case of Frénet-continuity for curves, to a connection matrix with a similar structure.

§1. Introduction

In this paper, we present the notion of F^p-contact between two surfaces at a common point. This notion is introduced from a very intuitive point of view: F^p-contact means that, at the considered common point the two surfaces have the same tangent plane and that, in any direction of this tangent plane, they have similar ways of moving away from their successive tangent subspaces.

We show that F^p-contact is characterized by a connection matrix whose block-structure is similar to that obtained in the case of F^p-continuity for curves. F^p-contact is also compared with G^p-contact (i.e., C^p-contact for a convenient reparametrization).

§2. Frénet-contact of Order p

Consider a surface (S) in $\mathbb{R}^d$, defined by a regular C^p-parametrization X, and a given point $m_0 = X(a,b)$ on (S). The linear subspace spanned by all the partial derivatives of order less or equal to i of X at (a,b) will be denoted by L_i and called the ith linear tangent subspace of (S) at m_0 $(1 \leq i \leq p)$.

Let us fix a unit vector $\vec{v}$ in L_1 (i.e., the linear tangent plane spanned by $\partial_1 X(a,b)$ and $\partial_2 X(a,b)$). We now consider the normal section of (S) in the direction $\vec{v}$, that is to say the curve (c) obtained by intersecting (S) with the affine hyperplane passing through m_0 and the direction of which is the linear

Mathematical Methods in Computer Aided Geometric Design II
Tom Lyche and Larry L. Schumaker (eds.), pp. 413–418.

ISBN 0-12-460510-9.

subspace $L_1^\perp \oplus \mathbb{R}\vec{v}$. We suppose that, near m_0, the arc-length parametrization of (c) with origin at m_0 is defined by:

$$x(s) = X(u(s), v(s)), \tag{1}$$

and that the orientation of (c) is given by $\vec{v}$, which means in particular that

$$x(0) = m_0, \quad x'(0) = \vec{v}. \tag{2}$$

When we move away from m_0 on (S) in direction $\vec{v}$, that is to say along the curve (c), we obtain a point $m = x(s)$, for which s represents the algebraic distance covered on (c) from m_0 up to m. We are interested in evaluating, for s close to zero, the error which is made when confusing m with its orthogonal projection onto the successive affine tangent subspaces $m_0 + L_i$. In other words, we want to calculate the quantities:

$$\vec{e}_i(s) = \overrightarrow{m_0m} - P_{L_{i-1}}(\overrightarrow{m_0m}) = P_{L_{i-1}^\perp}(\overrightarrow{m_0m}), \tag{3}$$

for $2 \leq i \leq p$. Let us use the Taylor expansion of $\overrightarrow{m_0m} = x(s) - x(0)$ up to order i, i.e.:

$$\overrightarrow{m_0m} = sx'(0) + \ldots + \frac{s^{i-1}}{(i-1)!}x^{(i-1)}(0) + \frac{s^i}{i!}x^{(i)}(0) + \vec{o}(s^i). \tag{4}$$

For each $j = 1, \ldots, p$, the jth derivative of x at 0 can be written as follows:

$$\begin{aligned} x^{(j)}(0) &= u^{(j)}(0)\partial_1 X(a,b) + v^{(j)}(0)\partial_2 X(a,b) \\ &\quad + \ldots + \sum_{k=0}^{j} C_j^k u'(0)^{j-k} v'(0)^k \partial_{1^{j-k}2^k} X(a,b) \end{aligned}. \tag{5}$$

Up to order $i-1$, each of these derivatives belongs to L_{i-1}; so, from (3) and (4), we have

$$\vec{e}_i(s) = \frac{s^i}{i!} P_{L_{i-1}^\perp}(x^{(i)}(0)) + \vec{o}(s^i).$$

Moreover, by (5), the previous equality leads to

$$\vec{e}_i(s) = \frac{s^i}{i!} \sum_{k=0}^{i} C_i^k u'(0)^{i-k} v'(0)^k G_{1^{i-k}2^k} + \vec{o}(s^i), \tag{6}$$

where, for $0 \leq k \leq i$

$$G_{1^{i-k}2^k} = P_{L_{i-1}^\perp}\left(\partial_{1^{i-k}2^k} X(a,b)\right). \tag{7}$$

By associating any vector $\ell_1 \partial_1 X(a,b) + \ell_2 \partial_2 X(a,b)$ of L_1 with the quantity

$$\sum_{k=0}^{i} C_i^k \ell_1^{i-k} \ell_2^k G_{1^{i-k}2^k},$$

we define an homogeneous mapping of degree i, denoted by $f_i : L_1 \longrightarrow \mathbb{R}^d$. By (1) and (2), we have

$$\vec{v} = x'(0) = u'(0)\partial_1 X(a,b) + v'(0)\partial_2 X(a,b);$$

thus, the sum in the right hand side of (6) appears to be $f_i(\vec{v})$. Thus, (6) finally provides us the following easy formula:

$$\vec{e}_i(s) = \frac{s^i}{i!} f_i(\vec{v}) + \vec{o}(s^i). \tag{8}$$

Now suppose that $m_0 = \Xi(\alpha, \beta)$, where Ξ is a given regular parametrization of another C^p-surface (Σ) in $\mathbb{R}^d$. Moreover, suppose that, at m_o, (Σ) has the same tangent plane as (S), that is to say, with evident notations, that $\Lambda_1 = L_1$. In the same way as before, we denote by (γ) the normal section of (Σ) in direction $\vec{v}$, and we suppose that the orientation of (γ) is given by $\vec{v}$. If μ represents the point obtained by covering on (γ) the same algebraic distance s from m_0, the error $\vec{\epsilon}_i(s)$ which is made when confusing μ with its orthogonal projection on the $(i-1)$th affine tangent subspace $m_0 + \Lambda_{i-1}$ of (Σ) at m_0 is similarly given by

$$\vec{\epsilon}_i(s) = \frac{s^i}{i!} \phi_i(s) + \vec{o}(s^i). \tag{9}$$

Here $\phi_i : \Lambda_1 = L_1 \longrightarrow \mathbb{R}^d$ is the homogeneous mapping of degree i defined by

$$\phi_i\left(\lambda_1 \partial_1 \Xi(\alpha,\beta) + \lambda_2 \partial_2 \Xi(\alpha,\beta)\right) = \sum_{k=0}^{i} C_i^k \lambda_1^{i-k} \lambda_2^k \Gamma_{1^{i-k}2^k}, \tag{10}$$

with, for $0 \le k \le i$

$$\Gamma_{1^{i-k}2^k} = P_{\Lambda_{i-1}^{\perp}}\left(\partial_{1^{i-k}2^k} \Xi(\alpha,\beta)\right).$$

The notion of geometric contact at m_o we are going to introduce is intuitively "similar ways of moving away from the successive tangent subspaces in any direction of the common tangent plane". More precisely, we give the following definition:

Definition. *Two C^p surfaces (S) and (Σ) in $\mathbb{R}^d$, are said to have a F^p-contact at $m_0 \in (S) \cap (\Sigma)$ if and only if the two following conditions are satisfied:*
(i) (S) and (Σ) have the same tangent plane $\Lambda_1 = L_1$ at m_o,
(ii) whatever the unit vector $\vec{v}$ may be in L_1, for each $i = 2, \ldots, p$,

$$\vec{\epsilon}_i(s) = \vec{e}_i(s) + \vec{o}(s^i). \tag{11}$$

Let us introduce the following notation: for each $i = 1, \ldots, p$, $\partial^i X(a,b)$ will denote the vector-valued column matrix composed of all the partial derivatives of order i of X at (a,b), i.e.

$$\partial^i X(a,b) = \begin{pmatrix} \partial_{1^i} X(a,b) \\ \partial_{1^{i-1}2} X(a,b) \\ \vdots \\ \partial_{2^i} X(a,b) \end{pmatrix}.$$

First, observe that condition (i) means the existence of a regular 2×2 matrix A_1 such that

$$\partial^1 \Xi(\alpha,\beta) = A_1 \cdot \partial^1 X(a,b). \tag{12}$$

From now on we shall suppose that (i) is satisfied. We shall consider a given i $(2 \leq i \leq p)$; for such an integer, we can obviously deduce from (8) and (9) that equality (11) is satisfied if and only if $\phi_i = f_i$. As f_i and ϕ_i are two homogeneous mappings of degree i, this latter equality may as well be replaced by $\beta_i = b_i$, where b_i (resp. β_i): $(L_1)^i = (\Lambda_1)^i \to \mathbb{R}^d$ is the linear blossom of f_i (resp. ϕ_i); in other words, b_i is the unique mapping defined on $(L_1)^i$ which is symmetric, linear with respect to each variable, and whose restriction to the diagonal of $(L_1)^i$ coincides with f_i. As a basis of L_1, we choose $\{\partial_1\Xi(\alpha,\beta), \partial_2\Xi(\alpha,\beta)\}$; then the equality between β_i and b_i is clearly ensured if and only if β_i and b_i coincide on each element

$$(\underbrace{\partial_1\Xi(\alpha,\beta), \ldots, \partial_1\Xi(\alpha,\beta)}_{i-k}, \underbrace{\partial_2\Xi(\alpha,\beta), \ldots, \partial_2\Xi(\alpha,\beta)}_{k})$$

of $(L_1)^i$ $(0 \leq k \leq i)$. Now, the image of such an element by β_i is $\Gamma_{1^{i-k}2^k}$; thus, using (12) and the properties of b_i yields the following characterization:

$$\beta_i = b_i \quad \Longleftrightarrow \begin{pmatrix} \Gamma_{1^i} \\ \Gamma_{1^{i-1}2} \\ \vdots \\ \Gamma_{2^i} \end{pmatrix} = A_1^{[i]} \cdot \begin{pmatrix} G_{1^i} \\ G_{1^{i-1}2} \\ \vdots \\ G_{2^i} \end{pmatrix}, \tag{13}$$

where the symbolic power $A_1^{[i]}$ denotes the $(i+1) \times (i+1)$ matrix constructed from $A_1 = \begin{pmatrix} a_{11} & a_{12} \\ a_{21} & a_{22} \end{pmatrix}$ as follows: for $0 \leq k \leq i$, its $(k+1)$th row is composed of the coefficients of $(a_{11} + a_{12})^{i-k} \times (a_{21} + a_{22})^k$ written in the increasing order with respect to the total degree in (a_{12}, a_{22}).

Finally, we can conclude from (13) that condition (ii) is satisfied if and only if, for each $i = 2, \ldots, p$,

$$\partial^i \Xi(\alpha,\beta) - A_1^{[i]} \cdot \partial^i X(a,b) \in L_{i-1} = \Lambda_{i-1}. \tag{14}$$

We thus obtain the following result.

Theorem 1. *Two C^p-surfaces (S) and (Σ) have a F^p-contact at a common point $m_0 = X(a,b) = \Xi(\alpha,\beta)$ if and only if there exists a regular 2×2 matrix A_1 such that*

$$\begin{pmatrix} \partial^1 \Xi(\alpha,\beta) \\ \partial^2 \Xi(\alpha,\beta) \\ \vdots \\ \partial^p \Xi(\alpha,\beta) \end{pmatrix} = \begin{pmatrix} A_1 & 0 & \cdots & \\ & A_1^{[2]} & \cdots & \\ \vdots & \vdots & \ddots & \vdots \\ & & \cdots & A_1^{[p]} \end{pmatrix} \begin{pmatrix} \partial^1 X(a,b) \\ \partial^2 X(a,b) \\ \vdots \\ \partial^p X(a,b) \end{pmatrix}. \tag{15}$$

§3. Comparison Between F^p-contact and G^p-contact

Again, we consider two C^p-surfaces in $\mathbb{R}^d$, (S) and (Σ), assumed to have a common point $m_0 = X(a,b) = \Xi(\alpha,\beta)$. Another notion of geometric contact of order p at m_0 can be defined between (S) and (Σ) by requiring the existence (at least locally) of reparametrizations which have a C^p-contact at m_0 (i.e., for which all the partial derivatives up to order p are equal at m_0). As it is a natural extension of the well known notion of G^p-continuity for curves, this notion will be called " G^p-contact at m_0".

Just as for curves, the latter definition easily yields the following characterization:

Theorem 2. *Two C^p surfaces (S) and (Σ) have a G^p-contact at a common point $m_0 = X(a,b) = \Xi(\alpha,\beta)$ if and only if there exist a regular 2×2 matrix B_1, and, for each $i = 2, \ldots, p$, a $(i+1) \times 2$ matrix B_i such that:*

$$\begin{pmatrix} \partial^1 \Xi(\alpha,\beta) \\ \partial^2 \Xi(\alpha,\beta) \\ \vdots \\ \partial^p \Xi(\alpha,\beta) \end{pmatrix} = \begin{pmatrix} B_1 & 0 & \cdots & \cdots & 0 \\ B_2 & B_1^{[2]} & \cdots & \cdots & \\ B_3 & B_{32} & B_1^{[3]} & \cdots & \vdots \\ \vdots & \vdots & \vdots & \ddots & \vdots \\ B_p & B_{p2} & \cdots & B_{p,p-1} & B_1^{[p]} \end{pmatrix} \begin{pmatrix} \partial^1 X(a,b) \\ \partial^2 X(a,b) \\ \vdots \\ \partial^p X(a,b) \end{pmatrix},$$

where the B_{ij} $(2 \le j < i \le p)$ can be calculated from $B_i, \ldots, B_p$ by the very relations obtained when iterating the chain rule with respect to two variables.

Now, Theorems 1 and 2 allow us to compare the two notions we have introduced. In the first case, all the $\frac{p(p-1)}{2}$ matrices appearing under the block diagonal $(B_1, B_1^{[2]}, \ldots, B_1^{[p]})$ of the connection matrix are free matrices while, in the second case, only $B_2, \ldots, B_p$ are free. So G^p-contact always implies F^p-contact, and the two notions are equivalent as soon as $p \le 2$.

References

1. Barsky B. A., and T. D. DeRose, Geometric continuity of parametric curves: Three equivalent characterizations, CG& A **9** (1989), 60–68.
2. Barsky B. A., and T. D. DeRose, Geometric continuity of parametric curves: Construction of geometrically continuous splines. I.E.E.E. Computer Graphics and Applications (1990), 60–68.

3. Boehm W., Visual continuity, Computer Aided Design **20** (1988), 307–311.
4. Dyn N., and C. A. Micchelli, Piecewise polynomial spaces and geometric continuity of curves, Num. Math. **54** (1988), 319–337.
5. Farin G., *Curves and Surfaces for Computer Aided Geometric Design*, Academic Press, Boston, 1988.
6. Goldman R. N., and B. A. Barsky, On Beta-continuous functions and their application to the construction of geometrically continuous curves and surfaces, in *Mathematical Methods in Computer Aided Geometric Design*, T. Lyche and L.L. Schumaker (eds.), Academic Press, New York, 1989, 299–311.
7. Goodman T. N. T., Constructing piecewise rational curves with Frenet frame continuity, Computer Aided Geometric Design **7** (1990), 15–31.
8. Gregory J. A., and J. M. Hahn, Geometric continuity and convex combination patches, Computer Aided Geometric Design **4** (1987), 79–89.
9. Gregory J. A., Geometric continuity, in *Mathematical Methods in Computer Aided Geometric Design*, T. Lyche and L.L. Schumaker (eds.), Academic Press, New York, 1989, 353–371.
10. Hagen H., and H. Pottmann, Curvature continuous triangular interpolants, in *Mathematical Methods in Computer Aided Geometric Design*, T. Lyche and L.L. Schumaker (eds.), Academic Press, New York, 1989, 373–384.
11. Hohmeyer M. E., and B. A. Barsky, Rational continuity: Parametric, geometric, and Frénet frame continuity of rational curves, ACM Transactions on Graphics, **8** (1989), 335–359.
12. Hoschek J., and D. Lasser, *Grundlagen der geometrischen Daten-verarbeitung*, Teubner, 1989.
13. Lasser D., and M. Eck, Bézier representation of geometric spline curves; a general concept and the quintic case, Technical report NPS-53-88-004, Department of Mathematics, Naval Postgraduate School, Monterey, CA, 1988.
14. Mazure M. L., Geometric continuity and Frénet continuity of parametric curves, RR 846-M, Mars 1991, IMAG, Université Joseph Fourier, Grenoble.
15. Pottmann H., Projectively invariant classes of geometric continuity. Computer Aided Geometric Design **6** (1989), 307–322.

Marie-Laurence MAZURE
LMC-IMAG
Université Joseph Fourier,
B.P. 53X
38041 GRENOBLE (France)

On Non-Parametric Constrained Interpolation

B. H. Ong and K. Unsworth

Abstract. Given an ordered set of data points lying in a plane on one side of one or more given curves, schemes are presented for generating a non-parametric C^1 planar curve which interpolates the given data points and lies on the same side of these constraint curves as the data points. Both straight lines and quadratic curves are considered as constraint curves. The interpolating curve is represented by a piecewise rational cubic. A default curve is first generated and the curve segments which have crossed over any of the constraint curves are subsequently modified. In the case of a quadratic constraint curve, two methods of modification are discussed, one by changing the weights of the rational cubic segments of the interpolating curve and the other by the addition of new interpolation points. In the case of a straight line constraint, modification may be achieved by just changing the weights. A number of examples are presented.

§1. Introduction

In [2], Goodman et al consider parametric piecewise rational G^2 cubic interpolation of planar data points, in which the data points all lie on one side of one or more given straight lines, and in which the interpolating curve is constrained to lie on the same side of the given lines as these data points.

In this paper, we consider a similar problem for non-parametric C^1 interpolation. In particular, given planar data points and a value for the first derivative at each data point, we shall describe a scheme which interpolates these data with a piecewise rational cubic curve, with the property that both the curve and the data points lie on the same side of the given constraint curve(s). Both straight lines and quadratic curves are considered as constraint curves. We shall consider rational cubics of the form

$$R(x) = \frac{aw_0(1-x)^3 + bw_1(1-x)^2x + cw_2(1-x)x^2 + dw_3x^3}{w_0(1-x)^3 + w_1(1-x)^2x + w_2(1-x)x^2 + w_3x^3}, \quad 0 \le x \le 1, \tag{1.1}$$

Mathematical Methods in Computer Aided Geometric Design II
Tom Lyche and Larry L. Schumaker (eds.), pp. 419–430.

ISBN 0-12-460510-9.

where the weights, w_i, satisfy $w_i > 0, i = 0, \ldots, 3$, thus ensuring a positive denominator. Without loss of generality we assume $w_0 = 1$. In addition we assume that the values of the first derivative at $(0, a)$ and $(1, d)$ are given by D_1 and D_2 respectively. Hence in order for $R(x)$ to interpolate these derivatives we must have

$$b = a + D_1/w_1, \qquad c = d - D_2 w_3/w_2. \tag{1.2}$$

A C^1 curve which interpolates the given data is first constructed by joining each consecutive pair of given data points with a rational cubic curve segment with some initial (default) values for the weights. The curve segments which cross over any of the constraint curves are subsequently modified. In the case of a straight line constraint (§2), the weights may be adjusted in order to modify a curve segment. The case of a quadratic constraint is discussed in §3. Here, modification of a curve segment may also be achieved in certain cases by simply adjusting the weights, however in certain situations it is necessary to add an extra data point with the possibility of subsequent adjustment of the weights. Examples for both types of constraints are presented in §4.

The generation and subsequent modification of any one curve segment may be done locally, i.e. independent of any other segment, hence in the discussions below it suffices to consider just one segment. Moreover we are able to assume that the end points of such a segment are $(0, a), (1, d)$, since the transformation to the unit x-interval is always possible by the use of an appropriate linear change of variable. The end-point derivatives and constraint curve(s) expressions are transformed accordingly.

Central to the discussion that follows is the following theorem, quoted from [2].

Theorem 1. *Let*

$$P(x) = A(1-x)^3 + 3Bx(1-x)^2 + 3Cx^2(1-x) + Dx^3, \quad 0 \le x \le 1,$$

where $A, D > 0$, *and* $B < 0$ *and/or* $C < 0$. *If* $P(x) < 0$ *for some* $x \in (0,1)$ *[resp.* $P(x) = 0$ *for only one point in* $(0,1)$*], then*

$$C^2 > BD, \quad B^2 > AC \tag{1.3}$$

and

$$3B^2C^2 + 6ABCD - 4(AC^3 + B^3D) - A^2D^2 > 0 \quad [resp. = 0] \tag{1.4}$$

Moreover if (1.4) holds, then (1.3) also holds and $P(x) < 0$ *for some* $x \in (0,1)$ *[resp.* $P(x) = 0$ *for only one point in* $(0,1)$*].*

The above theorem will be used to determine whether a curve segment crosses the given constraint curve and to find the values of the new weights so that the modified curve segment just touches it.

§2. ight Line as Constraint Curve

We discuss the ca which the given data points all lie above a given line: similar argume hold for cases in which the data lie below the line.

Suppose e given data points $(0, a)$ and $(1, d)$ lie above the given straight li $mx + k$, $0 \leq x \leq 1$. Then $a - k > 0$ and $d - m - k > 0$, and we the interpolating curve segment $R(x)$ to lie above, or just touc (x): that is, we require

$$R(x) - l(x) \geq 0, \quad 0 \leq x \leq 1.$$

hat the linear expression x can be represented as a rational cubic by the right hand side of (1.1), with

$$a = 0, \qquad b = 1/w_1, \qquad c = (w_1 - 1)/w_2, \qquad d = 1,$$

d (assuming $w_0 = 1$)

$$1 + w_2 = w_1 + w_3. \tag{2.1}$$

Hence in this section, we shall consider a rational cubic $R_l(x)$ of the form (1.1), (1.2) which also satisfies (2.1), and in which $w_1, w_2, 1 + w_2 - w_1 > 0$. We then have

$$R_l(x) - l(x) = \frac{A(1-x)^3 + B(1-x)^2 x + C(1-x)x^2 + Dx^3}{(1-x)^3 + w_1(1-x)^2 x + w_2(1-x)x^2 + (1 + w_2 - w_1)x^3}, \quad 0 \leq x \leq 1,$$

where

$$\begin{aligned} A &= a - k, \\ B &= (a - k)w_1 + D_1 - m, \\ C &= (d - k - D_2)w_2 - (m - D_2)(w_1 - 1), \\ D &= (d - k - m)(1 + w_2 - w_1). \end{aligned} \tag{2.2}$$

Note that $A, D > 0$, since it is assumed that the given data points lie above the given line. Clearly if $B \geq 0$ and $C \geq 0$, then for $0 \leq x \leq 1$, $R_l(x) - l(x) \geq 0$ since the denominator is always positive. But if $B < 0$ and/or $C < 0$ then we can use the criterion as stated in Tand in 1 to determine whether $R_l(x)$, with the default values for the weights w_1 and w_2, crosses the line $l(x)$. If $R_l(x)$ does cross it, then we modify the curve $R_l(x)$ by increasing the weights w_1 and w_2 so that the new $R_l(x)$ just touches the given line. This is always possible by the following observation.

Suppose that the initial weights w_1 and w_2 of $R_l(x)$ are each increased by the addition of a positive quantity v, and we denote the resulting expression by $\tilde{R}_l(x, v)$. Note that by keeping the increase in w_1 and w_2 equal, the weight w_3 will remain unchanged. Clearly $\tilde{R}_l(0, v) = a$ and $\tilde{R}_l(1, v) = d$, and in addition

$$\tilde{R}_l(x, v) - [a(1-x) + dx] = \frac{x(x-1)(\alpha x - a - D_1 + d)}{1 + (w_1 - 3)x + (3 - 2w_1 + w_2)x^2 + vx(1-x)},$$

where

$$\alpha = (a - d)(2 + w_2 - w_1) + D_1 + D_2(1 + \ldots$$

Thus for fixed $x \in [0, 1]$, $|\tilde{R}_l(x, v) - [a(1 - x) + dx]| \to$ … $v \to \infty$, and the set of curve segments generated for differen… another at the two end points and at most one other point x …nically as

The unique v needed so that the modified curve segment …et one the constraint line can be found by solving equation (1.4) for v wh… and D are as in (2.2), but with w_1 and w_2 replaced by $w_1 + v$ an… respectively.

The arguments above extend easily to the case of more than one … straint line. More precisely, suppose $n(> 1)$ constraint lines are given a… the initial rational curve segment crosses more than one of them. Then with respect to each constraint line, l_i, which it crosses, the necessary increase v_i for w_1 and w_2 may be calculated as above. The maximum of all such values is taken to be the amount by which each of the weights w_1 and w_2 should be increased.

An alternative way of modifying a curve segment is to add a knot to the segment. Observe that there are exactly two points of intersection between the rational cubic segment and the given straight line. The added knot could be chosen as a point on the constraint line between these two points of intersection. The derivative at the knot is chosen to be the value of the gradient of the given line. The initial curve segment is replaced by two new curve segments, one which joins the the left data point to the knot while the other joins the knot to the right data point. The initial values of the weights w_1 and w_2 of the two new segments are the same as those of the initial segment. They are subsequently increased or decreased, if necessary, so that each new segment just touches the given line at the knot. We shall not discuss the details here as the idea used is similar to the one discussed later when considering a quadratic constraint curve. However, it should be noted that in this case it is more difficult to extend the method to the case in which there is more than one constraint line, as it is not obvious where a knot (or knots) should be placed.

We should mention at this stage that in [1], Delbourgo and Gregory have considered piecewise rational cubic interpolation with the weights satisfying $w_0 = w_3 = 1$ and $w_1 = w_2$. This class of rational cubics satisfies (2.1). It was shown in [1] that if w_1 is sufficiently large, then a shape preserving curve which preserves both the monotonicity and convexity of the data is obtained. Thus by taking the weights w_1 and w_2 of $R_l(x)$ to be equal, and with w_1 sufficiently large, the additional shape preserving property can be attained.

§3. Quadratic Curve as Constraint Curve

We now assume that the data points lie on one side of a given quadratic curve. It suffices, as before, to consider the case in which the data points lie above this curve. As the scheme described below is also a local scheme, it is still only necessary to consider just one rational cubic segment. We may assume that this segment interpolates two consecutive data points which have x coordinates equal to 0 and 1 respectively.

Now the expression x^2, $0 \le x \le 1$, can be expressed in the form of a rational cubic as given by the right hand side of (1.1) with $a, b = 0, c = 1/w_2, d = 1$ and satisfying conditions (2.1) and

$$3 + w_2 = 2w_1. \tag{3.1}$$

From (2.1) and (3.1) we obtain

$$w_2 = 2w_1 - 3 \quad \text{and} \quad w_3 = w_1 - 2.$$

This leads us to consider the subset of rational cubics $R_q(x)$ which have the form (1.1), (1.2), where (2.1) and (3.1) are both satisfied, and where $w_1 > 2$. That is,

$$R_q(x) = R_q^n(x)/R_q^d(x), \quad 0 \le x \le 1, \tag{3.2}$$

where

$$\begin{aligned} R_q^n(x) =& a(1-x)^3 + (aw_1 + D_1)(1-x)^2 x + \\ &+ [d(2w_1 - 3) - D_2(w_1 - 2)](1-x)x^2 + d(w_1 - 2)x^3, \end{aligned}$$

and

$$R_q^d(x) = (1-x)^3 + w_1(1-x)^2 x + (2w_1 - 3)(1-x)x^2 + (w_1 - 2)x^3.$$

Note that $R_q^d(x)$ reduces to $1 + (w_1 - 3)x$. Writing the quadratic constraint curve as $Q(x) = px^2 + mx + k$, we have

$$R_q(x) - Q(x) = \frac{A(1-x)^3 + B(1-x)^2 x + C(1-x)x^2 + Dx^3}{1 + (w_1 - 3)x}, \quad 0 \le x \le 1,$$

where

$$\begin{aligned} A &= a - k, \\ B &= (a - k)w_1 + D_1 - m, \\ C &= [2(d - k) - D_2 - m]w_1 + 2D_2 + m - p - 3(d - k), \\ D &= (d - k - m - p)(w_1 - 2). \end{aligned} \tag{3.3}$$

As earlier we note that $A, D > 0$. It is useful to observe the limiting behaviour of $R_q(x)$ in each of the cases $w_1 \to 2^+$ and $w_1 \to \infty$. Let $\tilde{R}_q(x, w_1)$ be given by (3.2). Then for $x \in [0, 1]$, as $w_1 \to \infty$,

$$\tilde{R}_q(x, w_1) \to Q_u(x),$$

and as $w_1 \to 2^+$,

$$\tilde{R}_q(x, w_1) \to Q_l(x),$$

where

$$Q_u(x) = a(1-x)^2 + (2d - D_2)(1-x)x + dx^2,$$
$$Q_l(x) = a(1-x)^2 + (2a + D_1)(1-x)x + dx^2,$$

and will be referred to as the "limiting" quadratics. These two quadratics interpolate the data points (0,a) and (1,d) and each reduces to the straight line joining the data points when $D_1 = D_2 = d - a$.

Moreover for $w_1 \geq 2$, $0 \leq x < 1$,

$$\tilde{R}_q(x, w_1) - Q_u(x) = \frac{x(x-1)^2[2(a-d) + D_1 + D_2]}{1 + (w_1 - 3)x}.$$

Thus all of the curve segments, which are generated using different values for $w_1 (> 2)$, lie between the two limiting quadratics $Q_u(x)$ and $Q_l(x)$, and in general they only meet each other at $x = 0$ and $x = 1$. The exception is the case in which $2(a-d) + D_1 + D_2 = 0$, in which case all of the curves coincide.

We now note the conditions for the quadratic expression

$$F(x) = A(1-x)^2 + B(1-x)x + Cx^2, \quad 0 \leq x \leq 1,$$

with $A, C > 0$ but $B < 0$, to be non-negative. It is easy to see that, for $0 \leq x \leq 1$,

$$F(x) > 0 \quad [\text{resp.} \quad F(x) \geq 0 \text{ and } F(x) = 0 \text{ for only one } x \in (0,1)],$$

$$\textit{if and only if} \quad 4AC - B^2 > 0 \quad [\text{resp.} \quad 4AC - B^2 = 0].$$

We can use this result to decide whether the limiting quadratics cross the given quadratic $Q(x)$ by considering the difference between $Q_u(x)$ and $Q(x)$ and the difference between $Q_l(x)$ and $Q(x)$.

The rational cubic segment, as given by (3.2) with a default value for w_1, can be tested to see if it crosses the given quadratic $Q(x)$ by using the criterion specified in Theorem 1 on the numerator of (3.2). Suppose the curve segment crosses the given quadratic curve. Then we would like to modify this rational cubic segment so that it just touches the constraint curve either by just changing the value of the weight w_1, or by adding a knot when changing the weight fails to "rectify" the curve.

As noted above, all of the curve segments which are generated using different values for w_1 lie between the two limiting quadratics. Thus the behaviour of the limiting quadratics with respect to the given quadratic constraint curve determines when it is possible to "rectify" the curve by just changing the weight w_1. More precisely, if the limiting quadratic $Q_u(x)$ does not cross the given quadratic, then the curve segment can be rectified by increasing w_1; if the limiting quadratic $Q_l(x)$ does not cross the given quadratic, then the curve segment can be rectified by decreasing w_1. In each case, the new unique value of w_1 which allows the resulting modified curve segment to just touch the given constraint curve, is found by solving (1.4) for w_1 with A, B, C, D given by (3.3).

We shall now assume that the two limiting quadratics are not identically equal, but that they both cross the given quadratic. In this case, an extra data point (knot) is added and a curve segment which crosses the given quadratic curve is replaced by two new curve segments, each connecting the knot to one of the original end points. The location of this new data point must now be calculated.

Suppose $Q_u(x)$ and $Q_l(x)$ intersect the given quadratic at $x = u_1, u_2$ and $x = l_1, l_2$ respectively, where $u_1 < u_2$ and $l_1 < l_2$. Then either $[l_1, l_2] \subset [u_1, u_2]$ or vice versa. The knot is then chosen to lie on the constraint curve and located at $x = x_0$ where x_0 is the mid-point of the smaller of the two intervals $[l_1, l_2]$ and $[u_1, u_2]$. The derivative at the knot is chosen to be that of the constraint curve at x_0, based upon the observation that when any two smooth curves touch, then at the point of contact the derivatives of the two curves are equal. The choice of the location of the knot stems from the the observation that in the process of "lifting" the rational curve segment (which initially crosses the given quadratic) so that it just touches the constraint curve, the two points of intersection eventually coalesce into just one point of contact. In addition this choice of derivative and knot location eases any further curve modification, since we can rectify the curve by adjusting, if necessary, the weight w_1 of each of the new segments.

Consider for example the new segment which extends from $(0, a)$ to the knot. Suppose that this segment, with the default value for w_1, crosses the given quadratic curve. We shall modify it by changing w_1. With an appropriate linear transformation we may assume, as before, that the representation $R_1(x)$ of this segment be defined over the interval $[0, 1]$ instead of $[0, x_0]$. With respect to this transformation, suppose the constraint curve becomes $Q_1(x) = p_1x^2 + m_1x + k_1$, and let the transformed derivative, corresponding to the derivative D_1 at the left data point, be $\tilde{D}_1$. Then

$$R_1(x) - Q_1(x) = \frac{a_1(1-x)^3 + b_1(1-x)^2x + c_1(1-x)x^2 + d_1x^3}{1 + (w_1 - 3)x}, \quad 0 \le x \le 1,$$

and by our choice of knot location, and derivative value at the knot, we obtain

$$\begin{aligned} c_1 &= d_1 = 0, \\ a_1 &= a - k_1, \\ b_1 &= (a - k_1)w_1 + \tilde{D}_1 - m_1, \end{aligned}$$

where $a_1 > 0$, since the given data point lies above the constraint curve. It is easy to see that for $0 \le x \le 1$, $R_1(x) - Q_1(x) \ge 0$ if and only if $b_1 \ge 0$. Hence w_1 can be increased until $b_1 = 0$.

Similarly, consider the new segment extending from the knot to $(1, d)$. Suppose, again using a suitable linear transformation, that $R_2(x)$ represents this rational cubic segment, with $\tilde{D}_2$ representing the transformed derivative corresponding to the derivative D_2 at the right data point, and with the constraint curve given by $Q_2(x) = p_2x^2 + m_2x + k_2$, $0 \le x \le 1$. Then

$$R_2(x) - Q_2(x) = \frac{a_2(1-x)^3 + b_2(1-x)^2x + c_2(1-x)x^2 + d_2x^3}{1 + (w_1 - 3)x}, \quad 0 \le x \le 1,$$

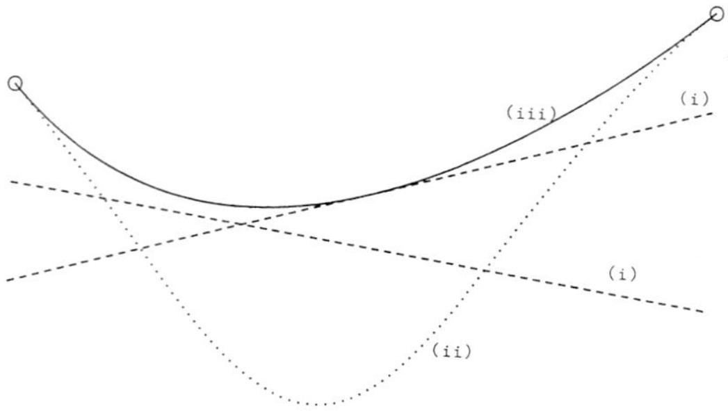

Table 1 for example 1

x	0.0	1.0
y	1.0	1.5

First derivatives: -4.0, 3.0.

Equations of lines :

$$y = x$$
$$y = -0.5x + 0.5$$

Figure 1. Default curve crosses two constraint lines.

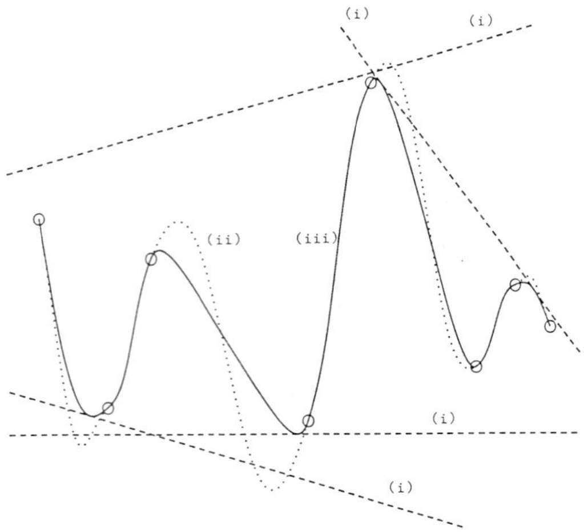

First derivatives : default.

Table 2 for example 2

x	-2.5	-1.0	0.0	3.5	5.0	7.3	8.2	9.0
y	1.0	-0.4	0.7	-0.5	2.0	-0.1	0.5	0.2

Equations of lines :

$$y = -0.6$$
$$y = -0.1x - 0.6$$
$$y = -0.45x + 4.35$$
$$y = 0.09x + 1.62$$

Figure 2. Case of four constraint lines.

where

$$\begin{aligned}
a_2 &= b_2 = 0, \\
d_2 &= d - k_2 - m_2 - p_2, \\
c_2 &= [2(d - k_2) - \tilde{D}_2 - m_2]w_1 + 2\tilde{D}_2 + m_2 - p_2 - 3(d - k_2),
\end{aligned}$$

where $d_2 > 0$, since the given data point lies above the constraint curve. Again for $0 \le x \le 1$, $R_2(x) - Q_2(x) \ge 0$ if and only if $c_2 \ge 0$, and w_1 can be either increased or decreased as necessary until $c_2 = 0$. Note that in the case in which the value of w_1 is decreased the new value of w_1 remains larger than 2. For, suppose $2(d-k_2) - \tilde{D}_2 - m_2 > 0$, then w_1 may be increased until $c_2 = 0$. However, if $2(d - k_2) - \tilde{D}_2 - m_2 < 0$, then

$$\begin{aligned}
2\tilde{D}_2 + m_2 - p_2 - 3(d - k_2) &> 2\tilde{D}_2 + m_2 + k_2 + m_2 - d - 3(d - k_2) \\
&= -2[2(d - k_2) - \tilde{D}_2 - m_2] \\
&> 0.
\end{aligned}$$

Thus w_1 may be decreased until $c_2 = 0$, and the new value for w_1 is

$$[2\tilde{D}_2 + m_2 - p_2 - 3(d - k_2)]/[m_2 + \tilde{D}_2 - 2(d - k_2)]$$

which is greater than 2. Finally, if $2(d - k_2) - \tilde{D}_2 - m_2 = 0$, then

$$2\tilde{D}_2 + m_2 - p_2 - 3(d - k_2) = d - k_2 - m_2 - p_2 > 0.$$

Then $c_2 > 0$ and no modification to this right segment is necessary.

§5. Numerical Results

We shall now describe some numerical examples for the scheme discussed above. The following notation has been adopted for the figures. The given data points are indicated by '0', the constraint curves are drawn with bold broken lines and marked as '(i)', the initial (default) curves are drawn in light broken lines and marked as '(ii)', the modified curves in full lines and marked as '(iii)', the limiting quadratics $Q_l(x)$ and $Q_u(x)$ in light dotted lines with the former ones being marked as '(iv)' and the latter ones marked as '(v)'.

The first example (Figure 1) comprises one default curve segment which crosses both constraint lines. The increase, v, in the weights w_1 and w_2 is calculated with respect to each of the given lines and the larger value for v is chosen. Hence the modified curve touches one line and stays above the other. Figure 2 shows a default curve with four curve segments crossing one or more of the four constraint lines. The derivative chosen at each of the data points is calculated by using the arithmetic mean of the adjacent chord gradients.

In the remaining examples the constraint curves are quadratic curves. Figure 3 shows an initial curve segment which crosses the constraint curve. The limiting quadratic curve $Q_l(x)$ also crosses the constraint curve but the

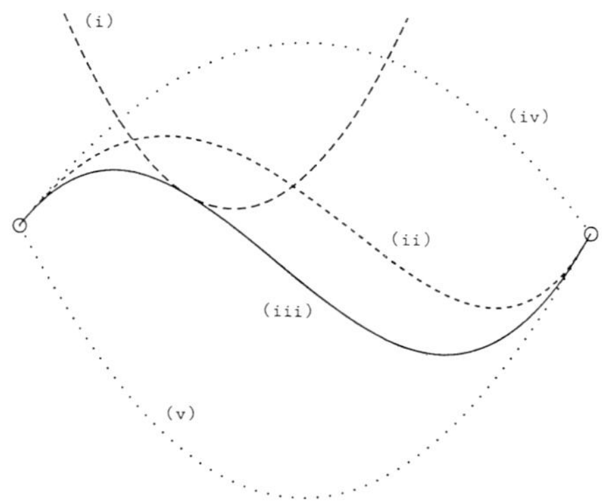

Table 3 for example 3

x	0.0	1.0
y	0.0	0.0

First derivatives : 7.0, 10.0.

Equation of quadratic curve : $y = 20x^2 - 15x + 3$

Figure 3. Modification by adjustment of weight.

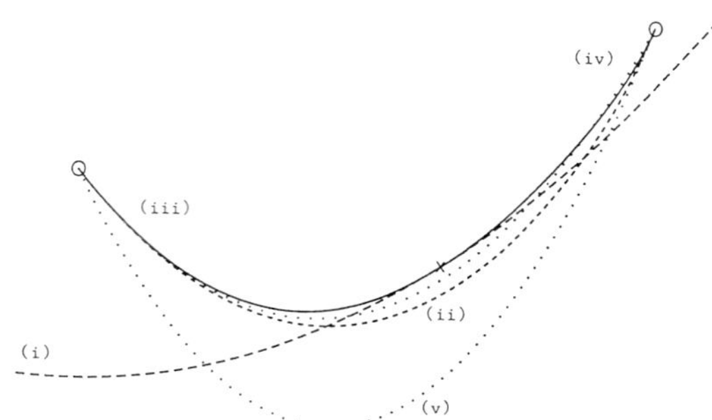

Table 4 for example 4

x	0.0	1.0
y	1.0	1.5

First derivatives : -2.35, 4.83.

Equation of quadratic curve : $y = x^2 + 0.3$

Figure 4. Modification by addition of new data point.

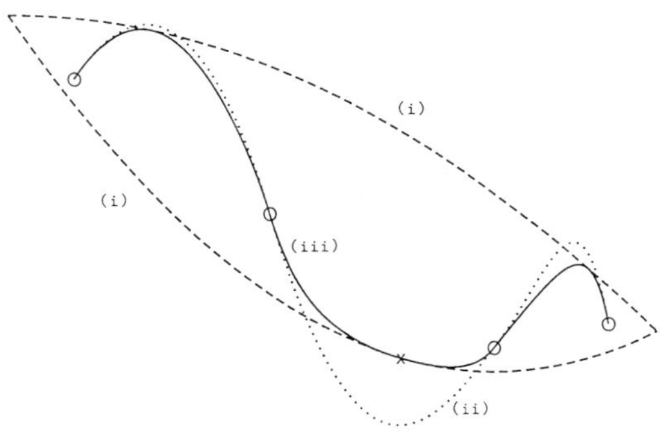

Table 5 for example 5

x	0.4	1.6	3.0	3.7
y	3.2	1.5	-0.2	0.1

First derivatives : 3.0, -7.0, 2.5, -12.0.

Equations of quadratic curves :

$y = -0.25x^2 + 4$

$y = 0.5x^2 - 3x + 4$

Figure 5. Case of two quadratic constraint curves.

other limiting quadratic, $Q_u(x)$, does not. Thus the modified curve segment is obtained by increasing the weight w_1. In Figure 4, as both of the limiting quadratics cross the constraint curve, modification of the curve segment requires the addition of a knot which splits the original segment into two new segments. The new right segment does not require any change in the initial value of the weight w_1, this being equal to the weight of the original segment, while the new left segment experiences a small increase in the value of its weight w_1. More precisely, the weight of the initial curve segment and that of the new right segment is 2.1 while for the new left segment the weight is 2.13. The last example (Figure 5) shows two constraint curves as the boundary of a closed region. The initial curve consists of three curve segments, each crossing one of the constraint curves, but not both. Each of the first and third segments can be modified by decreasing the value of w_1, but the second segment requires the addition of a knot. More precisely, the initial value of w_1 for each segment is 2.5 while the values of w_1 for the modified first and third segments are 2.07 and 2.23 respectively. The two new segments, to the left and to the right of the additional knot and which replace the original second segment, have w_1 equal to 4.50 and 2.35 respectively.

§6. Conclusions

In this paper, we have considered schemes for generating a C^1 non-parametric piecewise rational cubic interpolating curve, which is constrained to lie on one side of one or more given constraint curves. The motivation behind the work was to further develop the ideas of [2], by considering more complex constraint curves. Hence both straight line and quadratic constraint curves have been considered. By considering non-parametric curves, it was found that both types of constraints could be treated using ideas similar to those described in [2]. In the straight line case, the interpolating curve can be forced to satisfy the constraints by adjustment of the weights, while in the quadratic curve case addition of an extra data point may be necessary. Both approaches are entirely local. More complex constraints for both parametric and non-parametric cases remain as projects for the future.

Acknowledgments. This research was performed while the first author was visiting the Department of Mathematics and Computer Science, University of Dundee, on leave from Universiti Sains Malaysia. She wishes to acknowledge the financial support, in the form of a Royal Society research grant, which made this visit possible.

References

1. Delbourgo, R. and J. A. Gregory, Shape preserving piecewise rational interpolation, SIAM J. Sci. Statist. Comput. **6** (1985), 967–976.
2. Goodman, T. N. T., B. H. Ong, and K. Unsworth, Constrained interpolation using rational cubic splines, in *Nurbs for Curve and Surface Design*, G. Farin (ed.), SIAM, Philadelphia, 1991, 59–74.

K. Unsworth
Department of Mathematics and Computer Science
University of Dundee
Dundee DD1 4HN
Scotland.
unsworth@mcs.dund.ac.uk

B. H. Ong
School of Mathematical Sciences
Universiti Sains Malaysia
11800 Penang, Malaysia
obh@cs.usm.my

Tensor Product Slices

R. Pfeifle, R. Bartels, R. Goldman

Abstract. We introduce a variant of tensor product splines, called the *tensor product slice*, which can be used to create surfaces with multi-sided patches.

§1. Introduction

By restricting the domain of a tensor product spline in more than two variables to an affine plane, surface patches with non-rectangular geometries can be obtained. An example is shown in Figure 1.

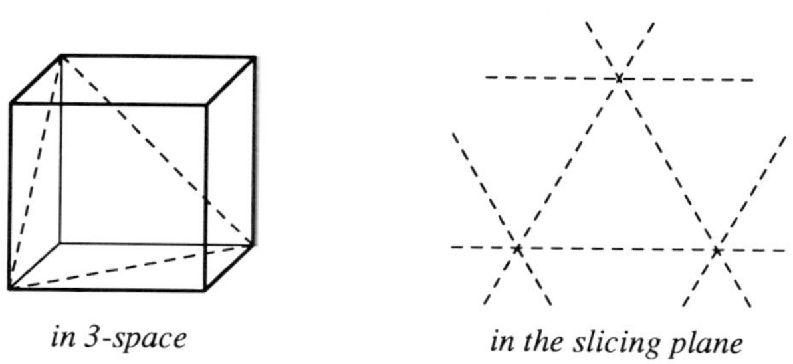

Figure 1. Creating a Triangular Domain by Slicing a Cube.

Here a uniform tensor product spline volume is restricted to the affine plane that intersects the domain cubes in a regular triangular lattice. The image of the spline volume restricted to this plane is a parametric surface, composed of triangular patches. The surface inherits the continuity of the volume.

Mathematical Methods in Computer Aided Geometric Design II
Tom Lyche and Larry L. Schumaker (eds.), pp. 431–440.

ISBN 0-12-460510-9.

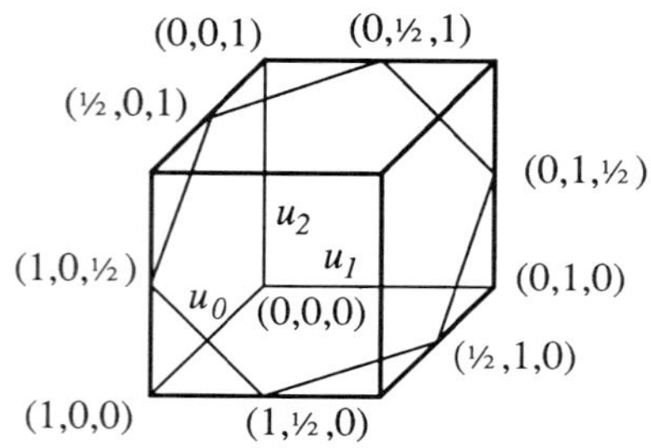

Figure 2. A hexagonal domain.

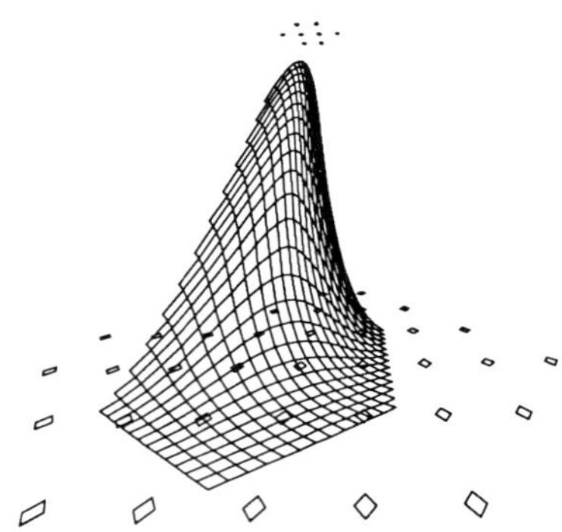

Figure 3. A hexagonal patch. Control vertices appear as diamonds.

Slices whose domains do not produce a regular tiling of the plane also may be useful. For example, Figure 3 presents a patch whose domain, a regular hexagon, is formed as shown in Figure 2.
As the dimension of the tensor product is increased, the number of sides that a domain may possess is also increased. Further geometries may be obtained by restricting any of the above domains to a polygon; e.g., a octagon could be cut from the interior of the hexagon in Figure 2.

A trivial restriction of a tensor product domain is one that holds all variables constant save two. The resulting surface is isomorphic to a standard tensor product surface. Thus, slice surfaces subsume all types of tensor product surfaces.

Other approaches to multi-sided patches have been proposed. Our approach most closely resembles Loop and DeRose's S-patches [3, 4]. However, these authors use a polygonal domain embedded in a high dimensional Bézier simplex, while we use a planar or polygonal domain embedded in a high dimensional cube. Our method produces integral surfaces from integral splines and rational surfaces from rational splines, whereas an S-patch will always be rational when the number of its sides is greater than four. For certain regular domain geometries (e.g., a tiling of the plane with triangles), we can form multi-patch C^k surfaces.

Section 2 gives basic definitions. Section 3 mentions important properties that slices inherit from their tensor product parents. Section 4 introduces the idea of *dimension raising*. Section 5 provides an optimal ordering for carrying out affine recurrences on a tensor product slice. Section 6 covers the conversion of tensor product slices to Bézier form. Section 7 discusses the joining of a slice to an existing surface, and Section 8 concludes with a summary of progress in implementing and applying the results of this work.

§2. Tensor Products and Slices

Consider splines of degree d, defined on an interval $[a,b]$, having knots $\tau_0 \leq \cdots \leq \tau_d \leq a \leq \tau_{d+1} \leq \cdots \leq b \leq \tau_{m+1} \leq \cdots \leq \tau_{m+d+1}$, with the usual restriction that $\tau_i < \tau_{i+d+1}$ for all i. We use $\{B_{i,d}(u) : i = 0, \ldots, m\}$ to designate the basis for these splines that is nonnegative, has compact support, and partitions unity. Let $[a_0, b_0], \ldots, [a_n, b_n]$ be real intervals, let $\mathcal{D} = [a_0, b_0] \times [a_1, b_1] \times \cdots \times [a_n, b_n]$, and let $B^j_{i_j,d_j}(u_j)$ be basis functions in the variable $u_j \in [a_j, b_j]$. If $\mathbf{u} = (u_0, u_1, \ldots, u_n) \in \mathcal{D}$ and $P_{i_0,i_1,\ldots,i_n} \in \mathbb{R}^\beta$ for $j = 0, \ldots, n$, $i_j = 0, \ldots, m_j$, then $F : \mathcal{D} \to \mathbb{R}^\beta$, given by

$$F(\mathbf{u}) = \sum_{i_0=0}^{m_0} \sum_{i_1=0}^{m_1} \cdots \sum_{i_n=0}^{m_n} P_{i_0,i_1,\ldots,i_n} B^0_{i_0,d_0}(u_0) B^1_{i_1,d_1}(u_1) \cdots B^n_{i_n,d_n}(u_n)$$

is an $n+1$ dimensional tensor product of the basis functions $B^j_{i_j,d_j}(u_j)$ over the domain $\mathcal{D}$ with the control vertices $P_{i_0,i_1,\ldots,i_n}$. A useful notation for certain purposes is $\sum_I P_I B_{I,D}(\mathbf{u})$.

If $\mathbf{c}_0 = (c_{00}, \ldots, c_{n0})$ and $\mathbf{c}_1 = (c_{01}, \ldots, c_{n1})$ are linearly independent vectors in $\mathbb{R}^{n+1}$, and if $\mathbf{c}_2 = (c_{02}, \ldots, c_{n2})$ is any point in $\mathbb{R}^{n+1}$, then the set $\{\mathbf{u} = v_0\mathbf{c}_0 + v_1\mathbf{c}_1 + \mathbf{c}_2 : v_0, v_1 \in \mathbb{R}\}$ is an affine plane in $n+1$ dimensions. For notational convenience, we may compress this into $\mathbf{u} = \mathbf{C}\mathbf{v}$ where $\mathbf{C}$ is the $(n+1) \times 3$ matrix having $\mathbf{c}_j$ as its j-th column and $\mathbf{v}$ is the column vector with components v_0, v_1, 1. The affine plane is defined entirely by $\mathbf{C}$.

Definition: *Let F be an $n+1$ dimensional tensor product over domain $\mathcal{D}$ and let $\mathbf{C}$ define an affine plane having a nonempty intersection with the interior of $\mathcal{D}$. The tensor product* **slice** *of F is the set of points of the form $F(\mathbf{C}\mathbf{v})$.*

Definition: *A tensor product* **cell** *is any set $[\tau_{\kappa_0}, \tau_{\kappa_0+1}] \times \cdots \times [\tau_{\kappa_n}, \tau_{\kappa_n+1}]$ having interior.*

Definition: *A* **slice segment** *is any intersection with interior of the affine plane with a cell. A* **slice patch** *is the image of a slice segment under the mapping $F(\mathbf{C}\mathbf{v})$.*

Example: *Using the linear Bernstein polynomials as basis and*

$$\begin{bmatrix} u_0 \\ u_1 \\ u_2 \end{bmatrix} = \begin{bmatrix} 1 & 0 & 0 \\ 0 & 1 & 0 \\ -1 & -1 & \frac{3}{2} \end{bmatrix} \begin{bmatrix} v_0 \\ v_1 \\ 1 \end{bmatrix}$$

results in the slice patch over the slice segment of Figure 2 given by

$$\begin{aligned} &- P_{0,0,0}(1-v_0)(1-v_1)(\frac{1}{2}-v_0-v_1) + P_{0,0,1}(1-v_0)(1-v_1)(\frac{3}{2}-v_0-v_1) \\ &- P_{0,1,0}(1-v_0)v_1(\frac{1}{2}-v_0-v_1) + P_{0,1,1}(1-v_0)v_1(\frac{3}{2}-v_0-v_1) \\ &- P_{1,0,0}v_0(1-v_1)(\frac{1}{2}-v_0-v_1) + P_{1,0,1}v_0(1-v_1)(\frac{3}{2}-v_0-v_1) \\ &- P_{1,1,0}v_0v_1(\frac{1}{2}-v_0-v_1) + P_{1,1,1}v_0v_1(\frac{3}{2}-v_0-v_1). \end{aligned}$$

In general, the functions $B^j_{i_j,d_j}(u_j) = B^j_{i_j,d_j}(v_0c_{j0} + v_1c_{j1} + c_{j2})$ will be linearly dependent. Dependency can be removed by reducing the slice patches to bivariate Bézier form.

Definition: *A slice patch given as*

$$\sum_{i_0=\kappa_0-d_0}^{\kappa_0} \cdots \sum_{i_n=\kappa_n-d_n}^{\kappa_n} P_{i_0,\ldots,i_n} B_{i_0,d_0}(u_0)\cdots B_{i_n,d_n}(u_n) \ ,$$

where the B_{i_j,d_j} may be any basis, is said to be in **tensor product form**. *The same slice patch given as*

$$\sum_{i_0=0}^{\delta_0} \sum_{i_1=0}^{\delta_1} L_{i_0,i_1} B_{i_0,\delta_0}(v_0) B_{i_1,\delta_1}(v_1) \ ,$$

where B_{i_j,δ_j} is the i_j-th Bernstein polynomial of degree δ_j, $j = 0, 1$, is said to be in **Bézier assimilated form**. *In particular, for the example of Figure 2, this becomes*

$$\sum_{i=0}^{3} \sum_{j=0}^{3} L_{i,j} B_{i,3}(v_0) B_{j,3}(v_1)$$

with $L_{2,2}$ given by

$$\frac{P_{0,0,0}}{6} - \frac{P_{0,0,1}}{18} + \frac{2\,P_{0,1,0}}{9} + \frac{2\,P_{1,0,0}}{9} + \frac{2\,P_{1,1,0}}{9} + \frac{2\,P_{1,1,1}}{9} \ .$$

(Indeed, in this case, the slice can be represented as a quadratic, but a cubic is more representative of trilinear slices, so a coefficient of the cubic representation is given.)

§3. Generally Inherited Properties

Slicing preserves many tensor product properties. For example, any algorithm that evaluates the original tensor product can be used to evaluate the slice; the slice of a tensor product with C^k continuity will be C^k continuous; polynomials of degree d remain polynomials of degree d (the sliced polynomial has fewer variables, but they are of higher degree).

As for the basis functions, summation to one over the full domain implies summation to one over the partial domain, so slices are affine invariant. Basis nonnegativity ensures that the slice will lie within the convex hull of the $P_{i_0,i_1,\ldots,i_n}$. If some control vertex is interpolated by a tensor product at a point in the slice domain, then that vertex is interpolated by the tensor product slice.

§4. Dimension Raising

The shape of each slice segment depends upon the number of cell boundaries of $\mathcal{D}$ intersected by the affine plane. Increasing the dimension increases the potential number of cell boundaries that may be intersected, allowing more complex slice segments. To raise the dimension by one, a new variable and basis, u_{n+1} and $B^{n+1}_{i_{n+1},d_{n+1}}$, are introduced, making use of partition of unity:

$$\begin{aligned} F(\mathbf{u}) &= \sum_I P_I B_{I,D}(\mathbf{u}) = \sum_I P_I B_{I,D}(\mathbf{u}) \times \sum_{i_{n+1}=0}^{m_{n+1}} B^{n+1}_{i_{n+1},d_{n+1}}(u_{n+1}) \\ &= \sum_{\overline{I}} P_I B_{\overline{I},\overline{D}}(\overline{\mathbf{u}}) = \sum_{\overline{I}} \overline{P}_{\overline{I}} B_{\overline{I},\overline{D}}(\overline{\mathbf{u}}) = F(\overline{\mathbf{u}}), \end{aligned}$$

where

$$\begin{aligned} &\overline{I} = i_0 \cdots i_{n+1} \ , \ \overline{D} = d_0 \cdots d_{n+1} \ , \ \overline{\mathbf{u}} = (u_0 \, , \ldots , u_{n+1}) \\ &\overline{P}_{i_0,i_1,\ldots,i_{n+1}} = P_{i_0,i_1,\ldots,i_n} \text{ for } i_{n+1} = 0 \, , \ldots , m_{n+1}. \end{aligned}$$

§5. Ordering of Affine Recurrences

A tensor product value is given, recursively, by $\sum_{i_\ell=0}^{m_\ell} \mathbf{P}_{i_\ell} B^\ell_{i_\ell,d_\ell}(u_\ell)$ where $\mathbf{P}_{i_\ell}$ represents summation in the dimensions $\{0,\ldots,n\}\backslash\{\ell\}$. With many basis functions the sum can be evaluated by forming a sequence of affine combinations of the $\mathbf{P}_{i_\ell}$ (*cf.* the de Casteljau and de Boor algorithms [2 p. 27]). The value computed is independent of the order in which dimensions are processed, but the total number of affine combinations required may depend on the ordering.

For the algorithms mentioned above, the cost $\Delta(d)$ of computing the value of a point on a curve of degree d by a sequence of affine combinations is consistent with the inequality $\frac{\Delta(d+k)}{\Delta(d)} \geq \frac{d+k}{d}$ for $k > 0$. If the dimension ordering is given by the permutation $\pi : \{0,\ldots,n\} \to \{\pi(0),\ldots,\pi(n)\}$ and the degree in dimension ℓ is d_ℓ, the cost of finding a surface point is:

$$\mu_{n+1}(d_{\pi(0)},\ldots,d_{\pi(n)}) = \sum_{i=0}^{n} \Delta(d_{\pi(i)}) \prod_{r=0}^{i-1} (d_{\pi(r)}+1) \ .$$

It can be shown [5] by induction on n, using only the inequality satisfied by Δ, that μ_{n+1} is minimized over all possible orderings when $d_{\pi(0)} \geq \cdots \geq d_{\pi(n)}$.

Many transformations of splines also involve recursively applied affine modifications to control vertices. Among such transformations are Bézier degree raising and subdivision, and the insertion of knots at a single location in B-spline representations. They should be carried out in order of decreasing dimension.

§6. Conversion to Bézier Assimilated Form via Blossoming

Blossoms have been studied extensively by Ramshaw [6]. In using them for this section we restrict our attention to a single cell of a tensor product and the corresponding segment of a slice.

Definition: *Let $\mathcal{T} = T_0^{d_0} \times T_1^{d_1} \times \cdots \times T_n^{d_n}$, where $T_\ell^{d_\ell}$ is $T_\ell \times T_\ell \times \cdots \times T_\ell$ d_ℓ times, and T_ℓ is a knot interval of the ℓ-th basis function, $[\tau_{\kappa_\ell}, \tau_{\kappa_\ell+1}]$. Let $(u_{0,0}, \ldots, u_{0,d_0-1}; u_{1,0}, \ldots, u_{1,d_1-1}; \cdots; u_{n,0}, \ldots, u_{n,d_n-1}) \in \mathcal{T}$, where semi-colons are used to separate variables in different T_ℓ. Then $f : \mathcal{T} \to \mathbb{R}$, is* **symmetric in dimension** *ℓ, if, for any permutation σ*

$$f(\cdots; u_{\ell,0}, \ldots, u_{\ell,d_\ell-1}; \cdots) = f(\cdots; u_{\ell,\sigma(0)}, \ldots, u_{\ell,\sigma(d_\ell-1)}; \cdots)$$

Definition: *let $F(u_0, u_1, \ldots, u_n)$ be a tensor product on the cell $T_0 \times \cdots \times T_n$. The* **blossom** *of F is the unique polynomial $f(\cdots; u_{\ell,0}, \ldots, u_{\ell,d_\ell-1}; \cdots)$ that*

- *is affine in each parameter u_{i_ℓ,j_ℓ} individually,*
- *is symmetric in each dimension ℓ,*
- *reproduces F in the cell as its* **diagonal***:*
  $F(u_0, \ldots, u_n) = f_F(u_0, \ldots, u_0; \ldots; u_n, \ldots, u_n)$

The Bézier control vertices of the tensor product are readily obtained from its blossom. Let

$$B_{i_\ell,d_\ell}(u_\ell; \tau_{\kappa_\ell}; \tau_{\kappa_\ell+1}) = \binom{d_\ell}{i_\ell} \left(\frac{\tau_{\kappa_\ell+1} - u_\ell}{\tau_{\kappa_\ell+1} - \tau_{\kappa_\ell}} \right)^{d_\ell - i_\ell} \left(\frac{u_\ell - \tau_{\kappa_\ell}}{\tau_{\kappa_\ell+1} - \tau_{\kappa_\ell}} \right)^{i_\ell}$$

and let

$$F(\cdots, u_\ell, \cdots) = \cdots \sum_{i_\ell=0}^{d_\ell} \cdots P_{i_0, \ldots, i_n} \cdots B_{i_\ell,d_\ell}(u_\ell; \tau_{\kappa_\ell}; \tau_{\kappa_\ell+1}) \cdots$$

Then

$$f_F(\cdots; \tau_{\kappa_\ell,\ell}, \ldots, \tau_{\kappa_\ell,\ell}, \tau_{\kappa_\ell+1,\ell}, \ldots, \tau_{\kappa_\ell+1,\ell}; \cdots) = P_{i_0,i_1, \ldots, i_n}$$

when, in dimension ℓ of the blossom, the number of variables that have been set to $\tau_{\kappa_\ell,\ell}$ is i_ℓ and the number set to $\tau_{\kappa_\ell+1,\ell}$ is $d_\ell - i_\ell$. Evaluating f_F, at least when F is expressed in terms of B-splines or Bernstein polynomials, is as easy as evaluating F itself. In the univariate case, the recursive evaluation scheme for F is invoked using a different value of $u = u_0$ at each level of recursion in the de Boor or de Casteljau algorithm, one value for each instance of u_0 in the blossom. The multivariate case is carried out this way in each variable position separately.

We can use values of f_F as in [1] to provide the Bézier assimilated form of any slice. Note that u_ℓ, which is equal to $v_0 c_{\ell 0} + v_1 c_{\ell 1} + c_{\ell 2}$, can be represented by $\sum_{j=0}^{1} \sum_{k=0}^{1} p_{\ell jk} B_{j,1}(v_0; 0; 1) B_{k,1}(v_1; 0; 1)$, where $p_{\ell 00} = c_{\ell 2}$, $p_{\ell 01} = c_{\ell 1} + c_{\ell 2}$, $p_{\ell 10} = c_{\ell 0} + c_{\ell 2}$, $p_{\ell 11} = c_{\ell 0} + c_{\ell 1} + c_{\ell 2}$. So, with respect to dimension ℓ,

$$\begin{aligned} F(\cdots, u_\ell, \cdots) &= f_F(\cdots; \underbrace{u_\ell, \ldots, u_\ell}_{d_\ell \text{ times}}; \cdots) \\ &= \sum_{j_1=0}^{1} \cdots \sum_{j_{d_\ell}=0}^{1} B_{j_1,1}(v_0; 0; 1) \cdots B_{j_{d_\ell},1}(v_0; 0; 1) \\ &\qquad \times \sum_{k_1=0}^{1} \cdots \sum_{k_{d_\ell}=0}^{1} B_{k_1,1}(v_1; 0; 1) \cdots B_{k_{d_\ell},1}(v_1; 0; 1) \\ &\qquad\qquad \times f_F(\cdots; p_{\ell j_1 k_1}, \ldots, p_{\ell j_{d_\ell} k_{d_\ell}}; \cdots) \end{aligned}$$

The summations can be written in multi-index form as

$$\begin{aligned} &\sum_{J_\ell} [B_{0,1}(v_0; 0; 1)]^{z(J_\ell)} [B_{1,1}(v_0; 0; 1)]^{o(J_\ell)} \\ &\qquad \times \sum_{K_\ell} [B_{0,1}(v_1; 0; 1)]^{z(K_\ell)} [B_{1,1}(v_1; 0; 1)]^{o(K_\ell)} \\ &\qquad\qquad \times f_F(\cdots; p_{\ell j_1 k_1}, \ldots, p_{\ell j_{d_\ell} k_{d_\ell}}; \cdots) \end{aligned}$$

where $J_\ell = (j_1, \ldots, j_{d_\ell})$ and $K_\ell = (k_1, \ldots, k_{d_\ell})$ run through all 2^{d_ℓ} 0-1 patterns of length d_ℓ . $z(J_\ell)$ gives the number of zeros in J_ℓ and $o(J_\ell)$ gives the number of ones, and similarly for $z(K_\ell)$, $o(K_\ell)$. Such products can be rephrased in terms of higher degree Bernstein polynomials using the product rule [2].

The remaining term $f_F(\cdots; p_{\ell j_1 k_1}, \ldots, p_{\ell j_{d_\ell} k_{d_\ell}}; \cdots)$ will have the form

$$f_F(\cdots; \underbrace{p_{\ell 00}, \ldots, p_{\ell 00}}_{zz(J_\ell, K_\ell)}, \underbrace{p_{\ell 01}, \ldots, p_{\ell 01}}_{zo(J_\ell, K_\ell)}, \underbrace{p_{\ell 10}, \ldots, p_{\ell 10}}_{oz(J_\ell, K_\ell)}, \underbrace{p_{\ell 11}, \ldots, p_{\ell 11}}_{oo(J_\ell, K_\ell)}; \cdots)$$

where $zz(J_\ell, K_\ell)$ counts the number of positions at which J_ℓ and K_ℓ both have a zero, $zo(J_\ell, K_\ell)$ counts the number of positions at which J_ℓ has a zero and K_ℓ has a one, and so on.

The conversion to Bézier assimilated form, from this result forward, is mechanical, involving logical operations on 0-1 bit patterns to collect products of $B_{i,1}$ functions that have a common B_{j,d_ℓ} result. The example of Figure 2 serves as an easy case, since $f_F(\mathbf{u}) = F(\mathbf{u})$. In this example u_0 has $p_{000} = 0$,

$p_{001} = 0$, $p_{010} = 1$, $p_{011} = 1$; u_1 has $p_{100} = 0$, $p_{101} = 1$, $p_{110} = 0$, $p_{111} = 1$, and u_1 has $p_{200} = \frac{3}{2}$, $p_{201} = \frac{1}{2}$, $p_{210} = \frac{1}{2}$, $p_{211} = -\frac{1}{2}$. Thus

$$\begin{aligned} F(u_0, u_1, u_2) = \cdots + B_{j_1,1}(v_0; 0; 1) B_{j_2,1}(v_0; 0; 1) B_{j_3,1}(v_0; 0; 1) \\ B_{k_1,1}(v_1; 0; 1) B_{k_2,1}(v_1; 0; 1) B_{k_3,1}(v_1; 0; 1) \\ F(p_{0j_1k_1}, p_{1j_2k_2}, p_{2j_3k_3}) + \cdots \end{aligned}$$

a representative term of which would be

$$\begin{aligned} B_{1,1}(v_0; 0; 1) B_{0,1}(v_0; 0; 1) B_{1,1}(v_0; 0; 1) \\ B_{1,1}(v_1; 0; 1) B_{1,1}(v_1; 0; 1) B_{0,1}(v_1; 0; 1) \\ F(p_{010}, p_{111}, p_{210}) \end{aligned}$$

which equals

$$\begin{aligned} B_{1,1}(v_0; 0; 1)^2 B_{0,1}(v_0; 0; 1) \\ B_{1,1}(v_1; 0; 1)^2 B_{0,1}(v_1; 0; 1) \\ F(p_{010}, p_{111}, p_{210}) \end{aligned}$$

or

$$\left(\frac{1}{18} P_{1,1,0} + \frac{1}{18} P_{1,1,1} \right) B_{2,3}(v_0; 0; 1) B_{2,3}(v_1; 0; 1) \, .$$

Several other terms produce factors of $B_{2,3}(v_0; 0; 1) B_{2,3}(v_1; 0; 1)$, all combining to produce the coefficient $L_{2,2}$ given earlier for this example.

§7. Joining a Slice Patch to an Existing Surface

The issues in joining a patch to several consistent patches are given in [4]. With adjustments in notation, we confine ourselves here to indicating the form of necessary derivatives (required by (A0)–(A3) and (8) in [4]).

A polygon in the affine plane $\mathbf{Cv}$ will have perimeter determined by the points $\mathbf{p}_0, \ldots, \mathbf{p}_k \in \{(v_0, v_1)\}$; e.g., Figure 2 has $\mathbf{p}_0 = (\frac{1}{2}, 0)$, $\mathbf{p}_1 = (0, \frac{1}{2})$, $\mathbf{p}_2 = (0, 1)$, $\mathbf{p}_3 = (\frac{1}{2}, 1)$, $\mathbf{p}_4 = (1, \frac{1}{2})$, and $\mathbf{p}_5 = (1, 0)$. The edges of the patch will be given by the points of the form $\mathbf{p}(\alpha_j) = \mathbf{p}_j + \alpha_j \mathbf{e}_j$, where the edge vector $\mathbf{e}_j$ is $\mathbf{p}_{j+1} - \mathbf{p}_j$. In terms of the affine plane, edges appear as $\mathbf{u}(\alpha_j) = \mathbf{Cv}(\alpha_j) = \alpha_j \mathbf{C}_{01} \mathbf{e}_j + \mathbf{r}_j$, where $\mathbf{r}_j = \mathbf{C}_{01} \mathbf{p}_j + \mathbf{c}_2$ and $\mathbf{C}_{01} = [\mathbf{c}_0, \mathbf{c}_1]$.

Let $S_i(\alpha_i, \beta_i), \ldots, S_{i+r}(\alpha_{i+r}, \beta_{i+r})$ be adjacent, G^1 compatible surfaces parameterized so that $S_j(0,1) = \mathbf{p}_j$ and $S_j(1,1) = \mathbf{p}_{j+1}$ for $j = i, \ldots, i+r$, and so that $S_j(\alpha_j, 1)$ provides the boundary along which we wish to join F. We must arrange that $F(\mathbf{Cv}(\alpha_j)) = S_j(\alpha_j, 1)$ for $0 \le \alpha_j \le 1$, that

$$D_{\mathbf{t}_j} F(\mathbf{Cv}(\alpha_j)) = \mu_j(\alpha_j) \frac{\partial S_j(\alpha_j, 1)}{\partial \alpha_j} + \nu_j(\alpha_j) \frac{\partial S_j(\alpha_j, 1)}{\partial \beta_j}$$

for polynomials μ_j and ν_j of sufficiently high degrees and some vector $\mathbf{t}_j$ independent of $\mathbf{e}_j$; for example, $\mathbf{t}_j = \mathbf{e}_{j-1}$ [4], and that

$$D_{\mathbf{t}_{j-1}} D_{\mathbf{t}_j} F(\mathbf{Cv}(\alpha_j)) = D_{\mathbf{t}_j} D_{\mathbf{t}_{j-1}} F(\mathbf{Cv}(\alpha_j))$$

when $\alpha_j = 0$.

Note that the derivative of F along edge $\mathbf{t}_j$ will be found from

$$\begin{aligned}\frac{\partial F}{\partial \alpha_j} &= \sum_{\ell} \frac{\partial F}{\partial u_\ell}\frac{\partial u_\ell}{\partial \alpha_j}\\ &= \sum_{\ell} \frac{\partial F}{\partial u_\ell}\left[c_{\ell 0}e_{j0} + c_{\ell 1}e_{j1}\right]\\ &= \sum_{\lambda=0}^{1} e_{j\lambda} \sum_{\ell} c_{\ell\lambda}\frac{\partial F}{\partial u_\ell}\end{aligned}$$

with

$$\begin{aligned}&\frac{\partial}{\partial u_\ell} \sum_{i_0=0}^{m_0} \cdots \sum_{i_\ell=0}^{m_\ell} \cdots \sum_{i_n=0}^{m_n} P_{i_0,\ldots,i_\ell,\ldots,i_n}\\ &\qquad \times B_{i_0,d_0}(u_0)\cdots B_{i_\ell,d_\ell}(u_\ell)\cdots B_{i_n,d_n}(u_n)\\ &\quad = \sum_{I_\ell} d_\ell E_\ell(P_I) B_{I_\ell,D_\ell}(\mathbf{u}),\end{aligned}$$

where E_ℓ is the forward difference operator in the ℓ-th index.

We may combine the above and summarize as follows: to attach F to S_j, we must arrange that

$$\sum_I P_I B_{I,D}(\mathbf{Cv}(\alpha_j)) = S_j(\alpha_j, 1)$$

$$\begin{aligned}&\sum_{\lambda=0}^{1} e_{\lambda j-1} \sum_{\ell=0}^{n} c_{\ell\lambda}\left[\sum_{I_\ell} d_\ell E_\ell(P_I) B_{I_\ell,D_\ell}(\mathbf{Cv}(\alpha_j))\right] =\\ &\qquad\qquad = \mu_j(\alpha_j)\frac{\partial S_j(\alpha_j,1)}{\partial \alpha_j} + \nu_j(\alpha_j)\frac{\partial S_j(\alpha_j,1)}{\partial \beta_j}\end{aligned}$$

The forms for $D_{\mathbf{t}_{j-1}} D_{\mathbf{t}_j} F(\mathbf{Cv}(\alpha_j))$ are similar.

If the basis functions are not rational, the left and right side of these equations are polynomials in α_j. The tensor product control points, P_I, and the coefficients of the polynomials $\mu_j(\alpha_j)$, $\nu_j(\alpha_j)$ are the unknowns to be found. The equations are linear in these unknowns. The equations can be made underdetermined by selecting the degrees of the bases, the degrees of μ_j, ν_j, and the dimension of the tensor product to be sufficiently high.

§8. Progress

Some implementation has been carried out to investigate these ideas more fully. A project at the University of Waterloo's Computer Graphics Laboratory has been devoted to the establishment of C++ classes out of which prototype, spline modelers can be constructed. A student using these tools has completed an interactive display tool for slice surfaces. We expect to expand this display tool to provide interactive manipulations for exploring C^k surfaces composed of triangular patches.

Acknowledgments. Supported by NSERC, ITRC, GM, DEC, and SGI.

References

1. DeRose, T., Goldman, R., Hagen, H., and Mann, S., Composition algorithms via blossoming: theory, applications, and implementation, Technical Report 91-05-04, Department of Computer Science and Engineering, University of Washington, Seattle, Washington 98195, 1991.
2. Farin, G., *Curves and Surfaces for Computer Aided Geometric Design* (second edition), Academic Press, Boston, 1990.
3. Loop, C., and DeRose, T., A multisided generalization of Bézier surfaces, Transactions on Graphics **8** (1989), 204–234.
4. Loop, C., and DeRose, T., Generalized B-spline surfaces of arbitrary topology, Computer Graphics **24** (1990), 347–356.
5. Pfeifle, R., Bartels, R., and Goldman, R., *Tensor product slices*, Technical Report CS-89-56, Computer Science Department, University of Waterloo, Waterloo, Ontario N2L 3G1, Canada, 1989.
6. Ramshaw, L., Blossoms are polar forms, Computer Aided Geom. Design **6** (1989), 323–358.

Ronald F. Pfeifle and Richard H. Bartels
University of Waterloo
Department of Computer Science
Waterloo, Ontario, Canada N2L 3G1
rhbartels@waterloo.edu
rfpfeifle@waterloo.edu

Ronald N. Goldman
Rice University
Department of Computer Science
Houston, Texas, U.S.A. 77251
rng@rice.edu

Construction of Smooth Surfaces by Piecewise Tensor Product Polynomials

A. R. M. Piah

Abstract. The aim of this paper is to construct surfaces with tangent plane continuity defined by tensor product polynomials over a mesh of rectangles, by subdividing the mesh and keeping the continuity conditions across mesh lines fixed.

§1. Introduction

Various attempts have been made to define a *geometrically smooth simple closed surface* whose parametric representation is similar locally to that of box splines. Most have used the notion of *subdivision* to construct surfaces by a process of recursive subdivision, see for instance Ball and Storry [1], Catmull and Clark [3], and Doo and Sabin [4]. Recently Goodman [5] suggested an approach to generate *geometrically smooth closed surfaces* by defining a biquadratic spline on a *simple quadrilateral mesh*. Höllig and Mögerle [8] generalized this idea to arbitrary networks of rectangular polynomial patches. Lee and Majid [9,10] followed by using piecewise bicubic and biquartic splines on an arbitrary polyhedral mesh.

In this paper, following the idea of Goodman [5] we construct a *geometrically smooth simple closed surface* defined by tensor product polynomials over a mesh of rectangles. We subdivide the mesh but always keep the continuity conditions across mesh lines fixed. The advantage of keeping the continuity conditions across mesh lines fixed is that the resulting functions can be added and still produce a geometrically smooth surface, thus allowing one to approximate data by such a surface by first approximating on a course mesh and then successively subdividing the mesh and approximating the error, where this is significant. Such applications will be considered in another paper.

We begin by describing briefly what is meant by a *simple quadrilateral mesh* (SQM) and our way of subdividing the mesh in Section 2. In Section 3

Mathematical Methods in Computer Aided Geometric Design II
Tom Lyche and Larry L. Schumaker (eds.), pp. 441–455.

ISBN 0-12-460510-9.

we briefly touch on the continuity conditions that we require. A new SQM, M is obtained upon *subdivision* of a cube as prescribed in Section 4 and upon further *subdivision* as in Section 5. We consider the space V(M) of all functions on M whose restrictions to each face are certain polynomials and which satisfy our continuity conditions across the edges. We construct a basis of positive functions (there are some slightly negative functions for the case in Section 5) with small support which sum to 1 and are thus useful for the design of surfaces. We also show that V(M) (or subspace of V(M) for the case in Section 5) is suitable for interpolating data at the midpoints of the faces of M.

§2. Simple Quadrilateral Meshes

A closed polyhedral mesh with quadrilateral faces such that at every vertex, either three or four edges meet, is called a *simple quadrilateral mesh* (SQM). A vertex where only three edges meet is called *extraordinary* or *singular*. A face is called *singular* if it contains a singular vertex. The minimum number of faces occurs when all the vertices are singular, which is obtained by a cube (see Goodman [5]).

A new SQM is obtained by subdividing a cube as follows. A new vertex is added on each edge and in the interior of each face of the cube. Each face is then divided into 4 new faces by joining the new vertices on the boundary of the face to the new vertex in the interior of the face.

§3. Continuity Conditions

Let M be a SQM and on each face of M, local coordinates u, v are taken so that the face has the form $\{(\mathrm{u},\mathrm{v}) : 0 \le \mathrm{u} \le 1,\ 0 \le \mathrm{v} \le 1\}$. We shall construct a vector space V(M) of functions on M whose restrictions to each face of M are tensor product polynomials in u and v. This shall be constructed so that any surface $s : \mathrm{M} \to \mathbb{R}^3$, whose components lie in V(M) and such that on each face

$$\frac{\partial s}{\partial \mathrm{u}} \times \frac{\partial s}{\partial \mathrm{v}} \neq 0,$$

is guaranteed to be G^1 . To ensure this, we require that the elements of V(M) satisfy certain continuity conditions across the edges of M.

We will choose the simplest and most natural conditions which are sufficient for our purpose. Suppose that a function f in V(M) coincides with a polynomial p on a face $\{(\mathrm{u},\mathrm{v}) : 0 \le \mathrm{u} \le 1,\ 0 \le \mathrm{v} \le 1\}$ and with a polynomial q on a face $\{(\tilde{\mathrm{u}},\tilde{\mathrm{v}}) : 0 \le \tilde{\mathrm{u}} \le 1,\ 0 \le \tilde{\mathrm{v}} \le 1\}$, where the faces have a common edge :

$$(\mathrm{u},0) = (\tilde{\mathrm{u}},1), \quad 0 \le \mathrm{u} = \tilde{\mathrm{u}} \le 1. \tag{1}$$

We also require

$$\mathrm{p}(t,0) = \mathrm{q}(t,1), \quad 0 \le t \le 1, \tag{2}$$

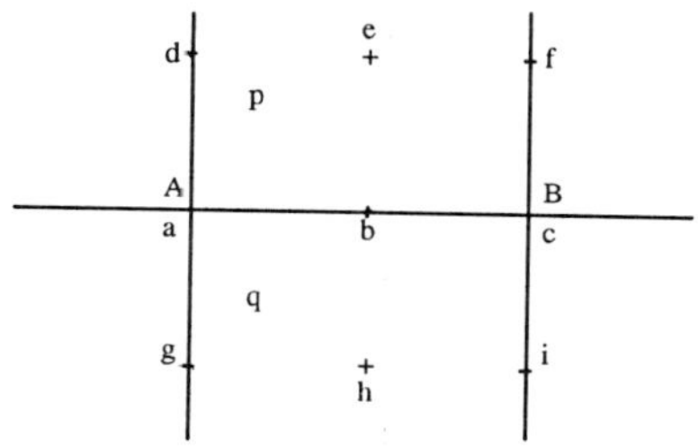

Figure 1. Between two vertices.

and

$$p_v(t,0) = q_v(t,1) + l(t)p_u(t,0), \quad 0 \le t \le 1, \tag{3}$$

for some polynomial l . We wish that whenever possible, this third condition will reduce to C^1 continuity, that is $l \equiv 0$.

Let us begin by examining two faces of a cube which are adjacent. We denote a vertex of the cube by A_1 and one of the other vertices joining to it by an edge, by B_1. Both A_1 and B_1 are singular. Along all such edges, we require $l(A_1) = -1$ and $l(B_1) = 1$ and we choose l to be linear, so that $l(t) = \gamma(1-t) + \delta t$, where $\gamma, \delta \in \mathbb{R}$ such that $-1 \le \gamma,\ \delta \le 1$. Along all other edges, we have C^1 continuity. We now construct the space V(M), where M is a SQM with 24 faces and 96 faces respectively. This is the case after a cube has been subdivided once and twice respectively using the scheme mentioned earlier. The space where M is a cube has been dealt with by Goodman [5].

§4. SQM upon Subdivision of a Cube

Upon *subdivision* of a cube, we will have 24 new faces which are all singular. A singular vertex is always joined by an edge to a nonsingular vertex, but a nonsingular vertex may also be joined by an edge to a nonsingular one.

In this case we assume that any function f in V(M) coincides with a biquadratic polynomial on any face $\{(u,v) : 0 \le u \le 1,\ 0 \le v \le 1\}$. Using the usual Bézier representation, this has the form

$$p(u,v) = \sum_{i,j=0}^{2} a_{ij} \binom{2}{i} \binom{2}{j} u^i (1-u)^{2-i} v^j (1-v)^{2-j},\ 0 \le u \le 1,\ 0 \le v \le 1. \tag{4}$$

Next, we consider how to match f on two adjacent faces of M. Suppose that on one face, f is given by (4) and on an adjacent face it is given by

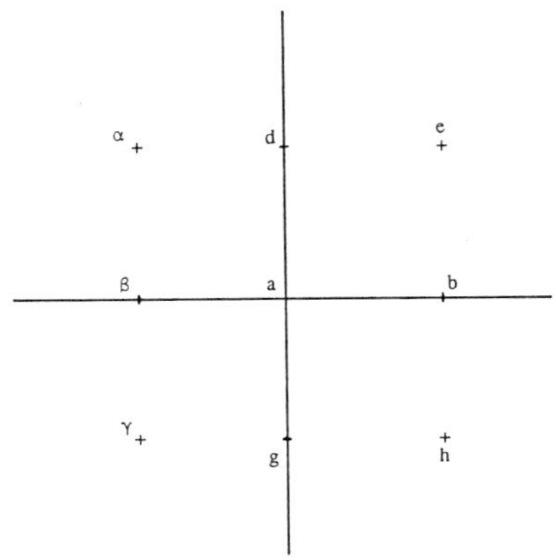

Figure 2. At a vertex.

$$q(\tilde{u},\tilde{v}) = \sum_{i,j=0}^{2} \tilde{a}_{ij}\binom{2}{i}\binom{2}{j}\tilde{u}^i(1-\tilde{u})^{2-i}\tilde{v}^j(1-\tilde{v})^{2-j},\ 0 \le \tilde{u} \le 1,\ 0 \le \tilde{v} \le 1. \tag{5}$$

Condition (2) says that $a_{i0} = \tilde{a}_{i2}$, for $i = 0, 1$ and 2. For simplicity, we shall write $a_{00} = a$, $a_{10} = b$, $a_{20} = c$, $a_{01} = d$, $a_{11} = e$, $a_{21} = f$, $\tilde{a}_{01} = g$, $\tilde{a}_{11} = h$, and $\tilde{a}_{21} = i$ (see Figure 1).

Across any edge joining two nonsingular vertices, we assume that l is identically zero. Thus, it suffices to look at only one of the 24 faces. Referring to Figure 1, suppose that A is a singular vertex (where $t = 0$) and B is nonsingular (where $t = 1$). Since B is the midpoint of an edge joining A to another singular vertex and l is linear, we require that l(B)=0. Thus, putting $l(t) = -(1-t),\ 0 \le t \le 1$ in (3), and using the notation as in Figure 1, we have from (4) and (5):

$$3a = b + d + g, \tag{6}$$

$$5b = 2e + 2h + c, \tag{7}$$

and

$$2c = f + i. \tag{8}$$

Similarly, if A is nonsingular and B is singular we have

$$2a = d + g, \tag{9}$$

$$5b = 2e + 2h + a, \tag{10}$$

$$3c = b + i + f. \tag{11}$$

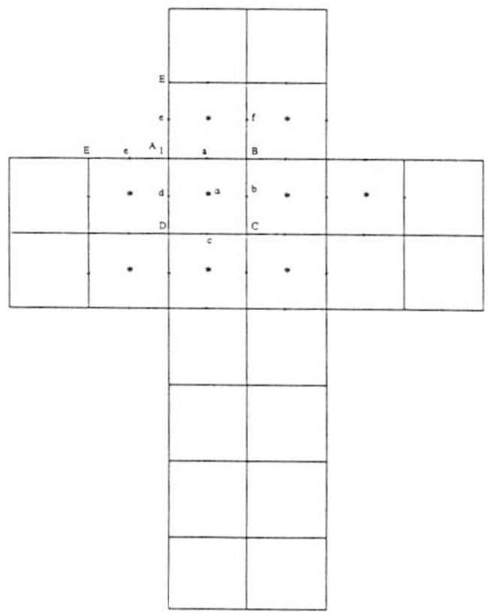

Figure 3.
Bézier coefficients of a singular face.

Thus any element of V(M) is described by all of the Bézier coefficients on all of the faces, namely one at each vertex, one in the interior of each edge and one in the interior of each face, while these coefficients are subject to the conditions (6), (7) and (8) for each edge. At any singular vertex, we have one condition (corresponding to (6)), which says that the coefficient at the vertex is the average of three adjoining edge coefficients.

We now consider the conditions at a nonsingular vertex. Denoting the vertex coefficient by a and the adjacent edge coefficients by b,d,g and β as in Figure 2, the conditions are

$$2a = b + \beta = d + g. \tag{12}$$

Recalling the edge conditions (7) and (10) together with the conditions

$$2d = \alpha + e, \tag{13}$$

and

$$2g = \gamma + h, \tag{14}$$

we have $5\beta = 2\alpha + 2\gamma + a$ and $5b = 2e + 2h + a$, that is $5(b + \beta) = 2(e + \alpha) + 2(h + \gamma) + 2a$, which will reduce to $2a = d + g$. This shows that the two conditions in (12) are not independent and in fact (12) reduces to only one condition. Thus the conditions on the Bézier coefficients can be described by one condition at each vertex and one condition (condition (7)) on each edge.

Theorem 1. *dim V(M) = 24.*

Proof: The total number of Bézier coefficients is 98. But the number of independent conditions is at most 74. So dim V(M) $\geq 98 - 74 = 24$.

We prove equality by showing that all the coefficients are determined by the 24 coefficients in the interiors of the faces. So we assume these coefficients are zero, and deduce that the other coefficients are also zero (refer to Figure 3).

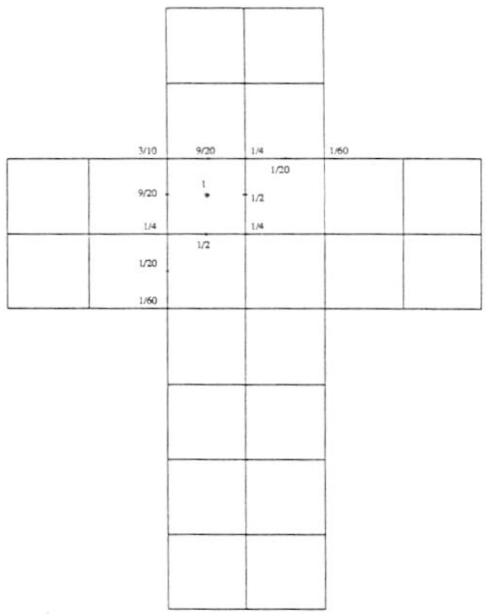

Figure 4.
A basis function when a face coefficient = 1.

By the conditions of the form (12), (13) and (14) we see that all the coefficients along the edge joining two nonsingular vertices are zero. This in turn will lead to the coefficient on the edge joining nonsingular and singular vertices to be zero (condition 7), and the coefficient of a singular vertex will then be zero (condition 6). ■

We now return to Figure 3. Consider $\alpha = 1$ and all the other face coefficients equal to zero. By using conditions (6) to (14) appropriately, we have $\mathrm{b} = \mathrm{c} = 1/2$. We also have $\mathrm{f} = 0$, $\mathrm{C} = \mathrm{D} = 1/4$. With $3\mathrm{A}_1 = \mathrm{a} + \mathrm{e} + \mathrm{d}$, $5\mathrm{a} = 2 + \mathrm{B}$ and $2\mathrm{B} = \mathrm{b} + \mathrm{f}$, we have $\mathrm{B} = 1/4$, $\mathrm{a} = \mathrm{d} = 9/20$, $\mathrm{e} = 0$ and so $\mathrm{A}_1 = 3/10$.

Thus we have constructed a function such that the coefficient for that face is 1 and the other face coefficients are zero (see Figure 4). For each of the other 23 faces we have a corresponding function. These 24 functions sum to 1. Thus we have constructed a basis for V(M) of positive functions that sum to one.

These functions can be used to design a closed G^1 surface using the following standard procedure. Denote the functions by S_i , $i = 1, \ldots, 24$. Given any points M_i in $\mathbb{R}^3$, $i = 1, \ldots, 24$, we can consider the parametrically defined surface

$$s = \sum_{i=1}^{24} \mathrm{M}_i \mathrm{S}_i. \tag{15}$$

Then the surface lies in the convex hull of $M_1, \ldots, M_{24}$ and we can alter the shape of the surface by moving the points M_i.

We can also consider an interpolation problem with these basis functions.

Theorem 2. *Let c_1, ..., c_{24} respectively be the points at the centres of the 24 faces. Then given a_1,..., a_{24} in $\mathbb{R}$, there exists a unique element f of V(M) satisfying $f(c_i) = a_i$, $i = 1, 2, \ldots, 24$.*

Proof: Let S_1 be the function defined as above with Bézier coefficient 1 in the face with centre c_1 . Since the basis functions sum to one and are all positive, it suffices to show that S_1 when evaluated at c_1 will have a value larger than 1/2.

Evaluating S_1 at $u = v = 1/2$ will give us the value 177/320 which is larger than 1/2. Therefore we have a system of linear equations which is strictly diagonally dominant, thus there exists a unique solution. ∎

From this theorem, we can take points $N_i = (x_i, y_i, z_i)$ in $\mathbb{R}^3$, $i = 1, \ldots, 24$ such that there will exist (A_j, B_j, C_j) in $\mathbb{R}^3$, $j = 1, \ldots, 24$ such that

$$\sum_{j=1}^{24} (A_j, B_j, C_j) S_j(c_i) = (x_i, y_i, z_i), \;\; i = 1, \ldots, 24$$

giving a surface of the form (15) interpolating the points $N_1, \ldots, N_{24}$.

§5. Further Subdivision of A Cube

Upon further subdivision, we get M with 96 faces. There are 24 singular faces and the rest are nonsingular. We assume that any function f in V(M) will coincide with a certain polynomial on a face. On any nonsingular face that is not adjacent to a singular one, f will coincide with a biquadratic polynomial of the form (4). However, for faces adjacent to a singular face, f will coincide with a function of the form

$$\begin{aligned} q(u, v) = &\sum_{i=0}^{3}\sum_{j=0}^{1} a_{ij} \binom{3}{i}\binom{3}{j} u^i(1-u)^{3-i}v^j(1-v)^{3-j} \\ &+ 3v^2(1-v)\{a_{02}(1-u)^2 + 2Au(1-u) + a_{32}u^2\} \\ &+ v^3\{a_{03}(1-u)^2 + 2Bu(1-u) + a_{33}u^2\}. \end{aligned} \tag{16}$$

Finally on any of the 24 singular faces, f will coincide with a bicubic polynomial, which, using the Bézier representation, has the form

$$r(u, v) = \sum_{i,j=0}^{3} a_{ij} \binom{3}{i}\binom{3}{j} u^i(1-u)^{3-i}v^j(1-v)^{3-j}. \tag{17}$$

We now consider matching f on two adjacent faces of M. Suppose that both faces are also adjacent to singular faces, and on one face f is given by (16) and on an adjacent face it is given by

$$\begin{aligned} \tilde{q}(\tilde{u}, \tilde{v}) = &\sum_{i=0}^{3}\sum_{j=2}^{3} \tilde{a}_{ij} \binom{3}{i}\binom{3}{j} \tilde{u}^i(1-\tilde{u})^{3-i}\tilde{v}^j(1-\tilde{v})^{3-j} \\ &+ 3\tilde{v}(1-\tilde{v})^2\{\tilde{a}_{01}(1-\tilde{u})^2 + 2C\tilde{u}(1-\tilde{u}) + \tilde{a}_{31}\tilde{u}^2\} \\ &+ (1-\tilde{v})^3\{\tilde{a}_{00}(1-\tilde{u})^2 + 2D\tilde{u}(1-\tilde{u}) + \tilde{a}_{30}\tilde{u}^2\}, \end{aligned} \tag{18}$$

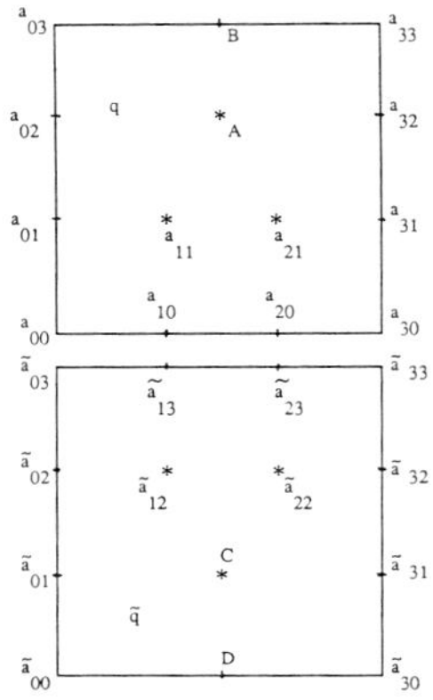

Figure 5.
Matching on adjacent faces, both adjacent to singular faces.

(see Figure 5). The same conditions (1), (2) and (3) will again be used. Condition (2) says that

$$a_{i0} = \tilde{a}_{i3}, \quad i = 0, 1, 2, 3.$$

Putting

$$l(t) = \gamma t + \delta(1-t), \quad 0 \le t \le 1$$

in (3) and using the notation as in Figure 5, we have upon simplification :

$$a_{01} - a_{00} = a_{00} - \tilde{a}_{02} + \delta(a_{10} - a_{00}), \tag{19}$$

$$a_{11} - a_{10} = a_{10} - \tilde{a}_{12} + 2\delta(a_{20} - a_{10})/3 + \gamma(a_{10} - a_{00})/3, \tag{20}$$

$$a_{21} - a_{20} = a_{20} - \tilde{a}_{22} + 2\gamma(a_{20} - a_{10})/3 + \delta(a_{30} - a_{20})/3, \tag{21}$$

$$a_{31} - a_{30} = a_{30} - \tilde{a}_{32} + \gamma(a_{30} - a_{20}). \tag{22}$$

Let us now match f across an edge where l is identically zero. Suppose we assume that on one face f is given by (16) and on an adjacent face it is given by

$$\begin{aligned} \mathrm{s}(\tilde{\mathrm{u}}, \tilde{\mathrm{v}}) = &\sum_{i=0}^{3}\sum_{j=0}^{1} \tilde{a}_{ij} \binom{3}{i}\binom{3}{j} \tilde{\mathrm{u}}^i (1-\tilde{\mathrm{u}})^{3-i} \tilde{\mathrm{v}}^j (1-\tilde{\mathrm{v}})^{3-j} \\ &+ 3\tilde{\mathrm{v}}^2(1-\tilde{\mathrm{v}})\{\tilde{a}_{02}(1-\tilde{\mathrm{u}})^2 + 2\tilde{A}\tilde{\mathrm{u}}(1-\tilde{\mathrm{u}}) + \tilde{a}_{32}\tilde{\mathrm{u}}^2\} \\ &+ \tilde{\mathrm{v}}^3\{\tilde{a}_{03}(1-\tilde{\mathrm{u}})^2 + 2\tilde{B}\tilde{\mathrm{u}}(1-\tilde{\mathrm{u}}) + \tilde{a}_{33}\tilde{\mathrm{u}}^2\}, \end{aligned} \tag{23}$$

we obtain :

$$2a_{30} = a_{20} + \tilde{a}_{10}, \tag{24}$$

$$2a_{31} = a_{21} + \tilde{a}_{11}, \tag{25}$$

$$2a_{32} = A + \tilde{A}, \tag{26}$$

$$2a_{33} = B + \tilde{B}. \tag{27}$$

Similarly, by considering the matching of f where on one face f is given by (17) and on an adjacent face it is given by (23), we have (24), (25) and

$$5a_{32} = 3a_{22} + 2\tilde{A}, \tag{28}$$

$$5a_{33} = 3a_{23} + 2\tilde{B}. \tag{29}$$

Matching f on two adjacent faces with f given by (16) on one face and on an adjacent face by

$$\mathrm{m}(\tilde{\mathrm{u}}, \tilde{\mathrm{v}}) = \sum_{i,j=0}^{?} \tilde{a}_{ij} \binom{2}{i} \binom{2}{j} \tilde{\mathrm{u}}^i (1 - \tilde{\mathrm{u}})^{2-i} \tilde{\mathrm{v}}^j (1 - \tilde{\mathrm{v}})^{2-j}, \tag{30}$$

will lead to

$$5a_{03} = 3a_{02} + 2\tilde{a}_{01}, \tag{31}$$

$$5B = 3A + 2\tilde{a}_{11}, \tag{32}$$

$$5a_{33} = 3a_{32} + 2\tilde{a}_{21}. \tag{33}$$

Finally, for singular faces adjacent to each other where on one face f has the form (17) and on the other it has the form

$$\tilde{\mathrm{r}}(\tilde{\mathrm{u}}, \tilde{\mathrm{v}}) = \sum_{i,j=0}^{3} \tilde{a}_{ij} \binom{3}{i} \binom{3}{j} \tilde{\mathrm{u}}^i (1 - \tilde{\mathrm{u}})^{3-i} \tilde{\mathrm{v}}^j (1 - \tilde{\mathrm{v}})^{3-j}, \tag{34}$$

matching f across their common edge will give us the same equations (19)–(22).

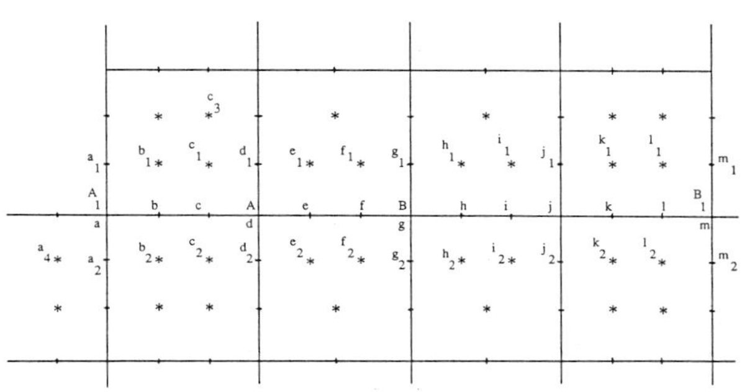

Figure 6. Faces above and below an edge joining singular vertices.

Now, let us consider the 8 faces that lie just above and below the edge joining singular vertices A_1 and B_1. For simplicity, let us denote the Bézier coefficients as in Figure 6. Using the appropriate values for γ and δ in equations (19)–(22) so that $l(A_1)$ = -1 and $l(B_1) = 1$, let us consider the conditions at the two nonsingular vertices A and B. At A, we have

$$6c = 3c_1 + 3c_2 - b + d, \tag{35}$$

$$3d = 2d_1 + 2d_2 - c, \tag{36}$$

$$5d = 2d_1 + 2d_2 + e, \tag{37}$$

$$7e = 3e_1 + 3e_2 + f. \tag{38}$$

From (24) and (25) we also have

$$2d_1 = c_1 + e_1, \tag{39}$$

$$2d = c + e, \tag{40}$$

$$2d_2 = c_2 + e_2. \tag{41}$$

Thus, from (35) and (38) we obtain

$$11d = 6d_1 + 6d_2 - b + f - e. \tag{42}$$

But from (36) and (37), we obtain

$$11d = 6d_1 + 6d_2 - 2c + e. \tag{43}$$

Therefore, we have two conditions

$$2d = c + e = d_1 + d_2 + (e - c)/4 \tag{44}$$

at A. These conditions will reduce to one condition if and only if

$$2(e - c) = f - b. \tag{45}$$

From (24) and (25), we have

$$2g_1 = f_1 + h_1, \tag{46}$$

$$2g = f + h, \tag{47}$$

$$2g_2 = f_2 + h_2, \tag{48}$$

at B. We also have

$$13f = 6f_1 + 6f_2 + g, \tag{49}$$

$$2g = g_1 + g_2, \tag{50}$$

$$13h = 6h_1 + 6h_2 + g. \tag{51}$$

(49) and (51) give $13(f + h) = 6(f_1 + h_1) + 6(f_2 + h_2) + 2g$, which by using (46), (47) and (48) and upon simplification, gives $2g = g_1 + g_2$. Thus, conditions (47) and (50) reduce to one condition and are not independent.

From above, we have two conditions at a nonsingular vertex adjoining a singular vertex and only one condition at any other vertex.

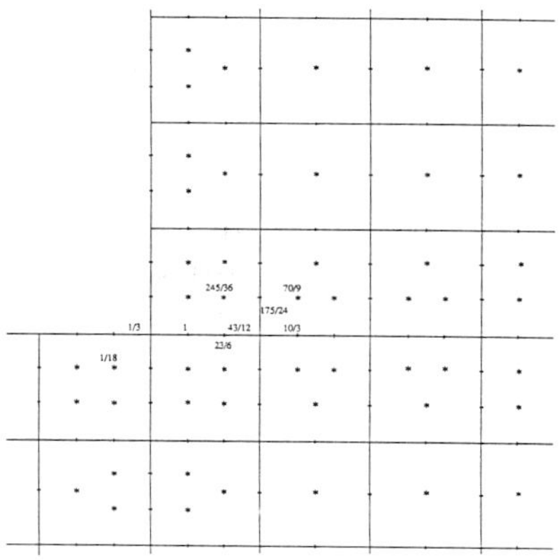

Figure 7.
Construction of a function associated with a singular face with one of the edge coefficients adjacent to singular vertex = 1.

Theorem 3. *If V(M) is defined as in this section where M is a SQM with 96 faces, then dim V(M) $\geq$ 240.*

Proof: Let e, v and f respectively denote the total number of all the Bézier coefficients in the interior of the edges, at the vertices and in the interior of the faces of M. Thus the number of Bézier coefficients is $e + v + f$. But the number of independent conditions is at most $e + v + 24$ and so

dim V(M) $\geq (e + v + f) - (e + v + 24) = f - 24 = 264 - 24 = 240$. ■

What we would like to do is construct a function associated with each face such that it is positive with small support and the 96 functions sum to 1. From the above theorem, it is obvious that we have many degrees of freedom to do this. We start by using the constraint of the form (45) to reduce the two conditions at the vertex to one condition. Let us first construct a function associated with a singular face. Suppose we let one of the boundary coefficients around the singular vertex equal 1 and the other two equal zero. Using the notation as in Figure 6, this means that $b = 1$, $a_1 = a_2 = 0$. From (19) with $\delta = -1$, that is

$$3a = a_1 + a_2 + b, \tag{52}$$

which implies $a = 1/3$. From (20) with $\delta = -1$ and $\gamma = -1/2$, we have

$$15b = 6b_1 + 6b_2 + 4c - a \tag{53}$$

and with $b_1 = b_2 = 0$ we obtain $c = 23/6$. Using (35) with $c_2 = 0$ will give us $c_1 = 245/36$. With $f_1 = f_2 = h_1 = h_2 = 0$, we get $f = g = 0$ from (49) and (50). Using (39), (40) and (41) with $e_2 = 0$, we have from (38) that $e = 10/3$, $e_1 = 70/9$, $d = 43/12$ and $d_1 = 175/24$. Similarly, with all the other face coefficients equal zero we get $a_4 = 1/18$ (see Figure 7). (Note that only the non-zero coefficients are stated in Figure 7 and the following figures.)

We proceed further by adding corresponding coefficients in Figure 7 and its symmetrical counterpart and then divide by 2. Call this case 1.

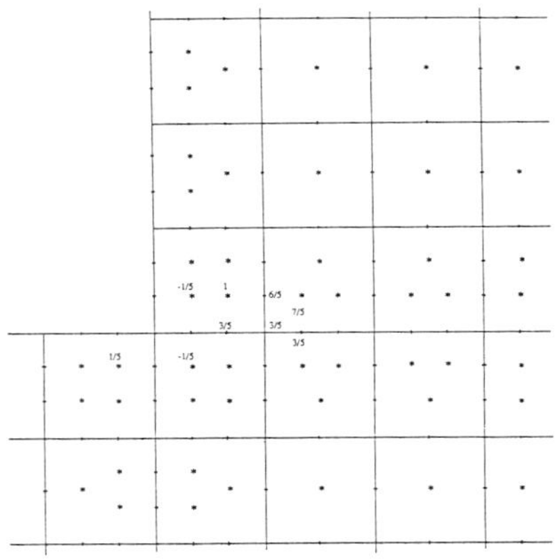

Figure 8.
Construction of a function associated with a singular face with one of the face coefficients = 1.

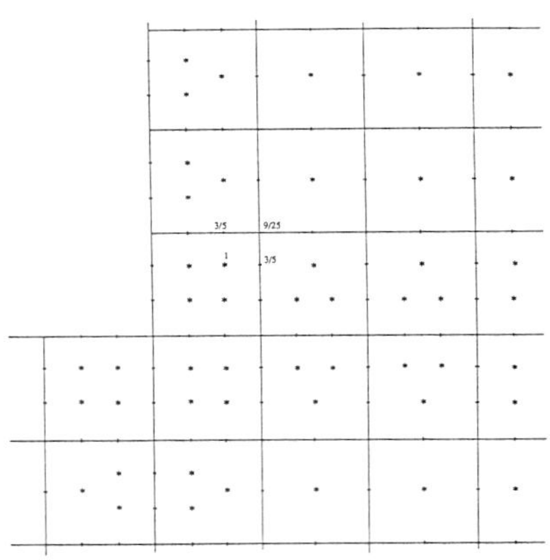

Figure 9.
Construction of a function associated with a singular face with the face coefficient farthest from singular vertex = 1.

Let us refer back to Figure 6, where this time we assume $c_1 = 1$ and all the coefficients adjacent to singular vertex equal zero. This will give us $a = 0$. By following a similar approach to that above with $c_2 = e_2 = f_1 = f_2 = h_1 = h_2 = 0$, we get coefficients as in Figure 8. As was done in case 1, we add corresponding coefficients in Figure 8 and its symmetrical counterpart and divide by 2. Then we multiply every coefficient by 85/18. We now subtract these resulting coefficients from corresponding ones produced earlier from case 1. Call this case 2.

By putting $c_3 = 1$ in Figure 6 and all the other face coefficients and the adjacent coefficients to singular vertices equal zero we get Figure 9. Now, from case 2 and Figure 9, as in Figure 10, the desired function associated with a singular face is obtained. This function has a support of 8 squares.

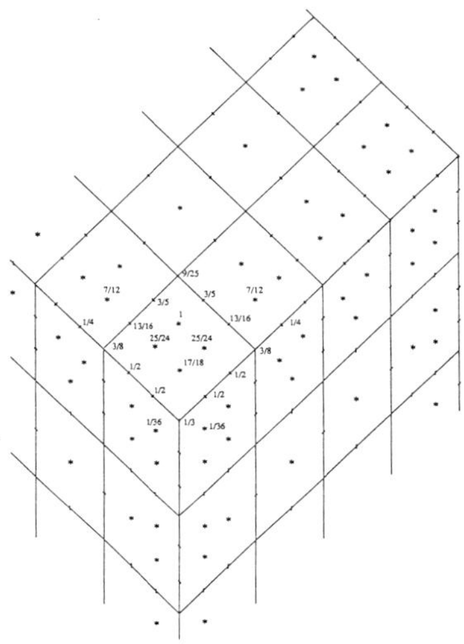

Figure 10. A function associated with a singular face.

For a nonsingular face which is adjacent to a singular face, by using a similar approach, we can construct a function associated with the face, as given in Figure 11 which has a support of 9 squares. This function is slightly negative with a minimum value of -0.00037 at the point u = 0.143 and v = 0.269. These values are obtained using the relevant routines E04KAF or E04KCF in the NAG Fortran Library (NAG Ltd. 1990). Altogether, there are 48 functions of this type. Finally for a face that is not adjacent to a singular face, we have a function which is associated with the face as illustrated in Figure 12. This function also has a support of 9 squares. Thus, we have constructed functions which will form a basis for the required subspace of V(M). Except for the case, as in Figure 11, where it is slightly negative, all the other functions are positive. Furthermore, the functions sum to 1.

The functions that we have constructed can be used to design a closed G^1 surface using the same standard procedure we have mentioned in the previous section. Thus given any points M_i in $\mathbb{R}^3$, $i = 1, \ldots, 96$ the surface will lie in the convex hull of $\mathrm{M}_1, \ldots, \mathrm{M}_{96}$ and moving the points M_i will alter the shape of the surface. Let us denote these basis functions by S_i, $i = 1, \ldots, 96$.

They are also useful in constructing interpolating surfaces.

Theorem 4. *For $i = 1, \ldots, 96$ let c_i respectively be the centroid of the face i of M. Then given $(\mathrm{x}_i, \mathrm{y}_i, \mathrm{z}_i)$ in $\mathbb{R}^3$, $i = 1, \ldots, 96$ there exists a unique set of points $(\alpha_j, \beta_j, \gamma_j)$ in $\mathbb{R}^3$, $j = 1, \ldots, 96$ such that*

$$\sum_{j=1}^{96} (\alpha_j, \beta_j, \gamma_j) S_j(c_i) = (\mathrm{x}_i, \mathrm{y}_i, \mathrm{z}_i), \ \ i = 1, \ldots, 96.$$

Proof: Let F_1 be a singular face with centre c_1 and F_2 be its adjacent face

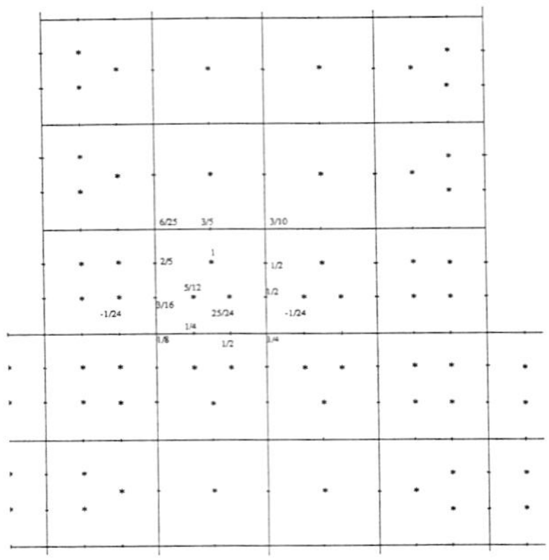

Figure 11.
A function associated with a nonsingular face adjacent to a singular face.

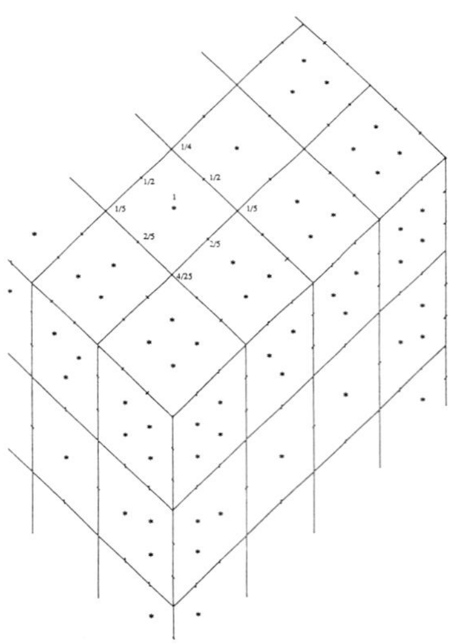

Figure 12.
A function associated with a nonsingular face away from an edge joining two singular vertices.

with centre c_2. Let F_3 be a nonsingular face with centre c_3 which is adjacent to F_2 but have no common edge with a singular face.

Let S_1 be the basis function that is associated with F_1,S_2 with F_2 and S_3 with F_3. Since all the basis functions sum to 1, it suffices to show that S_i when evaluated at c_i, $i = 1, 2, 3$ will have a value larger than 1/2, and we also need to show that all basis functions are positive at any centroid of their respective supports. Evaluating each of the functions S_i at the respective centroids of the faces F_i, $i = 1, 2, 3$ give us $S_1(c_1) = 31301/38400$, $S_2(c_2) = 15261/25600$ and $S_3(c_3)$=841/1600 which are clearly larger than 1/2. Since S_1 and S_3 are positive it is obvious that they must be positive at any centroid within their respective supports. For S_2 we need only show that it is positive at c_1 and c_3

(see Figure 11). Evaluating S_2 at both centroids, we obtain $S_2(c_1)=701/25600$ and $S_2(c_3) = 199/2560$.

Therefore we have a system of linear equations which is strictly diagonally dominant, and thus there exists a unique solution. ■

Acknowledgements. The author wishes to thank Dr. T. N. T. Goodman and Dr. K. Unsworth for their constructive criticism of the first version of this paper. This work was made possible by a study grant, Academic Staff Higher Education Scheme from Universiti Sains Malaysia.

References

1. Ball, A. A. and D. J. Storry, Conditions for tangent plane continuity over recursively generated B-spline surfaces, ACM ACM Trans. Graphics **7** (1988), 83–102.
2. Boehm, W., G. Farin, and J. Kahman, A survey of curve and surface methods in CAGD, Computer Aided Geom. Design **1** (1984), 1–60.
3. Catmull, E. and J. Clark, Recursively generated B-spline surface on arbitrary topological meshes, Comput. Aided Design **10** (1978), 350–355.
4. Doo, D. and M. A. Sabin, Behaviour of recursive subdivision surfaces near extraordinary points, Comput. Aided Design **10** (1978), 356–360.
5. Goodman, T. N. T., Closed surfaces defined from biquadratic splines, Constr. Approx. **7** (1991), 149–160.
6. Goodman, T. N. T., and S. L. Lee, Geometrically continous surface defined parametrically from piecewise polynomials, in *The Mathematics of Surfaces II*, R. Martin, (ed.), Clarendon, Oxford, 1987, 343–361.
7. Gregory, J. A. and J. M. Hahn, Geometric continuity and convex combination patches, Computer Aided Geom. Design **3** (1987), 79–89.
8. Höllig, K. and H. Mögerle, G-splines, Computer Aided Geom. Design **7** (1990), 197–207.
9. Lee, S. L. and A. A. Majid, Closed smooth piecewise bicubic surfaces, ACM ACM Trans. Graphics , to appear.
10. Lee, S. L. and A. A. Majid, Smooth piecewise biquartic surfaces on arbitrary polyhedral mesh, preprint.
11. NAG Ltd., NAG Fortran Library Manual, Mark 14, NAG Ltd. Wilkinson House, Jordan Hill Road, Oxford, 1990.
12. Nasri, A. H., Polyhedral subdivision methods for free-form surfaces, ACM ACM Trans. Graphics **6** (1987), 29–73.
13. Wick, J. J. van, Bicubic patches for approximating non-rectangular control -point meshes, Computer Aided Geom. Design **3** (1986), 1–13.

A. Rahni M. Piah
Department of Mathematics and Computer Science
The University of Dundee
Dundee DD1 4HN, Scotland.

The Virtues of Cyclides in CAGD

Mike Pratt

Abstract. This paper summarises the current state of research into the uses of cyclides in CAGD, gives an account of their virtues and limitations, and suggests directions for future work. Space only allows a cursory treatment of recent published research, and references are given throughout to original papers giving fuller details. The main message is that cyclides are surfaces with important virtues; they have low algebraic and parametric rational polynomial degree, and they have a simple geometric characterisation permitting their use in an intuitive way for the design of simple blends. However, they appear not to provide quite enough flexibility for free-form surface design, which prompts a search for appropriate generalisations with more degrees of freedom.

§1. Surface Representations

We start with a brief review of the major characteristics of surface representations as used in CAGD, namely parametric, implicit and geometric (the last in a sense to be defined shortly). Some practical CAD systems also use *procedural* representations, in which surfaces are not represented by formulae but in terms of defining data and a set of computational procedures operating on it. The *chord-height blend* surface in the PDGS system developed by the Ford Motor Co. provides a good example [15]. However, such definitions are very nonstandard and give rise to many problems both from a mathematical and from a practical engineering point of view. They will therefore be ignored in what follows.

§1.1. Parametric Representation

Parametric representations in CAGD conventionally use polynomial or rational functions. The choice is one of computational convenience; the rationals are the most complex functions which can be evaluated exactly (in principle)

Mathematical Methods in Computer Aided Geometric Design II
Tom Lyche and Larry L. Schumaker (eds.), pp. 457–473.

ISBN 0-12-460510-9.

using the four basic arithmetic operations of addition, subtraction, multiplication and division. Commercial systems differ in the maximum degree implemented, which ranges between two and twenty. Parametric representations may of course also be based on other classes of functions, as witness do Carmo [8] or any other textbook on differential geometry. A simple example is the well-known sine/cosine representation of the circle; this is exact, requires very little defining data and gives a parametrisation proportional to arc-length, which is convenient for many applications. The advantages of such representations should be more widely considered, and this is a point which will be returned to later. Parametric formulations of geometry in CAGD usually take the form of linear combinations of scalar basis functions having vector-valued coefficients (this is true also for rational representations provided we are prepared to work in the appropriate projective space). The advantage is that the application of a transformation leads to minimal changes in the representation of geometry; the defining coefficients may be transformed independently whilst the basis functions remain unchanged.

Polynomial and rational parametric surface geometry also has significant disadvantages. Firstly, it is difficult for traditionally trained draughtsmen and designers to use, since particularly for surface design it forces them to work three-dimensionally. The Bézier and B-spline formulations, based on the use of control points, seem to be best for the designer, but even so a total of 48 scalar values must somehow be provided for the definition of even a bicubic surface patch. It is not easy to shield the user from all this complexity.

The second disadvantage is a related one. Sederberg [25] first pointed out to the CAGD community that polynomial and rational curve and surface representations can in principle be recast in implicit algebraic form. A four-sided polynomial patch of degree n has algebraic degree $2n^2$, so that even the lowly bicubic patch has algebraic degree 18. This is some indication as to why it requires a lot of defining data, why it is sometimes not easy to control and why its intersection curves with other surfaces can be remarkably complex and difficult to compute.

§1.2. Implicit Representation

An implicit surface equation has the form $f(x, y, z) = 0$. In the CAGD context the only class of implicit surfaces which has been studied in any detail is that of *algebraic* surfaces, for which the function f is a polynomial in x, y and z. The algebraic surfaces of degree two are the quadrics. The *natural quadrics* (the plane, circular cylinder, right circular cone and the sphere) are widely used in solid modelling, but the more general forms such as the elliptic hyperboloid have so far found little application in CAGD, and there is insufficient freedom in such surfaces to allow the construction, for example, of smooth free-form surfaces out of quadric patches. The cubic algebraic surfaces are more promising. Sederberg [26] has pointed out that these are the lowest degree algebraic surfaces from which tangent-plane continuous piecewise free-form surfaces may be constructed. They are also the highest degree of such surfaces which generally admit of a rational parametrisation, a fact which is useful, for example, in the computation of intersection curves [23].

The advantage of working with low-degree algebraic surfaces is that we can obtain good freedom for design whilst avoiding the complexity which results from geometric computations involving parametric surfaces which, as pointed out earlier, have an algebraic degree much higher than their parametric degree. Algorithms for computing with low-degree algebraic surfaces can therefore be made robust and relatively fast, both valuable virtues in the CAGD context.

The chief problem of working with these surfaces is that of providing the CAD system user with suitable 'handles' for their definition and manipulation. It is certainly not possible to predict, for example, what will be the effect of a particular choice of coefficients in an algebraic representation. Sederberg [24] has suggested a design technique based on a variation of the Bézier approach to curve and surface definition, but this has not yet been developed to a stage where it is suitable for use by the mathematically unsophisticated user of a practical design system. Bajaj [1] gives a good review of some of the problems to be solved in the use of algebraic surfaces for CAGD.

§1.3. Geometric Representation

The term *geometric representation* was coined by Goldman [13] for a method of describing commonly used curves and surfaces in terms of a few salient parameters. For example, an infinite right circular cylinder is completely specified by (a) a code denoting the nature of the surface, (b) a straight line, the axis of the cylinder, and (c) a scalar giving the radius. The straight line may be defined in various ways, for example in terms of a point and a direction. The first item, the code, is essential to ensure the correct interpretation of the ensuing data. This type of representation is conceptually very simple. It has the disadvantage that different kinds of curves and surfaces all have their own distinctive data requirements, so that no 'unified' formulation is possible. On the other hand, the information stored in a geometric representation is often in the precise form required by application programs making use of the geometry, for example in the generation of manufacturing data. Furthermore, the existence of a simple geometric representation usually implies that a designer can gain a good intuitive grasp of the nature of a surface and the ways it may be handled in a CAD environment. However, this type of representation is not particularly well suited to many of the geometric interrogations and computations which are often required in a CAD context.

§1.4. Ideal Surfaces for CAGD

As we have seen, each type of surface representation has its advantages and disadvantages for CAGD. In order to obtain all the advantages but avoid most of the disadvantages we would like to identify surface types with the following characteristics:

1.) Having the necessary flexibility for free-form surface design when used in a piecewise manner.
2.) Having a low-degree parametric representation for compatibility with existing CAD systems and data exchange standards.
3.) Having a low-degree algebraic representation allowing the use of rapid and robust geometric algorithms.
4.) Having a simple geometric representation providing the unsophisticated CAD user with 'handles' for easy surface creation and manipulation.

It may be noted that even the widely-used parametric bicubic surface fails on the third and fourth of these criteria. With these aims in mind, the cyclide surfaces will now be introduced. It will be shown that they satisfy three of the conditions listed above, and only just fail to satisfy the remaining one.

§2. The Dupin Cyclides

Early in the last century, at the age of 17, the French mathematician and naval architect Dupin defined a class of surfaces which he named *cyclides* [9]. These were subsequently further investigated by Maxwell [18] in the context of a problem in electromagnetic theory, and by several other writers from a purely mathematical point of view. Darboux [5], rather unfortunately, used the term 'cyclide' to refer to a wider class of surfaces, and we here restrict attention to the Dupin cyclides. After the end of the 19th century, interest in the cyclides declined for many years, though they are mentioned as surfaces having noteworthy properties in several differential geometry texts dating from some eighty years ago. One such book is that of Forsyth [12], which gives equations for the cyclides in both algebraic and parametric forms, as follows:

Algebraic:

$$(x^2 + y^2 + z^2 - \mu^2 + b^2)^2 = 4(ax - c\mu)^2 + 4b^2y^2\ ;$$

Parametric (for $0 \le \theta < 2\pi, 0 \le \psi < 2\pi$):

$$x(\theta, \psi) = \frac{\mu(c - a\cos\theta\cos\psi) + b; 2\cos\theta}{a - c\cos\theta\cos\psi}\ ,$$

$$y(\theta, \psi) = \frac{b\sin\theta(a - \mu\cos\psi)}{a - c\cos\theta\cos\psi}\ ,$$

$$z(\theta, \psi) = \frac{b\sin\psi(c\cos\theta - \mu)}{a - c\cos\theta\cos\psi}\ .$$

Here x, y, z are cartesian coordinates, while a, b, c and μ are constants, only three of which are independent since $c^2 = a^2 - b^2$. In the parametric form the parameters are θ and ψ. We see that algebraically the cyclides are generally surfaces of the fourth degree (quartics). A particular member of the family is distinguished by values of the three constants a, c, μ, and it will be shown that these have a simple geometric interpretation. Furthermore, although the parametric form is expressed in terms of trigonometric functions of the parameters, the standard substitution for these functions in terms of tangents of half-angles leads to a rational biquadratic formulation. This possibility was noted by Martin [16], and the specific details of the rational Bézier form were given in an earlier paper by the present author [22]. It should be noted that the cyclides satisfy Items 2, 3 and 4 of the list of desirable surface properties given in Section 1.4. The extent to which they satisfy Item 1 will be discussed later.

§2.1. Geometry of the Cyclides

A useful review of cyclide geometry is given in [3]. The more important points are summarised here. The simplest form is the *ring cyclide*, which resembles a torus, though the minor radius varies around the central hole. There are two mutually perpendicular planes of symmetry, in which the cross-sections consist of two circles whose radii are simple functions of a, c and μ, as shown in Figure 1. In fact, by comparison with the torus, a is analogous to the major radius, μ to the minor radius and c is a measure of the asymmetry of the surface about the third coordinate plane ($c = 0$ gives a standard torus). There are also two self-intersecting forms of the cyclide, arising respectively when the cross-sectional circles in the x, y-plane and the z, x-plane intersect. These are the *horned cyclide* and the *spindle cyclide*. All three forms are illustrated in Figure 2. There is yet another form of the cyclide in which some of the circles degenerate to straight lines; this is not a closed surface, and its algebraic equation reduces to cubic degree.

It is important to note that the three constants defining a particular cyclide have a straightforward geometric interpretation which may be easily grasped by an engineering designer. Another significant point is that if the radii of the cross-sectional circles in the two planes of symmetry are known the cyclide is completely determined; this fact proves very useful in CAGD applications, as will be shown. Further notable properties of the cyclides are listed below:

1.) Both families of lines of curvature on a cyclide are circles. Forsyth [12] shows that in his parametrisation given earlier these are also the isoparametric lines of the surface. The cross-sectional circles in the planes of symmetry are lines of curvature.
2.) The constant distance parallel offset of a cyclide is also a cyclide, differing from the original only in the value of the constant μ. This is proved in [22].
3.) The planes of the circular lines of curvature of either family all intersect in a common line.
4.) Along any line of curvature on a cyclide the angle ϕ between the surface normal and the principal normal to the curve is constant. This is a consequence of Joachimstal's Theorem [8]. It follows that there is a right

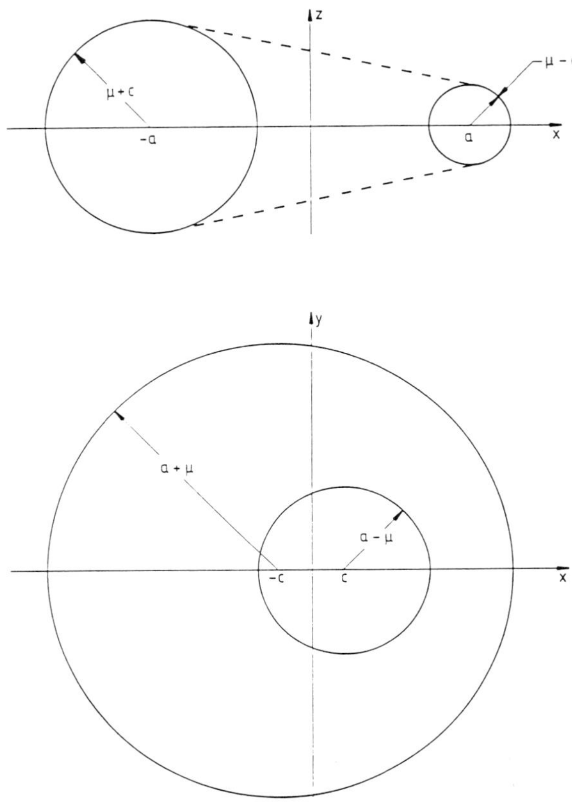

Figure 1. Cross-sections of a Dupin cyclide in its planes of symmetry.

circular cone, with semi-vertex angle ϕ, tangent to a cyclide around any of its lines of curvature (see Figure 3). The existence of these tangent cones is fundamental to the use of cyclides as blending surfaces, as will be explained later.

5.) With regard to inversion in a sphere, the inverse of a Dupin cyclide is also a Dupin cyclide provided the centre of inversion does not lie on the original surface [14]. A use of this property will be noted later in §3.4.

Fischer [10] has published a book containing photographs of 19th-century plaster-of-paris models of cyclides and many other surfaces. A modern treatment of the analytic geometry of cyclides may be found in Fladt and Baur [11].

§2.2. Cyclide Patches and Piecewise Cyclide Surfaces

The renaissance of interest in cyclides came in the early 1980's at Cambridge University in the course of some research into novel methods of computer aided

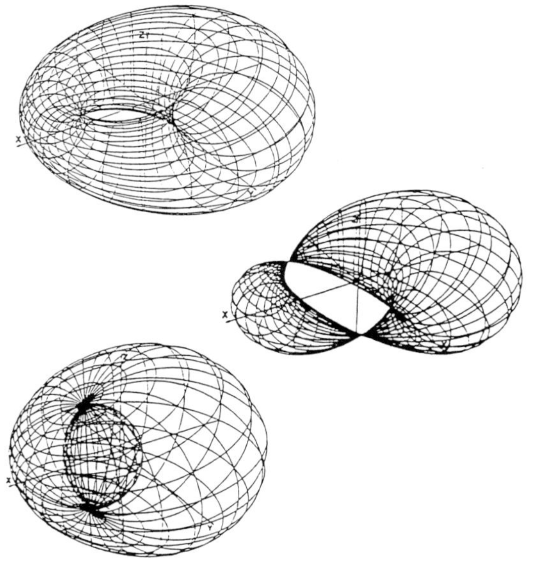

Figure 2. The ring, horned and spindle forms of the Dupin cyclide.

design initiated by A. W. Nutbourne. The idea was to create surfaces having curvature characteristics prescribed to some extent in advance by starting with a network of curves which were required to be lines of curvature of the surface generated. The initial choice of curve type was piecewise circular, since this was expected to lead to a comparatively simple geometrical analysis. The corresponding surface type has a piecewise definition, each patch being bounded by four circular arcs which are also lines of curvature of the patch surface. It was soon realised that such patches are regions of Dupin cyclides.

Martin [16] has given the name *principal patches* to all such surface regions bounded by lines of curvature. He developed methods for the construction of cyclide patches and proved the important result that the four corners of a cyclide patch are coplanar, lying on a circle (the simplest proof uses quaternions). Another obvious restriction, since the two lines of curvature through any point are mutually orthogonal, is that adjacent patch boundaries meet at right angles. Subsequently de Pont [7] extended Martin's work, studying intersections involving cyclides, inter-patch blending, local modifications and various generalisations of the cyclide patch. The most promising of these is the *cyclide Q-patch*, an assembly of four cyclide patches which together possess degrees of freedom for design similar to those of the conventional parametric bicubic patch. Sharrock [27] later examined the requirements for a practical design system based on the use of cyclide patches.

He studied the construction of three-dimensional networks of piecewise circular arcs as frameworks for smooth assemblies of cyclide patches. He also experimented with optimisation techniques for ensuring surface 'fairness'.

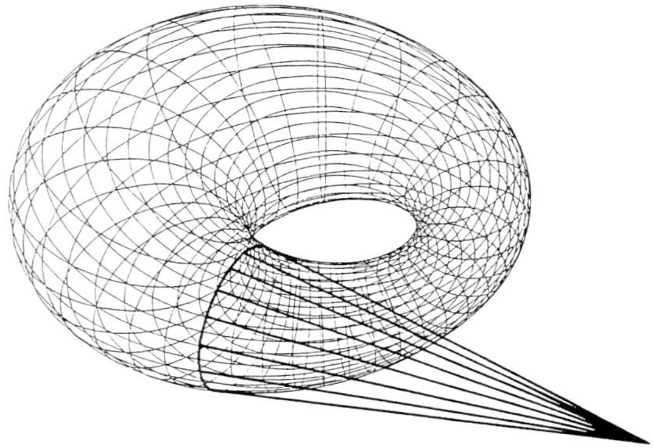

Figure 3. The tangent cone to a line of curvature on a cyclide.

Much of the work mentioned is summarised in [17].

A single cyclide patch is bounded by four lines of curvature of the cyclide to which it belongs. In [16] it is shown that under the parallel offset operation the circular lines of curvature on an original cyclide offset in their plane of definition to become lines of curvature on the offset cyclide. Thus the offset of a cyclide patch is another cyclide patch. Consider now a G^1 assembly of two cyclide patches sharing a common boundary arc. If both patches are offset simultaneously the offset boundary arc remains common to both offset patches, and the lines of curvature in the transverse direction remain G^1 continuous. In fact the same property extends to G^1 arrays of any number of cyclide patches; in short, the parallel offset property of the cyclide extends to piecewise defined G^1 surfaces composed of cyclide patches.

The conclusion reached by the Cambridge team is that there is not quite sufficient geometric freedom in the cyclide patch to provide adequate flexibility for free-form design. This was confirmed by McLean [19], who developed a method for the construction of piecewise cyclide surfaces in terms of the spherical representations [8] of its individual patches. The problem stems from the requirement mentioned earlier that patch corners must lie on a circle. This has the effect of decreasing design freedom as a surface is created by building onto an initially defined patch. Since it appears that only slightly more flexibility is needed for effective free-form design, it will be worthwhile to search for generalisations providing more geometric freedom. Some possibilities will be mentioned later.

Before leaving the topic of cyclide patches a few words should be said about their Bézier representation. As already mentioned this is rational biquadratic. Since each patch boundary is a circular arc the positions of the eight boundary control points are clearly at the patch corners and at the intersection of the corner tangent vectors; these are easily determined from the geometry of the arcs. The position of the central control point is less immediately clear. Investigation shows that the tangent planes at the four patch corners meet in a common point, and this is the location of the desired control point. This was first proved by Degen [6] in a more general context. Computational formulae for the control points and weights are given by Pratt [22]. If conventional Bézier patches are used, each boundary arc must turn through less than 180 degrees, so that a minimum of nine patches is needed to represent a complete cyclide. Boehm [2] and Mellado [20] have independently shown that the use of 'control points at infinity', as advocated by Piegl [21], allows the use of patch boundaries which turn through a full 180 degrees, so that the complete cyclide can be represented by just four patches. Even so, the amount of defining data is still quite large compared with the three constants needed for the geometric representation, though admittedly this will require an additional transformation matrix to give a cyclide in a general position and orientation. Note also that the original, trigonometric parametrisation given by Forsyth has the virtue of representing the complete cyclide without resort to piecewise methods. It has the further advantage that all isoparametric circles are uniformly parametrised with respect to arc-length. This is not immediately apparent, but is simply verified. It is a useful property in many applications, and one which does not hold for the Bézier formulation. These facts support the point made earlier that parametrisations based on functions other than polynomials may have practical advantages which deserve further exploration.

§3. Cyclides in Solid Modelling

Solid modelling has now been under development for more than twenty years. Early systems of this type could create models of objects bounded by planes, circular cylinders, cones, spheres and (in some cases) tori. All of these except the last are quadric surfaces. The tori are quartics, and they therefore fitted rather uneasily into an overall framework which was largely based on computations involving quadrics. In some cases the general quadrics were also implemented to obtain better 'geometric coverage' of engineering parts, but they were found to give little added practical advantage. There are several clues here to the way forward. Firstly the useful quadrics are the *natural quadrics*, all of whose lines of curvature are straight lines or circles. These surfaces, but not the more general quadrics, are all limiting cases of the cyclide. Reference to Figure 1 indicates that the cylinder results when $c = 0$, μ is finite and a is infinite. If μ also becomes infinite the cylinder unrolls to give a plane. The cone is a limiting case of the horned cyclide in which the cone vertex is a self-intersection point of the cyclide. The sphere occurs (twice) as the limiting case of the spindle cyclide when the cross-sectional circles in both planes of symmetry coincide. The torus is also a special case of the cyclide for which $c = 0$, as already noted. It therefore seems natural to seek improvement in the range of shapes which can be modelled by implementing the general cyclides in solid modellers. Early investigations show that the cyclides are very useful in extending the primary function of the torus in solid modelling, the blending of sharp edges and corners.

§3.1. Cyclide Blending - Symmetric Cases

The uses of cyclides as blending surfaces have been studied by Dutta [4], Boehm [2] and the present writer [22]. The first reference analyses the problem of approximate blending using piecewise cyclides, while the other two are concerned with exact blending. In all cases it is tangent-plane or G^1 continuity which is sought in constructing the blends. Only the exact blending case will be discussed here.

The fundamental property needed for cyclide blending is Property 4 from Section 2.1, which states that around any line of curvature of a cyclide there exists a right circular tangent cone. Then if we can find a surface which blends smoothly into a cone around one of its circular meridians, it will also blend smoothly into any cyclide tangent to the cone around that circle. The initial problem, then, is how to construct blends to cones. Two subcases arise, (a) the blend between a cone and a sphere, and (b) the blend between two cones. In what follows right circular cylinders may be regarded as special cases of cones.

The cone/sphere blend is easy to construct. We may work in the plane of symmetry of the configuration, which contains the axis of the cone and the centre of the sphere. This will also be a plane of symmetry of the blending cyclide. First choose a meridian of the cone to act as the first blend boundary, which will intersect the plane of symmetry in two points. Next, working in that plane, construct two circles each tangent to the cone at one of these points and also tangent to the third circle which is the cross-section of the sphere (see Figure 4). These circles form the cross-section of the blending cyclide; the three constants a, c and μ defining it may be found from their centres and radii (see Figure 1). The cyclide is tangent to both the cone and the sphere around a circle on each surface by virtue of the symmetry of the situation. Note that one degree of freedom is available to the designer, the choice of the first blend boundary, but that everything else follows automatically. The method also works when the sphere degenerates to a plane. The salient property shared by spheres and planes is that at points on *any* circle on either surface, whether a line of curvature or not, there is a constant angle ϕ between the principal curve normal and the normal to the surface.

Blending between two cones is more fundamental, because it is the basis of blending between two general cyclides. It is also less straightforward. Sabin has shown that two cones possess a cyclide blend similar to that described in the last paragraph provided they are both tangent along meridional circles to a common sphere. A proof is given in [22]. This condition clearly requires the axes of both cones to pass through the centre of the sphere, so that they define a plane of symmetry. For this case the generation of the blending cyclide is much as previously. A first blend boundary is chosen, and circles are constructed in the plane of symmetry tangent to the first cone where the

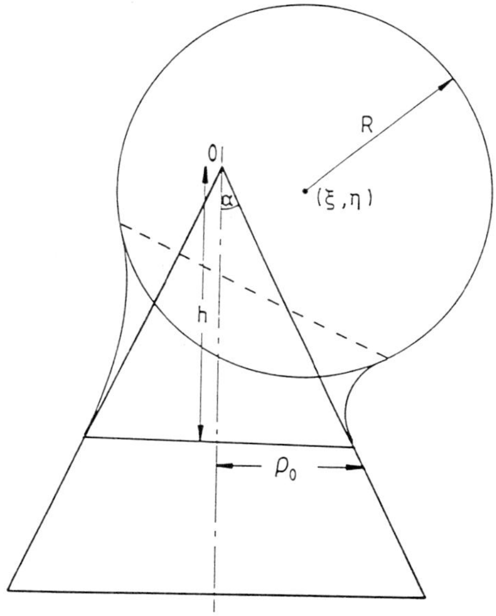

Figure 4. Construction of cone/sphere blend.

meridian intersects this plane, and also tangent to the second cone. If Sabin's condition applies, the points of tangency on the second cone will lie on a common meridian, which will form the second boundary of the blend. As in the cone/sphere blend there is one degree of freedom, the choice of the first blend boundary. As shown in Figure 5, the inscribed sphere condition permits us to construct four different possible blends between two cones.

More difficult is the case where the cone axes intersect (so that there is a plane of symmetry) but the inscribed sphere condition does not hold. The present writer has shown [22] that a cyclide blend may still be constructed, but that it must be built up from parts of two cyclides. Thus the general cone/cone blend is a *double-cyclide blend* rather than a *simple cyclide blend* as described for the two previous cases. Suppose that a meridional blend boundary has been chosen on both of the cones to be blended. Then a third, intermediate cone may be constructed which shares one tangent sphere with the first cone and another with the second. Then simple cyclide blends may be constructed between each of the original cones and the intermediate cone. The resulting configuration is G^1 continuous, but it can be improved upon. The two blend boundaries lying on the intermediate cone are both meridional circles, and by suitable adjustment of the blend boundaries on the two original cones they may be made to coincide. The result is a composite blend made up of two cyclides. Analysis [22] shows that there remains one degree of freedom, the choice of position of the intermediate blend boundary. The designer may

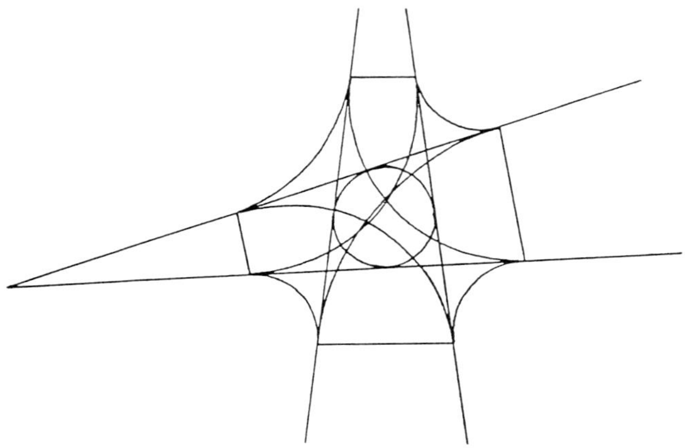

Figure 5. Four possible simple cyclide blends between a pair of cones.

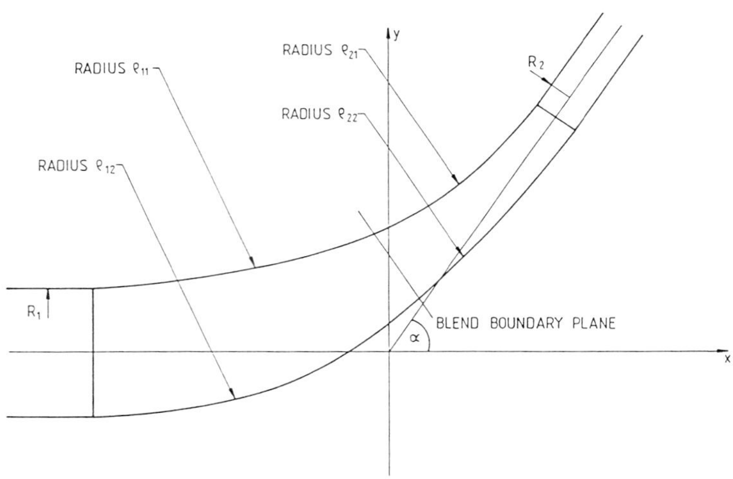

Figure 6. Double-cyclide blend between two cylinders.

be absolved from making this choice if a suitable default is implemented. A convenient choice is to require equal angles between the axes of the first and intermediate cone and those of the intermediate and second cone. This results in a smooth change in axial direction and a 'fair' blend. A double-cyclide blend is illustrated in Figure 6.

Cone/cone blending is the key to blending between cyclides in general, and in particular between the surfaces traditionally used in solid modelling, which are all special cases of cyclides. Cyclide blending as described above therfore provides a useful extension to blending capability in solid modelling. However, all the blends so far discussed have circular boundaries and require the surfaces being blended to share a common plane of symmetry. Such blends occur quite frequently in the design of engineering objects, but these are nevertheless severe restrictions. In the following sections it will be shown how they may be at least partially overcome.

§3.2. Cyclide Blending - Asymmetric Cases

First we examine the possibility of blending in asymmetric situtations. It is easy to see that this is possible. Consider the symmetric double-cyclide blend shown in Figure 6. The intermediate blend boundary is a circle lying on the intermediate cone. Suppose now that the part of the configuration to the left of the intermediate blend boundary is held fixed while the part to its right is rotated about the axis of the intermediate cone. The axis of the right-hand cylinder is thereby rotated out of the plane of symmetry and an asymmetric configuration results. Since both parts of the composite blend surface remain tangent to the intermediate cone, G^1 continuity of the blend is maintained. The question is, given the inverse situation where the two original cylinders have skew axes, how can the details of a double-cyclide blend between them be computed? An answer is sketched below.

In fact asymmetric blends may be created in more general ways than this, since given any simple or double-cyclide blend a similar rotation may be performed about the axis of any one of the complete circular lines of curvature lying on the blend surface; the use of the intermediate blend boundary as above is merely one of infinitely many possibilities.

Before leaving the topic of asymmetric blends a few details will be given about a practical procedure for their computation. Denote the axes of two cones to be blended by L_1, L_2, and suppose that these are skew lines. Now choose points P_1, P_2 on each of these lines respectively, and consider the effect of rotating about P_1P_2. Assume that this line is not the common perpendicular, since in that case a special situation arises which allows little freedom for blending. Suppose, in particular, that L_1 is not perpendicular to P_1P_2. Rotation of L_2 about P_1P_2 sweeps out a double cone (reducing to a plane if the lines are perpendicular) which intersects L_1 at two (possibly coincident) points. This indicates that a rotation can be found which brings the two cone axes into coplanarity. Taking a local origin of coordinates at P_1, with L_1 as the x-axis and P_1P_2 lying in the x, y-plane, it is a simple matter to compute the effect of a rotational transformation through an angle γ on the direction vector of the axis L_2. Then we require the z-component of the transformed vector to be zero, which immediately gives an expression for $tan\gamma$. Rotation by γ about P_1P_2 then gives a symmetric configuration, which may be blended using the double-cyclide method given earlier, with the intermediate cone having P_1P_2 as its axis. Finally, a reverse rotation $-\gamma$ is applied to the cone whose axis is L_2, together with its attached part of the composite blend surface.

§3.3. Cyclide Blending - Multi-surface Blends

The cyclide blends so far discussed provide smooth transitions between two surfaces, and are restricted to circular blend boundaries. Methods have also been found for blending more than two surfaces simultaneously. The simplest, due to Dutta and Srinivas and at present unpublished, is by using a sphere as a transitional element and reducing the problem to a set of two-surface blends. For example, suppose we have three cylindrical pipes of differing diameters and we wish them to have a smooth junction. This may be achieved by choosing a sphere of radius larger than any of the pipes and blending each pipe onto it using the simple method of Section 3.1. Some care is needed to avoid overlaps between the blends, but otherwise this approach is very straightforward.

The other multi-surface blending method is due to Boehm [2]. He notes that, in some cases when two cones have a common inscribed sphere (which it will be recalled guarantees that they possess a simple cyclide blend), they are also sandwiched between two common tangent planes. The specific cases for which this is true are those where the cones do not include each others' vertices. In such a case, if a blending cyclide is constructed, then the two planes are also tangent to the cyclide around lines of curvature. This permits the construction of a blended configuration of two cones, a cyclide and two planar regions, and allows blending in more general situations than those discussed in the earlier sections, as Boehm shows in his paper. He also demonstrates a more sophisticated method than that described above for blending between three cones using spherical surfaces.

One interesting aspect of the use of planar and spherical regions in the construction of blends is that it enables the blending of configurations which exhibit umbilics. It is known that cyclides in general have no umbilical points; the only exceptions are the sphere and the plane, for which all points are umbilical. The insertion of a region from one of these surfaces can therefore be used to introduce an umbilic.

§3.4. Cyclide Intersections

The computation of cyclide blends as described earlier does not require the calculation of intersections, but a practical solid modelling system must be capable of creating intersection curves between any of the surface types implemented. Methods are therefore needed for handling intersections between cyclides (which include all the 'traditional' solid modelling surface types). It was pointed out by de Pont [7] that we may make use of the alternative algebraic and parametric formulations of the cyclide. If the parametric expressions for the cartesian coordinates of a point on one cyclide are substituted into the algebraic equation for another cyclide the result can be expressed as an equation relating the surface parameters of the first cyclide, an expression of the intersection curve in parameter space. This equation has overall algebraic degree eight, but is only quartic in each of the variables. It may therefore in principle be solved exactly for one of the parameters, for each of a sequence of specified values of the other.

Johnstone [14] has recently given an alternative method, based on inversion (see Item 5 in the list of cyclide properties in Section 2.1). He considers the intersection between a cyclide and what he calls a *ringed surface*, the envelope of the general spatial motion of a variable radius circle. The overall intersection is found in terms of the intersection points of the cyclide

with particular circles from the ringed surface. In detail, Johnstone's method involves a sequence of inversions, but the main principle is to invert each chosen circle into a straight line. The intersection problem then becomes a sequence of line/cyclide problems, each expressible in terms of a quartic equation of one of the surface parameters.

§4. Summary and Conclusions

The intention of this paper has been to review the virtues of the Dupin cyclides as surfaces for practical shape modelling. They have low degree when regarded as either algebraic or rational parametric surfaces, and this brings with it a freedom from pathological behaviour and the possibility of fast, robust geometric computations. They also have a simple geometric representation which makes them easy for a designer to comprehend. It is easy to pass between the various representations when one or other is most convenient for some specific purpose. The cyclides have been shown to provide a very useful extension of 'geometric coverage' in solid modelling, primarily by their utility as blending surfaces in a range of restricted but commonly occurring situations. There remains scope for further research in this area to extend the range of techniques so far discovered. However, the geometrical freedom in the cyclides has proved marginally insufficient to allow their effective use in free-form surface modelling. For this reason it has been suggested that appropriate generalisations be sought which retain the benefits of the cyclides as far as possible whilst providing the necessary additional flexibility.

References

1. Bajaj, C. L., Geometric modelling with algebraic surfaces, in *The Mathematics of Surfaces III*, D. C. Handscomb (ed.), Oxford University Press, 1989.
2. Boehm, W., On cyclides in geometric modelling, Computer Aided Geom. Design **7** (1990), 243–255.
3. Chandru, V., Dutta, D. and Hoffmann, C. M., On the geometry of Dupin cyclides, The Visual Computer **5** (1989), 277-290.
4. Chandru, V., Dutta, D. and Hoffmann, C. M., Variable radius blending using Dupin cyclides, in *Geometric Modeling for Product Engineering*, M. J. Wozny, J. Turner and K. Preiss (eds.), North-Holland Publ. Co., 1990.

5. Darboux, G., *La théorie générale des surfaces*, Gauthier-Villars, Paris, 1887.
6. Degen, W. L. F., Generalised cyclides for use in computer aided geometric design, in A. D. Bowyer (ed.), *The Mathematics of Surfaces IV*, Oxford University Press (in preparation).
7. de Pont, J. J., Essays on the cyclide patch, PhD dissertation, Cambridge University Engineering Dept., 1984.
8. do Carmo, M. P., *Differential Geometry of Curves and Surfaces*, Prentice-Hall, 1976.
9. Dupin, C., *Applications de géometrie et de méchanique*, Bachelier, Paris, 1822.
10. Fischer, G. (ed.), *Mathematische Modelle*, Vieweg & Sohn, Braunschweig & Wiesbaden, 1986.
11. Fladt, A. and Baur, K. *Analytische Geometrie spezieller Flächen und Raumkurven*, Vieweg & Sohn, Braunschweig & Wiesbaden, 1975.
12. Forsyth, A. R., *Lectures on the Differential Geometry of Curves and Surfaces*, Cambridge University Press, 1912.
13. Goldman, R. N., Two approaches to a computer model for quadric surfaces, IEEE Comp. Graph. Appl. **3** (1983), 21–24.
14. Johnstone, J. K., A New Intersection Algorithm using Circle Decomposition, preprint from 2nd SIAM Conf. on Computer Aided Geometric Design, Nov. 1991, Tempe, AZ.
15. Lachance, M. A., Data proliferation in the exchange of surfaces, in *The Mathematics of Surfaces II*, R. R. Martin (ed.), Oxford University Press, 1987, 475–490.
16. Martin, R. R., Principal patches for computational geometry, PhD dissertation, Cambridge University Engineering Dept., 1982.
17. Martin, R. R., J. de Pont, and T. J. Sharrock, Cyclide surfaces in computer aided design, in *The Mathematics of Surfaces*, J. A. Gregory (ed.), Oxford University Press, 1986, 253–267.
18. Maxwell, J. C., On the cyclide, Quart. J. Pure & Applied Math. **IX** (1868) 111–126.
19. McLean, D., A method of generating surfaces as a composite of cyclide patches, Comput. J. **28** (1985), 433–438.
20. Mellado, M., Computational experiments on cyclide surfaces, MSc thesis, Cranfield Institute of Technology, 1989.
21. Piegl, L., Infinite control points - a method of representing surfaces of revolution using boundary data, IEEE Comp. Graph. Appl. **7** (1987), 45–55.
22. Pratt, M. J., Cyclides in computer aided geometric design, Computer Aided Geom. Design **7** (1990), 221–242.
23. Pratt, M. J. and A. D. Geisow, Surface/surface intersection problems, in *The Mathematics of Surfaces*, J. A. Gregory (ed.), Oxford University Press, 1986, 117–142.
24. Sederberg, T. W., Piecewise algebraic surface patches, Computer Aided Geom. Design **2**, 53–59.

25. Sederberg, T. W., Anderson, D. C. and Goldman, R. N., Implicit representation of parametric curves and surfaces, Computer Vision, Graphics & Image Processing **28**, 72–84.
26. Sederberg, T. W. and J. P. Snively, Parametrization of cubic algebraic surfaces, in *The Mathematics of Surfaces II*, R. R. Martin (ed.), Oxford University Press, 1987, 299–319.
27. Sharrock, T. J., Surface design with cyclide patches, PhD dissertation, Cambridge University Engineering Dept., 1985.

Mike Pratt
Applied Mathematics & Computing Group
School of Mechanical Engineering
Cranfield Institute of Technology
Cranfield, Bedford MK43 0AL, ENGLAND

Simple Surfaces Have No Simple C^1 Parametrization

Hartmut Prautzsch and Wilfried Trump

Abstract. Euler's formula is used to show that surfaces diffeomorphic to spheres, half, and quarter spheres cannot be parametrized by a "simple" C^1 map.

§1. Introduction

Closed C^1 surfaces can be modeled by triangular patches. For example, in [2], Chapter 18.10, p. 327, it is shown how to represent an octant of a sphere as a rational quartic Bézier triangle. On composing a sphere with it, one obtains a so called G^1 parametrization of the sphere.

Definition. *We will say that two adjacent triangular patches* $\mathbf{p}$ *and* $\mathbf{q}$ *have a simple* C^1 *joint if the corresponding parameter domains* P *and* Q *are triangles such that after an affine mapping of* Q*, if necessary, the quadrilaterals spanned by the tangent vectors* $D_{\mathbf{u}}\mathbf{p}$, $D_{\mathbf{v}}\mathbf{p} = D_{\mathbf{v}'}\mathbf{q}$, *and* $D_{\mathbf{u}'}\mathbf{q}$ *at every common boundary point are affine images of* $P \cup Q$*, as shown in Figure 1.*

This condition for a simple C^1 joint is stated in [2], Chapter 18.6, for polynomial Bézier triangles. It is widely used in CAGD and referred to as *the* C^1 condition whereas the concept of tangent plane continuity leads to so-called G^1 conditions.

While the simple C^1 condition is usually preferred over a G^1 condition, it is well known that it is impossible to model a closed surface with genus 0 with triangular patches such that adjacent patches have a simple C^1 joint. For this reason G^1 surface schemes have been developed in, *e.g.*, [3,5]. Herron also shows that a sphere cannot be modelled with one (subdivided) triangular patch such that the simple C^1 condition is satisfied.

Mathematical Methods in Computer Aided Geometric Design II
Tom Lyche and Larry L. Schumaker (eds.), pp. 475–480.

ISBN 0-12-460510-9.

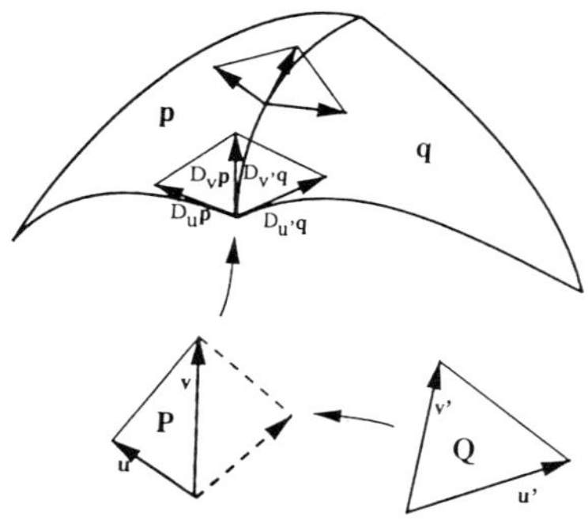

Figure 1. The simple C^1 joint.

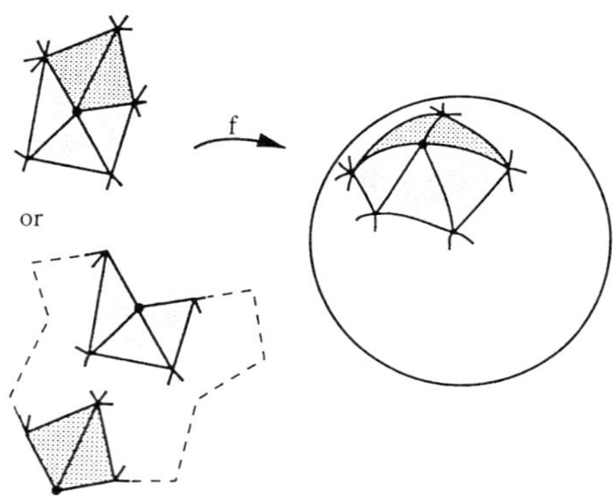

Figure 2. Mapping a subdomain onto the surface.

Here, we give a proof for the general case of arbitrary many triangular patches. The proof is extended in Section 3 allowing two further conclusions.

The results of this paper could also be obtained from some complex theorems of differential or algebraic topology. (One idea, for example, is to use vector fields and the *Poincaré–Brower theorem*, also called *the Porcupine Theorem*, (cf. [1], p. 116) for spherical surfaces.) Here a more elementary proof is presented using only the *Euler–Poincaré formula*.

§2. Simple Closed Surfaces

From the above definition we observe that a closed surface composed of regular triangular patches with simple C^1 joints can be parametrized by a regular C^1 map f defined on a triangulated domain (see Figure 2).

Then each vertex on the surface corresponds to either one interior vertex or to two or more boundary vertices of the triangulated domain.

Let V_I be the number of vertices corresponding to an interior point of the domain and V be the total number of all vertices. Then

$$2(V - V_I) + \gamma \tag{1}$$

is the number of all boundary domain vertices, where γ is some positive number. Furthermore, let F be the number of all triangular patches on the surface. It also equals the number of triangles in the domain. Then the total sum of all angles in the domain triangulation is given by

$$F \cdot 180°,$$

or equivalently by

$$V_I \cdot 360° + (2(V - V_I) + \gamma - 2) \cdot 180°,$$

which implies

$$F = 2V + \gamma - 2. \tag{2}$$

On introducing E, the number of all boundary curves of all triangular patches on the surface, we have two further identities: (1) Euler's formula for simple closed surfaces (*i.e.*, surfaces with genus 0) is (see, *e.g.*, [4])

$$V - E + F = 2. \tag{3}$$

(2) Since each triangular patch has three boundary curves and since each boundary curve belongs to two patches, we get

$$E = \frac{3}{2}F. \tag{4}$$

Equations (2–4) imply that $\gamma = -2$, which is a contradiction to above. Hence, a simple closed surface has no simple regular C^1 parametrization.

§3. Simple Open Surfaces

There is a straightforward generalization of the above proof to simple surfaces with a boundary like a half or a quarter sphere, see Figure 3. These surfaces can be obtained by cutting one hole into a simple closed surface.

As in the previous section we assume that this surface is composed of regular triangular patches with simple C^1 joints. Then it is parametrized by a regular C^1 map f defined on a triangular domain such that the boundary vertices of the surface correspond to at least one boundary vertex of the domain triangulation.

Note that if a surface boundary vertex is associated with only one domain boundary vertex, the corresponding angles on the surface and the domain are

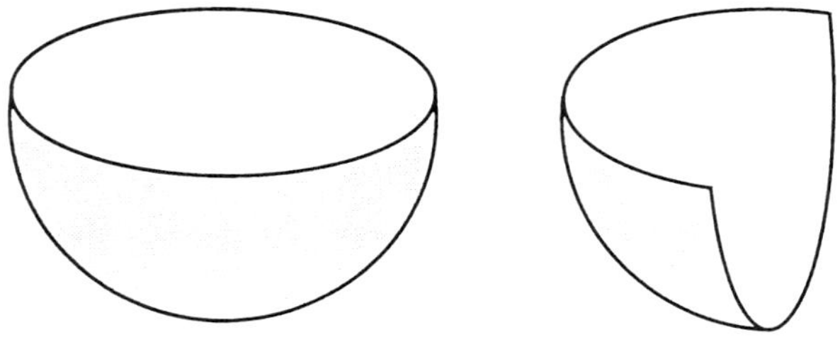

Figure 3. A half and a quarter of a sphere.

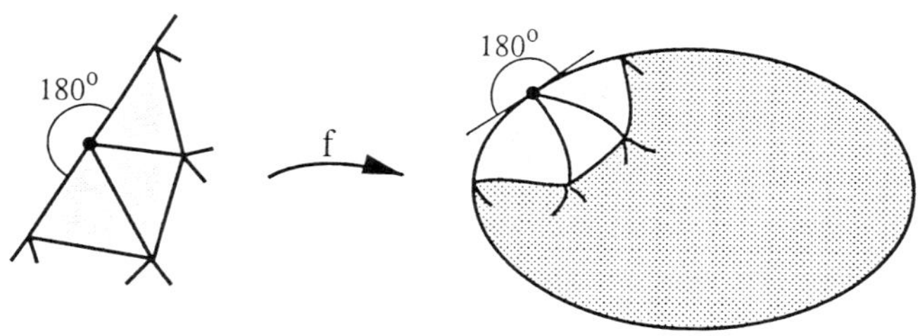

Figure 4. A 180° boundary angle.

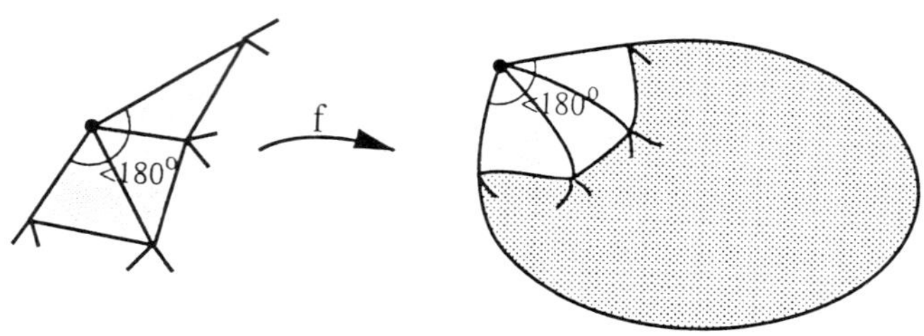

Figure 5. A convex boundary angle.

both 180° or both convex (< 180°) or both concave (> 180°); see Figures 4 and 5.

Since the boundary on the surface is a closed curve, its preimage is also closed or consists of disjoint parts on the boundary of the domain. If the preimage is closed, it must have at least three convex corners.

Therefore, if the surface boundary has no more than two convex corners, then there must be one boundary vertex on the surface with two preimages. Thus the number of all boundary vertices in the domain is given by

$$V_B + \delta + 2(V - V_B - V_I) + \gamma,$$

where V, V_B, γ are as in the previous section and V_B is the number of all surface boundary vertices and $V_B + \gamma$ the number of all their preimages. Hence $\delta \geq 0$ and if the surface boundary has less than three convex corners, even $\delta \geq 1$. Note that (1) is obtained for $V_B = \delta = 0$.

As a generalization of (3) we obtain

$$F = 2V - V_B + \gamma + \delta - 2,$$

while Euler's formula for this kind of surface is given by

$$V - E + F = 1,$$

and the relation between E and F is

$$E = \frac{3}{2}F + \frac{1}{2}V_B.$$

Together, these three identities entail that $\gamma + \delta = 0$, *i.e.*, $\gamma = \delta = 0$. Hence we conclude

Corollary. *One cannot model a simple connected surface with one closed boundary curve with triangular regular C^1 patches such that adjacent patches have a simple C^1 joint unless the boundary has at least three convex corners.*

Figure 6 shows a surface which has no simple C^1 parametrization.

Figure 6. A surface without simple C^1 parametrization.

§4. Concluding Remarks

It should be mentioned that in the above proofs the simple C^1 condition is needed only at the vertices of the triangular patches. Also we did not assume polynomial or rational patches. Furthermore, the results apply also to surfaces composed of arbitrary n-sided patches since these patches could be decomposed into triangular subpatches.

References

1. Aleksandrov, P. S., *Combinatorial Topology, Vol. 3*, Graylock Press, Albany, N. Y., 1960.
2. Farin, G., *Curves and Surfaces for Computer Aided Geometric Design*, Second Edition, Academic Press, San Diego, 1990.
3. Herron, G., Smooth closed surfaces with discrete triangular interpolants, Computer Aided Geom. Design **2** (1985), 297–306.
4. Hirsch, M. W., *Differential Topology*, Springer-Verlag, New York, 1976.
5. Piper, B., Visually smooth interpolation with triangular Bézier patches in *Geometric Modelling: Algorithms and New Trends*,G. Farin (ed.), SIAM, Philadelphia, 1987, 221–233.

Hartmut Prautzsch and Wilfried Trump
University of Karlsruhe
Am Fasanengarten 5
7500 Karlsruhe, Germany
prau@ira.uka.de, trump@ira.uka.de

Some Tools for Quasi-Interpolation on Cardinal Grids

Christophe Rabut

Abstract. We give in this paper some tools to study the properties of quasi-interpolation on cardinal grids (i.e. the mapping which, for a given function B, defines from some data $(jh, y_j)_{j\in\mathbb{Z}^d}$, the function $\sigma = \sum_{j\in\mathbb{Z}^d} y_j\, B(\bullet/h - j))$. We first attempt to give a proper definition of "quasi-interpolant" and relative notions, we then show that the "transfer function", the "interpolating power", and the "filtering power" of a quasi-interpolant are very convenient tools to analyse the properties of a quasi-interpolant; at last, we give some ways to define a "better" quasi-interpolant B_a from a given quasi-interpolant B. Examples are given, mainly for polynomial and polyharmonic splines

Introduction

Given a function B from $\mathbb{R}^d$ to $\mathbb{R}$, a vector $y = (y_j)_{j\in\mathbb{Z}^d}$, and a non-negative value h, we study in this paper the relation between y and the function σ defined (when the summation is convergent) by

$$\sigma = \sum_{j\in\mathbb{Z}^d} y_j\, B(\bullet/h - j). \qquad (1)$$

Under adequate conditions on B, the function σ is "close to" the "data points" $(jh, y_j)_{j\in\mathbb{Z}^d}$, i.e. $\forall j \in \mathbb{Z}^d \;,\quad \sigma(jh) \simeq y_j$, where the symbol $\simeq$ means "is not very different from", in a sense which has to be precised. In such a case, it is quite common to call B a *quasi-interpolant*, and σ the *quasi-interpolation of* $(jh, y_j)_{j\in\mathbb{Z}^d}$ *with respect to* B.

The aim of this paper is first to give main tools to handle such "quasi-interpolation" and to easily derive main properties of σ from properties of y and B, second to give some way to derive "better" quasi-interpolants, say B_a from a given quasi-interpolant B (i.e. to derive a new function B such $\sigma(jh)$, defined by (1) is nearer the data), and third to show some examples for which quasi-interpolation seems particularly interesting.

Mathematical Methods in Computer Aided Geometric Design II
Tom Lyche and Larry L. Schumaker (eds.), pp. 481–496.

ISBN 0-12-460510-9.

§1. What is a Quasi-interpolant?

1.1 Notations

Multi-integers

We use standard notations for multi-integers $\alpha, \beta \in \mathbb{Z}^d$:
$|\alpha| = \sum_{j=1}^{d} \alpha_j$; $\alpha! = \prod_{j=1}^{d} \alpha_j$; $x^\alpha = \prod_{j=1}^{d} x_j^{\alpha_j}$; $\mathrm{D}^\alpha f = \partial^\alpha f / \partial x^{|\alpha|}$;
$\beta \leq \alpha \iff \forall j \in [1, d]\,,\ \beta_j \leq \alpha_j$; $\binom{\alpha}{\beta} = \frac{\alpha!}{\beta!\,(\alpha-\beta)!}$

Sample, with step h, of a function

For any function f from $\mathbb{R}^d$ into $\mathbb{R}$, we call the vector f_h defined by

$$\forall j \in \mathbb{Z}^d\,, \quad (f_h)_j = f(jh). \tag{2}$$

the sample of f with step h. As a particular case, if $h = 1$, f_1 is the vector whose j^{th} component is $f(j)$.

Convolutions and hybrid convolutions

We shall use $*$ for convolutions:

Discrete convolution of two vectors: For two vectors u and v such that the summation in (3) is convergent, we define

$$\forall k \in \mathbb{Z}^d\,, \quad (u * v)_k = \sum_{j \in \mathbb{Z}^d} u_j\, v_{k-j}. \tag{3}$$

Continuous convolution of two functions: For two functions f and g such that the summation in (4) is convergent, we define

$$f * g = \int_{\mathbb{R}^d} f(x)\, g(\bullet - x) dx. \tag{4}$$

Hybrid convolution of a vector and a function: For a vector u and a function f such that the summation in 5) is convergent, we call the function defined by

$$u * f = \sum_{j \in \mathbb{Z}^d} u_j\, f(\bullet - j) \tag{5}$$

the *hybrid convolution of u and f.*

We shall write $*_h$ for *convolutions of step h*:

Hybrid convolution, with step h, of a vector and a function: Under the same condition as before, we define the *hybrid convolution with step h, of u and f* by

$$u *_h f = (u * f)(\bullet/h) = \sum_{j \in \mathbb{Z}^d} u_j\, f(\bullet/h - j). \tag{6}$$

Discrete convolution, with step h, of two functions: Finally, for two functions f and g such that the summation in (7) is convergent, we define the *discrete convolution, with step h* to be the hybrid convolution, with step h, of the vector f_h (defined from f by (2)) and the function g, i.e. the function defined by

$$f *_h g = f_h *_h g = \sum_{j \in \mathbb{Z}^d} f(jh)\, g(\bullet/h - j). \tag{7}$$

Note that this "discrete convolution of step h" is essentially non-symetric, i.e. $f *_h g$ is usually different from $g *_h f$.

Fourier transforms

We shall write $\mathcal{F}$ for the Fourier transforms:

Fourier transform of a function: if $f \in L^1(\mathbb{R}^d)$, $\mathcal{F}f$ is the function defined by

$$\mathcal{F}f = \int_{\mathbb{R}^d} f(x)\, e^{-2i\pi x \bullet} dx. \tag{8}$$

Fourier transform of a vector: if $u \in \ell^1(\mathbb{R}^d)$, we call (continous) Fourier Transform of the vector u, the function from $\mathbb{R}^d$ into $\mathbb{R}$ defined by

$$\mathcal{F}u = \sum_{j \in \mathbb{Z}^d} u_j\, e^{-2i\pi j \bullet} dx. \tag{9}$$

Discrete Fourier transform, with step h, of a function: if $f \in L^1(\mathbb{R}^d)$, we call discrete Fourier transform, with step h, of the function f, and we note $\mathcal{F}_h f$, or $\mathcal{T}f$ when $h = 1$, the Fourier transform of the vector f_h (as defined by (2) and (9)), taken at the value $h\bullet$, i.e.:

$$\mathcal{F}_h f = (\mathcal{F} f_h)(h\bullet) = \sum_{j \in \mathbb{Z}^d} f(jh)\, e^{-2i\pi jh\bullet}. \tag{10}$$

$$\mathcal{T}f = \mathcal{F}_1 f = \sum_{j \in \mathbb{Z}^d} f(j)\, e^{-2i\pi j\bullet}. \tag{11}$$

A few properties linking $*$, $*_h$, $\mathcal{F}$, and $\mathcal{T}$

In addition to the usual properties involving convolutions and Fourier transform, we can easily prove the following properties, for vectors u and functions f and g:

$$\mathcal{F}(u * f) = \mathcal{F}u \cdot \mathcal{F}f \tag{12}$$

$$\mathcal{F}(u *_h f) = h\ \mathcal{F}u \cdot \mathcal{F}f \tag{13}$$

$$\mathcal{F}(f *_h g) = h\ \mathcal{F}_h f \cdot \mathcal{F}f \tag{14}$$

$$(u *_h f)_h = u * f_1 \tag{15}$$

$$\mathcal{T}(u * f) = \mathcal{F}u \cdot \mathcal{T}f \tag{16}$$

A few properties using norms

$\|\bullet\|_1$ (resp. $\|\bullet\|_2$, resp. $\|\bullet\|_\infty$)denotes as well the L^1-norm as the ℓ^1-norm (resp. the L^2-norm and the ℓ^2-norm , resp. the L^∞-norm and the ℓ^∞-norm). Norms of periodic functions are taken over a single period. Thanks to Parseval's formula, we have

$$\|u\|_2 = \|\mathcal{F}u\|_2\,, \tag{17}$$

and so

$$\|f_1\|_2 = \|\mathcal{T} f\|_2 \,. \tag{18}$$

Furthermore,

$$\|\mathcal{F} u\|_\infty \leq \|u\|_1 \,. \tag{19}$$

1.2 Definitions and Terminology

In the literature, the notions of "quasi-interpolation " and of "quasi-interpolant " are quite commonly used, but, as far as we know, have never been precisely defined. We propose here some definitions which try to cover the general use of these terms.

Quasi-interpolation

Quasi-interpolation of a vector y, with respect to a function B, is defined by

$$\sigma = y * B \tag{20}$$

Quasi-interpolation, with step h, of a vector y, with respect to a function B, is defined by

$$\sigma_h = y *_h B \tag{21}$$

As this type of approximation of data $(jh, y_j)_{j \in \mathbb{Z}^d}$ has first been defined by I. J. Schoenberg (in one dimension) in his 1946 paper [6], so we also propose to call this "*Schoenberg's approximation (with step h) of y with respect B*".

In the sequel, $\mathcal{V}$ is a set of vectors (for example $\mathcal{V} = \mathbb{P}_k(\mathbb{Z}^d)$, i.e. the set of vectors y such there exists a polynomial p, of degree at most k, enjoying $\forall j \in \mathbb{Z}^d$, $y_j = p(j)$), and $\mathcal{K}$ is a set of function (for example $\mathcal{K} = \mathbb{P}_k(\mathbb{R}^d)$, i.e. the set of polynomials, with variable in $\mathbb{R}^d$, of total degree at most k).

$\mathcal{V}$-interpolant

A function B is said to be "$\mathcal{V}$-interpolant" if and only if for any y in $\mathcal{V}$,

$$(y * B)_1 = y \ ,$$

i.e. if and only if $\forall k \in \mathbb{Z}^d$, $\sum_{j \in \mathbb{Z}^d} y_j B(k - j) = y_k$.

A function L is said to be an interpolant if and only if for any vector y, $\sum_{j \in \mathbb{Z}^d} y_j \, L(k-j) = y_k$, i.e. if and only if $L(0) = 1$, $\forall j \in \mathbb{Z}^d - \{0\}$, $L(j) = 0$ (L is then called " cardinal function" by some authors, "Lagrangean function" by others). When L is a combination of translates of B (i.e. $L = \sum_{j \in \mathbb{Z}^d} a_j \, B(\bullet - j)$), L is called a "B-interpolant".

$\mathcal{K}$-reproducing

A function B is said to be "$\mathcal{K}$-reproducing" if and only if , for any f in $\mathcal{K}$,

$$f * B = f \ ,$$

i.e. if and only if $\sum_{j \in \mathbb{Z}^d} f(j) B(\bullet - j) = f$.

$\mathcal{K}$-quasi-interpolant

The expression "quasi-interpolant" is often used and needs be precisely defined. We think that the first notion usually required is that for some class of vectors y the function $y *_h B$ "almost interpolates" the data $(jh, y_j)_{j\in\mathbb{Z}^d}$. In these terms, thinks are far too imprecise to be usable; we think that by the notion "almost interpolates", we usually require that if the vector y is the sample, with step h, of a function f (i.e. if there is a function f such that $y = f_h$), then the vector $(y *_h B)_h$ (i.e. the sample, with step h, of the function $y *_h B$) should be all the nearer to y than h is small; for technical reasons, this will be required only on bounded sets of the variable. This can be expressed by the following:

A function B is said to be $\mathcal{K}$-quasi-interpolating if and only if , for any f in $\mathcal{K}$,

$$\forall N \in \mathbb{N}\,, \quad \sup_{\substack{k\in\mathbb{Z}^d \\ |kh|\le N}} \|(f *_h B)(kh) - f(kh)\| \xrightarrow[h\to 0]{} 0$$

A function B is said to be quasi-interpolating if and only if it is $\mathcal{K}$-quasi-interpolating for some set $\mathcal{K}$.

$\mathcal{K}$-quasi-reproducing

At last, there is another important property: we may require that, for some class of functions f, the function $f *_h B$ is in some neighbourhood of the function f. As for quasi-interpolation, we will express such a property as a limit, when h tends to zero: the notion of "quasi-interpolation" is, in other words, an asymptotical (when h goes to zero) interpolation, and we want to define "quasi-reproduction" which should be a sort of asymptotical reproduction. This can be expressed by the following:

A function B is said to be $\mathcal{K}$-quasi-reproducing if and only if , for any f in $\mathcal{K}$, and for any bounded set Ω in $\mathbb{R}^d$,

$$\sup_{x\in\Omega} |(f *_h B)(x) - f(x)| \xrightarrow[h\to 0]{} 0$$

A function B is said to be quasi-reproducing if and only if it is $\mathcal{K}$-quasi-reproducing for some set $\mathcal{K}$.

1.3 Examples

In one dimension, let $B(x) = (x+1)_+ - 2x_+ + (x-1)_+$; in two dimensions, let $B(x) = (1 + x_1 + x_2 - 2(x_1)_+ - 2(x_2)_+)_+$. The associated B-approximation are piecewise linear interpolations, they are $\mathbb{R}^{\mathbb{Z}^d}$-interpolating, $\mathbb{P}_1$-reproducing, $\mathcal{C}^0$-quasi-reproducing.

In one dimension, let $B(x) = \frac{1}{(2m-1)!}\sum_{j=-m}^{m}(-1)^{m+j}\binom{2m}{m+j}(x-j)_+^{2m-1}$, we get the usual polynomial B-approximation , which is $\mathbb{P}_1$-reproducing, $\mathcal{C}^0$-quasi-reproducing.

In d dimensions, let us define $B^h_{d,m} = h^d \Delta^m_h v_{d,m}$, where Δ_h is the discrete version (with step h) of the Laplacean operator Δ ($\Delta_h = \sum_{j=1}^d f(\bullet + he_j) - 2f + f(\bullet - he_j)$, (here e_j is the jth basis vector), $\Delta^m_h = \Delta_h \ldots \Delta_h$ and $v_{d,m}$ is a function such that $\Delta^m v_{d,m} = Dirac$. $B^h_{d,m}$ is then a "polyharmonic B-spline " as defined in [3,5]; polyharmonic B-approximation is $\mathbb{P}_1$-reproducing, $\mathcal{C}^0$-quasi-reproducing.

In d dimensions, let define B such that $\mathcal{F}B(\omega) = \frac{T(\omega)}{q(\omega)}$ where T is a 2π-periodic function whose unic zero in $[-\pi, \pi]^d$ is $\omega = 0$, q is a homogeneous function whose unic zero in $\mathbb{R}^d$ is $\omega = 0$, T and q being such that $\mathcal{F}B(\omega) = 1 + \mathcal{O}(\|\omega\|_\infty^{n+1})$ for some $n \in \mathbb{N}_*$. Then B-approximation is $\mathbb{P}_n$-reproducing. With such B's, we can build multiresolution analysis and wavelets, as shown in [2].

§2. B-approximation Viewed as a Digital Filter

Let $\sigma = y *_h B$; in this section, we only look at values $(\sigma(jh))_{j \in \mathbb{Z}^d}$: we use σ_h (as defined by (2) and (21)), i.e. the sample, with step h, of the B-approximation , with step h, of the vector y, and we study the mapping $f_B : y \rightarrow \sigma_h$; we show that this mapping is a (digital) filter and we give main tools about this digital filter: transfer function, interpolating power, and filtering power.

2.1 Digital Filter Associated with a B-approximation

Notations Let $r \in \mathbb{R}_+$. We define Y_r to be the set of vectors satisfying

$$|y_j| = \mathcal{O}\left(\|j\|_\infty^r\right) \quad \text{when } \|j\|_\infty \rightarrow \infty. \tag{22}$$

We write $|\bullet|_r$ for the norm on Y_r defined by

$$\forall y \in Y_r\ , \quad |y|_r = \max\left(|y_0|\ , \sup_{j \in \mathbb{Z}^d - \{0\}} \frac{|y_j|}{\|j\|_\infty^r}\right). \tag{23}$$

Theorem 2.1. *Let $r \in \mathbb{R}_+$; let $r' \in \mathbb{R}$ such that $r' > r + d$. Suppose B is a function from $\mathbb{R}^d$ to $\mathbb{R}$ such that $B_h \in Y_{-r'}$, and f_B is the mapping from Y_r to Y_r defined by*

$$\forall y \in Y_r\ , \quad f_B(y) = (y *_h B)_h\ . \tag{24}$$

Then f_B is a linear function, continous for the norm $|\bullet|_r$, and shift-invariant (i.e. for any $k \in \mathbb{Z}^d$ and any vector y, $f_B(T_k(y)) = T_k(f_B(y))$, where $T_k(u)$ is defined for any vector u by $(T_k(u))_j = u_{j+k}$ for any $j \in \mathbb{Z}^d$).

So f_B is called a (digital) "filter" (see [1]).

Proof: Linearity and shift-invariance are easy to prove; so let us prove continuity; writing b instead of B_1, we get

$$\begin{aligned}
\|j\|_\infty^{-r} |(b*y)_j| &= \|j\|_\infty^{-r} \left(|b_0 y_j| + |b_j y_0| + \sum_{k\neq 0,j} |b_k y_{j-k}| \right) \\
&\leq \|j\|_\infty^{-r} |b_0 y_j| + \|j\|_\infty^{-r-r'} \|j\|_\infty^{r'} |b_j y_0| \\
&\quad + \sum_{k\neq 0,j} \frac{\|j-k\|_\infty}{\|j\|_\infty^r \|k\|_\infty^r} \frac{|b_k|}{\|k\|_\infty^{-r'}} \frac{|y_{j-k}|}{\|j-k\|_\infty^r} \frac{1}{\|k\|_\infty^{r'-r}} \\
&\leq |b|_{-r'} \|y\|_r + \|j\|_\infty^{-r-r'} |b|_{-r'} |y|_r \\
&\quad + \left(\sum_{k\neq 0,j} \frac{\|j-k\|_\infty^r}{\|j\|_\infty^r \|k\|_\infty^r} \frac{1}{\|k\|_\infty^{r'-r}} \right) |b|_{-r'} |y|_r \\
&\leq A\, |b|_{-r'} \cdot |y|_r,
\end{aligned}$$

where $A = 2 + \left(\sup_{j,k\neq 0} \frac{\|j-k\|_\infty^r}{\|j\|_\infty^r \|k\|_\infty^r}\right) \left(\sum_{k\neq 0} \frac{1}{\|k\|_\infty^{r'-r}}\right)$. ■

Definition. *f_B is called the (digital) filter associated to the B-approximation.*

2.2 Transfer Function Associated with a B-approximation

Theorem 2.2. *Let r, r' and B_h satisfy the hypothesis of Theorem 2.1. Let $\omega \in \mathbb{R}$, and $\mathrm{e}_{h\omega}$ be the vector defined by*

$$\forall j \in \mathbb{Z}^d, \quad (\mathrm{e}_{h\omega})_j = e^{2i\pi jh\omega}. \tag{25}$$

Then $\mathrm{e}_{h\omega}$ is an eigenvector of f_B, the associated eigenvalue being

$$(\mathcal{T}B)(\omega) = \sum_{j\in\mathbb{Z}^d} B(j)\, e^{-2i\pi jh\omega}. \tag{26}$$

$\mathcal{T}B$ is called the transfer function of the filter f_B, or transfer function associated to the B-approximation , or simply transfer function of B.

Proof:

$$\begin{aligned}
(f(\mathrm{e}_{h\omega}))_k &= \sum_{j\in\mathbb{Z}^d} a_j (\mathrm{e}_{h\omega})_{k-j} = \sum_{j\in\mathbb{Z}^d} a_j\, e^{2i\pi(k-j)h\omega} \\
&= \left(\sum_{j\in\mathbb{Z}^d} a_j\, e^{-2i\pi jh\omega} \right) e^{2i\pi kh\omega} = (\mathcal{F}a)(\omega) \cdot (\mathrm{e}_{h\omega})_j\,. \quad ■
\end{aligned}$$

Remark

As a direct consequence of this, $|\mathcal{T}B(\omega)|$ is the coefficient of the amplification (or of the attenuation) of a (co)sinusoidal signal with frequency ω; $\arg(\mathcal{T}B(\omega))$ is the change of phase of the same signal.

2.3 Interpolating Power and Filtering Power

The whole shape of $\mathcal{T}B$ is important and often of high interest; however, in order to know rapidly whether a B-approximation is far from the interpolation or not (and in order to get some error majoration), it is also of interest to have a single number which should measure some "distance" of a B-approximation to the interpolation. In the same way it is of interest to have a number measuring the level of filtering of a B-approximation. This is the aim of the two following definitions (we remind the reader that $\|f\|_2$ denotes the L^2-norm if f is a function, the ℓ^2-norm if f is a vector).

Definition 2.3. *Let ε be the vector defined by $\varepsilon_0 = 1$, $\varepsilon_j = 0$ if $j \in \mathbb{Z}^d - \{0\}$. We define the interpolating power and filtering power of a B-approximation to be the quantities*

$$\overline{B} = 1 - \|B_1 - \varepsilon\|_2 = 1 - \|\mathcal{T}B - 1\|_2 \qquad (27)$$

$$\underline{B} = 1 - \|B_1\|_2 = 1 - \|\mathcal{T}B\|_2 \,. \qquad (28)$$

The equivalence of the two expressions proposed above for $\overline{B}$, as well as for $\underline{B}$ is directly due to (16), (17), and the obvious relation $\mathcal{F}\varepsilon = 1$.

Remark

Obviously, it is equivalent to say:

- B is an interpolant
- $\mathcal{T}B \equiv 1$
- $\overline{B} = 1$.

Then, we have $\underline{B} = 0$. The closer $\overline{B}$ is to 1, the closer the B-approximation is to the interpolation. Conversely, the closer $\underline{B}$ is to 1, the more filtering the B-approximation provides.

§3. Properties

3.1 Interpolation of Polynomials

Theorem 3.1. *Let B be a function, $\mathcal{T}B$ the transfer function associated to the B-approximation ; let $k \in \mathbb{N}$. It is equivalent to say (D^α is the derivative of order α; ε has been defined in 2.3)*

i. *B is $\mathbb{P}_k$-interpolating (i.e. $\forall y \in \mathbb{P}_k(\mathbb{Z}^d)$, $(y * B)_1 = y$).*

ii. $\forall \alpha \in \mathbb{N}^d$, $|\alpha| \leq k \Longrightarrow \sum_{j\in\mathbb{Z}^d} j^\alpha B(j) = \varepsilon_\alpha$.

iii. $\forall \alpha \in \mathbb{N}^d \ , \quad |\alpha| \leq k \Longrightarrow (\mathrm{D}^\alpha \mathcal{T} B)(0) = \varepsilon_\alpha$.

Proof: i. $\Rightarrow$ ii. For any $\alpha \in \mathbb{N}^d$ such that $|\alpha| \leq k$, let choose y such that $\forall j \in \mathbb{Z}^d$, $y_j = (-j)^\alpha$. Then $(y*B)_1 = \sum_{j\in\mathbb{Z}^d} y_j B(\bullet - j) = \sum_{j\in\mathbb{Z}^d} (-j)^\alpha B(\bullet - j)$ $= \sum_{j\in\mathbb{Z}^d} j^\alpha B(\bullet + j)$. So, $\varepsilon_0 = y_0 = (y * B)_1(0) = \sum_{j\in\mathbb{Z}^d} j^\alpha B(j)$.
ii. $\Rightarrow$ i. Taking linear combinations of $\sum_{j\in\mathbb{Z}^d} j^\alpha B(-j)$ for various α, we have, for any $n \in \mathbb{N}$ such that $n \leq k$: $\forall i \in \mathbb{Z}^d$, $\sum_{j\in\mathbb{Z}^d}(i-j)^n B(j) = i^n$, which is, by linear combinations for various $n \leq k$: $\forall p \in \mathbb{P}_k$, $\sum_{j\in\mathbb{Z}^d} p(i-j)\, B(j) = p(i)$, which is i.
ii. $\Leftrightarrow$ iii. $\mathcal{T}B(\omega) = \sum_{j\in\mathbb{Z}^d} B(j)\, e^{-2i\pi j\omega}$, whence
$\mathrm{D}^\alpha \mathcal{T}B(0) = (-2i\pi h)^\alpha \sum_{j\in\mathbb{Z}^d} j^\alpha B(j)$. ■

Remarks

Polynomial reproduction is then a consequence of the properties of interpolation with translates of B: if B is $\mathbb{P}_k$-interpolating and if the interpolation with functions of the form $\sum_{j\in\mathbb{Z}^d} a_j B(\bullet - j)$ is $\mathbb{P}_\ell$-reproducing, then B is $\mathbb{P}_n$-reproducing, with $n = \min(k, l)$.

We generally require $\mathcal{T}B(0) = 1$ (which is the same as $\sum_{j\in\mathbb{Z}^d} B(j) = 1$), as this is equivalent to $\mathbb{P}_0$-interpolation, and so to $\mathbb{P}_0$-reproduction when interpolation is $\mathbb{P}_0$-reproducing.

3.2 Error Bounds

Theorem 3.2. *With the hypothesis of Theorem 2.1 (ε is defined in 2.3), we have:*

i. $y \in \ell^1 \implies \|(y *_h B)_h - y\|_2 \leq (1 - \overline{B})\, \|y\|_1$.
ii. $y \in \ell^2 \implies \|(y *_h B)_h - y\|_2 \leq \|\mathcal{T}B - 1)\|_\infty \|y\|_2$.
iii. $y \in \ell^\infty \implies \|(y *_h B)_h - y\|_\infty \leq \|B_1 - \varepsilon\|_1 \|y\|_\infty$.

Proof: i. By using the definition of $(y*B)$, then (17), (12), (25), and finally (19), we get :

$$\begin{aligned}\|(y*B) - y\|_2^2 &= \|(B_1 - \varepsilon) * y\|_2^2 = \|\mathcal{F}[(B_1 - \varepsilon) * y]\|_2^2 \\ &= \|\mathcal{F}(B_1 - \varepsilon)\cdot \mathcal{F}y\|_2^2 \leq \|\mathcal{F}(B_1 - \varepsilon)\|_2^2\ \|\mathcal{F}y\|_\infty^2 \\ &\leq (1 - \overline{B})^2\ \|y\|_1^2 .\end{aligned}$$

ii. Starting as in i., we also have (using (17)):

$$\begin{aligned}\|(y*B) - y\|_2^2 &= \|\mathcal{F}(B_1 - \varepsilon)\cdot \mathcal{F}y\|_2^2 \leq \|\mathcal{F}(B_1 - \varepsilon)\|_\infty^2\ \|\mathcal{F}y\|_2^2 \\ &\leq \|\mathcal{T}B - 1\|_\infty^2 \cdot \|y\|_2^2\end{aligned}$$

iii.
$$\begin{aligned}\left|((y *_h B)_h - y)_k\right| &= |((B_1 - \varepsilon) * y)_k| = \left|\sum_{j\in\mathbb{Z}^d} (B(j) - \varepsilon_j)\, y_{k-j}\right| \\ &\leq \left(\sum_{j\in\mathbb{Z}^d} |B(j) - \varepsilon_j|\right)\cdot \|y\|_\infty = \|B_1 - \varepsilon\|_1\ \|y\|_\infty .\end{aligned}$$ ■

Remark 3.2
As we can see, a B-approximation is close to the interpolation when $\overline{B}$ is close to 1, i.e. when the whole function $\mathcal{T}B$ is close to 1; whereas polynomial interpolation (and, as a consequence of it, as we saw in last the remark, polynomial reproduction) is dependent on the properties of $\mathcal{T}B$ in the *neighbourhood of* 0. This is the reason why we claim that *polynomial reproduction* -as well as polynomial interpolation- *is not a good criterion* to evaluate how far a B-approximation is from interpolation.

Finer error bounds are available for B-approximation when B is compactly supported, as is the case for (tensor product of) univariate polynomial B-splines. Details are given in [3].

3.3 Fourier Transform

Theorem 3.3. *With the hypothesis of Theorem 2.1, let* $\sigma = y * B$. *Then*

$$\mathcal{F}\sigma = (\mathcal{F}y) \cdot (\mathcal{F}B) \tag{29}$$

Proof: Direct consequence of (12). ■

This theorem is important as it shows the behaviour of σ (in terms of its Fourier transform), not only for integer values (at approximation points), but for all values in $\mathbb{R}^d$.

§4. Modified Approximants

The aim of this section is to define a new approximant, say B_a, such that the B_a-approximation is "nearer" the interpolation than the B-approximation is. We require that the B_a-approximation of a vector y should be in the same vectorial space as the B-approximation of y (i.e. the closure of the space generated by h-integer translates of B), and that it should be quite easily computed.

4.1 Definition and Properties of Modified Approximants

Definition 4.1. *With the hypothesis of Theorem 2.1, let* $a \in Y_r$ *(usually* a *is furthermore symetric and compactly supported). Let*

$$B_a = a * B = \sum_{j \in \mathbb{Z}^d} a_j \, B(\bullet - j). \tag{30}$$

We suppose in the sequel that $(B_a)_h \in Y_{-r'}$.

Theorem 4.1. *Let* a *and* B_a *be defined as above; let* $y \in Y_r$ *and* $\sigma_a = y * B_a$*; let*

$$Y_a = a * y. \tag{31}$$

Then

i. $$\sigma_a = Y_a * B. \tag{32}$$

ii. $$\mathcal{T}B_a = (\mathcal{F}a)\cdot(\mathcal{T}B). \tag{33}$$

iii. $$\mathcal{F}(y * B_a) = (\mathcal{F}y)\cdot(\mathcal{F}a)\cdot(\mathcal{T}B). \tag{34}$$

Proof: i.

$$\begin{aligned}\sigma_a &= \sum_{j\in\mathbb{Z}^d} y_j \left(\sum_{k\in\mathbb{Z}^d} a_k\, B(\bullet - j - k)\right)\\ &= \sum_{j\in\mathbb{Z}^d}\sum_{k\in\mathbb{Z}^d} y_j\, a_k\, B(\bullet - j - k)\\ &= \sum_{\ell\in\mathbb{Z}^d}\left(\sum_{j\in\mathbb{Z}^d} y_j\, a_{\ell-j}\right) B(\bullet - \ell) \;=\; Y_a * B.\end{aligned}$$

ii. is a direct consequence of (16) and (30).
iii. is a direct consequence of (29) and (33). ■

Remark
i. means that in order to get the B_a-approximation of a vector y, we can first compute Y_a (i.e. do a pre-treatment of data y), and then do the B-approximation of Y_a. So B_a-approximation may be considered as a (modified) B-approximation.

The introduction of B_a gives us, via the vector a, a lot of degrees of freedom in order to give any desired shape to the approximant B_a (which is given, in terms of the function by (32), in terms of its Transfer function by (33)). Main properties of the associated B_a-approximation are given by Theorem 4.1. We can now choose a in order to obtain specific properties for B_a-approximation , such as, for example, $\mathbb{P}_k$-reproduction.

We now propose and compare various criteria in order to define suitable vectors a. We want that B_a-approximation should be as near the interpolation as possible, i.e. that, for some classes of vectors y, $(y * B_a)_1$ should be as small as possible.

4.2 $\mathbb{P}_k$-interpolating criterion

As we saw in Remark 3.2, $\mathbb{P}_k$-interpolating is not a good criterion to evaluate how far a B_a-approximation is from interpolation. However, as the $\mathbb{P}_k$-interpolating property is quite commonly examined for approximation, we now examine how to choose a in order to get a B_a-approximation which should be $\mathbb{P}_k$-interpolating for the highest possible k.

In the sequel, we denote by $P_{d,k}$ the dimension of the vector space $\mathbb{P}_k$, and we write b instead of B_1 (i.e. $\forall j \in \mathbb{Z}^d\,,\; b_j = B(j)$).

Theorem 4.2. *With the hypothesis of Theorem 4.1, let $k \in \mathbb{N}$ and suppose the vector a satisfies the following relation for any $\alpha \in \mathbb{N}^d$ such that $|\alpha| \leq k$:*

$$\sum_{j\in\mathbb{Z}^d} a_j \left(\sum_{\ell\in\mathbb{Z}^d} \sum_{\beta\leq\alpha} (j+\ell)^\alpha\, j^\beta\, \ell^{\alpha-\beta}\, b_\ell \right) = \varepsilon_\alpha. \tag{35}$$

Then the B_a-approximation is $\mathbb{P}_k$-interpolating .

Proof: Using (33) and Theorem 3.1, we only need to prove that (35) is equivalent to

$$\mathrm{D}^\alpha(\mathcal{F}a \cdot \mathcal{T}B)(0) = \varepsilon_\alpha. \tag{36}$$

Now, using

$$\mathrm{D}^\alpha(f\cdot g) = \sum_{\beta\leq\alpha} \binom{\alpha}{\beta} \mathrm{D}^\beta f\, \mathrm{D}^{\alpha-\beta},$$

$$\mathrm{D}^\beta(\mathcal{F}a)(0) = \sum_{j\in\mathbb{Z}^d} a_j (2i\pi)^{|\beta|} j^\beta a_j$$

$$\mathrm{D}^{\alpha-\beta}\mathcal{T}B(0) = \sum_{\ell\in\mathbb{Z}^d} (-2i\pi)^{|\alpha-\beta|} \ell^{\alpha-\beta}\, b_\ell,$$

we see that (35) is equivalent to (36) (the constant factors drop out as they are equal to one, or $\varepsilon_\alpha = 0$). ■

Remarks

Theorem 4.2 gives a way to build a $\mathbb{P}_k$-interpolating quasi-interpolant from any function B such that $B_h \in Y_{-r'}$ where $r' > d$ (i.e. when B is sufficiently decaying at ∞); as a consequence of it, if we have a $\mathbb{P}_0$-reproducing quasi-interpolant B, we can build a $\mathbb{P}_k$-interpolating quasi-interpolant as a linear combination of $P_{d,k}$ translates of B, by merely solving a linear system of order $P_{d,k}$.

Other ways to build $\mathbb{P}_k$-interpolating quasi-interpolants are shown in [3], but, if B is not compactly supported, they are combinations of an *infinite* number of translates of B.

4.3 Approximant with maximum interpolating power

As shown in Theorem 3.2i, it is of interest to maximize $\overline{B}_a$ in order to minimize the error $\|(y *_h B)_h - y\|$. The aim of this section is to show how to determine a (bounded supported) vector a such that the B_a-approximation is (eventually $\mathbb{P}_k$-reproducing) as near interpolation as possible.

Theorem 4.3. *With the hypothesis of Theorem 4.1, let $k \in \mathbb{N}$. Let J be a finite subset of $\mathbb{Z}^d$. Let V_J be the set of vectors whose support is in J (i.e. $a \in V_J$ if and only if $\forall j \notin J$, $a_j = 0$). Then*

i.) *Among all vectors in V_J, the one which maximizes the interpolating power of B_a does enjoy*

$$\forall j \in J\ , \quad (a * b * b)_j = b_j. \tag{37}$$

We now suppose card $J \geq P_{d,k}$.

ii.) *Among all vectors in* V_J *and such that the* B_a*-approximation is* $\mathbb{P}_k$*-interpolating, the one which maximizes the interpolating power of* B_a *does enjoy*

$$\begin{cases} \forall j \in J \ , \quad (a * b * b)_j + \displaystyle\sum_{|\alpha| \leq k} \lambda_\alpha \left(\sum_{\ell \in \mathbb{Z}^d} (j+\ell)^\alpha \, b_\ell \right) = b_j \\ \forall \alpha \in \mathbb{N}^d \ , \quad |\alpha| \leq k \implies \displaystyle\sum_{j \in J} a_j \left(\sum_{\ell \in \mathbb{Z}^d} (j+\ell)^\alpha \, b_\ell \right) = \varepsilon_\alpha, \end{cases} \tag{38}$$

where $(\lambda_\alpha)_{|\alpha| \leq k}$ *are Lagrange coefficients.*

Proof: i. We have $(B_a)_1 = a * b$, and so, using (27), maximizing $\overline{B}_a$ is minimizing $E = \|a{*}b - \varepsilon\|_2^2$, which is equivalent to $\forall j \in J \ , \ \frac{\partial E}{\partial a_j} = 0$, which is (37).
ii. Using Theorem 4.2, we have to minimize $E = \|a{*}b - \varepsilon\|_2^2$ with the constraint $|\alpha| \leq k \Rightarrow$ (35); using i. and the Lagrange's method, we obtain (38). ■

4.4 Truncated basis criterion

Now, we propose another way to choose suitable a; this one does not give an optimal result (the optimal result is given by the maximum interpolating power), but we can obtain a without solving linear systems, and in some cases we can easily get error majorations.

Let L be a B-interpolant, written in the form $L = \sum_{j \in \mathbb{Z}^d} \ell_j \, B(\bullet - j)$. Then, for some $j \subset \mathbb{Z}^d$, we propose to choose a such that

$$\forall j \in J \ , \quad a_j = \ell_j \quad ; \quad \forall j \in \mathbb{Z}^d - J \ , \quad a_j = 0. \tag{39}$$

It is possible too to modify some non-zero a_j's such that the B_a-approximation is $\mathbb{P}_k$-reproducing for some k.

Let us mention that the vector ℓ is quite easily known, or by the relation

$$\forall j \in \mathbb{Z}^d \ , \quad \ell_j = \int_{[0,1]^d} \frac{1}{(\mathcal{T}B)(\omega)} \, e^{2i\pi j\omega} \, d\omega \ , \tag{40}$$

or better by iterative methods when B is at least $\mathbb{P}_0$-reproducing (for more details, see [3]).

§5. Examples

5.1 Polynomial Splines

Here are values of transfer functions, interpolating and filtering power for the lowest degrees (for even degrees, we use the symetrical B-spline):

degree 2: $\mathcal{T}B(\omega) = (3 + \cos 2\pi\omega)/4 = 1 - (\sin^2 \pi\omega)/2$
$\overline{B} = 1 - \sqrt{6}/8 \simeq 0.69 \quad ; \quad \underline{B} = 1 - \sqrt{38}/8 \simeq 0.23$

degree 3: $\mathcal{T}B(\omega) = (2+\cos 2\pi\omega)/3 = 1 - 2(\sin^2 \pi\omega)/3$
$\overline{B} = 1 - \sqrt{6}/6 \simeq 0.59$; $\underline{B} = 1 - \sqrt{2}/2 \simeq 0.29$

degree 4: $\mathcal{T}B(\omega) = (115 + 76\cos 2\pi\omega + \cos 4\pi\omega)/192$
$= 1 - 5(\sin^2 \pi\omega)/2 + (\sin^4 \pi\omega)/24$
$\overline{B} = 1 - \sqrt{35270}/384 \simeq 0.51$; $\underline{B} = 1 - \sqrt{64454}/384 \simeq 0.34$

degree 5: $\mathcal{T}B(\omega) = (133 + 26\cos 2\pi\omega + \cos 4\pi\omega)/160$
$= 1 - \sin^2 \pi\omega + 2(\sin^4 \pi\omega)/15$
$\overline{B} = 1 - \sqrt{4270}/120 \simeq 0.46$; $\underline{B} = 1 - \sqrt{5710}/120 \simeq 0.37$

As illustrated by Figure 1, the higher the degree of the B-spline is, the lower $B(\omega)$ is for any ω in $\mathbb{R}$, and, as a consequence of it, the lower $\overline{B}$ and the higher $\underline{B}$ are, all this means that the higher the degree of the B-spline is the less interpolating and the more filtering the B-approximation is, as it is well known.

For cubic splines, if B is the usual B-spline with knots -2, -1, 0, 1, 2, the $\mathbb{P}_k$-reproducing approximant with smallest support is $B_a = a * B$, with $\forall j \in \mathbb{Z} - [-k\,,\,k]\,,\ a_j = 0$, and:

$$\forall j \in [-k\,,\,k]\,,\quad a_j = (-1)^j (1/6)^k \sum_{i=|j|}^{n} \binom{2i}{i+j}. \tag{36}$$

B_a can also be written in the very simple way

$$B_a = \sum_{j=o}^{k} (-1/6)^j\, \delta^{2j} B\ , \tag{37}$$

where $\delta^{2j}B$ is defined by $\delta^2 B = B(\bullet+1) - B + B(\bullet-1)$ and $\delta^{2j}B = \delta^2(\delta^{2j-2}B)$.

As a consequence of it, the cubic spline interpolant can be written

$$L = \sum_{j\in\mathbb{N}} (-1/6)^j\, \delta^{2j} B. \tag{38}$$

Furthermore, we have

$$y * B_a - y = (-1/6)^{k+1}\, \delta^{2k+2} y\ , \tag{39}$$

(here $\delta^2 y$ is defined by $(\delta^2 y)_j = y_{j+1} - 2y + y_{j-1}$ for any j in $\mathbb{Z}^d$), and so

$$\|y * B_a - y\|_\infty \le (2/3)^{k+1}\, \|y\|_\infty\,. \tag{40}$$

Proofs and more details are given in [3,III-2.3.1].

The Lagrangean cubic spline can also be written

$$L = \sqrt{3} \sum_{j\in\mathbb{Z}} (\sqrt{3}-2)^{|j|}\, B(\bullet - j)\ ,$$

and so, to obtain a $\mathbb{P}_0$-reproducing, truncated basis, quasi-interpolant , with support $[-k-2\,,\,k+2]$, we propose the following values:
$\forall j \in [-k+1\,,\,k-1]\,,\ a_j = \sqrt{3}(\sqrt{3}-2)^{|j|}$; $a_k = a_{-k} = \left(1 - \sum_{j=-k+1}^{k-1} a_j\right)/2$
then

$$\|y * B_a - y\|_\infty \le 2\,(2-\sqrt{3})^k\, \|y\|_\infty\,. \tag{41}$$

Comparing (40) and (41) shows a much faster convergence, when k increases, with truncated basis criterion than with $\mathbb{P}_k$-reproduction criterion.

5.2 Polyharmonic Splines

For polyharmonic B-splines, as defined in Section 1.3, we have

$$\mathcal{F}B^h_{d,m} = h^d \frac{\|\sin \pi h\bullet\|^{2m}}{\|\pi h\bullet\|^{2m}} \tag{42}$$

($\sin x$ is the vector whose j^{th} component is $\sin x_j$; $\|\bullet\|$ is the Euclidean norm),

$$\mathcal{T}B^h_{d,m} = \sum_{j\in\mathbb{Z}^d} B^h_{d,m}(jh)\cos 2\pi h\omega = \sum_{j\in\mathbb{Z}^d}\left(\frac{\|\sin \pi h\bullet\|}{\|\pi h\bullet - \pi j\|}\right)^{2m}. \tag{43}$$

Furthermore, $\overline{B^h_{2,2}} \simeq 0.63$, $\underline{B^h_{2,2}} \simeq 0.31$, $\overline{B^h_{2,3}} \simeq 0.39$, and $\underline{B^h_{2,3}} \simeq 0.51$. Details and other values of interpolating and filtering power are given in [3,5].

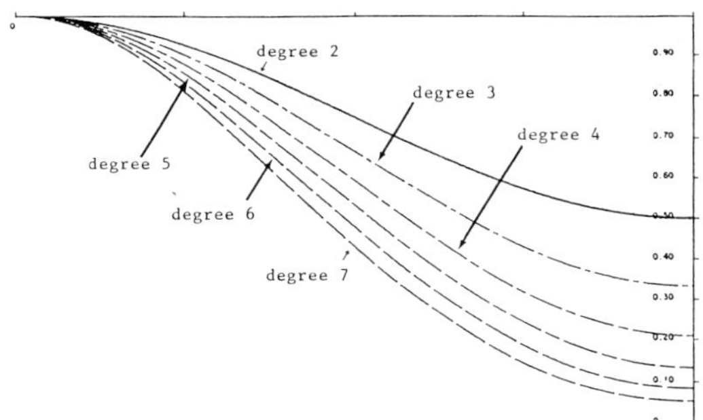

Figure 1. Transfer functions of polynomial splines.

5.3 Figures

Figure 1 shows transfer functions of B-approximation for polynomial splines of degree 2 to 7. Figure 2 and 3 show $\mathcal{T}B^h_{2,2}$ and $\mathcal{T}B^h_{2,3}$. At last, Figure 4 shows the curve obtained by B-approximation and B_a-approximation ($\mathbb{P}_7$-interpolating) for cubic splines and data extracted from the Oseberg ship data (the squares are the retained data, the "+" are the Y_a); as we can see, the B_a-approximation is very near the interpolation of the data points.

Conclusion

The transfer function is very useful for giving a complete view of the properties of a B-approximation; interpolating and filtering power are very convenient to have a quick glance on them. From any quasi-interpolant B, it is quite easy to derive other quasi-interpolants which are nearer the interpolating function than B is. Many criteria are possible; polynomial interpolation (or polynomial reproduction) is quite a poor criterion, whereas maximising the interpolating power is a very efficient one, and the truncated basis is very efficient and very convenient.

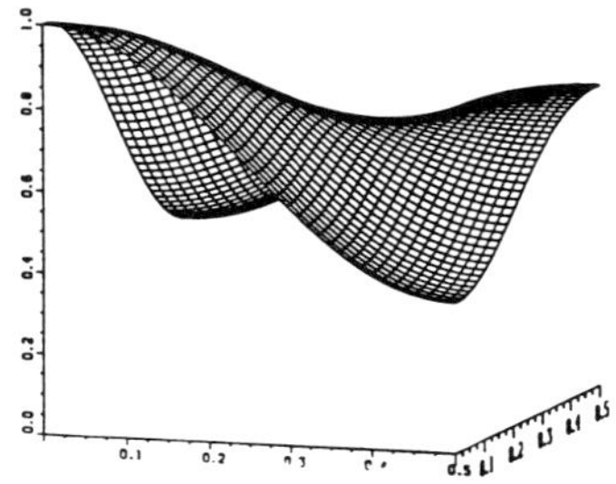

Figure 2. $\mathcal{T}B^h_{2,2}$.

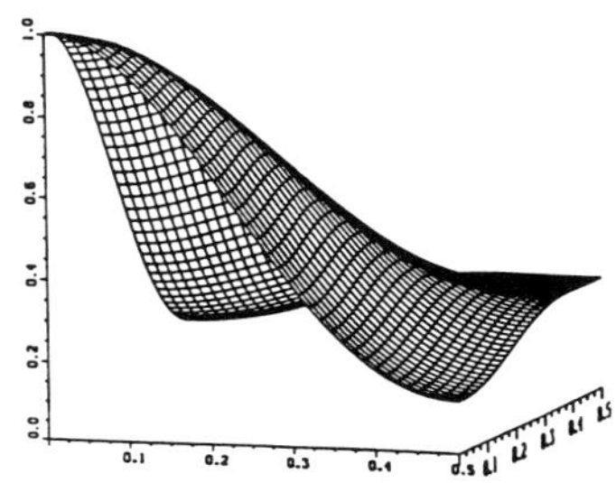

Figure 3. $\mathcal{T}B^h_{2,3}$.

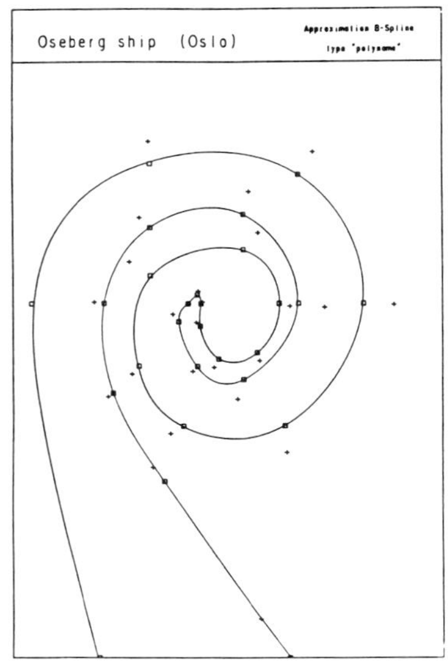

Figure 4. Oseberg Ship.

References

1. Hamming, R. W., *Digital Filters*, Prentice Hall, Englewood cliffs, NJ, 1977.
2. Micchelli, C. A., Christophe Rabut, and Florencio Utreras, Using the refinement equation for the construction of pre-wavelets. III: Elliptic splines, Numerical Algorithms **1** 1991, 331–352.
3. Rabut, C., B-splines polyharmoniques cardinales : interpolation, quasi-interpolation , filtrage., Thèse d'Etat, Toulouse University, 1990.
4. Rabut, C., How to build quasi-interpolants. Application to polyharmonic B-splines, *Curves and Surfaces*, P.-J. Laurent, A. Le Méhauté, and L. L. Schumaker (eds.), Academic Press, New York, 1991, 391–402.
5. Rabut, C., Elementary m-harmonic cardinal B-splines, Numerical Algorithms **2** (1992), to appear.
6. Schoenberg, I. J., Contributions to the problem of approximation of equidistant data by analytic functions..., Quarterly Applied Mathematics IV, 1946, 45–99 and 112–137.

Discrete Bézier Curves and Surfaces

Paul Sablonnière

Abstract. This paper is a short survey on discrete Bézier curves and surfaces expressed in various discrete Bernstein bases. Among the algorithms presented here, a subdivision algorithm is of interest. We also discuss different notions of discrete C^k continuity between two curves.

§1. Introduction

In this paper are given some definitions and algorithms concerning discrete Bézier curves (DBC) and surfaces (DBS). They are expressed in discrete Bernstein bases (DBB) of polynomials defined on intervals, squares and triangles (with obvious extensions to hypercubes or simplices). Discrete splines were studied in various papers by Mangasarian and Schumaker [6,7], by Schumaker [12], and Lyche [5].

More recently, Barry [1] and Goldman [3] introduced Polya curves and urn models: many algorithms given here are probably particular cases of those given in their work. However, it seemed interesting to write them down explicitly for direct use.

Here is an outline of the paper: in Section 2 definitions and main properties of DBB on $[0, N]$ are given. In Sections 3, 4 and 5, one finds a discrete de Casteljau (DDC) algorithm, algorithms for finite differences, degree elevation and degree reduction, and algorithms for subdivision and extension. In Section 6 we propose three definitions of C^k continuity between two DBC. As an example, we give the explicit construction of various discrete C^2 cubic B-splines. In Sections 7 and 8, we try to extend the above results to DBS on squares and triangles. The subdivision algorithm for tensor-product DBS is detailed. In Section 9, we describe some discrete Bernstein quasi-interpolants (DBQI) corresponding to those defined by the author in the continuous case [9,10]. Finally in Section 10, following ideas developed by Ciesielski in the continuous case [2], a dual basis of the DBB of $\mathbb{P}_n[0, N]$ is introduced and

Mathematical Methods in Computer Aided Geometric Design II
Tom Lyche and Larry L. Schumaker (eds.), pp. 497–515.

ISBN 0-12-460510-9.

the condition number of this latter is computed (see e.g. [4] for this notion). We also give a formula connecting finite differences and divided differences via discrete Bernstein polynomials. We hope that this short survey will help the interested reader in developing new applications of DBC and DBS which could be competitive with the classical continuous ones.

§2. Discrete Bernstein Basis

Definition. *The discrete Bernstein basis (DBB) of degree $n \le N$ on $[0, N]$ is the family of polynomials*

$$b_i^n(N,x) = \binom{x}{i}\binom{N-x}{n-i} \Big/ \binom{N}{n} \quad (0 \le i \le n \le N)$$

They form a basis of $P_n[0, N]$.

Properties:

(DB1) $b_i^n(N,x) \ge 0$ for $x \in [0, N] \cap \mathbb{N}$

(DB2) $\sum_{i=0}^n b_i^n(N,x) = 1$ for all $x \in \mathbb{R}$

(DB3) More generally, setting $e_0(x) = x^{(0)} = 1$ and for $r \ge 1$: $e_r(x) = x^{(r)} = x(x-1)\cdots(x-r+1)$, we have the following representation:

$$e_r(x) = \left(\sum_{i=r}^n e_r(i) b_i^n(N,x)\right) \frac{e_r(N)}{e_r(n)}.$$

In particular, for $r = 1$ and 2:

$$e_1(x) = x = \frac{N}{n}\sum_{i=r}^n i b_i^n(N,x),$$

$$e_2(x) = x(x-1) = \frac{N(N-1)}{n(n-1)}\sum_{i=2}^n i(i-1) b_i^n(N,x)$$

(DB4) **Recurrence relations**

(i) $b_i^n(N,x) = \dfrac{x-i+1}{N-n+1} b_{i-1}^{n-1}(N,x) + \dfrac{N-n+1-x+i}{N-n+1} b_i^{n-1}(N,x)$

(ii) $b_i^n(N,x) = \dfrac{n-i+1}{n+1} b_i^{n+1}(N,x) + \dfrac{i+1}{n+1} b_{i+1}^{n+1}(N,x).$

(DB5) **Finite differences**

$\Delta b_i^n(N,x) = \dfrac{n}{N}[b_{i-1}^{n-1}(N-1,x) - b_i^{n-1}(N-1,x)]$ for $0 \le x \le N-1$, where $\Delta f(x) = f(x+1) - f(x)$.

Proof: (DB1) is obvious. (DB2) is Vandermonde's convolution identity

$$\sum_{i=0}^n \binom{x}{i}\binom{N-x}{n-i} = \binom{N}{n} \text{ for } 0 \le x \le N.$$

(DB3) is proved by using generating functions

$$(1+t)^{N-x} = \sum_{j\geq 0} \binom{N-x}{j} t^j. \tag{1}$$

By successive derivation with respect to the variable t of

$$(1+t)^x = \sum_{j\geq 0} \binom{x}{j} t^j,$$

we obtain

$$x^{(k)}(1+t)^{x-k}t^k = \sum_{i\geq k} i^{(k)} \binom{x}{i} t^i. \tag{2}$$

By multiplying (1) (with $x = k$) by $x^{(k)}t^k$ and (1) by (2), we get

$$x^{(k)}(1+t)^{N-k}t^k = \sum_{\ell\geq k} x^{(k)} \binom{N-k}{\ell-k} t^\ell = \sum_{\ell\geq k} \left[\sum_{i+j=\ell} i^{(k)} \binom{x}{i} \binom{N-x}{j} \right] t^\ell.$$

Now, by identification, we obtain the result

$$\binom{N-k}{n-k} x^{(k)} = \sum_{i+j=n} i^{(k)} \binom{x}{i} \binom{N-x}{j}.$$

(DB4) and (DB5) are proved directly from the definition of the DBB. These formulas are valid if we adopt the classical convention $b_i^n = 0$ for $i < 0$ and $i > n$. ■

§3. Algorithms for Discrete Bézier Curves

Definition. *A discrete Bézier curve (DBC) of degree n in $\mathbb{R}^d$ is defined for $x \in [0, N]$ by $B_n(x) = B_nP(x) = \sum_{i=0}^{n} P_i b_i^n(N, x)$, where $P = \{P_i, 0 \leq i \leq n\}$ is a given set of $n+1$ points, called the B-polygon (or control polygon) associated with the curve.*

DDC Algorithm (discrete de Casteljau):

(i) $P_{i,0}(x) = P_i$ for $0 \leq i \leq n$

(ii) for $1 \leq k \leq n$, and $0 \leq i \leq n-k$:

$$P_{i,k}(x) = (1-\beta_i)P_{i,k-1}(x) + \beta_i P_{i+1,k-1}(x), \tag{A1}$$

where $\beta_i = (x-i)/(N-n+k)$

(iii) $P_{0,n}(x) = B_n(x)$.

Proof: Using (DB4i) for all values of k, it is easily proved that

$$B_n(x) = \sum_{i=0}^{n-k} P_{i,k}(x) b_i^{n-k}(N, x).$$

In particular, for $k = n$ we obtain the final result. ■

Remark: β_i is positive for $x \geq i$ and $1-\beta_i$ is positive for $x \leq N-n+k+i$. Therefore, the above linear combinations *are not all convex ones*, contrary to those of the classical de Casteljau algorithm.

FD Algorithm (finite differences):

(i) $P_i^{(0)} = P_i$ for $0 \leq i \leq n$

(ii) for $1 \leq k \leq n$, and $0 \leq i \leq n-k$,

$$P_i^{(k)} = \frac{n-k+1}{N-k+1}(P_{i+1}^{(k-1)} - P_i^{(k-1)}) \tag{A2}$$

(iii) for $0 \leq x \leq N-k$,

$$\Delta^k B_n(x) = \sum_{i=0}^{n-k} P_i^{(k)} b_i^{n-k}(N-k, x).$$

Proof: We proceed by induction on k, using (DB5). In particular, we obtain the finite differences of B_n at $x=0$ and $x=N$:

$$\begin{cases} \Delta^k B_n(0) = \frac{n^{(k)}}{N^{(k)}}\Delta^k P_0 \\ \nabla^k B_n(N) = \Delta^k B_n(N-n) \quad = \frac{n^{(k)}}{N^{(k)}}\Delta^k P_{n-k} = \frac{n^{(k)}}{N^{(k)}}\nabla^k P_{n-k} \end{cases} \tag{3}$$

where $\nabla f(x) = f(x) - f(x-1)$ and $\nabla P_i = P_i - P_{i-1}$. ■

DE Algorithm (degree elevation):

(i) $P_i^{[0]} = P_i$ for $0 \leq i \leq n$.

(ii) for $1 \leq k \leq N-n$ and $0 \leq i \leq n+k$

$$P_i^{[k]} = \frac{i}{n+k}P_{i-1}^{[k-1]} + \frac{n+k-i}{n+k}P_i^{[k-1]} \tag{A3}$$

(iii) for $x \in [0, N]$,

$$B_n P(x) = \sum_{i=0}^{n+k} P_i^{[k]} b_i^{n+k}(N, x).$$

In particular, $P_i^{[N-n]} = B_n(i)$ for $0 \leq i \leq N$.

Proof: We proceed by induction, using (DB4ii). Since $P_i^{[k]}$ is a convex combination of $P_{i-1}^{[k-1]}$ and $P_i^{[k-1]}$, we deduce from this algorithm *a geometrical proof of the variation diminishing property of discrete Bézier curves.* Let $h(v) = a^T v + b = 0$ be the equation of some affine hyperplane H of $\mathbb{R}^d (v, a \in \mathbb{R}^d, b \in \mathbb{R})$. Then the number of sign changes of the sequence $\{h(B_n P(x)), x \in [0, N] \cap \mathbb{N}\}$ is not greater than the number of sign changes of the sequence $\{h(P_i), 0 \leq i \leq n)$. This is due to the fact that the first sequence is deduced from the second one by a succession of $N-n$ applications of algorithm (A3), which is "corner cutting". In other words, the variation of the discrete curve w.r.t. H is not greater than the variation of its control polygon. ■

LDR Algorithm (left degree reduction):
Given $B_{n+k}P(x) = \sum_{i=0}^{n+k} P_i b_i^{n+k}(N,x)$, use degree elevation (algorithm DE) to find $B_n L(x) = \sum_{i=0}^{n} L_i b_i^n(N,x)$ such that $L_i^{[k]} = P_i$ for $0 \le i \le n$. In other words, since

$$\begin{array}{rclcl} \Delta^r B_{n+k}P(0) & = & \frac{(n+k)^{(r)}}{N^{(r)}}\Delta^r P_0 & = & \frac{(n+k)^{(r)}}{N^{(r)}}\Delta^r L_0^{(k)} \\ & = & \Delta^r B_{n+k} L^{[k]}(0) & = & \Delta^r B_n L(0) \ \text{ for } 0 \le r \le n, \end{array}$$

the two curves $B_{n+k}P$ and $B_n L$ coincide on the first $n+1$ points $x = 0, 1, \ldots, n$ of $[0, N]$.

(i) $L_i^{[k]} = P_i$ for $0 \le i \le n$

(ii) for $r = k, k-1, \ldots, 1$ and $0 \le i \le n$

$$L_i^{[r-1]} = \frac{n+r}{n+r-i} L_i^{[r]} - \frac{i}{n+r-i} L_{i-1}^{[r-1]} \tag{A4}$$

(iii) $L_i = L_i^{[0]}$ for $0 \le i \le n$.

Proof: Obvious from (A3) by computing $P_i^{[k-1]}$ in terms of $P_i^{[k]}$ and $P_{i-1}^{[k-1]}$. ■

In a symmetric way, we obtain the

RDR Algorithm (right degree reduction):
Given $B_{n+k}P(x) = \sum_{i=0}^{n+k} P_i b_i^{n+k}(x)$, use degree elevation to find $B_n R(x) = \sum_{i=0}^{n} R_i b_i^n(x)$ such that

$$R_i^{[k]} = P_i \quad \text{for} \quad k \le i \le n+k.$$

In other words, the two curves coincide on the last $n+1$ points $x = N - n, \ldots, N$ of [0, N].

(i) $R_i^{[k]} = P_i$ for $k \le i \le n+k$

(ii) for $s = k, k-1, \ldots, 1$ and $s \le i \le n+s$,

$$R_{i-1}^{[s-1]} = \frac{n+s}{i} R_i^{[s]} - \frac{n+s-i}{i} R_i^{[s-1]} \tag{A5}$$

(iii) $R_i = R_i^{[0]}$ for $0 \le i \le n$.
Application: verify the continuity of two curves (see §6).

§4. Subdivision Algorithm

Let $B_n P(x) = \sum_{i=0}^{n} P_i b_i^n(N,x)$ be a DBC of degree n, defined on $[0, N]$. For $N = N_1 + N_2$, we want to compute the two representations of $B_n P$ on $[0, N_1]$ and $[N_1, N]$:

$$B_n P_L(x) = \sum_{i=0}^{n} L_i b_i^n(N_1, x), \qquad x \in [0, N_1]$$

$$B_n P_R(x) = \sum_{i=0}^{n} R_i b_i^n(N_2, x - N_1), \qquad x \in [N_1, N].$$

LS Algorithm (left subdivision):

(i) $L_i^{(0)} = P_i$ for $0 \leq i \leq n$

(ii) for $1 \leq k \leq n$ and $k \leq i \leq n$

$$L_i^{(k)} = \frac{N_2 + i - 2k + 1}{N - k + 1} L_{i-1}^{(k-1)} + \frac{N_1 + i - k}{N - k + 1} L_i^{(k-1)} \tag{A6}$$

(iii) $L_i = L_i^{(i)}$ for $0 \leq i \leq n$.

RS Algorithm (right subdivision):

(i) $R_i^{(0)} = P_i$ for $0 \leq i \leq n$

(ii) for $1 \leq k \leq n$ and $k \leq i \leq n$

$$R_{n-i}^{(k)} = \frac{N_2 - i + k}{N - k + 1} R_{n-i}^{(k-1)} + \frac{N_1 + 1 + i - 2k}{N - k + 1} R_{n+1-i}^{(k-1)} \tag{A7}$$

(iii) $R_i = R_i^{(n-i)}$ for $0 \leq i \leq n$.

Remark: When $N_1 = N_2$, setting $\alpha_{ik} = \frac{N_1 - i + k}{N - k + 1}$, the two formulas (A6) and (A7) become respectively

$$L_i^{(k)} = (1 - \alpha_{ik}) L_{i-1}^{(k-1)} + \alpha_{ik} L_i^{(k-1)}$$

$$R_{n-i}^{(k)} = \alpha_{ik} R_{n-i}^{(k-1)} + (1 - \alpha_{ik}) R_{n-i+1}^{(k-1)}$$

for $1 \leq k \leq n$ and $k \leq i \leq n$. Therefore, the same coefficients α_{ik} are used in both algorithms and can be computed once for all, for a fixed integer N. *Moreover they do not depend on n and they are positive.* It is possible to prove the variation diminishing property of DBC by using this algorithm which is also "corner cutting".

Proof of LS: Obviously, we have $P_L(0) = L_0 = P_0 = P(0)$. Then, for $k = 1$, $L_i^{(1)} - L_{i-1}^{(0)} = \frac{N_1 - i + 1}{N} (L_i^{(0)} - L_{i-1}^{(0)})$ for $1 \leq i \leq n$. In particular, for $i = 1$,

$$\Delta P_L(0) = \frac{n}{N_1} (L_1^{(1)} - L_0^{(0)}) = \frac{n}{N} (P_1 - P_0) = \Delta P(0).$$

Let us define the operators D and Δ by

$$DL_{i-1}^{(k-1)} = L_i^{(k)} - L_{i-1}^{(k-1)}$$
$$\Delta L_{i-1}^{(k-1)} = L_i^{(k-1)} - L_{i-1}^{(k-1)}.$$

Then, (A6) gives, for $1 \leq k \leq n$ and $k \leq i \leq n$,

$$DL_{i-1}^{(k-1)} = \frac{N_1 - i + k}{N - k + 1} \Delta L_{i-1}^{(k-1)}.$$

Now, suppose that we have already proved that for some $0 \le r \le n$, for all $0 \le s \le r$ and $0 \le i \le n - s$,

$$D^s L_i^{(0)} = \sum_{j=0}^{s} (-1)^j \binom{s}{j} L_{s+i-j}^{(s-j)} = \frac{(N_1 - i)^{(s)}}{N^{(s)}} \Delta^s L_i. \tag{4}$$

For $i = 0$, this is equivalent to $\Delta^s P_L(0) = \Delta^s P(0)$ for $0 \le s \le r$. If this is true for $0 \le r \le n$, the two polynomial $B_n P$ and $B_n P_L$ are identical and the proof is complete.

Now we have to prove identity (4) for $s = r + 1$, or equivalently that $\Delta^{r+1} P_L(0) = \Delta^{r+1} P(0)$. We shall verify the following equality in two ways:

$$D^{r+1} L_i^{(0)} = \frac{N_1 - i}{N - r} D^r L_{i+1}^{(0)} - \frac{N_1 - i - r}{N - r} D^r L_i^{(0)}. \tag{5}$$

First, we substitute in (5) the second expression of $D^r L_{i+1}^{(0)}$ and $D^r L_i^{(0)}$ given by (4) $(s = r)$ to get

$$\begin{aligned} D^{r+1} L_i^{(0)} &= \frac{(N_1 - i)}{N - r} \frac{(N_1 - i - 1)^{(r)}}{N^{(r)}} \Delta^r L_{i+1} - \frac{(N_1 - i - r)}{N - r} \frac{(N_1 - i)}{N^{(r)}} \Delta^r L_i \\ &= \frac{(N_1 - i)^{(r+1)}}{N^{(r+1)}} (\Delta^r L_{i+1} - \Delta^r L_i) = \frac{(N_1 - i)^{(r+1)}}{N^{(r+1)}} \Delta^{r+1} L_i \end{aligned} \tag{6}$$

Now we substitute in (5) the first expression of $D^r L_{i+1}^{(0)}$ and $D^r L_i^{(0)}$ given by (4) $(s = r)$ to get

$$D^{r+1} L_i^{(0)} = \sum_{j=0}^{r} (-1)^j \binom{r}{j} \left[\frac{N_1 - i}{N - r} L_{r+i+1-j}^{(r-j)} - \frac{N_1 - i - r}{N - r} L_{r+i-j}^{(r-j)} \right]. \tag{7}$$

By the LS algorithm we obtain

$$L_{r+i+1-j}^{(r-j)} = \frac{N - r + j}{N_1 - i} L_{r+i+1-j}^{(r+i-j)} - \frac{N_2 - r + i + j}{N_1 - i} L_{r+i-j},$$

which we substitute in the brackets of (7) to get

$$\begin{aligned} D^{r+1} L_i^{(0)} &= \sum_{j=0}^{r} (-1)^j \binom{r}{j} \left[\frac{N - r + j}{N - r} L_{r+i+1-j}^{(r+1-j)} - \frac{N - 2r + j}{N - r} L_{r+1-j}^{(r-j)} \right] \\ &= \sum_{j=0}^{r+1} (-1)^j \binom{r+1}{j} L_{r+i+1-j}^{(r+1-j)}, \end{aligned} \tag{8}$$

which is true by definition. Thus, finally, the two expressions (6) and (8) are identical and give equality (4) for $s = r + 1$. ■

Cost of the algorithm:

Assuming that $N = 2^q n$, there are q steps in the global subdivision algorithm computing all values of $B_n P$ on $[0, N]$ and the cost is $1 + 2 + \cdots + 2^{q-1} = 2^q - 1$ subdivisions (right or left). Therefore, the total number of convex combinations of type (A6) or (A7) is $n_S = n(n+1)(2^q - 1) \sim nN$. For the DDC algorithm, this number is $n_C = \frac{1}{2}n(n+1)(N-1) \sim \frac{1}{2}n^2 N$, and so $\frac{n_c}{n_s} \sim \frac{1}{2}n$ for N large. This is confirmed by numerical examples.

Evaluation by Horner type algorithms costs about nN multiplications, i.e., two times less than subdivision. But the subdivision algorithm involves only convex combinations and is numerically more stable. It would be interesting to compare their respective advantages in practical examples.

§5. Extension Algorithms

Let $B_n L(x) = \sum_{i=0}^n L_i b_i^n(N_1, x)$ be a DBC of degree n, defined on $[0, N_1]$, we want to extend it to a DBC defined on $[0, N_1 + N_2]$ and to compute its restriction on $[N_1, N]$ which we denote by $B_n P(x) = \sum_{i=0}^n P_i b_i^n(N, x)$ and $B_n R(x) = \sum_{i=0}^n R_i b_i^n(N_2, x - N_1)$, respectively.

This is the converse of the subdivision algorithm: the formulas are deduced from (A6) with some changes of indices.

RE Algorithm (right extension):

(i) $P^{(0)} = L_i, 0 \le i \le n$

(ii) for $1 \le k \le n$ and $k \le i \le n$

$$\begin{cases} P_i^{(k)} &= (1 - \lambda_{i,k})P_{i-1}^{(k-1)} + \lambda_{i,k} P^{(k-1)} \\ \text{with } \lambda_{i,k} &= (N - i + k)/(N_1 - k + 1) \end{cases} \tag{A8}$$

(iii) $P_i = P_i^{(i)}$ for $0 \le i \le n$.

In a symmetric way, we deduce from RS an algorithm computing P from R and then L from P by left subdivision.

LE Algorithm (left extension):

(i) $P_i^{(0)} = R_i, 0 \le i \le n$

(ii) for $1 \le k \le n$ and $k \le i \le n$

$$\begin{cases} P_{n-i}^{(k)} &= \mu_{ik} P_{n-i}^{(k-1)} + (1 - \mu_{ik}) P_{n-i+1}^{(k-1)} \\ \text{with } \mu_{ik} &= (N - i + k)/(N_2 - k + 1) \end{cases} \tag{A9}$$

(iii) $P_i = P_i^{(n-i)}$ for $0 \le i \le n$

(iv) Apply LS algorithm (left subdivision) to get L from P.

Remark. Comparing (A8) and (A6), resp. (A9) and (A7)), we see that *the RE algorithm* (resp. LE) *coincides with the LS algorithm* (resp. RS) *by permuting* N *and* N_1 (resp. N and N_2). This is obvious, since they simply express the equality of finite differences of order not greater than n of both polynomials at $x = 0$ (resp. at $x = N$).

§6. Joining two Discrete Bézier Curves

Given two DBC $L(x) = \sum_{i=0}^{n} L_i b_i^n(N_1, x+N_1)$ and $R(x) = \sum_{i=0}^{n} R_i b_i^n(N_2, x)$, defined respectively on $[-N_1, 0]$ and $[0, N_2]$, we want to join them "smoothly" at $x = 0$. There is a variety of ways for realizing this condition. We give three of them.

6.1. $C^{k,\ell}$ Continuity

For $0 \le k \le n$ and $0 \le \ell \le k$, we define *$C^{k,\ell}$ continuity of L and R at $x = 0$* by the fact that *both curves coincide at the $k+1$ integer points $x = \ell - k, \ell - k + 1, \dots, \ell$*. There are $k - \ell + 1$ points in $[-N_1, 0]$ and $\ell + 1$ ones in $[0, N_2]$.

One possible method for verifying this condition is to extend L to R^* on $[0, N_2]$, R to L^* on $[-N_1, 0]$ and to check that

$$L_i = L_i^* \text{ for } i = n, n-1, \dots, n-k+\ell$$
$$R_i = R_i^* \text{ for } i = 0, 1, \dots, n-k+\ell.$$

According to the results of § 5, the cost is $2n(n+1)$ linear combinations. This cost is reduced to $n(n+1)$ in the extreme cases $\ell = 0$ or $\ell = k$ because we have only to extend L or R respectively.

6.2. Equality of Forward and Backward Differences

For classical (continuous) Bézier curves L and R, defined e.g. on $[-\alpha, 0]$ and $[0, \beta]$, the equality of derivatives up to order k is equivalent to

$$\alpha^{-r}\nabla^r L_n = \beta^{-r}\Delta^r R_0 \qquad 0 \le r \le n. \tag{9}$$

A possible extension to DBC is to define C^k continuity by equality of finite differences at $x = 0$:

$$\nabla^r L(0) = \Delta^r R(0) \qquad r = 0, 1, \dots, k. \tag{10}$$

Using relations (3) of § 3 (algorithm FD), this is equivalent to

$$\frac{1}{N_1^{(r)}}\nabla^r L_n = \frac{1}{N_2^{(r)}}\Delta^r R(0) \qquad (0 \le r \le k),$$

which is very close to (9), especially for large values of N_1, N_2 and small values of r.

To compute $\{R_i, 0 \le i \le k\}$ from $\{L_{n-i}, 0 \le i \le k\}$, one can use the following algorithm:

(i) Setting $L_i^{(0)} = L_i$ for $n - k \le i \le n$, then for $r = 1, \dots, k$ and $i = n-k, \dots, n-r$ compute $L_i^{(r)} = \Delta L_i^{(r-1)} = L_{i+1}^{(r-1)} - L_i^{(r-1)}$

(ii) Setting $R_i^{(0)} = L_{n-i}^{(i)} \times N_2^{(i)}/N_1^{(i)}$ for $0 \le i \le k$, then for $r = 1, \dots, k$ and $i = n, n-1, \dots, r$, compute $R_i^{(r)} = R_{i-1}^{(r-1)} + R_i^{(r-1)}$

(iii) $R_i = R_i^{(i)}$ for $0 \le i \le k$.

Therefore, the algorithmic cost is $(k+1)(k+3)$ arithmetic operations. For $k = 1, 2$ and 3, we obtain

$$
\begin{aligned}
&R_0 = L_n, \qquad R_1 = (1+a_1)L_n - a_1 L_{n-1}\\
&R_2 = (1+2a_1+a_1a_2)L_n - 2a_1(1+a_2)L_{n-1} + a_1a_2L_{n-2}\\
&R_3 = (1+3a_1(1+a_2)+a_1a_2a_3)L_n - 3a_1(1+2a_2+a_2a_3)L_{n-1}\\
&\quad + 3a_1a_2(1+a_3)L_{n-2} - a_1a_2a_3L_{n-3},
\end{aligned}
$$

where

$$a_1 = N_2/N_1, \quad a_2 = (N_2-1)/(N_1-1), \quad a_3 = (N_2-2)/(N_1-2).$$

In particular, for $\alpha = \beta$ and $N_1 = N_2$, we see that (9) and (11) are both equivalent to

$$\nabla^r L_n = \Delta^r R_0 \qquad 0 \le r \le n.$$

This is equivalent to C^k-continuity of continuous Bézier curves on equal intervals. Thus the classical de Casteljau algorithm can be applied in this case, and we obtain R from L by the following formulae:

(i) $R_i^{(0)} = L_i$ for $n-k \le i \le n$

(ii) $R_i^{(r)} = -R_{i-1}^{(r-1)} + 2R_i^{(r-1)}$ for $1 \le r \le k$ and $n-k+1 \le i \le n$.

6.3. Joining Curve of Degree k

A third possibility is the following one: L and R are *k-smoothly joined* at $x = 0$ iff there exists a *polynomial curve J of degree k which coincides with L on the $k+1$ last integer points of* $[-N_1, 0]$ *and with R on the $k+1$ first integer points of* $[0, N_2]$.

The curve J has the two following representations:

$$J^L(x) = \sum_{i=0}^{k} J_i^L b_i^k(N_1, N_1+x) \qquad x \in [-N_1, 0]$$

$$J^R(x) = \sum_{i=0}^{k} J_i^R b_i^k(N_2, x) \qquad x \in [0, N_2].$$

By degree elevation, J^L becomes a DBC curve of degree n whose $k+1$ last coefficients coincide with those of L. Similary, J^R becomes a DBC of degree n whose $k+1$ first coefficients coincide with those of R. Practically, to obtain R from L, we follow the following procedure:

(1) by right degree reduction, compute J^L from L: the cost is $k(n-k)$ linear combinations.

(2) by right extension and right subdivision, compute J^R from J^L: the cost is $k(k+1)$ linear combinations.

(3) by left degree elevation, compute the first $k+1$ B-coefficients of R from J^R: the cost is $k(n-k)$ linear combinations.

The cost of this algorithm is $k(2n-k+1)$ linear combinations.

6.4. Application

Construction of a cubic B-spline of class C^2 on the interval $[-2N, 2N]$ subdivided into four intervals of length N.

1: $C^{2,1}$ continuity. We seek four polynomials

$$\begin{array}{ll} A(x) = \sum_{i=0}^{3} \alpha_i b_i^3(N, x+2N) & x \in [-2N, -N] \\ B(x) = \sum_{i=0}^{3} \beta_i b_i^3(N, x+N) & x \in [-N, 0] \\ C(x) = \sum_{i=0}^{3} \gamma_i b_i^3(N, x) & x \in [0, N] \\ D(x) = \sum_{i=0}^{3} \delta_i b_i^3(N, x-N) & x \in [N, 2N] \end{array}$$

which satisfy the following conditions:

(i) $A(-2N-1) = A(2N) = A(-2N+1) = 0$

(ii) $B(x) = A(x)$ for $x = -N-1, -N, -N+1$

(iii) $C(x) = B(x)$ for $x = -1, 0, 1$

(iv) $D(x) = C(x)$ for $x = N-1, N, N+1$

(v) $D(2N-1) = D(2N) = D(2N+1) = 0$.

By (i) and (v), we obtain $\alpha_0 = \alpha_1 = \delta_2 = \delta_3 = 0$ and

$$0 = A(-2N-1) = \sum_{i=0}^{3} \alpha_i b_i^3(N, -1) = \frac{6}{N(N-1)(N-2)}[(N+1)\alpha_2 - \alpha_3]$$

$$0 = D(2N+1) = \sum_{i=0}^{3} \delta_i b_i^3(N, N+1) = \frac{6}{N(N-1)(N-2)}[-\delta_0 + (N+1)\delta_1].$$

Therefore, $\alpha_3 = (N+1)\alpha_2$ and $\delta = (N+1)\delta_1$. By (ii), (iii) and (iv), we obtain successively: $\beta_0 = \alpha_3, \gamma_0 = \beta_3, \delta_0 = \gamma_3$, $3(N-1)(N-2)\alpha_2 + (N-1)(N-2)(N-3)\alpha_3 = (N+1)N(N-1)\beta_0 - 3(N+1)N\beta_1 + 6(N+1)\beta_2 - 6\beta_3 - 6\alpha_0 + 6(N+1)\alpha_1 - 3(N+1)N\alpha_2 + (N+1)N(N-1)\alpha_3 = (N-1)(N-2)(N-3)\beta_0 + 3(N-1)(N-2)\beta_1$ and similar relations for (β, γ) and (γ, δ).

We have 7 unknowns $\{\alpha_2, \beta_1, \beta_2, \gamma_0, \gamma_1, \gamma_2, \delta_1\}$ satisfying 6 equations. The solutions are

$$\begin{aligned} &\alpha_3 = \beta_0 = \gamma_3 = \delta_0 = (N+1)\alpha_2 \\ &\beta_1 = \gamma_2 = \alpha_2(N+1)(2N-1)/(N-1) \\ &\beta_2 = \gamma_1 = 2\alpha_2(2N^2 - N + 1)/(N-1) \\ &\beta_3 = \gamma_0 = 2\alpha_2(2N^2+1)/(N-1). \end{aligned}$$

Normalizing this B-spline to get a partition of unity gives $2\beta_0 + \gamma_0 = \beta_1 + \gamma_1 + \delta_1 = 1$, *i.e.*

$$\begin{aligned} &\alpha_2 = (N-1)/6N^2 \\ &\alpha_3 = \beta_0 = \gamma_3 = \delta_0 = (N^2-1)/6N^2 \\ &\beta_1 = \gamma_2 = (N+1)(2N-1)/6N^2 \\ &\beta_2 = \gamma_1 = (2N^2 - N + 1)3N^2. \end{aligned}$$

When $N \to +\infty$, these B-coefficients tends to those of the continuous cubic B-spline.

2: $C^{2,2}$ continuity. Now the conditions are the following:

(i) $A(x) = 0$ for $x = -2N, -2N+1, -2N+2$

(ii) $B(x) = A(x)$ for $x = -N, -N+1, -N+2$

(iii) $C(x) = B(x)$ for $x = 0, 1, 2$

(iv) $D(x) = C(x)$ for $x = N, N+1, N+2$

(v) $D(x) = 0$ for $x = 2N, 2N+1, 2N+2$.

Solving the associated system of linear equations and normalizing gives

$$\begin{aligned}
&\alpha_0 = \alpha_1 = \alpha_2 = \delta_3 = 0\\
&\alpha_3 = \beta_0 = (N-1)(N-2)/6N^2\\
&\beta_1 = (N-1)^2/3N^2, \beta_2 = (2N-1)(N-1)/3N^2\\
&\gamma_0 = \beta_3 = 2(N^2-1)/3N^2\\
&\gamma_1 = (2N^2+N-2)/3N^2, \gamma_2 = (N^2+3N-2)/3N^2\\
&\delta_0 = \gamma_3 = (N+1)(N+2)/6N^2\\
&\delta_1 = (N+1)/3N^2, \delta_2 = 1/3N^2.
\end{aligned}$$

When N tends to infinity, we have

$$\begin{aligned}
\lim \beta_0 &= \lim \delta_0 = \frac{1}{6}\\
\lim \beta_1 &= \lim \gamma_2 = \frac{1}{3}\\
\lim \beta_2 &= \lim \gamma_0 = \lim \gamma_1 = \frac{2}{3}.
\end{aligned}$$

3: Equality of forward and backward differences. In this case, we obtain of course the same B-coefficients as in the continuous case, i.e.

$$\alpha = \frac{1}{6}(0,0,0,1), \quad \beta = \frac{1}{6}(1,2,4,4), \quad \gamma = \frac{1}{6}(4,4,2,1), \quad \delta = \frac{1}{6}(1,0,0,0).$$

4: Joining parabola. In this case, we obtain

$$\alpha = (0,0,0,\beta_0) \quad \text{and} \quad \delta = (\delta_0, 0, 0, 0).$$

Let A^* be the parabola coinciding with A at the 3 points $x = -N-2, -N-1, -N$. Since $A(x) = \beta_0 b_3^3(N, x+2N)$, we obtain by degree reduction:

$$A^*(x) = \beta_0 b_0^2(N, x+2N) - \frac{1}{2}\beta_0 b_1^2(N, x+2N) + \beta_0 b_2^2(N, x+2N).$$

By extension to $[-2N, 0]$ and subdivision, we obtain A^* on the latter interval:

$$A^*(x) = \beta_0 \left[b_0^2(N, x+N) + \frac{5N+1}{2(N-1)} b_1^2(N, x+N) + \frac{7N-1}{N-1} b_2^2(x, x+N) \right].$$

Now, by degree elevation, we obtain the first coefficients of B:

$$\beta_1 = \frac{2N}{(N-1)}\beta_0 \quad \text{and} \quad \beta_2 = \frac{4N}{(N-1)}\beta_0.$$

Similarly, by symmetry and C^2 continuity at $x = 0$, we obtain the value $\gamma_0 = \beta_3 = \frac{2N+1}{3N}$. By normalization, we have $2\beta_0 + \gamma_0 = 1$, and finally

$$\beta_0 = \frac{N-1}{6N}, \beta_0 = \frac{1}{3}, \beta_2 = \frac{2}{3}, \beta_3 = \frac{2N+1}{3N}.$$

When N tends to infinity, we have

$$\lim \beta_0 = \frac{1}{6}, \quad \lim \beta_1 = \frac{1}{3}, \quad \lim \beta_2 = \lim \gamma_0 = \frac{2}{3}.$$

§7. Rectangular Discrete Bézier Surfaces

There is no theoretical difficulty to extend the above algorithms to tensor product surfaces. Let us only describe the subdivision algorithm for a discrete Bézier (DBS) defined on the rectangle $[0, M] \times [0, N]$:

$$B_{mn}P(x, y) = \sum_{i=0}^{m}\sum_{j=0}^{n} P_{ij} b_i^m(M, x) b_j^n(N, y).$$

This algorithm computes the B-coefficients of the four polynomials A, B, C, D which coincide with $B_{mn}P$ on the rectangles $[0, M_1]\times[0, N_1]$, $[M_1, M]\times[0, N_1]$, $[0, M_1] \times [N_1, N]$ and $[M_1, M] \times [N_1, N]$ respectively.

$$A(x, y) = \sum_{i=0}^{m}\sum_{j=0}^{n} A_{ij} b_i^m(M_1, x) b_j^n(N_1, y)$$

$$B(x, y) = \sum_{i=0}^{m}\sum_{j=0}^{n} B_{ij} b_i^m(M_2, x - M_1) b_j^n(N_1, y)$$

$$C(x, y) = \sum_{i=0}^{m}\sum_{j=0}^{n} C_{ij} b_i^m(M_1, x) b_j^n(N_2, y - N_1)$$

$$D(x, y) = \sum_{i=0}^{m}\sum_{j=0}^{n} D_{ij} b_i^m(M_2, x - M_1) b_j^n(N_2, y - N_1)$$

First step (left and right subdivisions in x):
For each $j = 0, 1, \ldots, n$ do

(i) $L_{ij}^{(0)} = R_{ij}^{(0)} = P_{ij} \qquad 0 \le i \le m$

(ii) for $k = 1, 2, \ldots, m$ and $k \le i \le m$

$$L_{ij}^{(k)} = (1 - \alpha_i)L_{i-1,j}^{(k-1)} + \alpha_i L_{i,j}^{(k-1)}$$

$$R_{m-ij}^{(k)} = \beta_i R_{m-i,j}^{(k-1)} + (1 - \beta_i)R_{m-i+1,j},$$

where $\alpha_i = (M_1 - i + k)/(M - k)$ and $\beta_i = (M_2 - i + k)/(M - k)$.

(iii) $L_{ij} = L_{ij}^{(i)}$ and $R_{ij} = R_{ij}^{(m-i)}, \qquad 0 \le i \le m$

B - Coefficients of the normalized B-spline with different C^2-continuity conditions

α	$C^{2,1}$	$C^{2,2}$	$\nabla\Delta$	JC
α_0	0	0	0	0
α_1	0	0	0	0
α_2	$\frac{N-1}{6N^2}$	0	0	0
$\beta_0 = \alpha_3$	$\frac{N^2-1}{6N^2}$	$\frac{(N-1)(N-2)}{6N^2}$	$\frac{1}{6}$	$\frac{N-1}{6N}$
β_1	$\frac{(N+1)(2N-1)}{6N^2}$	$\frac{(N-1)^2}{3N^2}$	$\frac{1}{3}$	$\frac{1}{3}$
β_2	$\frac{2N^2-N+1}{3N^2}$	$\frac{(2N-1)(N-1)}{3N^2}$	$\frac{2}{3}$	$\frac{2}{3}$
$\gamma_0 = \beta_3$	$\frac{2N^2+1}{3N^2}$	$\frac{(2N^2-1)}{3N^2}$	$\frac{2}{3}$	$\frac{2N+1}{3N}$
γ_1	$\frac{2N^2-N+1}{3N^2}$	$\frac{2N^2+N-2}{3N^2}$	$\frac{2}{3}$	$\frac{2}{3}$
γ_2	$\frac{(N+1)(2N-1)}{6N^2}$	$\frac{N^2+3N-2}{3N^2}$	$\frac{1}{3}$	$\frac{1}{3}$
$\delta_0 = \gamma_0$	$\frac{N^2-1}{6N^2}$	$\frac{(N-1)(N-2)}{6N^2}$	$\frac{1}{6}$	$\frac{N-1}{6N}$
δ_1	$\frac{N-1}{6N^2}$	$\frac{N+1}{3N^2}$	0	0
δ_2	0	$\frac{1}{3N^2}$	0	0
δ_3	0	0	0	0

Second step (left and right subdivisions in y):
For each $i = 0, 1, \ldots, m$ do

(i)
$$\left.\begin{aligned} A_{ij}^{(0)} &= C_{ij}^{(0)} = L_{ij} \\ B_{ij}^{(0)} &= D_{ij}^{(0)} = R_{ij} \end{aligned}\right\} \text{ for } 0 \le j \le n$$

(ii) for $\ell = 1, 2, \ldots, n$ and $\ell \le j \le n$

$$\begin{aligned} A_{ij}^{(\ell)} &= (1-\gamma_j)A_{ij-1}^{(\ell-1)} + \gamma A_{ij}^{(\ell-1)} \\ C_{i,n-j}^{(\ell)} &= \delta_j C_{n,n-j}^{(\ell-1)} + (1-\delta_j)C_{i,n-j+1}^{(\ell-1)} \end{aligned}$$

$$\begin{aligned} B_{ij}^{(\ell)} &= (1-\gamma_j)B_{ij-1}^{(\ell-1)} + \gamma B_{ij}^{(\ell-1)} \\ D_{i,n-j}^{(\ell)} &= \delta_j C_{i,n-j}^{(\ell-1)} + (1-\delta_j)D_{i,n-j+1}^{(\ell-1)}, \end{aligned}$$

where $\gamma_j = (N_1 - j + \ell)/(N - \ell)$ and $\delta_j = (N_2 - j + \ell)/(N - \ell)$.

(iii)
$$\left.\begin{aligned} A_{ij} &= A_{ij}^{(0)}, B_{ij} = B_{ij}^{(j)} \\ C_{ij} &= C_{i,j}^{(n-j)}, D_{ij} = D_{i,j}^{(n-j)} \end{aligned}\right\} \text{ for } 0 \le j \le n$$

Cost of the algorithm: $(n+1)m(m+1)$ linear combinations (LC) for the first step and $(2m+1)n(n+1)$ ones for the second step, i.e.: $n_S = (m+1)(n+1)(n+2m)LC$ if we perform subdivisions for x first and then for y. If we exchange the roles of variables, we get $n'_S = (n+1)(m+1)(m+2n)LC$. Therefore the first method (resp. the second) should be used when $m < n$ (resp. $n < m$).

Global cost ($m = n, M = N = 2^q n$): The total number of subdivisions is $1+4+\cdots+4^{q-1} = \frac{1}{3}(4^q - 1) = (N^2 - n^2)/3n^2$, and the total number of LC is $n_S = (n+1)^2(N^2 - n^2)/n$. Using the tensor product form of DDC algorithm yields $n_C = \frac{1}{2}n(n+1)(N-1)(N+n)LC$, and therefore we have

$$\frac{n_C}{n_S} = \frac{n^2(N-1)}{2(N+1)(N-m)} \sim \frac{n}{2} \quad \text{for } N \text{ large.}$$

Proof: Write $P(x,y) = \sum_{j=0}^{m} P_j(x) b_j^n(N,y)$, where $P_j(x) = \sum_{i=0}^{m} P_{ij} b_i^m(M,x)$. In the first step, for each $0 \le j \le n$,

$$P_j(x) = \sum_{i=0}^{m} L_{ij} b_i^m(M_1, x) \quad \text{for } x \in [0, M_1]$$

$$P_j(x) = \sum_{i=0}^{m} R_{ij} b_i^m(M_2; x - M_1) \quad \text{for } x \in [M_1, M]$$

whose coefficients are computed by subdivision in x. Next, for $x \in [0, M_1]$, we write

$$P(x,y) = \sum_{j=0}^{n} \left\{ \sum_{i=0}^{m} L_{ij} b_i^m(M_1, x) \right\} b_j^n(N, y)$$

$$= \sum_{i=0}^{m} \left\{ \sum_{j=0}^{n} L_{ij} b_i^n(N, y) \right\} b_i^m(M_1, x).$$

Now, for $y \in [0, N_1]$, we have

$$P(x,y) = \sum_{i=0}^{m} \left\{ \sum_{j=0}^{n} A_{ij} b_i^n(N_1, y) \right\} b_i^m(M_1, x),$$

and for each fixed $0 \le i \le m$, the array $\{A_{ij}, 0 \le j \le n\}$ is deduced from the array $\{L_{ij}, 0 \le j \le n\}$ by the subdivision algorithm in y.

Similar observations lead to the computation of coefficients of B, C and D. ■

§8. Triangular Discrete Bézier Surfaces

Let $S_N = \{x = (x_1, x_2, x_3) \in [0, N]^3 \cap \mathbb{N}^3 : |x| = x_1 + x_2 + x_3 = N\}$ denote the discrete triangle with $(N+1)(N+2)/2$ integer points. The DBB of $\mathbb{P}_n(S_N)$ is given by

$$b_i^n(N, x) = \binom{x}{i} \Big/ \binom{N}{n} = \binom{x_1}{i_1} \binom{x_2}{i_2} \binom{x_3}{i_3} \Big/ \binom{N}{n}$$

for $i = (i_1, i_2, i_3) \in S_N, x \in S_N$, and $\binom{s}{\alpha} = \frac{s^{(\alpha)}}{\alpha!} = s(s-1)\cdots(s-\alpha+1)/\alpha!$.

A triangular DBS is defined by $B_n P(x) = \sum_{i \in S_n} P_i b_i^n(N, x)$, where $x \in S_N$ and $P = \{P_i, i \in S_n\}$ is the control net of the discrete surface. There is no special difficulty to extend to these DBS the de Casteljau, finite difference, and degree elevation algorithms. For subdivision, there are at least three possible extensions corresponding to the partition of the original triangle into two, three or four triangles. Similarly, C^k continuity between two triangular DBS can be defined in different ways corresponding to those given in § 6. These notions will be developed in another paper, due to space limitation here.

§9. Discrete Bernstein Quasi-interpolants

The discrete version of the classical Bernstein quasi-interpolants is given by

$$\mathcal{D}_{n,N} f(x) = \sum_{i=0}^{n} f(ip) b_i^n(N, x), x \in [0, N],$$

where $p = N/n$ and $f \in \mathbb{R}^{[0,N]}$ (see also Stancu [13]).

From result of § 1, one verifies that $\mathcal{D}_{n,N}e_0 = e_0$ and $\mathcal{D}_{n,N}e_1 = e_1$, i.e., $\mathcal{D}_{n,N}$ is exact on $\mathbb{P}_1$. More generally, let $\{v_k, 0 \le k \le n\}$ be the Newton basis associated with $\{0, p, \ldots, np = N\}$, then one can prove that for $k = 0, 1, \ldots, n$

$$\mathcal{D}_{n,N}v_k = \lambda_k(n,N)e_k \quad \text{with} \quad \lambda_k(n,N) = p^k n^{(k)}/N^{(k)}.$$

Therefore, $\mathcal{D}_{n,N}$ is an isomorphism of $\mathbb{P}_n[0,N]$ whose eigenvalues are the $\lambda_k(n,N)$. Obviously, $\mathcal{D}_{n,N}$ is a positive operator and $\|\mathcal{D}_{n,N}\|_\infty = 1$. Moreover, $\mathcal{D}_{n,N}$ preserves all convexities of f and is variation (and total variation) diminishing. Using ideas developed recently by the author in the continuous case [9,10], one can define a sequence of discrete Bernstein quasi-interpolants (DBQI) $\mathcal{D}_{n,N}^{(k)}, 0 \le k \le n$, which are exact on $\mathbb{P}_k[0,N]$.

$$\mathcal{D}_{n,N}^{(0)} = \mathcal{D}_{n,N}^{(1)} = \mathcal{D}_{n,N}$$
$$\mathcal{D}_{n,N}^{(2)}f(x) = \mathcal{D}_{n,N}f(x) + \left(\frac{1}{p} - 1\right)\frac{x(N-x)}{2(n-1)}\delta^2\mathcal{D}_n f(x),$$

where $\delta^2 f(x) = \nabla\Delta f(x) = f(x+1) - 2f(x) + f(x-1)$

$$\mathcal{D}_{n,N}^{(3)}f(x) = \mathcal{D}_{n,N}^{(2)}f(x) + \left(\frac{1}{p} - 1\right)\left(\frac{1}{p} - 2\right)\frac{x(N-x)}{6(n-1)(n-2)}$$
$$\times \{(N-1-x)\Delta\delta^2\mathcal{D}_{n,N}f(x) - (x-1)\nabla\delta^2\mathcal{D}_{n,N}f(x)\}.$$

More generally, for $0 \le k \le n$,

$$\mathcal{D}_{n,N}^{(k)}f(x) = \mathcal{D}_{n,N}^{(k-1)}f(x) + \left(\frac{1}{p} - 1\right)^{(k)}$$
$$\frac{n}{k!n^{(k)}}\left\{\sum_{i=1}^{k-1} a_i x^{(i)}(N-x)^{(k-i)}\nabla^i\Delta^{k-i}\mathcal{D}_{n,N}f(x)\right\},$$

where the $k-1$ unknown coefficients $\{a_i, 1 \le i \le k-1\}$ are solutions of a linear system of order $k-1$ with a triangular matrix and can be computed by formal calculus. As in the continuous case, it is possible to give upper bounds of $\|\mathcal{D}_{n,N}^{(k)}\|_\infty$ independent of n and N, for any $k \ge 0$ fixed.

Similar results can also be proved for Hahn-Bernstein quasi-interpolants defined and studied in [8], whose eigenfunctions are Hahn polynomials (see also [11] for applications of degree elevation and subdivision algorithms to this class of discrete orthogonal polynomials). The above results will be detailed elsewhere.

§10. Dual Basis, Condition Number, Finite Differences

Following ideas developed by Ciesielski [2] in the continuous case, one can define a dual basis of the DBB of $\mathbb{P}_n[0,N]$ in the following way. Let $t_n(N,x) =$

$\sum_{i=0}^{n}(-1)^i \binom{n}{i} b_i^n(N,x)$ be the discrete Chebyshev polynomial of degree $n \le N$. These polynomials are orthogonal with respect to the discrete scalar product

$$\langle f, g\rangle = \frac{1}{N+1}\sum_{x=0}^{N} f(x)g(x),$$

i.e., they satisfy $\langle t_i(N,\cdot), t_j(N,\cdot)\rangle = 0$ for $i \neq j$. Let $\frac{1}{c_n^2} = \|t_n(N,\cdot)\|^2 = \langle t_n(N\cdot), t_n(N,\cdot)\rangle$ and define by induction on n, for $0 \le i \le n$,

$$\overline{b}_i^n(N,\cdot) = \left(1 - \frac{i}{n}\right)\overline{b}_i^{n-1}(N,\cdot) + \frac{i}{n}\overline{b}_i^{n-1}(N,\cdot) + (-1)^i \binom{n}{i} c_n^2 t_n(N,\cdot).$$

Then it can be proved that the polynomials $\{\overline{b}_i^n(N,\cdot), 0 \le i \le n\}$ form a basis of $\mathbb{P}_n[0,N]$ which is orthogonal to the DBB, i.e.

$$\langle b_i^n(N,\cdot), \overline{b}_j^n(N,\cdot)\rangle = 0 \quad \text{for} \quad i \neq j.$$

Following the same technique as the one given by Ciesielski, it can be deduced from this fact that the condition number (see [4] for the definition) of the DBB of $\mathbb{P}_n[0,N]$ is $\mathcal{O}(2^n)$ for N large enough. The detailed proof will be given elsewhere.

Another interesting property of the DBB is the following formula which makes the connection between finite differences of order $n+1$ of a given function on *consecutive* integer points of $[0, N+n+1]$ and divided differences on *extreme* points of this interval. For $0 \le k \le n$, we have

$$\frac{1}{N+1}\sum_{x=0}^{N} b_k^n(N,x)\Delta^{n+1}f(x) = n![k,\ldots,n,N+1,\ldots,N+k+1]f,$$

the last quantity being the divided difference of order $n+1$ of f at the given points. The proof of this formula can be made by induction on n using Leibniz formula for finite differences. The extension to tensor products is straightforward, but it would also be interesting to extend it to triangles.

References

1. Barry, P. J., Urn models, recursive curve schemes, and computer aided geometric design, Ph. D. dissertation, University of Utah, Salt Lake City, 1987.
2. Ciesielski, Z., The degenerate B-splines as a basis in the space of algebraic polynomials, Annales Polonici Mathematici XLVI (1985), 71–79.
3. Goldman, R. N., Polya's urn model and computer aided geometric design, SIAM J. on Algebraic and Discrete Methods **6** (1985), 1–28.
4. Gautschi, W., Questions of numerical condition related to polynomials, in *Studies in numerical analysis*, G. Golub (ed), The Mathematical Association of America, 1984, 140–177.

5. Lyche, T., Discrete polynomial spline approximation methods, thesis, University of Texas, Austin, 1975.
6. Mangasarian, O. L., and L. L. Schumaker, Discrete splines via mathematical programming, SIAM J. Control **9** (1971), 174–183.
7. Mangasarian, O. L., and L. L. Schumaker, Best summation formulae and discrete splines, SIAM J. Numerical Analysis **10** (1973), 448–459.
8. Sablonnière, P., Hahn polynomials as eigenvectors of positive operators, in *Second International Symposium on orthogonal polynomials and their applications, Segovia, 1986*, Monografias de la Academia de Ciencias de Zaragoza, 1988, 139–146.
9. Sablonnière, P., Bernstein quasi-interpolants on $[0,1]$, in *Multivariate approximation theory* IV, C.K. Chui, W. Schempp, K. Zeller (eds.), ISNM, vol. 90, Birkhäuser Verlag, Basel, 1989, 287–294.
10. Sablonnière, P., Bernstein type quasi-interpolants, in *Curves and Surfaces*, P.-J. Laurent, A. Le Méhauté, and L. L. Schumaker (eds.), Academic Press, New York, 1991, 421–426.
11. Sablonnière, P., Discrete Bernstein basis and Hahn polynomials, in *VII Simposium sobre polinomios ortogonales y aplicaciones*. Granada, Sept. 23-27, 1991. Publication LANS n° 37 ; INSA de Rennes.
12. Schumaker, L. L., Constructive aspects of discrete polynomial spline functions, in *Approximation Theory*, G.G. Lorentz ed., Academic Press, New York, 1973, 469–476.
13. Stancu, D. D., Approximation of functions by a new class of linear polynomial operators, Rev. Roum. Math. Pures et Appl., **13** (1968), 1173–1194.

Paul Sablonnière
Laboratoire LANS,
INSA de Rennes
20, Avenue des Buttes de Coesmes
35043 Rennes Cedex
France

Rational Geometric Curve Interpolation

R. Schaback

Abstract. This survey covers geometric (i.e. parametrization–independent) curve interpolation methods by piecewise rational functions with geometric continuity. It includes shape–preserving properties, orders of accuracy, and provides a number of examples.

§1. Introduction

A *classical* (non–parametric) univariate interpolation problem is based on two different kinds of data: given *ordinates*

$$y_0, y_1, \ldots, y_{N-1} \in \mathbb{R}^d, \quad d, N \in \mathbb{N} \tag{1}$$

are to be interpolated at given *abscissae*

$$a \le t_0 < t_1 < \cdots < t_{N-1} \le b, \quad a, b \in \mathbb{R} \tag{2}$$

by a function f_N from a set $F_N \subseteq C[a,b]$ of (possibly vector–valued) functions on $[a,b] \subset \mathbb{R}$ in the sense that the ordinate values y_i are attained by f_N on the abcissae, i.e.:

$$f_N(t_i) = y_i, \quad 0 \le i \le N-1. \tag{3}$$

In Computer–Aided Design applications, however, only the ordinates or *positions* (1) are given, and there is no information on the abscissae or *parameters* (2). The interpolation conditions still have the form (3), but the parameters t_i have to be determined from the positions alone (*geometric* interpolation). If the data

$$y_i = f(s_i), \quad 0 \le i \le N-1$$

are sampled from a curve

$$f : [a,b] \to \mathbb{R}^d$$

Mathematical Methods in Computer Aided Geometric Design II
Tom Lyche and Larry L. Schumaker (eds.), pp. 517–535.

ISBN 0-12-460510-9.

for unknown values

$$0 \leq s_0 < s_1 < \cdots < s_{N-1} \leq b$$

of the parameter s of f, then the positions y_i are considered as being independent of changes of the parametrization. The geometric information contained in the range $R = f([a,b])$ is the only available source for the determination of the unknown parameters t_i or s_i.

This survey will concentrate on *geometric* curve interpolation. Recent accounts of the state-of-the-art of nonparametric univariate interpolation are by Gregory [10], Schmidt [26] and Späth [27].

In view of the well-known negative results on polynomial interpolation, the sets F_N of interpolating curves are usually chosen as spaces of *piecewise* polynomial or rational functions. Each piece should be defined with as little dependence on parametrization as possible, and the local connections between pieces should also not depend on parametrization. The latter is achieved by the notion of GC^k (i.e. geometric C^k) continuity of two curves

$$f[a,b] \to \mathbb{R}^d, \quad g : [b,c] \to \mathbb{R}^d$$

at their common parameter b. It is required that there should exist strictly monotonic C^k reparametrization functions

$$\varphi : [a,b] \to [\alpha, \beta], \quad \psi : [b,c] \to [\beta, \gamma]$$

such that the functions $f \circ \varphi^{-1}$ and $g \circ \psi^{-1}$ coincide up to their k-th derivative at β.

Local polynomial or rational curve pieces in *Bernstein–Bézier representation* have the advantage of being at least independent of linear or rational reparametrizations. They involve $n+1$ vectors $b_0, \ldots, b_n$ called *control points* and represent a vector-valued polynomial as

$$BB[b_0, \ldots, b_n](t) := \sum_{j=0}^{n} \beta_{j,n}(t) b_j \tag{4}$$

of degree $\leq n$, blending the control points b_j by Bernstein polynomials

$$\beta_{j,n}(t) := \binom{n}{j} \left(\frac{b-t}{b-a}\right)^{n-j} \left(\frac{t-a}{b-a}\right)^{j}, \quad 0 \leq j \leq n.$$

Rational pieces in Bernstein–Bézier representation additionally use $n+1$ positive weights $w_0, w_1, \ldots, w_n$ to define

$$R[w_0, b_0, \ldots, w_n, b_n](t) := \frac{BB[w_0 b_0, \ldots, w_n b_n](t)}{BB[w_0, \ldots, w_n](t)}. \tag{5}$$

Basic facts on polynomial Bernstein–Bézier representations (4) can be found in any textbook on Computer–Aided Design, e.g. Farin [3] or Hoschek & Lasser [12], while a comprehensive account of rational curves and surfaces is given by Farin in [2].

One of the standard properties of rational Bernstein–Bézier representations (5) for $n \geq 1$ is that by a suitable reparametrization two of the weights can be chosen to be equal to one. For rational quadratics, usually $w_0 = w_2 = 1$ leaves w_1 to cover the whole range of conics by

$$\begin{array}{lcll} w_1 & < & 1 & : \text{ elliptic curve,} \\ w_1 & = & 1 & : \text{ parabolic curve,} \\ w_1 & > & 1 & : \text{ hyperbolic curve.} \end{array}$$

Details can be found in the survey of Lee [13] on rational Bézier representations of conics. For rational cubics it is somewhat more convenient to use $w_1 = w_2 = 1$, leaving w_0 and w_3 as free weights with the special cases

$$\begin{array}{ccccll} & & w_0 & = & w_3 & : \text{ denominator of degree } \leq 2, \\ 1 & = & w_0 & = & w_3 & : \text{ constant denominator, polynomial cubic.} \end{array}$$

Unfortunately, there is no survey of the special geometric properties of Bernstein–Bézier representations of rational cubics. From the available sources on rational curves we note without further explanation that successive subdivision of rational curves takes successive means of weights, causing all weights to tend to one, if two of the weights are normalized to be exactly one. Thus some precautions to fixed choices of weights have to be made; for sufficiently dense data the interpolation methods using rational pieces should allow for variable weights tending to one.

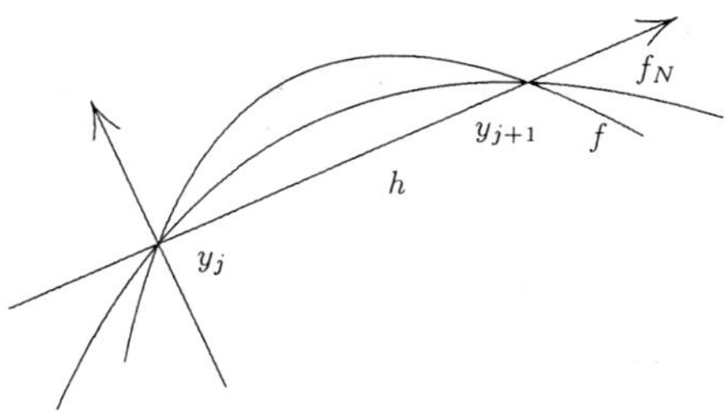

Figure 1. Error between curve and interpolant.

To measure the error of an interpolant f_N to a curve f that supplied the interpolation data, usually f is assumed to be smooth and regular in the sense

$$f'(t) \neq 0 \text{ for all } t \in [a, b].$$

Then the behavior of f and f_N between two interpolation points $y_j \neq y_{j+1}$ is considered (see Figure 1).

If the data sample is dense enough in the sense that

$$h := \max_j \|y_{j+1} - y_j\|$$

is sufficiently small, the chord $y_j y_{j+1}$ defines a local coordinate system, along which the curve f can be uniquely reparametrized by a monotonic function φ_j. The interpolant f_N must be proven to be regularly reparametrizable between y_j and y_{j+1} by a monotonic function ψ_j. For instance, if f and f_N are regular, then on the unique hyperplane $H(t)$ perpendicular to $y_{j+1} - y_j$ and containing the point

$$y(t) = (1-t)y_j + ty_{j+1},\ t \in [0,1]$$

there are two uniquely defined points $f(\varphi_j(t))$ and $f_N(\psi_j(t))$, and the local error between y_j and y_{j+1} can be measured as

$$\max_{0\le t\le 1} \|f(\varphi_j(t)) - f_N(\psi_j(t))\|,$$

if h is sufficiently small. Taking another maximum over j, the overall error can be evaluated. If the error has the behavior $\mathcal{O}(h^m)$ for $h \to 0$, the interpolation method is said to be of *order* m or of *accuracy* h^m.

This survey will focus on geometric rational curve interpolation in $\mathbb{R}^d$ for dimensions $d = 2$ or $d = 3$ with geometric GC^k continuity for $k = 1$ or $k = 2$, but some detours into piecewise polynomial cases will be necessary. There is a vast literature on methods with additional "tension" parameters which can be used for local shape control in design applications. This survey excludes these methods and concentrates on interpolants which are completely determined by the given data. The reader is tacitly assumed to be familiar with the basics of Bernstein–Bézier representations. Because a number of existing interpolation methods treat variations of two–point Hermite interpolation problems, these will be considered first, being followed later by Lagrange interpolation problems. Some examples will conclude the survey.

§2. Two–point GC^2 Hermite Interpolation

In contrast to classical derivatives, the notions of *tangent direction*

$$r(t) := f'(t)/\|f'(t)\|_2, \quad t \in [a,b] \tag{6}$$

and *curvature*

$$\kappa(t) = \|f'(t) \times f''(t)\|_2/\|f'(t)\|_2^3, \quad t \in [a,b] \tag{7}$$

of smooth curves $f : [a,b] \to \mathbb{R}^d$ are independent of the parametrization. Thus they are ideal candidates for specifying parametrization–independent Hermite interpolation data. To make (6) and (7) well–defined, the underlying curve should be regular and the classical cross–product is employed for $d = 3$, while for $d = 2$ we use

$$\begin{pmatrix} x_1 \\ y_1 \end{pmatrix} \times \begin{pmatrix} x_2 \\ y_2 \end{pmatrix} = x_1 y_2 - x_2 y_1$$

to define *signed* curvature as

$$\kappa(t) = (f'(t) \times f''(t))/\|f'(t)\|_2^3, \quad t \in [a, b].$$

A two–point GC^1 Hermite interpolation problem will specify two positions $y_0 \neq y_1$ together with two tangent directions r_0, r_1 to be attained by the interpolant at y_0, y_1. In $\mathbb{R}^2$, two curvature values for a two–point GC^2 Hermite interpolation problem can be simply specified as two (signed) real numbers κ_0 and κ_1, while in $\mathbb{R}^3$ one should specify two vectors $\kappa_0\eta_0$ and $\kappa_1\eta_1$ with positive curvature values κ_0 and κ_1. Here, η_0 and η_1 are (normalized) *binormal* vectors, being orthogonal to the osculating planes and to the tangent directions r_0 and r_1 at y_0 and y_1, respectively.

If the data y_0, y_1, r_0 and r_1 lie on a line L in $\mathbb{R}^2$, while the curvature data, illustrated in Figure 2 by small circular arcs at y_0 and y_1, require different signs of curvature, then any polynomial or rational interpolant in Bernstein–Bézier representation will in general need at least six control points of the form

$$\begin{array}{lllllll} b_0 & = & y_0, & b_5 & = & y_1 \\ b_1 & = & y_0 + \alpha_0 r_0 & b_4 & = & y_1 - \alpha_1 r_1 \\ b_2 & = & y_0 + \beta_0 \eta_0 & b_3 & = & y_0 - \beta_1 \eta_1 \end{array}$$

with positive real constants α_i, β_i and vectors η_0, η_1 not parallel to L (see Figure 2 for the $\mathbb{R}^2$ situation, but a similar construction with two disjoint osculating planes will suffice for $\mathbb{R}^3$). This proves

Theorem 1. *A general strategy to solve a two–point GC^2 Hermite interpolation problem by a rational or polynomial function in Bernstein–Bézier representation will require a degree of at least five.*

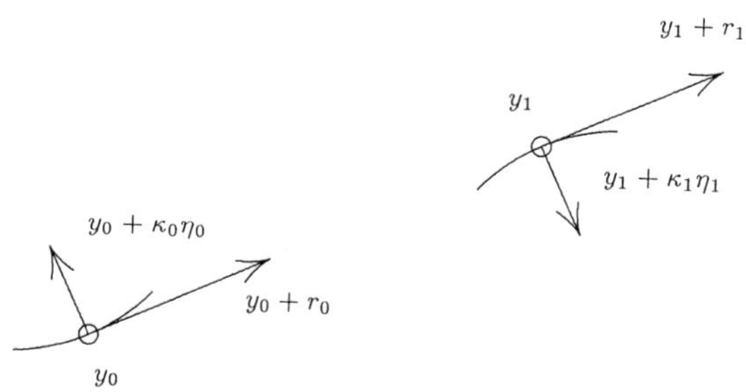

Figure 2. A special two–point GC^2 Hermite problem in $\mathbb{R}^2$.

The same degree would suffice to produce a C^2 solution, if the parameters t_0, t_1 were known together with $f^{(j)}(t_i)$ for $i = 0,1$ and $j = 0,1,2$. But a simple comparison of the dimension of the problem and the degrees of freedom gives a quite different result. In $\mathbb{R}^3$, the problem dimensions are

$$\begin{array}{lcll} 3 & + & 3 & \text{for positions } y_0, y_1 \\ 2 & + & 2 & \text{for tangent directions } r_0, r_1 \text{ with } \|r_i\|_2 = 1 \\ 2 & + & 2 & \text{for } \kappa_i\eta_i \text{ values with } \eta_i \perp r_i,\ i = 0,1 \end{array}$$

summing up to 14, while the degrees of freedom are

$$\begin{array}{ll} 18 & \text{for polynomial quintics} \\ 15 & \text{for polynomial quartics} \\ 14 & \text{for rational cubics} \end{array}$$

in Bernstein–Bézier representation.

Thus the degree five interpolant, as required by Theorem 1 for the general case, has at least four superfluous degrees of freedom. However, no method is known that handles the general case by a quintic polynomial and makes a clever use of its free parameters under special circumstances. A method based on polynomial quartics,]s given by Peters 16], can only handle a a restricted class of subproblems and still will have one degree of freedom in a number of cases.

However, there is no additional degree of freedom if rational cubics are used, though they will not be able to handle the general problem. In [11], Höllig gives a solution and proves

Theorem 2. *If data are sampled from a smooth and regular curve in $\mathbb{R}^3$ with nonvanishing curvature and torsion, there is a unique solution to the GC^2 Hermite interpolation problem with accuracy $\mathcal{O}(h^6)$, if h is sufficiently small.*

Höllig solves the problem via a linear system. A direct and explicit solution can be given by writing the interior control points b_1 and b_2 as

$$\begin{array}{rcll} b_{1+2i} & = & b_{3i} + (-1)^i \delta_i r_i, & i = 0,1, \\ \delta_i & = & \|b_{1+2i} - b_{3i}\|_2 > 0, & i = 0,1, \end{array}$$

and observing that the biorthogonal vector η_i must be perpendicular to both $b_1 - b_{3i}$ and $b_2 - b_{3i}$, yielding

$$\delta_{1-i} = \frac{\eta_i^T (y_1 - y_0)}{\eta_i^T r_{1-i}}, \quad i = 0,1 \tag{8}$$

independent of the prescribed curvature. Positivity of the solution requires $y_1 - y_0$ and r_{1-i} to be in the same half-space defined by the osculating plane at y_i.

If the data are not coplanar and if the above condition is not satisfied, then a cubic solution cannot exist. Since we can resort to an $\mathbb{R}^2$ method for coplanar data, equations (8) characterize the solvability of the problem by cubics. Curvature continuity is achieved by choosing $w_1 = w_2 = 1$ and determining w_0 and w_3 directly from

$$\kappa_i \eta_i = \frac{2}{3} w_{3i} (-1)^i r_i \times (y_1 - y_0 - \delta_{1-i} r_{1-i}), \quad i = 0, 1 \tag{9}$$

(see Höllig [11]), which is possible if the data are not coplanar. Equations (8) and (9) provide an easy implementation of the method. However, there still is no satisfactory embedding of this special case into the general situation.

In $\mathbb{R}^2$, the dimensions of a two–point GC^2 Hermite interpolation problem are

2	+	2	for positions y_0, y_1,
1	+	1	for tangent directions r_0, r_1 with $\|r_i\|_2 = 1$,
1	+	1	for (scalar) curvature values

which sum up to 8, and the degrees of freedom are

6	for polynomial quadratics,
7	for rational quadratics,
8	for polynomial cubics,
10	for rational cubics
10	for polynomial quartics,
12	for polynomial quintics.

Again, there is an excess of 4 degrees of freedom in the polynomial quintics, which are required by Theorem 1 to handle the general case. The "natural" interpolants for a class of subproblems will consist of piecewise polynomial cubics, and this case is treated by de Boor, Höllig, and Sabin in [1]. They show

Theorem 3. *If the data are sampled from a smooth and regular curve in $\mathbb{R}^2$ with nonvanishing curvature, there is a unique solution to the GC^2 Hermite interpolation problem with accuracy $\mathcal{O}(h^6)$, if h is small enough.*

The underlying system is a special case of (9) for planar data, i.e.:

$$\frac{3}{2} \delta_i^2 \kappa_i = c_i - c_0 \delta_{1-i}, \qquad i = 0, 1 \tag{10}$$

with

$$\begin{aligned} \eta c_i &= (-1)^i r_i \times (y_1 - y_0) \\ \eta c &= r_0 \times r_1 \end{aligned}$$

for $\eta = \eta_0 = \eta_1 \perp r_0, r_1, y_1 - y_0$, if everything is written in $\mathbb{R}^3$ for compatibility with (9). These two quadratic equations with two unknowns may have several or no solutions, and it is quite surprising that for $h \to 0$ there are three coalescing positive solutions, if the data are sampled from a smooth and regular curve with nonvanishing curvature.

Unfortunately, the method does not work for general h, even if the above assumption holds. For h small enough, it even works for general data sets, but it might produce up to three local solutions. Furthermore, the accuracy may go down to $\mathcal{O}(h^4)$ in cases with zeros of curvature. In [24] Theorem 3 is extended to the case of curves which may have simple zeros of curvature. As is shown in [1] by an example, the accuracy necessarily must go down to $\mathcal{O}(h^4)$ near *double* zeros of curvature. Thus for $h \to 0$ the behavior of the method is fully characterized, leaving its feasibility and numerical treatment for coarse data as an open question.

The approach of Goodman, Unsworth and others [5,6,9] uses rational cubics to interpolate planar two–point Hermite data with GC^2 continuity. Counting the degrees of freedom yields two free parameters per interval which the cited papers fix in a way that maintains convex interpolants to convex data and preserves circular arcs.

To be more specific, equations (9) and (10) in this case become

$$\frac{3}{2}\,\kappa_i\delta_i^2 = w_{3i}(c_i - c\delta_{1-i}), \quad i = 0, 1$$

and are satisfied for each choice of δ_0, δ_i by determination of w_{3i}. This leaves δ_0 and δ_1 as free parameters, and in [5] Goodman suggests

$$\delta_0 = \frac{2\|y_1 - y_0\||\sin\alpha_1|}{2\lambda|\sin\alpha_1| + (1-\lambda)\|y_1 - y_0\|\cdot|\kappa_1| + 2|\sin(\alpha_0+\alpha_1)|}$$

$$\delta_1 = \frac{2\|y_1 - y_0\||\sin\alpha_0|}{2\mu|\sin\alpha_0| + (1-\mu)\|y_1 - y_0\|\cdot|\kappa_0| + 2|\sin(\alpha_0+\alpha_1)|}$$

with parameters $\lambda, \mu \in (0, 1)$ and for angles

$$\begin{array}{llcl} \alpha_0 & \text{between} & y_1 - y_0 & \text{and} & r_0, \\ \alpha_1 & \text{between} & r_1 & \text{and} & y_1 - y_0, \end{array}$$

respectively. This choice is made for convex data and will yield a convex interpolant, while

$$\delta_0 = \gamma\|y_1 - y_0\|, \qquad \delta_1 = \delta\|y_1 - y_0\|$$

is suggested with $0 < \gamma, \delta < \gamma + \delta < 1$ for the other cases. Note that $\gamma = \delta = 1/3$ holds in the $h/3$–rule given by (11). So far there is no order of accuracy known for this method. Its advantage is its generality, providing a widely applicable GC^2 solution for planar data.

Goodman and Unsworth give an interactive extension in [9] which allows linear pieces to be inserted, and provide a detailed formulation of the algorithm together with a series of numerical examples. The paper [6] of Goodman, Ong, and Unsworth additionally allows to keep the interpolant away from a number of given lines which restrict the convex solution in the plane.

In [7], Goodman and Unsworth provide a GC^n interpolation method for planar data using piecewise polynomials of degree $\leq n + 1$. It solves the Lagrange problem via a Hermite problem with estimated tangent data and zero curvature data at the breakpoints. Though it is convexity–preserving and allows general data, its curvature properties restrict it to be of at most second order.

§3. Two-point GC^1 Hermite Interpolation

Here the given data only consist of two positions y_0, y_1 and correspondent tangent directions r_0, r_1 with normalization $\|r_i\|_2 = 1$. If the rays $y_i + (-1)^i \alpha r_i$ for $\alpha \geq 0$ do not intersect, a GC^1 solution will necessarily need at least four control points (see Figure 3).

Theorem 4. *A general strategy to solve a two–point GC^1 Hermite interpolation problem by a rational or polynomial function in Bernstein–Bézier representation will require a degree of at least three.*

Figure 3. A special two–point GC^1 Hermite problem.

All existing methods to solve this problem have additional degrees of freedom, which can be used, for instance,

1. to increase smoothness or
2. for certain reproduction properties or
3. for higher accuracy or
4. for additional interpolation conditions.

If the problem is posed in $\mathbb{R}^3$ with non-coplanar data, no partial solution by a quadratic method can exist because it would be confined to a plane defined by its three control points. Cubic polynomials will have two excess degrees of freedom, and no strategy seems to be known that makes clever use of this fact. The methods of [1] show that the simple $h/3$-rule

$$\begin{aligned} h &= \|y_1 - y_0\| \\ b_1 &= y_0 + \frac{h}{3} r_0, \quad b_2 = y_1 - \frac{h}{3} r_1 \\ b_0 &= y_0, \quad b_3 = y_1 \end{aligned} \tag{11}$$

constructs a polynomial cubic GC^1 interpolant with $\mathcal{O}(h^4)$ accuracy. It works in any dimension and seems to be rather robust.

Rational cubics in $\mathbb{R}^3$ have 4 excess degrees of freedom and were considered by Piegl in [18]. When written in the form (5), Piegl's interpolant y_P gets weights (1, 1/3 , 1/3 ,1), because it has the denominator $(1-t)^2 + t^2$.

If compared to a two–point cubic Hermite interpolant y_H, the condition $y_H(\ 1/2\ ) = y_P(\ 1/2\ )$ is imposed, leading to control points

$$\begin{aligned} b_0 &= y_0, & b_3 &= y_1 \\ b_1 &= y_0 + \frac{h}{2}\, r_0, & b_2 &= y_1 - \frac{h}{2}\, r_1 \end{aligned}$$

for (5) with $h = \|y_1 - y_0\|$ and weights as above. Now, if the data are sampled from a smooth curve f with arclength parametrization in the sense

$$\begin{aligned} y_0 &= f(0), & y_1 &= f(t) \\ r_0 &= f'(0), & r_1 &= f'(t), \end{aligned}$$

then curvature of Piegl's interpolant y_P at zero will be

$$\frac{8}{h^3}\left(2 \cdot \frac{h}{2}\, r_0 \times (f(t) - \frac{h}{2}\, f'(t) - f(0))\right) = \mathcal{O}(h)$$

because of

$$y_P'(0) = b_1 - b_0, \quad y_P''(0) = 2(b_2 - b_0), \quad \text{and} \quad t = h + \mathcal{O}(h^3)$$

for $h \to 0$. Thus Piegl's interpolant will have small curvature at the breakpoints for dense data. Its order of convergence cannot exceed two. Nevertheless, there may be visually pleasing results for coarse data.

In $\mathbb{R}^2$, polynomial cubics again have two additional degrees of freedom, which were used by de Boor, Höllig, and Sabin in [1] to increase smoothness to GC^2 and accuracy to $\mathcal{O}(h^6)$, if curvature data are known. The approach of [23] shows how to generate such auxiliary data, and then Theorem 3 can be carried over to the GC^1 case.

Rational or polynomial quadratics become feasible for planar data. A quadratic Bernstein-Bézier polynomial yields a GC^1 interpolant to Hermite data from convex curves by simply constructing the missing control point b_1 by intersection of tangents (see Figure 4). The methods of [1] easily supply a proof of

Theorem 5. *If the data are sampled from a smooth and regular curve in* $\mathbb{R}^2$ *with nonvanishing curvature, there is a unique solution to the* GC^1 *Hermite interpolation problem with accuracy* $\mathcal{O}(h^4)$.

Going to rational quadratics, there will be an excess degree of freedom in the choice of the weight w_1, which controls the type of the conic represented by the rational quadratic. Since three points and two tangents uniquely define a conic (this is a classical result of projective geometry), an additional data point $y_{1/2}$ within the triangle $y_0\ b_1\ y_1$ can be prescribed and there still will be a unique solution. For an explicit construction and further details see [25].

Both methods work wherever the data allow the formation of triangles like in Figures 4 and 5, and they give visually pleasing results. Note that the method using rational quadratics will reproduce any kind of conic.

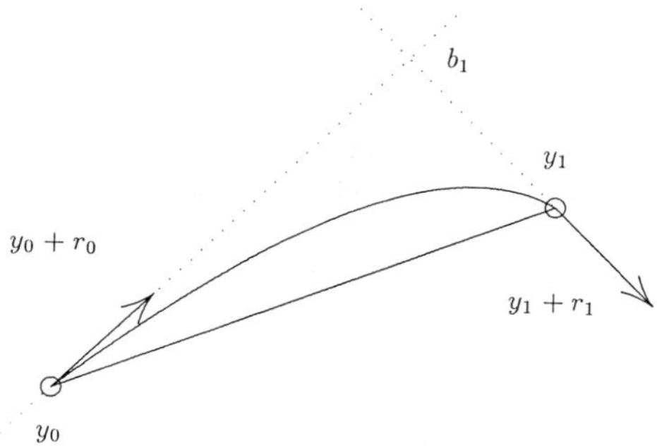

Figure 4. Quadratic GC^1 Hermite Interpolant.

§4. Lagrange Interpolation via Hermite Interpolation

It is very convenient to use two-point Hermite interpolation sequentially on large data sets, because the methods are purely local, the complexity grows linearly with the sample size, and one can work on a data stream with only local storage, calculating secondary information (derivatives, bending energy etc.) "on the fly". But then one needs good estimates for derivative data. As a rule of thumb, which is made precise in [23], a Hermite interpolation method of final accuracy h^m requires accuracy h^{m-j} for its j-th derivative input. Unfortunately, for purely positional data the lack of information on the parametrization makes it hard to design methods of high order which construct data like tangent directions or values of curvature or torsion with sufficiently high accuracy.

Interpolating positional data $y_i \in \mathbb{R}^d$ at successive chordlength abscissae t_i with

$$t_{i+1} - t_i = \|y_{i+1} - y_i\|_2$$

cannot yield approximations of order more than three in general, but if chordlength is taken with respect to a fixed center y_ℓ, then

$$t_i = t_\ell + \|y_i - y_\ell\|_2 \operatorname{sgn}(i - \ell), \qquad i \text{ around } \ell$$

does not spoil approximation orders, but still requires elimination of the parametrization effects. A detailed analysis of this approach can be found in [23] together with methods for GC^{k-1} interpolation by piecewise polynomials of degree $\leq 2k - 1$ with accuracy $\mathcal{O}(h^{2k})$ for arbitrary k, provided that the data are sampled from a sufficiently smooth and regular curve.

§5. Direct Lagrange Interpolation in $\mathbb{R}^3$

This is a field in which still a lot of work must be done, though it is most important for applications. To see why this case is difficult to handle, we again compare the number of parameters of the problem with the degrees of freedom of certain families of interpolating functions. The simplest conceivable and useful piecewise polynomial interpolant (apart from the trivial case of classical Lagrange interpolation at estimated parameter values) seems to be a GC^2 piecewise cubic. Given purely positional data $y_0, \ldots, y_{N-1} \in \mathbb{R}^3$, we are left with two free control points per section, if we place the breakpoints at the "interior" data points $y_1, \ldots, y_{N-2}$. Then GC^1 continuity enforces collinearity of three control points around each interior data point, yielding $2(N-2)$ scalar conditions. Furthermore, GC^2 continuity requires coplanarity of five points around these data points, plus a scalar curvature continuity condition on each osculating plane. Again, this amounts to $2(N-2)$ scalar conditions, which still leaves $6(N-1) - 4(N-2) = 2N+2$ unused degrees of freedom. No algorithm is known which makes a clever use of these degrees of freedom. For instance, additional data points $y_{i+1/2}$ for $i = 0, \ldots, N-2$ might be prescribed in each section joining y_i and y_{i+1}. If the local parameter $t_{i+1/2}$ with

$$f_N(t_{i+1/2}) = y_{i+1/2}, \quad 0 \le i \le N-1$$

is not prescribed, this will fix $2(N-1)$ degrees of freedom, allowing two additional boundary tangent directions at y_0 and y_{N-1} to be specified. No solution to this problem is known so far.

Polynomial quadratics can only form GC^1 interpolants, because the osculating planes will be piecewise constant. Still they will have $N+1$ degrees of freedom, and again there seems to be no method available in the literature. Going over to rational pieces does not improve the situation.

§6. Direct Lagrange Interpolation in $\mathbb{R}^2$

In contrast to the situation in $\mathbb{R}^3$, where there is no straightforward way to formulate a well–posed problem for a suitable family of interpolating functions, there are several possibilities to handle *planar* Lagrange data. Polynomial quadratics can form a GC^2 interpolant to $y_0, y_1, \ldots, y_{N-1}$ in $\mathbb{R}^2$ with breakpoints at $y_1, \ldots, y_{N-2}$, because the $N-1$ additional control points $b_1, \ldots, b_{N-1}$ needed to form interpolating Bernstein-Bézier pieces $BB[y_{i-1}, b_i, y_i]$ for $i = 1, \ldots, N-1$ yield $2N-2$ degrees of freedom which are enough to account for the $2(N-2)$ conditions for GC^2 continuity at the $N-2$ interior points. This leaves two degrees of freedom to be used for boundary conditions in the form of prescribed tangent directions at y_0 and y_{N-1}. However, a GC^2 interpolant by piecewise quadratics cannot have a zero of curvature. Thus the interpolation can work for convex data only.

This can be characterized in terms of the angles γ_i between the lines $y_{i-1}\,y_i$ and $y_i\,y_{i+1}$ for $1 \le i \le N-2$ plus γ_0 and γ_{N-1} being defined via the boundary conditions: the lines with angles γ_0 at y_0 and γ_{N-1} at y_{N-1} should be tangent to the solution at y_0 and y_{N-1}, respectively.

In [20] the following is proven:

Theorem 6. *For*

$$\begin{array}{lcccl} 0 & < & \gamma_i & < & \pi, \qquad 0 \le i \le N-1 \\ 0 & < & \gamma_{i-1} + \gamma_i & < & \pi, \qquad 1 \le i \le N \end{array}$$

there exists a GC^2 interpolant. The interpolant is unique, if

$$\begin{array}{lcccll} 0 & < & \gamma_i & < & \pi/2, & 0 \le i \le N-1, \\ 0 & < & \gamma_{i-1} + \gamma_i & < & \pi/2, & 1 \le i \le N. \end{array}$$

If the data are a sufficiently dense sample from a smooth curve with nonvanishing curvature, there is a unique interpolant with $\mathcal{O}(h^4)$ accuracy.

The solution can be calculated by solving a nonlinear system of equations with a tridiagonal and diagonally dominant Jacobian, using a stepsize-controlled Newton–Raphson iteration.

Rational quadratics introduce additional $N-1$ degrees of freedom, because they have one free weight per interval. Consequently, they possibly can interpolate an additional point per interval. As a variation of the situation of Figure 5, there now are additional angles δ_i^+ and δ_i^- as shown in Figure 6, which now requires an odd number of data points $y_0, y_1, \ldots, y_{2N}$.

Again, the data must be "convex", i.e.: there should be a smooth and regular curve with nonvanishing curvature which interpolates the data. Then [25] proves

Theorem 7. *For angles γ_i, δ_i^- and δ_i^+ as in Figure 6, being positive and satisfying*

$$0 < \delta_i^+ + \delta_i^- < \gamma_i < \pi/2,$$

there is a unique GC^2 interpolant by piecewise conics of varying type. The interpolation has accuracy $\mathcal{O}(h^5)$ if the data are a dense sample from a smooth and regular curve with nonvanishing curvature.

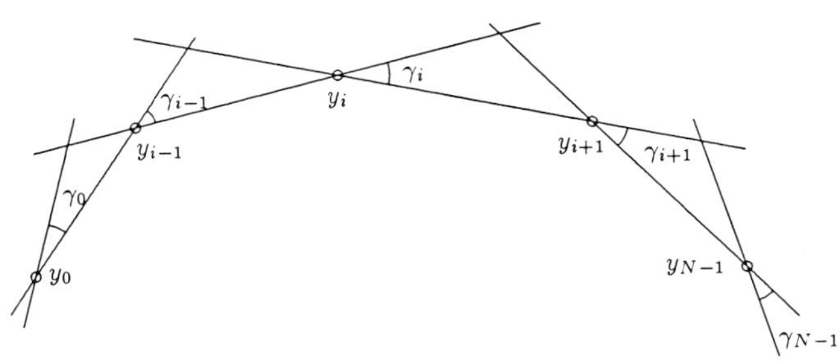

Figure 5. Quadratic Lagrange GC^2 Interpolation Data.

Both methods of this section have a serious drawback: they cannot handle data with inflection points. For the GC^2 piecewise quadratic interpolant, the paper [21] shows how to add cubic pieces to cope with zeros of curvature or collinear data. Some remarks in [25] do the same for general conics, but a proof of the accuracy of a "mixed" method still is missing (see [19] for a "mixed" nonparametric interpolation which also lacks a convergence analysis).

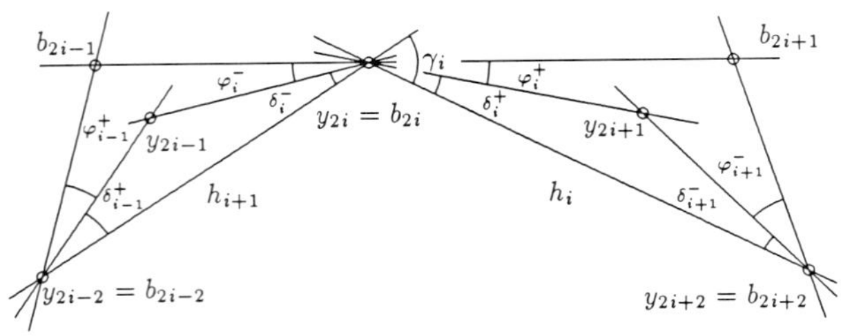

Figure 6. Quadratic Rational Lagrange GC^2 Interpolation Data.

§7. Examples

This section shows some experimental results for the following methods:

- $H3$ Höllig's method of piecewise rational cubic GC^2 Hermite interpolation of $\mathbb{R}^3$ data [11].
- $h/3$ The $h/3$ method for piecewise cubic GC^1 Hermite interpolation of $\mathbb{R}^3$ data.
- BHS DeBoor, Höllig, and Sabin's method of piecewise cubic GC^2 Hermite interpolation of $\mathbb{R}^2$ data [187].
- $GU2$ Goodman's and Unsworth's method of convexity–preserving piecewise rational cubic GC^2 Hermite interpolation of $\mathbb{R}^2$ data [5], [6], [9].
- Sk Piecewise polynomial GC^{k-1} Lagrange interpolation of degree $\leq 2k$ to $\mathbb{R}^3$ data [23].
- $SH2$ Convexity–preserving piecewise rational quadratic GC^1 Hermite interpolation of $\mathbb{R}^2$ data [25].
- $SL2$ Convexity–preserving piecewise rational quadratic GC^2 Lagrange interpolation of $\mathbb{R}^2$ data [25].

The algorithms of [18] and [7] were not included here, because they systematically produce much too small values of curvature.

We start with 9 equidistant points on a semi–circle with radius one, and generate (exact) tangent and curvature information from local five–point interpolation by conics. Then the semi–circle will be reproduced by methods $GU2$, $SH2$, and $SL2$. All methods nicely reproduce the circular arc within plot precision, and some even produce a constant–looking curvature plot. Therefore we only include plots that contain nontrivial information (see Figures 7 to 9).

The second example in taken from [23] and comprises 33 points on a twisted logarithmic spiral in $\mathbb{R}^3$ (see Figure 10) with nonvanishing curvature and torsion, as required for Höllig's method. Plots for the method Sk are in the cited paper, while the behavior of Höllig's method is shown in Figures 11 and 12.

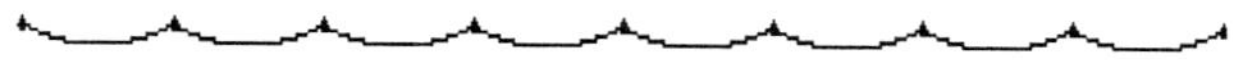

Figure 7. Curvature of $h/3$ Interpolant.

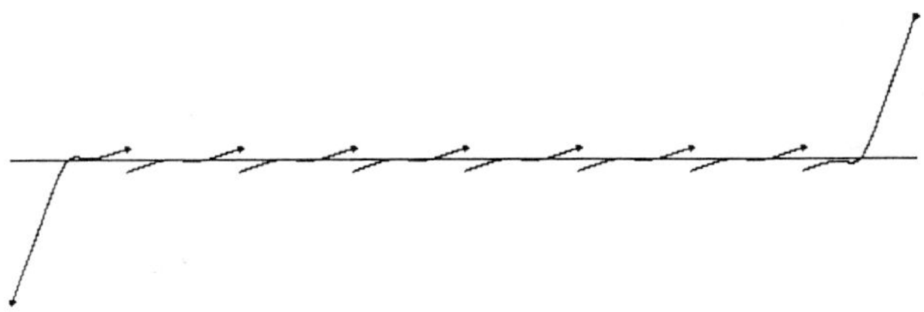

Figure 8. Derivative of curvature of $S3$ interpolant of order 6.
Hermite data generated as in [23]

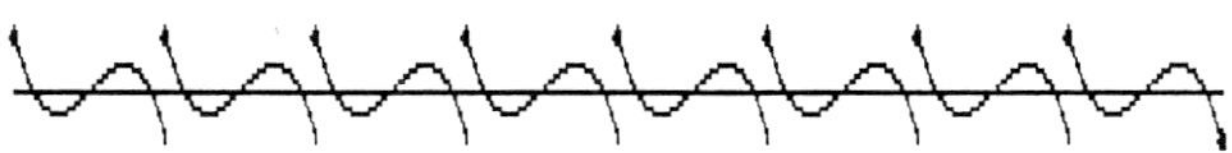

Figure 9. Derivative of curvature of BHS interpolant of order 6.

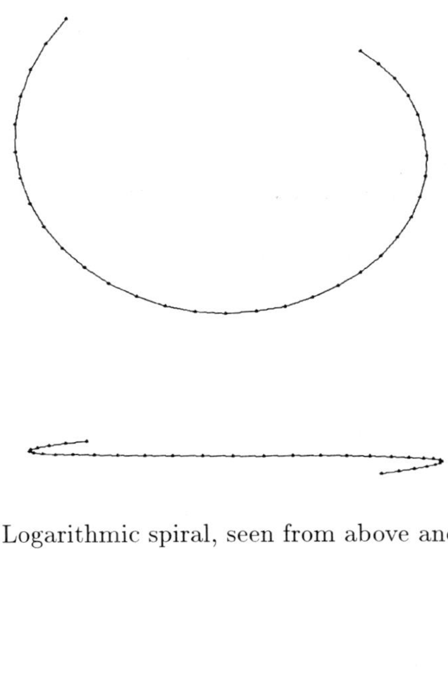

Figure 10. Logarithmic spiral, seen from above and from the side.

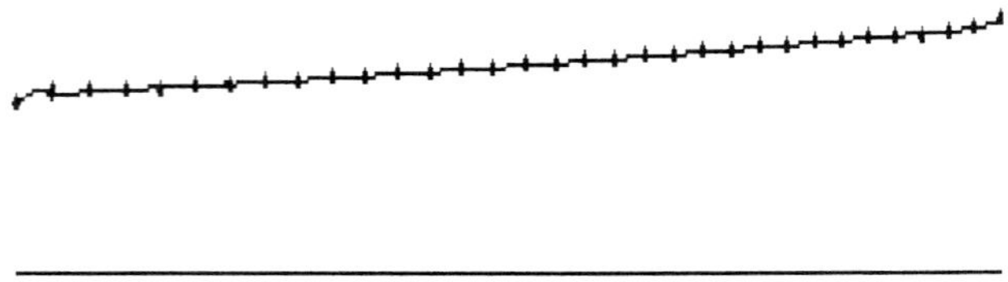

Figure 11. Derivative of curvature of Höllig's cubic rational interpolant.

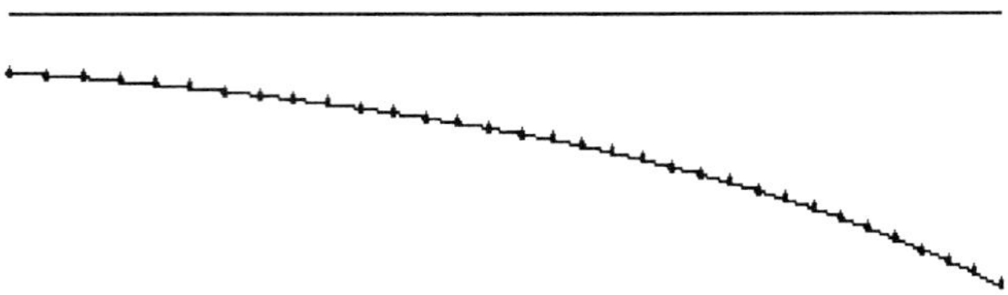

Figure 12. Torsion of Höllig's cubic rational interpolant.

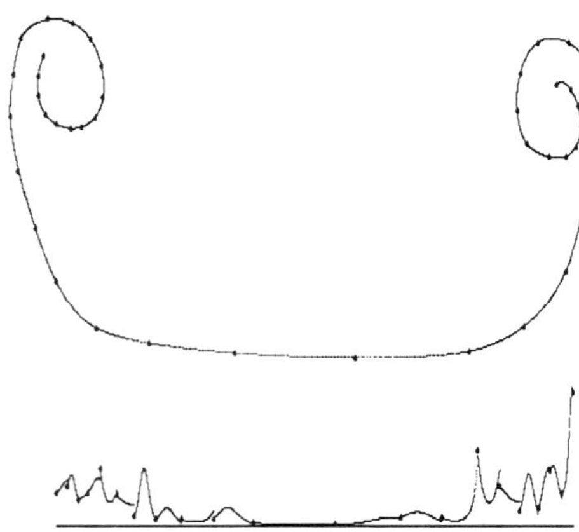

Figure 13. $SH2$ interpolant, plus its curvature.

The last example shows a manually entered coarse data set in $\mathbb{R}^2$, its form being inspired by the keel of the Oseberg ship. All methods give essentially the same shape of the curve, but the curvature plots show some of the differences.

In all cases the necessary Hermite data were generated by local five–point conic interpolation, the derivatives being cleared from parametrization effects using the methods of [23]. This is why the discontinuous curvature plots still look reasonable: the order of approximation is still high because of the good quality of data estimates. The method of deBoor, Höllig, and Sabin also produced a nice–looking plot of the interpolant, but the curvature values were

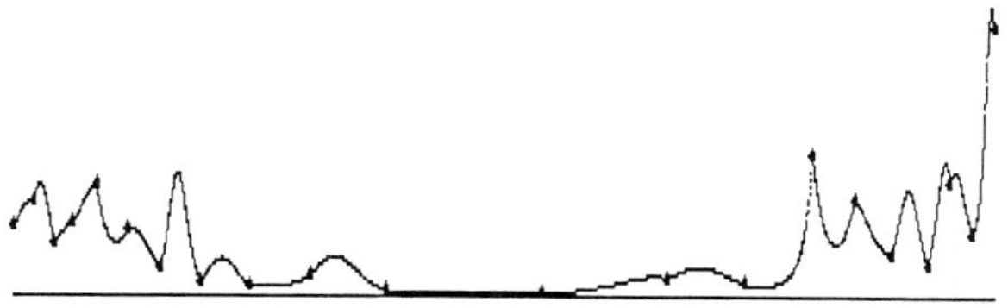

Figure 14. Curvature of $SL2$ interpolant.

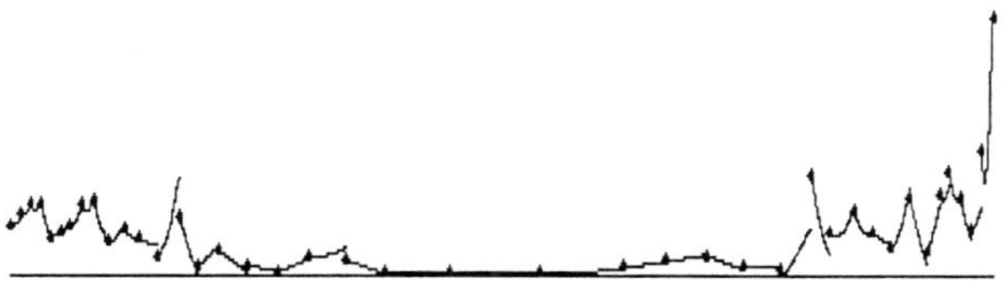

Figure 15. Curvature of $h/3$–rule interpolant.

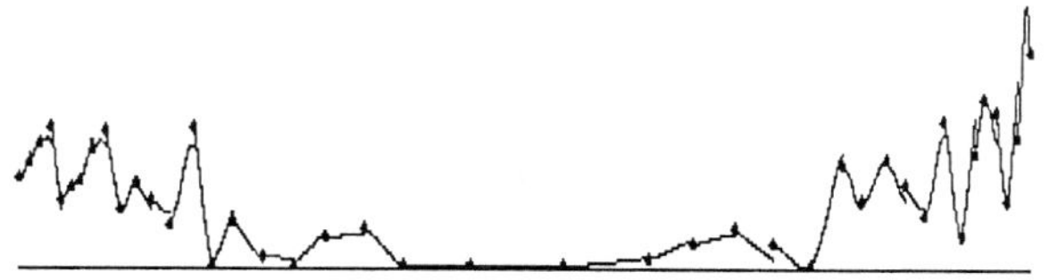

Figure 16. Curvature of fourth–order $S2$ interpolant.

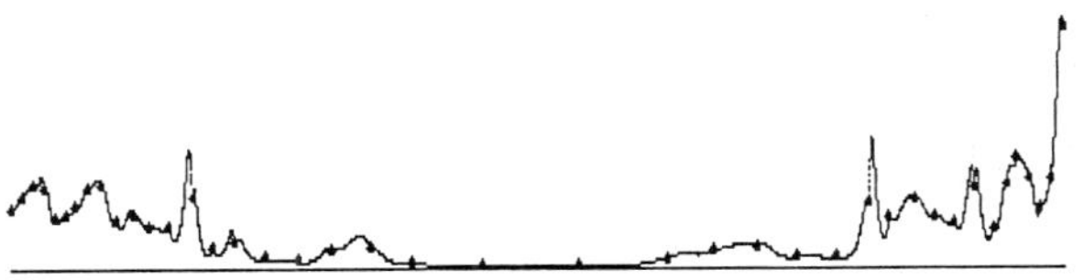

Figure 17. Curvature of $GU2$ interpolant.

bad, because in several intervals there was no GC^2 solution, and we had to insert pieces using the $h/3$ rule. Note that the quadratic rational methods have only about half the number of breakpoints.

References

1. deBoor, C., K. Höllig, and M. Sabin, High accuracy geometric Hermite interpolation, Computer Aided Geom. Design **4** (1987), 269–278.
2. Farin, G. Rational curves and surfaces, in *Mathematical Methods in Computer Aided Geometric Design*, Tom Lyche and L. L. Schumaker (eds.), Academic Press, New York, 1989, 215–238.
3. Farin, G. *Curves and Surfaces for Computer-Aided Geometric Design*, Academic Press, San Diego, 1988.
4. Foley, T. A., Goodman, T. N. T., and K. Unsworth, An algorithm for shape preserving parametric interpolating curves with GC^2 continuity, *Mathematical Methods in Computer Aided Geometric Design*, Tom Lyche and L. L. Schumaker (eds.), Academic Press, New York, 1989, 249–259.
5. Goodman, T. N. T., Shape preserving interpolation by parametric rational cubic splines, *Numerical Mathematics Singapore 1988*, R. P. Agarwal, Y. M. Chow, and S. J. Wilson (eds.), International Series of Numerical Mathematics, Vol. 86, Birkhäuser, Basel, 1988, 149–158.
6. Goodman, T. N. T., B. H. Ong, and K. Unsworth, Constrained interpolation using rational cubic splines, Univ. of Dundee, Computer Sciences Report CS90/01, 1990.
7. Goodman, T. N. T., and K. Unsworth, Shape preserving interpolation by parametrically defined curves, SIAM J. Numer. Anal. **25** (1988), 1453–1465.
8. Goodman, T. N. T., and K. Unsworth, Shape preserving interpolation by curvature continuous parametric curves, Computer Aided Geom. Design **5** (1988), 323–340.
9. Goodman, T. N. T., and K. Unsworth, An algorithms for generating shape preserving parametric interpolating curves using rational cubic splines, Univ. of Dundee, Computer Science Report CS 89/01, 1989.
10. Gregory, J. A., Shape preserving rational spline interpolation, in *Rational Approximation and Interpolation*, P. R. Graves–Morris, E. B. Saff, and R. S. Varga (eds.), Lecture Note 1105, Springer, Berlin, 1984.
11. Höllig, K., Algorithms for rational spline curves, ARO-Report 88–1, 1988, 287–300.
12. Hoschek, K., and D. Lasser, Grundlagen der geometrischen Datenverarbeitung, Teubner, Stuttgart, 1989.
13. Lee, E. T., The rational Bézier representation for conics in Geometric Modeling, in *Geometric Modeling: Algorithms and New Trends*, G. Farin (ed.), SIAM, Philadelphia, 1987, 3–19.
14. McConalogue, D. J., A quasi–intrinsic scheme for passing a smooth curve through a discrete set of points, Computer J. **13** (1970), 392–396.

15. McLaughlin, H. W., Shape preserving planar interpolation: an algorithm, IEEE Comp. Graph. Appl. **3** (1983), 58–67.
16. Peters, J., Local generalized Hermite interpolation by quartic C^2 space curves, ACM Trans. Graphics **8** (1989), 235–242.
17. Piegl, L., Interactive data interpolation by rational Bézier curves, IEEE Comp. Graph. Appl. **7** (1987), 45–58.
18. Piegl, L., Hermite– and Coons-like interpolants using rational Bézier approximation form with infinite control points, Comput. Aided Design **20** (1988), 2–10.
19. Schaback, R. Adaptive rational splines, Constr. Approx. **6** (1990), 167–179.
20. Schaback, R. Interpolation in $\mathbb{R}^2$ by piecewise quadratic visually C^2 Bezier polynomials, Computer Aided Geom. Design **6** (1989), 219–233.
21. Schaback, R. On global GC^2 convexity preserving interpolation of planar curves by piecewise Bézier polynomials, in *Mathematical Methods in Computer Aided Geometric Design*, Tom Lyche and L. L. Schumaker (eds.), Academic Press, New York, 1989, 539–547.
22. Schaback, R. Convergence of planar curve interpolation schemes, in *Approximation Theory VI*, C. K. Chui, L. L. Schumaker, and J. Ward (eds.), Academic Press, New York, 1989, 581–584.
23. Schaback, R. Geometric differentiation and high–accuracy curve interpolation,, Approximation Theory, Spline Functions, and Applications, S. P. Singh (ed.), Kluwer, to appear.
24. Schaback, R. Remarks on high accuracy geometric Hermite interpolation, preprint, Göttingen, 1989.
25. Schaback, R. Interpolation by piecewise conics of arbitrary type, preprint, Göttingen 1991.
26. Schmidt, J. W. On shape preserving spline interpolation: Existence theorems and determination of optimal splines, Banach Center Publications 22, Warsaw 1989.
27. Späth, H. Eindimensionale Spline–Algorithmen, Oldenbourg, München, 1990.

R. Schaback
Georg–August–Universität Göttingen
Lotzestrasse 16–18
D–3400 Göttingen, Germany
U0881@DGOGWDG5.Bitnet

Curvature Properties of Parametric Triangular Bézier Patches

G. Schmeltz

Abstract. A geometric condition is given for deciding whether a point on a polynomial or rational Bézier patch is elliptic, parabolic, or hyperbolic. For patches of second degree, it is proved that parabolic points always lie along straight lines in the parameter plane. Based on this result, a convexity condition is developed depending on the location of the zeroes of three cubic univariate polynomials whose coefficients are built from the Bézier points (and weights). Finally, polynomial patches of second degree are shown to be part of an affine deformation of one specific surface out of fourteen. The corresponding distributions of the sign of curvature are given.

§1. Introduction

Most people agree that a well designed surface should not have any extra bumps or wiggles — if at all possible, it should be convex. In any event, sign changes of the Gaussian curvature are undesired. This paper deals with criteria for deciding the sign of the Gaussian curvature of triangular Bézier patches. After setting up some notation and definitions in Section 1, Section 2 gives a condition for a patch vertex. Section 3 contains the global criterion valid for degree 2 patches, and Section 4 gives a complete list of all polynomial degree 2 surfaces which is needed for the proof of the necessary and sufficient condition in Section 3.

For the forthcoming analysis, it is useful to change the notion of a triangular Bézier surface slightly as follows. A polynomial triangular Bézier *surface* is a map from some parameter plane Ω into $\mathbb{R}^3$ that can be written in the form

$$X(\omega) = \sum_{i+j+k=n} \frac{n!}{i!j!k!} u^i v^j w^k P_{ijk}\,, \qquad n,\, i, j, k \in \mathbb{N}_0, \quad P_{ijk} \in \mathbb{R}^3\,, \quad (1)$$

Mathematical Methods in Computer Aided Geometric Design II
Tom Lyche and Larry L. Schumaker (eds.), pp. 537–548.

ISBN 0-12-460510-9.

where u, v, w are the barycentric coordinates of ω with respect to some *reference triangle* $\triangle UVW \subset \Omega$ such that $\omega = uU + vV + wW$ and $u + v + w = 1$. According to this definition, a change of the reference triangle will not change the surface but merely its representation by control points. By a *patch*, I mean the restriction of the surface to the triangle $u, v, w \geq 0$, which of course changes with U, V, and W. Commonly, both terms "surface" and "patch" are used for what I call a patch.

Equation (1) can be rewritten as

$$\begin{aligned} X(\omega) &= \sum_{i=0}^{n} \frac{n!}{i!(n-i)!} u^i (1-u)^{n-i} Q_i \\ Q_i &= \sum_{j+k=n-i} \frac{(n-i)!}{j!\,k!} \left(\frac{v}{v+w}\right)^j \left(\frac{w}{v+w}\right)^k P_{ijk} \,. \end{aligned} \tag{2}$$

Figure 1 illustrates this and introduces some suggestive notation for the control points near vertex $X(U)$.

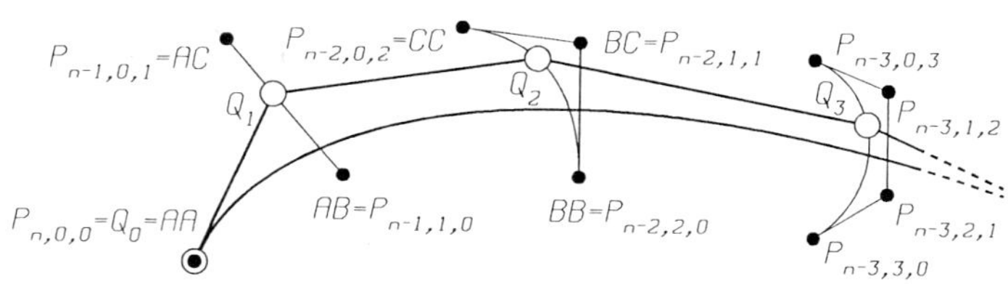

Figure 1. Image of the straight line $v/(v+w) = const$ in Ω.

For rational triangular Bézier surfaces, the control points have to be taken from $\mathbb{R}^4$, and so the range of the map X will be in $\mathbb{R}^4$ as well. The resulting surface is then projected by

$$\begin{array}{rccc} \pi : & \mathbb{R}^4 \setminus (\{0\} \times \mathbb{R}^3) & \to & \{1\} \times \mathbb{R}^3 \\ & (h, x, y, z) & \mapsto & (1, \frac{x}{h}, \frac{y}{h}, \frac{z}{h}) \,. \end{array} \tag{3}$$

§2. The Criterion for a Vertex

Let $X : \Omega \to \mathbb{R}^3$ be a C^2-manifold and $\omega_0 \in \Omega$.

Definition 1. *The surface point $X(\omega_0)$ as well as its parameter value ω_0 is called singular if some directional derivative vanishes there. Otherwise it is called regular.*

Curvature is only defined at regular points. The Gaussian curvature is given in terms of the *normal curvatures* κ_N which are by definition

$$\kappa_N(\omega_0, T) = N \cdot \frac{d^2}{ds^2} X(\omega(s))\big|_{s=0}, \tag{4}$$

where N is the unit normal vector at ω_0, $\omega(s)$ is a curve in Ω passing through ω_0 at $s = 0$ with tangent direction T, and s is the arc length of $X(\omega(s))$. As indicated, this value depends only on ω_0 and T. The *Gaussian curvature* K of X at parameter point ω_0 is simply the product of the minimal and maximal normal curvature, the extrema taken over all directions T:

$$K(\omega_0) = \min\{\kappa_N(\omega_0, T) | T \in \mathbb{R}^2 \setminus \{0\}\} \cdot \max\{\kappa_N(\omega_0, T) | T \in \mathbb{R}^2 \setminus \{0\}\}.$$

Now, κ_N depends continuosly on T. Therefore, $K > 0$ iff κ_N has no zeroes, $K = 0$ iff κ_N has a zero but does not change sign, and $K < 0$ iff κ_N changes sign (when varying T). This corresponds to $X(\omega_0)$ being an *elliptic*, *parabolic*, or *hyperbolic point* of the surface respectively. Hence, only the sign of the normal curvatures is of interest, and this gives rise to some simplifications in (4): The normal vector need not be of unit length and can be replaced by the cross product of any two tangent vectors. There is no further need to use arc length parametrization.

The sign of the normal curvature at vertex $X(U)$ of a polynomial Bézier patch is given by

$$\begin{aligned} \operatorname{sgn} \kappa_N(U, T) &= \operatorname{sgn} \det \left(D_1 X \quad D_2 X \quad \tfrac{d^2}{dt^2} X(U + tT)\big|_{t=0} \right) \\ &= \operatorname{sgn} \det \left(AB - AA \quad AC - AA \quad Q_2 - 2Q_1 + Q_0 \right) \\ &= \operatorname{sgn} \det \begin{pmatrix} 1 & 1 & 1 & 1 \\ AA & AB & AC & Q_2 \end{pmatrix}, \end{aligned} \tag{5}$$

where $D_1 X$ and $D_2 X$ denote the directional derivatives of U along the edges of the reference triangle; the other symbols have been explained in Figure 1.

The matrices are made up of column vectors; an additional row of 1's was supplied in the last line. The points Q_0 and Q_1 disappear because they lie in the tangent plane spanned by AA, AB, and AC.

Note that the last determinant is six times the volume of the tetrahedron with vertices AA, AB, AC, and Q_2. Its sign indicates on which side of the tangent plane at $X(U)$ the point Q_2 is located. Since the direction T is arbitrary, Q_2 is a general point on the Bézier curve defined by BB, BC, and CC, and thus we get the following geometric criterion:

Theorem 1. *The surface X is elliptic (parabolic or hyperbolic) at the regular vertex AA iff the Bézier curve defined by BB, BC, and CC misses (touches or intersects) the plane spanned by AA, AB, and AC.*

Discussion:

- If AA is singular, the control points AA, AB, and AC do not span a plane.
- The Bézier curve must be viewed without the usual restriction to the parameter interval $[0, 1]$. So it is a whole (possibly degenerate) parabola instead of just a segment.

- A Bézier curve that is parallel to the tangent plane must be considered touching it at infinity and so belongs to the parabolic case. This exception is a consequence of the fact that for T parallel to the line VW, $v+w=0$ would hold which is impossible in Equation (2). ■

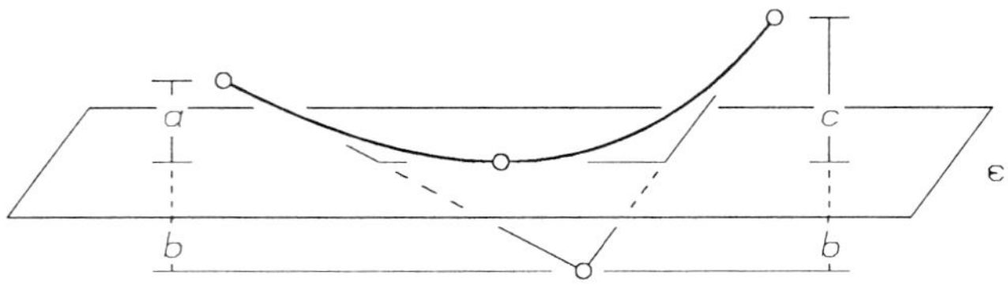

Figure 2. The parabola touches the plane ε iff $\frac{a}{b} = \frac{b}{c}$.

The criterion is similar to the one used in [1] for triangular patches on quadrics to determine the type of the quadric. Let us now look at a method to compute the relative locations of the plane and parabola in question. Figure 2 shows a parabola touching a plane subdivided at the point of contact. Obviously this is a correct instance of de Casteljau's algorithm exactly if the plane divides the two edges of the original control polygon in the same ratio $\frac{a}{b} = \frac{b}{c}$. In the case of inequality, "$<$" means intersecting and "$>$" means missing the plane. Since the distance of a point from the tangent plane at AA is proportional to a determinant like the one in (5), we get the following algebraic version of the geometric criterion in Theorem 1 :

Theorem 2. *The surface X is elliptic/parabolic/hyperbolic at the regular vertex AA iff*

$$\begin{aligned}\Phi = &\left[\det\begin{pmatrix}1 & 1 & 1 & 1\\ AA & AB & AC & BC\end{pmatrix}\right]^2\\ &- \det\begin{pmatrix}1 & 1 & 1 & 1\\ AA & AB & AC & BB\end{pmatrix}\cdot\det\begin{pmatrix}1 & 1 & 1 & 1\\ AA & AB & AC & CC\end{pmatrix} \gtreqless 0\,.\end{aligned}\tag{6}$$

It is clear that all the determinants — and thus Φ — are zero if AA is singular.

In the rational case, X has to be replaced by $\pi\circ X$. With $'$ denoting some directional derivative, (3) implies

$$(\pi\circ X)' = \frac{1}{h}X' + \left(\frac{1}{h}\right)'X \quad\text{and}\quad (\pi\circ X)'' = \frac{1}{h}X'' + \left(\frac{1}{h}\right)'X' + \left(\frac{1}{h}\right)''X\,. \tag{7}$$

Note that the first coordinate is always zero here. Formula (5) now reads

$$\begin{aligned}\operatorname{sgn}\kappa_N(U,T) &= \operatorname{sgn}\det\left(D_1(\pi\circ X) \quad D_2(\pi\circ X) \quad \tfrac{d^2}{dt^2}(\pi\circ X(U+tT))\big|_{t=0}\right)\\ &= \operatorname{sgn}\det\left(\tfrac{1}{h}X \quad \tfrac{1}{h}D_1X \quad \tfrac{1}{h}D_2X \quad \tfrac{1}{h}\tfrac{d^2}{dt^2}X(U+tT)\big|_{t=0}\right)\\ &= \operatorname{sgn}\det\left(AA \quad AB-AA \quad AC-AA \quad Q_2-2Q_1+Q_0\right)\\ &= \operatorname{sgn}\det\left(AA \quad AB \quad AC \quad Q_2\right)\,.\end{aligned}$$

In the first line the determinant is to be understood with the row of zeroes removed. The second line is verified by expansion with respect to the first row and column manipulations using (7). Finally the four factors $\frac{1}{h}$ are removed because they are irrelevant to the sign. The last determinant can be geometrically interpreted as the oriented distance of the point Q_2 from the 3-space spanned by AA, AB, AC, and the origin.

It is now obvious that Theorems 1 and 2 are still true in the rational setting with the following minor changes observed:

- The tangent plane at AA (spanned by AA, AB, and AC) is to be replaced by the 3-space spanned by AA, AB, AC, and the origin.
- The additional rows of ones that had been supplied in the polynomial case is now superfluous because the control points have already four components.

§3. The Global Criterion

Up to now, only a criterion for determining the sign of the Gaussian curvature at a vertex has been given. But by first appropriately subdividing the patch, any point on the surface can be made a vertex. When the point in question is chosen to depend on two parameters so that it is actually an arbitrary point on the surface, a global criterion results. Φ then becomes some bivariate polynomial whose sign is equal to the sign of the Gaussian curvature. I will not pursue that further; the results would be similar to those in [2,4]. The drawback is that the sign changes of a bivariate polynomial in a whole region are somewhat hard to calculate. Therefore, I will now turn to the degree 2 case, where an additional insight justifies concentrating on the curvature on the patch boundaries. Note that for second degree, the control points BB, BC and CC define a boundary curve.

Definition 2. *A straight line in the parameter domain* Ω

- *with constant tangent plane along that line and*
- *vanishing normal curvature on and in the direction of the line*
- *mapping to parabolic and singular points only,*

will be called a parabolic or p-line. To exclude some degenerate situations, a p-line should have at least one regular point on it.

Theorem 3. *Let* $X(\omega)$ *be a (regular) parabolic point of a (polynomial or rational) surface of second degree. Then there is a p-line through* ω.

The proof will be given for the polynomial case. Almost nothing has to be changed to cover the rational case as well.

Proof: Since $X(\omega)$ is parabolic, there is a direction T with $\kappa_N(\omega, T) = 0$. Let $L = \{\omega + tT \mid t \in \mathbb{R}\}$, and choose $U = \omega$ and $V \in L \backslash \{U\}$. Theorem 1 now implies that the parabola defined by BB, BC, and CC touches the plane spanned by AA, AB, and AC at the point BB. But this means that the control points AA, AB, AC, BB, and BC are all in the same plane. It follows that $BB = X(V)$ is parabolic (or singular), the tangent plane at BB is the same as in $AA = X(U)$, and the direction of vanishing normal curvature is T. Since V was arbitrary on L, the proof is complete. ■

Consider again the expression Φ. Given a degree 2 surface, it depends on the choice of the barycentric coordinate system U, V, W; and it tells about the curvature sign at $X(U)$. Now define three functions $\Phi_i(t)$ telling about the curvature sign on the three boundary lines of the patch with reference triangle U_0, V_0, W_0:

$$\begin{array}{lll} \Phi_1(t) & \text{based on} & (1-t)V_0 + tW_0, U_0, V_0 \\ \Phi_2(t) & \text{based on} & (1-t)W_0 + tU_0, V_0, W_0 \\ \Phi_3(t) & \text{based on} & (1-t)U_0 + tV_0, W_0, U_0 \end{array} \tag{8}$$

Note that $t = 0$ is not allowed since this would make two vertices equal. These definitions enable us to formulate the following theorem:

Theorem 4. *For a (polynomial or rational) patch of second degree, $\forall t \in (0,1] : \Phi_1(t), \Phi_2(t) > 0$ is a sufficient, and $\forall t \in (0,1] : \Phi_1(t), \Phi_2(t), \Phi_3(t) \geq 0$ is a necessary convexity condition.*

Proof: This is more or less a direct consequence of Theorems 2 and 3. In the sufficiency part, it is legal to focus on the two boundary segments, because any change in curvature sign must be on a p-line, and every such p-line must hit two of the three edges of triangle U, V, W. But this would contradict the positiveness of Φ_1 and Φ_2; so no sign change can occur. ■

The formula for Φ_1 is derived by substituting the control points which belong to the new reference triangle (see (8)) into (6).

$$\begin{aligned} \Phi_1(t) = t^4 \cdot \Big[& \Big(-\det(AB\,BB\,BC\,AC)^2 - \det(AA\,AB\,BB\,BC) \cdot \det(AB\,BB\,BC\,CC)\Big) \\ & \qquad \cdot (1-t)^3 \\ & + \Big(\det(AA\,AB\,BB\,BC) \cdot \det(BB\,BC\,CC\,AC) \\ & \quad - 2\det(AB\,BB\,BC\,AC) \cdot \det(AB\,BB\,CC\,AC) + \det(AB\,BB\,BC\,AC)^2 \\ & \quad + (-\det(AA\,BB\,BC\,AC) - \det(AA\,AB\,BB\,CC) + \det(AA\,AB\,BB\,BC)) \\ & \qquad \cdot \det(AB\,BB\,BC\,CC)\Big) \cdot t(1-t)^2 \\ & + \Big(\det(AB\,BC\,CC\,AC)^2 - 2\det(AB\,BB\,CC\,AC) \cdot \det(AB\,BC\,CC\,AC) \\ & \quad - \det(AA\,BC\,CC\,AC) \cdot \det(AB\,BB\,BC\,CC) \\ & \quad + (-\det(AA\,BC\,CC\,AC) + \det(AA\,BB\,CC\,AC) + \det(AA\,AB\,BC\,CC)) \\ & \qquad \cdot \det(BB\,BC\,CC\,AC)\Big) \cdot t^2(1-t) \\ & + \Big(\det(AA\,BC\,CC\,AC) \cdot \det(BB\,BC\,CC\,AC) - \det(AB\,BC\,CC\,AC)^2\Big) \cdot t^3\Big]. \end{aligned}$$

In the polynomial case, the matrices must be filled up with a row of ones. The factor t^4 is due to the degeneration of the triangle U, V, W when $t = 0$.

With Theorem 4, the convexity of a patch of second degree can be decided in most cases by determining whether or not the univariate cubic polynomials $\Phi_i(t)/t^4$ have roots in the unit interval. Alas, it seems impossible to interpret these polynomials or their coefficients geometrically. The next section will prove that for polynomial degree 2 patches, the necessary condition is also sufficient. This will be done by describing all possible distributions of curvature sign, thus eliminating the possibility of a patch with $K \geq 0$ on the boundary but $K < 0$ somewhere in the interior. You will also see that the opposite ($K \leq 0$ on the boundary but $K > 0$ somewhere in the interior) does happen.

§4. Classification of Polynomial Degree Two Surfaces

Let's first state and prove a few lemmas on p- and s-lines (to be defined later) that are useful for classifying the degree 2 patches. They are here proved for the polynomial case, but their proofs can easily be changed to cover the rational case as well. (The notion of s-lines would have to be changed a little.)

Lemma 1. *If two p-lines L_1 and L_2 meet in a regular point, then the patch is planar.*

Proof: Let $L_1 \cap L_2 = \{U\}$, $V \in L_1$, and $W \in L_2$. Then the Bézier curve defined by BB, BC, and CC must touch the tangent plane spanned by AA, AB, and AC at both points BB and CC. Thus it is part of the plane which, hence, contains all the control points. (Regularity is needed to make the three control points really span a plane.) ■

Lemma 2. *If there are two parallel p-lines L_1 and L_2, then the surface is a parabolic cylinder (or a plane).*

Proof: Let $U, V \in L_1$, $W \in L_2$ and U and W regular. Then $W \in L_2$ implies plane AA, AB, BB parallel to plane BC, CC, AC, and $U, V \in L_1$ implies that points AA, AB, BB, BC, and AC are all in the same plane. This is only possible if either all control points are contained in one plane or $AA - AB$, $AB - BB$, and $AC - BC$ are all parallel to each other. In the former case, the patch is planar. In the latter case, since any directional derivative in the $(U–V)$-direction is a convex combination of the difference vectors given above, all those derivative vectors are parallel to each other. This forces the surface to be a cylinder. ■

Because p-lines meet in singular points, I will now have a closer look at singular points.

Definition 3. *For every singular point $X(\omega)$ with exactly one vanishing directional derivative, consider the straight line L through ω with that direction. If L maps to a doubly traced half line with turning point $X(\omega)$, then L will be called a singular or s-line and $X(\omega)$ a simple singular point.*

Discussion: "Very singular" points:

- If there are two independent directions of vanishing derivative, then all derivatives are zero. Choosing $U = \omega$ implies $X(U) = AA = AB = AC$, and the construction in Figure 1 shows that every straight line through ω maps to a half line. Thus in this case, the surface is a (parabolic) cone (or even a plane).

- The line L in the preceding definition must map to some degenerate parabola because ω is not only singular to the surface but also to the line. However, it may map to a single point instead of a half line. Choosing $U = \omega$ and $U, V \in L$ makes $X(U) = AA = AB = BB = X(V)$. Then any derivative in $(U - V)$-direction is parallel to $AC - BC$, since $AA - AB = 0$ and $AB - BB = 0$. — The surface is a parabolic cylinder (or a plane). ■

Note that along an s-line, the surface always intersects itself. If on the other hand, two points in the parameter plane Ω map to the same surface point, the straight line joining them must be an s-line.

Lemma 3. *Let $X(U)$ be a simple singular point and L the corresponding s-line. Then*

$$\begin{array}{ll} V \in L & \text{iff } AA = AB\,, \\ W \in L & \text{iff } AA = AC\,, \\ VW \parallel L & \text{iff } AB = AC\,. \end{array}$$

Proof: Compute the vanishing directional derivative in terms of the Bézier points. ■

Now I turn to the central theorem of this section.

Theorem 5. *Any polynomial patch of second degree is part of an affine deformation of one of the fourteen surfaces given below.*

It may seem surprising that up to an affine transformation, there is only a finite number of degree 2 surfaces; so let's first count degrees of freedom: An affine map in 3-space fixes 4 of the 6 control points of a patch. But I claimed that there is a finite number of surfaces, not patches. The remaining 2 control points give rise to 6 degrees of freedom which can be fixed by choosing a reference triangle U, V, W in Ω.

No formal proof of Theorem 5 will be given. Instead the general ideas of the proof are presented, and some of the details will be mentioned along with the list of surfaces. The remaining details will be included in my dissertation [3]. The main line of thought is the following: First, some assumptions are made on the surface concerning the existence and location of s- and p-lines. Since the surface is independent of the triangle U, V, W, w.l.o.g. the vertices can be chosen as singular points, on p-lines, and the like. Then, four affinely independent control points are chosen, and — if necessary — de Casteljau's algorithm is used to move the other two control points into prespecified positions within the affine coordinate system provided by them. This shows that

— up to an affine transformation — only one surface fulfills the assumptions I began with.

I now start the classification with the surfaces with "very singular" points which have already been discussed.

I. Simple surfaces.
   1. The plane.
   2. The parabolic cylinder.
   3. The (parabolic) cone.

For the plane and the cylinder, there are also other parametrizations. For all other surfaces, parametrization (as a polynomial Bézier surface of degree 2) is unique up to a change of the reference triangle. The remaining surfaces are assumed to be non-planar.

II. Surfaces with at least two simple singular points with s-lines S_1 and S_2. Lemma 3 implies that some control points coincide and only 4 different control points are left when the reference triangle is chosen appropriately.
   – $S_1 \parallel S_2$ $\Longrightarrow$ **2.**
   4. $S_1 \cap S_2 =$ a regular point.
   5. $S_1 \cap S_2 =$ a singular point.

We now look at surfaces with one simple singular point. To show that no case is missing, the surfaces with more singular points appear here once more. Let AA be the singular point, and $AB = AC$. Let g be the straight line joining AA and AB, and Π the plane spanned by BB, BC, and CC (If they do not span a plane, we again get a cylinder.). The three control points BB, BC, and CC define a parabola in Π.

III. Surfaces with at least one simple singular point and $g \parallel \Pi$.
   6. $g \nparallel$ axis of parabola.
   7. $g \parallel$ axis of parabola.

If $g \nparallel \Pi$, choose a coordinate system so that Π is parallel to the x-y-plane. Then either V and W can be moved so that $AB = AC \in \Pi$, or the z-components of AA, AB, and Π are equally spaced (try de Casteljau's algorithm !).

IV. Surfaces with at least one simple singular point, $AB \in \Pi$.
   8. $g \cap \Pi$ inside the parabola
   – $g \cap \Pi \in$ parabola $\Longrightarrow$ **5.**
   – $g \cap \Pi$ outside the parabola $\Longrightarrow$ **4.**

V. Surfaces with at least one simple singular point, z-components of AA, AB, and Π equally spaced.
   9. $g \cap \Pi$ inside the parabola
   10. $g \cap \Pi \in$ parabola
   – $g \cap \Pi$ outside the parabola $\Longrightarrow$ **5.**

Surface **10.** is known as "Whitney's Umbrella".

VI. Surfaces without singular points. It turns out that there are only the following four possibilities (neglecting planes and cylinders) :

11. A surface with one p-line.
12. The elliptic paraboloid.
13. The hyperbolic paraboloid.
14. A ruled surface with one family of lines.

That all surfaces differ is obvious from the different distributions of curvature sign (see Figure 3). Among these surfaces, only **4** and **8** occur generically.

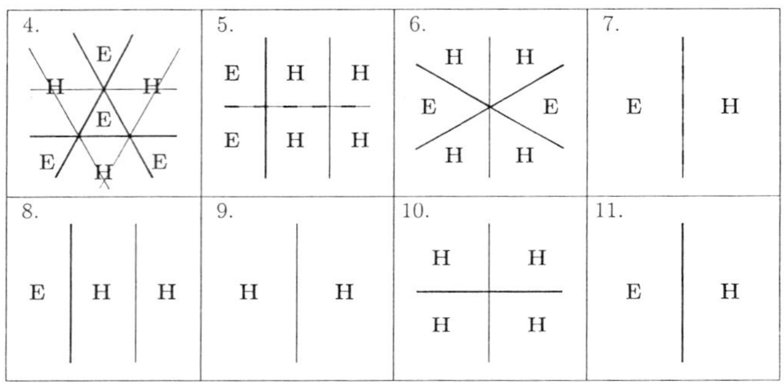

Figure 3. p-lines, s-lines, and curvature sign for degree two surfaces.

Figure 3 sketches the parameter domains Ω of surfaces **4**–**11**. Bold lines are p-lines, and thin lines are s-lines; broken bold and thin lines are both p- and s-lines at the same time. The elliptic and hyperbolic regions are also indicated. The p- and s-lines and curvature signs were computed from representatives of each class using MACSYMA. Surfaces **1**–**3** consist of parabolic points only, while the paraboloids **12** and **13** are entirely elliptic and hyperbolic, respectively. Surface 14 is also hyperbolic everywhere.

In order to give an impression of how complicated patches of such a low degree can be, Figure 4 shows Surfaces **4**–**11**. The pictures of Surfaces **4**, **5**, **7**, **8**, and **11** show single triangular patches. Pictures **6** and **10** are both composed of four patches; with the p-lines as boundaries in Picture **6**. In Picture **9** you see two patches with the s-line as common boundary curve. But even where several patches are used, this is only to emphasize the symmetry of the surfaces, which would be hidden by a triangular parameter domain. Every picture shows part of one polynomial surface whose parameter domain is the whole plane Ω.

The thick lines in the interior of the patches are the images of p- and s-lines. You can tell s-lines from p-lines, if you remember that singular lines map to straight lines of self-intersection. You may also try to visualize the planes containing the images of the parabolic lines; they are tangent to the surface at every point of the p-line.

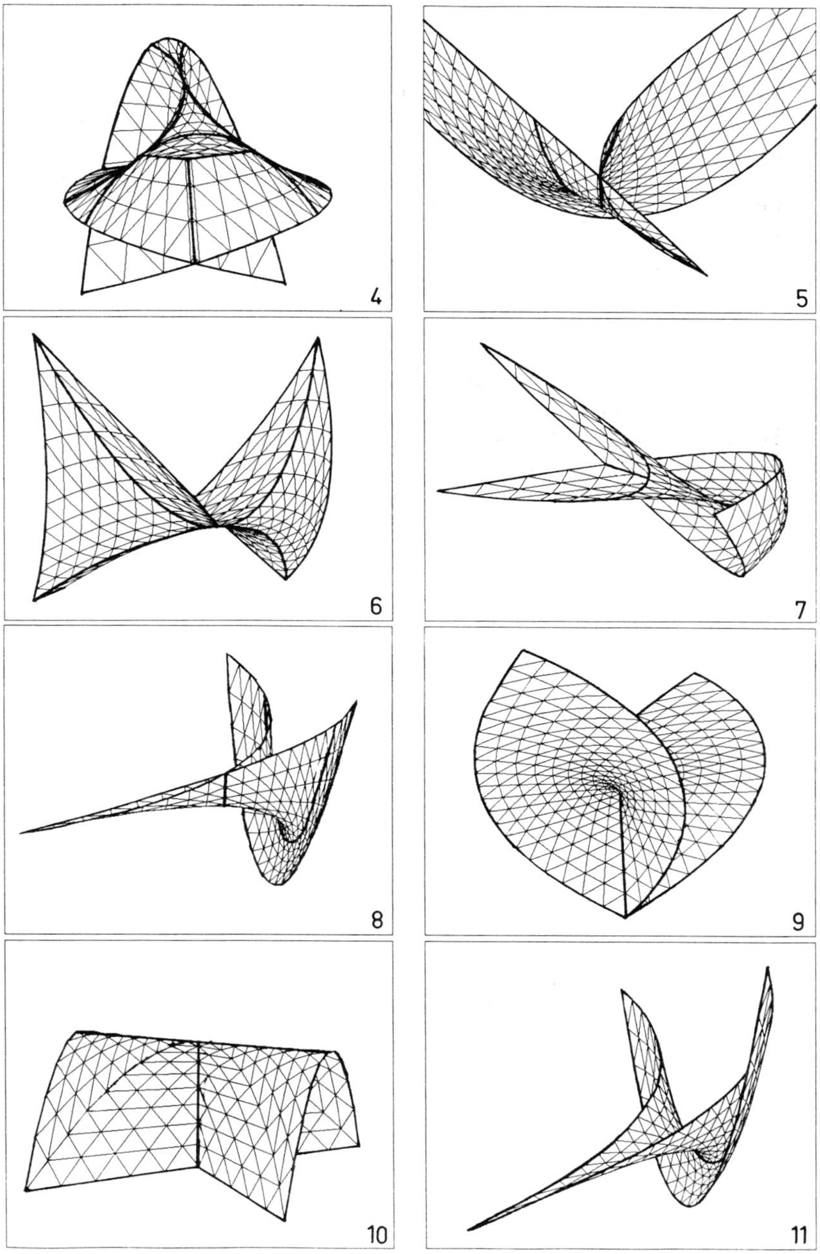

Figure 4. Degree two surfaces.

References

1. Böhm, W., and D. Hansford, Parametric representation of quadric surfaces, Math. Meth. and Num. Anal., to appear.
2. Cao, Y., and X. Hua, The convexity of quadratic parametric triangular Bernstein-Bézier surfaces, Computer Aided Geom. Design **8** (1991), 1–6.
3. Schmeltz, G., Variationsreduzierende Kurvendarstellungen und Krümmungskriterien für Bézierflächen, dissertation, Technische Hochschule Darmstadt, 1992.
4. Zhou, C.-Z., On the convexity of parametric Bézier triangular surfaces, Computer Aided Geom. Design **7** (1991), 459–463.

Gerd Schmeltz
Technische Hochschule Darmstadt
Schloßgartenstraße 7
W-6100 Darmstadt, Germany
xbr3dc56@ddathd21.bitnet

Offsets of Polynomial Bézier Curves: Hermite Approximation with Error Bounds

Thomas W. Sederberg and David B. Buehler

Abstract. A method is presented for determining the error bound in approximating the offset of a polynomial Bézier curve using Hermite interpolation. The method is based on interval arithmetic and polynomials in Bernstein form. In particular, we show how an offset of a Bézier curve can itself be represented as a Bézier curve in which one of the control points is a rectangular interval. The size of this interval control point bounds the approximation error.

§1. Introduction

The offset $\mathbf{P}^\rho(t)$ of a parametric curve $\mathbf{P}(t)$ is given by the equation

$$\mathbf{P}^\rho(t) = \mathbf{P}(t) + \rho\mathbf{N}(t), \tag{1}$$

where ρ is the offset radius, and $\mathbf{N}(t)$ is the normal function

$$\mathbf{N}(t) = \frac{(-y'(t), x'(t))}{\sqrt{x'^2(t) + y'^2(t)}}$$

as shown in Figure 1.

Offset curves are so fundamental to computer aided geometric design that a review here is not necessary. For a good overview, see [2].

An unfortunate property of offset curves is that unless $x'^2(t) + y'^2(t)$ is a perfect square, the offset of a Bézier curve cannot be expressed exactly as a rational Bézier curve [2]. This fact has led to several papers which propose methods for approximating the offset of a Bézier or B-spline curve

Mathematical Methods in Computer Aided Geometric Design II

Tom Lyche and Larry L. Schumaker (eds.), pp. 549–558.

ISBN 0-12-460510-9.

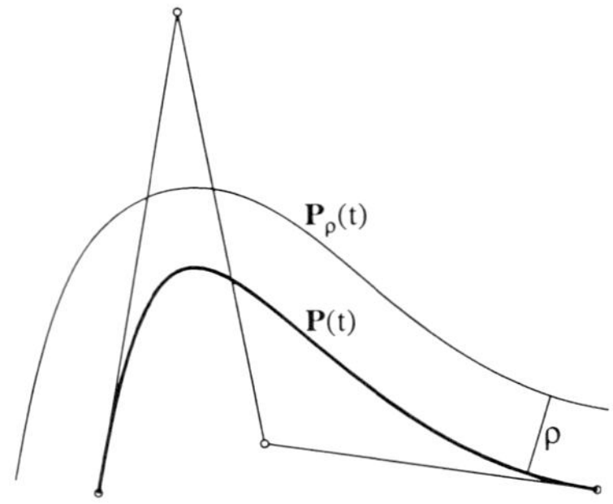

Figure 1. Offset curve.

[1,5,6,7,8,11,14]. This paper addresses the problem of computing an error bound for a Hermite approximation of the offset of a polynomial Bézier curve.

Section 2 reviews interval arithmetic and interval Bézier curves. Section 3 discusses how to derive an interval Bézier curve which contains the offset of a given Bézier curve. Section 4 reviews pertinent details from [12] on how a rational Bézier curve can be bounded by an interval polynomial Bézier curve. Section 5 provides a numerical example, and Section 6 comments on the tightness and computational expense of the error bound.

§2. Interval Bézier Curves

Recall a few concepts from interval arithmetic [9,10]. The scalar interval $[a, b]$ denotes a closed set of real values:

$$[a,b] = \{ t \mid a \leq t \leq b \}.$$

Given two intervals $[a, b]$ and $[c, d]$, binary arithmetic operations are defined:

$$[a,b] \star [c,d] = \{ x \star y \mid x \in [a,b] \text{ and } y \in [c,d] \}, \quad \star \in \{+,-,\times,/\}.$$

Specifically, one can verify that

$$[a,b] + [c,d] = [a+c, b+d], \quad [a,b] - [c,d] = [a-d, b-c],$$

$$[a,b] \times [c,d] = [\min(ac,ad,bc,bd), \max(ac,ad,bc,bd)],$$

$$[a,b] / [c,d] = [a,b] \cdot [1/d, 1/c],$$

where division is usually defined only for denominator intervals that do not contain 0. A single real value can be treated as a *degenerate* interval, $a = [a, a]$, so that arithmetic operations can be applied to mixed operands (e.g., $a + [b,c] = [a+b, a+c]$). A single symbol enclosed in square parentheses, such as $[u] = [a, b]$, will also be used to denote an interval.

$[q](t) = \sum_{k=0}^{n} [q_k, r_k] B_k^n(t)$ denotes an interval polynomial in Bernstein form. An interval Bézier curve [13] is written $[\mathbf{Q}](t) = \sum_{k=0}^{n} [\mathbf{Q}_k] B_k^n(t)$, where the interval control points $[\mathbf{Q}_k]$ are rectangular (possibly degenerate) intervals of the form

$$[\mathbf{Q}_k] = \{ (x, y) \mid x \in [a_k, b_k] \text{ and } y \in [c_k, d_k] \}.$$

For a given t, $[\mathbf{Q}](t)$ is the region defined by

$$[\mathbf{Q}](t) = \{\mathbf{Q} = \sum_{k=0}^{n} \hat{\mathbf{Q}}_k B_k^n(t) | \hat{\mathbf{Q}}_k \in [\mathbf{Q}_k], \ k = 0, \ldots, n\}.$$

$[\mathbf{Q}](t)$ is always a rectangular interval. Figure 2 shows a sample interval Bézier curve for which the middle control point is an interval. Twenty interval values of $[\mathbf{Q}](t)$ are also plotted.

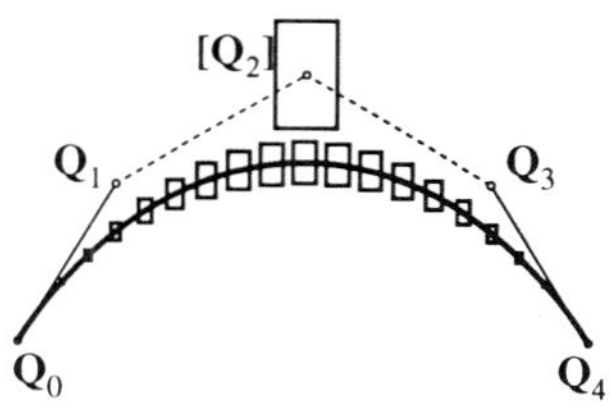

Figure 2. Interval Bézier curve.

Our representation of an offset curve will be in this form — an interval Bézier curve $[\tilde{\mathbf{P}}^\rho](t)$ of even degree n with a single interval control point $[\mathbf{P}_{\frac{n}{2}}]$ with $[\tilde{\mathbf{P}}^\rho](t) \supseteq \mathbf{P}^\rho(t)$. Thus, a Hermite approximation of $\mathbf{P}^\rho(t)$ can be obtained by replacing the interval control point of $[\tilde{\mathbf{P}}^\rho](t)$ with its midpoint, and the error of that Hermite approximation is bounded by the size of the interval control point. Denoting the Hermite approximation by $\tilde{\mathbf{P}}^\rho(t)$, we have the error bound

$$|\tilde{\mathbf{P}}^\rho(t) - \mathbf{P}^\rho(t)| \leq |[\mathbf{P}_{\frac{n}{2}}]| B_{\frac{n}{2}}^n(t).$$

Figure 3 shows a quartic Bézier curve and its offset approximated in this form. The figure on the left involves a single approximating curve, and the figure on the right uses two approximating curves The error reduction suggested here after a single subdivision is in keeping with the rate of convergence of Hermite approximation of $O(h^n)$. Thus, the size of the interval control points on the right are roughly $\left(\frac{1}{2}\right)^4$ the size of the interval control point in the single approximating curve.

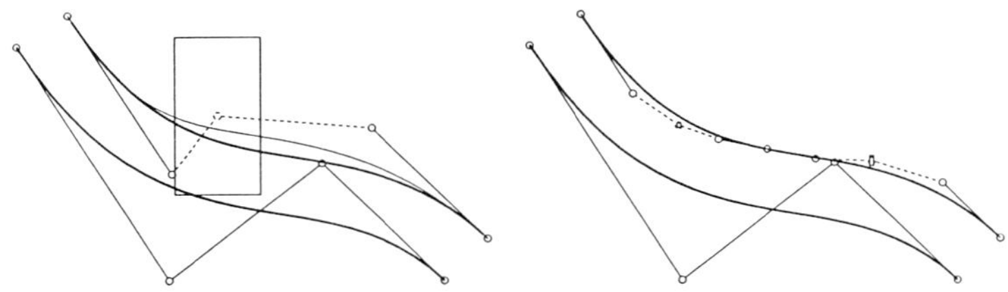

Figure 3. Interval Bézier curve.

§3. Hermite Approximation of Offset Curves

We now overview how to compute an interval Bézier curve $[\tilde{\mathbf{P}}^{\rho}](t)$ which contains the offset $\mathbf{P}^{\rho}(t)$ of a given Bézier curve $\mathbf{P}(t)$. Only the middle control point of $[\tilde{\mathbf{P}}^{\rho}](t)$ is an interval (the rest are points), and the size of the interval indicates the tightness of the approximation. $[\tilde{\mathbf{P}}^{\rho}](t)$ is computed in these steps:

---

0. Choose the degree of the subsequent Hermite offset approximation, an even integer n, and degree elevate the given curve $\mathbf{P}(t)$ to degree n.
1. Compute
$$w(t) = x'^2(t) + y'^2(t).$$
2. Find $[a(t), b(t)] \supseteq \sqrt{w(t)}$, $t \in [0,1]$ i.e., $a(t) \le \sqrt{w(t)} \le b(t)$.
3. Compute a polynomial interval Bézier curve $[\tilde{\mathbf{N}}](t)$ which satisfies:
$$[\tilde{\mathbf{N}}](t) \supseteq \frac{(-y'(t), x'(t))}{[a(t), b(t)](t)} \supseteq \frac{(-y'(t), x'(t))}{\sqrt{x'^2(t) + y'^2(t)}} \equiv \mathbf{N}(t)$$
4. Compute
$$[\tilde{\mathbf{P}}^{\rho}](t) \equiv \mathbf{P}(t) + \rho[\tilde{\mathbf{N}}](t) \supseteq \mathbf{P}(t) + \rho\mathbf{N}(t) \equiv \mathbf{P}^{\rho}(t). \tag{2}$$

Now for the details. In Step 1, $w(t)$ must be degree $2n$. If $\mathbf{P}(t)$ has control points (x_i, y_i), $i = 0, \ldots, n$, we can compute the coefficients of $w(t) = \sum_{i=0}^{2n} w_i B_i^{2n}(t)$ as

$$w_i = \frac{n^2}{\binom{2n}{i}} \sum_{\substack{j+k+l=i \\ j,k \in [0,\ldots,n-1] \\ l \in [0,1,2]}} [(x_{j+1}-x_j)(x_{k+1}-x_k)+(y_{j+1}-y_j)(y_{k+1}-y_k)]\binom{n-1}{j}\binom{n-1}{k}\binom{2}{l}.$$

In Step 2, to find $[a(t), b(t)] \supseteq \sqrt{w(t)}$, $t \in [0,1]$, note that $w(t) \ge 0$. Thus if $a^2(t) \le w(t)$ then $a(t) \le \sqrt{w(t)}$, and if $b^2(t) \ge w(t)$ and $b(t) \ge a(t)$, then

$b(t) \geq \sqrt{w(t)}$ (all for $t \in [0,1]$). Our problem thus reduces to finding degree n polynomials whose squares are upper and lower bounds for $w(t)$:

$$a(t) = \sum_{i=0}^{n} a_i B_i^n(t), \text{ and } b(t) = \sum_{i=0}^{n} b_i B_i^n(t).$$

This is done by solving for the a_i and b_i so that $a^2(t) \leq w(t) \leq b^2(t)$. Writing

$$a^2(t) = \sum_{i=0}^{2n} \hat{a}_i B_i^{2n}(t), \quad b^2(t) = \sum_{i=0}^{2n} \hat{b}_i B_i^{2n}(t),$$

where

$$\hat{a}_i = \sum_{\substack{j+k=i \\ j,k\in[0,\ldots,n]}} \frac{\binom{n}{j}\binom{n}{k} a_j a_k}{\binom{2n}{i}}, \quad \hat{b}_i = \sum_{\substack{j+k=i \\ j,k\in[0,\ldots,n]}} \frac{\binom{n}{j}\binom{n}{k} b_j b_k}{\binom{2n}{i}},$$

we need $\hat{a}_i \leq w_i \leq \hat{b}_i$, $i = 0, \ldots 2n$. We can always solve $a_i = b_i$, $i \neq \frac{n}{2}$, from the *equalities* $\hat{a}_i = w_i = \hat{b}_i$, $i = 0, \ldots, \frac{n}{2} - 1, \frac{3n}{2} + 1, \ldots, 2n$ as follows:

$$a_0 = b_0 = \sqrt{w_0} \quad (w_0 \text{ is always non} - \text{negative}) \tag{3}$$

$$a_1 = b_1 = w_1 / a_0 \text{ if } n \geq 4. \tag{4}$$

$$a_i = b_i = \frac{\binom{2n}{i} w_i - \sum_{j=1}^{i-1} \binom{n}{j}\binom{n}{i-j} a_j a_{i-j}}{2\binom{n}{i} a_0}, \quad i = 1, \ldots, \frac{n}{2} - 1 \tag{5}$$

$$a_n = b_n = \sqrt{w_{2n}} \quad (w_{2n} \text{ is always non} - \text{negative}) \tag{6}$$

$$a_{n-1} = b_{n-1} = w_{2n-1} / a_n \text{ if } n \geq 4. \tag{7}$$

$$a_i = b_i = \binom{2n}{n+j} \frac{w_{n+i} - \sum_{j=n-1}^{i+1} \binom{n}{j}\binom{n}{n+i-j} a_j a_{n+i-j}}{2\binom{n}{i} a_n}, \quad i = n-1, \ldots, \frac{n}{2} + 1. \tag{8}$$

The middle coefficients $a_{\frac{n}{2}} \leq b_{\frac{n}{2}}$ are chosen to satisfy the inequalities $\hat{a}_i \leq w_i \leq \hat{b}_i$, $i = \frac{n}{2}, \ldots, \frac{3n}{2}$. To do this, compute the values

$$\underbrace{c_i = \frac{\binom{2n}{i} w_i - \sum_{\substack{j+k=i \\ j,k\in[0,\ldots,\frac{n}{2}-1,\frac{n}{2}+1,\ldots,n]}} \binom{n}{j}\binom{n}{k} a_j a_k}{2\binom{n}{n/2}\binom{n}{i-n/2} a_{i-n/2}}}_{i\in\{\frac{n}{2},\ldots,n-1,n+1,\ldots\frac{3n}{2}\}}, \quad c_n = \frac{\binom{2n}{n} w_n - \sum_{\substack{j=0 \\ j\neq\frac{n}{2}}}^{n} \binom{n}{j}^2 a_{n-j} a_j}{\binom{n}{d/2}^2}. \tag{9}$$

Then, $a_{\frac{n}{2}}$ is the largest member of the set

$$A = \{x | x^2 \leq c_n;\ x \leq c_i \text{ if } a_{i-n/2} \geq 0;\ x \geq c_i \text{ if } a_{i-n/2} \leq 0\}, \tag{10}$$

where $i \in \{\frac{n}{2}, \ldots, n-1, n+1, \ldots \frac{3n}{2}\}$ and $b_{\frac{n}{2}}$ is the smallest member of the set

$$B = \{x | x^2 \geq c_n;\ x \geq c_i \text{ if } a_{i-n/2} \geq 0;\ x \leq c_i \text{ if } a_{i-n/2} \leq 0\}. \tag{11}$$

Note that for some $w(t)$, at least one of these sets may be empty in which case we cannot solve for $[a(t), b(t)] \supseteq \sqrt{w(t)}$. However, such cases can always be subdivided until a piecewise solution is possible. When a cusp exists on $\mathbf{P}(t)$, then $w(t) = 0$ at the cusp and the set A is normally empty. The solution is to subdivide $\mathbf{P}(t)$ at the cusp.

In Step 3 we interpret $\frac{(-y'(t),x'(t))}{[a(t),b(t)]}$ as a rational Bézier curve of degree n for which the middle weight $[a_{d/2}, b_{d/2}]$ is an interval value. A discussion of how to compute the interval Bézier curve $[\tilde{\mathbf{N}}](t)$ which contains $\frac{(-y'(t),x'(t))}{[a(t),b(t)](t)}$ is presented in Section 4, and Section 5 runs through some examples.

§4. Polynomial Approximation of Rational Bézier Curves

This section builds on the result from [12] that any rational Bézier curve can be expressed exactly as a *hybrid* curve — a polynomial Bézier curve with one "moving control point" which is itself a rational Bézier curve. The current paper requires a slight modification of that result, because the middle weight is an interval. For simplicity of explanation, we begin this section by assuming that $a_{d/2} = b_{d/2}$, that is, that all weights are zero-width intervals and $a(t) \equiv b(t)$. This actually happens when $w(t)$ is a perfect square.

Consider the rational curve

$$\mathbf{N}(t) = \frac{(-y'(t), x'(t))}{a(t)} = \frac{\sum_{i=0}^{n} \mathbf{N}_i B_i^n(t)}{\sum_{i=0}^{n} a_i B_i^n(t)}.$$

If the control points of the given curve $\mathbf{P}(t)$ in (1) are (x_i, y_i), then the control points $\mathbf{N}_i$ are based on the degree elevated derivative of $\mathbf{P}(t)$:

$$\mathbf{N}_i = ((i-n)y_{i+1} + (n-2i)y_i + iy_{i-1}, (n-i)x_{i+1} + (2i-n)x_i - ix_{i-1})\,. \tag{12}$$

$\mathbf{N}(t)$ can be expressed exactly as a hybrid curve:

$$\mathbf{H}(t) = \sum_{i=0,\ i \neq \frac{n}{2}}^{n} \mathbf{H}_i B_i^n(t) + \mathbf{M}(t) B_{\frac{n}{2}}^n(t) \equiv \mathbf{N}(t),$$

where $\mathbf{M}(t) = \frac{\sum_{i=0}^{n} a_i \mathbf{M}_i B_i^n(t)}{\sum_{i=0}^{n} a_i B_i^a(t)}$ is the moving control, and a_i are the same weights as in $\mathbf{N}(t)$.

Figure 4 shows a sample quartic rational Bézier curve $\mathbf{N}(t)$ next to a hybrid representation of the same curve. The hybrid curve has control points

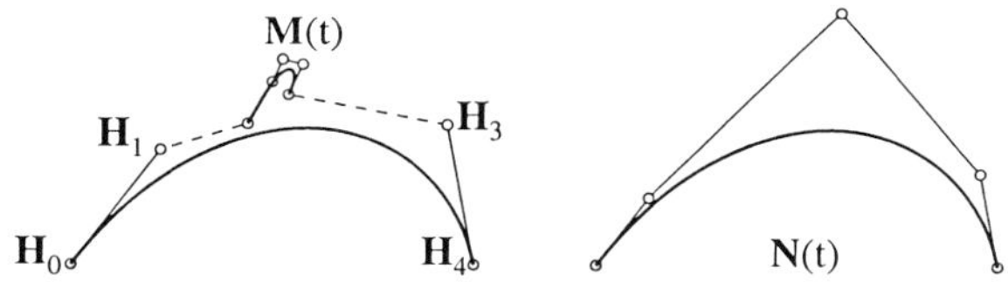

Figure 4. Hybrid representation of rational curve.

$\mathbf{H}_0$, $\mathbf{H}_1$, $\mathbf{M}(t)$, $\mathbf{H}_3$, $\mathbf{H}_4$. The moving control point $\mathbf{M}(t)$ is a quartic rational Bézier curve. In order to evaluate this hybrid curve at some value t, first evaluate $\mathbf{M}(t)$ and then evaluate at the same parameter value t the polynomial Bézier curve with control points $\mathbf{H}_0$, $\mathbf{H}_1$, $\mathbf{M}(t)$, $\mathbf{H}_3$, $\mathbf{H}_4$. This hybrid curve is an exact representation of, not an approximation to, $\mathbf{N}(t)$.

This is the idea used in solving

$$[\tilde{\mathbf{N}}](t) \equiv \sum_{i=0, i\neq \frac{n}{2}}^{n} \tilde{\mathbf{N}}_i B_i^n(t) + [\tilde{\mathbf{N}}]_{\frac{n}{2}} B_{\frac{n}{2}}^n(t) \supseteq \frac{\sum_{i=0}^{n} \mathbf{N}_i B_i^n(t)}{\sum_{i=0}^{n} [a]_i B_i^n(t)} \supseteq \mathbf{N}(t),$$

where $[a]_{\frac{n}{2}} = [a_{\frac{n}{2}}, b_{\frac{n}{2}}]$ is an interval and all other weights are non-intervals. Adapting the development in [12], we can solve for the control points $\tilde{\mathbf{N}}_i$ and the interval control point $[\tilde{\mathbf{N}}]_{\frac{n}{2}}$ as follows:

$$\tilde{\mathbf{N}}_0 = \mathbf{N}_0 / a_0 \tag{13}$$

$$\tilde{\mathbf{N}}_1 = \frac{\mathbf{N}_0 + \mathbf{N}_1 - a_1 \tilde{\mathbf{N}}_0}{a_0} \tag{14}$$

$$\tilde{\mathbf{N}}_i = \frac{\mathbf{N}_0}{a_0} + \frac{\sum_{j=1}^{\min(n,i)} \binom{n}{j}\binom{n}{i-j}(\mathbf{N}_j - a_j \tilde{\mathbf{N}}_{i-j}]}{a_0 \binom{n}{i}}. \tag{15}$$

For $i = p + r, \ldots, m + 1 + r$,

$$\tilde{\mathbf{N}}_n = \mathbf{N}_n / a_n \tag{16}$$

$$\tilde{\mathbf{N}}_{n-1} = \frac{\mathbf{N}_n + \mathbf{N}_{n-1} - a_{n-1} \tilde{\mathbf{N}}_n}{a_n} \tag{17}$$

$$\tilde{\mathbf{N}}_{i-n} = \frac{\mathbf{N}_n}{a_n} + \frac{\sum_{j=\max(0,i-n)}^{n-1} \binom{n}{j}\binom{n}{i-j}(\mathbf{N}_j - a_j \tilde{\mathbf{N}}_{i-j})}{a_n \binom{n}{i-n}}. \tag{18}$$

The interval control point $[\tilde{\mathbf{N}}]_{\frac{n}{2}}$ is taken to be the smallest rectangular interval which contains all

$$[\mathbf{M}]_i = \sum_{j=\max(0,i-\frac{n}{2})}^{\min(i+\frac{n}{2},n)} \frac{\binom{n}{j}\binom{n}{i+\frac{n}{2}-j}(\mathbf{N}_j - [a]_j \tilde{\mathbf{N}}_{i+\frac{n}{2}-j})}{[a]_i \binom{n}{\frac{n}{2}}\binom{n}{i}}; \quad i = 0, \ldots, n. \tag{19}$$

That is, choose $[\tilde{\mathbf{N}}]_{\frac{n}{2}}$ so that $[\tilde{\mathbf{N}}]_{\frac{n}{2}} \subseteq [\mathbf{M}]_i$, $i = 0, \ldots, n$.

§5. Numerical Example

We illustrate this process by approximating an offset with $\rho = 1$ for the curve in Figure 5.

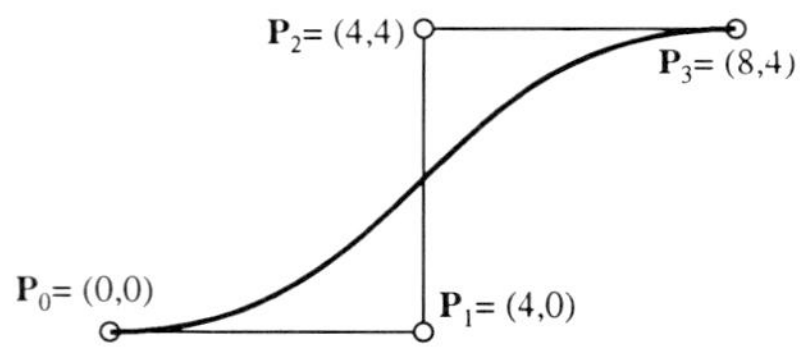

Figure 5. Sample cubic Bézier curve.

Following the steps in Section 3,

0. Choose the degree $n = 4$. Degree elevation produces $\mathbf{P}_0 = (0,0)$, $\mathbf{P}_1 = (3,0)$, $\mathbf{P}_2 = (4,2)$, $\mathbf{P}_3 = (5,4)$, $\mathbf{P}_4 = (8,4)$.
1. For this curve, $x'(t) = 12B_0^3 + 4B_1^3 + 4B_2^3 + 12B_3^3$ and $y'(t) = 0B_0^3 + 8B_1^3 + 8B_2^3 + 0B_3^3$, and after degree elevation, $w(t) = x'^2(t) + y'^2(t) = 144B_0^8 + 72B_1^8 + \frac{432}{7}B_2^8 + 72B_3^8 + \frac{2736}{35}B_4^8 + 72B_5^8 + \frac{432}{7}B_6^8 + 72B_7^8 + 144B_8^8$.
2. Compute $a_0 = b_0 = \sqrt{144} = 12$, $a_1 = b_1 = w_1/a_0 = 72/12 = 6$, $a_4 = b_4 = 12$, $a_3 = b_3 = 6$. To determine the interval $[a_2, b_2]$, first compute the values $c_2 = 8$, $c_3 = 12$, $c_4 = 112$, $c_5 = 12$, and $c_6 = 8$. Then a_2 is the largest number in the set $\{x | x^2 \leq 112;\ x \leq 8;\ x \leq 12\}$, or $a_2 = 8$. b_2 is the smallest number in the set $\{x | x^2 \geq 112;\ x \geq 8;\ x \geq 12\}$, or $b_2 = 12$.
3. To find $[\tilde{\mathbf{N}}](t)$, first compute from (12) $\mathbf{N}_0 = (0,12)$, $\mathbf{N}_1 = (-6,6)$, $\mathbf{N}_2 = (-8,4)$, $\mathbf{N}_3 = (-6,6)$, $\mathbf{N}_4 = (0,12)$. From (13)-(18), $\tilde{\mathbf{N}}_0 = \mathbf{N}_0/a_0 = (0,12)/12 = (0,1)$; $\tilde{\mathbf{N}}_1 = \frac{\mathbf{N}_0 + \mathbf{N}_1 - a_1\tilde{\mathbf{N}}_0}{a_0} = \frac{(0,12)+(-6,6)-6(0,1)}{12} = (-0.5,1)$; likewise $\tilde{\mathbf{N}}_4 = (0,1)$; and $\tilde{\mathbf{N}}_3 = (-.5,1)$. From (19), $[\tilde{\mathbf{N}}]_2 = ([-2,-2/3],[-1/3,3/4])$.
4. From equation (4), $[\tilde{\mathbf{P}}^\rho](t) \equiv \mathbf{P}(t) + \rho[\tilde{\mathbf{N}}](t)$, compute using $\rho = 1$: $\tilde{\mathbf{P}}_0^\rho = (0,0) + (0,1) = (0,1)$; $\tilde{\mathbf{P}}_1^\rho = (3,0) + (-.5,1) = (2.5,1)$; $[\tilde{\mathbf{P}}]_2^\rho = (4,2) + ([-2,-2/3],[-1/3,3/4]) = ([2,10/3],[5/3,11/4])$; $\tilde{\mathbf{P}}_3^\rho = (5,4) + (-.5,1) = (4.5,5)$; $\tilde{\mathbf{P}}_4^\rho = (8,4) + (0,1) = (8,5)$;

The resulting offset curve is shown in Figure 6.

The approximate offset curve is obtained by replacing the interval control point $[\tilde{\mathbf{P}}]_2^\rho = ([2,10/3],[2/3,11/4])$ with its midpoint $\tilde{\mathbf{P}}_2^\rho = (8/3,53/24)$. The approximation error is then defined by the width and heighth of the interval:

$$[\tilde{\mathbf{P}}^\rho](t) - \mathbf{P}^\rho(t) \subseteq ([-2/3,2/3],[-13/24,13/24])B_2^2(t).$$

Since $B_2^2(1/2) = 3/8$, the x error is bounded by $\pm 2/3 \times 3/8 = \pm 1/4$ and the error in y is bounded by $\pm 13/24 \times 3/8 = \pm 13/64$.

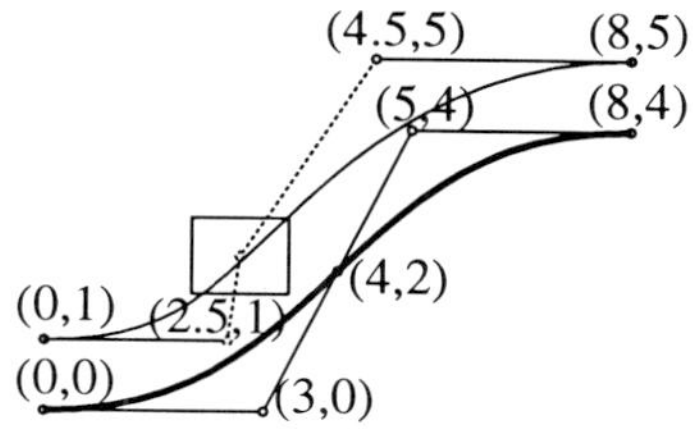

Figure 6. Sample cubic Bézier curve.

§6. Discussion

The primary contribution of this paper is an error bound for a Hermite approximation of an offset curve. As a byproduct of computing the error bound, the algorithm also produces the control points of a Hermite approximation of any even degree not less than the degree of the initial curve.

The cost of computing the Hermite approximation and its error bound is about 530 floating point operations for a degree four curve. The alternative to computing an error bound, which is to *approximate* the maximum error by sampling several discrete points along the curve, would allow roughly ten discrete samples for the computational expense of computing the guaranteed error bound.

There are two ways to measure how tight the error bound is. One way is to explore how much the interval control point $[\tilde{\mathbf{P}}]^{\rho}_{\frac{n}{2}}$ could be shrunk without violating

$$[\tilde{\mathbf{P}}^{\rho}](t) \supseteq \mathbf{P}^{\rho}(t), \quad t \in [0,1].$$

For the approximation in Figure 6, the interval control point could be shrunk by a factor of 0.11 in x and 0.40 in y. Alternately, the tightness of the error bound can be measured by comparing the *maximum* error bound $|[\tilde{\mathbf{P}}]^{\rho}_{\frac{n}{2}} B^{n}_{\frac{n}{2}}(\frac{1}{2})|$ with the actual maximum error $\max_{t\in[0.1]} |\tilde{\mathbf{P}}^{\rho}(t) - \mathbf{P}^{\rho}(t)|$. For the approximation in Figure 6, this ratio is 7.5.

In Figure 6, if the approximation is performed with two quartic Hermite curves, the dimensions of the largest interval control point drop to 0.15×0.12; that control point could safely be shrunk by 0.25 in x and 0.6 in y, and the ratio of the maximum error bound to the actual maximum error is 5.2. If the approximation is performed with four quartic Hermite curves, the dimensions of the largest interval control point drop to 0.003×0.007; that control point could safely be shrunk by 0.88 in x and 0.34 in y, and the ratio of the maximum error bound to the actual maximum error is 3.5.

Acknowledgments. This research was sponsored in part by the National Science Foundation under grant DMC-8657057.

References

1. Coquillart, S., Computing offsets of B-spline curves, Comput. Aided Design **19** (1987), 305–309.
2. Farouki, R. T., and C. A. Neff, Analytic properties of plane offset curves, Computer Aided Geom. Design **7** (1990), 83–99.
3. Farouki, R. T., and T. Sakkalis, Pythagorean hodographs", IBM Journal of Research and Development **34** (1990), 736-752.
4. Farouki, R. T., Pythagorean-hodograph curves in practical use, in *Geometry Processing for Design and Manufacturing*, R.E. Barnhill (ed.), SIAM, 1991, to appear.
5. Hoschek, J., Offset curves in the plane, Comput. Aided Design **17** (1985), 77–82.
6. Hoschek, J., Spline approximation of offset curves, Comput. Aided Geom. Design **5** (1988), 33–40.
7. Hoschek, J., and N. Wissel, Optimal approximate conversion of spline curves and spline approximation of offset curves, Comput. Aided Design **20** (1988), 475–483.
8. Klass, R., An offset spline approximation for plane cubic splines, Comput. Aided Design **15** (1983), 297–299.
9. Moore, R. E., *Interval Analysis*, Prentice–Hall, Englewood Cliffs, NJ., 1966.
10. Mudur, S. P., and P. A. Koparkar, Interval methods for processing geometric objects, IEEE Comp. Graph. Appl. **4** (2) (1984), 7–17.
11. Pham, B., Offset approximation of uniform B-splines, Comput. Aided Design **20** (1988), 471–474.
12. Sederberg, T. W., and M. Kakimoto, Approximating rational curves with polynomial curves, in *NURBS for Curve and Surface Design*, G. Farin (ed.), SIAM, 1991, 149–158.
13. Sederberg, T. W., and R. T. Farouki, Approximation by interval Bézier curves, preprint.
14. Tiller, W. and E. G. Hanson, Offsets of two-dimensional profiles, IEEE Comp. Graph. Appl. **4** (1984), 36–46.

Thomas W. Sederberg and David B. Buehler
Brigham Young University
368 Clyde Building
Provo, Utah 84602 USA
tom@spock.byu.edu

Representing Piecewise Polynomials as Linear Combinations of Multivariate B-Splines

H.-P. Seidel

Abstract. The classical Curry-Schoenberg Theorem is generalized to the multivariate setting. It is shown that every C^k-continuous piecewise polynomial F of degree n over an arbitrary triangulation $T = \{\Delta(I) \mid I \in \mathcal{J}\}$ of $\mathbb{R}^s$ can be written as a linear combination of the normalized multivariate B-splines $\{N_\lambda^I\}_{I \in \mathcal{J}, |\lambda|=n}$ over an appropriate knot sequence $\mathcal{K}$ as constructed in [4], where every vertex of the given triangulation T appears as knot of multiplicity $m = n - k$. The coefficients $c_{I,\lambda}$ in the resulting representation $F(u) = \sum_{I,\lambda} N_\lambda^I(u) c_{I,\lambda}$ are given as $c_{I,\lambda} = f_I(t_{i_0,0}, \ldots, t_{i_0,\lambda_0-1}, \ldots, t_{i_s,0}, \ldots, t_{i_s,\lambda_s-1})$, where f_I is the polar form of the restriction F_I of F to the simplex $\Delta(I) = [t_{i_0}, \ldots, t_{i_s}]$.

§1. Introduction

The classical Curry-Schoenberg Theorem [2,5] states that every C^k-continuous piecewise polynomial F of degree n on the real line can be written as a linear combination of the normalized B-splines $\{N_i^n\}_{i \in \mathbb{Z}}$ over an appropriate knot sequence $\mathcal{K} = \{t_i\}_{i \in \mathbb{Z}}$, where every breakpoint appears as a knot of multiplicity $m = n - k$. Furthermore [9,11], the coefficients $\mathbf{c}_i$ in the resulting representation $F(u) = \sum_i N_i^n(u)\mathbf{c}_i$ are given as

$$\mathbf{c}_i = f_i(t_{i+1}, \ldots, t_{i+n}),$$

where f_i is the polar form of the restriction F_i of F to the interval $\Delta_i = [t_i, t_{i+1}]$.

In this paper we generalize this result to the multivariate setting. More specifically, we show (cf. Theorem 6) that every C^k-continuous piecewise polynomial F of degree n over an arbitrary triangulation $T = \{\Delta(I) \mid I \in \mathcal{J}\}$ of

Mathematical Methods in Computer Aided Geometric Design II
Tom Lyche and Larry L. Schumaker (eds.), pp. 559–566.

ISBN 0-12-460510-9.

$\mathbb{R}^s$ can be written as a linear combination of the normalized multivariate B-splines $\{N_\lambda^I\}_{I\in\mathcal{J},|\lambda|=n}$ over an appropriate knot sequence $\mathcal{K}$ as constructed in [4], where every vertex of the given triangulation T appears as a knot of multiplicity $m = n - k$. Furthermore, the coefficients $\mathbf{c}_{I,\lambda}$ in the resulting representation $F(\mathbf{u}) = \sum_{I,\lambda} N_\lambda^I(\mathbf{u})\mathbf{c}_{I,\lambda}$ are given as

$$\mathbf{c}_{I,\lambda} = f_I(\mathbf{t}_{i_0,0}, \ldots, \mathbf{t}_{i_0,\lambda_0-1}, \ldots, \mathbf{t}_{i_s,0} \cdots \mathbf{t}_{i_s,\lambda_s-1}),$$

where f_I is the polar form of the restriction F_I of F to the simplex $\Delta(I) = [\mathbf{t}_{i_0}, \ldots, \mathbf{t}_{i_s}]$. Thus our result extends the polynomial identity (3.12) in [4] for the representation of polynomials from polynomials to piecewise polynomials.

The paper is organized as follows. Section 2 discusses some basic results about polar forms that are needed in our proof. Section 3 recalls the basic setup and the main properties of the new multivariate B-spline space constructed in [4]. Finally, Section 4 states and proves the main theorem.

§2. Polynomials and Polar Forms

P. de Faget de Casteljau [6] and L. Ramshaw [9] observed that the study of polynomials and piecewise polynomials can be simplified through the use of polar forms. The relationship between polynomials and polar forms can be phrased as follows [9]:

Theorem 1 (Blossoming Principle). *Polynomials $F : \mathbb{R}^s \to \mathbb{R}^t$ of degree n, symmetric multiaffine maps $f : (\mathbb{R}^s)^n \to \mathbb{R}^t$, and symmetric multilinear maps $f_* : (\mathbb{R}^{s+1})^n \to \mathbb{R}^t$ are equivalent. In particular, given a map of either type, unique maps of the other two types exist that satisfy the identity $F(\mathbf{u}) = f(\mathbf{u}, \ldots, \mathbf{u}) = f_*(\bar{\mathbf{u}}, \ldots, \bar{\mathbf{u}})$. In this situation f and f_* are called the multiaffine and multilinear polar form or blossom of F. Furthermore, the k-th directional derivative of F with respect to vectors $\xi_1, \ldots, \xi_k \in \mathbb{R}^s$ is given as*

$$D_{\xi_1,\ldots,\xi_k} F(\mathbf{u}) = \frac{n!}{(n-k)!} f_*(\bar{\mathbf{u}}, \ldots, \bar{\mathbf{u}}, \hat{\xi}_1, \ldots, \hat{\xi}_k), \tag{1}$$

where $\bar{\mathbf{u}} = (\mathbf{u}, 1) \in \mathbb{R}^{s+1}$ and $\hat{\xi} = (\xi, 0) \in \mathbb{R}^{s+1}$.

Proof: We start with existence of f_*. Since $\mathcal{B}_0 = \{\bar{\mathbf{o}}, \hat{\mathbf{e}}_1, \ldots, \hat{\mathbf{e}}_s\}$ is a basis of $\mathbb{R}^{s+1}$, any symmetric multilinear map $f_* : (\mathbb{R}^{s+1})^n \to \mathbb{R}^t$ is completely defined by the $\binom{n+s}{s}$ values $f_*(\bar{\mathbf{o}}, \ldots, \bar{\mathbf{o}}, \hat{\mathbf{e}}_1 \ldots, \hat{\mathbf{e}}_1, \ldots, \hat{\mathbf{e}}_s \ldots, \hat{\mathbf{e}}_s)$. Setting

$$f_*(\underbrace{\bar{\mathbf{o}}, \ldots, \bar{\mathbf{o}}}_{\lambda_0}, \underbrace{\hat{\mathbf{e}}_1, \ldots, \hat{\mathbf{e}}_1}_{\lambda_1}, \ldots, \underbrace{\hat{\mathbf{e}}_s, \ldots, \hat{\mathbf{e}}_s}_{\lambda_s}) = \frac{\lambda_0!}{n!} \frac{\partial^{\lambda_1+\cdots+\lambda_s} F(\mathbf{o})}{\partial_{u_1}^{\lambda_1} \cdots \partial_{u_s}^{\lambda_s}} \tag{2}$$

we thus obtain a multilinear map f_*, and writing $\bar{\mathbf{u}} = \bar{\mathbf{o}} + u_1\hat{\mathbf{e}}_1 + \cdots + u_s\hat{\mathbf{e}}_s$

we see that f_* satisfies

$$f_*(\bar{\mathbf{u}}, \ldots, \bar{\mathbf{u}}) =$$

$$= \sum_{|\lambda|=n} \binom{n}{\lambda_0 \ldots \lambda_s} f_*(\underbrace{\bar{\mathbf{o}}, \ldots, \bar{\mathbf{o}}}_{\lambda_0}, \underbrace{\hat{\mathbf{e}}_1, \ldots, \hat{\mathbf{e}}_1}_{\lambda_1}, \ldots, \underbrace{\hat{\mathbf{e}}_s, \ldots, \hat{\mathbf{e}}_s}_{\lambda_s}) u_1^{\lambda_1} \cdots u_s^{\lambda_s}$$

$$= \sum_{|\lambda|=n} \frac{\partial^{\lambda_1+\ldots+\lambda_s} F(\mathbf{o})}{\partial_{u_1}^{\lambda_1} \cdots \partial_{u_s}^{\lambda_s}} \frac{u_1^{\lambda_1} \cdots u_s^{\lambda_s}}{\lambda_1! \cdots \lambda_s!} = F(\mathbf{u}),$$

where the last step follows from Taylor's Theorem.

The above argument also shows that any symmetric multiaffine map f_* satisfying $f_*(\bar{\mathbf{u}}, \ldots, \bar{\mathbf{u}}) = F(\mathbf{u})$ must satisfy (2), hence f_* is unique. Finally, (1) follows from the fact that the differential operator D_ξ is linear in ξ. Together with the well-known fact that symmetric multiaffine maps and symmetric multilinear maps are equivalent to each other this completes the proof of Theorem 1. ■

Using the identity $\bar{\mathbf{u}} + \hat{\xi} = \overline{\mathbf{u} + \xi}$, equation (1) directly implies the following corollary (Prop. 8.8 in [9]) which expresses the continuity conditions between polynomials in terms of polar forms.

Corollary 2 (C^k-Conditions at a Point). *Let $F : \mathbb{R}^s \to \mathbb{R}^t$ and $G : \mathbb{R}^s \to \mathbb{R}^t$ be two polynomials of degree n, and let $\mathbf{u} \in \mathbb{R}^s$. Then F and G are C^k-continuous at $\mathbf{u}$ if and only if*

$$f(\mathbf{u}, \ldots, \mathbf{u}, \mathbf{v}_1, \ldots, \mathbf{v}_k) = g(\mathbf{u}, \ldots, \mathbf{u}, \mathbf{v}_1, \ldots, \mathbf{v}_k) \tag{3}$$

for arbitrary $\mathbf{v}_1, \ldots, \mathbf{v}_k \in \mathbb{R}^s$. ■

By polarizing (3) we obtain the following slightly more general version (Prop. 21.1 in [9]) of Corollary 2:

Corollary 3 (C^k-Conditions along a Simplex). *Let $F : \mathbb{R}^s \to \mathbb{R}^t$ and $G : \mathbb{R}^s \to \mathbb{R}^t$ be two polynomials of degree n, and let $\Lambda \subseteq \mathbb{R}^s$ be a (possibly degenerate) simplex. Then F and G are C^k-continuous along Λ if and only if*

$$f(\mathbf{u}_1, \ldots, \mathbf{u}_{n-k}, \mathbf{v}_1, \ldots, \mathbf{v}_k) = g(\mathbf{u}_1, \ldots, \mathbf{u}_{n-k}, \mathbf{v}_1, \ldots, \mathbf{v}_k) \tag{4}$$

for arbitrary $\mathbf{u}_1, \ldots, \mathbf{u}_{n-k} \in \Lambda$ and $\mathbf{v}_1, \ldots, \mathbf{v}_k \in \mathbb{R}^s$. ■

It is cumbersome to indicate the number of arguments in a polar form by underbracing. This notational inconvenience can easily be overcome by switching from additive to *multiplicative* notation. This means that from now on we will write

$$f(\mathbf{u}_1 \cdots \mathbf{u}_n) \quad \text{instead of} \quad f(\mathbf{u}_1, \ldots, \mathbf{u}_n). \tag{5}$$

If all arguments coincide, our notation then simplifies to

$$f(\mathbf{u}^n) \quad \text{instead of} \quad f(\underbrace{\mathbf{u},\ldots,\mathbf{u}}_{n}). \tag{6}$$

Terms with several multiple arguments can then be written, e.g., as

$$f({\mathbf{u}_0}^{\lambda_0}\cdots{\mathbf{u}_s}^{\lambda_s}) \quad \text{instead of} \quad f(\underbrace{\mathbf{u}_0,\ldots,\mathbf{u}_0}_{\lambda_0},\ldots,\underbrace{\mathbf{u}_s,\ldots,\mathbf{u}_s}_{\lambda_s}). \tag{7}$$

If nothing else, multiplicative notation avoids the writing of brackets and indices, and also saves some space. A deeper mathematical justification for this notation, using tensors, is given in [9].

§3. The New B-spline Scheme

Using polar forms, a new multivariate B-spline space has recently been constructed in [4]. The construction is based on the observation that the recurrence relations for simplex splines [8] and for B-patches [12] agree under proper selection of the knots and under proper renormalization.

The construction in [4] proceeds as follows: Let $T = \{\Delta(I) = [\mathbf{t}_{i_0},\ldots,\mathbf{t}_{i_s}] \mid I = (i_0,\ldots,i_s) \in \mathcal{J} \subseteq \mathbb{Z}_+^{s+1}\}$ define a triangulation of $\mathbb{R}^s$ (or of a bounded domain $D \subset \mathbb{R}^s$). This means that $D = \cup_{I\in\mathcal{J}}\Delta(I)$, while for any two $I, J \in \mathcal{J}$, $\Delta(I)\cap\Delta(J)$ is empty or is a common point or face of $\Delta(I)$ and $\Delta(J)$. In particular, any $(s-1)$-dimensional face of some $\Delta(I) \in T$ is the common edge of exactly two simplices $\Delta(I)$ and $\Delta(J)$, say.

In order to construct simplex splines of degree n over the triangulation T we first assign a cloud of knots $\mathbf{t}_{i,0},\ldots,\mathbf{t}_{i,n}$ to every vertex $\mathbf{t}_i$ of the triangulation in such a way that $\mathbf{t}_{i,0} = \mathbf{t}_i$ and such that the regions Ω_n^I defined by

$$\Omega_n^I := \operatorname{int}(\cap_{|\lambda|=n}\Omega_\lambda^I) \tag{8}$$

with

$$\Omega_\lambda^I := \cap_{\mu\le\lambda}\Delta_\mu^I\,, \quad \Delta_\mu^I = [\mathbf{t}_{i_0,\mu_0},\ldots,\mathbf{t}_{i_s,\mu_s}] \tag{9}$$

satisfy $\mathrm{vol}_s(\Omega_n^I) \ne 0$. Loosely speaking, this condition states that the clouds of knots associated with different vertices of a simplex $\Delta(I)$ are all separated from each other. In particular, all simplices $\Delta_\mu^I = [\mathbf{t}_{i_0,\mu_0},\ldots,\mathbf{t}_{i_s,\mu_s}]$ then have the same orientation.

With the knot net $\mathcal{K} = \{\mathbf{t}_{i,l} \mid i \in \mathbb{Z}, l = 1,\ldots,n\}$ in place, we then consider the simplex splines [8]

$$M_\lambda^I(\mathbf{u}) = M(\mathbf{u}|V_\lambda^I)\,, \quad I \in \mathcal{J}, |\lambda| = n \tag{10}$$

with

$$V_\lambda^I = \{\mathbf{t}_{i_0,0},\ldots,\mathbf{t}_{i_0,\lambda_0},\ldots,\mathbf{t}_{i_s,0},\ldots,\mathbf{t}_{i_s,\lambda_s}\}. \tag{11}$$

It is shown in [4] that under the above conditions

$$B_\lambda^I(\mathbf{u}) = |d(W_\lambda^I)|\, M(\mathbf{u}|V_\lambda^I) \quad \text{for all } \mathbf{u} \in \Omega_n^I, \tag{12}$$

where $B_\lambda^I(\mathbf{u})$ are the normalized B-patch blending functions as defined in [12]. In fact, the recipe for choosing the sets V_λ^I according to (11) precisely stems from this observation. It is perhaps worthwile to point out that this knot selection rule is completely different from another rule as developed earlier in [3,7], and subsequently used, e.g., in [1,13]. The above identity (12) suggests introducing the *normalized B-splines*

$$N_\lambda^I(\mathbf{u}) := |d(W_\lambda^I)|\, M(\mathbf{u}|V_\lambda^I). \tag{13}$$

It follows from the properties of B-patches [12] that the normalized B-splines $\{N_\lambda^I\}_{I,\lambda}$ are linearly independent and locally linearly independent over every subdomain of Ω_n^I, and it is shown in [4] that they also form a global partition of unity. These normalized B-splines are used as blending functions in the new B-spline scheme.

Any linear combination

$$F(\mathbf{u}) = \sum_{\substack{I \in \mathcal{J} \\ |\lambda|=n}} N_\lambda^I(\mathbf{u})\, \mathbf{c}_{I,\lambda}\,, \quad \mathbf{c}_{I,\lambda} \in \mathbb{R}^t \tag{14}$$

then is a piecewise polynomial of degree n that lies inside the *convex hull* $[\{\mathbf{c}_{I,\lambda}\}_{I,\lambda}]$ of its control points. The relationship between F and its control points is *affinely invariant*, and movement of a single control point $\mathbf{c}_{I,\lambda}$ only affects the surface *locally* on the simplex $\Delta(I)$ and on its neighbours.

One of the main results in [4] states that the B-spline space as developed in this section allows to represent polynomials. It had been shown in [12] that every polynomial F on $\mathbb{R}^s$ can be represented as a B-patch, and that B-patches can be evaluated recursively by a generalization of the univariate de Boor algorithm to the multivariate setting. The results in [4] are based upon studying to what extent this recursive evaluation persists for linear combinations of B-splines. The essential observation is formulated in the following Lemma (Lemma 3.1 in [4]):

Lemma 4 (Recursive Evaluation). *Suppose the sequence $\{c_{I,\lambda}\}_{I\in\mathcal{J},|\lambda|=n}$ has the following property. For any two adjacent simplices $\Delta(I)$ and $\Delta(J) \in T$ whose common face is*

$$[\mathbf{t}_{i_0,0},\ldots,\mathbf{t}_{i_{p-1},0},\mathbf{t}_{i_{p+1},0},\ldots,\mathbf{t}_{i_s,0}] = [\mathbf{t}_{j_0,0},\ldots,\mathbf{t}_{j_{q-2},0},\mathbf{t}_{j_q,0},\ldots,\mathbf{t}_{j_s,0}] \tag{15}$$

and for any λ, μ such that

$$\begin{aligned} \mu_j &= \lambda_j, \quad j = 0,\ldots,p-1\ , \ j = q,\ldots,s \\ \mu_{q-1} &= \lambda_p = 0 \\ \mu_j &= \lambda_{j+1}, \quad j = p,\ldots,q-2. \end{aligned} \tag{16}$$

one has

$$\mathbf{c}_{I,\lambda} = \mathbf{c}_{J,\mu}. \tag{17}$$

Then

$$\sum_{\substack{I \in \mathcal{J} \\ |\lambda|=n}} N_\lambda^I(\mathbf{u})\, \mathbf{c}_{I,\lambda} = \sum_{\substack{I \in \mathcal{J} \\ |\lambda|=n-1}} N_\lambda^I(\mathbf{u})\, \mathbf{c}_{I,\lambda}^{(1)}(\mathbf{u}) \tag{18}$$

where

$$\mathbf{c}_{I,\lambda}^{(1)}(\mathbf{u}) = \sum_{j=0}^{s} \tau_{\lambda,j}^I(\mathbf{u})\, \mathbf{c}_{I,\lambda+\mathbf{e}_j}. \;\blacksquare \tag{19}$$

Using Lemma 4, the following polynomial identity has been proved in Theorem 3.1 of [4]:

Theorem 5 (Polynomial Representation). *Let $F : \mathbb{R}^s \to \mathbb{R}^t$ be a polynomial of degree n. Then F can be represented as a linear combination of the normalized B-splines $\{N_\lambda^I\}_{I\in\mathcal{J},|\lambda|=n}$, and we get $F(\mathbf{u}) = \sum_{I,\lambda} N_\lambda^I(\mathbf{u})\; \mathbf{c}_{I,\lambda}$ with*

$$\mathbf{c}_{I,\lambda} = f(\mathbf{t}_{i_0,0} \cdots \mathbf{t}_{i_0,\lambda_0-1} \cdots \mathbf{t}_{i_s,0} \cdots \mathbf{t}_{i_s,\lambda_s-1}), \tag{20}$$

where f denotes the polar form of F. $\blacksquare$

§4. Piecewise Polynomials as Linear Combinations of B-splines

In this section we extend the polynomial identity (20) from polynomials to piecewise polynomials and prove the following multivariate analogue to the classical Curry-Schoenberg Theorem:

Theorem 6 (Piecewise Polynomial Representation). *Let F be a C^k-continuous piecewise polynomial of degree n over a given triangulation $T = \{\Delta(I) \mid I \in \mathcal{J}\}$ of $\mathbb{R}^s$. Assume that every vertex $\mathbf{t}_i$ of the triangulation T has multiplicity $m = n - k$, i.e., $\mathbf{t}_i = \mathbf{t}_{i,0} = \ldots = \mathbf{t}_{i,m-1}$. Then F can be represented as a linear combination of the normalized B-splines $\{N_\lambda^I\}_{I\in\mathcal{J},|\lambda|=n}$, and we get $F(\mathbf{u}) = \sum_{I,\lambda} N_\lambda^I(\mathbf{u})\; \mathbf{c}_{I,\lambda}$ with*

$$\mathbf{c}_{I,\lambda} = f_I(\mathbf{t}_{i_0,0} \cdots \mathbf{t}_{i_0,\lambda_0-1} \cdots \mathbf{t}_{i_s,0} \cdots \mathbf{t}_{i_s,\lambda_s-1}), \tag{21}$$

where f_I denotes the polar form of the restriction F_I of F to the simplex $\Delta(I)$.

Proof: We proceed by induction on k and start with $k = -1$. In this situation we have to show that C^{-1}-continuous, i.e., discontinuous, piecewise polynomials of arbitrary degree n can be represented as linear combinations of B-splines with $(n+1)$-fold knots $\mathbf{t}_{i,0} = \cdots = \mathbf{t}_{i,n}$. However, in this situation the B-spline representation agrees with the piecewise Bézier representation, and the assertion is trivial.

Suppose that the statement holds for $k-1$. Let F be a C^k-continuous piecewise polynomial, and let $\mathbf{w} \in \mathbb{R}^s$ be arbitrary. We consider the piecewise polynomial $F_{\mathbf{w}}$ defined by $F_{\mathbf{w}}(\mathbf{u}) = f_I(\mathbf{u}^{n-1}\mathbf{w})$ for $\mathbf{u} \in \Delta(I)$. By definition, $F_{\mathbf{w}}$ is a piecewise polynomial of degree $n-1$ over T.

We claim that $F_{\mathbf{w}}$ is C^{k-1}-continuous: Consider a point $\mathbf{u}$ that lies on a common face of the simplices $\Delta(I)$ and $\Delta(J)$. According to Corollary 2, the C^k-continuity of F at $\mathbf{u}$ is equivalent to the identity $f_I(\mathbf{u}^{n-k}\mathbf{v}_1 \cdots \mathbf{v}_k) = f_J(\mathbf{u}^{n-k}\mathbf{v}_1 \cdots \mathbf{v}_k)$ for all $\mathbf{v}_1, \ldots, \mathbf{v}_k \in \mathbb{R}^s$. Let $f_{\mathbf{w}_I}$ be the polar form of the restriction $F_{\mathbf{w}_I}$ of $F_{\mathbf{w}}$ to the simplex $\Delta(I)$. Then

$$\begin{aligned} f_{\mathbf{w}_I}(\mathbf{u}^{n-k}\mathbf{v}_1 \cdots \mathbf{v}_{k-1}) &= f_I(\mathbf{u}^{n-k}\mathbf{v}_1 \cdots \mathbf{v}_{k-1}\mathbf{w}) \\ &= f_J(\mathbf{u}^{n-k}\mathbf{v}_1 \cdots \mathbf{v}_{k-1}\mathbf{w}) \\ &= f_{\mathbf{w}_J}(\mathbf{u}^{n-k}\mathbf{v}_1 \cdots \mathbf{v}_{k-1}) \end{aligned}$$

for all $\mathbf{v}_1, \ldots, \mathbf{v}_{k-1} \in \mathbb{R}^s$, which in turn is equivalent to the C^{k-1}-continuity of $F_{\mathbf{w}}$ at $\mathbf{u}$.

Using the induction hypothesis and the fact that the polar form $f_{\mathbf{w}_I}$ is affine in each argument, we see that $F_{\mathbf{w}}$ satisfies

$$\begin{aligned} F_{\mathbf{w}}(\mathbf{u}) &= \sum_{\substack{I \in \mathcal{J} \\ |\lambda| = n-1}} N_\lambda^I(\mathbf{u})\; f_{\mathbf{w}_I}(\mathbf{t}_{i_0,0} \cdots \mathbf{t}_{i_0,\lambda_0-1} \cdots \mathbf{t}_{i_s,0} \cdots \mathbf{t}_{i_s,\lambda_s-1}) \\ &= \sum_{\substack{I \in \mathcal{J} \\ |\lambda| = n-1}} N_\lambda^I(\mathbf{u})\; f_I(\mathbf{t}_{i_0,0} \cdots \mathbf{t}_{i_0,\lambda_0-1} \cdots \mathbf{t}_{i_s,0} \cdots \mathbf{t}_{i_s,\lambda_s-1}\mathbf{w}) \\ &= \sum_{\substack{I \in \mathcal{J} \\ |\lambda| = n-1}} N_\lambda^I(\mathbf{u}) \sum_{j=0}^{s} \tau_{\lambda,j}(\mathbf{w})\; \mathbf{c}_{I,\lambda+\mathbf{e}_j} \end{aligned}$$

for all $\mathbf{u} \in \mathbb{R}^s$.

In order to complete the proof we use Lemma 4. Let $\Delta(I)$ and $\Delta(J)$ be two adjacent simplices with common face

$$\Lambda = [\mathbf{t}_{i_0,0}, \ldots, \mathbf{t}_{i_{p-1},0}, \mathbf{t}_{i_{p+1},0}, \ldots, \mathbf{t}_{i_s,0}] = [\mathbf{t}_{j_0,0}, \ldots, \mathbf{t}_{j_{q-2},0}, \mathbf{t}_{j_q,0}, \ldots, \mathbf{t}_{j_s,0}],$$

and let λ, μ be such that the conditions (16) of Lemma 4 are satisfied. Since every vertex $\mathbf{t}_{i,0} = \cdots = \mathbf{t}_{i,m-1}$ has multiplicity m, the sequence

$$\begin{aligned} &(\underbrace{\mathbf{t}_{i_0,0}, \ldots, \mathbf{t}_{i_0,\lambda_0-1}}_{\lambda_0}, \ldots, \underbrace{\mathbf{t}_{i_{p-1},0}, \ldots, \mathbf{t}_{i_{p-1},\lambda_{p-1}-1}}_{\lambda_{p-1}}, \\ &\quad \underbrace{\mathbf{t}_{i_{p+1},0}, \ldots, \mathbf{t}_{i_{p+1},\lambda_{p+1}-1}}_{\lambda_{p+1}}, \ldots, \underbrace{\mathbf{t}_{i_s,0}, \ldots, \mathbf{t}_{i_s,\lambda_s-1}}_{\lambda_s}) \end{aligned}$$

contains at least m knots on Λ. According to Corollary 3, the C^k-continuity of F along Λ thus implies

$$\begin{aligned} \mathbf{c}_{I,\lambda} &= f_I(\mathbf{t}_{i_0,0} \cdots \mathbf{t}_{i_{p-1},\lambda_{p-1}-1}\mathbf{t}_{i_{p+1},0} \cdots \mathbf{t}_{i_s,\lambda_s-1}) \\ &= f_J(\mathbf{t}_{i_0,0} \cdots \mathbf{t}_{i_{p-1},\lambda_{p-1}-1}\mathbf{t}_{i_{p+1},0} \cdots \mathbf{t}_{i_s,\lambda_s-1}) = \mathbf{c}_{J,\mu}, \end{aligned}$$

and the points $\mathbf{c}_{I,\lambda} = \mathbf{c}_{J,\mu}$ satisfy assumption (17) of Lemma 4. Using $F_{\mathbf{u}}(\mathbf{u}) = F(\mathbf{u})$ we therefore obtain

$$\begin{aligned} F(\mathbf{u}) =& F_{\mathbf{u}}(\mathbf{u}) = \sum_{\substack{I\in\mathcal{J} \\ |\lambda|=n-1}} N_\lambda^I(\mathbf{u}) \sum_{j=0}^{s} \tau_{\lambda,j}(\mathbf{u})\mathbf{c}_{I,\lambda+\mathbf{e}_j} \\ =& \sum_{\substack{I\in\mathcal{J} \\ |\lambda|=n-1}} N_\lambda^I(\mathbf{u})\ \mathbf{c}_{I,\lambda}^{(1)}(\mathbf{u}) = \sum_{\substack{I\in\mathcal{J} \\ |\lambda|=n}} N_\lambda^I(\mathbf{u})\ \mathbf{c}_{I,\lambda}, \end{aligned}$$

and the proof of Theorem 6 is complete. ■

References

1. Auerbach, S., R. H. J. Gmelig Meyling, M. Neamtu, and H. Schaeben, Approximation and geometric modeling with simplex B-splines associated with irregular triangles, Computer Aided Geom. Design **8** (1991), 67–87.
2. Curry, H. B., and I. J. Schoenberg, On pólya frequency functions. IV: The fundamental spline functions and their limits, Journal d'Analyse Math. **17** (1966), 71–107.
3. Dahmen, W., and C. A. Micchelli, On the linear independence of multivariate B-splines I. Triangulations of simploids, SIAM J. Numer. Anal. **19** (1982), 993–1012.
4. Dahmen, W., C. A. Micchelli, and H.-P. Seidel, Blossoming begets B-Splines built better by B-Patches, Research Report RC 16261(#72182), IBM Research Division, Yorktown Heights, 1990; also Math. Comp. , to appear.
5. de Boor, C., *A Practical Guide to Splines*, Springer, New York, 1978.
6. de Casteljau, P., *Formes à Pôles*, Hermes, Paris, 1985.
7. Höllig, K., Multivariate splines, SIAM J. Numer. Anal. **19** (1982), 1013–1031.
8. Micchelli, C. A., On a numerically efficient method for computing with multivariate B-splines, in *Multivariate Approximation Theory*, W. Schempp and K. Zeller, (eds.), Birkhäuser, Basel, 1979, 211–248.
9. Ramshaw, L., Blossoming: A connect-the-dots approach to splines, Technical report, Digital Systems Research Center, Palo Alto, 1987.
10. Ramshaw, L., Blossoms are polar forms, Computer Aided Geom. Design **6** (1989), 323–358.
11. Seidel, H.-P., A new multiaffine approach to B-splines, Computer Aided Geom. Design **6** (1989), 23–32.
12. Seidel, H.-P., Symmetric recursive algorithms for surfaces: B-patches and the de Boor algorithm for polynomials over triangles, Constr. Approx. **7** (1991), 257–279.
13. Traas, C. R., Practice of bivariate quadratic simplicial splines, in *Computation of Curves and Surfaces*, W. Dahmen, M. Gasca, and C.A. Micchelli (eds.), Kluwer Academic Publishers, Dordrecht, 1990, 383–422.

An Explicit Derivation of Discretely Shaped Beta-spline Basis Functions of Arbitrary Order

Gadiel Seroussi and Brian A. Barsky

Abstract. Beta-splines are a class of splines with applications in the construction of curves and surfaces for computer-aided geometric design. One of the salient features of the Beta-spline is that the curves and surfaces thus constructed are *geometrically continuous*, a more general notion of continuity than the one used in ordinary B-splines. The basic building block for Beta-splines of order k is a set of *Beta-polynomials* of degree $k-1$, which are used to form the Beta-spline basis functions. The coefficients of the Beta-polynomials are functions of certain *shape parameters* $\beta_{s;i}$. In this paper, we present a symbolic derivation of the Beta-polynomials as polynomials over the field K_n of real rational functions in the indeterminates $\beta_{s;i}$. We show, constructively, the existence and uniqueness of Beta-polynomials satisfying the design objectives of geometric continuity, minimum spline order, invariance under translation, and linear independence, and we present an explicit symbolic procedure for their computation. The derivation presented here, and the resulting procedure, are valid for the general case of discretely-shaped Beta-splines of arbitrary order, over uniform knot sequences. By extending the field K_n with symbolic indeterminates z_s representing the lengths of the parametric intervals, the result can be generalized to discretely-shaped Beta-splines over non-uniform knot sequences, as shown in Seroussi & Barsky (1991).

§1. Introduction

Beta-splines are a class of piecewise polynomial splines with applications in the construction of curves and surfaces for computer graphics and computer-aided geometric design. The original cubic Beta-spline was first developed in Barsky (1981). Beta-splines combine the power of a general mathematical formulation with an intuitive specification mechanism based on *shape parameters*. The shape parameters arise from the extra degrees of freedom that are liberated by constructing curves and surfaces using a more relaxed form of continuity

Mathematical Methods in Computer Aided Geometric Design II
Tom Lyche and Larry L. Schumaker (eds.), pp. 567–584.

ISBN 0-12-460510-9.

called *geometric continuity* (Barsky & DeRose, 1984). The widely used B-splines (de Boor, 1972; de Boor, 1978; Riesenfeld, 1973; Schumaker, 1981) are a special case of Beta-splines, obtained by a specific setting of the shape parameters.

Subsequent to the original development of the Beta-spline, various aspects of this representation have been pursued, and numerous articles on Beta-splines and geometric continuity have appeared in the literature. For a survey of work in this area, including recent results and extensive bibliographies, the reader is invited to consult some of the tutorial articles and books on the subject, such as Barsky (1981), Barsky (1988), Barsky (1989) and Barsky & DeRose (1990) for Beta-splines, and such as Barsky & DeRose (1989), Barsky & DeRose (1990), Gregory (1989), Herron (1987) and Höllig (1986) for geometric continuity. In addition, Bartels *et al.* (1987) is a comprehensive reference for both subjects.

Most of the results in the literature deal with *cubic* Beta-splines (order $k = 4$). Generalization of Beta-splines to higher order was first discussed in Barsky & DeRose (1984), where the conditions for geometric continuity of arbitrary order, called the *Beta-constraints*, were developed. The existence of Beta-splines of arbitrary order satisfying these constraints was established in Dyn & Micchelli (1988) and in Goodman (1985). An explicit method for the computation of basis functions in the case of *uniformly-shaped* Beta-splines (where the same set of shape parameters is used in all parts of the curve) is presented in Dyn *et al.* (1987), together with explicit examples for cubic and quartic basis functions. However, there has been no general procedure for deriving the more general case of *discretely-shaped* Beta-splines of arbitrary order (where different sets of shape parameters are used in different parts of the curve, as will be precisely defined in Sections 2 and 3), except on a case-by-case basis. In a recent report (Seidel, 1990), explicit geometric constructions and knot insertion algorithms for Beta-splines of arbitrary order are provided by computing the representation of a Beta-spline basis function as a piecewise Bézier polynomial (Bézier, 1970). These results, as well as the results of Dyn *et al.* (1987), Dyn & Micchelli (1988) and Goodman (1985), are based on the total positivity of certain connection matrices related to the Beta-constraints, and use a classical approximation-theoretic approach to Beta-splines.

The basic building block for Beta-splines of order k is a set of *Beta-polynomials* of degree $k - 1$, which are used to form the Beta-spline basis functions. The coefficients of the Beta-polynomials are symbolic functions of certain shape parameters $\beta_{s;i}$, for ranges of s and i to be specified elsewhere in the paper. The main result of this paper is a purely symbolic derivation of Beta-spline basis functions of arbitrary order. Our derivation treats geometric continuity as a set of symbolic constraints on the coefficients of the Beta-polynomials. The latter are regarded as polynomials over the field $\mathbf{K}_n$ of real rational functions in the *indeterminates* $\beta_{s;i}$, and all computations and proofs are carried out over that field. We prove the existence and uniqueness of Beta-polynomials of arbitrary order, and we present explicit expressions for their computation. The method used for the derivation is similar to the one used in Lempel & Seroussi (1990) for spline bases in more general function spaces.

The remainder of this paper is organized as follows. In Section 2 we describe the basic geometric setting for the construction of piecewise continuous parametric curves. We then present, first, the concept of parametric continuity on which traditional representations such as B-splines are based. Then, we define the more relaxed notion of geometric continuity, and present the Beta-constraints as the conditions for geometric continuity of a parametric curve. In Section 3 we formally define the Beta-polynomials and Beta-splines, and set four design objectives to be satisfied by them: geometric continuity, minimum spline order, invariance under translation, and linear independence. It is later shown in the paper that these design objectives uniquely determine the Beta-polynomials. In Section 4, we transform the design objectives into a set of linear equations in the coefficients of the Beta-polynomials. In Section 5 we explicitly solve these equations, and present a symbolic procedure for the computation of Beta-polynomials. Finally, in Section 6 we present some concluding remarks.

The derivation in Sections 4 and 5, and the resulting symbolic procedure, are valid for the general case of discretely-shaped Beta-splines over uniform knot sequences. By further extending the field $\mathbf{K}_n$ with indeterminates z_s representing the lengths of the parametric intervals, the result can be generalized to discretely-shaped Beta-splines over non-uniform knot sequences. This generalization is not presented here, for the sake of brevity. Also, most of the lemmas and theorems in the paper are presented without proof. A full account can be found in Seroussi & Barsky (1991).

§2. Geometric Continuity and Beta-constraints

Let k and m be positive integers, with $m \geq k$, and let $\mathbf{V}_1, \mathbf{V}_2, \ldots, \mathbf{V}_m$ be a sequence of points in $\mathbb{R}^d$. The following describes a standard way of constructing an *order k, piecewise polynomial, parametric curve* $\mathbf{q}(u)$ in $\mathbb{R}^d$: Let $u_0 < u_1 < \ldots < u_{m-k+1}$ be a monotonically increasing sequence of real numbers (known as *knots*). (A discussion of *monotonically non-decreasing* knot sequences can be found in Seroussi & Barsky (1991)). Then $\mathbf{q}(u)$ is defined over the domain $u_0 \leq u < u_{m-k+1}$ as a concatenation of *polynomial curve segments* $\mathbf{q}_1(u)$, $\mathbf{q}_2(u)$, ..., $\mathbf{q}_{m-k+1}(u)$ as follows:

$$\mathbf{q}(u) = \mathbf{q}_s(u), \qquad u_{s-1} \leq u < u_s, \quad 1 \leq s \leq m-k+1, \tag{2.1}$$

where

$$\mathbf{q}_s(u) = \sum_{i=0}^{k-1} \mathbf{V}_{s+i} b_{s;i}(u - u_{s-1}), \qquad u_{s-1} \leq u \leq u_s, \quad 1 \leq s \leq m-k+1. \tag{2.2}$$

Here, the $b_{s;i}(u)$ are scalar-valued polynomials of degree at most $k-1$ in u. Notice that in (2.2) we define $\mathbf{q}_s(u)$ over the closed interval $[u_{s-1}, u_s]$, even though in (2.1) we use it only over the semi-open interval $[u_{s-1}, u_s)$. We use a semicolon in the subscript of b to emphasize the distinction between the segment index s and the element index i. We shall use this convention throughout this paper for polynomials, vectors and matrices. For example, $A_{s;ij}$ will denote the $(i,j)^{th}$ entry of a matrix A_s, related to the segment with index s.

The points $\mathbf{V}_i$ are called *control vertices*, and they form the *control polygon* of $\mathbf{q}(u)$. Strictly speaking, $\mathbf{q}(u)$ is just one of many possible parametrizations of the represented curve. We assume that $\mathbf{q}(u)$ is a *regular parametrization*; i.e. its first derivative vector never vanishes. Clearly, being built of polynomial segments, $\mathbf{q}(u)$ is continuous and infinitely differentiable at any internal parametric value u in the range $u_{s-1} < u < u_s$, for $1 \leq s \leq m-k+1$, but potential discontinuities may occur at the knots, where the curve segments meet. We say that $\mathbf{q}(u)$ is n^{th} *degree parametrically continuous*, denoted C^n, at $u = u_s$, if

$$\mathbf{q}_{s+1}^{(j)}(u)|_{u=u_s} = \mathbf{q}_s^{(j)}(u)|_{u=u_s}, \qquad 0 \leq j \leq n, \quad 1 \leq s \leq m-k, \tag{2.3}$$

where $g^{(j)}(u)|_{u=x}$ denotes the jth derivative of a function $g(u)$ with respect to u, evaluated at $u = x$. In this case, we also say that $\mathbf{q}_s(u)$ *meets* $\mathbf{q}_{s+1}(u)$ with n^{th} degree parametric continuity at $u = u_s$.

The concept of parametric continuity is used in computer-aided geometric design to capture the notion of "smoothness" of a piecewise parametric curve. In particular, the widely used B-splines (see, for instance, de Boor, 1978; Bartels *et al.*, 1987) form a basis for a very general family of parametrically continuous curves. However, this notion of continuity is sometimes too restrictive, as it reflects a property of the specific parametrization used, rather than of the curve itself.

We now define a more general notion of continuity for parametric curves, called *geometric continuity*, which was first presented in the computer aided geometric design literature in Barsky & DeRose (1984); a similar concept called *contact of order n* was described in the German geometry literature in Geise (1962) and Scheffers (1910).

We say that $\mathbf{q}(u)$ is n^{th} *degree geometrically continuous*, denoted G^n, at $u = u_s$ if there exists a continuous, n times differentiable *reparametrization* function $u = f_s(\tilde{u})$, $f_s : [\tilde{u}_{s-1}, \tilde{u}_s] \rightarrow [u_{s-1}, u_s]$, such that at the end points of the interval, we have $f_s(\tilde{u}_{s-1}) = u_{s-1}$, $f_s(\tilde{u}_s) = u_s$, and $\mathbf{q}_s(f_s(\tilde{u}))$ meets $\mathbf{q}_{s+1}(u)$ with n^{th} degree parametric continuity at $u = u_s$; i.e., we have

$$\mathbf{q}_{s+1}^{(j)}(u)|_{u=u_s} = \mathbf{q}_s^{(j)}(f_s(\tilde{u}))|_{\tilde{u}=\tilde{u}_s}, \qquad 0 \leq j \leq n, \quad 1 \leq s \leq m-k, \tag{2.4}$$

where the jth derivative at the righthand side of (2.4) is taken with respect to $\tilde{u}$. In computer-aided geometric design applications, the reparametrization function f_s is required to be *orientation preserving*; i.e. $f_s^{(1)}(\tilde{u}) > 0$. This guarantees that the new parametrization is regular, and also, that $\mathbf{q}_s(f_s(\tilde{u}))$, $\tilde{u}_{s-1} \le \tilde{u} \le \tilde{u}_s$, traces the same curve in $\mathbb{R}^d$ as $\mathbf{q}_s(u)$, $u_{s-1} \le u \le u_s$, without backtracking. Notice also that in practical applications, we do not need to actually compute the reparametrized curve segment. Instead, the curve is manipulated using the original parametrization, and the existence of a parametrically continuous reparametrization guarantees geometric continuity, and hence, the "smoothness" of the curve.

Let

$$\beta_{s;j} = f_s^{(j)}(\tilde{u})|_{\tilde{u}=\tilde{u}_s}, \qquad 1 \le j \le n,$$

and let $\boldsymbol{\beta}_s = (\beta_{s;1}\ \beta_{s;2}\ \cdots\ \beta_{s;n})$. Then, using the chain rule for derivatives to expand the righthand side of (2.4), the latter is transformed into

$$\mathbf{q}_{s+1}^{(j)}(u_s) = \sum_{r=0}^{j} M_{s;jr}\mathbf{q}_s^{(r)}(u_s), \qquad 0 \le j \le n, \tag{2.5}$$

where the coefficients $M_{s;jr}$ are polynomials in $\beta_{s;1}$, $\beta_{s;2}$, ..., $\beta_{s;n}$ derived from the application of the chain rule. These polynomials are given by *Faa di Bruno's formulas*, as presented on page 50 of Knuth (1973):

$$M_{s;jr} = \sum_{\substack{k_1+k_2+\ldots+k_j=r \\ k_1+2k_2+\ldots+jk_j=j \\ k1,k2,\ldots,k_j\ge 0}} \frac{j!}{k_1!(1!)^{k_1}k_2!(2!)^{k_2}\cdots k_j!(j!)^{k_j}}\beta_{s;1}^{k_1}\beta_{s;2}^{k_2}\cdots\beta_{s;j}^{k_j} \tag{2.6}$$

for $0 \le j \le n$, $0 \le r \le n$. (In the above equation, we assume that empty sums are equal to zero. Also, the range of r has been extended to $0 \le r \le n$, but it is easily verified that $M_{s;jr} = 0$ for $r > j$). Define the $(n+1) \times (n+1)$ matrix $M_s(\boldsymbol{\beta}_s) = (M_{s;jr})_{0\le j\le n, 0\le r\le n}$. Then, $M_s(\boldsymbol{\beta}_s)$ is lower triangular, and it has the general form

$$M_s(\boldsymbol{\beta}_s) = \begin{pmatrix} 1 & 0 & 0 & 0 & \cdots & 0 \\ 0 & \beta_{s;1} & 0 & 0 & \cdots & 0 \\ 0 & \beta_{s;2} & \beta_{s;1}^2 & 0 & \cdots & 0 \\ 0 & \beta_{s;3} & 3\beta_{s;1}\beta_{s;2} & \beta_{s;1}^3 & \cdots & 0 \\ \vdots & \vdots & \vdots & \vdots & & \vdots \\ 0 & \beta_{s;n} & M_{s;n2} & M_{s;n3} & \cdots & \beta_{s;1}^n \end{pmatrix}. \tag{2.7}$$

$M_s(\boldsymbol{\beta}_s)$ is called a *Beta-connection matrix*. The following properties of $M_s(\boldsymbol{\beta}_s)$ are readily derived from (2.6):

(M1) $M_{s;00} = 1$, $M_{s;0r} = M_{s;j0} = 0$ for $1 \le j \le n$, $1 \le r \le n$.

(M2) $M_{s;j1} = \beta_{s;j}$ for $1 \le j \le n$.

(M3) $M_{s;jj} = \beta_{s;1}^j$ for $0 \le j \le n$.

(M4) $M_{s;jr}$ does not depend on n, as long as $j \le n$ and $r \le n$. For example, the 4×4 matrix at the upper left corner of $M_s(\boldsymbol{\beta}_s)$ in (2.7) is the Beta-connection matrix for $n = 3$.

(M5) $M_s([1\ 0\ 0\ \ldots\ 0]) = I$, the identity matrix of order $n+1$.

In the remainder of the paper, we shall denote the Beta-connection matrix by M_s, the dependence on $\boldsymbol{\beta}_s$ being understood from the context. Using the definition of M_s, equation (2.5) can be rewritten in matrix form as

$$\begin{pmatrix} \mathbf{q}_{s+1}^{(0)}(u_s) \\ \mathbf{q}_{s+1}^{(1)}(u_s) \\ \vdots \\ \mathbf{q}_{s+1}^{(n)}(u_s) \end{pmatrix} = M_s \begin{pmatrix} \mathbf{q}_{s}^{(0)}(u_s) \\ \mathbf{q}_{s}^{(1)}(u_s) \\ \vdots \\ \mathbf{q}_{s}^{(n)}(u_s) \end{pmatrix}. \tag{2.8}$$

These are known as the *Beta-constraints* (Barsky & DeRose, 1984; Barsky & DeRose, 1989; Goldman & Barsky, 1989; Goodman, 1985) for n^{th} degree geometric continuity at $u = u_s$. Notice that, by property (M5), when $\boldsymbol{\beta}_s = (1\ 0\ 0\ \ldots\ 0)$, (2.8) reduces to the usual parametric continuity constraints.

Let $z_s = u_s - u_{s-1}$. We shall now assume that the knots u_s are *uniformly spaced*, and that the parametric intervals have unit length, namely, $z_s = 1$ for $1 \le s \le m-k+1$. The more general case of non-uniform knot spacing is covered in Seroussi & Barsky (1991).

§3. Beta-splines

Under the uniform knot spacing assumption, the curve segment equation (2.2) can be rewritten as follows:

$$\mathbf{q}_s(u) = \sum_{i=0}^{k-1} \mathbf{V}_{s+i} b_{s;i}(u - u_{s-1}), \quad u_{s-1} \le u \le u_{s-1}+1 = u_s, \ 1 \le s \le m-k+1, \tag{3.1}$$

where $b_{s;0}(u)$, $b_{s;1}(u)$, ..., $b_{s;k-1}(u)$ are polynomials of degree at most $k-1$ in u. Notice that in (3.1), $b_{s;i}(u)$ is evaluated in the interval $0 \le u < 1$. The main result of this paper is the derivation of explicit symbolic expressions for polynomials $b_{s;i}(u)$ satisfying the following *design objectives*:

(i) G^n **Continuity**. For a given continuity degree $n \ge 0$, and for any choice of control vertices $\mathbf{V}_1, \mathbf{V}_2, \ldots, \mathbf{V}_m$, the Beta-constraints (2.8) are satisfied.

(ii) **Minimum spline order**. The order of the curve is the least integer k for which objective (i) can be satisfied.

(iii) **Shape preservation under translation**. For any fixed translation vector $\mathbf{Q} \in \mathbb{R}^d$, the following holds:

$$\mathbf{Q} + \mathbf{q}_s(u) = \sum_{i=0}^{k-1} (\mathbf{Q} + \mathbf{V}_{s+i}) b_{s;i}(u - u_{s-1}), \quad u_{s-1} \le u < u_s, \ 1 \le s \le m-k+1.$$

Hence, a translation of all control vertices by a fixed vector results in the translation of all points of the curve by the same vector.

(iv) **Linear independence**. The polynomials $b_{s;0}(u), b_{s;1}(u), \ldots, b_{s;k-1}(u)$ are linearly independent and thus, they form a basis for the k-dimensional linear space of polynomials of degree at most $k-1$ in u.

Polynomials $b_{s;i}(u)$ satisfying objectives (i)-(iv) will be called *Beta-polynomials*, and the curves constructed according to (2.1)-(2.2) will be called *Beta-spline curves*. The parameters $\beta_{s;i}$ are referred to as *shape parameters*. Beta-polynomials are also known in the literature as *Beta-spline basis segments* (Bartels *et al.*, 1987), and the corresponding piecewise functions $F_s(u)$, defined over the domain $[u_0, u_{m-k+1}]$ by

$$F_s(u) = \begin{cases} 0, & u < u_{s-k} \text{ or } u \geq u_s \\ b_{s-i;i}(u - u_{s-i-1}), & u_{s-i-1} \leq u < u_{s-i}, \quad k-1 \geq i \geq 0, \end{cases}$$

for $1 \leq s \leq m$, are known as the *Beta-spline basis functions* for the given knot sequence. This is justified by the fact that (2.1) and (2.2) are equivalent to

$$\mathbf{q}(u) = \sum_{s=1}^{m} \mathbf{V}_s F_s(u).$$

The relation between the Beta-polynomials and the basis functions is depicted in Figure 1.

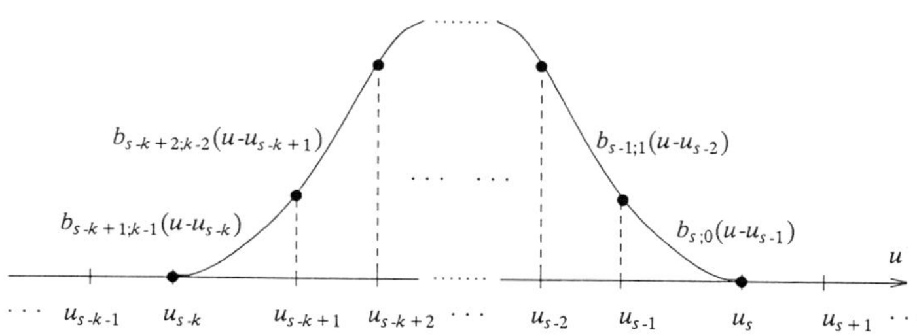

Figure 1. Basis function $F_s(u)$.

As we shall see, design objectives (i)-(iv) uniquely determine the Beta-polynomials. The existence of Beta-polynomials, and other properties of Beta-splines of arbitrary order, had been established in Dyn & Micchelli (1988) and Goodman (1985), based on the total positivity of the matrix M_s. The main result of this paper is a purely symbolic and constructive proof of the existence and uniqueness of Beta-polynomials of arbitrary order, and the derivation of explicit expressions for their symbolic computation.

We now present the formal algebraic framework for our symbolic derivation. While the $b_{s;i}(u)$ are polynomials in u, their coefficients are, in general, functions of parameters $\beta_{l;1}, \beta_{l;2}, \ldots, \beta_{l;n}$, for integers l in a range to be determined in the derivation. More formally, for integers $r \geq 0$ and $n \geq 1$, let

$$\mathbf{K_{r;n}} = \mathbb{R}(\beta_{-r;1}, \cdots, \beta_{-r;n}, \beta_{-r+1;1}, \cdots, \beta_{-r+1;n}, \cdots \quad \cdots, \beta_{r;1}, \cdots, \beta_{r;n})$$

denote the field of symbolic rational functions over $\mathbb{R}$, in the indeterminates $\beta_{l;i}$ for $-r \leq l \leq r$ and $1 \leq i \leq n$. Also, let

$$\mathbf{K}_n = \cup_{r=0}^{\infty} \mathbf{K}_{r;n}$$

denote the field of rational functions in the indeterminates $\beta_{l;i}$ for $1 \leq i \leq n$ and all integers l. Then, for $0 \leq i \leq k-1$, $b_{s;i}(u)$ will be a polynomial of degree at most $k-1$ with coefficients in $\mathbf{K}_n$. In the remainder of the paper, we deal with polynomials, vectors and matrices with coefficients in $\mathbf{K}_n$ and, unless explicitly stated otherwise, algebraic properties such as linear independence and nonsingularity will be understood to be defined over that field (for a treatment of extension fields and rational function fields see, for instance, Herstein, 1975).

A particular case of interest occurs when we make the substitution $\beta_{s;i} = \beta_i$ for $1 \leq i \leq n$ and all s; i.e. we use the same set of parameters $\boldsymbol{\beta} = (\beta_1 \ \beta_2 \ \ldots \ \beta_n)$ for the Beta-constraints at all the knots. The splines thus constructed are called *uniformly-shaped Beta-splines*, while the splines in the general case described above are referred to as *discretely-shaped Beta-splines* (Bartels *et al.*, 1987). Making a further restriction, we obtain another case of interest by setting $\beta_{s;1} = 1$, and $\beta_{s;2} = \beta_{s;3} = \ldots = \beta_{s;n} = 0$ for all s. In this case, the Beta-splines become the usual uniform B-splines, and $\mathbf{q}(u)$ is a uniform B-spline curve. If the uniform knot spacing assumption is removed, we obtain the non-uniform B-splines.

§4. Beta-constraints Revisited

We now proceed to transform the design objectives (i)-(iv) to a set of algebraic equations which will later be solved for the polynomials $b_{s;i}(u)$. In the derivation, we will ignore end conditions, namely, we will derive the Beta-polynomials $b_{s;i}(u)$ for a generic segment $\mathbf{q}_s(u)$ "far away" from the ends of the curve (or, in a different interpretation, we will assume that $b_{s;i}(u)$ and all the related parameters are defined for all integers s). End conditions are discussed in Seroussi & Barsky (1991).

Let $b_{s;ij}$ denote the coefficients of $b_{s;i}(u)$; i.e.

$$b_{s;i}(u) = \sum_{i=0}^{k-1} b_{s;ij} u^j, \qquad 0 \leq i \leq k-1,$$

and define the $k \times k$ matrix $B_s = (b_{s;ij})_{0 \leq i,j \leq k-1}$. Also let $\mathbf{b}_{s;i}$ be the vector $\mathbf{b}_{s;i} = (b_{s;i0} \ b_{s;i1} \ \cdots \ b_{s;i,k-1})$. Thus, we have

$$\begin{pmatrix} b_{s;0}(u) \\ b_{s;1}(u) \\ \vdots \\ b_{s;k-1}(u) \end{pmatrix} = \begin{pmatrix} \mathbf{b}_{s;0} \\ \mathbf{b}_{s;1} \\ \vdots \\ \mathbf{b}_{s;k-1} \end{pmatrix} \begin{pmatrix} 1 \\ u \\ \vdots \\ u^{k-1} \end{pmatrix} = B_s \begin{pmatrix} 1 \\ u \\ \vdots \\ u^{k-1} \end{pmatrix}. \tag{4.1}$$

Consider the jth order Beta-constraint (2.5), and substitute the expressions for $\mathbf{q}_s^{(j)}(u)$ and $\mathbf{q}_{s+1}^{(j)}(u)$ derived from (3.1) into (2.5). Recalling that $u_s - u_{s-1} = 1$ and that $M_{s;jr} = 0$ for $r > j$, we obtain

$$\sum_{i=0}^{k-1} \mathbf{V}_{s+i+1} b_{s+1;i}^{(j)}(0) - \sum_{r=0}^{n} M_{s;jr} \sum_{i=0}^{k-1} \mathbf{V}_{s+i} b_{s;i}^{(r)}(1) = 0, \qquad 0 \le j \le n. \tag{4.2}$$

(We assume that the geometric continuity degree n is given, and we have yet to establish a relation between n and k).

Since (4.2) must hold for all choices of control vertices, the coefficient of $\mathbf{V}_{s+i}$ in the lefthand side of (4.2) must be identically zero for $0 \le i \le k$. This leads to the following system of equations, equivalent to the Beta-constraints (2.8). To emphasize the link with (4.2), we label each equation with the control vertex whose coefficient is being equated to zero.

$$\mathbf{V}_s : \quad -\sum_{r=0}^{n} M_{s;jr} b_{s;0}^{(r)}(1) = 0, \quad 0 \le j \le n, \tag{4.3.a}$$

$$\mathbf{V}_{s+i+1} : \quad b_{s+1;i}^{(j)}(0) - \sum_{r=0}^{n} M_{s;jr} b_{s;i+1}^{(r)}(1) = 0, \quad 0 \le i \le k-2, \quad 0 \le j \le n, \tag{4.3.b}$$

$$\mathbf{V}_{s+k} : \quad b_{s+1;k-1}^{(j)}(0) = 0, \quad 0 \le j \le n. \tag{4.3.c}$$

Before we further investigate equations (4.3), we present a few more definitions. Let $S(k,u)$ be the $k \times k$ matrix with $(i,j)^{th}$ entry defined by

$$S_{ij}(k,u) := \frac{d^j(u^i)}{du^j} = \binom{i}{j} j! u^{i-j}, \qquad 0 \le i,j \le k-1. \tag{4.4}$$

It follows immediately from (4.4) that $S(k,u)$ is a lower triangular matrix, with main diagonal entries of the form $S_{ii}(k,u) = i!$ for $0 \le i \le k-1$. Let $S_j(k,u)$ denote the jth column of $S(k,u)$, $0 \le j \le k-1$, and let

$$\overline{S}(k,u) := [S_0(k,u) \; S_1(k,u) \; \cdots \; S_{k-2}(k,u)] \tag{4.5}$$

denote the $k \times (k-1)$ matrix obtained by deleting the last column from $S(k,u)$. For any row vector $\mathbf{v} = (v_0 \; v_1 \; \cdots \; v_r)$, let $v(u)$ denote the polynomial $v(u) = \sum_{i=0}^{r} v_i u^i$, and, conversely, given a polynomial $v(u)$, let $\mathbf{v}$ denote its vector of coefficients. It follows from (4.4) that for any polynomial $v(u)$, of degree $k-1$, we have

$$v^{(j)}(u) = \mathbf{v}S_j(k,u). \tag{4.6}$$

Lemma 1. *$S(k,u)$ is nonsingular for all u.*

We are now ready to address the "minimum spline order" design objective (ii).

Theorem 1. *The spline order k must satisfy $k \geq n+2$.*

Proof: Assume, by contradiction, that $k \leq n+1$. Then, it follows from (4.3.c) that $b^{(j)}_{s;k-1}(0) = 0$ for $0 \leq j \leq k-1$. Together with (4.6), this implies that $\mathbf{b}_{s;k-1}S(k,0) = \mathbf{0}$. Since, by Lemma 1, $S(k,0)$ is nonsingular, we must have $\mathbf{b}_{s;k-1} = \mathbf{0}$, contradicting the linear independence of $\mathbf{b}_{s;0}, \mathbf{b}_{s;1}, \ldots, \mathbf{b}_{s;k-1}$ required by design objective (iv). Hence, we must have $k \geq n+2$. ■

In the sequel, we assume $k = n+2$, thus meeting the "minimum spline order" requirement. Clearly, this also implies that $k \geq 2$.

We now resume our investigation of the Beta-constraints (4.3). Using (4.6), we observe that, for $0 \leq i \leq k-1$ and $0 \leq j \leq n$, we have

$$\sum_{r=0}^{n} M_{s;jr} b^{(r)}_{s;i}(u) = \sum_{r=0}^{n} M_{s;jr}\mathbf{b}_{s;i}S_r(k,u) = \mathbf{b}_{s;i}\sum_{r=0}^{n} M_{s;jr}S_r(k,u).$$

Setting $k = n+2$, and recalling the definition of $\overline{S}(k,u)$ in (4.5), it follows from the last equation that

$$\sum_{r=0}^{k-2} M_{s;jr} b^{(r)}_{s;i}(u) = \mathbf{b}_{s;i}\overline{S}(k,u)M^T_{s;j}, \qquad 0 \leq i \leq k-1, \quad 0 \leq j \leq k-2, \tag{4.7}$$

where $M_{s;j}$ is the jth row of M_s, and superscript T denotes transposition. Using (4.6) and (4.7), the Beta-constraints (4.3) can be rewritten as follows:

$$\mathbf{b}_{s;0}\overline{S}(k,1)M^T_{s;j} = 0, \qquad 0 \leq j \leq k-2, \tag{4.8.a}$$

$$\mathbf{b}_{s+1;i}S_j(k,0) - \mathbf{b}_{s;i+1}\overline{S}(k,1)M^T_{s;j} = 0, \qquad 0 \leq i \leq k-2, \quad 0 \leq j \leq k-2, \tag{4.8.b}$$

$$\mathbf{b}_{s+1;k-1}S_j(k,0) = 0, \qquad 0 \leq j \leq k-2. \tag{4.8.c}$$

Equation (4.8.a) represents $k-1$ scalar equations, each involving one column $M^T_{s;j}$. These can be combined into one vector equation involving the matrix M^T_s. A similar transformation can be applied to (4.8.b) and (4.8.c). Thus, we obtain the following equivalent version of the Beta-constraints:

$$\mathbf{b}_{s;0}\overline{S}(k,1)M_s^T = \mathbf{0}, \tag{4.9.a}$$

$$\mathbf{b}_{s+1;i}\overline{S}(k,0) - \mathbf{b}_{s;i+1}\overline{S}(k,1)M_s^T = \mathbf{0}, \qquad 0 \le i \le k-2, \tag{4.9.b}$$

$$\mathbf{b}_{s+1;k-1}\overline{S}(k,0) = \mathbf{0}. \tag{4.9.c}$$

The zero vectors at the right hand sides of (4.9.a)-(4.9.c) are of dimension $k-1$.

At this point, we can also address design objective (iii), "shape preservation under translation". As is well known (see, for example, page 191 of Bartels *et al.*, 1987), objective (iii) is satisfied if and only if the Beta-polynomials sum to unity. In vector notation, this implies

$$\sum_{i=0}^{k-1} \mathbf{b}_{s;i} = (1\ 0\ \cdots\ 0). \tag{4.9.d}$$

§5. Solving for the Beta-polynomials

We now solve equations (4.9) for the vectors $\mathbf{b}_{s;i}$. Let $\overline{\mathbf{b}}_{s;i}$ denote the vector obtained by deleting the last entry of $\mathbf{b}_{s;i}$; i.e.

$$\overline{\mathbf{b}}_{s;i} = (b_{s;i0}\ b_{s;i1}\ \cdots\ b_{s;i,k-2}), \qquad 0 \le i \le k-1,$$

and let $\gamma_{s;i}$ denote the last entry of $\mathbf{b}_{s;i}$; i.e.

$$\gamma_{s;i} = b_{s;i,k-1}, \qquad 0 \le i \le k-1.$$

Let $\mathbf{y}(k,u)$ denote the last row of the matrix $\overline{S}(k,u)$ defined in (4.5). It follows from (4.4) and (4.5) that

$$\overline{S}(k,u) = \begin{pmatrix} S(k-1,u) \\ \mathbf{y}(k,u) \end{pmatrix}. \tag{5.1}$$

Now, using the definitions of $\overline{\mathbf{b}}_{s;i}$ and $\gamma_{s;i}$, and (5.1), it follows from (4.9.c) (with index s decremented by one) that

$$\overline{\mathbf{b}}_{s;k-1}S(k-1,0) + \gamma_{s;k-1}\mathbf{y}(k,0) = \mathbf{0}.$$

Since $\mathbf{y}(k,0) = \mathbf{0}$, and $S(k-1,0)$ is nonsingular, this implies that we must have

$$\overline{\mathbf{b}}_{s;k-1} = \mathbf{0}. \tag{5.2}$$

One consequence of (5.2) is that $b_{s;k-1}(u)$ is always of the form $\gamma_{s;k-1}u^{k-1}$ for some $\gamma_{s;k-1} \in \mathbf{K}_n$. Using, again, (5.1) and the fact that $\mathbf{y}(k,0) = \mathbf{0}$, it follows from (4.9.b) that

$$\begin{aligned}\overline{\mathbf{b}}_{s+1;i}S(k-1,0) =&\mathbf{b}_{s;i+1}\overline{S}(k,1)M_s^T \\ =&\overline{\mathbf{b}}_{s;i+1}S(k-1,1)M_s^T + \gamma_{s;i+1}\mathbf{y}(k,1)M_s^T.\end{aligned} \tag{5.3}$$

Let A_s denote the $(k-1) \times (k-1)$ matrix over $\mathbf{K}_n$ defined by

$$A_s = S(k-1,1)M_s^T S(k-1,0)^{-1}. \tag{5.4}$$

Also, define the $(k-1)$-dimensional vector

$$\mathbf{a}_s = \mathbf{y}(k,1)M_s^T S(k-1,0)^{-1}. \tag{5.5}$$

Multiplying both sides of (5.3) to the right by $S(k-1,0)^{-1}$, and using (5.4) and (5.5), we obtain

$$\overline{\mathbf{b}}_{s+1;i} = \overline{\mathbf{b}}_{s;i+1}A_s + \gamma_{s;i+1}\mathbf{a}_s.$$

Now, decrementing the index s by one yields

$$\overline{\mathbf{b}}_{s;i} = \overline{\mathbf{b}}_{s-1;i+1}A_{s-1} + \gamma_{s-1;i+1}\mathbf{a}_{s-1}, \qquad 0 \le i \le k-2. \tag{5.6}$$

Equation (5.6) can be used recursively $k-i-1$ times, yielding

$$\begin{aligned}\overline{\mathbf{b}}_{s;i} =&\overline{\mathbf{b}}_{s-(k-i-1);k-1}A_{s-(k-i-1)}A_{s-(k-i-2)}\cdots A_{s-1} \\ &+\gamma_{s-(k-i-1);k-1}\mathbf{a}_{s-(k-i-1)}A_{s-(k-i-2)}A_{s-(k-i-3)}\cdots A_{s-1} + \cdots + \\ &+\gamma_{s-3;i+3}\mathbf{a}_{s-3}A_{s-2}A_{s-1} + \gamma_{s-2;i+2}\mathbf{a}_{s-2}A_{s-1} + \gamma_{s-1;i+1}\mathbf{a}_{s-1}.\end{aligned} \tag{5.7}$$

Applying (5.2) to index $s-(k-i-1)$ instead of s, we have $\overline{\mathbf{b}}_{s-(k-i-1);k-1} = \mathbf{0}$. Hence, the first term on the righthand side of (5.7) vanishes. To abbreviate notation, let $A(s;j)$ denote the matrix

$$A(s;j) = \begin{cases} I, & j = 0 \\ A_{s-j+1}A_{s-j+2}\cdots A_s, & j \ge 1. \end{cases} \tag{5.8}$$

Then, (5.7) can be rewritten as

$$\overline{\mathbf{b}}_{s;i} = \sum_{j=0}^{k-i-2} \gamma_{s-j-1;i+j+1}\mathbf{a}_{s-j-1}A(s-1;j), \qquad 0 \le i \le k-2. \tag{5.9}$$

Define

$$\overline{B}_s = \begin{pmatrix} \overline{\mathbf{b}}_{s;0} \\ \overline{\mathbf{b}}_{s;1} \\ \vdots \\ \overline{\mathbf{b}}_{s;k-2} \end{pmatrix}. \tag{5.10}$$

$\overline{B}_s$ is the $(k-1) \times (k-1)$ matrix at the upper left corner of B_s. Together with (5.2) and the definition of $\gamma_{s;i}$, we have

$$B_s = \begin{pmatrix} & & & & \gamma_{s;0} \\ & & & & \gamma_{s;1} \\ & & \overline{B}_s & & \\ & & & & \vdots \\ & & & & \gamma_{s;k-2} \\ 0 & 0 & \cdots & 0 \quad 0 & \gamma_{s;k-1} \end{pmatrix}. \tag{5.11}$$

Define also the $(k-1) \times (k-1)$ matrices

$$\Gamma_s = \begin{pmatrix} \gamma_{s-1;1} & \gamma_{s-2;2} & \cdots & \gamma_{s-k+2;k-2} & \gamma_{s-k+1;k-1} \\ \gamma_{s-1;2} & \gamma_{s-2;3} & \cdots & \gamma_{s-k+2;k-1} & 0 \\ \vdots & \vdots & & \vdots & \vdots \\ \gamma_{s-1;k-2} & \gamma_{s-2;k-1} & \cdots & 0 & 0 \\ \gamma_{s-1;k-1} & 0 & \cdots & 0 & 0 \end{pmatrix} \tag{5.12}$$

and

$$E_s = \begin{pmatrix} \mathbf{a}_{s-1} \\ \mathbf{a}_{s-2}A(s-1;1) \\ \mathbf{a}_{s-3}A(s-1;2) \\ \vdots \\ \mathbf{a}_{s-k+1}A(s-1;k-2) \end{pmatrix}. \tag{5.13}$$

Then, (5.9) can be rewritten in matrix form as

$$\overline{B}_s = \Gamma_s E_s. \tag{5.14}$$

Notice that E_s is defined in terms of the known vectors $\mathbf{a}_{s-j-1}$ and matrices $A(s-1;j)$, $0 \le j \le k-2$. Hence, to make (5.14) an explicit expression for the matrix $\overline{B}_s$, we need to determine the entries of Γ_s, namely $\gamma_{s-j-1;i+j+1}$ for $0 \le i \le k-2$, $0 \le j \le k-i-2$. If we also determine the parameters $\gamma_{s;i}$ for $0 \le i \le k-1$ then, by (5.11), the matrix B_s of coefficients of the Beta-polynomials will be completely determined.

Equation (5.14) was derived from (4.9.b)-(4.9.c). We still need to take into account the constraints imposed by (4.9.a) and (4.9.d). To achieve this goal, we need the result presented in the following lemma, a proof of which can be found in Seroussi & Barsky (1991), based on symbolic substitution arguments.

Lemma 2. *Let E_s be the matrix defined in (5.13). Then, E_s is nonsingular over $\mathbf{K}_n$.*

We are now ready to address constraint (4.9.d). It follows from (5.11) and (5.14) that

$$\sum_{i=0}^{k-1} \mathbf{b}_{s;i} = \left[\mathbf{1}\overline{B}_s \quad \sum_{j=0}^{k-1} \gamma_{s;j}\right] = \left[\mathbf{1}\Gamma_s E_s \quad \sum_{j=0}^{k-1} \gamma_{s;j}\right], \tag{5.15}$$

where $\mathbf{1}$ is a row vector of $(k-1)$ ones, and $[\mathbf{v} \quad x]$ denotes the vector obtained by appending x to $\mathbf{v}$. Hence, for (4.9.d) to hold, we must have

$$\mathbf{1}\Gamma_s E_s = (1\ 0\ \cdots\ 0), \tag{5.16}$$

and

$$\sum_{j=0}^{k-1} \gamma_{s;j} = 0. \tag{5.17}$$

By Lemma 2, (5.16) is equivalent to

$$\mathbf{1}\Gamma_s = (1\ 0\ \cdots\ 0)E_s^{-1}. \tag{5.18}$$

Let

$$\mathbf{w}_s := (w_{s;0} w_{s;1} \cdots w_{s;k-2}) = (1\ 0\ \cdots\ 0)E_s^{-1}. \tag{5.19}$$

Then, (5.16) is equivalent to

$$\mathbf{1}\Gamma_s = \mathbf{w}_s. \tag{5.20}$$

Recalling the definition of Γ_s in (5.12), equation (5.20) is equivalent to

$$\sum_{i=0}^{k-j-2} \gamma_{s-j-1;i+j+1} = w_{s;j}, \qquad 0 \le j \le k-2. \tag{5.21}$$

Equation (5.21) can be used to solve for the parameters $\gamma_{s-j;j}$. By direct application of (5.21), one can readily verify that

$$\gamma_{s-j;j} = \begin{cases} w_{s;j-1} - w_{s+1;j}, & 1 \le j \le k-2 \\ w_{s;k-2}, & j = k-1. \end{cases} \tag{5.22}$$

Translated to matrix form, (5.22) gives an expression for Γ_s. Consider the $(k-2) \times (k-2)$ matrix

$$W_s = \begin{pmatrix} w_{s+1;1} & w_{s+1;2} & \cdots & w_{s+1;k-2} & 0 \\ w_{s+2;2} & w_{s+2;3} & \cdots & 0 & 0 \\ \vdots & \vdots & \vdots & \vdots & \vdots \\ w_{s+k-3;k-3} & w_{s+k-3;k-2} & \cdots & 0 & 0 \\ w_{s+k-2;k-2} & 0 & \cdots & 0 & 0 \end{pmatrix}. \tag{5.23}$$

Then, it follows from (5.12) and (5.22) that

$$\Gamma_s = \begin{pmatrix} \mathbf{w}_s \\ W_s \end{pmatrix} - \begin{pmatrix} W_s \\ \mathbf{0} \end{pmatrix}. \tag{5.24}$$

Since E_s, E_s^{-1}, and $\mathbf{w}_s$ are known for all s, equation (5.22) uniquely determines the parameters $\gamma_{s;j}$, for all s and $1 \le j \le k-1$. Hence, the matrices Γ_s and $\overline{B}_s$ in (5.14), are uniquely determined, and so is the last column of B_s in (5.11), except for the first row entry $\gamma_{s;0}$. The latter is determined by (5.17), which dictates

$$\gamma_{s;0} = -\sum_{j=1}^{k-1} \gamma_{s;j}. \tag{5.25}$$

The preceding discussion is summarized in the following lemma.

Lemma 3. *There is one and only one solution B_s to equations (4.9.b)-(4.9.d). The solution is given by equations (5.11)-(5.14), (5.19), and (5.22)-(5.25).*

Notice that the solution B_s was uniquely determined from (4.9.b)-(4.9.d), without taking into account (4.9.a). Hence, the system (4.9) appears to be over-determined. However, the following lemma shows this not to be the case.

Lemma 4. *The solution B_s of Lemma 3 is consistent with (4.9.a).*

Finally, the following lemma establishes the linear independence of the Beta-polynomials, as required by design objective (iv).

Lemma 5. *B_s is nonsingular.*

The proofs of Lemmas 4 and 5 are presented in Seroussi & Barsky (1991). To summarize the results of this section, we present the following theorem.

Theorem 2. *There is one and only one set of Beta-polynomials $b_{s;0}(u)$, $b_{s;1}(u)$, $\ldots$, $b_{s;k-1}(u)$, with coefficients in $\mathbf{K}_n$, satisfying the design objectives (i)-(iv). The following procedure outlines the symbolic computation of the matrix of coefficients B_s of these polynomials:*

Procedure $\beta 1$: Computation of Beta-polynomials for uniform knot spacing.

1. Compute M_s, using either the chain rule for derivatives or (2.6).
2. Compute $S(k-1,0)$ and $S(k-1,1)$ using (4.4), and A_s using (5.4). Compute $\mathbf{a}_s$ using (5.5). Notice that once A_s and $\mathbf{a}_s$ are computed for a given index s, A_t and $\mathbf{a}_t$ are obtained for any index t by substituting $\beta_{t;j}$ for $\beta_{s;j}$ in A_s (respectively $\mathbf{a}_s$) for $1 \le j \le n$, without having to repeat the computation in (5.4) (respectively (5.5)). Notice also that A_s and $\mathbf{a}_s$ depend only on $\boldsymbol{\beta}_s$.
3. Compute E_s using (5.8) and (5.13).
4. Compute $\mathbf{w}_s = (1\ 0\ \cdots\ 0)E_s^{-1}$. Once $\mathbf{w}_s$ is known for a given index s, $\mathbf{w}_t$ is obtained, for any t, by symbolically substituting $\beta_{l+t-s;j}$ for $\beta_{l;j}$ in $\mathbf{w}_s$, for all l and $1 \le j \le n$.
5. Compute Γ_s using (5.23) and (5.24).
6. Compute $\overline{B}_s$ using (5.14).

7. Obtain $\gamma_{s;1}, \gamma_{s;2}, \ldots, \gamma_{s;k-1}$ from the first column of Γ_{s+1} (see (5.12)), and $\gamma_{s;0}$ from (5.25).
8. Compute B_s using (5.11).

For the sake of clarity, Procedure $\beta 1$ follows the derivation leading to Theorem 2 in a straightforward manner, and is not necessarily optimal from a computational point of view. The procedure was programmed in the symbolic manipulation system Mathematica^{TM}, Release 1.5 (Mathematica is a trademark of Wolfram Research Inc.), and Beta-polynomials for various orders were computed. Some examples of the resulting expressions can be found in Seroussi & Barsky (1991).

§6. Conclusion

We have presented a symbolic derivation of discretely-shaped Beta-spline bases of arbitrary order, over uniform knot sequences. The derivation led to explicit procedures for the symbolic computation of the Beta-polynomials and, hence, of the Beta-spline basis functions. The results can be generalized to the case of non-uniform knot sequences, as discussed in Seroussi & Barsky (1991). The latter report also shows how certain properties of Beta-splines, such as local control and the equivalence between uniform and non-uniform knot spacings, follow naturally from the symbolic framework. Other properties involving inequalities, such as the convex hull property, or the variation-diminishing property are less amenable to symbolic treatment, and are better dealt with using classical approximation theory methods (Dyn & Micchelli, 1988; Goodman, 1985). It should be noted, however, that the Beta-spline basis functions derived in this paper exist and are unique for all values of the shape parameters that do not make the denominators of the rational functions in the matrix B_s vanish, including values of the shape parameters for which the Beta-connection matrix M_s is not totally positive and hence, for which the results of Dyn & Micchelli (1988) and Goodman (1985) do not guarantee the variation-diminishing or convex hull properties.

The derivation and proofs leading to Theorem 2 and Procedure $\beta 1$ were carried out for geometric continuity based on the Beta-constraints (2.8) defined by the Beta-connection matrix $M_s(\boldsymbol{\beta}_s)$. However, a closer examination of the proofs reveals that only the following properties of the connection matrix M_s are required:

1. The existence of a substitution of the symbolic indeterminates which transforms M_s into an identity matrix (this substitution is used in the proofs of Lemmas 2, 3 and 5), and
2. The fact that the first column of M_s is $(1\ 0\ \ldots\ 0)^T$, which implies that the first row of A_s is $(1\ 0\ \ldots\ 0)$ (a necessary condition for the consistency proof in Lemma 4).

Therefore, the results presented in this paper are valid also for a more general class of connection matrices including, in particular, Frenet-frame connection matrices (Dyn & Micchelli, 1988; Seidel, 1990) of the form

$$C_s = \begin{pmatrix} 1 & 0 & 0 & 0 & \cdots \\ 0 & c_{s;1,1} & 0 & 0 & \cdots \\ 0 & c_{s;2,1} & c_{s;1,1}^2 & 0 & \cdots \\ 0 & c_{s;3,1} & c_{s;3,2} & c_{s;1,1}^3 & \cdots \\ \vdots & \vdots & \vdots & \vdots & \vdots \end{pmatrix}.$$

Acknowledgments. The second author was supported in part by the Hewlett-Packard Laboratories and by the National Science Foundation under grant number CCR-9015874.

References

1. Barsky, B.A., The Beta-spline: A local representation based on shape parameters and fundamental geometric measures, Ph.D. Thesis, University of Utah, Salt Lake City, Utah, December, 1981.
2. Barsky, B.A., *Computer Graphics and Geometric Modeling Using Beta-splines*, Springer-Verlag, Heidelberg, 1988.
3. Barsky, B.A., Author's comment: update on β-splines, letter to the editor, Computer-Aided Design **21** (1989), 472.
4. Barsky, B.A., and T. D. DeRose, Geometric continuity of parametric curves, Technical Report No. UCB/CSD 84/205, Computer Science Division, Electrical Engineering and Computer Sciences Department, University of California, Berkeley, October, 1984.
5. Barsky, B.A., and T. D. DeRose, Geometric continuity of parametric curves: Three equivalent characterizations, IEEE Computer Graphics and Applications **9** (1989), 60–68.
6. Barsky, B.A., and T. D. DeRose, Geometric continuity of parametric curves: Constructions of geometrically continuous splines, IEEE Computer Graphics and Applications **10** (1990), 60–68.
7. Bartels, R.H., J.C. Beatty, and B.A. Barsky, *An Introduction to Splines for Use in Computer Graphics and Geometric Modeling*, Morgan Kaufmann Publishers, Inc., San Mateo, California, 1987.
8. Bézier P.E., *Emploi des machines à commande numérique*, Masson et Cie., Paris, 1970. Translated by A. Robin Forrest and Anne F. Pankhurst as *Numerical Control - Mathematics and Applications*, John Wiley & Sons, New York, 1972.
9. de Boor, C., On calculating with B-splines, Journal of Approximation Theory, **6** (1972), 50–62.
10. de Boor, C., *A Practical Guide to Splines*, Springer-Verlag, New York, 1978.
11. Dyn, N., E. Edelman, and C.A. Micchelli, On locally supported basis functions for the representation of geometrically continuous curves, Analysis **7** (1987), 313–341.
12. Dyn, N., and C.A. Micchelli, Piecewise polynomial spaces and geometric continuity of curves, Numerische Mathematik, **54** (1988), 319–333.

13. Geise, G., Ueber beruehrende Kegelschnitte einer ebenen Kurve, Zeitschrift für Angewandte Mathematik und Mechanik **42** (1962), 297–304.
14. Goldman, R.N., and B. Barsky, On β-continuous functions and their application to the construction of geometrically continuous curves and surfaces, in *Mathematical Methods in Computer Aided Geometric Design*, T. Lyche, Tom and L. Schumaker (eds.), Academic Press, Boston, 1989, 299–311.
15. Goodman, T.N.T., Properties of Beta-splines, Journal of Approximation Theory **44** (1985), 132–153.
16. Goodman, T.N.T., and K. Unsworth, Manipulating shape and producing geometric continuity in Beta-spline curves, IEEE Computer Graphics and Applications **6** (1986), 50–56.
17. Gregory, J.A., Geometric continuity, in *Mathematical Methods in Computer Aided Geometric Design*, T. Lyche and L. Schumaker (eds.), Academic Press, Boston, 1989, 353–371.
18. Herron, G., Techniques for visual continuity, in *Geometric Modeling: Algorithms and New Trends*, Gerald Farin (ed.), SIAM, Philadelpia, 1987, 163–174.
19. Herstein, I.N., *Topics in Algebra*, John Wiley & Sons, New York, 1975.
20. Höllig, K., Geometric continuity of spline curves and surfaces, Technical Report No. 645, Computer Science Department, University of Wisconsin, Madison, 1986.
21. Knuth, D.E., *The Art of Computer Programming: Volume 1, Fundamental Algorithms*, Addison-Wesley Publishing Company, Reading, Massachusetts, 1973.
22. Lempel, A., and G. Seroussi, Systematic derivation of spline bases, Technical Report No. HPL-90-34, Hewlett-Packard Laboratories, May, 1990.
23. Riesenfeld, R.F., Applications of B-spline approximation to geometric problems of computer-aided design, Ph.D. Thesis, Syracuse University, Syracuse, N.Y., 1973. Available as Tech. Report No. UTEC-CSc-73-126, Department of Computer Science, University of Utah.
24. Scheffers, G., *Anwendung der Differential- und Integralrechnung auf die Geometrie, Bd. I: Einfuehrung in die Theorie der Kurven in der ebene und im Raume*, B.G. Teubner-Verlag, Berlin and Leipzig, 1910.
25. Schumaker, L.L., *Spline Functions: Basic Theory*, John Wiley and Sons, Inc., New York, 1981.
26. Seidel, H.P., Geometric constructions and knot insertion for geometrically continuous spline curves of arbitrary degree, Research Report No. CS-90-24, University of Waterloo, Waterloo, Ontario, June, 1990.
27. Seroussi, G. and B. A. Barsky, A symbolic derivation of Beta-splines of arbitrary order, Technical Report No. HPL-91-87(R.1), Hewlett-Packard Laboratories, June 1991. Also, Technical Report UCB/CSD 91/633, Computer Science Division, Electrical Engineering and Computer Sciences Department, University of California, Berkeley.

Discrete Convolution Schemes

Yohanes Stefanus and Ronald Goldman

Abstract. The Bernstein basis functions $\{B_k^n(t)\}$ can be generated by taking the discrete convolution of the sequence $\langle(1-t), t\rangle$ with itself $(n-1)$-times. The convolution blending functions $\{C_k^n(t)\}$ are generated by replacing the parameter t by an arbitrary linear function $X_j(t)$ in each factor of the discrete convolution. Under certain mild restrictions on these linear functions, the polynomials $C_k^n(t)$ form a basis for the space of polynomials of degree n. We investigate how this convolution basis extends many of the properties of the Bernstein basis, and we show that as generalizations of Bézier curves, the convolution curves $\sum_k P_k C_k^n(t)$ have many interesting and desirable properties for computer aided geometric design.

§1. Introduction

The theory of Bernstein polynomials and Bézier curves plays a central role in computer aided geometric design (CAGD). Bézier curves and surfaces are the mathematical basis of many CAGD systems and they also serve as a major paradigm for the development of new schemes for curves and surfaces [1].

The focus of this paper is a generalization of the Bézier scheme developed by extending the underlying *convolution* structure of the Bernstein basis functions. The highlights are:

- The new convolution scheme generalizes the Bézier scheme.
- Matrices whose entries are convolution functions provide the transformations from the Pólya representation to the Bézier representation and from the Bézier representation to the B-spline representation.
- The convolution functions are a special case of the discrete B-splines.
- The convolution scheme is closely related to the *blossoming* technique.
- A convolution curve is the locus of a B-spline control point of a Bézier curve.

In Section 2 we briefly review the concept of discrete convolution. We then show in Section 3 that the Bernstein basis has a convolution structure.

Mathematical Methods in Computer Aided Geometric Design II

Tom Lyche and Larry L. Schumaker (eds.), pp. 585–596.

ISBN 0-12-460510-9.

In Section 4 the convolution basis functions are defined in three ways — by a recurrence, from generating functions, and through blossoming. We present the basic properties of the convolution basis functions in Section 5. In Sections 6 to 9 we discuss the identities and formulas for the convolution basis functions listed in Table 1. Then we describe the properties of convolution curves in Section 10. Finally, in Section 11 we provide a brief summary of our results.

§2. Discrete Convolution

A sequence $G = \langle g_0, g_1, g_2, \ldots \rangle$ can conveniently be represented as a power series in an auxiliary variable z,

$$G(z) \;=\; g_0 + g_1 z + g_2 z^2 + \cdots \;=\; \sum_{k\geq 0} g_k z^k.$$

$G(z)$ is referred to as the **generating function** for the sequence G. When G is a finite sequence, the generating function $G(z)$ is a polynomial in z.

Definition 1. *Given two sequences $A = \langle a_0, a_1, a_2, \ldots \rangle$ and $B = \langle b_0, b_1, b_2, \ldots \rangle$, the convolution sequence $C = \langle c_0, c_1, c_2, \ldots \rangle$ obtained from A and B is defined by*

$$c_k = \sum_{i=0}^{k} a_i b_{k-i}. \tag{1}$$

We shall write $C = A * B$ to denote that C is the convolution of A and B. It is not difficult to verify the following basic properties of convolution.

- The generating function of the convolution of two sequences is equal to the product of their generating functions
- $A * B = B * A$ (commutativity)
- $A * (B * C) = (A * B) * C$ (associativity)
- Let the derivative of a sequence of functions A be the corresponding sequence of derivatives; i.e.,

$$A' = \langle a_0, a_1, a_2, \ldots \rangle' = \langle a_0{}', a_1{}', a_2{}', \ldots \rangle.$$

Then for n sequences $A_1, A_2, \ldots, A_n$, we have

$$(A_1 * A_2 * \cdots * A_n)' = $$
$$A_1{}' * A_2 * \cdots * A_n \;+\; A_1 * A_2{}' * \cdots * A_n \;+\; \cdots \;+\; A_1 * A_2 * \cdots * A_n{}'. \tag{2}$$

§3. Bernstein Basis as Discrete Convolution

In this section we examine the convolution structure of the sequence of Bernstein basis functions. It is this structure that we shall preserve when we generalize the Bernstein basis to the convolution basis in the next section.

The sequence of Bernstein basis functions of degree n is

$$B^n(t) = \langle B_0^n(t), B_1^n(t), \ldots, B_n^n(t)\rangle,$$

where $B_k^n(t) = \binom{n}{k} t^k (1-t)^{n-k}$ for $k = 0, 1, \ldots, n$ and $\binom{n}{k} = \frac{n!}{k!(n-k)!}$. The polynomials $B_k^n(t)$ satisfy the recurrence [9]

$$B_k^n(t) = (1-t)B_k^{n-1}(t) + tB_{k-1}^{n-1}(t) \tag{3}$$

with $B_0^0(t) \equiv 1$.

Using the recurrence relation, we can easily verify by induction on n that the sequence $B^n(t)$ is the n-fold convolution of the sequence $\langle 1-t, t\rangle$, where the 0-fold and the 1-fold convolution are defined as the sequence $\langle 1\rangle$ and the sequence $\langle 1-t, t\rangle$ respectively. In other words,

$$B^n(t) = \underbrace{\langle 1-t, t\rangle * \cdots * \langle 1-t, t\rangle}_{n}.$$

This convolution structure can be used to simplify the analysis of the Bernstein basis functions. For example, by (2) and the commutativity of convolution:

$$\begin{aligned}\langle \frac{\mathrm{d}}{\mathrm{d}t}B_0^n(t), \frac{\mathrm{d}}{\mathrm{d}t}B_1^n(t), \ldots, \frac{\mathrm{d}}{\mathrm{d}t}B_n^n(t)\rangle \\ &= n(\langle -1, 1\rangle * \underbrace{\langle 1-t, t\rangle * \cdots * \langle 1-t, t\rangle}_{n-1}) \\ &= n\langle -1, 1\rangle * B^{n-1}(t).\end{aligned} \tag{4}$$

From this identity we can immediately conclude that

$$\frac{\mathrm{d}}{\mathrm{d}t}B_k^n(t) = n[B_{k-1}^{n-1}(t) - B_k^{n-1}(t)].$$

The generating function for the sequence $\langle 1-t, t\rangle$ is clearly $(1-t+tz)$. Hence, by the multiplicative property of generating functions, the generating function for the sequence $\langle B_0^n(t), B_1^n(t), \ldots, B_n^n(t)\rangle$ is $(1-t+tz)^n$.

§4. Convolution Basis Functions

In this section we construct the convolution basis functions $C_k^n(t)$, a generalization of the Bernstein basis functions $B_k^n(t), k = 0, 1, \ldots, n$. There are three ways to define the convolution basis functions: by a recurrence, from generating functions, and through blossoming.

4.1 Recurrence

To generate the $C_k^n(t)$, we replace the parameter t in the recursion formula for $B_k^n(t)$ by n functions of t, $X_1(t), \ldots, X_n(t)$, where a different function $X_k(t)$ appears on each level of the recursion. Thus, as a generalization of (3), the recursion formula for $C_k^n(t)$ is

$$C_k^n(t) \;=\; (1 - X_n(t))\, C_k^{n-1}(t) \;+\; X_n(t)\, C_{k-1}^{n-1}(t) \tag{5}$$

with $C_0^0(t) \equiv 1$.

If $X_1(t), \ldots, X_n(t)$ are linear, the polynomials $C_k^n(t)$ have degree n. In Section 5 we will prove that under certain mild restrictions on these linear functions the polynomials $C_k^n(t)$ form a basis for the vector space of polynomials of degree n. From now on, unless explicitly stated otherwise, we assume that $X_k = a_k t + b_k$ where a_k and b_k are constants and $a_k \neq 0$. Sometimes we shall use the notation

$$C_k^n(X_1, \ldots, X_n)$$

to make explicit the argument functions. For instance, for $n = 3$ we have:

- $C_0^3(t) = (1 - X_1)(1 - X_2)(1 - X_3)$
- $C_1^3(t) = (1-X_1)(1-X_2)X_3 + (1-X_1)X_2(1-X_3) + X_1(1-X_2)(1-X_3)$
- $C_2^3(t) = (1 - X_1)X_2X_3 + X_1(1 - X_2)X_3 + X_1X_2(1 - X_3)$
- $C_3^3(t) = X_1X_2X_3$

As a generalization of the convolution structure of Bernstein basis functions, the sequence $C^n(t) = \langle C_0^n(t), C_1^n(t), \ldots, C_n^n(t)\rangle$ is the convolution of the n sequences

$$\langle 1 - X_k, X_k\rangle, \qquad k = 1, \ldots, n;$$

that is,

$$C^n(t) = \langle 1 - X_1, X_1\rangle * \langle 1 - X_2, X_2\rangle * \cdots * \langle 1 - X_n, X_n\rangle.$$

We can easily verify this result from (5) by induction on n. This justifies the name "convolution" for the functions $C_k^n(t)$.

4.2 Generating Functions

The generating function for the sequence $\langle 1-X_k, X_k\rangle$ is simply $(1-X_k+X_k z)$. Therefore, by the multiplicative property of generating functions, the sequence $\langle C_0^n(t), C_1^n(t), \ldots, C_n^n(t)\rangle$ can be generated from

$$C(z) \;=\; \prod_{k=1}^{n} (1 - X_k + X_k z), \tag{6}$$

that is,

$$\sum_{i=0}^{n} C_i^n(t) z^i \;=\; \prod_{k=1}^{n} (1 - X_k + X_k z), \tag{7}$$

or equivalently,

$$C_k^n(t) = \frac{1}{k!} C^{(k)}(z)\mid_{z=0} .$$

4.3 Blossoms of Bernstein Basis Functions

In Section 4.1 we established that the sequence $\langle C_0^n(t), C_1^n(t), \ldots, C_n^n(t)\rangle$ is equal to $\langle 1 - X_1, X_1\rangle * \cdots * \langle 1 - X_n, X_n\rangle$. Here we want to focus on three properties of $C_k^n(t)$ that follow from this convolution structure.

(1) $C_k^n(t)$ is a symmetric function with respect to $X_1, \ldots, X_n$, since convolution is commutative.

(2) $C_k^n(t)$ reduces to the Bernstein basis function $B_k^n(t)$ when $X_1 = X_2 = \cdots = X_n = t$, since $\langle B_0^n(t), \ldots, B_n^n(t)\rangle$ is equal to

$$\underbrace{\langle 1 - t, t\rangle * \cdots * \langle 1 - t, t\rangle}_{n}.$$

(3) $C_k^n(t)$ is multiaffine with respect to $X_1, \ldots, X_n$, since each X_i appears at most to the first power.

But these properties correspond exactly to those of a *blossom* [8,15]. (The *blossom* of a degree n polynomial $p(t)$ is the unique symmetric multiaffine polynomial $\mathcal{B}[p](u_1, \ldots, u_n)$ for which $\mathcal{B}[p](t, \ldots, t) = p(t)$.) Hence, the convolution basis functions can be obtained by blossoming the Bernstein basis functions and evaluating the blossoms off the diagonal, i.e.,

$$C_k^n(t) = \mathcal{B}[B_k^n(x)](X_1, \ldots, X_n). \tag{8}$$

§5. Basic Properties of the Convolution Basis Functions

5.1 Polynomial Basis

In this subsection we shall prove that under some mild conditions on the functions $X_1, \ldots, X_n$, the polynomials $C_k^n(t)$ form a basis for the space of polynomials of degree n. The approach is to look at *change of basis* to and from the monomial form.

Theorem 1. *The set of convolution polynomials $\{C_k^n(t) \mid k = 0, 1, \ldots, n\}$ is a basis for the space of polynomials of degree n if and only if the elementary symmetric functions $s_j(a_1, \ldots, a_n) \neq 0$ for $j = 1, \ldots, n$.*

Proof: Start with the identity for Bernstein basis functions [9]:

$$\sum_{k=j}^{n} \binom{k}{j} B_k^n(t) = \binom{n}{j} t^j.$$

Blossoming both sides with respect to t, evaluating at $X_1, \ldots, X_n$, and applying (8), we obtain

$$\sum_{k=j}^{n} \binom{k}{j} C_k^n(t) = s_j(X_1, \ldots, X_n). \tag{9}$$

The right hand side is a polynomial of degree j in t with the leading term $s_j(a_1,\dots,a_n)\,t^j$. Thus, if $s_j(a_1,\dots,a_n)\neq 0$, then the degree of the polynomial $\sum_{k=j}^{n}\binom{k}{j}C_k^n(t)$ is exactly equal to j. Hence $\{\sum_{k=j}^{n}\binom{k}{j}C_k^n(t)\mid j=0,\dots,n\}$ is a basis for the space of polynomials of degree n, and consequently so is $\{C_k^n(t)\mid k=0,\dots,n\}$.

Conversely, suppose that $s_p(a_1,\dots,a_n)=0$, for some $p\le n$. Then for $j=0,\dots,p$ the right hand side of (9) provides $p+1$ polynomials in t which are linearly dependent since t^p is missing. Let the j-th polynomial be denoted by $P_j(t)$. Then there exist constants K_j not all equal to zero such that

$$\sum_{j=0}^{p} K_jP_j(t)=0.$$

By (9), this corresponds to

$$(C_0^n\ \cdots\ C_n^n)\begin{pmatrix}1 & 0 & \cdots & 0\\ 1 & \binom{1}{1} & \cdots & 0\\ 1 & \binom{2}{1} & \cdots & 0\\ \vdots & \vdots & \vdots & \vdots\\ 1 & \binom{n}{1} & \cdots & \binom{n}{p}\end{pmatrix}\begin{pmatrix}K_0\\ K_1\\ \vdots\\ K_p\end{pmatrix}=0.$$

But

$$\begin{pmatrix}1 & 0 & \cdots & 0\\ 1 & \binom{1}{1} & \cdots & 0\\ 1 & \binom{2}{1} & \cdots & 0\\ \vdots & \vdots & \vdots & \vdots\\ 1 & \binom{n}{1} & \cdots & \binom{n}{p}\end{pmatrix}\begin{pmatrix}K_0\\ K_1\\ \vdots\\ K_p\end{pmatrix}=\begin{pmatrix}K_0\\ K_0+\binom{1}{1}K_1\\ K_0+\binom{2}{1}K_1+\binom{2}{2}K_2\\ \vdots\\ K_0+\binom{n}{1}K_1+\cdots+\binom{n}{p}K_p\end{pmatrix}$$

is not zero, because if this vector is zero then all the K_j are zero, which is not the case. Thus when $s_p(a_1,\dots,a_n)=0$, $C_0^n(t),\dots,C_n^n(t)$ are linearly dependent and hence do not form a basis. ■

5.2 Symmetry with Respect to Arguments

By the commutativity of convolution, $C_k^n(t)$ does not depend on the order of $X_1,\dots,X_n$, i.e., it is a symmetric function over $X_1,\dots,X_n$.

5.3 Partition of Unity

If z is set to 1, (7) becomes $\sum_{i=0}^{n}C_i^n(t)\;=\;1$.

5.4 Nonnegativity

Let r_k be the root of X_k and s_k be the root of $1-X_k$. If a_k are all positive, let L be the maximum of $\{r_k\mid k=1,\dots,n\}$ and U be the minimum of $\{s_k\mid k=1,\dots,n\}$; if a_k are all negative, let L be the maximum of $\{s_k\mid k=1,\dots,n\}$ and U be the minimum of $\{r_k\mid k=1,\dots,n\}$. Then $[L,U]$ is the largest interval where all the functions $C_0^n(t), \ldots, C_n^n(t)$ are nonnegative. The interval $[L, U]$ is vacuous if $L > U$.

Notice that if the a_k are not all of the same sign, then the interval $[L, U]$ is vacuous. And since the case where all a_k are negative is the mirror image of the case where all a_k are positive, we can assume without loss of generality that from now on, unless stated explicitly otherwise, all a_k are positive.

As will be discussed later, this nonnegativity is very important for constructing curves with certain desirable properties from these convolution basis functions. Therefore, $[L, U]$ is the interval of interest.

5.5 Total Positivity

A matrix is called **totally positive** if all its minors of every order are nonnegative. The total positivity of the collocation matrix for a set of basis functions implies that the basis functions satisfy Descartes' Law of Signs [11,16].

We conjecture that the convolution collocation matrix is totally positive over the parameter interval $[L, U]$. Unfortunately, this conjecture for arbitrary degree is still an open problem. The case of degree $n = 1$ is trivial. A detailed proof for degrees $n = 2$ and $n = 3$ can be found in [17].

5.6 Unimodality

If the convolution basis functions satisfy Descartes' Law of Signs in the interval $[L, U]$, then in this interval they are *unimodal* in t, i.e., they have a single local maximum.

The polynomials $\{C_k^n(t) \mid k = 0, 1, \ldots, n\}$ are also unimodal in k, i.e., for any fixed $t \in [L, U]$ there is an index k such that

$$C_0^n(t) \leq \cdots \leq C_{k-1}^n(t) \leq C_k^n(t) \geq C_{k+1}^n(t) \geq \cdots \geq C_n^n(t).$$

This property can be verified by induction on n using (5).

§6. Identities for the Convolution Basis Functions

Many identities involving the Bernstein basis functions are known [9]. Equation 8 allows us to extend these identities to the convolution basis functions by the blossoming technique. Table 1 summarizes these results. Each entry consists of two parts: the *first part* for the Bernstein basis functions, the *second part* for the convolution basis functions. Many of these identities can also be derived directly from the the generating functions presented earlier. Below we shall elaborate on some of these formulas. In the table, $\hat{X}$ means X is not present; π denotes a permutation function; $\sum_{\{\sigma_1,\ldots,\sigma_j\}}$ means summing over all j-subsets $\{\sigma_1, \ldots, \sigma_j\}$ of the set $\{1, 2, \ldots, n\}$ where the complements are $\{\sigma_{j+1}, \ldots, \sigma_n\}$; and Per denotes the *permanent* [14] of a matrix.

§7. Marsden's Identity and Convolution

Entry 10a in Table 1 is Marsden's Identity [13] for the Bernstein basis functions. If the X_i in identity 10b are constants, the left hand side is essentially a

	Table 1
1.a	$\sum_{k=0}^{n} B_k^n(t) = 1$
b	$\sum_{k=0}^{n} C_k^n(t) = 1$
2.a	$\sum_{k=0}^{n}(-1)^k B_k^n(t) = (1-2t)^n$
b	$\sum_{k=0}^{n}(-1)^k C_k^n(t) = \prod_{i=1}^{n}(1-2X_i)$
3.a	$B_k^n(t) = \binom{n}{k} t^k (1-t)^{n-k}$
b	$C_k^n(t) = \sum_{\{\sigma_1,\ldots,\sigma_k\}} X_{\sigma_1}\cdots X_{\sigma_k}(1-X_{\sigma_{k+1}})\cdots(1-X_{\sigma_n})$
4.a	$\sum_{k=j}^{n}\binom{k}{j} B_k^n(t) = \binom{n}{j} t^j$
b	$\sum_{k=j}^{n}\binom{k}{j} C_k^n(t) = \sum_{\{\sigma_1,\ldots,\sigma_j\}} X_{\sigma_1}\cdots X_{\sigma_j}$
5.a	$\frac{\mathrm{d}B_k^n(t)}{\mathrm{d}t} = n(B_{k-1}^{n-1}(t) - B_k^{n-1}(t))$
b	$\frac{\partial}{\partial X_j} C_k^n(X_1,\ldots,X_n)$ $= C_{k-1}^{n-1}(X_1,\ldots,\hat{X}_j,\ldots,X_n) - C_k^{n-1}(X_1,\ldots,\hat{X}_j,\ldots,X_n)$ $= C_k^n(X_1,\ldots,\underbrace{X_j}_{=1},\ldots,X_n) - C_k^n(X_1,\ldots,\underbrace{X_j}_{=0},\ldots,X_n)$
6.a	$tB_k^n(t) = \frac{(k+1)}{(n+1)} B_{k+1}^{n+1}(t)$
b	$\sum_{i=1}^{n+1} X_i C_k^n(X_1,\ldots,\hat{X}_i,\ldots,X_{n+1}) = (k+1)C_{k+1}^{n+1}(t)$
7.a	$(1-t)B_k^n(t) = \frac{(n+1-k)}{(n+1)} B_k^{n+1}(t)$
b	$\sum_{i=1}^{n+1}(1-X_i)C_k^n(X_1,\ldots,\hat{X}_i,\ldots,X_{n+1}) = (n+1-k)C_k^{n+1}(t)$
8.a	$B_k^n(t) = \frac{(n+1-k)}{(n+1)} B_k^{n+1}(t) + \frac{(k+1)}{(n+1)} B_{k+1}^{n+1}(t)$
b	$\sum\limits_{i=1}^{n+1} C_k^n(X_1,\ldots,\hat{X}_i,\ldots,X_{n+1}) = (n+1-k)C_k^{n+1}(t) + (k+1)C_{k+1}^{n+1}(t)$
9.a	$B_j^n(rt) = \sum_{k=j}^{n} B_j^k(r) B_k^n(t)$
b	$C_j^n(rX_1,\ldots,rX_n) = \sum_{k=j}^{n} B_j^k(r) C_k^n(t)$
c	$C_j^n(r_1 t,\ldots,r_n t) = \sum\limits_{k=j}^{n}\Big[\sum\limits_{\{\sigma_1,\ldots,\sigma_k\}} C_j^k(r_{\sigma_1},\ldots,r_{\sigma_k})/\binom{n}{k}\Big] B_k^n(t)$
10.a	$(t-x)^n = \sum_{k=0}^{n}(-1)^k B_{n-k}^n(t) B_k^n(x)/\binom{n}{k}$
b	$\prod_{i=1}^{n}(X_i - x) = \sum_{k=0}^{n}(-1)^k C_{n-k}^n(X_1,\ldots,X_n) B_k^n(x)/\binom{n}{k}$
c	$\prod_{i=1}^{n}(t - U_i) = \sum_{k=0}^{n}(-1)^k B_{n-k}^n(t) C_k^n(U_1,\ldots,U_n)/\binom{n}{k}$
d	$\frac{1}{n!}\mathrm{Per}(X_i - U_j) = \sum_{k=0}^{n}(-1)^k C_{n-k}^n(X_1,\ldots,X_n) C_k^n(U_1,\ldots,U_n)/\binom{n}{k}$
11.a	$\sum_{r=0}^{n}(-1)^{n-r}[B_i^n(t)]^{(r)}[B_{n-j}^n(t)]^{(n-r)}/\mathcal{K}_{nj} = \delta_{ij}$
b	$\sum\limits_{r=0}^{n}(-1)^{n-r}\binom{n}{r}\sum\limits_{\text{all }\pi}\frac{\partial^r C_i^n(X_1,\ldots,X_n)}{\partial X_{\pi(1)}\cdots\partial X_{\pi(r)}}\frac{\partial^{n-r} C_{n-j}^n(X_1,\ldots,X_n)}{\partial X_{\pi(r+1)}\cdots\partial X_{\pi(n)}}/\mathcal{K}_{nj} = \delta_{ij}$ where $\mathcal{K}_{nj} \equiv (-1)^{(n-j)} n! \binom{n}{n-j}$

Pólya basis function [1,3]. Thus, up to a constant, the convolution polynomials $C^n_{n-k}(X_1, \ldots, X_n)$ provide the transformation from the Pólya representation to the Bézier representation. identity 10c is similar. Identity 10d, obtained by applying the blossoming technique to both sides of identity 10a twice, provides a polynomial time algorithm for computing the permanent $\mathrm{Per}(X_i - U_j)$. See [18] for further details.

§8. Relation to the Discrete B-splines

The **Principle of Duality** [2] states that *transformations between B-splines are equivalent to transformations between corresponding Pólya bases in the opposite direction.* Thus, the convolution transformation from Pólya bases to Bernstein bases given by identity 10c of Table 1 also gives the transformation from Bernstein bases to B-splines. That is,

$$B^n_j(x) = \sum_{k=0}^{n} N^n_k(x) C^n_j(t_{k+1}, \ldots, t_{k+n}) \tag{10}$$

where $N^n_k(x)$ is the B-spline [4,6] with support $[t_k, t_{k+n+1}]$. Another way to derive (10) is by employing the dual functional property of the blossom for B-splines [8,15] — that is, if $p(t) = \sum_k P_k N^n_k(t)$, then $P_k = \mathcal{B}[p](t_{k+1}, \ldots, t_{k+n})$ — together with (8). In any case, Equation 10 says that $C^n_j(t_{k+1}, \ldots, t_{k+n})$ is a special case of a discrete B-spline [5,12].

§9. Dual Functionals

Formula 11a in Table 1 defines the de Boor-Fix dual functionals [7] for the Bernstein basis functions. Formula 11b defines the corresponding dual functionals for the convolution basis functions with the same condition on the a's as in Theorem 1. These dual functionals give us a general method for representing any polynomial basis in terms of the convolution basis. See [17] for further details.

§10. Convolution Curves

A convolution curve $C(t)$ of degree n is a free-form curve defined by a sequence of control points $P = (P_0, \ldots, P_n)$ blended by the convolution basis functions $C^n_0(t), \ldots, C^n_n(t)$. That is,

$$C(t) = \sum_{k=0}^{n} P_k C^n_k(t) \qquad t \in [L, U]. \tag{11}$$

Desirable properties for curves in CAGD are formulated in [10]. As a generalization of the Bézier curves, convolution curves inherit many desirable properties for CAGD.

10.1 Basic Properties

Polynomial
Since each $C_k^n(t)$ is a polynomial in t, the curve $C(t)$ is also a polynomial in t. Moreover, let $p(t)$ be a polynomial of degree $d \leq n$. If the control points lie uniformly on $p(t)$, that is, if $P_k = p\big(L + k(U-L)/n\big)$ for $k = 0, \ldots, n$, then the curve $C(t)$ is a polynomial of degree d. This property can be proved by induction on d using (4.b) in Table 1.

Affine Invariance
This property follows from the Partition of Unity property of the convolution basis functions.

Convex Hull
This property holds since the convolution basis functions are nonnegative over the interval $[L, U]$ and they have the Partition of Unity property.

Nondegenerate
Since the blending functions $C_k^n(t)$ form a basis, the convolution curve $C(t)$ is nondegenerate, i.e., the only time it collapses to a point is when all the control points are at that same point.

Unimodality
In Section 5.6 we point out that the convolution basis $\{C_k^n(t) \mid k = 0, 1, \ldots, n\}$ is unimodal in k. This property implies that for each point $C(t)$ on a convolution curve there is one control point P_k which has the biggest influence on $C(t)$, and the influence of P_j on the value of $C(t)$ decreases monotonically as j recedes from k.

Subdivision
Since the blending functions $C_k^n(t)$ form a polynomial basis, there exists a subdivision algorithm. For the case where $X_i = a_i t$ for $i = 1, \ldots, n$, a practical subdivision algorithm can be obtained from (9.b) in Table 1. See [17] for further details.

de Casteljau-type Evaluation Algorithm
This property follows from the recursion formula given in Section 4.1.

Extra Control Parameters
The parameters a_k and b_k in each $X_k(t)$ provide extra control over the shape of a convolution curve. For example,

- if one of the a-parameters is relatively much larger than the others, the resulting curve will appear to be linear since effectively only one $X_k(t)$ (the one with the largest a-parameter) varies with t;
- if one of the a-parameters is relatively much smaller than the others, the resulting curve will appear to be of degree $n-1$ since effectively one of the $X_k(t)$ (the one with the smallest a-parameter) becomes a constant;
- the parameters a_k and b_k can be chosen to optimize some function attributed to the curve.

Variation Diminishing
This property holds for the cases where the convolution basis functions satisfy Descartes' Law of Signs (see Section 5.5) [11,16].

Geometric Relationship with Bézier Curves
By the dual functional property of a blossom (see Section 8), the convolution curve

$$\sum_k C_k^n(t)P_k = \mathcal{B}[\sum_k B_k^n(t)P_k](X_1(t),\ldots,X_n(t))$$

can be viewed as the locus of a B-spline control point of the Bézier curve $\sum_k B_k^n(t)P_k$ as we vary the knots $X_1(t),\ldots,X_n(t)$.

Generalization to Surfaces
Once we have a curve scheme, we can always define a surface scheme by the standard tensor product construction [4]. Triangular patches which are more complicated may also be generated by blossoming the bivariate Bernstein polynomials [9,10] and evaluating off the diagonal.

§11. Summary

By replacing the parameter t in the recurrence for the Bernstein basis functions $B_k^n(t)$ by n linear functions (with some mild restrictions), we obtained the convolution basis functions $C_k^n(t)$. This generalization preserves the convolution structure of the Bernstein basis functions. Moreover, each convolution basis function is the blossom of the corresponding Bernstein basis function evaluated off the diagonal at those linear functions. This formulation is a powerful tool for deriving many identities and formulas for the convolution basis functions from known identities and formulas for the Bernstein basis functions, including Marsden's Identity and the de Boor-Fix dual functionals.

The convolution polynomials appear in many important contexts, such as the computation of a certain permanent and the transformations from the Pólya representation to the Bézier representation and from the Bézier representation to the B-spline representation. As a generalization of the Bézier scheme, the convolution scheme inherits many interesting and desirable properties for CAGD.

Acknowledgements. The first author has been supported in part by the Second University Development Project Fellowship of the Republic of Indonesia.

References

1. Barry, P. J., Urn models, recursive curve schemes, and Computer Aided Geometric Design, Ph.D. dissertation, Department of Mathematics, University of Utah, 1987.
2. Barry, P. J., and R. N. Goldman, What is the natural generalization of a Bézier curve?, in *Mathematical Methods in Computer Aided Geometric Design*, Tom Lyche and L. L. Schumaker (eds.), Academic Press, New York, 1989, 71–85.
3. Barry, P. J., and R. N. Goldman, Interpolation and approximation of curves and surfaces using Pólya polynomials, Comp. Graphics and Image Proc. **53** (1991), 137–148.

4. Bartels, R. H., J. C. Beatty and B. A. Barsky, *An Introduction to Splines for Use in Computer Graphics and Geometric Modeling*, Morgan Kaufmann, Los Altos, California, 1987.
5. Cohen, E., T. Lyche and R. Riesenfeld, Discrete B-splines and subdivision techniques in Computer Aided Geometric Design and Computer Graphics, Comp. Graphics and Image Proc. **14** (1980), 87–111.
6. de Boor, C., *A Practical Guide to Splines*, Applied Mathematical Sciences Volume 27, Springer-Verlag, New York, 1978.
7. de Boor, C., and G. Fix, Spline approximation by quasi-interpolants, J. Approx. Th. **8** (1973), 19–45.
8. de Casteljau, P., *Shape Mathematics and CAD*, Kogan Page, London, 1986.
9. Farin, G., *Curves and Surfaces for Computer Aided Geometric Design: A Practical Guide*, Academic Press, San Diego, California, 1990.
10. Goldman, R. N., Polya's urn model and Computer Aided Geometric Design, SIAM J. Alg. Disc. Meth. **6** (1985), 1–28.
11. Karlin, S., *Total Positivity*, Vol.1, Stanford University Press, Stanford, California, 1968.
12. Lyche, T., Discrete polynomial spline approximation methods, Ph.D. dissertation, Department of Mathematics, University of Texas, Austin, 1975.
13. Marsden, M. J., An identity for spline functions with applications to variation-diminishing spline approximation, J. Approx. Th. **3** (1970), 7–49.
14. Minc, H., *Permanents*, Addison-Wesley, London, 1978.
15. Ramshaw, L., Blossoming: a connect-the-dots approach to splines, Digital Equipment Corporation, SRC Report, June 21, 1987.
16. Schumaker, L. L., *Spline Functions: Basic Theory*, John Wiley & Sons, New York, 1981.
17. Stefanus, Y., Blossoming off the diagonal, Ph.D. dissertation, Department of Computer Science, University of Waterloo, Waterloo, Ontario, Canada, 1991.
18. Stefanus, Y., and R. Goldman, Blossoming Marsden's Identity, submitted for publication.

Yohanes Stefanus
CGL, Dept. of Computer Science, University of Waterloo
Waterloo, Ontario, Canada N2L-3G1
lystefan@watcgl.waterloo.edu

Ronald Goldman
Dept. of Computer Science, Rice University, P. O. Box 1892
Houston, Texas 77251 USA
rng@rice.edu

A Method for Removing the Singularities from Gregory Surfaces

Kenji Ueda

Abstract. The bicubic Gregory patch is mathematically equivalent to the bicubic Coons patch with Gregory's square. A method to convert a family of Gregory patches to rational Bézier surfaces which have no singularity is described. The method is mathematically an approximation of Gregory patches based on an approximation of the rational blending function by a rational Bézier function which has an implanted minute error.

§1. Introduction

Many surfaces in geometric and solid modeling are constructed by means of a network of rectangular patches. Bézier surfaces, Coons surfaces and B-spline surfaces are popular surfaces. The bicubic Gregory patch [4,5] is one such surface patch. This patch is developed by applying Gregory's square [1,9], which is a solution of the incompatibility [3,8] at the corner of the bicubic Coon's patch [7], to the Bézier patch. In this surface, a Bézier point is expressed as a rational blending function of two independent sub-control points like the twist terms in Gregory's square, and the cross boundary derivatives on four sides can be defined independently. It is easy to join two Gregory patches smoothly, because of the independence.

Local smooth surface interpolation methods including the Gregory patch have been discussed and classified [1,2,12], and a Gregory patch for rational boundary curves [6] and a C^2 Gregory patch [11] have been proposed. In [14] the properties of Gregory surfaces are discussed and several Gregory surfaces are surveyed.

Such Gregory surfaces are not standard surfaces like Bézier surfaces or B-spline surfaces, therefore it is often required to convert the Gregory surface into one of the standard surfaces. The Gregory surface has rational surface

Mathematical Methods in Computer Aided Geometric Design II
Tom Lyche and Larry L. Schumaker (eds.), pp. 597–606.

ISBN 0-12-460510-9.

properties and is convertible to a standard rational surface, e.g., a rational Bézier surface [13]. In this paper, we discuss a general procedure for converting Gregory surfaces to rational Bézier surface by representing rational blending functions by rational Bézier functions.

However, the converted surface has singularities, i.e. zero weights, at the four corners due to the singularities of the rational functions of the control points. Generally a rational surface having singularities is not permitted. For example, the weights of a rational B-spline surface must be non-zero real numbers in IGES (Initial Graphics Exchange Specification) files, and the libraries of graphic workstations usually cannot render NURBS surfaces with zero weights.

A method to convert a Gregory surface to a rational Bézier surface which has no singularity is presented. The method is mathematically an approximation of the Gregory patch based on an approximation of the rational blending function by a rational Bézier function which has an implanted minute error.

The conversion process is based on the Bernstein basis function, and therefore only calculations among binomial coefficients is required. This method is a general procedure and applicable to various Gregory patches, e.g., the rational boundary Gregory patch, the C^2 Gregory patch, etc.

§2. Gregory's Square

The bicubic Coons patch is defined in Hermite interpolation form as follows:

$$\mathbf{S}(u,v) = \begin{bmatrix} \mathrm{H}_0^3(u) \\ \mathrm{H}_1^3(u) \\ \mathrm{H}_2^3(u) \\ \mathrm{H}_3^3(u) \end{bmatrix}^T \begin{bmatrix} \mathbf{S}(0,0) & \mathbf{S}_v(0,0) & \mathbf{S}_v(0,1) & \mathbf{S}(0,1) \\ \mathbf{S}_u(0,0) & \mathbf{S}_{uv}(0,0) & \mathbf{S}_{uv}(0,1) & \mathbf{S}_u(0,1) \\ \mathbf{S}_u(1,0) & \mathbf{S}_{uv}(1,0) & \mathbf{S}_{uv}(1,1) & \mathbf{S}_u(1,1) \\ \mathbf{S}(1,0) & \mathbf{S}_v(1,0) & \mathbf{S}_v(1,1) & \mathbf{S}(1,1) \end{bmatrix} \begin{bmatrix} \mathrm{H}_0^3(v) \\ \mathrm{H}_1^3(v) \\ \mathrm{H}_2^3(v) \\ \mathrm{H}_3^3(v) \end{bmatrix} \tag{1}$$

where $\mathrm{H}_i^3(t)$ are cubic Hermite polynomials. Cubic Hermite polynomials are expressed using Bernstein polynomials, $\mathrm{B}_i^n(t) = \binom{n}{i}(1-t)^{n-i}t^i$.

$$\begin{bmatrix} \mathrm{H}_0^3(t) \\ \mathrm{H}_1^3(t) \\ \mathrm{H}_2^3(t) \\ \mathrm{H}_3^3(t) \end{bmatrix} = \begin{bmatrix} 1 & 1 & 0 & 0 \\ 0 & \frac{1}{3} & 0 & 0 \\ 0 & 0 & -\frac{1}{3} & 0 \\ 0 & 0 & 1 & 1 \end{bmatrix} \begin{bmatrix} \mathrm{B}_0^3(t) \\ \mathrm{B}_1^3(t) \\ \mathrm{B}_2^3(t) \\ \mathrm{B}_3^3(t) \end{bmatrix} \tag{2}$$

The surface is converted to a patch of cubic Bézier form by basis conversion. The Bézier points of the converted surface have the following values:

$$\begin{aligned} \mathbf{P}_{0,0} &= \mathbf{S}(0,0) \\ \mathbf{P}_{0,1} &= \mathbf{S}(0,0) + \frac{1}{3}\mathbf{S}_v(0,0) \\ \mathbf{P}_{1,0} &= \mathbf{S}(0,0) + \frac{1}{3}\mathbf{S}_u(0,0) \\ \mathbf{P}_{1,1} &= \mathbf{S}(0,0) + \frac{1}{3}\mathbf{S}_u(0,0) + \frac{1}{3}\mathbf{S}_v(0,0) + \frac{1}{9}\mathbf{S}_{uv}(0,0). \end{aligned} \tag{3}$$

Æ where are twist incompatibility problems in the bicubic Coons patch, i.e., $\mathbf{S}_{uv} \neq \mathbf{S}_{vu}$ at the four corners. Gregory's square is a solution of this problem, and is expressed as follows:

$$\begin{bmatrix} \mathbf{S}(0,0) & \mathbf{S}_v(0,0) & \mathbf{S}_v(0,1) & \mathbf{S}(0,1) \\ \mathbf{S}_u(0,0) & \frac{u\mathbf{S}_{uv}(0,0)+v\mathbf{S}_{vu}(0,0)}{u+v} & \frac{u\mathbf{S}_{uv}(0,1)+(1-v)\mathbf{S}_{vu}(0,1)}{u+(1-v)} & \mathbf{S}_u(0,1) \\ \mathbf{S}_u(1,0) & \frac{(1-u)\mathbf{S}_{uv}(1,0)+v\mathbf{S}_{vu}(1,0)}{(1-u)+v} & \frac{(1-u)\mathbf{S}_{uv}(1,1)+(1-v)\mathbf{S}_{vu}(1,1)}{(1-u)+(1-v)} & \mathbf{S}_u(1,1) \\ \mathbf{S}(1,0) & \mathbf{S}_v(1,0) & \mathbf{S}_v(1,1) & \mathbf{S}(1,1) \end{bmatrix}. \tag{4}$$

This method replaces the constant twist terms in (1) by variable twists.

§3. Gregory Surfaces

The bicubic Gregory patch is the result of an application of Gregory's square to the Bézier patch. It is obtained by converting the basis function to the Bernstein polynomial of the bicubic Coons patch having Gregory's square. The surface is defined as follows:

$$\mathbf{S}(u,v) = \sum_{i=0}^{3}\sum_{j=0}^{3} \mathrm{B}_i^3(u)\mathrm{B}_j^3(v)\mathbf{P}_{i,j}(u,v) \quad (0 \le u,v \le 1) \tag{5}$$

$$\begin{aligned} \mathbf{P}_{i,j}(u,v) &= \mathbf{P}^{\mathbf{v}}_{i,j} = \mathbf{P}^{\mathbf{u}}_{i,j} \equiv \mathbf{P}_{i,j} \qquad (i=0,3 \;\; j=0,3) \\ \mathbf{P}_{1,1}(u,v) &= \frac{u\mathbf{P}^{\mathbf{v}}_{1,1} + v\mathbf{P}^{\mathbf{u}}_{1,1}}{u+v} \\ \mathbf{P}_{1,2}(u,v) &= \frac{u\mathbf{P}^{\mathbf{v}}_{1,2} + (1-v)\mathbf{P}^{\mathbf{u}}_{1,2}}{u+(1-v)} \\ \mathbf{P}_{2,1}(u,v) &= \frac{(1-u)\mathbf{P}^{\mathbf{v}}_{2,1} + v\mathbf{P}^{\mathbf{u}}_{2,1}}{(1-u)+v} \\ \mathbf{P}_{2,2}(u,v) &= \frac{(1-u)\mathbf{P}^{\mathbf{v}}_{2,2} + (1-v)\mathbf{P}^{\mathbf{u}}_{2,2}}{(1-u)+(1-v)}, \end{aligned} \tag{6}$$

where $\mathbf{P}^{\mathbf{u}}_{i,j}$ and $\mathbf{P}^{\mathbf{v}}_{i,j}$ are the sub-control points of the Bézier points $\mathbf{P}_{i,j}(u,v)$. For example, the meaning of $\mathbf{P}_{1,1}(u,v)$ is expressed by the following relation:

$$\begin{aligned} &\mathbf{S}(0,0) + \frac{1}{3}\mathbf{S}_u(0,0) + \frac{1}{3}\mathbf{S}_v(0,0) + \frac{1}{9}\frac{u\mathbf{S}_{uv}(0,0) + v\mathbf{S}_{vu}(0,0)}{u+v} \\ =&\frac{u}{u+v}\left(\mathbf{S}(0,0) + \frac{1}{3}\mathbf{S}_u(0,0) + \frac{1}{3}\mathbf{S}_v(0,0) + \frac{1}{9}\mathbf{S}_{uv}(0,0)\right) + \\ &\frac{v}{u+v}\left(\mathbf{S}(0,0) + \frac{1}{3}\mathbf{S}_u(0,0) + \frac{1}{3}\mathbf{S}_v(0,0) + \frac{1}{9}\mathbf{S}_{vu}(0,0)\right) \\ =&\frac{u\mathbf{P}^{\mathbf{v}}_{1,1} + v\mathbf{P}^{\mathbf{u}}_{1,1}}{u+v}. \end{aligned} \tag{7}$$

In the bicubic Gregory patch the variable control points are expressed by rational blending functions of two sub-control points similar to the twist terms of Gregory's square. Surfaces having similar properties to the bicubic Gregory patch are proposed to interpolate to rational boundary curves [5], or to interpolate more smoothly [11]. These surfaces are called *Gregory surfaces.*

To solve the incompatibility at a corner, it is not necessary to use rational functions in the bicubic Gregory patch. For example, all variable control points may expressed [10] as the following function:

$$\mathbf{P}_{i,j}(u,v) = \frac{(1-u)u\mathbf{P}^{\mathbf{v}}_{i,j} + (1-v)v\mathbf{P}^{\mathbf{u}}_{i,j}}{(1-u)u + (1-v)v}. \tag{8}$$

To resolve the incompatibility at a corner (u^*, v^*), the functions must satisfy the following property:

$$g(u^*, v) = V, \qquad g(u, v^*) = U. \tag{9}$$

This property is inconsistent with the value of $g(u^*, v^*)$, and causes the singularity at a corner (u^*, v^*). At the corner, however, the value and the derivatives of the Gregory patch can be calculated only using boundary curve data. In the calculation of Gregory surface it is necessary to pay attention to this singularity.

Several functions are proposed to blend two control points or two surfaces [2,10]. We call rational functions which satisfy a property similar to (9), *Gregory functions.* The form of a Gregory function depends on the continuity condition at the boundary [14].

§4. Gregory Functions as Rational Bézier Functions

A family of Gregory functions is simply expressed using Bernstein polynomials $\mathrm{B}_i^n(t)$. Most proposed Gregory functions are included in this family. Furthermore, the functions can be transformed easily to rational Bézier form.

$$\begin{aligned}\frac{uV + vU}{u+v} &= \frac{\mathrm{B}_1^1(u)V + \mathrm{B}_1^1(v)U}{\mathrm{B}_1^1(u) + \mathrm{B}_1^1(v)} \\ &= \frac{\begin{bmatrix}\mathrm{B}_0^1(u) & \mathrm{B}_1^1(u)\end{bmatrix}\begin{bmatrix}0 & U \\ V & V+U\end{bmatrix}\begin{bmatrix}\mathrm{B}_0^1(v) \\ \mathrm{B}_1^1(v)\end{bmatrix}}{\begin{bmatrix}\mathrm{B}_0^1(u) & \mathrm{B}_1^1(u)\end{bmatrix}\begin{bmatrix}0 & 1 \\ 1 & 2\end{bmatrix}\begin{bmatrix}\mathrm{B}_0^1(v) \\ \mathrm{B}_1^1(v)\end{bmatrix}}\end{aligned} \tag{10}$$

$$\begin{aligned}\frac{(1-u)uV + (1-v)vU}{(1-u)u + (1-v)v} &= \frac{2(1-u)uV + 2(1-v)vU}{2(1-u)u + 2(1-v)v} \\ = \frac{\mathrm{B}_1^2(u)V + \mathrm{B}_1^2(v)U}{\mathrm{B}_1^2(u) + \mathrm{B}_1^2(v)} &= \frac{\begin{bmatrix}\mathrm{B}_0^2(u) \\ \mathrm{B}_1^2(u) \\ \mathrm{B}_2^2(u)\end{bmatrix}^T \begin{bmatrix}0 & U & 0 \\ V & V+U & V \\ 0 & U & 0\end{bmatrix}\begin{bmatrix}\mathrm{B}_0^2(v) \\ \mathrm{B}_1^2(v) \\ \mathrm{B}_2^2(v)\end{bmatrix}}{\begin{bmatrix}\mathrm{B}_0^2(u) \\ \mathrm{B}_1^2(u) \\ \mathrm{B}_2^2(u)\end{bmatrix}^T \begin{bmatrix}0 & 1 & 0 \\ 1 & 2 & 1 \\ 0 & 1 & 0\end{bmatrix}\begin{bmatrix}\mathrm{B}_0^2(v) \\ \mathrm{B}_1^2(v) \\ \mathrm{B}_2^2(v)\end{bmatrix}}.\end{aligned} \tag{11}$$

A Family of Gregory functions $g(u,v)$ is represented by the following rational Bézier form:

$$g(u,v) = \frac{\mathrm{B}_p^{p+q}(u)V + \mathrm{B}_r^{r+s}(v)U}{\mathrm{B}_p^{p+q}(u) + \mathrm{B}_r^{r+s}(v)} = \frac{\displaystyle\sum_{k=0}^{p+q}\sum_{l=0}^{r+s} \mathrm{B}_k^{p+q}(u)\mathrm{B}_l^{r+s}(v)M^{k,l}}{\displaystyle\sum_{k=0}^{p+q}\sum_{l=0}^{r+s} \mathrm{B}_k^{p+q}(u)\mathrm{B}_l^{r+s}(v)W^{k,l}}, \tag{12}$$

where

$$[M^{k,l}] = \begin{matrix} p\left\{\begin{matrix}\\ \\ \\ \end{matrix}\right. \\ \\ q\left\{\begin{matrix}\\ \\ \\ \end{matrix}\right. \end{matrix} \left[\begin{array}{ccc|c|ccc} \multicolumn{3}{c}{\overbrace{\hphantom{0 \ \ldots \ 0}}^{r}} & & \multicolumn{3}{c}{\overbrace{\hphantom{0 \ \ldots \ 0}}^{s}} \\ 0 & \ldots & 0 & U & 0 & \ldots & 0 \\ \vdots & \ddots & \vdots & U & \vdots & \ddots & \vdots \\ 0 & \ldots & 0 & U & 0 & \ldots & 0 \\ V & \ldots & V & V+U & V & \ldots & V \\ 0 & \ldots & 0 & U & 0 & \ldots & 0 \\ \vdots & \ddots & \vdots & U & \vdots & \ddots & \vdots \\ 0 & \ldots & 0 & U & 0 & \ldots & 0 \end{array}\right] \quad \begin{pmatrix} 0 \le p,q,r,s \\ 0 < p+q \\ 0 < r+s \end{pmatrix}, \tag{13}$$

$$[W^{k,l}] = \begin{bmatrix} 0 & \ldots & 0 & 1 & 0 & \ldots & 0 \\ \vdots & \ddots & \vdots & 1 & \vdots & \ddots & \vdots \\ 0 & \ldots & 0 & 1 & 0 & \ldots & 0 \\ 1 & \ldots & 1 & 2 & 1 & \ldots & 1 \\ 0 & \ldots & 0 & 1 & 0 & \ldots & 0 \\ \vdots & \ddots & \vdots & 1 & \vdots & \ddots & \vdots \\ 0 & \ldots & 0 & 1 & 0 & \ldots & 0 \end{bmatrix}.$$

It is visually clear that a zero sub-matrix in the corner of matrices of (13) causes the singularity at a corner of a Gregory function.

Representing Gregory functions by rational Bézier functions, each control point of a Gregory surface has the same form as a rational Bézier patch. This means that a control point is calculated in the same manner as for a rational Bézier patch. However, there exists singular points at the corners of the control point.

§5. Conversion of Gregory Surfaces to Rational Bézier Surfaces

Representing Gregory functions in rational Bézier form, Gregory surfaces are also defined as the following general form:

$$\mathbf{S}(u,v) = \frac{\displaystyle\sum_{i=0}^{m}\sum_{j=0}^{n} \mathrm{B}_i^m(u)\mathrm{B}_j^n(v)\mathbf{Q}_{i,j}(u,v)}{\displaystyle\sum_{i=0}^{m}\sum_{j=0}^{n} \mathrm{B}_i^m(u)\mathrm{B}_j^n(v)w_{i,j}(u,v)}, \tag{14}$$

where

$$\mathbf{Q}_{i,j}(u,v)=\frac{\sum_{k=0}^{p}\sum_{l=0}^{q}\mathrm{B}_k^p(u)\mathrm{B}_l^q(v)\mathrm{w}^{k,l}w_{i,j}^{k,l}\mathbf{P}_{i,j}^{k,l}}{\sum_{k=0}^{p}\sum_{l=0}^{q}\mathrm{B}_k^p(u)\mathrm{B}_l^q(v)\mathrm{w}^{k,l}},$$
$$w_{i,j}(u,v)=\frac{\sum_{k=0}^{p}\sum_{l=0}^{q}\mathrm{B}_k^p(u)\mathrm{B}_l^q(v)\mathrm{w}^{k,l}w_{i,j}^{k,l}}{\sum_{k=0}^{p}\sum_{l=0}^{q}\mathrm{B}_k^p(u)\mathrm{B}_l^q(v)\mathrm{w}^{k,l}}. \tag{15}$$

In this form, all denominators of the functions of the control points must be the same. At this point all control points of the Gregory patch have singularities. In the case all control points have the same weight like the bicubic Gregory patch, the weight of the control points are all one, i. e., $w_{i,j}(u,v)\equiv 1$.

In order to represent the bicubic Gregory patch in the form (15), the control points must be transformed to have the denominator $\{u+v\}\{u+(1-v)\}\{(1-u)+v\}\{(1-u)+(1-v)\}$, which is the least common multiplier of the denominators of the blending function for the control points in the bicubic Gregory patch. For example, the control points $\mathbf{P}_{0,0}(u,v)$ and $\mathbf{P}_{1,1}(u,v)$ must be transformed to have the following biquartic rational function forms:

$$\begin{aligned}\mathbf{P}_{0,0}(u,v)&=\mathbf{P}_{0,0}\cdot\frac{u+v}{u+v}\cdot\frac{u+(1-v)}{u+(1-v)}\cdot\frac{(1-u)+v}{(1-u)+v}\cdot\frac{(1-u)+(1-v)}{(1-u)+(1-v)}\\ \mathbf{P}_{1,1}(u,v)&=\frac{u\mathbf{P}_{1,1}^{\mathbf{v}}+v\mathbf{P}_{1,1}^{\mathbf{u}}}{u+v}\cdot\frac{u+(1-v)}{u+(1-v)}\cdot\frac{(1-u)+v}{(1-u)+v}\cdot\frac{(1-u)+(1-v)}{(1-u)+(1-v)},\end{aligned} \tag{16}$$

$\left[\mathbf{P}_{1,1}^{k,l}\right]$ and other components of its rational Bézier form are shown below:

$$\begin{bmatrix}\mathbf{0} & \mathbf{P}_{1,1}^{\mathbf{u}} & \mathbf{P}_{1,1}^{\mathbf{u}} & \mathbf{P}_{1,1}^{\mathbf{u}} & \mathbf{0}\\ \mathbf{P}_{1,1}^{\mathbf{v}} & \frac12\mathbf{P}_{1,1}^{\mathbf{u}}+\frac12\mathbf{P}_{1,1}^{\mathbf{v}} & \frac34\mathbf{P}_{1,1}^{\mathbf{u}}+\frac14\mathbf{P}_{1,1}^{\mathbf{v}} & \frac67\mathbf{P}_{1,1}^{\mathbf{u}}+\frac17\mathbf{P}_{1,1}^{\mathbf{v}} & \mathbf{P}_{1,1}^{\mathbf{u}}\\ \mathbf{P}_{1,1}^{\mathbf{v}} & \frac14\mathbf{P}_{1,1}^{\mathbf{u}}+\frac34\mathbf{P}_{1,1}^{\mathbf{v}} & \frac12\mathbf{P}_{1,1}^{\mathbf{u}}+\frac12\mathbf{P}_{1,1}^{\mathbf{v}} & \frac47\mathbf{P}_{1,1}^{\mathbf{u}}+\frac37\mathbf{P}_{1,1}^{\mathbf{v}} & \frac35\mathbf{P}_{1,1}^{\mathbf{u}}+\frac25\mathbf{P}_{1,1}^{\mathbf{v}}\\ \mathbf{P}_{1,1}^{\mathbf{v}} & \frac17\mathbf{P}_{1,1}^{\mathbf{u}}+\frac67\mathbf{P}_{1,1}^{\mathbf{v}} & \frac37\mathbf{P}_{1,1}^{\mathbf{u}}+\frac47\mathbf{P}_{1,1}^{\mathbf{v}} & \frac12\mathbf{P}_{1,1}^{\mathbf{u}}+\frac12\mathbf{P}_{1,1}^{\mathbf{v}} & \frac12\mathbf{P}_{1,1}^{\mathbf{u}}+\frac12\mathbf{P}_{1,1}^{\mathbf{v}}\\ \mathbf{0} & \mathbf{P}_{1,1}^{\mathbf{v}} & \frac25\mathbf{P}_{1,1}^{\mathbf{u}}+\frac35\mathbf{P}_{1,1}^{\mathbf{v}} & \frac12\mathbf{P}_{1,1}^{\mathbf{u}}+\frac12\mathbf{P}_{1,1}^{\mathbf{v}} & \mathbf{0}\end{bmatrix}$$

$$\left[w_{1,1}^{k,l}\right]=\begin{bmatrix}0&1&1&1&0\\1&1&1&1&1\\1&1&1&1&1\\1&1&1&1&1\\0&1&1&1&0\end{bmatrix}\quad\left[\mathrm{w}^{k,l}\right]=\begin{bmatrix}0&\frac12&\frac56&\frac12&0\\ \frac12&\frac78&\frac76&\frac78&\frac12\\ \frac56&\frac76&\frac{13}{9}&\frac76&\frac56\\ \frac12&\frac78&\frac76&\frac78&\frac12\\ 0&\frac12&\frac56&\frac12&0\end{bmatrix} \tag{17}$$

The Gregory surface $\mathbf{S}(u,v)$ of degree $m\times n$ is converted to the surface of rational Bézier patch of degree $(m+p)\times(n+q)$:

$$\mathbf{S}(u,v) = \frac{\displaystyle\sum_{i=0}^{m}\sum_{j=0}^{n} \mathrm{B}_i^m(u)\mathrm{B}_j^n(v)\left(\sum_{k=0}^{p}\sum_{l=0}^{q} \mathrm{B}_k^p(u)\mathrm{B}_l^q(v)\mathrm{w}^{k,l} w_{i,j}^{k,l}\mathbf{P}_{i,j}^{k,l}\right)}{\displaystyle\sum_{i=0}^{m}\sum_{j=0}^{n} \mathrm{B}_i^m(u)\mathrm{B}_j^n(v)\left(\sum_{k=0}^{p}\sum_{l=0}^{q} \mathrm{B}_k^p(u)\mathrm{B}_l^q(v)\mathrm{w}^{k,l} w_{i,j}^{k,l}\right)}$$
$$\Rightarrow \frac{\displaystyle\sum_{i=0}^{m+p}\sum_{j=0}^{n+q} \mathrm{B}_i^{m+p}(u)\mathrm{B}_j^{n+q}(v)\bar{w}_{i,j}\bar{\mathbf{P}}_{i,j}}{\displaystyle\sum_{i=0}^{m+p}\sum_{j=0}^{n+q} \mathrm{B}_i^{m+p}(u)\mathrm{B}_j^{n+q}(v)\bar{w}_{i,j}}. \tag{18}$$

The control points of the converted surface are obtained by the following relation:

$$\begin{aligned}
&\sum_{i=0}^{m}\sum_{j=0}^{n} \mathrm{B}_i^m(u)\mathrm{B}_j^n(v)\left(\sum_{k=0}^{p}\sum_{l=0}^{q} \mathrm{B}_k^p(u)\mathrm{B}_l^q(v)V_{i,j}^{k,l}\right)\\
&=\sum_{i=0}^{m}\sum_{j=0}^{n}\sum_{k=0}^{p}\sum_{l=0}^{q} \mathrm{B}_{i+k}^{m+p}(u)\mathrm{B}_{j+l}^{n+q}(v)\frac{\binom{m}{i}\binom{p}{k}\binom{n}{j}\binom{q}{l}}{\binom{m+p}{i+k}\binom{n+q}{j+l}}V_{i,j}^{k,l} \qquad (19)\\
&=\sum_{i=0}^{m+p}\sum_{j=0}^{n+q} \mathrm{B}_i^{m+p}(u)\mathrm{B}_j^{n+q}(v)\bar{V}_{i,j}^{k,l},
\end{aligned}$$

where

$$\bar{V}_{i,j}^{k,l} = \sum_{k=\max(0,i-p)}^{\min(m,i)}\ \sum_{l=\max(0,j-q)}^{\min(n,j)} \frac{\binom{m}{k}\binom{p}{i-k}\binom{n}{l}\binom{q}{j-l}}{\binom{m+p}{i}\binom{n+q}{j}} V_{k,l}^{i-k,j-l}. \tag{20}$$

By this conversion, the bicubic Gregory patch is converted to a rational Bézier patch of degree 7×7. If the form of all control points of the bicubic Gregory patch are replaced by the form of (8), it is converted to a biquintic rational Bézier patch. Any Gregory surface is converted to a rational Bézier surface of higher degree. The degree of raising is dependent on the form of Gregory functions expressing its control points.

The converted rational Bézier surface also has singularities, i.e., zero weights at the four corners, due to the singularities of the rational functions for the control points. Conversely, a control point at a corner of a rational Bézier surface is not necessary to represent the surface for which the compatibility at the corner is corrected. While this is a reasonable consequence in the surface theory, the singularity at the corner is a serious problem for handling the Bézier surface.

§6. Removal of the Singurality of Gregory Functions

Rational Bézier surfaces converted from Gregory surfaces have singular points at the four corners. Generally rational surfaces having zero weights are not permitted. A method of removing the singularity is presented below. The method is based on an approximation of a Gregory function. A Gregory function $g(u,v)$ of (12) is replaced by the following function:

$$g_\epsilon(u,v) = \frac{\mathrm{B}_p^{p+q}(u)V + \mathrm{B}_r^{r+s}(v)U + \{1-\mathrm{B}_p^{p+q}(u)\}\{1-\mathrm{B}_r^{r+s}(v)\}(V+U)\epsilon}{\mathrm{B}_p^{p+q}(u) + \mathrm{B}_r^{r+s}(v) + \{1-\mathrm{B}_p^{p+q}(u)\}\{1-\mathrm{B}_r^{r+s}(v)\}2\epsilon}$$
$$= \frac{\sum_{k=0}^{p+q}\sum_{l=0}^{r+s} \mathrm{B}_k^{p+q}(u)\mathrm{B}_l^{r+s}(v)M_\epsilon^{k,l}}{\sum_{k=0}^{p+q}\sum_{l=0}^{r+s} \mathrm{B}_k^{p+q}(u)\mathrm{B}_l^{r+s}(v)W_\epsilon^{k,l}}. \tag{21}$$

The $[M_\epsilon^{k,l}]$ and $[W_\epsilon^{k,l}]$ in (21) are obtained by substituting $(V+U)\epsilon$ for each zero of $[M^{k,l}]$ and 2ϵ for each zero of $[W^{k,l}]$ in (13).

$$[M_\epsilon^{k.l}] = \begin{bmatrix} (V+U)\epsilon & \dots & (V+U)\epsilon & U & (V+U)\epsilon & \dots & (V+U)\epsilon \\ \vdots & \ddots & \vdots & U & \vdots & \ddots & \vdots \\ (V+U)\epsilon & \dots & (V+U)\epsilon & U & (V+U)\epsilon & \dots & (V+U)\epsilon \\ V & \dots & V & V+U & V & \dots & V \\ (V+U)\epsilon & \dots & (V+U)\epsilon & U & (V+U)\epsilon & \dots & (V+U)\epsilon \\ \vdots & \ddots & \vdots & U & \vdots & \ddots & \vdots \\ (V+U)\epsilon & \dots & (V+U)\epsilon & U & (V+U)\epsilon & \dots & (V+U)\epsilon \end{bmatrix}$$
$$[W_\epsilon^{k.l}] = \begin{bmatrix} 2\epsilon & \dots & 2\epsilon & 1 & 2\epsilon & \dots & 2\epsilon \\ \vdots & \ddots & \vdots & 1 & \vdots & \ddots & \vdots \\ 2\epsilon & \dots & 2\epsilon & 1 & 2\epsilon & \dots & 2\epsilon \\ 1 & \dots & 1 & 2 & 1 & \dots & 1 \\ 2\epsilon & \dots & 2\epsilon & 1 & 2\epsilon & \dots & 2\epsilon \\ \vdots & \ddots & \vdots & 1 & \vdots & \ddots & \vdots \\ 2\epsilon & \dots & 2\epsilon & 1 & 2\epsilon & \dots & 2\epsilon \end{bmatrix} \tag{22}$$

The function $g_\epsilon(u,v)$ has no singularity and does not affect the continuity condition at the boundary. At a singular point (u^*,v^*) the function has the value

$$g_\epsilon(u^*,v^*) = (V+U)/2. \tag{23}$$

By this modification, for example, the upper left part of $[\mathrm{w}^{k,l}]$ in (7) is changed as follows:

$$\begin{bmatrix} 0 & \frac{1}{2} & \frac{5}{6} \\ \frac{1}{2} & \frac{7}{8} & \frac{7}{6} \\ \frac{5}{6} & \frac{7}{6} & \frac{13}{9} \end{bmatrix} \Rightarrow \begin{bmatrix} 4\epsilon & \frac{1+5\epsilon+4\epsilon^2}{2} & \frac{5+8\epsilon+20\epsilon^2}{6} \\ \frac{1+5\epsilon+4\epsilon^2}{2} & \frac{7+20\epsilon+6\epsilon^2+8\epsilon^3}{8} & \frac{7+11\epsilon+8\epsilon^2+4\epsilon^3}{6} \\ \frac{5+8\epsilon+20\epsilon^2}{6} & \frac{7+11\epsilon+8\epsilon^2+4\epsilon^3}{6} & \frac{13+12\epsilon+16\epsilon^2+4\epsilon^3}{9} \end{bmatrix}. \tag{24}$$

Because a Gregory function is well approximated by selecting a very small value for ϵ, this method is sufficient for practical use. The rational Bézier surface, which is converted from the Gregory surface with this modification for its Gregory functions, has no zero weights.

§7. Conclusion

In this paper, we first discussed a technique for representing a Gregory function as a rational Bézier function. No special function is necessary to calculate the control points of Gregory surfaces. NURBS hardware is applicable for this purpose. Secondly, a general procedure for converting Gregory surfaces to rational Bézier surfaces is described. The procedure does not require any complex calculations, and is applicable to any kind of Gregory surface; e.g., rational boundary Gregory surface, etc. A Gregory surface of any degree can be converted to a rational Bézier surface of higher degree. The technique of representing a rational function by a rational Bézier function is generally applicable to convert a modified Bezier surface, whose Bézier points are expressed by rational functions, into a NURBS surface.

Finally, a method for removing the singularity from Gregory functions is presented. Using this method and the conversion procedure, Gregory surfaces are converted to rational Bézier surfaces having no singularity. Although this conversion is an approximation of a Gregory surface by a rational Bézier surface, this method is sufficient for practical use.

A rational Bézier surface converted mathematically from a Gregory surface has a singularity at the corner. Using the method described in this paper, Gregory surfaces can be well approximated by rational Bézier surfaces or NURBS surfaces with no singularity. All techniques and utilities for NURBS surfaces are then applicable to the surfaces derived from Gregory surfaces.

References

1. Barnhill, R. E., J. H. Brown and I. M. Klucewicz, A new twist in computer aided geometric design, Comp. Graphics and Image Proc. **8** (1978), 78–91.
2. Barnhill, R. E., Computer aided surface representation and design, in *Surfaces in CAGD*, R. E. Barnhill and W. Boehm (eds.), North-Holland, 1983, 1–24.
3. Barnhill, R. E., G. Farin, L. Fayard and H. Hagen, Twists, curvatures and surface interrogation, Comput. Aided Design **20** (1988), 341–346.
4. Chiyokura H. and F. Kimura, Design of solids with free-form surfaces, Computer Graphics **17(3)** (1983), 289–298.
5. Chiyokura H., *Solid Modelling with DESIGNBASE: Theory and Implementation*, Addison-Wesley, 1988.
6. Chiyokura H. and T. Takamura, K. Konno and T. Harada, G^1 surface interpolation over irregular meshes with rational curves, in *NURBS for Curve and Surface Design*, G. Farin (ed.), SIAM, 1991, 15–34.

7. Coons S. A., Surfaces for computer aided design of space forms, MIT Project MAC, TR-41, 1967.
8. Farin, G., *Curves and Surfaces for Computer Aided Geometric Design: A Practical Guide*, Academic Press, 1988.
9. Gregory, J. A., Smooth interpolation without twist constraints, in *Computer Aided Geometric Design*, R. E. Barnhill and R. F. Riesenfeld (eds.), Academic Press, 1974, 71–87.
10. Gregory, J. A., C^1 rectangular and non-rectangular surface patches, in *Surfaces in CAGD*, R. E. Barnhill and W. Boehm (eds.), North-Holland, 1983, 25–33.
11. Miura, K. T. and Wang K.-K., C^2 Gregory patch, in *EUROGRAPHICS '91*, F. H. Post and W. Barth (eds.), North-Holland, 1991, 481–492.
12. Peters, J., Local smooth surface interpolation: a classification, Computer Aided Geom. Design **7** (1990), 191–195.
13. Takamura, T., M. Ohta, H. Toriya and H. Chiyokura, A method to convert a Gregory patch and a rational boundary Gregory patch to a rational Bézier patch and its applications, in *CG International '90: Computer Graphics Around the World*, T. S. Chua and T. L. Kunii (eds.), Springer-Verlag, 1990, 543–562.
14. Ueda, K. and T. Harada, Generalization of a family of Gregory surfaces, in *Scientific Visualization of Physical Phenomena*, N. M. Patrikalakis (ed.), Springer-Verlag, 1991, 417–434.

Kenji Ueda
Software Research Center
RICOH Company, Ltd.
1-1-17, Koishikawa, Bunkyo-ku
Tokyo, 112, JAPAN
ueda@src.ricoh.co.jp

Bivariate Spline Approximation by Penalized Least Squares

N. Weyrich

Abstract. We discuss the results of using cubic C^1 splines on a type-1 partition to approximate an unknown function based on scattered data $z_i = f(x_i, y_i) + \varepsilon_i$ $(i = 1, \ldots, n)$. The approximation is constructed by the method of penalized least squares. The associated smoothing parameter is obtained using the method of generalized cross-validation and the Monte-Carlo method of Girard.

§1. Introduction

Suppose we are given scattered noisy data of the form

$$z_i = f(x_i, y_i) + \varepsilon_i \quad (i = 1, \ldots, n),$$

$$\mathrm{E}[\varepsilon_i] = 0 \quad (i = 1, \ldots, n),$$

$$\mathrm{E}[\varepsilon_i \varepsilon_j] = \sigma^2 \delta_{ij} \quad (i, j = 1, \ldots, n),$$

where E denotes the mathematical mean. Here we are assuming that the variance σ^2 of the errors ε_i is unknown, and that f is an unknown smooth function in the space

$$H^2(G) := \{g \; : \; G \to \mathrm{I\!R} \; ; \; \int\!\!\int_G \sum_{|\alpha| \le 2} D^\alpha g \, dx dy < \infty\},$$

where G is the smallest rectangle containing all given nodes (x_i, y_i) $(i = 1, \ldots, n)$ of the data.

To fit the function f, we propose to minimize the following combination of smoothness and goodness of fit:

Mathematical Methods in Computer Aided Geometric Design II
Tom Lyche and Larry L. Schumaker (eds.), pp. 607–614.

ISBN 0-12-460510-9.

$$J(g) := \frac{1}{n}\sum_{i=1}^{n}(g(x_i,y_i)-z_i)^2 + \lambda \int\int_G (g_{xx}^2 + g_{yy}^2 + 2g_{xy}^2)\,dxdy, \qquad (1)$$

where $\lambda > 0$ is a fixed *smoothing parameter*. Minimizing this quantity over an appropriate class of smooth functions leads to a smoothing thin plate spline; see e.g. [7,8]. These splines are non-polynomial functions. To get piecewise polynomial functions, we may simply restrict our minimization problem to such a space.

The idea of minimizing functionals like $J(g)$ over fixed finite dimensional spaces is called *penalized least squares*. For a detailed general treatment, see Golitschek and Schumaker [5]. The purpose of this paper is to present the results of an explicit penalized least squares fitting method based on splines on a type-1 triangulation.

To introduce our spline space, suppose Δ_{n_x,n_y} is a uniform type-1 triangulation of the rectangle G (see Figure 1). Now associated with this partition, we define the usual space of C^1 cubic splines

$$S_3^1(\Delta_{n_x,n_y}) := \{g \in C^1(G) \ : \ g|_{G_i} \in \Pi_3^2\},$$

where Π_3^2 is the space of polynomials of total degree 3, and G_i are the triangles of the type-1-triangulation.

It is known (cf. [6]) that

$$\dim S_3^1(\Delta_{n_x,n_y}) := m = 2(n_x+2)(n_y+2) - 5.$$

We have chosen this space because it has a basis consisting of basis functions with local support as developed by Fredrickson and Chui and Wang [1]. In particular, the space is spanned by suitably chosen translates of the two basis splines shown in Figure 2. We may represent each of the cubic pieces of these basis functions in Bernstein-Bézier form. The numbers shown in Figure 2 are the Bezier ordinates of the two types of basis splines.

The paper is organized as follows. In Section 2 we discuss how to solve the minimization problem, and explore the role of λ. In Section 3 we present an explicit algorithm, and in Section 4 we discuss some numerical experiments.

§2. Solution of the Problem

We denote the basis functions of $S_3^1(G)$ by B_i ($i = 1,\ldots,m$). Then any $g \in S_3^1(G)$ can be written as

$$g(x,y) = \sum_{i=1}^{m} g_i B_i(x,y), \qquad (g_i \in \mathbb{R}). \qquad (2)$$

We shall use the notation

$$\underline{z} := (z_1,\ldots,z_n)^T, \quad \underline{g} := (g_1,\ldots,g_m)^T,$$

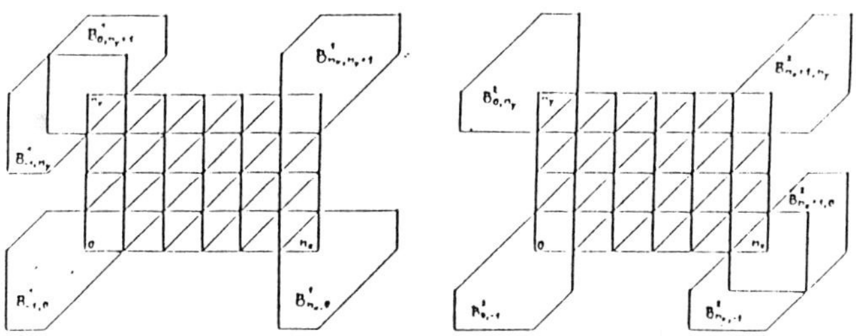

Figure 1. Type-1 triangulation and supports of the basis splines.

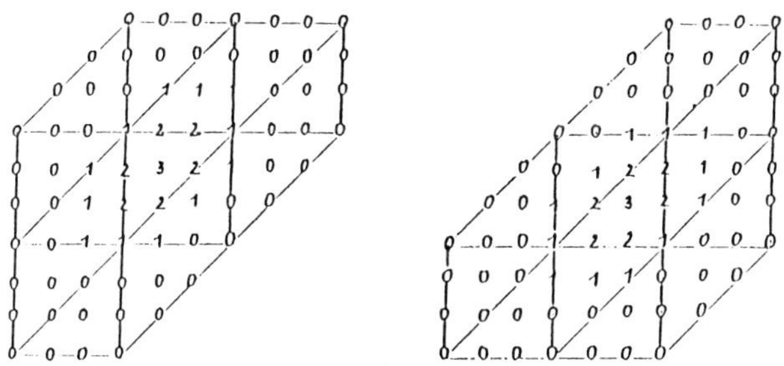

Figure 2. Local supports of the basis splines.

$$F := (f_{i,j})_{i=1,j=1}^{n\quad m}, \quad f_{i,j} := B_j(x_i, y_i),$$

$$M := (m_{i,j})_{i,j=1}^{m}, \quad m_{i,j} := \int\int_G (B_{i\,xx}B_{j\,xx} + B_{i\,yy}B_{j\,yy} + 2B_{i\,xy}B_{j\,xy})\,dxdy.$$

Now it is well-known (cf. [5]), that the problem

$$\text{minimize } J(g) \text{ over } g \in S_3^1(\Delta_{n_x,n_y}) \tag{3}$$

reduces to solving the linear system of m equations

$$(F^T F + n\lambda M)\underline{g}_{n\lambda} = F^T \underline{z}, \tag{4}$$

where

$$\underline{g}_{n\lambda} := (g_{n\lambda\,1}, \ldots, g_{n\lambda\,m})^T$$

is the vector of the spline coefficients.

The role of the smoothing parameter $\lambda > 0$ was discussed in [5]. It was shown there that the solution of (3) tends to the bivariate linear (regression) polynomial for $\lambda \to \infty$, and to the (not necessarily unique) least squares solution over the space of splines for $\lambda \to 0^+$, respectively.

For a given set of data, the best choice of λ will be some number lying in $(0, \infty)$. The (theoretically) optimal smoothing parameter λ_R is given by

$$\mathrm{E}[R(\lambda_R)] := \min\{\mathrm{E}[R(\lambda)] \; : \; \lambda > 0\},$$

where

$$R(\lambda) := \frac{1}{n} \sum_{i=1}^{n} (g_{n\lambda}(x_i, y_i) - f(x_i, y_i))^2$$

and E denotes the mathematical mean. In general we cannot compute the optimal parameter exactly because R depends on the unknown function f.

It was shown in [5] that under the statistical assumptions made here, there is an (asymptotically) optimal choice of smoothing parameter λ_V which can be calculated by the method of generalized cross-validation introduced by Craven and Wahba [2]. To find λ_V, we minimize $E[V(\lambda)]$, where

$$V(\lambda) := \frac{\frac{1}{n}\|(I - A)\underline{z}\|^2}{(\frac{1}{n}\mathrm{Tr}(I - A))^2}, \tag{5}$$

and A is the $(n \times n)$ influence matrix

$$A = F(F^T F + n\lambda M)^{-1} F^T. \tag{6}$$

As shown in [5], this is the matrix such that

$$A\underline{z} = (g_{n\lambda}(x_1, y_1), \ldots, g_{n\lambda}(x_n, y_n))^T \quad (\underline{z} \in \mathbb{R}^n). \tag{7}$$

To minimize $V(\lambda)$ over all positive λ, we use a search procedure which involves computing the value of $V(\lambda)$ for a number of choices of λ. If we proceed directly with the formula (5), this means evaluating the trace, which can be quite expensive. A better alternative is to use Girard's Monte-Carlo method [4].

§3. The Algorithm

this section we discuss an efficient algorithm for solving the penalized least ares problem using the space of cubic C^1 splines introduced above. First, ıote that the matrices $F^T F$ and M are positive semidefinite matrices. ıermore the matrix $F^T F + n\lambda M$ is regular, and consequently positive 'e. Thus, we can compute the vector of the spline coefficients $\underline{g}_{n\lambda}$ by Cholesky decomposition of $F^T F + n\lambda M$.

the evaluation of the optimal smoothing parameter, we have to cal- 'e generalized cross-validation function $V(\lambda)$. The numerator of (5) termined in the following way:

$$- A)\underline{z}\|^2 = \|\underline{z} - F\underline{g}_{n\lambda}\|^2 = \underline{z}^T \underline{z} - 2\underline{g}_{n\lambda} F^T \underline{z} + \underline{g}_{n\lambda} F^T F \underline{g}_{n\lambda}. \tag{8}$$

The evaluation of the denominator requires the evaluation of the inverse of the matrix $F^TF + n\lambda M$. Because of the high numerical cost, we use Girard's Monte-Carlo method [4] which works as follows:

i) Generate a vector $\underline{\omega}$ of n independent pseudo-random variables from the standard normal distribution.
ii) Replace $\frac{1}{n}\mathrm{Tr}(I - A)$ by $\frac{\underline{\omega}^T(I-A)\underline{\omega}}{\underline{\omega}^T\underline{\omega}}$.

Now let

$$\underline{d}_\omega := (d_{\omega_1}, \ldots, d_{\omega_m})^T := (F^TF + n\lambda M)^{-1}F^T\underline{\omega}.$$

Then using (7), we have

$$\frac{\underline{\omega}^T(I-A)\underline{\omega}}{\underline{\omega}^T\underline{\omega}} = 1 - \frac{\underline{\omega}^TA\underline{\omega}}{\underline{\omega}^T\underline{\omega}} = 1 - \frac{\underline{\omega}^TF\underline{d}_\omega}{\underline{\omega}^T\underline{\omega}} = 1 - \frac{\underline{d}_\omega^TF^T\underline{\omega}}{\underline{\omega}^T\underline{\omega}}, \tag{9}$$

and consequently we do not need to evaluate $(F^TF + n\lambda M)^{-1}$ to get the denominator. Indeed, we can calculate $\frac{\underline{\omega}^T(I-A)\underline{\omega}}{\underline{\omega}^T\underline{\omega}}$ using just a few operations and the Cholesky decomposition of $F^TF + n\lambda M$.

Algorithm. Evaluation of bivariate smoothing cubic splines.

(0) Input: z_i $(i = 1, \ldots, n)$ data,
x_i , y_i $(i = 1, \ldots, n)$ nodes of given data,
ω_i $(i = 1, \ldots, n)$ $N(0,1)$-distributed pseudo-random numbers,
n_x , n_y parameters of the type-1 triangulation of G.

(i) Evaluation of $\underline{z}^T\underline{z}$, $F^T\underline{z}$, $\underline{\omega}^T\underline{\omega}$, $F^T\underline{\omega}$, F^TF , M .

(ii) Find the minimizer of the generalized cross-validation function (for example by the method of golden section):
Cholesky decomposition of $F^TF + n\lambda M$,
evaluate $\underline{g}_{n\lambda}$ from $(F^TF + n\lambda M)\underline{g} = F^T\underline{z}$,
evaluate $\underline{d}_\omega$ from $(F^TF + n\lambda M)\underline{d} = F^T\underline{\omega}$,
evaluate $\underline{g}_{n\lambda}^TF^T\underline{z}$, $\underline{g}_{n\lambda}^TF^TF\underline{g}_{n\lambda}$, $\underline{d}_\omega^TF^T\underline{\omega}$,
evaluate $V(\lambda)$ using (8) and (9),
if λ is the minimizer, then go to step (iii),
else go to step (ii) with new λ .

(iii) Output: $\underline{g}_{n\lambda}$ vector of the spline coefficients.

§4. Numerical Results

We have tested our algorithm for the two functions

$$F_1(x,y) := \frac{1,25 + \cos(5,4y)}{6 + 6(3x-1)^2} \quad (0 \le x, y \le 1),$$

$$F_2(x,y) := \sin(2x) + 2\sin(y) + \sin(x+y) + 10 \quad (0 \le x, y \le 2\pi),$$

Figure 3. (a), (b).

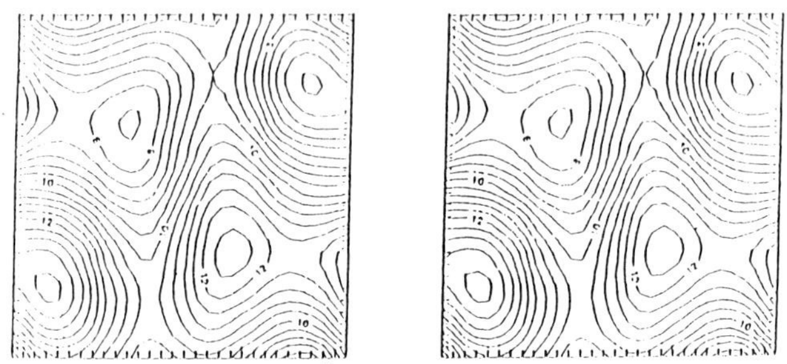

Figure 4. (a), (b).

and noisy data of the form $z_i = F_j(x_i, y_i) + \varepsilon_i$ $(i = 1, \dots, n\ ;\ j = 1, 2)$, where ε_i are pseudo-random numbers of nearly 10% of the variation of F_j $(j = 1, 2)$.

For the case $n_x = n_y = 5$ (i.e. partition of the domain G into 50 triangles), we have evaluated the smoothing parameter λ for several numbers of data by using the method of generalized cross-validation and the Monte-Carlo method of Girard (MCG), the method of generalized cross-validation without the Monte-Carlo method (GCV) and by minimizing the function $R(\lambda)$ (optimal). The results are shown in Tables 1 and 2. Note that in every case the optimal smoothing parameter is greater than its estimate.

In addition, we have considered the calculation of optimal choices for the parameters n_x, n_y describing the triangulation for the case of the function F_1 (# of data = 225). Wahba has suggested that to estimate these kinds of parameters, one can evaluate $V(\lambda_V)$, where λ_V is the minimizer of $V(\lambda)$, for various values n_x, n_y and then look for those parameters which produce the minimum of V. A comparison of the Tables 3–5 (λ_R is the minimizer of $R(\lambda)$) shows that this strategy does not work here. From Table 3 we can see that $n_x = 4$, $n_y = 5$ are the optimal parameters. The Figures 3 and 4 are plots of the isolines of the function F_2 (Figure 3 (a): data with errors of nearly 10%, Figure 3 (b): original data, Figure 4 (a): smoothing approximation with GCV, Figure 4 (b): smoothing approximation with MCG).

Smoothing parameter λ for function F_1			
# of data	MCG	GCV	optimal
100	0.03341	0.03238	0.05981
225	0.05631	0.03352	0.07028
400	0.04912	0.04198	0.09389
625	0.08796	0.06154	0.10945

Table 1

Smoothing parameter λ for function F_2			
# of data	MCG	GCV	optimal
100	0.00881	0.00968	0.01322
225	0.01877	0.01676	0.01997
400	0.01762	0.01861	0.02631
625	0.02839	0.02346	0.03775

Table 2

$R(\lambda_R) * 10^{-4}$						
$n_y \setminus n_x$	2	3	4	5	6	7
2	0.629	0.682	0.635	0.676	0.784	0.832
3	0.497	0.492	0.422	0.484	0.552	0.638
4	0.493	0.469	0.393	0.421	0.462	0.514
5	0.484	0.437	0.368	0.385	0.414	0.447
6	0.564	0.456	0.381	0.380	0.391	0.411
7	0.593	0.480	0.400	0.387	0.387	0.397

Table 3

$R(\lambda_V) * 10^{-4}$						
$n_y \setminus n_x$	2	3	4	5	6	7
2	0.637	0.686	0.636	0.678	0.786	0.834
3	0.499	0.496	0.423	0.488	0.556	0.649
4	0.504	0.470	0.395	0.423	0.470	0.528
5	0.485	0.438	0.369	0.389	0.431	0.470
6	0.567	0.458	0.382	0.381	0.400	0.423
7	0.596	0.481	0.402	0.396	0.422	0.428

Table 4

$V(\lambda_V) * 10^{-4}$						
$n_y \setminus n_x$	2	3	4	5	6	7
2	3.924	3.923	3.807	3.881	3.990	3.774
3	3.886	3.878	3.741	3.691	3.767	3.653
4	4.039	3.847	3.732	3.692	3.723	3.728
5	3.997	3.900	3.786	3.728	3.720	3.733
6	4.247	4.052	3.907	3.846	3.818	3.837
7	4.191	3.998	3.902	3.847	3.819	3.844

Table 5

References

1. Chui, C. K., and R. H. Wang, Spaces of bivariate cubic and quartic splines on type-1-triangulations, J. Math. Anal. Appl. **101** (1984), 540–554.
2. Craven, P., and G. Wahba, Smoothing noisy data with spline functions: estimating the correct degree of smoothing by the method of generalized cross-validation, Numer. Math. **31** (1979), 377–403.
3. Farin, G., Subsplines over triangles (German), dissertation, TU Braunschweig, 1979.
4. Girard, D., A fast Monte-Carlo cross-validation procedure for large least squares problems with noisy data, Numer. Math. **56** (1989), 1–23.
5. von Golitschek, M., and L. L. Schumaker, Data fitting by penalized least squares, in: *Algorithms for Approximation II* , J. Mason and M. Cox (eds.), Chapman and Hall, London, 1988, 210–227.
6. Schumaker, L. L., On the dimension of spaces of piecewise polynomials in two variables, in: *Multivariate Approximation Theory*, W. Schempp and K. Zeller (eds.), Birkhäuser Verlag, Basel, 1979, 396–412.
7. Utreras, F. I., On generalized cross-validation for multivariate smoothing spline functions, SIAM J. Sci. Statist. Comput. **8** (1988).
8. Wahba, G., *Spline Models for Observational Data*, SIAM, Philadelphia, Pennsylvania, 1990.

Acknowledgments. The author would like to thank Prof. L. L. Schumaker for his helpful comments.

Norman Weyrich
Universität Rostock, FB Mathematik
Universitätsplatz 1
O-2500 Rostock, Germany

Interpolation with g-splines

Rainer Zeifang

Abstract. We discuss interpolation of vertex data of a rectangular closed mesh in $\mathbb{R}^3$ with arbitrary Euler characteristic using g-splines. The mesh is essentially arbitrary. We merely assume that singular vertices, i. e. vertices with more or less than 4 edges, are separated by at least two edges. Our goal is to calculate a G^1 interpolating surface of low degree. We show that this interpolation problem is solvable with g-spline surfaces of degree three, if the order of the singular vertices is odd. If the order of the singular vertices is even, then solvability is attained with g-spline surfaces of degree four. We obtain these results by constructing Lagrange functions and by using the discrete Fourier transform for solving the systems of linear equations of the smoothness constraints. We compute interpolating surfaces by eliminating the degrees of freedom with least squares methods.

§1. Introduction

Modelling oriented closed surfaces of arbitrary genus p is a central problem in CAGD. Tensor product Bézier patches are currently most widely used in CAGD systems for free-form modelling. However, these patches can only model surfaces of genus $p = 1$. This fact is due to the Euler characteristic. If a closed rectangular mesh of genus p contains c vertices, e edges and f faces, then

$$c + f - e = 2(1 - p).$$

Because $2c = e$ and $2f = e$ for rectangular closed meshes, this leads to $p = 1$, which is a surface equivalent to a torus. Therefore, for arbitrary genus $p \neq 1$, we must admit meshes with so called *singular* vertices where more or less then 4 edges meet.

Using C^r conditions for tensor product patches along edges containing a singular vertex lead to singular surface points. We use geometric continuity (G^r) to solve this problem.

Mathematical Methods in Computer Aided Geometric Design II
Tom Lyche and Larry L. Schumaker (eds.), pp. 615–626.

ISBN 0-12-460510-9.

In Section 2 we recall the concept of geometric continuity following [1] and [4]. Based on this we recall the definition of g-splines of Höllig and Mögerle [5] which is a generalization of a spline space used by Goodman [2]. In Section 3 we calculate the linear constraints we need for modelling the surface. We use these constraints in Section 4 to show the existence of a solution of the interpolation problem. Applications of these results to free-form modelling are discussed in Section 5.

§2. The g-Spline Space

We consider networks of rectangular polynomial patches. We label the patches by $F := \{0, 1, \ldots, \sharp F - 1\}$, the boundary curves by $E := \{0, 1, \ldots \sharp E - 1\}$, and the vertices by $C := \{0, 1, \ldots, \sharp C - 1\}$. Then, the topological structure of the network can be described by a mapping

$$g : F \times \{1, 2, 3, 4\} \mapsto E$$

and a mapping

$$h : F \times \{1, 2, 3, 4\} \mapsto C.$$

For each $f \in F$,

$$\begin{aligned} g(f) &:= \{g(f,1), g(f,2), g(f,3), g(f,4)\} \\ h(f) &:= \{h(f,1), h(f,2), h(f,3), h(f,4)\} \end{aligned}$$

where $g(f)$ are the labels of the boundary curves and $h(f)$ are the labels of the vertices of the patch f, ordered counterclockwise with respect to the outward surface normals. Thus, two patches f and $\tilde{f}$ have a common boundary curve e if

$$e = g(f, \nu) = g(\tilde{f}, \tilde{\nu}) \tag{1}$$

for some indices $1 \leq \nu, \tilde{\nu} \leq 4$.

Each patch $f \in F$ is represented by a polynomial of coordinate degree $\leq d$, i. e., a parametrization in Bézier representation

$$(u, v) \mapsto p(u, v, f) := \sum_{i=0}^{d} \sum_{j=0}^{d} b_{i,j}^{f} B_i^d(u) B_j^d(v), \qquad (u, v) \in [0, 1]^2 \tag{2}$$

with the Bernstein polynomials B_i^d and Bézier coefficients $b_{i,j}^f \in \mathbb{R}^3$. It is assumed that these parametrizations are consistently oriented, i. e. that the outward surface normal is a multiple of $p_u \times p_v \neq 0$. Then, the labels $g(f)$ refer to the boundary curves

$$p([0, 1], 0, f), \quad p(1, [1, 0], f), \quad p([0, 1], 1, f), \quad p(0, [0, 1], f),$$

and the labels $h(f)$ refer to the vertices

$$p(0, 0, f), \quad p(1, 0, f), \quad p(1, 1, f), \quad p(0, 1, f).$$

For the description of the smoothness constraints, we consider two patches with labels $f < \tilde{f}$ which have a common boundary curve e. Denote by a_ν the affine mapping which rotates the unit square counterclockwise by an angle of $(\nu - 1) \cdot \pi/2$ about its center. Then, after a change of variables,

$$q(u,v) := p(A_\nu(u,v), f), \quad \tilde{q}(u,v) := p(A_{\tilde{\nu}+1 \bmod 4}(u,v), \tilde{f}), \tag{3}$$

with $\nu, \tilde{\nu}$ defined by (1), and we obtain

Definition 1. *Two patches $f, \tilde{f}$ with common boundary curve*

$$q([0,1],0) = \tilde{q}(0,[0,1])$$

join with geometric continuity of order r if there exist a 1 — 1 orientation preserving mapping

$$\phi(\cdot, e) := (\alpha, \beta) : \mathbb{R}^2 \to \mathbb{R}^2$$

with

$$\alpha(u,0) = u, \quad \beta(u,0) = 0,$$

such that the transversal (with respect to the common boundary curve) derivatives up to order r of the parametrizations $q \circ \phi$ and $\tilde{q}$ are continuous along the common boundary curve. This means that

$$\left(D_v^l q(\alpha(u,v), \beta(u,v))\right)_{|(u,v)=(t,0)} = (-1)^l \left(D_u^l \tilde{q}(u,v)\right)_{|(u,v)=(0,t)}, \tag{4}$$

$0 \le t \le 1, 0 \le l \le r$. The equations (4) are linear in q and $\tilde{q}$ for any choice of $\phi(\cdot, e)$ and hold simultaneously in each component. We call these equations G^r smoothness constraints, and for later reference we abbreviate them by

$$L(q, \tilde{q}, (\alpha, \beta), l) = 0 \qquad 0 \le l \le r. \tag{5}$$

With the above definition we define the g-spline space as in [5].

Definition 2. *A g-spline is a function*

$$s : \Omega := [0,1]^2 \times F \to \mathbb{R}$$

which satisfies the G^r smoothness constraints

$$L(s(A_\nu(\cdot), f), s(A_{\tilde{\nu}+1 \bmod 4}(\cdot), \tilde{f}, \phi(\cdot, e), l) = 0, \qquad 0 \le l \le r, \tag{6}$$

for any interior boundary curve e related to f and $\tilde{f}$ by (2). The linear space of all such functions s is denoted by $\mathcal{S}(d, r, g, \phi)$. A g-spline surface has a parametrization of the form (2) with coordinate functions in S.

For our purposes we have to classify the vertices of rectangular meshes.

Definition 3. *A rectangular oriented mesh is given by the mappings g and h.*

1. *A vertex c of the mesh is of order n if*
$$\sharp(h^{-1}(\{c\})) = n.$$
2. *A vertex of order $n \neq 4$ is called a singular vertex.*
3. *An edge e of the mesh is called a singular edge if it contains a singular vertex.*
4. *All the other edges are called regular edges.*
5. *A vertex of order 4 on a singular edge is called a G-vertex.*
6. *A vertex of order 4 on 4 regular edges is called a regular vertex.*
7. *A rectangular mesh in which every G-vertex is contained on exactly one singular edge is called a mesh with isolated singular vertices.*
8. *For meshes with isolated singular vertices every singular edge is uniquely related to a singular vertex of order n. These edges are called singular edges of order n.*

From now on we will only discuss meshes with isolated singular vertices. This assumption is not very restrictive. For an arbitrary rectangular mesh we obtain a mesh with isolated singular vertices by subdividing every patch into four patches.

On regular edges we want to maintain the standard C^1 conditions and choose $\phi = id$ along these edges. Along singular edges $\phi = id$ leads to a point on the surface with $p_u \times p_v = 0$ at a singular vertex, a so-called *singular surface point*. If we choose $\phi(u, v, e) = (u + \text{const}(1-u)v, v)$ along a singular edge e, we obtain a singular surface point at the G-vertex as shown in [5]. Therefore our choice for ϕ is

$$\phi(u, v, e) := \begin{cases} u + \lambda_n(1-u)^2 v, v) & \text{if } e \text{ singular of order } n, \\ (u, v) & \text{otherwise,} \end{cases} \tag{7}$$

where $\lambda_n = -2\cos(2\pi/n)$ and the choice of λ_n is motivated by symmetry considerations.

§3. G^1 Constraints

In order to calculate the G^1 constraints, we consider two patches q and $\tilde{q}$ with common boundary curve e according to (1) and parametrizations according to (3). For a regular edge e, i. e., $\phi(u, v) = (u, v)$ we obtain the usual C^1 constraints

$$\begin{aligned} b^{\nu}_{i,0} &= b^{\tilde{\nu}}_{0,i} \\ b^{\nu}_{i,1} - b^{\nu}_{i,0} &= -(b^{\tilde{\nu}}_{i,1} - b^{\tilde{\nu}}_{0,i}) \end{aligned} \qquad i = 0, \ldots, d. \tag{8}$$

For a singular edge e of order n, i. e., $\phi(u, v) = (u + \lambda_n(1-u)^2 v, v)$ we have to evaluate (4) and obtain

$$\begin{aligned} q(t, 0) &= \tilde{q}(0, t), \\ \lambda_n(1-u)^2 q_u(t, 0) + q_v(t, 0) &= -\tilde{q}_u(0, t). \end{aligned} \tag{9}$$

Some easy calculations with Bernstein polynomials lead us to the second part of the following

Theorem 4. *A g-spline surface with smoothness $r = 1$ and coordinate degree d defined over a rectangular mesh with isolated singular vertices given by the mappings g and h and the mapping*

$$\phi(u,v,e) := \begin{cases} (u + \lambda_n(1-u)^2 v, v) & \text{if } e \text{ singular of order } n, \\ (u,v) & \text{otherwise} \end{cases}$$

can be described by the following constraints on the Bézier coefficients:

1. *constraints at a regular edge*

$$\begin{aligned} b^{\nu}_{i,0} - b^{\tilde\nu}_{0,i} &= 0 \\ 2b^{\nu}_{i,0} - b^{\nu}_{i,1} - b^{\tilde\nu}_{1,i} &= 0 \end{aligned} \qquad i = 0,\ldots,d$$

2. *constraints at a singular edge of order n*

$$\begin{aligned} b^{\nu}_{i,0} - b^{\tilde\nu}_{0,i} &= 0 \qquad\qquad i = 0,\ldots,d \\ (2+\lambda_n)b^{\nu}_{0,0} - \lambda_n b^{\nu}_{1,0} - b^{\nu}_{0,1} - b^{\tilde\nu}_{1,0} &= 0 \\ \sum_{j=0}^{d} (-1)^j \binom{d}{j} b^{\nu}_{j,0} &= 0 \\ \binom{d}{i-1}(2b^{\nu}_{i-1,0} - b^{\nu}_{i-1,1} - b^{\tilde\nu}_{1,i-1}) + & \\ + \lambda_n \binom{d-1}{i} b^{\nu}_{i,0} + \sum_{j=i+1}^{d} (-1)^{i+j}\lambda_n \binom{d}{j} b^{\nu}_{j,0} &= 0 \qquad\qquad i = 2,\ldots,d-1 \\ 2b^{\nu}_{d-1,0} - b^{\nu}_{d-1,1} - b^{\tilde\nu}_{1,d-1} &= 0 \\ 2b^{\nu}_{d,0} - b^{\nu}_{d,1} - b^{\tilde\nu}_{1,d} &= 0. \end{aligned}$$

§4. A G^1 Interpolation Problem and its Solution

We can now formulate

Problem 5.

Given:
- *a g-spline space $\mathcal{S}(d,r,g,\phi)$, where*
  - *$d(f) = d$ for all $f = 0,\ldots,\sharp F - 1$*
  - *$r = 1$*
  - *ϕ is given by (7)*
  - *g describes a rectangular mesh with isolated singular vertices*
- *for every vertex $c \in C$ a data point $p_c (\in \mathbb{R}^3)$*

Find: *a g-spline surface p with coordinate functions $\in \mathcal{S}(d,r,g,\phi)$ which interpolates the given data points, i. e.*

$$x \in h^{-1}(\{c\}) \Rightarrow p(x) = p_c \qquad c = 0,\ldots,\sharp C - 1.$$

Our goal is to show the existence of solutions of Problem 5 for low degree d. In order to do this we construct Lagrange functions l_c

$$l_c : \Omega \to \mathbb{R}$$
$$l_c \in \mathcal{S}(d, r, g, \phi),$$

where

$$l_c(x) = \begin{cases} 1 & \text{if } x \in h^{-1}(\{c\}), \\ 0 & \text{if } x \in h^{-1}(C)\backslash h^{-1}(\{c\}). \end{cases} \tag{10}$$

Then, an interpolating g-spline surface p for arbitrary data $p_c(\in \mathbb{R}^3)$, $c = 0, \ldots, \sharp C - 1$ is represented by

$$p(u, v, f) = \sum_{c=0}^{\sharp C-1} p_c l_c(u, v, f)$$

for each patch $f \in F$ because of the linearity of $\mathcal{S}(d, r, g, \phi)$.

In Bézier representation the functions l_c are uniquely determined by the Bézier coefficients $b^f_{i,j}(\in \mathbb{R})$, $i, j = 0, \ldots, d$, $f = 0, \ldots, \sharp F - 1$. The Bézier coefficients $b^f_{i,j}$ are determined by solving the linear equations of Theorem 4.

Theorem 6. *With the assumption on the g-spline space $\mathcal{S}(d, 1, g, \phi)$ as described in Problem 5, the following holds:*

1. *If the mesh contains only singular vertices of odd order, then for every vertex c, there exists a Lagrange function $l_c \in \mathcal{S}(d, 1, g, \phi)$ with $d \geq 3$.*
2. *If the mesh contains singular vertices of arbitrary order (especially even), then for every vertex c exists a Lagrange function $l_c \in \mathcal{S}(d, 1, g, \phi)$ with $d \geq 4$.*

Proof: Because $\mathcal{S}(d, r, g, \phi) \in \mathcal{S}(\tilde{d}, r, g, \phi)$, if $d \leq \tilde{d}$ we remark that it is always sufficient to discuss the case of the lower degree d. We first prove part 1 for $d = 3$. We construct Lagrange functions for the vertices classified in Definition 3.

- **Regular vertex**
  This is the standard tensor product case and various C^1 interpolation methods are well known.
- **Singular vertex of order $n = 2m + 1$**
  Without loss of generality we label the patches containing the singular vertex counterclockwise by 0 to $n - 1$. In the following, all indices ν are to be taken mod$n$. The G^1 constraints

$$\begin{aligned} \lambda_n b^\nu_{1,0} + b^{\nu+1}_{1,0} + b^{\nu-1}_{1,0} &= \lambda_n + 2 \\ 3b^\nu_{1,0} - 3b^\nu_{2,0} &= 1 \\ 6b^\nu_{1,0} + \lambda_n b^\nu_{2,0} - 3b^\nu_{1,1} - 3b^{\nu-1}_{1,1} &= 0 \\ 2b^\nu_{2,0} - b^\nu_{2,1} - b^{\nu-1}_{1,2} &= 0 \\ b^\nu_{3,1} + b^{\nu-1}_{1,3} &= 0 \end{aligned} \tag{11}$$

along the edge $v = 0$ of the patch ν, $\nu = 0, \ldots, n-1$, are obtained by using the equations of Theorem 4 for the singular edge and taking into account the identities $b^{\nu}_{i,0} = b^{\nu-1}_{0,i}$ and the definition of Lagrange functions (10) $b_{0,0} = 1$ and $b^{\nu}_{3,0} = 0$. We solve the first three equations of (11) by using the discrete Fourier transform

$$\hat{z}_\mu := \frac{1}{n}\sum_{j=0}^{n-1} z_j \omega^{-j\mu} \qquad \omega := \exp(\frac{2\pi i}{n}). \tag{12}$$

With $y^\nu := (b^{\nu}_{1,0}, b^{\nu}_{1,1}, b^{\nu}_{2,0})'$ we obtain

$$A_0 y^\nu + A_1 y^{\nu-1} + A_{n-1} y^{\nu+1} = B,$$

where $B = (\lambda_n + 2, 1, 0)'$ and

$$A_0 = \begin{pmatrix} \lambda_n & 0 & 0 \\ 3 & 0 & -3 \\ 6 & -3 & \lambda_n \end{pmatrix}, A_1 = \begin{pmatrix} 1 & 0 & 0 \\ 0 & 0 & 0 \\ 0 & -3 & 0 \end{pmatrix}, A_{n-1} = \begin{pmatrix} 1 & 0 & 0 \\ 0 & 0 & 0 \\ 0 & 0 & 0 \end{pmatrix},$$

or equivalently the equations

$$\sum_{j=0}^{n-1} A_{k-j} y^k = B_k \qquad k = 0, \ldots, n-1,$$

where in addition $A_k = 0$ for $k = 2, \ldots, n-2$ and $B_k = B$ for $k = 0, \ldots, n-1$. By using the Fourier transform (12), we obtain

$$n\hat{A}_\mu \hat{y}_\mu = \hat{B}_\mu \qquad \mu = 0, \ldots, n-1.$$

Since $\lambda_n = -2\cos(2\pi/n) = -(\omega + \omega^{-1})$,

$$\hat{A}_\mu = \frac{1}{n}\begin{pmatrix} -\omega - \omega^{-1} + \omega^\mu + \omega^{-\mu} & 0 & 0 \\ 3 & 0 & -3 \\ 6 & -3 - 3\omega^{-\mu} & -\omega - \omega^{-1} \end{pmatrix},$$

$\mu = 0, \ldots, n-1$, and

$$\hat{B}_0 = \begin{pmatrix} -\omega - \omega^{-1} + 2 \\ 1 \\ 0 \end{pmatrix}, \qquad \hat{B}_\mu = 0 \quad \mu = 1, \ldots, n-1.$$

Taking into account that

$$\begin{aligned} -\omega - \omega^{-1} + \omega^\mu + \omega^{-\mu} = 0 &\Longleftrightarrow \mu = 1 \vee \mu = n-1, \\ -3 - 3\omega^{-\mu} = 0 &\Longleftrightarrow n \text{ even } \wedge \mu = n/2, \end{aligned} \tag{13}$$

we get the solutions

$$\hat{y}_0 = \begin{pmatrix} 1 \\ 1 - \frac{1}{9}(\omega + \omega^{-1}) \\ \frac{2}{3} \end{pmatrix}, \hat{y}_1 = r \begin{pmatrix} 1 \\ \frac{6-\omega-\omega^{-1}}{3+3\omega^{-1}} \\ 1 \end{pmatrix}, \hat{y}_{n-1} = s \begin{pmatrix} 1 \\ \frac{6-\omega-\omega^{-1}}{3+3\omega} \\ 1 \end{pmatrix}$$

and

$$\hat{y}_\mu = 0 \qquad \mu = 2, \ldots, n-2.$$

Remark 7. For n even the proof holds too. The solution only changes for $\nu = n/2$, where according to part 2 of (13), $-3 - 3\omega^{-n/2} = 0$. This leads to the additional solution

$$\hat{y}_{n/2} = t \begin{pmatrix} 0 \\ 1 \\ 0 \end{pmatrix}.$$

- **The G-vertex of order** $n = 2m + 1$:
At a G-vertex the G^1 constraints which come from the singular vertex (respectively the singular edge) strongly influence the Lagrange function. Therefore, as in the previous case we label the patches around the singular vertex counterclockwise by 0 to $n-1$ so that the patches containing the G-vertex have the labels 0 and $n-1$. We get the equations

$$\begin{aligned} \lambda_n b_{1,0}^\nu + b_{1,0}^{\nu+1} + b_{1,0}^{\nu-1} &= 0 \\ 3b_{1,0}^\nu - 3b_{2,0}^\nu &= -\delta_{\nu,0} \\ 6b_{1,0}^\nu + \lambda_n b_{2,0}^\nu - 3b_{1,1}^\nu - 3b_{1,1}^{\nu-1} &= \lambda_n \delta_{\nu,0} \\ 2b_{2,0}^\nu - b_{2,1}^\nu - b_{1,2}^{\nu-1} &= 0 \\ b_{3,1}^\nu + b_{1,3}^{\nu-1} &= 2\delta_{\nu,0} \end{aligned}$$

with the Kronecker-δ. We solve the first three equations as in the previous case using the same notations. Since the left sides of (13) are the same as in (11) we obtain the same A_k and $\hat{A}_k$ as before. The right hand side of (13) leads to

$$B_\nu = \delta_{\nu,0} \begin{pmatrix} 0 \\ -1 \\ \lambda_n \end{pmatrix}, \qquad \hat{B}_\nu = \frac{1}{n} \begin{pmatrix} 0 \\ -1 \\ -\omega - \omega^{-1} \end{pmatrix} \qquad \nu = 0, \ldots, n-1.$$

This gives the solutions

$$\hat{y}_\mu = \frac{1}{n} \begin{pmatrix} 0 \\ \frac{2(\omega+\omega^{-1})}{9(1+\omega^{-\mu})} \\ \frac{1}{3} \end{pmatrix} \qquad \mu = 0, \ldots, n-1, \quad \mu \neq 1, n-1,$$

$$\hat{y}_1 = \frac{1}{n} \begin{pmatrix} 0 \\ \frac{2(\omega+\omega^{-1})}{9(1+\omega^{-1})} \\ \frac{1}{3} \end{pmatrix} + r \begin{pmatrix} 1 \\ \frac{6-\omega-\omega^{-1}}{3+3\omega^{-1}} \\ 1 \end{pmatrix}$$

and

$$\hat{y}_{n-1} = \frac{1}{n}\begin{pmatrix} 0 \\ \frac{2(\omega+\omega^{-1})}{9(1+\omega)} \\ \frac{1}{3} \end{pmatrix} + s\begin{pmatrix} 1 \\ \frac{6-\omega-\omega^{-1}}{3+3\omega} \\ 1 \end{pmatrix}.$$

Remark 8. Here is where the proof fails for degree $d = 3$. We obtain at a G-vertex the system

$$\begin{pmatrix} -\omega-\omega^{-1}-2 & 0 & 0 \\ 3 & 0 & -3 \\ 6 & 0 & -\omega-\omega^{-1} \end{pmatrix}\hat{y}_{n/2} = \frac{1}{n}\begin{pmatrix} 0 \\ -1 \\ -\omega-\omega^{-1} \end{pmatrix}$$

which is not solvable.

This proves part 1 of Theorem 6. For the proof of part 2 of Theorem 6, the only interesting vertex is the G-vertex. The other two types of vertices are easy to handle because we have obtained solutions for a regular vertex for degree $d = 3$ and according to Remark 7 for a singular vertex.

- **The G-vertex of order** $n = 2m$:
  The system which has to be solved is

$$\begin{aligned} \lambda_n b_{1,0}^{\nu} + b_{1,0}^{\nu+1} + b_{1,0}^{\nu-1} &= 0 \\ -4b_{1,0}^{\nu} + 6b_{2,0}^{\nu} - 4b_{3,0}^{\nu} &= -\delta_{\nu,0} \\ 8b_{1,0}^{\nu} + 3\lambda_n b_{2,0}^{\nu} - 4\lambda_n b_{3,0}^{\nu} - 4b_{1,1}^{\nu} - 4b_{1,1}^{\nu-1} &= -\lambda_n\delta_{\nu,0} \\ 12b_{2,0}^{\nu} + \lambda_n b_{3,0}^{\nu} - 6b_{2,1}^{\nu} - 6b_{1,2}^{\nu-1} &= \lambda_n\delta_{\nu,0} \\ 2b_{3,0}^{\nu} - b_{3,1}^{\nu} - b_{1,3}^{\nu-1} &= 0 \\ b_{4,1}^{\nu} + b_{1,4}^{\nu-1} &= 2\delta_{\nu,0}. \end{aligned}$$

  This system is derived from part 2 of Theorem 4 with the same assumptions as above. We solve the first four equations of the system by using the discrete Fourier transform and $y^{\nu} := (b_{1,0}^{\nu}, b_{1,1}^{\nu}, b_{2,0}^{\nu}, b_{3,0}^{\nu})'$. Using the same notations as above, this leads us to the systems

$$n\hat{A}_{\nu}\hat{y}_{\nu} = \hat{B}_{\nu} \qquad \nu = 0,\ldots,n-1.$$

  where

$$\hat{A}_{\nu} = \frac{1}{n}\begin{pmatrix} -\omega-\omega^{-1}+\omega^{\nu}+\omega^{-\nu} & 0 & 0 & 0 \\ -4 & 0 & 6 & -4 \\ 8 & -4-4\omega^{-\nu} & -3\omega-3\omega^{-1} & 4\omega+4\omega^{-1} \end{pmatrix}$$

  and

$$\hat{B}_{\nu} = \frac{1}{n}\begin{pmatrix} 0 \\ -1 \\ \omega+\omega^{-1} \end{pmatrix},$$

$\nu = 0, \ldots, n-1$. The solutions are

$$\hat{y}_\nu = \frac{1}{n}\begin{pmatrix} 0 \\ \frac{-\omega-\omega^{-1}}{8(1+\omega^{-\nu})} \\ -\frac{1}{6} \\ 0 \end{pmatrix} + r_\nu \begin{pmatrix} 0 \\ \frac{\omega+\omega^{-1}}{2(1+\omega^{-\nu})} \\ \frac{2}{3} \\ 1 \end{pmatrix} \qquad \nu \neq 1, n-1, n/2,$$

$$\hat{y}_\nu = \frac{1}{n}\begin{pmatrix} 0 \\ \frac{-\omega-\omega^{-1}}{8(1+\omega^{-\nu})} \\ -\frac{1}{6} \\ 0 \end{pmatrix} + r_\nu \begin{pmatrix} 0 \\ \frac{\omega+\omega^{-1}}{2(1+\omega^{-\nu})} \\ \frac{2}{3} \\ 1 \end{pmatrix} + s_\nu \begin{pmatrix} 1 \\ 0 \\ 0 \\ 0 \end{pmatrix} \qquad \nu = 1, n-1,$$

and

$$\hat{y}_{n/2} = \frac{1}{n}\begin{pmatrix} 0 \\ 0 \\ 0 \\ \frac{1}{4} \end{pmatrix} + t \begin{pmatrix} 0 \\ 1 \\ 0 \\ 0 \end{pmatrix}.$$

The real parameters can be found by using the inverse Fourier transform. This finishes the proof. ■

§5. Interpolating Closed Surfaces

As we have seen in Section 4, there are still some degrees of freedom for G^1 interpolation over a closed oriented rectangular mesh with isolated singular vertices. In order to eliminate these degrees of freedom we can introduce some design criteria into the problem of interpolation. We use a method similar to [3] where it is done for C^1 interpolation with Bézier triangles.

Thinking of the G^1 constraints and the interpolation constraints as a system of equations

$$Bx = b,$$

where x contains the Bézier coefficients, and representing other design criteria also in form of linear equations

$$Ax = a,$$

leads to the least squares problem

$$\min_{Bx=b} \|Ax - a\|,$$

where $\|\cdot\|$ is the Euclidean norm. A solution can be obtained by solving

$$\begin{pmatrix} A'A & B' \\ B & 0 \end{pmatrix}\begin{pmatrix} x \\ \mu \end{pmatrix} = \begin{pmatrix} A'a \\ b \end{pmatrix}.$$

The simplest choice of a design criterion is to minimize the distance to a known surface, e. g. the C^0 interpolant to the vertex data. This idea was used

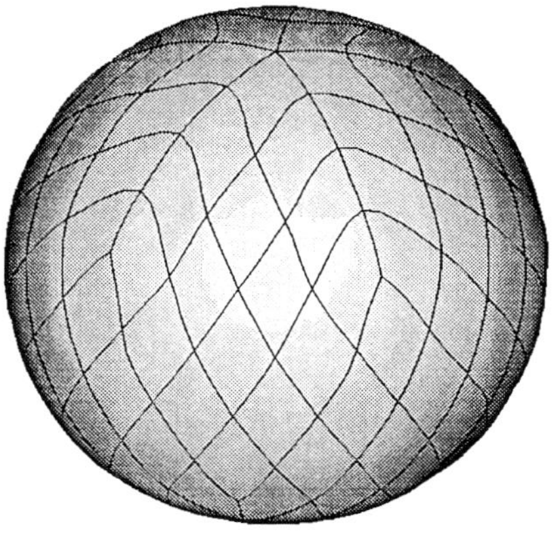

Figure 1. A sphere of degree 3 with singular vertices of order 3 and 5 .

by Grandine [3] for Bézier triangles. For Bézier rectangles the equivalent is the bilinear C^0 interpolant. The idea is that the G^1 surface should be a small perturbation of the C^0 interpolant. Our problem reduces to

$$\min_{Bx=b} \|x - a\|,$$

where a contains the coefficients of the bilinear interpolant,

$$\begin{aligned} b^f_{i,0} &= \frac{i}{d} b^f_{0,0} + \frac{d-i}{d} b^f_{d,0}, \\ b^f_{i,d} &= \frac{i}{d} b^f_{0,d} + \frac{d-i}{d} b^f_{d,d}, \\ b^f_{i,j} &= \frac{j}{d} b^f_{i,0} + \frac{d-j}{d} b^f_{i,d}, \end{aligned}$$

This method works well for degree $d = 3$. For $d = 4$, however, the method leads to sharp edges because it leaves the inner Bézier coefficients of the patches flat. We can resolve this difficulty by minimizing the distance to a cubic patch, i. e. by using for the system $Ax = a$ equations of degree reduction which can be written in the form

$$\begin{aligned} &\sum_{i=0}^{4} (-1)^i \binom{4}{i} b^f_{i,j} = 0 \qquad j = 0, \ldots, 4, \\ &\sum_{j=0}^{4} (-1)^j \binom{4}{j} b^f_{i,j} = 0 \qquad i = 0, \ldots, 4. \end{aligned}$$

Some examples are shown in the following figures.

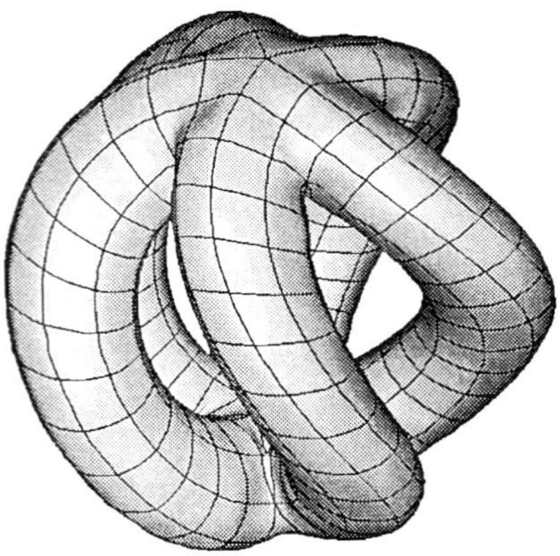

Figure 2. A surface of genus 4 and degree 4 with singular vertices of order 5 and 6.

Acknowledgments. I would like to thank Professor Klaus Höllig for many helpful hints and detailed comments on this paper.

References

1. DeRose, T. D., Geometric continuity: a parametrization independent measure of continuity for computer aided geometric design, PhD thesis, University of California at Berkeley, 1985.
2. Goodman, T. N. T., Closed surfaces defined from biquadratic splines, Constr. Approx. **7** (1991), 149–160.
3. Grandine, T. A., An iterative method for computing bivariate C^1 piecewise cubic polynomial interpolants, Technical Summary Report ♯29xx, MRC, Univ. Wisconsin, 1986.
4. Höllig, K., Geometric continuity, in *From curves to Surfaces*, SIGGRAPH, 1986.
5. Höllig, K., and H. Mögerle, G-splines, Computer Aided Geom. Design **7** (1990), 197–207.

Rainer Zeifang
Universität Stuttgart
Mathematisches Institut A
Pfaffenwaldring 57
7000 Stuttgart 80, Germany
zeifang@merkur.mathematik.uni-stuttgart.de